Springer Series on
SIGNALS AND COMMUNICATION TECHNOLOGY

SIGNALS AND COMMUNICATION TECHNOLOGY

Wireless Network Security
Y. Xiao, D.-Z. Du, X. Shen
ISBN 978-0-387-28040-0

Terrestrial Trunked Radio – TETRA
A Global Security Tool
P. Stavroulakis   ISBN 978-3-540-71190-2

Multirate Statistical Signal Processing
O.S. Jahromi   ISBN 978-1-4020-5316-0

Wireless Ad Hoc and Sensor Networks
A Cross-Layer Design Perspective
R. Jurdak   ISBN 978-0-387-39022-2

**Positive Trigonometric Polynomials
and Signal Processing Applications**
B. Dumitrescu   ISBN 978-1-4020-5124-1

Face Biometrics for Personal Identification
Multi-Sensory Multi-Modal Systems
R.I. Hammoud, B.R. Abidi, M.A. Abidi (Eds.)
ISBN 978-3-540-49344-0

**Cryptographic Algorithms
on Reconfigurable Hardware**
F. Rodríguez-Henríquez
ISBN 978-0-387-33883-5

**Ad-Hoc Networking
Towards Seamless Communications**
L. Gavrilovska   ISBN 978-1-4020-5065-7

Multimedia Database Retrieval
A Human-Centered Approach
P. Muneesawang, L. Guan
ISBN 978-0-387-25627-6

Broadband Fixed Wireless Access
A System Perspective
M. Engels; F. Petre
ISBN 978-0-387-33956-6

Acoustic MIMO Signal Processing
Y. Huang, J. Benesty, J. Chen
ISBN 978-3-540-37630-9

Algorithmic Information Theory
Mathematics of Digital Information
Processing
P. Seibt   ISBN 978-3-540-33218-3

Continuous-Time Signals
Y.S. Shmaliy   ISBN 978-1-4020-4817-3

Interactive Video
Algorithms and Technologies
R.I. Hammoud (Ed.)   ISBN 978-3-540-33214-5

Distributed Cooperative Laboratories
Networking, Instrumentation,
and Measurements
F. Davoli, S. Palazzo, S. Zappatore (Eds.)
ISBN 978-0-387-29811-5

Topics in Acoustic Echo and Noise Control
Selected Methods for the Cancellation
of Acoustical Echoes, the Reduction
of Background Noise, and Speech Processing
E. Hänsler, G. Schmidt (Eds.)
ISBN 978-3-540-33212-1

**EM Modeling of Antennas
and RF Components for Wireless
Communication Systems**
F. Gustrau, D. Manteuffel
ISBN 978-3-540-28614-1

**Orthogonal Frequency Division Multiplexing
for Wireless Communications**
Y. Li, G.L. Stuber (Eds.)
ISBN 978-0-387-29095-9

Advanced Man-Machine Interaction
Fundamentals and Implementation
K.-F. Kraiss   ISBN 978-3-540-30618-4

**The Variational Bayes Method
in Signal Processing**
V. Šmídl, A. Quinn   ISBN 978-3-540-28819-0

Voice and Speech Quality Perception
Assessment and Evaluation
U. Jekosch   ISBN 978-3-540-24095-2

**Circuits and Systems Based
on Delta Modulation**
Linear, Nonlinear and Mixed Mode Processing
D.G. Zrilic   ISBN 978-3-540-23751-8

Speech Enhancement
J. Benesty, S. Makino, J. Chen (Eds.)
ISBN 978-3-540-24039-6

Uwe Meyer-Baese

Digital Signal Processing with Field Programmable Gate Arrays

Third Edition

With 359 Figures and 98 Tables
Book with CD-ROM

 Springer

Dr. Uwe Meyer-Baese
Florida State University
College of Engineering
Department Electrical & Computer Engineering
Pottsdamer St. 2525
Tallahassee, Florida 32310
USA
E-Mail: *Uwe.Meyer-Baese@ieee.org*

Originally published as a monograph

Library of Congress Control Number: 2007933846

ISBN   978-3-540-72612-8   Springer Berlin Heidelberg New York

Springer is a part of Springer Science+Business Media

springer.com

© Springer-Verlag Berlin Heidelberg 2007

Typesetting: Data conversion by the author
Production: LE-TEX Jelonek, Schmidt & Vöckler GbR, Leipzig
Cover Design: WMXDesign GmbH, Heidelberg

Printed on acid-free paper   60/3180/YL   5 4 3 2 1   SPIN 12631506

To my Parents,

Anke and Lisa

Preface

Field-programmable gate arrays (FPGAs) are on the verge of revolutionizing digital signal processing in the manner that programmable digital signal processors (PDSPs) did nearly two decades ago. Many front-end digital signal processing (DSP) algorithms, such as FFTs, FIR or IIR filters, to name just a few, previously built with ASICs or PDSPs, are now most often replaced by FPGAs. Modern FPGA families provide DSP arithmetic support with fast-carry chains (Xilinx Virtex, Altera FLEX) that are used to implement multiply-accumulates (MACs) at high speed, with low overhead and low costs [1]. Previous FPGA families have most often targeted TTL "glue logic" and did not have the high gate count needed for DSP functions. The efficient implementation of these front-end algorithms is the main goal of this book.

At the beginning of the twenty-first century we find that the two programmable logic device (PLD) market leaders (Altera and Xilinx) both report revenues greater than US$1 billion. FPGAs have enjoyed steady growth of more than 20% in the last decade, outperforming ASICs and PDSPs by 10%. This comes from the fact that FPGAs have many features in common with ASICs, such as reduction in size, weight, and power dissipation, higher throughput, better security against unauthorized copies, reduced device and inventory cost, and reduced board test costs, and claim advantages over ASICs, such as a reduction in development time (rapid prototyping), in-circuit reprogrammability, lower NRE costs, resulting in more economical designs for solutions requiring less than 1000 units. Compared with PDSPs, FPGA design typically exploits parallelism, e.g., implementing multiple multiply-accumulate calls efficiency, e.g., zero product-terms are removed, and pipelining, i.e., each LE has a register, therefore pipelining requires no additional resources.

Another trend in the DSP hardware design world is the migration from graphical design entries to hardware description language (HDL). Although many DSP algorithms can be described with "signal flow graphs," it has been found that "code reuse" is much higher with HDL-based entries than with graphical design entries. There is a high demand for HDL design engineers and we already find undergraduate classes about logic design with HDLs [2]. Unfortunately *two* HDL languages are popular today. The US west coast and Asia area prefer Verilog, while US east coast and Europe more frequently use VHDL. For DSP with FPGAs both languages seem to be well suited, although some VHDL examples are a little easier to read because of the supported signed arithmetic and multiply/divide operations in the IEEE VHDL 1076-1987 and 1076-1993 standards. The gap is expected to disappear after approval of the Verilog IEEE standard 1364-1999, as it also includes signed arithmetic. Other constraints may include personal preferences, EDA library and tool availability, data types, readability, capability, and language extensions using PLIs, as well as commercial, business, and marketing issues, to name just a few [3]. Tool providers acknowledge today that both languages have to be supported and this book covers examples in both design languages.

We are now also in the fortunate situation that "baseline" HDL compilers are available from different sources at essentially no cost for educational use. We take advantage of this fact in this book. It includes a CD-ROM with Altera's newest `MaxPlusII` software, which provides a complete set of design tools, from a content-sensitive editor, compiler, and simulator, to a bitstream generator. All examples presented are written in VHDL and Verilog and should be easily adapted to other propriety design-entry systems. Xilinx's "Foundation Series," ModelTech's ModelSim compiler, and Synopsys FC2 or FPGA Compiler should work without any changes in the VHDL or Verilog code.

The book is structured as follows. The first chapter starts with a snapshot of today's FPGA technology, and the devices and tools used to design state-of-the-art DSP systems. It also includes a detailed case study of a frequency synthesizer, including compilation steps, simulation, performance evaluation, power estimation, and floor planning. This case study is the basis for more than 30 other design examples in subsequent chapters. The second chapter focuses on the computer arithmetic aspects, which include possible number representations for DSP FPGA algorithms as well as implementation of basic building blocks, such as adders, multipliers, or sum-of-product computations. At the end of the chapter we discuss two very useful computer arithmetic concepts for FPGAs: distributed arithmetic (DA) and the CORDIC algorithm. Chapters 3 and 4 deal with theory and implementation of FIR and IIR filters. We will review how to determine filter coefficients and discuss possible implementations optimized for size or speed. Chapter 5 covers many concepts used in multirate digital signal processing systems, such as decimation, interpolation, and filter banks. At the end of Chap. 5 we discuss the various possibilities for implementing wavelet processors with two-channel filter banks. In Chap. 6, implementation of the most important DFT and FFT algorithms is discussed. These include Rader, chirp-z, and Goertzel DFT algorithms, as well as Cooley–Tuckey, Good–Thomas, and Winograd FFT algorithms. In Chap. 7 we discuss more specialized algorithms, which seem to have great potential for improved FPGA implementation when compared with PDSPs. These algorithms include number theoretic transforms, algorithms for cryptography and errorcorrection, and communication system implementations.

The appendix includes an overview of the VHDL and Verilog languages, the examples in Verilog HDL, and a short introduction to the utility programs included on the CD-ROM.

Acknowledgements. This book is based on an FPGA communications system design class I taught for four years at the Darmstadt University of Technology; my previous (German) books [4, 5]; and more than 60 Masters thesis projects I have supervised in the last 10 years at Darmstadt University of Technology and the University of Florida at Gainesville. I wish to thank all my colleagues who helped me with critical discussions in the lab and at conferences. Special thanks to: M. Acheroy, D. Achilles, F. Bock, C. Burrus, D. Chester, D. Childers, J. Conway, R. Crochiere, K. Damm, B. Delguette, A. Dempster, C. Dick, P. Duhamel, A. Drolshagen, W. Endres, H. Eveking, S. Foo, R. Games, A. Garcia, O. Ghitza, B. Harvey, W. Hilberg, W. Jenkins, A. Laine, R. Laur, J. Mangen, J. Massey, J. McClellan, F. Ohl, S. Orr, R. Perry, J. Ramirez, H. Scheich, H. Scheid, M. Schroeder, D. Schulz, F. Simons, M. Soderstrand, S. Stearns, P. Vaidyanathan, M. Vetterli, H. Walter, and J. Wietzke.

I would like to thank my students for the innumerable hours they have spent implementing my FPGA design ideas. Special thanks to: D. Abdolrahimi, E. Allmann, B. Annamaier, R. Bach, C. Brandt, M. Brauner, R. Bug, J. Burros, M. Burschel, H. Diehl, V. Dierkes, A. Dietrich, S. Dworak, W. Fieber, J. Guyot, T. Hattermann, T. Häuser, H. Hausmann, D. Herold, T. Heute, J. Hill, A. Hundt, R. Huthmann, T. Irmler, M. Katzenberger, S. Kenne, S. Kerkmann, V. Kleipa, M. Koch, T. Krüger, H. Leitel, J. Maier, A. Noll, T. Podzimek, W. Praefcke, R. Resch, M. Rösch, C. Scheerer, R. Schimpf, B. Schlanske, J. Schleichert, H. Schmitt, P. Schreiner, T. Schubert, D. Schulz, A. Schuppert, O. Six, O. Spiess, O. Tamm, W. Trautmann, S. Ullrich, R. Watzel, H. Wech, S. Wolf, T. Wolf, and F. Zahn.

For the English revision I wish to thank my wife Dr. Anke Meyer-Bäse, Dr. J. Harris, Dr. Fred Taylor from the University of Florida at Gainesville, and Paul DeGroot from Springer.

For financial support I would like to thank the DAAD, DFG, the European Space Agency, and the Max Kade Foundation.

If you find any errata or have any suggestions to improve this book, please contact me at `Uwe.Meyer-Baese@ieee.org` or through my publisher.

Tallahassee, May 2001   *Uwe Meyer-Bäse*

Preface to Second Edition

A new edition of a book is always a good opportunity to keep up with the latest developments in the field and to correct some errors in previous editions. To do so, I have done the following for this second edition:

- Set up a web page for the book at the following URL:
  http://hometown.aol.de/uwemeyerbaese
  The site has additional information on DSP with FPGAs, useful links, and additional support for your designs, such as code generators and extra documentation.
- Corrected the mistakes from the first edition. The errata for the first edition can be downloaded from the book web page or from the Springer web page at www.springer.de, by searching for Meyer-Baese.
- A total of approximately 100 pages have been added to the new edition. The major new topics are:
  - The design of serial and array dividers
  - The description of a complete floating-point library
  - A new Chap. 8 on adaptive filter design
- Altera's current student version has been updated from 9.23 to 10.2 and all design examples, size and performance measurements, i.e., many tables and plots have been compiled for the EPF10K70RC240-4 device that is on Altera's university board UP2. Altera's UP1 board with the EPF10K20RC240-4 has been discontinued.
- A solution manual for the first edition (with more than 65 exercises and over 33 additional design examples) is available from Amazon. Some additional (over 25) new homework exercises are included in the second edition.

Acknowledgements. I would like to thank my colleagues and students for the feedback to the first edition. It helped me to improve the book. Special thanks to: P. Ashenden, P. Athanas, D. Belc, H. Butterweck, S. Conners, G. Coutu, P. Costa, J. Hamblen, M. Horne, D. Hyde, W. Li, S. Lowe, H. Natarajan, S. Rao, M. Rupp, T. Sexton, D. Sunkara, P. Tomaszewicz, F. Verahrami, and Y. Yunhua.

From Altera, I would like to thank B. Esposito, J. Hanson, R. Maroccia, T. Mossadak, and A. Acevedo (now with Xilinx) for software and hardware support and the permission to include datasheets and MaxPlus II on the CD of this book.

From my publisher (Springer-Verlag) I would like to thank P. Jantzen, F. Holzwarth, and Dr. Merkle for their continuous support and help over recent years.

I feel excited that the first edition was a big success and sold out quickly. I hope you will find this new edition even more useful. I would also be grateful, if you have any suggestions for how to improve the book, if you would e-mail me at `Uwe.Meyer-Baese@ieee.org` or contact me through my publisher.

Tallahassee, October 2003   *Uwe Meyer-Bäse*

Preface to Third Edition

Since FPGAs are still a rapidly evolving field, I am very pleased that my publisher Springer Verlag gave me the opportunity to include new developments in the FPGA field in this third edition. A total of over 150 pages of new ideas and current design methods have been added. You should find the following innovations in this third edition:

1) Many FPGAs now include embedded 18×18-bit multipliers and it is therefore recommended to use these devices for DSP-centered applications since an embedded multiplier will save many LEs. The Cyclone II EP2C35F672C6 device for instance, used in all the examples in this edition, has 35 18×18-bit multipliers.

2) MaxPlus II software is no longer updated and new devices such as the Stratix or Cyclone are only supported in Quartus II. All old and new examples in the book are now compiled with Quartus 6.0 for the Cyclone II EP2C35F672C6 device. Starting with Quartus II 6.0 integers are by default initialized with the smallest negative number (similar to with the ModelSim simulator) rather than zero and the verbatim 2/e examples will therefore not work with Quartus II 6.0. Tcl scripts are provided that allow the evaluation of all examples with other devices too. Since downloading Quartus II can take a long time the book CD includes the web version 6.0 used in the book.

3) The new device features now also allow designs that use many MAC calls. We have included a new section (2.9) on MAC-based function approximation for trigonometric, exponential, logarithmic, and square root.

4) To shorten the time to market further FPGA vendors offer intellectual property (IP) cores that can be easily included in the design project. We explain the use of IP blocks for NCOs, FIR filters, and FFTs.

5) Arbitrary sampling rate change is a frequent problem in multirate systems and we describe in Sect. 5.6 several options including B-spline, MOMS, and Farrow-type converter designs.

6) FPGA-based microprocessors have become an important IP block for FPGA vendors. Although they do not have the high performance of a custom algorithm design, the software implementation of an algorithm with a µP usually needs much less resources. A complete new chapter (9) covers many aspects from software tool to hard- and softcore µPs. A

complete example processor with an assembler and C compiler is developed.

7) A total of 107 additional problems have been added and a solution manual will be available later from www.amazon.com at a not-for-profit price.

8) Finally a special thank you goes to Harvey Hamel who discovered many errors that have been summarized in the errata for 2/e that is posted at the book homepage http://hometown.aol.de/uwemeyerbaese

Acknowledgements. Again many colleagues and students have helped me with related discussions and feedback to the second edition, which helped me to improve the book. Special thanks to:

P. Athanas, M. Bolic, C. Bentancourth, A. Canosa, S. Canosa, C. Chang, J. Chen, T, Chen, J. Choi, A. Comba, S. Connors, J. Coutu, A. Dempster, A. El-wakil, T. Felderhoff, O. Gustafsson, J. Hallman, H. Hamel, S. Hashim, A. Hoover, M. Karlsson, K. Khanachandani, E. Kim, S. Kulkarni, K. Lenk, E. Manolakos, F.Mirzapour, S. Mitra, W. Moreno, D. Murphy, T. Meiβner, K. Nayak, H. Ningxin, F.von Münchow-Pohl, H. Quach, S. Rao, S. Stepanov, C. Suslowicz, M. Unser J. Vega-Pineda, T. Zeh, E. Zurek

I am particular thankful to P. Thévenaz from EPFL for help with the newest developments in arbitrary sampling rate changers.

My colleagues from the ISS at RHTH Aachen I would like to thank for their time and efforts to teach me LISA during my Humboldt award sponsored summer research stay in Germany. Special thanks go to H. Meyr, G. Ascheid, R. Leupers, D. Kammler, and M. Witte.

From Altera, I would like to thank B. Esposito, R. Maroccia, and M. Phipps for software and hardware support and permission to include datasheets and Quartus II software on the CD of this book. From Xilinx I like to thank for software and hardware support of my NSF CCLI project J. Weintraub, A. Acevedo, A. Vera, M. Pattichis, C. Sepulveda, and C. Dick.

From my publisher (Springer-Verlag) I would like to thank Dr. Baumann, Dr. Merkle, M. Hanich, and C. Wolf for the opportunity to produce an even more useful third edition.

I would be very grateful if you have any suggestions for how to improve the book and would appreciate an e-mail to Uwe.Meyer-Baese@ieee.org or through my publisher.

Tallahassee, May 2007   *Uwe Meyer-Bäse*

Contents

Preface VII

Preface to Second Edition XI

Preface to Third Edition XIII

1. Introduction 1
   1.1 Overview of Digital Signal Processing (DSP) 1
   1.2 FPGA Technology 3
       1.2.1 Classification by Granularity 3
       1.2.2 Classification by Technology 6
       1.2.3 Benchmark for FPLs 7
   1.3 DSP Technology Requirements 10
       1.3.1 FPGA and Programmable Signal Processors 12
   1.4 Design Implementation 13
       1.4.1 FPGA Structure 18
       1.4.2 The Altera EP2C35F672C6 22
       1.4.3 Case Study: Frequency Synthesizer 29
       1.4.4 Design with Intellectual Property Cores 35
   Exercises 42

2. Computer Arithmetic 53
   2.1 Introduction 53
   2.2 Number Representation 54
       2.2.1 Fixed-Point Numbers 54
       2.2.2 Unconventional Fixed-Point Numbers 57
       2.2.3 Floating-Point Numbers 71
   2.3 Binary Adders 74
       2.3.1 Pipelined Adders 76
       2.3.2 Modulo Adders 80
   2.4 Binary Multipliers 82
       2.4.1 Multiplier Blocks 87
   2.5 Binary Dividers 91
       2.5.1 Linear Convergence Division Algorithms 93

2.5.2   Fast Divider Design............................ 98
2.5.3   Array Divider 103
2.6   Floating-Point Arithmetic Implementation 104
2.6.1   Fixed-point to Floating-Point Format Conversion..... 105
2.6.2   Floating-Point to Fixed-Point Format Conversion..... 106
2.6.3   Floating-Point Multiplication 107
2.6.4   Floating-Point Addition 108
2.6.5   Floating-Point Division 110
2.6.6   Floating-Point Reciprocal 112
2.6.7   Floating-Point Synthesis Results 114
2.7   Multiply-Accumulator (MAC) and Sum of Product (SOP) .. 114
2.7.1   Distributed Arithmetic Fundamentals 115
2.7.2   Signed DA Systems 118
2.7.3   Modified DA Solutions 120
2.8   Computation of Special Functions Using CORDIC 120
2.8.1   CORDIC Architectures 125
2.9   Computation of Special Functions using MAC Calls......... 130
2.9.1   Chebyshev Approximations 131
2.9.2   Trigonometric Function Approximation 132
2.9.3   Exponential and Logarithmic Function Approximation 141
2.9.4   Square Root Function Approximation 148
Exercises 154

3.   **Finite Impulse Response (FIR) Digital Filters** 165
3.1   Digital Filters... 165
3.2   FIR Theory... 166
3.2.1   FIR Filter with Transposed Structure 167
3.2.2   Symmetry in FIR Filters 170
3.2.3   Linear-phase FIR Filters 171
3.3   Designing FIR Filters 172
3.3.1   Direct Window Design Method...................... 173
3.3.2   Equiripple Design Method 175
3.4   Constant Coefficient FIR Design 177
3.4.1   Direct FIR Design 178
3.4.2   FIR Filter with Transposed Structure 182
3.4.3   FIR Filters Using Distributed Arithmetic............ 189
3.4.4   IP Core FIR Filter Design 204
3.4.5   Comparison of DA- and RAG-Based FIR Filters 207
Exercises 209

4.   **Infinite Impulse Response (IIR) Digital Filters** 215
4.1   IIR Theory 218
4.2   IIR Coefficient Computation 221
4.2.1   Summary of Important IIR Design Attributes 223
4.3   IIR Filter Implementation 224

4.3.1   Finite Wordlength Effects 228
4.3.2   Optimization of the Filter Gain Factor 229
4.4   Fast IIR Filter 230
4.4.1   Time-domain Interleaving 230
4.4.2   Clustered and Scattered Look-Ahead Pipelining 233
4.4.3   IIR Decimator Design.............................. 235
4.4.4   Parallel Processing 236
4.4.5   IIR Design Using RNS 239
Exercises 240

5.   **Multirate Signal Processing** 245
5.1   Decimation and Interpolation 245
5.1.1   Noble Identities 246
5.1.2   Sampling Rate Conversion by Rational Factor........ 248
5.2   Polyphase Decomposition............................... 249
5.2.1   Recursive IIR Decimator 254
5.2.2   Fast-running FIR Filter 254
5.3   Hogenauer CIC Filters 256
5.3.1   Single-Stage CIC Case Study 257
5.3.2   Multistage CIC Filter Theory 259
5.3.3   Amplitude and Aliasing Distortion 264
5.3.4   Hogenauer Pruning Theory 266
5.3.5   CIC RNS Design 272
5.4   Multistage Decimator 273
5.4.1   Multistage Decimator Design Using Goodman–Carey
        Half-band Filters 274
5.5   Frequency-Sampling Filters as Bandpass Decimators 277
5.6   Design of Arbitrary Sampling Rate Converters 280
5.6.1   Fractional Delay Rate Change 284
5.6.2   Polynomial Fractional Delay Design 290
5.6.3   B-Spline-Based Fractional Rate Changer 296
5.6.4   MOMS Fractional Rate Changer 301
5.7   Filter Banks 308
5.7.1   Uniform DFT Filter Bank 309
5.7.2   Two-channel Filter Banks 313
5.8   Wavelets 328
5.8.1   The Discrete Wavelet Transformation 332
Exercises 335

6.   **Fourier Transforms** 343
6.1   The Discrete Fourier Transform Algorithms 344
6.1.1   Fourier Transform Approximations Using the DFT ... 344
6.1.2   Properties of the DFT 346
6.1.3   The Goertzel Algorithm 349
6.1.4   The Bluestein Chirp-z Transform.................... 350

        6.1.5   The Rader Algorithm 353
        6.1.6   The Winograd DFT Algorithm...................... 359
    6.2   The Fast Fourier Transform (FFT) Algorithms 361
        6.2.1   The Cooley–Tukey FFT Algorithm 363
        6.2.2   The Good–Thomas FFT Algorithm................. 373
        6.2.3   The Winograd FFT Algorithm 375
        6.2.4   Comparison of DFT and FFT Algorithms 379
        6.2.5   IP Core FFT Design 381
    6.3   Fourier-Related Transforms 385
        6.3.1   Computing the DCT Using the DFT................ 387
        6.3.2   Fast Direct DCT Implementation 388
    Exercises 391

7. **Advanced Topics** 401
    7.1   Rectangular and Number Theoretic Transforms (NTTs) 401
        7.1.1   Arithmetic Modulo $2^b \pm 1$ 403
        7.1.2   Efficient Convolutions Using NTTs 405
        7.1.3   Fast Convolution Using NTTs 405
        7.1.4   Multidimensional Index Maps...................... 409
        7.1.5   Computing the DFT Matrix with NTTs............. 411
        7.1.6   Index Maps for NTTs............................. 413
        7.1.7   Using Rectangular Transforms to Compute the DFT .. 416
    7.2   Error Control and Cryptography 418
        7.2.1   Basic Concepts from Coding Theory 419
        7.2.2   Block Codes 424
        7.2.3   Convolutional Codes 428
        7.2.4   Cryptography Algorithms for FPGAs 436
    7.3   Modulation and Demodulation 453
        7.3.1   Basic Modulation Concepts 453
        7.3.2   Incoherent Demodulation 457
        7.3.3   Coherent Demodulation 463
    Exercises 472

8. **Adaptive Filters** 477
    8.1   Application of Adaptive Filter 478
        8.1.1   Interference Cancellation 478
        8.1.2   Prediction 479
        8.1.3   Inverse Modeling 479
        8.1.4   Identification................................. 480
    8.2   Optimum Estimation Techniques 481
        8.2.1   The Optimum Wiener Estimation 482
    8.3   The Widrow–Hoff Least Mean Square Algorithm 486
        8.3.1   Learning Curves................................ 493
        8.3.2   Normalized LMS (NLMS) 496
    8.4   Transform Domain LMS Algorithms 498

8.4.1   Fast-Convolution Techniques...................... 498
8.4.2   Using Orthogonal Transforms 500
8.5   Implementation of the LMS Algorithm 503
8.5.1   Quantization Effects 504
8.5.2   FPGA Design of the LMS Algorithm 504
8.5.3   Pipelined LMS Filters........................... 507
8.5.4   Transposed Form LMS Filter 510
8.5.5   Design of DLMS Algorithms 511
8.5.6   LMS Designs using SIGNUM Function 515
8.6   Recursive Least Square Algorithms 518
8.6.1   RLS with Finite Memory 521
8.6.2   Fast RLS Kalman Implementation................... 524
8.6.3   The Fast a Posteriori Kalman RLS Algorithm........ 529
8.7   Comparison of LMS and RLS Parameters 530
Exercises 532

9.   **Microprocessor Design** 537
9.1   History of Microprocessors.............................. 537
9.1.1   Brief History of General-Purpose Microprocessors 538
9.1.2   Brief History of RISC Microprocessors 540
9.1.3   Brief History of PDSPs 541
9.2   Instruction Set Design 544
9.2.1   Addressing Modes 544
9.2.2   Data Flow: Zero-,One-, Two- or Three-Address Design 552
9.2.3   Register File and Memory Architecture 558
9.2.4   Operation Support 562
9.2.5   Next Operation Location 565
9.3   Software Tools 566
9.3.1   Lexical Analysis................................. 567
9.3.2   Parser Development 578
9.4   FPGA Microprocessor Cores 588
9.4.1   Hardcore Microprocessors 589
9.4.2   Softcore Microprocessors 594
9.5   Case Studies 605
9.5.1   T-RISC Stack Microprocessors 605
9.5.2   LISA Wavelet Processor Design 610
9.5.3   Nios FFT Design 625
Exercises 634

References... 645

A.   Verilog Source Code 2001 661

B. VHDL and Verilog Coding 729

    B.1 List of Examples 731

    B.2 Library of Parameterized Modules (LPM) 733

        B.2.1 The Parameterized Flip-Flop Megafunction (lpm_ff) .. 733

        B.2.2 The Adder/Subtractor Megafunction 737

        B.2.3 The Parameterized Multiplier Megafunction
              (lpm_mult) 741

        B.2.4 The Parameterized ROM Megafunction (lpm_rom) ... 746

        B.2.5 The Parameterized Divider Megafunction
              (lpm_divide) 749

        B.2.6 The Parameterized RAM Megafunction (lpm_ram_dq) 751

C. Glossary 755

D. CD-ROM File: "1readme.ps" 761

Index 769

1. Introduction

This chapter gives an overview of the algorithms and technology we will discuss in the book. It starts with an introduction to digital signal processing and we will then discuss FPGA technology in particular. Finally, the Altera EP2C35F672C6 and a larger design example, including chip synthesis, timing analysis, floorplan, and power consumption, will be studied.

1.1 Overview of Digital Signal Processing (DSP)

Signal processing has been used to transform or manipulate analog or digital signals for a long time. One of the most frequent applications is obviously the *filtering* of a signal, which will be discussed in Chaps. 3 and 4. Digital signal processing has found many applications, ranging from data communications, speech, audio or biomedical signal processing, to instrumentation and robotics. Table 1.1 gives an overview of applications where DSP technology is used [6].

Digital signal processing (DSP) has become a mature technology and has replaced traditional analog signal processing systems in many applications. DSP systems enjoy several advantages, such as insensitivity to change in temperature, aging, or component tolerance. Historically, analog chip design yielded smaller die sizes, but now, with the noise associated with modern submicrometer designs, digital designs can often be much more densely integrated than analog designs. This yields compact, low-power, and low-cost digital designs.

Two events have accelerated DSP development. One is the disclosure by Cooley and Tuckey (1965) of an efficient algorithm to compute the discrete Fourier Transform (DFT). This class of algorithms will be discussed in detail in Chapter 6. The other milestone was the introduction of the programmable digital signal processor (PDSP) in the late 1970s, which will be discussed in Chap. 9. This could compute a (fixed-point) "multiply-and-accumulate" in only one clock cycle, which was an essential improvement compared with the "Von Neuman" microprocessor-based systems in those days. Modern PDSPs may include more sophisticated functions, such as floating-point multipliers, barrelshifters, memory banks, or zero-overhead interfaces to A/D and D/A converters. EDN publishes every year a detailed overview of available PDSPs

Table 1.1. Digital signal processing applications.

Area	DSP algorithm
General-purpose	Filtering and convolution, adaptive filtering, detection and correlation, spectral estimation and Fourier transform
Speech processing	Coding and decoding, encryption and decryption, speech recognition and synthesis, speaker identification, echo cancellation, cochlea-implant signal processing
Audio processing	hi-fi encoding and decoding, noise cancellation, audio equalization, ambient acoustics emulation, audio mixing and editing, sound synthesis
Image processing	Compression and decompression, rotation, image transmission and decompositioning, image recognition, image enhancement, retina-implant signal processing
Information systems	Voice mail, facsimile (fax), modems, cellular telephones, modulators/demodulators, line equalizers, data encryption and decryption, digital communications and LANs, spread-spectrum technology, wireless LANs, radio and television, biomedical signal processing
Control	Servo control, disk control, printer control, engine control, guidance and navigation, vibration control, power-system monitors, robots
Instrumentation	Beamforming, waveform generation, transient analysis, steady-state analysis, scientific instrumentation, radar and sonar

[7]. We will return in and Chap. 2 (p. 116) and Chap. 9 to PDSPs after we have studied FPGA architectures.

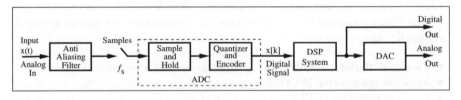

Fig. 1.1. A typical DSP application.

Figure 1.1 shows a typical application used to implement an analog system by means of a digital signal processing system. The analog input signal is feed through an analog anti aliasing filter whose stopband starts at half the sampling frequency f_s to suppress unwonted mirror frequencies that occur during the sampling process. Then the analog-to-digital converter (ADC) follows that typically is implemented with a sample-and-hold and a quantize (and encoder) circuit. The digital signal processing circuit perform then the steps that in the past would have been implemented in the analog system. We may want to further process or store (i.e., on CD) the digital processed data, or we may like to produce an analog output signal (e.g., audio signal) via a digital-to-analog converter (DAC) which would be the output of the equivalent analog system.

1.2 FPGA Technology

VLSI circuits can be classified as shown in Fig. 1.2. FPGAs are a member of a class of devices called field-programmable logic (FPL). FPLs are defined as programmable devices containing repeated fields of small logic blocks and elements[2]. It can be argued that an FPGA is an ASIC technology since FPGAs are application-specific ICs. It is, however, generally assumed that the design of a classic ASIC required additional semiconductor processing steps beyond those required for an FPL. The additional steps provide higher-order ASICs with their performance and power consumption advantage, but also with high nonrecurring engineering (NRE) costs. At 65 nm the NRE cost are about $4 million, see [8] . Gate arrays, on the other hand, typically consist of a "sea of NAND gates" whose functions are customer provided in a "wire list." The wire list is used during the fabrication process to achieve the distinct definition of the final metal layer. The designer of a *programmable* gate array solution, however, has full control over the actual design implementation without the need (and delay) for any physical IC fabrication facility. A more detailed FPGA/ASIC comparison can be found in Sect. 1.3, p. 10.

1.2.1 Classification by Granularity

Logic block size correlates to the *granularity* of a device that, in turn, relates to the effort required to complete the wiring between the blocks (routing channels). In general three different granularity classes can be found:

- Fine granularity (Pilkington or "sea of gates" architecture)
- Medium granularity (FPGA)
- Large granularity (CPLD)

Fine-Granularity Devices

Fine-grain devices were first licensed by Plessey and later by Motorola, being supplied by Pilkington Semiconductor. The basic logic cell consisted of a

---

[2] Called configurable logic block (CLB) by Xilinx, logic cell (LC) or logic elements (LE) by Altera.

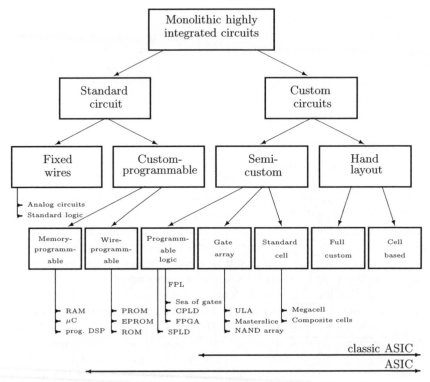

Fig. 1.2. Classification of VLSI circuits (©1995 VDI Press [4]).

single NAND gate and a latch (see Fig. 1.3). Because it is possible to realize any binary logic function using NAND gates (see Exercise 1.1, p. 42), NAND gates are called *universal* functions. This technique is still in use for gate array designs along with approved logic synthesis tools, such as ESPRESSO. Wiring between gate-array NAND gates is accomplished by using additional metal layer(s). For programmable architectures, this becomes a bottleneck because the routing resources used are very high compared with the implemented logic functions. In addition, a high number of NAND gates is needed to build a simple DSP object. A fast 4-bit adder, for example, uses about 130 NAND gates. This makes fine-granularity technologies unattractive in implementing most DSP algorithms.

Medium-Granularity Devices

The most common FPGA architecture is shown in Fig. 1.4a. A concrete example of a contemporary medium-grain FPGA device is shown in Fig. 1.5. The elementary logic blocks are typically small tables (e.g., Xilinx Virtex with 4- to 5-bit input tables, 1- or 2-bit output), or are realized with ded-

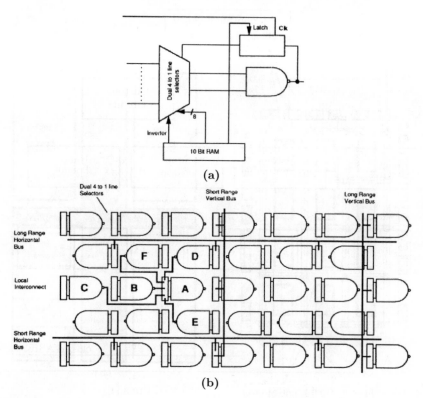

Fig. 1.3. Plessey ERA60100 architecture with 10K NAND logic blocks [9]. **(a)** Elementary logic block. **(b)** Routing architecture (©1990 Plessey).

icated multiplexer (MPX) logic such as that used in Actel ACT-2 devices [10]. Routing channel choices range from short to long. A programmable I/O block with flip-flops is attached to the physical boundary of the device.

Large-Granularity Devices

Large granularity devices, such as the complex programmable logic devices (CPLDs), are characterized in Fig. 1.4b. They are defined by combining so-called simple programmable logic devices (SPLDs), like the classic GAL16V8 shown in Fig. 1.6. This SPLD consists of a programmable logic array (PLA) implemented as an AND/OR array and a universal I/O logic block. The SPLDs used in CPLDs typically have 8 to 10 inputs, 3 to 4 outputs, and support around 20 product terms. Between these SPLD blocks wide busses (called programmable interconnect arrays (PIAs) by Altera) with short delays are available. By combining the bus and the fixed SPLD timing, it is possible to provide predictable and short pin-to-pin delays with CPLDs.

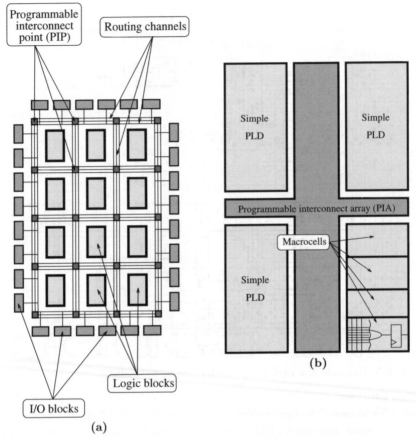

Fig. 1.4. (a) FPGA and (b) CPLD architecture (©1995 VDI Press [4]).

1.2.2 Classification by Technology

FPLs are available in virtually all memory technologies: SRAM, EPROM, E^2PROM, and antifuse [11]. The specific technology defines whether the device is *reprogrammable* or *one-time programmable*. Most SRAM devices can be programmed by a single-bit stream that reduces the wiring requirements, but also increases programming time (typically in the ms range). SRAM devices, the dominate technology for FPGAs, are based on static CMOS memory technology, and are re- and in-system programmable. They require, however, an external "boot" device for configuration. Electrically programmable read-only memory (EPROM) devices are usually used in a one-time CMOS programmable mode because of the need to use ultraviolet light for erasure. CMOS electrically erasable programmable read-only memory (E^2PROM) can be used as re- and in-system programmable. EPROM and E^2PROM have the advantage of a short setup time. Because the programming information is

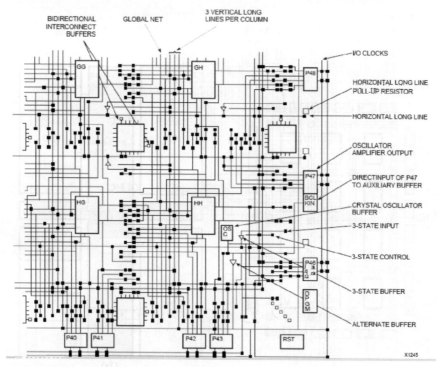

Fig. 1.5. Example of a medium-grain device (©1993 Xilinx).

not "downloaded" to the device, it is better protected against unauthorized use. A recent innovation, based on an EPROM technology, is called "flash" memory. These devices are usually viewed as "pagewise" in-system reprogrammable systems with physically smaller cells, equivalent to an E^2PROM device. Finally, the important advantages and disadvantages of different device technologies are summarized in Table 1.2.

1.2.3 Benchmark for FPLs

Providing objective benchmarks for FPL devices is a nontrivial task. Performance is often predicated on the experience and skills of the designer, along with design tool features. To establish valid benchmarks, the Programmable Electronic Performance Cooperative (PREP) was founded by Xilinx [12], Altera [13], and Actel [14], and has since expanded to more than 10 members. PREP has developed nine different benchmarks for FPLs that are summarized in Table 1.3. The central idea underlining the benchmarks is that each vendor uses its own devices and software tools to implement the basic blocks as many times as possible in the specified device, while attempting to maximize speed. The number of instantiations of the same logic block within

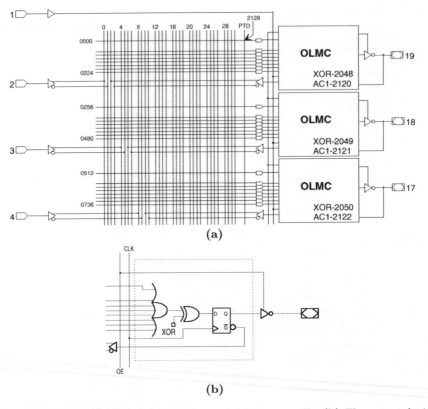

Fig. 1.6. The GAL16V8. **(a)** First three of eight macrocells. **(b)** The output logic macrocell (OLMC) (©1997 Lattice).

one device is called the *repetition rate* and is the basis for all benchmarks. For DSP comparisons, benchmarks five and six of Table 1.3 are relevant. In Fig. 1.7, repetition rates are reported over frequency, for typical Actel (A_k), Altera (o_k), and Xilinx (x_k) devices. It can be concluded that modern FPGA families provide the best DSP complexity and maximum speed. This is attributed to the fact that modern devices provide fast-carry logic (see Sect. 1.4.1, p. 18) with delays (less than 0.1 ns per bit) that allow fast adders with large bit width, without the need for expensive "carry look-ahead" decoders. Although PREP benchmarks are useful to compare equivalent gate counts and maximum speeds, for concrete applications additional attributes are also important. They include:

- Array multiplier (e.g., 18×18 bits)
- Embedded hardwired microprocessor (e.g., 32-bit RISC PowerPC)
- On-chip RAM or ROM (LE or large block size)
- External memory support for ZBT, DDR, QDR, SDRAM

Table 1.2. FPL technology.

Technology	SRAM	EPROM	E²PROM	Antifuse	Flash
Repro-grammable	✓	✓	✓	–	✓
In-system programmable	✓	–	✓	–	✓
Volatile	✓	–	–	–	–
Copy protected	–	✓	✓	✓	✓
Examples	Xilinx Spartan	Altera MAX5K	AMD MACH	Actel ACT	Xilinx XC9500
	Altera Cyclone	Xilinx XC7K	Altera MAX 7K		Cypress Ultra 37K

- Pin-to-pin delay
- Internal tristate bus
- Readback- or boundary-scan decoder
- Programmable slew rate or voltage of I/O
- Power dissipation
- Ultra-high speed serial interfaces

Some of these features are (depending on the specific application) more relevant to DSP application than others. We summarize the availability of some of these key features in Tables 1.4 and 1.5 for Xilinx and Altera, respectively. The first column shows the device family name. The columns $3 - 9$ show the (for most DSP applications) relevant features: (3) the support of fast-carry logic for adder or subtractor, (4) the embedded array multiplier of 18×18 bit width, (5) the on-chip RAM implemented with the LEs, (6) the on-chip kbit memory block of size larger of about 1-16 kbit,(7) the on-chip Mbit memory block of size larger of about 1 mega bit, (8) embedded microprocessor: IBM's PowerPC on Xilinx or the ARM processor available with Altera devices, and (9) the target price and availability of the device family. Device that are no longer recommended for new designs are classified as mature with m. Low-cost devices have a single $ and high price range devices have two $$.

Figure 1.8 summarizes the power dissipation of some typical FPL devices. It can be seen that CPLDs usually have higher "standby" power consumption. For higher-frequency applications, FPGAs can be expected to have a higher power dissipation. A detailed power analysis example can be found in Sect. 1.4.2, p. 27.

Table 1.3. The PREP benchmarks for FPLs.

Number	Benchmark name	Description
1	Data path	Eight 4-to-1 multiplexers drive a parallel-load 8-bit shift register (see Fig. 1.27, p. 44)
2	Timer/counter	Two 8-bit values are clocked through 8-bit value registers and compared (see Fig. 1.28, p. 45)
3	Small state machine	An 8-state machine with 8 inputs and 8 outputs (see Fig. 2.59, p. 159)
4	Large state machine	A 16-state machine with 40 transitions, 8 inputs, and 8 outputs (see Fig. 2.60, p. 161)
5	Arithmetic circuit	A 4-by-4 unsigned multiplier and 8-bit accumulator (see Fig. 4.23, p. 243)
6	16-bit accumulator	A 16-bit accumulator (see Fig. 4.24, p. 244)
7	16-bit counter	Loadable binary up counter (see Fig. 9.40, p. 642)
8	16-bit synchronous prescaled counter	Loadable binary counter with asynchronous reset (see Fig. 9.40, p. 642)
9	Memory mapper	The map decodes a 16-bit address space into 8 ranges (see Fig. 9.41, p. 643)

1.3 DSP Technology Requirements

The PLD market share, by vendor, is presented in Fig. 1.9. PLDs, since their introduction in the early 1980s, have enjoyed in the last decade steady growth of 20% per annum, outperforming ASIC growth by more than 10%. In 2001 the worldwide recession in microelectronics reduced the ASIC and FPLD growth essentially. Since 2003 we see again a steep increase in revenue for the two market leader. The reason that FPLDs outperformed ASICs seems to be related to the fact that FPLs can offer many of the advantages of ASICs such as:

- Reduction in size, weight, and power dissipation
- Higher throughput
- Better security against unauthorized copies
- Reduced device and inventory cost
- Reduced board test costs

without many of the disadvantages of ASICs such as:

Table 1.4. Xilinx FPGA family DSP features.

Family	Feature						
	Fast adder carry logic	Emb. mult. 18×18 bits	LE RAM	Kbit RAM	Mbit RAM	Emb. µP	Low cost/ mature
XC2000	−	−	−	−	−	−	m
XC3000	−	−	−	−	−	−	m
XC4000	✓	−	✓	−	−	−	m
Spartan-XL	✓	−	✓	−	−	−	$
Spartan-II	✓	−	✓	✓	−	−	$
Spartan-3	✓	✓	✓	✓	−	−	$
Virtex	✓	−	✓	✓	−	−	$$
Virtex-II	✓	✓	✓	✓	−	−	$$
Virtex-II Pro	✓	✓	✓	✓	−	✓	$$
Virtex-4-LX	✓	✓	✓	✓	−	−	$$
Virtex-4-SX	✓	✓	✓	✓	−	−	$$
Virtex-4-FX	✓	✓	✓	✓	−	✓	$$
Virtex-5	✓	✓	✓	✓	−	−	$$

Table 1.5. Altera FPGA family DSP features.

Family	Feature						
	Fast adder carry logic	Emb. mult. 18×18 bits	LE RAM	Kbit RAM	Mbit RAM	Emb. µP	Low cost/ mature
FLEX8K	✓	−	−	−	−	−	m
FLEX10K	✓	−	−	✓	−	−	m
APEX20K	✓	−	−	✓	−	−	m
APEX II	✓	−	−	✓	−	−	m
ACEX	✓	−	−	✓	−	−	m
Mercury	✓	−	−	✓	−	−	m
Excalibur	✓	−	−	✓	−	✓	m
Cyclone	✓	−	−	✓	−	−	$
Cyclone II	✓	✓	−	✓	−	−	$
Stratix	✓	✓	−	✓	✓	−	$$
Stratix II	✓	✓	−	✓	✓	−	$$

- A reduction in development time (rapid prototyping) by a factor of three to four
- In-circuit reprogrammability
- Lower NRE costs resulting in more economical designs for solutions requiring less than 1000 units

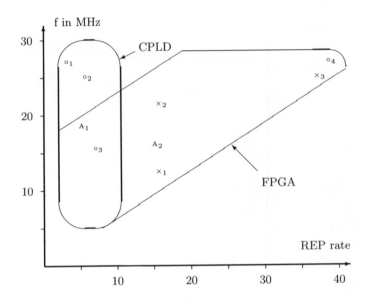

Fig. 1.7. Benchmarks for FPLs (©1995 VDI Press [4]).

CBIC ASICs are used in high-end, high-volume applications (more than 1000 copies). Compared to FPLs, CBIC ASICs typically have about ten times more gates for the same die size. An attempt to solve the latter problem is the so-called hard-wired FPGA, where a gate array is used to implement a verified FPGA design.

1.3.1 FPGA and Programmable Signal Processors

General-purpose programmable digital signal processors (PDSPs) [6, 15, 16] have enjoyed tremendous success for the last two decades. They are based on a reduced instruction set computer (RISC) paradigm with an architecture consisting of at least one fast array multiplier (e.g., 16×16-bit to 24×24-bit fixed-point, or 32-bit floating-point), with an extended wordwidth accumulator. The PDSP advantage comes from the fact that most signal processing algorithms are multiply and accumulate (MAC) intensive. By using a multistage pipeline architecture, PDSPs can achieve MAC rates limited only by the speed of the array multiplier. More details on PDSPs can be found in Chap. 9. It can be argued that an FPGA can also be used to implement MAC cells [17], but cost issues will most often give PDSPs an advantage, if the PDSP meets the desired MAC rate. On the other hand we now find many high-bandwidth signal-processing applications such as wireless, multimedia, or satellite transmission, and FPGA technology can provide more bandwidth through multiple MAC cells on one chip. In addition, there are several al-

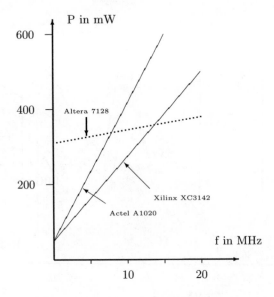

Fig. 1.8. Power dissipation for FPLs (©1995 VDI Press [4]).

gorithms such as CORDIC, NTT or error-correction algorithms, which will be discussed later, where FPL technology has been proven to be more efficient than a PDSP. It is assumed [18] that in the future PDSPs will dominate applications that require complicated algorithms (e.g., several `if-then-else` constructs), while FPGAs will dominate more front-end (sensor) applications like FIR filters, CORDIC algorithms, or FFTs, which will be the focus of this book.

1.4 Design Implementation

The levels of detail commonly used in VLSI designs range from a geometrical layout of full custom ASICs to system design using so-called set-top boxes. Table 1.6 gives a survey. Layout and circuit-level activities are absent from FPGA design efforts because their physical structure is programmable but fixed. The best utilization of a device is typically achieved at the gate level using register transfer design languages. Time-to-market requirements, combined with the rapidly increasing complexity of FPGAs, are forcing a methodology shift towards the use of intellectual property (IP) macrocells or mega-core cells. Macrocells provide the designer with a collection of predefined functions, such as microprocessors or UARTs. The designer, therefore, need only specify selected features and attributes (e.g., accuracy), and a

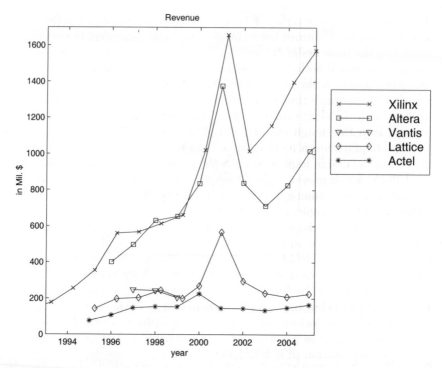

Fig. 1.9. Revenues of the top five vendors in the PLD/FPGA/CPLD market.

Table 1.6. VLSI design levels.

Object	Objectives	Example
System	Performance specifications	Computer, disk unit, radar
Chip	Algorithm	μP, RAM, ROM, UART, parallel port
Register	Data flow	Register, ALU, COUNTER, MUX
Gate	Boolean equations	AND, OR, XOR, FF
Circuit	Differential equations	Transistor, R, L, C
Layout	None	Geometrical shapes

synthesizer will generate a hardware description code or schematic for the resulting solution.

A key point in FPGA technology is, therefore, powerful design tools to

- Shorten the design cycle
- Provide good utilization of the device
- Provide synthesizer options, i.e., choose between optimization speed versus size of the design

A CAE tool taxonomy, as it applies to FPGA design flow, is presented in Fig. 1.10. The design entry can be graphical or text-based. A formal check that eliminates syntax errors or graphic design rule errors (e.g., open-ended wires) should be performed before proceeding to the next step. In the function extraction the basic design information is extracted from the design and written in a functional netlist. The netlist allows a first functional simulation of the circuit and to build an example data set called a testbench for later testing of the design with timing information. If the functional test is not passed we start with the design entry again. If the functional test is satisfactory we proceed with the design implementation, which usually takes several steps and also requires much more compile time then the function extraction. At the end of the design implementation the circuit is completely routed within our FPGA, which provides precise resource data and allows us to perform a simulation with all timing delay information as well as performance measurements. If all these implementation data are as expected we can proceed with the programming of the actual FPGA; if not we have to start with the design entry again and make appropriate changes in our design. Using the JTAG interface of modern FPGAs we can also directly monitor data processing on the FPGA: we may read out just the I/O cells (which is called a boundary scan) or we can read back all internal flip-flops (which is called a full scan). If the in-system debugging fails we need to return to the design entry.

In general, the decision of whether to work within a graphical or a text design environment is a matter of personal taste and prior experience. A graphical presentation of a DSP solution can emphasize the highly regular dataflow associated with many DSP algorithms. The textual environment, however, is often preferred with regard to algorithm control design and allows a wider range of design styles, as demonstrated in the following design example. Specifically, for Altera's Quartus II, it seemed that with text design more special attributes and more-precise behavior can be assigned in the designs.

Example 1.1: Comparison of VHDL Design Styles

The following design example illustrates three design strategies in a VHDL context. Specifically, the techniques explored are:
- Structural style (component instantiation, i.e., graphical netlist design)
- Data flow, i.e., concurrent statements
- Sequential design using PROCESS templates

The VHDL design file `example.vhd`[4] follows (comments start with --):

```
PACKAGE eight_bit_int IS    -- User-defined type
   SUBTYPE BYTE IS INTEGER RANGE -128 TO 127;
END eight_bit_int;

LIBRARY work;
USE work.eight_bit_int.ALL;

LIBRARY lpm;                        -- Using predefined packages
USE lpm.lpm_components.ALL;
```

---

[4] The equivalent Verilog code `example.v` for this example can be found in Appendix A on page 663. Synthesis results are shown in Appendix B on page 731.

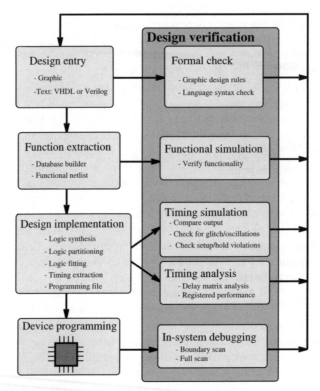

Fig. 1.10. CAD design circle.

```
LIBRARY ieee;
USE ieee.std_logic_1164.ALL;
USE ieee.std_logic_arith.ALL;

ENTITY example IS                        ------> Interface
  GENERIC (WIDTH : INTEGER := 8);   -- Bit width
  PORT (clk  :  IN STD_LOGIC;
        a, b :  IN BYTE;
        op1  :  IN STD_LOGIC_VECTOR(WIDTH-1 DOWNTO 0);
        sum  :  OUT STD_LOGIC_VECTOR(WIDTH-1 DOWNTO 0);
        d    :  OUT BYTE);
END example;

ARCHITECTURE fpga OF example IS

  SIGNAL  c, s       :  BYTE;        -- Auxiliary variables
  SIGNAL  op2, op3   :  STD_LOGIC_VECTOR(WIDTH-1 DOWNTO 0);

BEGIN

  -- Conversion int -> logic vector
```

```
op2 <= CONV_STD_LOGIC_VECTOR(b,8);

add1: lpm_add_sub            ------> Component instantiation
   GENERIC MAP (LPM_WIDTH => WIDTH,
                LPM_REPRESENTATION => "SIGNED",
                LPM_DIRECTION => "ADD")
   PORT MAP (dataa => op1,
             datab => op2,
             result => op3);
reg1: lpm_ff
   GENERIC MAP (LPM_WIDTH => WIDTH )
   PORT MAP (data => op3,
             q => sum,
             clock => clk);

c <= a + b ;                 ------> Data flow style

p1: PROCESS                  ------> Behavioral style
BEGIN
   WAIT UNTIL clk = '1';
   s <= c + s;               ----> Signal assignment statement
END PROCESS p1;
d <= s;

END fpga;
```

1.1

To start the simulator[5] we first copy the file from the CD to the project directory and use File→Open Project to select the example project. Now select Simulator Tool under the Processing menu. A new window to control the simulation parameter will pop up. To perform a function simulation in Quartus II the Generate Functional Simulation Netlist button needs to be activated first by selecting Functional as Simulation mode. If successful we can proceed and start with the design implementation as shown in Fig. 1.10. To do this with the Quartus II compiler, we choose Timing as the Simulation mode. However, the timing simulation requires that all compilation steps (Analysis & Synthesis, Fitter, Assembler and Timing Analyzer) are first performed. After completion of the compilation we can then conduct a simulation with timing, check for glitches, or measure the Registered Performance of the design, to name just a few options. After all these steps are successfully completed, and if a hardware board (like the prototype board shown in Fig. 1.11) is available, we proceed with programming the device and may perform additional hardware tests using the read-back methods, as reported in Fig. 1.10. Altera supports several DSP development boards with a large set of useful prototype components including fast A/D, D/A, audio CODEC, DIP switches, single and 7-segment LEDs, and push

---

[5] Note that a more detailed design tool study will follow in section 1.4.3.

buttons. These development boards are available from Altera directly. Altera offers Stratix S25, Stratix II S60,and S80 and Cyclone II boards, in the $995-$5995 price range, which differs not only in FPGA size, but also in terms of the extra features, like number, precision and speed of A/D channels, and memory blocks. For universities a good choice will be the lowest-cost Cyclone II board, which is still more expensive than the UP2 or UP3 boards used in many digital logic labs, but has a fast A/D and D/A and a two-channel CODEC, and large memory bank outside the FPGA, see Fig. 1.11a. Xilinx on the other side has very limited direct board support; all boards for instance available in the university program are from third parties. However some of these boards are priced so low that it seems that these boards are not-for-profit designs. A good board for DSP purposes (with on-chip multipliers) is for instance offered by Digilent Inc. for only $99, see Fig. 1.11b. The board has a XC3S200 FPGA, flash, four 7-segment LEDs, eight switches, and four push buttons. For DSP experiments, A/D and D/A mounted on very small daughter boards are available for $19.95 each, so a nice DSP board can be built for only $138.90.

(a)   (b)

Fig. 1.11. Low-cost prototype boards: **(a)** Cyclone II Altera board. **(b)** Xilinx Nexsys board with ADC and DAC daughter boards.

1.4.1 FPGA Structure

At the beginning of the 21st century FPGA device families now have several attractive features for implementing DSP algorithms. These devices provide fast-carry logic, which allows implementations of 32-bit (nonpipelined) adders at speeds exceeding 300 MHz [1, 19, 20], embedded 18 × 18 bit multipliers, and large memory blocks.

Xilinx FPGAs are based on the elementary logic block of the early XC4000 family and the newest derivatives are called Spartan (low cost) and Virtex (high performance). Altera devices are based on FLEX 10K logic blocks and the newest derivatives are called Stratix (high performance) and Cyclone (low cost). The Xilinx devices have the wide range of routing levels typical of a FPGAs, while the Altera devices are based on an architecture with the wide busses used in Altera's CPLDs. However, the basic blocks of the Cyclone and Stratix devices are no longer large PLAs as in CPLD. Instead the devices now have medium granularity, i.e., small look-up tables (LUTs), as is typical for FPGAs. Several of these LUTs, called logic elements (LE) by Altera, are grouped together in a logic array block (LAB). The number of LEs in an LAB depends on the device family, where newer families in general have more LEs per LAB: Flex10K utilizes eight LEs per LAB, APEX20K uses 10 LEs per LAB and Cyclone II has 16 LEs per LAB.

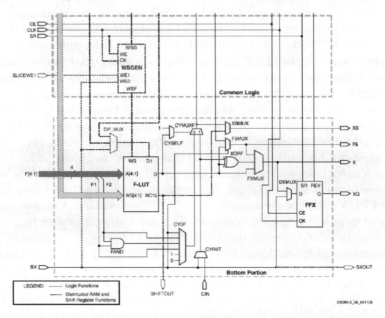

Fig. 1.12. Spartan-3 low portion of a slice/logic element (©2006 Xilinx).

Since the Spartan-3 devices are part of a popular DSP board offered by Digilent Inc., see Figure 1.11b, we will have a closer look at this FPGA family. The basic logic elements of the Xilinx Spartan-3 are called slices having two separate four-input one-output LUTs, fast-carry dedicated logic, two flip-flops, and some shared control signals. In the Spartan-3 family four slices are combined in a configurable logic blocks (CLB), having a total of eight four-input one-output LUTs, and eight flip-flops. Figure 1.12 shows the lower part of the left slice. Each slice LUT can be used as a 16×1 RAM or ROM. The dashed part is used if the slice is used to implement distributed memory or shift registers, and is only available in 50% of the slices. The Xilinx device has multiple levels of routing, ranging from CLB to CLB, to long lines spanning the entire chip. The Spartan-3 device also includes large memory block

Table 1.7. The Xilinx Spartan-3 family.

Device	Total 4-input LUTs	CLB	RAM blocks	DCM	Emb. mult. 18×18	Max. I/O mbit	Conf. file
XC3S50	1536	192	4	2	4	124	0.4
XC3S200	3840	480	12	4	12	173	1.0
XC3S400	7168	896	16	4	16	264	1.7
XC3S1000	15 360	1920	24	4	24	391	3.2
XC3S1500	26 624	3328	32	4	32	487	5.2
XC3S2000	40 960	5120	40	4	40	565	7.6
XC3S4000	55 296	6912	96	4	96	712	11.3
XC3S5000	66 560	8320	104	4	104	784	13.2

(18,432 bits or 16,384 bits if no parity bits are used) that can be used as single- or dual-port RAM or ROM. The memory blocks can be configure as $2^9 \times 32, 2^{10} \times 16, \ldots, 2^{14} \times 1$, i.e., each additional address bit reduces the data bit width by a factor of two. Another interesting feature for DSP purpose is the embedded multiplier in the Spartan-3 family. These are fast 18×18 bit signed array multipliers. If unsigned multiplication is required 17×17 bit multiplier can be implemented with this embedded multiplier. This device family also includes up to four complete clock networks (DCMs) that allow one to implement several designs that run at different clock frequencies in the same FPGA with low clock skew. Up to 13 Mbits configuration files size is required to program Spartan-3 devices. Tables 1.7 shows the most important DSP features of members of the Xilinx Spartan-3 family.

As an example of an Altera FPGA family let us have a look at the Cyclone II devices used in the low-cost prototyping board by Altera, see Fig. 1.11a. The basic block of the Altera Cyclone II device achieves a medium granularity using small LUTs. The Cyclone device is similar to the Altera 10K device used in the popular UP2 and UP3 boards, with increased RAM blocks memory size to 4 kbits, which are no longer called EAB as in Flex 10K or ESB as in the APEX family, bur rather M4K memory blocks, which better reflects their size. The basic logic element in Altera FPGAs is called a logic element (LE)[6] and consists of a flip-flop, a four-input one-output or three-input one-output LUT and a fast-carry logic, or AND/OR product term expanders, as shown in Fig. 1.13. Each LE can be used as a four-input LUT in the normal mode, or in the arithmetic mode, as a three-input LUT with an additional fast carry. Sixteen LEs are combined in a logic array block (LAB) in Cyclone II devices. Each row contains at least one embedded 18×18 bit multiplier and one M4K memory block. One 18×18 bit multiplier can also be used as two signed 9×9 bit multipliers, or one unsigned 17×17 bit multiplier. The M4K memory can be configured as $2^7 \times 32, 2^8 \times 16, \ldots, 4096 \times 1$ RAM or ROM. In addition one

---

[6] Sometimes also called logic cells (LCs) in a design report file.

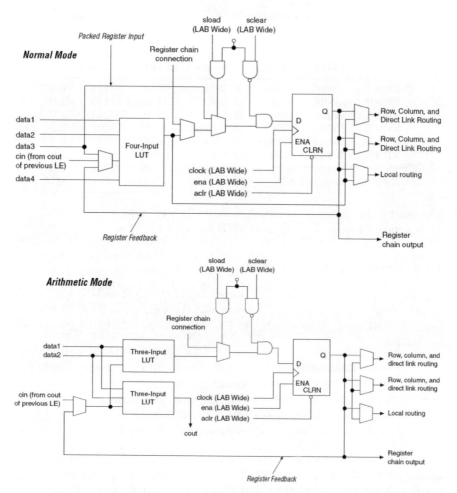

Fig. 1.13. Cyclone II logic cell (©2005 Altera).

parity bit per byte is available (e.g., 128×36 configuration), which can be used for data integrity. These M4Ks and LABs are connected through wide high-speed busses as shown in Fig. 1.14. Several PLLs are in use to produce multiple clock domains with low clock skew in the same device. At least 1 Mbits configuration files size is required to program the devices. Table 1.8 shows some members of the Altera Cyclone II family.

If we compare the two routing strategies from Altera and Xilinx we find that both approaches have value: the Xilinx approach with more local and less global routing resources is synergistic to DSP use because most digital signal processing algorithms process the data locally. The Altera approach, with wide busses, also has value, because typically not only are single bits

Table 1.8. Altera's Cyclone II device family.

Device	Total 4-input LUTs	RAM blocks M4K	PLLs/ clock networks	Emb. mul. 18×18	Max. I/O	Conf. file Mbits
EP2C5	4608	26	2/8	13	89	1.26
EP2C8	8256	36	2/8	18	85	1.98
EP2C20	18 752	52	4/16	26	315	3.89
EP2C35	33 216	105	4/16	35	475	6.85
EP2C50	50 528	129	4/16	86	450	9.96
EP2C70	68 416	250	4/16	150	622	14.31

processed in bit slice operations, but normally wide data vectors with 16 to 32 bits must be moved to the next DSP block.

1.4.2 The Altera EP2C35F672C6

The Altera EP2C35F672C6 device, a member of the Cyclone II family, which is part of the DSP prototype board provided through Altera's university program, is used throughout this book. The device nomenclature is interpreted as follows:

```
EP2C35F672C6
  |   |   |   |--> speed grade
  |   |   |-----> Package and pin number
  |   |---------> LEs in 1000
  |------------> Device family
```

Specific design examples will, wherever possible, target the Cyclone II device EP2C35F672C6 using Altera-supplied software. The enclosed Quartus II software is a fully integrated system with VHDL and Verilog editor, synthesizer, simulator, and bitstream generator. The only limitation in the web version is that not all pinouts of every devices are available. Because all examples are available in VHDL and Verilog, any other simulator may also be used. For instance, the device-independent ModelTech compiler has successfully been used to compile the examples using the synthesizable code for lpm functions on the CD-ROM provided by EDIF. The use of Xilinx ISE software is also discussed in appendix D.

Logic Resources

The EP2C35 is a member of the Altera Cyclone II family and has a logic density equivalent to about 35 000 logic elements (LEs). An additional 35 multipliers of size 18×18 bits (or twice this number if a size of 9×9 bit is used) are available. From Table 1.8 it can be seen that the EP2C35 device

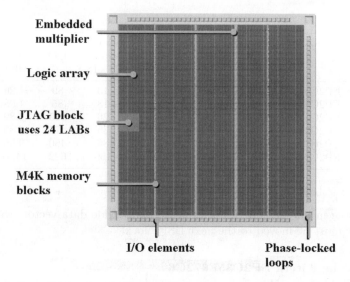

Embedded multiplier

Logic array

JTAG block uses 24 LABs

M4K memory blocks

I/O elements   **Phase-locked loops**

Fig. 1.14. Overall floorplan in Cyclone II devices.

has 33 216 basic logic elements (LEs). This is also the maximum number of implementable full adders. Each LE can be used as a four-input LUT, or in the arithmetic mode, as a three-input LUT with an additional fast carry as shown in Fig. 1.13. Sixteen LEs are always combined into a logic array block (LAB), see Fig. 1.15a. The number of LABs is therefore 33,216/16=2076. These 2076 LABs are arranged in 35 rows and 60 columns. In the left medium area of the device the JTAG interface is placed and uses the area of 24 LABs. This is why the total number of LABs in not just the product of rows × column, i.e., $35 \times 60 - 24 = 2100 - 24 = 2076$. The device also includes three columns of 4-kbit memory block (called M4K memory blocks, see Fig. 1.15b) that have the height of one LAB and the total number of M4Ks is therefore $3 \times 35 = 105$. The M4Ks can be configured as 128×36, 128×32, 256×18, 256×16, ...4096×1 RAM or ROM, where for each byte one parity bit is available. The EP2C35 also has one column of 18×18 bit fast array multipliers, that can also be configured as two 9×9 bit multipliers, see Fig. 1.16. Since there are 35 rows the number of multipliers is 35 for the 18×18 bit type or 70 of the 9×9 bit multiplier type. Figure 1.14 presents the overall device floorplan.

Routing Resources

All 16 LEs in a LAB share the same reset and clock signals. Each LE has a fan-out of 48 for fast local connection to neighboring LABs. The next level of routing are the R4 and C4 fast row and column local connections that allow wires to reach LABs at a distance of ±4 LABs, or 3 LABs and one embedded multiplier or M4K memory block. The longest connection available

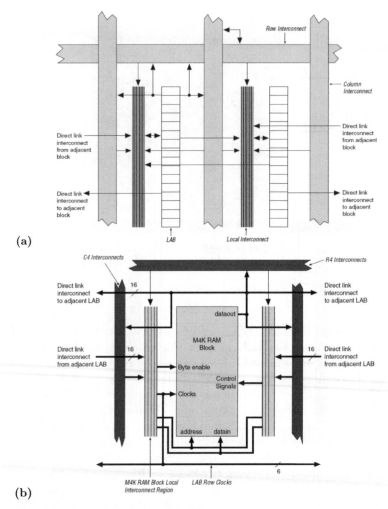

(a)

(b)

Fig. 1.15. Cyclone II resources: (a) logic array block structure (b) M4K memory block interface (© 2005 Altera [21]).

are R24 and C16 wires that allows 24 rows or 16 column LAB, respectively, to build connections that span basically the entire chip. It is also possible to use any combination of row and column connections, in case the source and destination LAB are not only in different rows but also in different columns. As we will see in the next section the delay in these connections varies widely and the synthesis tool tries always to place logic as close together as possible to minimize the interconnection delay. A 32 bit adder, for instance, would be best placed in two LABs in two rows one above the other, see Fig. 1.20, p. 33.

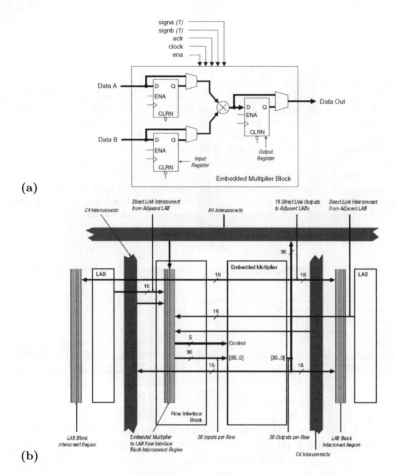

Fig. 1.16. Embedded multiplier (a) Architecture (b) LAB interface (© 2005 Altera [21]).

Timing Estimates

Altera's Quartus II software calculates various timing data, such as the Registered Performance, setup/hold time (t_{su}, t_h) and non-registered combination delay (t_{pd}). For a full description of all timing parameters, refer to the Timing Analysis Settings under EDA Tools Settings in the Assignments menu. To achieve optimal performance, it is necessary to understand how the software physically implements the design. It is useful, therefore, to produce a rough estimate of the solution and then determine how the design may be improved.

Example 1.2: Speed of an 32-bit Adder

Assume one is required to implement a 32-bit adder and estimate the design's maximum speed. The adder can be implemented in two LABs, each using the fast-carry chain. A rough first estimate can be done using the carry-in to carry-out delay, which is 71 ps for Cyclone II speed grade 6. An upper bound for the maximum performance would then be $32 \times 71 \times 10^{-12} = 2.272$ ns or 440 MHz. But in the actual implementation additional delays occur: first the interconnect delay from the previous register to the first full adder gives an additional delay of 0.511 ns. Next the first carry t_{cgen} must be generated, requiring about 0.414 ns. With the group of eight LEs each and in between the LAB we see from the floorplan that stronger drivers are used, requiring an additional 75-88 ps. Finally at the end of the carry chain the full sum bit needs to be computed (about 410 ps) and the setup time for the output register (84 ps) needs to be taken into account. The results are then stored in the LE register. The following table summarizes these timing data:

LE register clock-to-output delay	t_{co}	$=$	223 ps
Interconnect delay	t_{ic}	$=$	511 ps
Data-in to carry-out delay	t_{cgen}	$=$	414 ps
Carry-in to carry-out delay	$27 \times t_{\text{cico}}$	$=27 \times 71\,ps$	$=1917$ ps
8 bit LAB group carry-out delay	$2 \times t_{\text{cico8LAB}}$	$=2 \times 159\,ps$	$= 318$ ps
Same column carry out delay	$t_{\text{samecolumn}}$	$=$	146 ps
LE look-up table delay	t_{LUT}	$=$	410 ps
LE register setup time	t_{su}	$=$	84 ps
Total		$=$	4,022 ps

The estimated delay is 4.02 ns, or a rate of 248.63 MHz. The design is expected to use about 32 LEs for the adder and an additional 2×32 to store the input data in the registers (see also Exercise 1.7, p. 43).   1.2

If the two LABs used can not be placed in the same column next to each other then an additional delay would occur. If the signal comes directly from the I/O pins much longer delays have to be expected. For a 32 bit adder with data coming from the I/O pins the Quartus II `Timing Analyzer Tool` reports a propagation delay of 8.944 ns, much larger than the registered performance when the data comes directly from the register next to the design under test. Datasheets [21, Chap. 5] usually report the best performance that is achieved if I/O data of the design are placed in registers close to the design unit under test. Multiplier and block M4K (but not the adder) have additional I/O registers to enable maximum speed, see Fig. 1.16. The additional I/O registers are usually not counted in the LE resource estimates, since it is assumed that the previous processing unit uses a output register for all data. This may not always be the case and we have therefore put the additional register needed for the adder design in parentheses. Table 1.9 reports some typical data measured under these assumptions. If we compare this measured data with the delay given in the data book [21, Chap. 5] we notice that for some blocks Quartus II limits the upper frequency to a specific bound less than the delay in the data book. This is a conservative and more-secure estimate – the design may in fact run error free at a slightly higher speed.

Table 1.9. Some typical `Registered Performance` and resource data for the Cyclone II EP2C35F672C6.

Design	LE	M4K memory blocks	Multiplier blocks 9×9 bit	Registered Performance MHz
16 bit adder	16(+32)	−	−	369
32 bit adder	32(+64)	−	−	248
64 bit adder	64(+128)	−	−	151
ROM $2^9 \times 8$	−	1	−	260
RAM $2^9 \times 8$	−	1	−	230
9×9 bit multiplier	−	−	1	260
18×18 bit multiplier	−	−	2	260

Power Dissipation

The power consumption of an FPGA can be a critical design constraint, especially for mobile applications. Using 3.3 V or even lower-voltage process technology devices is recommended in this case. The Cyclone II family for instance is produced in a 1.2 V, 90-nm, low-k-dielectric process from the Taiwan ASIC foundry TSMC, but I/O interface voltages of 3.3 V, 2.5V, 1.8V and 1.5V are also supported. To estimate the power dissipation of the Altera device EP2C35, two main sources must be considered, namely:

1) Static power dissipation, $I_{\text{standby}} \approx 66\,\text{mA}$ for the `EP2C35F672C6`
2) Dynamic (logic, multiplier, RAM, PLL, clocks, I/O) power dissipation, I_{active}

The first parameter is not design dependent, and also the standby power in CMOS technology is generally small. The active current depends mainly on the clock frequency and the number of LEs or other resources in use. Altera provides an EXCEL work sheet, called `PowerPlay Early Power Estimator`, to get an idea about the power consumption (e.g., battery life) and possible cooling requirements in an early project phase.

For LE the dynamic power dissipation is estimated according to the proportional relation

$$P \approx I_{\text{dynamic}} V_{cc} = K \times f_{\max} \times N \times \tau_{\text{LE}} V_{cc}, \qquad (1.1)$$

where K is a constant, $f_{\max}$ is the operating frequency in MHz, N is the total number of logic cells used in the device, and τ_{LE} is the average percentage of logic cells toggling at each clock (typically 12.5%). Table 1.10 shows the results for power estimation when all resource of the EP2C35F672C6 are in use and a system clock of 100 MHz is applied. For less resource usage or lower system clock the data in (1.1) can be adjusted. If, for instance, a system clock is reduced from 100 MHz to 10 MHz then the power would be reduced to

Table 1.10. Power consumption estimation for the Cyclone II EP2C35F672C6.

Parameter	Units	Toggle rate (%)	Power mW
P_{static}			85
LEs 33216 @ 100 MHz	33216	12.5%	572
M4K block memory	105	50%	37
18 × 18 bit multiplier	35	12.5%	28
I/O cells (3.3V,24 mA)	475	12.5%	473
PLL	4		30
Clock network	33831		215
Total			1440

$85 + 1355/10 = 220.5 \, \text{mW}$, and the static power consumption would now be account for 38%.

Although the `PowerPlay` estimation is a useful tool in a project planing phase, it has its limitations in accuracy because the designer has to specify the toggle rate. There are cases when it become more complicated, such as for instance in frequency synthesis design examples, see Fig. 1.17. While the block RAM estimation with a 50% toggle may be accurate, the toggle rate of the LEs in the accumulator part is more difficult to determine, since the LSBs will toggle at a much higher frequency than the MSBs, since the accumulators produce a triangular output function. A more-accurate power estimation can be made using Altera's `PowerPlay Power Analyzer Tool` available from the `Processing` menu. The `Analyzer` allows us to read in toggle data computed from the simulation output. The simulator produces a "Signal Activity File" that can be selected as the input file for the `Analyzer`. Table 1.11 shows a comparison between the power estimation and the power analysis.

Table 1.11. Power consumption for the design shown in Fig. 1.17 for a Cyclone II EP2C35F672C6.

Parameter	Estimation 12.5% toggle rate power/mW	Analysis toggle rate measured power/mW
Static	79.91	80.02
Dynamic	5.09	6.68
I/O	50.60	83.47
Total	135.60	170.17

We notice a discrepancy of 20% between estimation and analysis. The analysis however requires a complete design including a testbench, while the estimation may be done at an early phase in the project.

The following case study should be used as a detailed scheme for the examples and self-study problems in subsequent chapters.

1.4.3 Case Study: Frequency Synthesizer

The design objective in the following case study is to implement a classical frequency synthesizer based on the Philips PM5190 model (circa 1979, see Fig. 1.17). The synthesizer consists of a 32-bit accumulator, with the eight most significant bits (MSBs) wired to a SINE-ROM lookup table (LUT) to produce the desired output waveform. A *graphical* solution, using Altera's Quartus II software, is shown in Fig. 1.18, and can be found on the CD-ROM as `book3e/vhdl/fun_graf.bdf`. The equivalent HDL text file `fun_text.vhd` and `fun_text.v` implement the design using component instantiation. In the following we walk through all steps that are usually performed when implementing a design using Quartus II:

1) Compilation of the design
2) Design results and floor plan
3) Simulation of the design
4) A performance evaluation

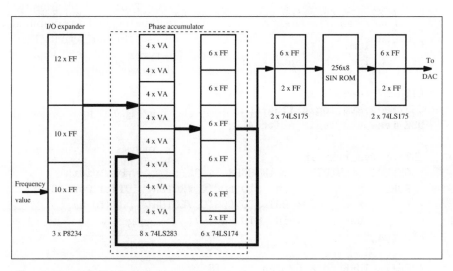

Fig. 1.17. PM5190 frequency synthesizer.

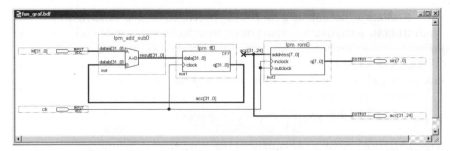

Fig. 1.18. Graphical design of the frequency synthesizer.

Design Compilation

To check and compile the file, start the Quartus II Software and select
File→Open Project or launch File→New Project Wizard if you do not
have a project file yet. In the project wizard specify the project directory you
would like to use, and the project name and top-level design as fun_text.
Then press Next and specify the HDL file you would like to add, in our case
fun_text.vhd. Press Next again and then select the device EP2C35F672C6
from the Cyclone II family and press Finish. If you use the project file from
the CD the file fun_text.qsf will already have the correct file and device
specification. Now select File→Open to load the HDL file. The VHDL design[7] reads as follows:

```
--  A 32-bit function generator using accumulator and ROM

LIBRARY lpm;
USE lpm.lpm_components.ALL;

LIBRARY ieee;
USE ieee.std_logic_1164.ALL;
USE ieee.std_logic_arith.ALL;

ENTITY fun_text IS
  GENERIC ( WIDTH   : INTEGER := 32);      -- Bit width
  PORT ( M        : IN  STD_LOGIC_VECTOR(WIDTH-1 DOWNTO 0);
         sin, acc : OUT STD_LOGIC_VECTOR(7 DOWNTO 0);
         clk      : IN  STD_LOGIC);
END fun_text;

ARCHITECTURE fpga OF fun_text IS
```

---

[7] The equivalent Verilog code fun_text.v for this example can be found in Appendix A on page 664. Synthesis results are shown in Appendix B on page 731.

```
    SIGNAL s, acc32 : STD_LOGIC_VECTOR(WIDTH-1 DOWNTO 0);
    SIGNAL msbs     : STD_LOGIC_VECTOR(7 DOWNTO 0);
                                        -- Auxiliary vectors
BEGIN

    add1: lpm_add_sub                -- Add M to acc32
       GENERIC MAP ( LPM_WIDTH => WIDTH,
                     LPM_REPRESENTATION => "SIGNED",
                     LPM_DIRECTION => "ADD",
                     LPM_PIPELINE => 0)
       PORT MAP ( dataa => M,
                  datab => acc32,
                  result => s );

    reg1: lpm_ff                     -- Save accu
       GENERIC MAP ( LPM_WIDTH => WIDTH)
       PORT MAP ( data => s,
                  q => acc32,
                  clock => clk);

    select1: PROCESS (acc32)
               VARIABLE i : INTEGER;
             BEGIN
             FOR i IN 7 DOWNTO 0 LOOP
                 msbs(i) <= acc32(31-7+i);
             END LOOP;
             END PROCESS select1;

    acc <= msbs;

    rom1: lpm_rom
       GENERIC MAP ( LPM_WIDTH => 8,
                     LPM_WIDTHAD => 8,
                     LPM_FILE => "sine.mif")
       PORT MAP ( address => msbs,
                  inclock => clk,
                  outclock => clk,
                  q => sin);

END fpga;
```

The object LIBRARY, found early in the code, contains predefined modules and definitions. The ENTITY block specifies the I/O ports of the device and generic variables. Using component instantiation, three blocks (see labels add1, reg1, rom1) are called as subroutines. The select1 PROCESS construct is used to select the eight MSBs to address the ROM. Now start the compiler tool (it has a little factory symbol) that can be found under the Processing menu. A window similar to the one shown in Fig. 1.19 pops up. You can now see the four steps envolved in the compilation, namely: Analysis & Synthesis, Fitter, Assembler and Timing Analyzer. Each of these steps has four small buttons. The first arrow to the right starts the processing, the second "equal-sign with a pen" is used for the assignments, the third is used to display the report files, and the last button starts an additional function, such as the hierarchy of the project, timing closure floorplan, programmer, or timing summary. To optimize the design for speed, click on the assignment symbol (equal-sign with a pen) in the Analysis & Synthesis step in the compiler window. A new window pops up that allows you to specify the Optimization Technique as Speed, Balanced or Area. Select the option Speed, and leave the other synthesis options unchanged. To set the target speed of the design click on the assignment button of the Timing Analyzer section and set the Default required fmax to 260 MHz. Note that you can also find all the assignments also under EDA Tools Settings under the Assignment menu. Next, start the Analysis & Synthesis by clicking on the right arrow in the compiler window or with <Ctrl+K> or by selecting Start Analysis & Synthesis in the Start item in the Processing menu. The compiler checks for basic syntax errors and produces a report file that lists resource estimation for the design. After the syntax check is successful, compilation can be started by pressing the large Start button in the lower left corner of the compiler tool window, or by pressing <Ctrl+L>. If all compiler steps were successfully completed, the design is fully implemented. Press the Report button in the compiler window to get a flow summary report that should show 32 LEs and 2048 memory bits use. Check the memory initialization file sine.mif, containing the sine table in offset binary form. This file was generated using the program sine3e.exe included on the CD-ROM under book3e/util. Figure 1.19 summarizes all the processing steps of the compilation, as shown in the Quartus II compiler window.

Floor Planing

The design results can be verified by clicking on the 4. button (i.e., Timing Closure Floorplan or opening the Tool→Chip Editor) to get a more-detailed view of the chip layout. The Chip Editor view is shown in Fig. 1.20. Use the Zoom in button (i.e., the ± magnifying glass) to produce the screen shown in Fig. 1.20. Zoom in to the area where the LAB and an M4K are highlighted in a different color. You should then see the two LABs used by the accumulation highlighted in blue and the M4K block highlighted in green. In addition several I/O cell are also highlighted in brown. Click on the Critical Path Setting[8] button and use the slider to select graph display.

---

[8] Note, as with all MS Window programs, just move your mouse over a button (no need to click on the button) and its name/function will be displayed.

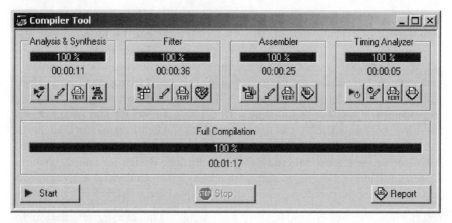

Fig. 1.19. Compilation steps in Quartus II.

You should then see in the Chip Editor a blue line that shows the worst-case path running from the first to the last bit of the accumulator. Now click on the Bird's Eye View button on the left menu buttons and an additional window will pop up. Now select the Coordinate option, i.e., the last entry in the Main Window options. You may also try out the connection display. First select for instance the M4K block and then press the button Generate Fan-In Connections or Generate Fan-Out Connections several times and more and more connections will be displayed.

Simulation

To simulate, open the Simulator Tool under the Processing menu. Under simulation mode you can select between Functional, Timing and Timing using fast Timing Model. The Functional simulation requires the func-

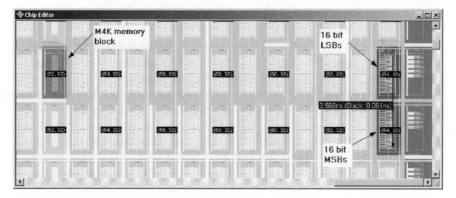

Fig. 1.20. Floorplan of the frequency synthesizer design.

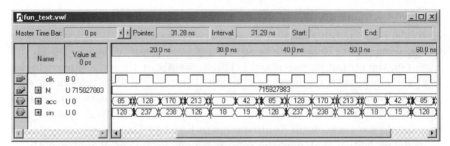

Fig. 1.21. VHDL simulation of frequency synthesizer design.

tional netlist to be generate first; this takes additional time, but is much faster than the full compilation of the design. The Timing simulation requires that you first make a full compile of the design as described in the previous section. You should then click the Open button and the waveform window will pop up. Notice also that 24 new buttons on the left have been added. Move your mouse (without clicking) over the buttons to become familiar with their name and function. If you have copied the waveform file fun_text.vwf from the CD into the project directory you will see a simulation with timing loaded. If you start with an empty waveform Quartus II helps you to assemble a waveform file. Double click the space under the Name section to insert a node or bus. Click the Node Finder button and select in the Node Finder window as Filter the entry Pins: all. Now press List and all I/O pins are listed. Note that the I/O pins are available with both functional and timing simulation, but internal signals may not all be available in the netlist generated. Now select all bus signals, i.e., acc, clk, M, and sin, and press the single arrow to the right button. Now the four signals should be listed in the Selected Nodes window. If you selected only one bus signal repeat the selection by pressing the arrow with the other signals too. After all four signals are listed in the right window, press OK. Now the signal should be listed in the waveform window. Sort the signal according to Fig. 1.21, i.e., list first input control signals like clk followed by input data signal(s) M. In the last place in the list put the output data (acc and sin). To move a signal make sure that the arrow is selected from the left menu buttons. Then select the signal you would like to move and hold down the mouse button while moving the signal. When all signals are arranged correctly we can start defining the input signals. Select first the clk signal and then press the clock symbol from the left menu buttons. As the period select $1/260\,\mathrm{MHz} = 3.846\,\mathrm{ns}$. Next set $M = 715\,827\,883$ ($M = 2^{32}/6$), so that the period of the synthesizer is six clock cycles long. After we have specified the input signals we are ready to simulate. The default simulation time is always $1\mu s$. You can change the default value by selecting End Time under the Edit menu. You may set it to about 60 ns to display the first 15 clock cycles. You may want to start with a functional simulation first and then proceed with the timing simulation. Select `Functional Simulation` and do not forget to generate the functional netlist first. You can then press the `Start` button in the `Simulator Tool` window. You should see the waveforms without delays, i.e., the output signals should change exactly at the same time samples that the clock has a rising edge. You can then proceed with the timing simulation. Remember to conduct a full compile first if you have not done so already. Then press the `Start` button in the `Simulator Tool` window and you should see a waveform result similar to Fig. 1.21 that shows a simulation with delay. Notice that now the output signals no longer change exactly at the rising edge and that the signals also need some time to settle down, i.e., become stable. Make sure that the period is of length 6, both in the `accu` as well in the `sin` signal. Notice that the ROM has been coded in binary offset (i.e., zero = 128). This is a typical coding used in D/A converters and is used in the D/A converter of the Cyclone II DSP prototype board. When complete, change the frequency so that a period of eight cycles occurs, i.e., ($M = 2^{32}/8$), and repeat the simulation.

Performance Analysis

To initiate a performance analysis, select the `Timing Analyzer Tool` under the `Processing` menu. Usually the `Registered Performance` is the most important measurement. For a combination circuit (only) the propagation delay t_{pd} should be monitored. You can change the goals in the `EDA Tools Setting` under the `Assignment` menu. Note that it not sufficient to set the synthesis options to `Speed`; if you do not specify a requested `Default Required fmax` the synthesis results will most likely not reach the best performance that can be achieved with the device.

In order to display timing data a full compile of the design has to be done first. The `Registered Performance` is then displayed using a speed meter, while the other timing data like t_{pd} are shown in table form. The result for `Registered Performance` should be similar to that shown in Fig. 1.22. You can also list the worst-case path. Select 1 in `Number of path to list` and press the `List Paths` button. The path is shown as information in the message window. Pressing the plus sign, expand the information to see the full path detail. The path information of each node includes interconnect delay, cell delay, the LAB cell location with x and y coordinates, and the local signal name. You can verify this data using the `Chip Editor` described in the previous "Floor Planning" section.

This concludes the case study of the frequency synthesizer.

1.4.4 Design with Intellectual Property Cores

Although FPGAs are known for their capability to support rapid prototyping, this only applies if the HDL design is already available and sufficiently

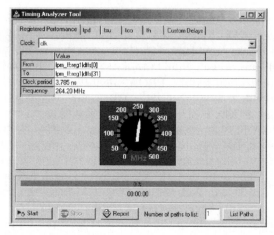

Fig. 1.22. Register performance of frequency synthesizer design.

tested. A complex block like a PCI bus interface, a pipelined FFT, an FIR filter, or a μP may take weeks or even months in development time. One option that allows us to essentially shorten the development time is available with the use of a so-called intellectual property (IP) core. These are predeveloped (larger) blocks, where typical standard blocks like numeric controlled oscillators (NCO), FIR filters, FFTs, or microprocessors are available from FPGA vendors directly, while more-specialized blocks (e.g., AES, DES, or JPEG codec, a floating-point library, or I2C bus or ethernet interfaces) are available from third-party vendors. On a much smaller scale we have already used IP blocks. The library of parameterize modules (LPM) blocks we used in the `example` and `fun_text` designs are parameterized cores, where we could select, for instance, bitwidth and pipelining that allow fast design development. We can use the LPM blocks and configure a pipelined multiplier or divider or we can specify to use memory blocks as CAM, RAM, ROM or FIFO. While this LPM blocks are free in the Quartus II package the larger more-sophisticated blocks may have a high price tag. But as long as the block meets your design requirement it is most often more cost effective to use one of these predefined IP blocks.

Let us now have a quick look at different types of IP blocks and discuss the advantages and disadvantages of each type [22, 23, 24]. Typically IP cores are divided into three main forms, as described below.

Soft Core

A *soft core* is a behavioral description of a component that needs to be synthesized with FPGA vendor tools. The block is typically provided in a hardware description language (HDL) like VHDL or Verilog, which allows easy modification by the user, or even new features to be added or deleted before synthesis for a specific vendor or device. On the downside the IP block may also require more work to meet the desired size/speed/power requirements. Very few of the blocks provided by FPGA vendors are available in this form, like the Nios microprocessor from Altera or the PICO blaze microprocessor by Xilinx. IP protection for the FPGA vendor is difficult to achieve since the block is provided as synthesizable HDL and can quite easily be used with a competing FPGA tool/device set or a cell-based ASIC. The price of third-party FPGA blocks provided in HDL is usually much higher than the moderate pricing of the parameterized core discussed next.

Parameterized Core

A *parameterized* or firm core is a structural description of a component. The parameters of the design can be changed before synthesis, but the HDL is usually not available. The majority of cores provided by Altera and Xilinx come in this type of core. They allow certain flexibility, but prohibit the use of the core with other FPGA vendors or ASIC foundries and therefore offers better IP protection for the FPGA vendors than soft cores. Examples of parameterized cores available from Altera and Xilinx include an NCO, FIR filter compiler, FFT (parallel and serial) and embedded processors, e.g., Nios II from Altera. Another advantage of parameterized cores is that usually a resource (LE, multiplier, block RAMs) is available that is most often correct within a few percent, which allows a fast design space exploration in terms of size/speed/power requirements even before synthesis. Testbenches in HDL (for ModelSim simulator) that allow cycle-accurate modeling as well as C or MATLAB scripts that allow behavior-accurate modeling are also standard for parameterized cores. Code generation usually only takes a few seconds. Later in this section we will study an NCO parameterized core and continue this in later chapters (Chap. 3 on FIR filter and Chap. 6 on FFTs).

Hard Core

A *hard core* (fixed netlist core) is a physical description, provided in any of a variety of physical layout formats like EDIF. The cores are usually optimized for a specific device (family), when hard realtime constrains are required, like for instance a PCI bus interface. The parameters of the design are fixed, like a 16-bit 256-point FFT, but a behavior HDL description allows simulation and integration in a larger project. Most third-party IP cores from FPGA vendors and several free FFT cores from Xilinx use this core type. Since the layout is fixed, the timing and resource data provided are precise and do not depend on synthesis results. But the downside is that a parameter change is not possible, so if the FFT should have 12- or 24-bit input data the 16-bit 256-point FFT block can not be used.

IP Core Comparison and Challenges

If we now compare the different IP block types we have to choose between design flexibility (soft core) and fast results and reliability of data (hard core). Soft cores are flexible, e.g., change of system parameters or device/process technology is easy, but may have longer debug time. Hard cores are verified in silicon. Hard cores reduce development, test, and debug time but no VHDL code is available to look at. A parameterized core is most often the best compromise between flexibility and reliability of the generated core.

There are however two major challenges with current IP block technology, which are pricing of a block and, closely related, IP protection. Because the cores are reusable vendor pricing has to rely on the number of units of IP blocks the customer will use. This is a problem known for many years in patent rights and most often requires long licence agreements and high penalties in case of customer misuse. FPGA-vendor-provided parameterized blocks (as well as the design tool) have very moderate pricing since the vendor will profit if a customer uses the IP block in many devices and then usually has to buy the devices from this single source. This is different with third-party IP block providers that do not have this second stream of income. Here the licence agreement, especially for a soft core, has be drafted very carefully.

For the protection of parameterized cores FPGA vendor use FlexLM-based keys to enable/disable single IP core generation. Evaluation of the parameterized cores is possible down to hardware verification by using time-limited programming files or requiring a permanent connection between the host PC and board via a JTAG cable, allowing you to program devices and verify your design in hardware before purchasing a licence. For instance, Altera's OpenCore evaluation feature allows you to simulate the behavior of an IP core function within the targeted system, verify the functionality of the design, and evaluate its size and speed quickly and easily. When you are completely satisfied with the IP core function and you would like to take the design into production, you can purchase a licence that allows you to generate non-time-limited programming files. The Quartus software automatically downloads the latest IP cores from Altera's website. Many third-party IP providers also support the OpenCore evaluation flow but you have to contact the IP provider directly in order to enable the OpenCore feature.

The protection of soft cores is more difficult. Modification of the HDL to make them very hard to read, or embedding watermarks in the high-level design by minimizing the extra hardware have been suggested [24]. The watermark should be robust, i.e., a single bit change in the watermark should not be possible without corrupting the authentication of the owner.

IP Core-Based NCO Design

Finally we evaluate the design process of an IP block in an example using the case study from the last section, but this time our design will use Altera's

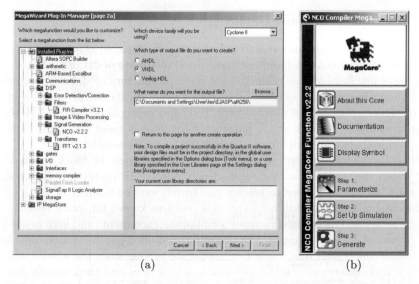

(a)   (b)

Fig. 1.23. IP design of NCO (**a**) Library element selection. (**b**) IP toolbench.

NCO core generator. The NCO compiler generates numerically controlled oscillators (NCOs) optimized for Altera devices. You can use the IP toolbench interface to implement a variety of NCO architectures, including ROM-based, CORDIC-based, and multiplier-based options. The MegaWizard also includes time- and frequency-domain graphs that dynamically display the functionality of the NCO based on the parameter settings. For a simple evaluation we use the graphic design entry. Open a new project and BDF file, then double click in the empty space in the BDF window and press the button MegaWizard Plug-In Manager. In the first step select the NCO block in the MegaWizard Plug-In Manager window, see Fig. 1.23a. The NCO block can by found in the Signal Generation group under the DSP cores. We then select the desired output format (AHDL, VHDL, or Verilog) and specify our working directory. Then the IP toolbench pops up (see Fig. 1.23b) and we have access to documentation and can start with step 1, i.e., the parametrization of the block. Since we want to reproduce the function generator from the last section, we select a 32-bit accumulator, 8 bit output precision, and the use of a large block of RAM in the parameter window, see Fig. 1.24. As expected for an 8 bit output we get about 50 dB sidelope suppression, as can be seen in the Frequency Domain Response plot in the lower part of the NCO window. Phase dithering will make the noise more equally distributed, but will require twice as many LEs. In the Implementation window we select Single Output since we only require one sine but no cosine output as is typical for I/Q receivers, see Chap. 7. The Resource Estimation provides as data 72 LEs, 2048 memory bits and one M4K block. After we are satisfied with our parameter selection we then proceed to step 2 to specify if we want

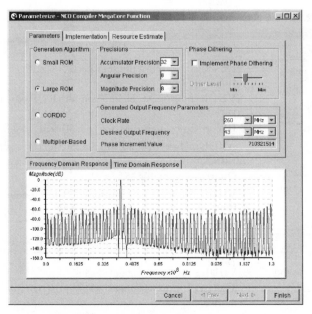

Fig. 1.24. IP parametrization of NCO core according to the data from the case study in the previous section.

to generate behavior HDL code, which speeds up simulation time. Since our block is small we deselect this option and use the full HDL generated code directly. We can now continue with step 3, i.e., `Generate` on the Toolbench. The listing in Table 1.12 gives an overview of the generated files.

We see that not only are the VHDL and Verilog files generated along with their component file, but MATLAB (bit accurate) and ModelTech (cycle accurate) testbenches are also provided to enable an easy verification path. We decide to instantiate our block in the graphical design and connect the input and outputs, see Fig. 1.25a. We notice that the block (outside that we have asked for) has some additional useful control signal, i.e., `reset`, `clken`, and `data_ready`, whose function is self-explanatory. All control signals are high active. We start with a `Functional` simulation first and then proceed (after a full compile) with the `Timing` simulation. With the full compile data available we can now compare the actual resource requirement with the estimate. The memory requirement and block RAM predictions were correct, but for the LEs with 86 LEs (actual) to 72 LEs (estimated) we observe a 17% error margin. Using the same value $M = 715\,827\,883$ as in our function generator (see Fig. 1.21, p. 34) we get a period of 6 in the output signal, as shown in Fig. 1.25b. We may notice a small problem with the IP block, since the output is a signed value, but our D/A converter expects unsigned (or more precisely binary offset) numbers. In a soft core we would be able to change the HDL code of the design, but in the parameterized core we do not have this option.

Table 1.12. IP file generation for the NCO core.

File	Description
nco.bsf	Quartus II symbol file for the IP core function variation
nco.cmp	VHDL component declaration for the IP core function variation
nco.html	IP core function report file that lists all generated files
nco.inc	AHDL include declaration file for the IP core function variation
nco.vec	Quartus vector file
nco.vhd	VHDL top-level description of the custom IP core function
nco.vho	VHDL IP functional simulation model
nco_bb.v	Verilog HDL black-box file for the IP core function variation
nco_inst.vhd	VHDL sample instantiation file
nco_model.m	MATLAB M-file describing a MATLAB bit-accurate model
nco_sin.hex	Intel Hex-format ROM initialization file
nco_st.v	Generated NCO synthesizable netlist
nco_tb.m	MATLAB Testbench
nco_tb.v	Verilog testbench
nco_tb.vhd	VHDL testbench
nco_vho_msim.tcl	ModelSim TCL Script to run the VHDL IP functional simulation model in the ModelSim simulation software
nco_wave.do	ModelSim waveform file

But we can solved this problem by attaching an adder with constant 128 to the output that make it an offset binary representation. The offset binary is not a parameter we could select in the block, and we encounter extra design effort. This is a typical experience with the parameterized cores – the core provide a 90% or more reduction in design time, but sometimes small extra design effort is necessary to meet the exact project requirements.

Exercises

Note: If you have no prior experience with the Quartus II software, refer to the case study found in Sect. 1.4.3, p. 29. If not otherwise noted use the

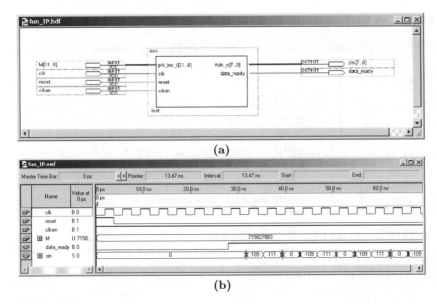

(a)

(b)

Fig. 1.25. Testbench for NCO IP design **(a)** Instantiation of IP block in graphical design. **(b)** Verification via simulation.

EP2C35F672C6 from the Cyclone II family for the Quartus II synthesis evaluations.

1.1: Use only two input NAND gates to implement a full adder:
(a) $s = a \oplus b \oplus c_{in}$
(Note: $\oplus$=XOR)
(b) $c_{out} = a \times b + c_{in} \times (a + b)$
(Note: +=OR; ×=AND)
(c) Show that the two-input NAND is *universal* by implementing NOT, AND, and OR with NAND gates.
(d) Repeat (a)-(c) for the two input NOR gate.
(e) Repeat (a)-(c) for the two input multiplexer $f = xs' + ys$.

1.2: (a) Compile the HDL file `example` using the Quartus II compiler (see p. 15) in the `Functional` mode. Start first the `Simulation Tool` under the `Processing` menu. Then select `Functional` as `Simulation mode` and press the button `Generate Functional Simulation Netlist`.
(b) Simulate the design using the file `example.vwf`.
(c) Compile the HDL file `example` using the Quartus II compiler with `Timing`. Perform a full compilation using the `Compiler Tool` under the `Processing` menu. Then select `Timing` as `Simulation mode` option in the the `Simulation Tool`.
(d) Simulate the design using the file `example.vwf`.
(e) Turn on the option `Check Outputs` in the simulator window and compare the functional and timing netlists.

1.3: (a) Generate a waveform file for `clk,a,b,op1` that approximates that shown in Fig. 1.26.

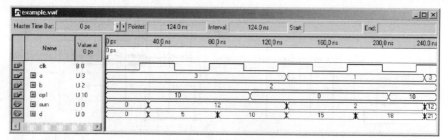

Fig. 1.26. Waveform file for Example 1.1 on p. 15.

(b) Conduct a simulation using the HDL code `example`.

(c) Explain the algebraic relation between `a,b,op1` and `sum,d`.

1.4: (a) Compile the HDL file `fun_text` with the synthesis optimization technique set to `Speed`, `Balanced` or `Area` that can be found in the `Analysis & Synthesis Settings` under `EDA Tool Settings` in the `Assignments` menu.

(b) Evaluate `Registered Performance` and the LE's utilization of the designs from (a). Explain the results.

1.5: (a) Compile the HDL file `fun_text` with the synthesis `Optimization Technique` set to `Speed` that can be found in the `Analysis & Synthesis Settings` under `EDA Tool Settings` in the `Assignments` menu.

For the period of the clock signal

(I) 20 ns,

(II) 10 ns,

(III) 5 ns,

(IV) 3 ns,

use the waveform file `fun_text.vwf` and enable

(b) `Setup and hold time violation detection`,

(c) `Glitch detection`, and

(d) `Check outputs`.

Select one option after the other and not all three at the same time. For `Check outputs` first make a `Functional Simulation`, then select `Check outputs`, and perform then `Timing` simulation. Under `Waveform Comparison Setting` select `sin` for comparison and deselect all other signals. Set the `Default comparison timing tolerance` to `<<default>>`, i.e., halve the clock period or the falling edge of clock. Click on the `Report` button in the `Simulation Tool` window if there are violation.

1.6: (a) Open the file `fun_text.vwf` and start the simulation.

(b) Select `File→Open` to open the file `sine.mif` and the file will be displayed in the `Memory editor`. Now select `File→Save As` and select `Save as type: (*.hex)` to store the file in Intel HEX format as `sine.hex`.

(c) Change the `fun_text` HDL file so that it uses the Intel HEX file `sine.hex` for the ROM table, and verify the correct results through a simulation.

1.7: (a) Design a 32-bit adder using the `LPM_ADD_SUB` macro with the Quartus II software.

(b) Measure the `Registered Performance` and compare the result with the data from Example 1.2 (p. 26).

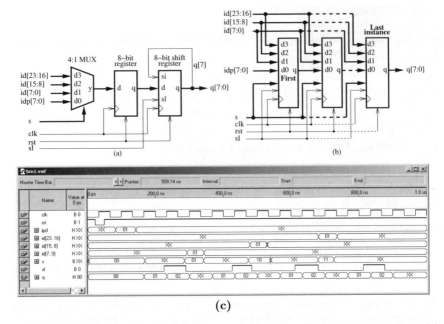

Fig. 1.27. PREP benchmark 1. **(a)** Single design. **(b)** Multiple instantiation. **(c)** Testbench to check the function.

1.8: **(a)** Design the PREP benchmark 1, as shown in Fig. 1.27a with the Quartus II software. PREP benchmark no. 1 is a data path circuit with a 4-to-1 8-bit multiplexer, an 8-bit register, followed by a shift register that is controlled by a shift/load input **sl**. For **sl=1** the contents of the register is cyclic rotated by one bit, i.e., $q(k) = q(k-1), 1 \le k \le 7$ and $q(0) <= q(7)$. The reset **rst** for all flip-flops is an asynchronous reset and the 8-bit registers are positive-edge triggered via **clk**, see the simulation in Fig. 1.27c for the function test.

(b) Determine the **Registered Performance** and the used resources (LEs, multipliers, and M2Ks/M4Ks) for a single copy. Compile the HDL file with the synthesis **Optimization Technique** set to **Speed**, **Balanced** or **Area**; this can be found in the **Analysis & Synthesis Settings** section under **EDA Tool Settings** in the **Assignments** menu. Which synthesis options are optimal in terms of size and **Registered Performance**?

Select one of the following devices:

(b1) EP2C35F672C6 from the Cyclone II family
(b2) EPF10K70RC240-4 from the Flex 10K family
(b3) EPM7128LC84-7 from the MAX7000S family

(c) Design the multiple instantiation for benchmark 5 as shown in Fig. 1.27b.

(d) Determine the **Registered Performance** and the used resources (LEs, multipliers, and M2Ks/M4Ks) for the design with the maximum number of instantiations of PREP benchmark 1. Use the optimal synthesis option you found in (b) for the following devices:

(d1) EP2C35F672C6 from the Cyclone II family
(d2) EPF10K70RC240-4 from the Flex 10K family
(d3) EPM7128LC84-7 from the MAX7000S family

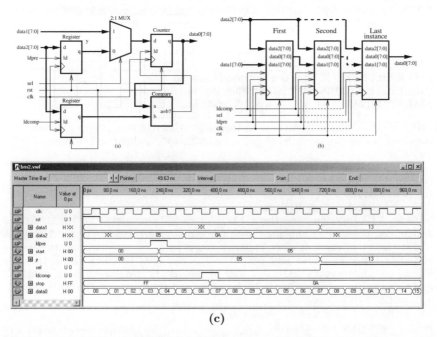

Fig. 1.28. PREP benchmark 2. **(a)** Single design. **(b)** Multiple instantiation. **(c)** Testbench to check the function.

1.9: (a) Design the PREP benchmark 2, as shown in Fig. 1.28a with the Quartus II software. PREP benchmark no. 2 is a counter circuit where 2 registers are loaded with start and stop values of the counter. The design has two 8-bit register and a counter with asynchronous reset **rst** and synchronous load enable signal (**ld, ldpre** and **ldcomp**) and positive-edge triggered flip-flops via **clk**. The counter can be loaded through a 2:1 multiplexer (controlled by the **sel** input) directly from the **data1** input or from the register that holds **data2** values. The load signal of the counter is enabled by the equal condition that compares the counter value **data** with the stored values in the **ldcomp** register. Try to match the simulation in Fig. 1.28c for the function test. Note there is a mismatch between the original PREP definition and the actual implementation: We can not satisfy, that the counter start counting after reset, because all register are set to zero and **ld** will be true all the time, forcing counter to zero. Also in the simulation testbench signal value have been reduced that simulation fits in a 1 μs time frame.

(b) Determine the **Registered Performance** and the used resources (LEs, multipliers, and M2Ks/M4Ks) for a single copy. Compile the HDL file with the synthesis **Optimization Technique** set to **Speed**, **Balanced** or **Area**; this can be found in the **Analysis & Synthesis Settings** section under **EDA Tool Settings** in the **Assignments** menu. Which synthesis options are optimal in terms of size and **Registered Performance**?

Select one of the following devices:

(b1) EP2C35F672C6 from the Cyclone II family

(b2) EPF10K70RC240-4 from the Flex 10K family

(b3) EPM7128LC84-7 from the MAX7000S family

(c) Design the multiple instantiation for benchmark 2 as shown in Fig. 1.28b.

(d) Determine the `Registered Performance` and the used resources (LEs, multipliers, and M2Ks/M4Ks) for the design with the maximum number of instantiations of PREP benchmark 2. Use the optimal synthesis option you found in (b) for the following devices:

(d1) EP2C35F672C6 from the Cyclone II family

(d2) EPF10K70RC240-4 from the Flex 10K family

(d3) EPM7128LC84-7 from the MAX7000S family

1.10: Use the Quartus II software and write two different codes using the structural (use only one or two input basic gates, i.e., `NOT`, `AND`, and `OR`) and behavioral HDL styles for:

(a) A 2:1 multiplexer

(b) An XNOR gate

(c) A half-adder

(d) A 2:4 decoder (demultiplexer)

Note for VHDL designs: use the `a_74xx` Altera SSI component for the structural design files. Because a component identifier can not start with a number Altera has added the `a_` in front of each 74 series component. In order to find the names and data types for input and output ports you need to check the library file `libraries\vhdl\altera\MAXPLUS2.VHD` in the Altera installation path. You will find that the library uses `STD_LOGIC` data type and the names for the ports are `a_1`, `a_2`, and `a_3` (if needed).

(e) Verify the function of the design(s) via

(e1) A `Functional` simulation.

(e2) The `RTL viewer` that can be found under the `Netlist Viewers` in the `Tools` menu.

1.11: Use the Quartus II software language templates and compile the HDL designs for:

(a) A tri-state buffer

(b) A flip-flop with all control signals

(c) A counter

(d) A state machine with asynchronous reset

Open a new HDL text file and then select `Insert Template` from the `Edit` menu.

(e) Verify the function of the design(s) via

(e1) A `Functional` simulation

(e2) The `RTL viewer` that can be found under the `Netlist Viewers` in the `Tools` menu

1.12: Use the `search` option in Quartus II software help to study HDL designs for:

(a) The 14 counters, see `search`→`implementing sequential logic`

(b) A manually specifying state assignments, `Search`→`enumsmch`

(c) A latch, `Search`→`latchinf`

(d) A one's counter, `Search`→`proc`→`Using Process Statements`

(e) A implementing CAM, RAM & ROM, `Search`→`ram256x8`

(f) A implementing a user-defined component, `Search`→`reg24`

(g) Implementing registers with clr, load, and preset, `Search`→`reginf`

(h) A state machine, `Search`→`state_machine`→`Implementing...`

Open a new project and HDL text file. Then `Copy/Paste` the HDL code, save and compile the code. Note that in VHDL you need to add the STD_LOGIC_1164 IEEE library so that the code runs error free.

(i) Verify the function of the design via

(i1) A `Functional` simulation

(i2) The RTL viewer that can be found under the Netlist Viewers in the Tools menu

1.13: Determine if the following VHDL identifiers are valid (true) or invalid (false).
(a) VHSIC   (b) h333   (c) A_B_C
(d) XyZ   (e) N#3   (f) My-name
(g) BEGIN   (h) A__B   (i) ENTITI

1.14: Determine if the following VHDL string literals are valid (true) or invalid (false).
(a) B"11_00"   (b) O"5678"   (c) O"0_1_2"
(d) X"5678"   (e) 16#FfF#   (f) 10#007#
(g) 5#12345#   (h) 2#0001_1111_# (i) 2#00_00#

1.15: Determine the number of bits necessary to represent the following integer numbers.
(a) INTEGER RANGE 10 TO 20;
(b) INTEGER RANGE -2**6 TO 2**4-1;
(c) INTEGER RANGE -10 TO -5;
(d) INTEGER RANGE -2 TO 15;
Note that ** stand for the power-of symbol.

1.16: Determine the error lines (Y/N) in the VHDL code below and explain what is wrong, or give correct code.

VHDL code	Error (Y/N)	Give reason
LIBRARY ieee; /* Using predefined packages */		
ENTITY error is		
PORTS (x: in BIT; c: in BIT;		
Z1: out INTEGER; z2 : out BIT);		
END error		
ARCHITECTURE error OF has IS		
SIGNAL s ; w : BIT;		
BEGIN		
w := c;		
Z1 <= x;		
P1: PROCESS (x)		
BEGIN		
IF c='1' THEN		
x <= z2;		
END PROCESS P0;		
END OF has;		

1.17: Determine the error lines (Y/N) in the VHDL code below, and explain what is wrong, or give correct code.

48   1. Introduction

VHDL code	Error (Y/N)	Give reason
LIBRARY ieee; /* Using predefined packages */		
USE altera.std_logic_1164.ALL;		
ENTITY srhiftreg IS		
GENERIC (WIDTH : POSITIVE = 4);		
PORT(clk, din : IN STD_LOGIC;		
dout : OUT STD_LOGIC);		
END;		
ARCHITECTURE a OF shiftreg IS		
COMPONENT d_ff		
PORT (clock, d : IN std_logic;		
q : OUT std_logic);		
END d_ff;		
SIGNAL b : logic_vector(0 TO witdh-1);		
BEGIN		
d1: d_ff PORT MAP (clk, b(0), din);		
g1: FOR j IN 1 TO width-1 GENERATE		
d2: d-ff		
PORT MAP(clk => clock,		
din => b(j-1),		
q => b(j));		
END GENERATE d2;		
dout <= b(width);		
END a;		

1.18: Determine for the following process statements
(a) the synthesized circuit and label I/O ports
(b) the cost of the design assuming a cost 1 per adder/subtractor
(c) the critical (i.e., worst-case) path of the circuit for each process. Assume a delay
of 1 for an adder or subtractor.

```
-- QUIZ VHDL2graph for DSP with FPGAs
LIBRARY ieee; USE ieee.std_logic_1164.ALL;
USE ieee.std_logic_arith.ALL;
USE ieee.std_logic_unsigned.ALL;

ENTITY qv2g IS
 PORT(a, b, c, d : IN  std_logic_vector(3 DOWNTO 0);
      u, v, w, x, y, z : OUT std_logic_vector(3 DOWNTO 0));
END;
ARCHITECTURE a OF qv2g IS BEGIN

  P0: PROCESS(a, b, c, d)
  BEGIN
    u <= a + b - c + d;
  END PROCESS;

  P1: PROCESS(a, b, c, d)
  BEGIN
    v <= (a + b) - (c - d);
  END PROCESS;

  P2: PROCESS(a, b, c)
```

```
BEGIN
  w <= a + b + c;
  x <= a - b - c;
END PROCESS;

P3: PROCESS(a, b, c)
VARIABLE t1 :  std_logic_vector(3 DOWNTO 0);
BEGIN
  t1 := b + c;
  y <= a + t1;
  z <= a - t1;
END PROCESS;
END;
```

(a)

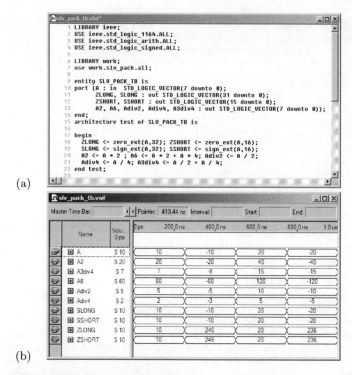

(b)

Fig. 1.29. STD_LOGIC_VECTOR package testbench. (a) HDL code. (b) Functional simulation result.

1.19: (a) Develop a functions for zero- and sign extension called zero_ext(ARG,SIZE) and sign_ext(ARG,SIZE) for the STD_LOGIC_VECTOR data type.
(b) Develop "∗" and "/" function overloading to implement multiply and divide operation for the STD_LOGIC_VECTOR data type.
(c) Use the testbench shown in Fig. 1.29 to verify the correct functionality.

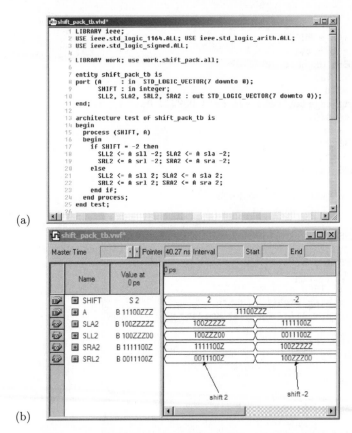

Fig. 1.30. STD_LOGIC_VECTOR shift library testbench. **(a)** HDL code. **(b)** Functional simulation result.

1.20: **(a)** Design a function library for the STD_LOGIC_VECTOR data type that implement the following operation (defined in VHDL only for the bit_vector data type):
(a) srl　　**(b)** sra　　**(c)** sll　　**(d)** sla
(e) Use the testbench shown in Fig. 1.30 to verify the correct functionality. Note the high impedance values Z that are part of the STD_LOGIC_VECTOR data type but are not included in the bit_vector data type. A left/right shift by a negative value should be replaced by the appropriate right/left shift of the positive amount inside your function.

1.21: Determine for the following PROCESS statements the synthesized circuit type (combinational, latch, D-flip-flop, or T-flip-flop) and the function of a, b, and c, i.e., clock, a-synchronous set (AS) or reset (AR) or synchronous set (SS) or reset (SR). Use the table below to specify your classification.

```
LIBRARY ieee; USE ieee.std_logic_1164.ALL;

ENTITY quiz IS
  PORT(a, b, c : IN  std_logic;
```

```
      d          : IN  std_logic_vector(0 TO 5);
      q          : BUFFER std_logic_vector(0 TO 5));
END quiz;
ARCHITECTURE a OF quiz IS BEGIN
  P0: PROCESS (a)
  BEGIN
    IF rising_edge(a) THEN
      q(0) <= d(0);
    END IF;
  END PROCESS P0;

  P1: PROCESS (a, d)
  BEGIN
    IF a= '1' THEN   q(1) <= d(1);
             ELSE   q(1) <= '1';
    END IF;
  END PROCESS P1;

  P2: PROCESS (a, b, c, d)
  BEGIN
    IF a = '1' THEN q(2) <= '0';
    ELSE IF rising_edge(b) THEN
           IF c = '1' THEN q(2) <= '1';
                      ELSE q(2) <= d(1);
           END IF;
         END IF;
    END IF;
  END PROCESS P2;

  P3: PROCESS (a, b, d)
  BEGIN
    IF a = '1' THEN q(3) <= '1';
    ELSE IF rising_edge(b) THEN
           IF c = '1' THEN q(3) <= '0';
                      ELSE q(3) <= not q(3);
           END IF;
         END IF;
    END IF;
  END PROCESS P3;

  P4: PROCESS (a, d)
  BEGIN
    IF a = '1' THEN q(4) <= d(4);
    END IF;
  END PROCESS P4;

  P5: PROCESS (a, b, d)
  BEGIN
    IF rising_edge(a) THEN
      IF b = '1' THEN q(5) <= '0';
                 ELSE q(5) <= d(5);
      END IF;
    END IF;
```

END PROCESS P5;

Process	Circuit type	CLK	AS	AR	SS	SR
P0						
P1						
P2						
P3						
P4						
P5						
P6						

1.22: Given the following MATLAB instructions,

```
a=-1:2:5
b=[ones(1,2),zeros(1,2)]
c=a*a'
d=a.*a
e=a'*a
f=conv(a,b)
g=fft(b)
h=ifft(fft(a).*fft(b))
```

determine a-h.

2. Computer Arithmetic

2.1 Introduction

In computer arithmetic two fundamental design principles are of great importance: number representation and the implementation of algebraic operations [25, 26, 27, 28, 29]. We will first discuss possible number representations, (e.g., fixed-point or floating-point), then basic operations like adder and multiplier, and finally efficient implementation of more difficult operations such as square roots, and the computation of trigonometric functions using the CORDIC algorithm or MAC calls.

FPGAs allow a wide variety of computer arithmetic implementations for the desired digital signal processing algorithms, because of the physical bit-level programming architecture. This contrasts with the programmable digital signal processors (PDSPs), with the fixed multiply accumulator core. Careful choice of the bit width in FPGA design can result in substantial savings.

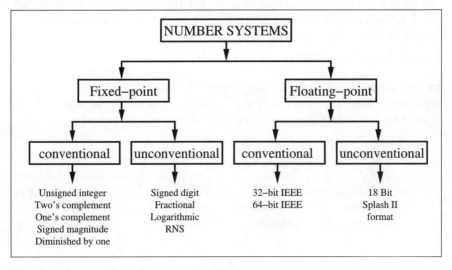

Fig. 2.1. Survey of number representations.

2.2 Number Representation

Deciding whether fixed- or floating-point is more appropriate for the problem must be done carefully, preferably at an early phase in the project. In general, it can be assumed that fixed-point implementations have higher speed and lower cost, while floating-point has higher dynamic range and no need for scaling, which may be attractive for more complicated algorithms. Figure 2.1 is a survey of conventional and less conventional fixed- and floating-point number representations. Both systems are covered by a number of standards but may, if desired, be implemented in a proprietary form.

2.2.1 Fixed-Point Numbers

We will first review the fixed-point number systems shown in Fig. 2.1. Table 2.1 shows the 3-bit coding for the 5 different integer representations.

Unsigned Integer

Let X be an N-bit unsigned binary number. Then the range is $[0, 2^N - 1]$ and the representation is given by:

$$X = \sum_{n=0}^{N-1} x_n 2^n, \tag{2.1}$$

where x_n is the nth binary digit of X (i.e., $x_n \in [0,1]$). The digit x_0 is called the least significant bit (LSB) and has a relative weight of unity. The digit x_{N-1} is the most significant bit (MSB) and has a relative weight of 2^{N-1}.

Signed-Magnitude (SM)

In signed-magnitude systems the magnitude and the sign are represented separately. The first bit x_{N-1} (i.e., the MSB) represents the sign and the remaining $N - 1$ bits the magnitude. The representation becomes:

$$X = \begin{cases} \sum_{n=0}^{N-2} x_n 2^n & X \geq 0 \\ -\sum_{n=0}^{N-2} x_n 2^n & X < 0. \end{cases} \tag{2.2}$$

The range of this representation is $[-(2^{N-1} - 1), 2^{N-1} - 1]$. The advantage of the signed-magnitude representation is simplified prevention of overflows, but the disadvantage is that addition must be split depending on which operand is larger.

Two's Complement (2C)

An N-bit two's complement representation of a signed integer, over the range $[-2^{N-1}, 2^{N-1} - 1]$, is given by:

$$X = \begin{cases} \sum_{n=0}^{N-2} x_n 2^n & X \geq 0 \\ -2^{N-1} + \sum_{n=0}^{N-2} x_n 2^n & X < 0. \end{cases} \qquad (2.3)$$

The two's complement (2C) system is by far the most popular signed numbering system in DSP use today. This is because it is possible to add several signed numbers, and as long as the final sum is in the N-bit range, we can ignore *any* overflow in the arithmetic. For instance, if we add two 3-bit numbers as follows

$$
\begin{array}{rcl}
3_{10} & \longleftrightarrow & 0\ 1\ 1_{2C} \\
-2_{10} & \longleftrightarrow & 1\ 1\ 0_{2C} \\
1_{10} & \longleftrightarrow & 1.\ 0\ 0\ 1_{2C}
\end{array}
$$

the overflow can be ignored. All computations are modulo 2^N. It follows that it is possible to have intermediate values that can not be correctly represented, but if the final value is valid then the result is correct. For instance, if we add the 3-bit numbers $2 + 2 - 3$, we would have an intermediate value of $010 + 010 = 100_{2C}$, i.e., -4_{10}, but the result $100 - 011 = 100 + 101 = 001_{2C}$ is correct.

Two's complement numbers can also be used to implement modulo 2^N arithmetic without any change in the arithmetic. This is what we will use in Chap. 5 to design CIC filters.

One's Complement (1C)

An N-bit one's complement system (1C) can represent integers over the range $[-(2^{N-1} + 1), 2^{N-1} - 1]$. In a one's complement code, positive and negative numbers have bit-by-bit complement representations including for the sign bit. There is, in fact a redundant representation of zero (see Table 2.1). The representation of signed numbers in a 1C system is formally given by:

$$X = \begin{cases} \sum_{n=0}^{N-2} x_n 2^n & X \geq 0 \\ -2^{N-1} + 1 + \sum_{n=0}^{N-2} x_n 2^n & X < 0. \end{cases} \qquad (2.4)$$

For example, the three-bit 1C representation of the numbers -3 to 3 is shown in the third column of Table 2.1.

From the following simple example

$$3_{10} \longleftrightarrow \quad 0 \ 1 \ 1_{1C}$$
$$-2_{10} \longleftrightarrow \quad 1 \ 0 \ 1_{1C}$$
$$1_{10} \longleftrightarrow \quad 1. \quad 0 \ 0 \ 0_{1C}$$
$$\text{Carry} \qquad \hookrightarrow \rightarrow \rightarrow 1_{1C}$$
$$1_{10} \longleftrightarrow \quad 0 \ 0 \ 1_{1C}$$

we remember that in one's complement a "carry wrap-around" addition is needed. A carry occurring at the MSB must be added to the LSB to get the correct final result.

The system can, however, efficiently be used to implement modulo $2^N - 1$ arithmetic without correction. As a result, one's complement has specialized value in implementing selected DSP algorithms (e.g., Mersenne transforms over the integer ring $2^N - 1$; see Chap. 7).

Diminished One System (D1)

A diminished one (D1) system is a biased system. The positive numbers are, compared with the 2C, diminished by 1. The range for $(N+1)$-bit D1 numbers is $[-2^{N-1}, 2^{N-1}]$, excluding 0. The coding rule for a D1 system is defined as follows:

$$X = \begin{cases} \sum_{n=0}^{N-2} x_n 2^n + 1 & X > 0 \\ -2^{N-1} + \sum_{n=0}^{N-2} x_n 2^n & X < 0 \\ 2^N & X = 0. \end{cases} \qquad (2.5)$$

From adding two D1 numbers

$$3_{10} \longleftrightarrow \qquad 0 \ 1 \ 0_{D1}$$
$$-2_{10} \longleftrightarrow \qquad 1 \ 1 \ 0_{D1}$$
$$1_{10} \longleftrightarrow \quad 1. \qquad 0 \ 0 \ 0_{D1}$$
$$\text{Carry} \qquad \hookrightarrow \boxed{\times - 1} \rightarrow 0_{D1}$$
$$1_{10} \longleftrightarrow \qquad 0 \ 0 \ 0_{D1}$$

we see that, for D1 a complement and add of the *inverted* carry must be computed.

D1 numbers can efficiently be used to implement modulo $2^N + 1$ arithmetic without any change in the arithmetic. This fact will be used in Chap. 7 to implement Fermat NTTs in the ring $2^N + 1$.

Bias System

The biased number system has a bias for all numbers. The bias value is usually in the middle of the binary range, i.e., bias $= 2^{N-1} - 1$. For a 3-bit system, for instance the bias would be $2^{3-1} - 1 = 3$. The range for N-bit biased numbers is $[-2^{N-1} - 1, 2^{N-1}]$. Zero is coded as the bias. The coding rule for a biased system is defined as follows:

Table 2.1. Conventional coding of signed binary numbers.

Binary	2C	1C	D1	SM	Bias
011	3	3	4	3	0
010	2	2	3	2	−1
001	1	1	2	1	−2
000	0	0	1	0	−3
111	−1	−0	−1	−3	4
110	−2	−1	−2	−2	3
101	−3	−2	−3	−1	2
100	−4	−3	−4	−0	1
1000	−	−	0	−	−

$$X = \sum_{n=0}^{N-1} x_n 2^n - \text{bias}. \qquad (2.6)$$

From adding two biased numbers

$$
\begin{aligned}
3_{10} &\longleftrightarrow 1\,1\,0_{\text{bias}} \\
+(-2_{10}) &\longleftrightarrow 0\,0\,1_{\text{bias}} \\
4_{10} &\longleftrightarrow 1\,1\,1_{\text{bias}} \\
-\text{bias} &\longleftrightarrow 0\,1\,1_{\text{bias}} \\
1_{10} &\longleftrightarrow 1\,0\,0_{\text{bias}}
\end{aligned}
$$

we see that, for each addition the bias needs to be subtracted, while for every subtraction the bias needs to be added.

Bias numbers can efficiently be used to simplify comparison of numbers. This fact will be used in Sect. 2.2.3 (p. 71) for coding the exponent of floating-point numbers.

2.2.2 Unconventional Fixed-Point Numbers

In the following we continue the review of number systems according to Fig. 2.1 (p. 53). The unconventional fixed-point number systems discussed in the following are not as often used as for instance the 2C system, but can yield significant improvements for particular applications or problems.

Signed Digit Numbers (SD)

The signed digit (SD) system differs from the traditional binary systems presented in the previous section in the fact that it is ternary valued (i.e., digits have the value $\{0, 1, -1\}$, where -1 is sometimes denoted as $\bar{1}$).

SD numbers have proven to be useful in carry-free adders or multipliers with less complexity, because the effort in multiplication can typically be estimated through the number of nonzero elements, which can be reduced by using SD numbers. Statistically, half the digits in the two's complement coding of a number are zero. For an SD code, the density of zeros increases to two thirds as the following example shows:

Example 2.1: SD Coding

Consider coding the decimal number $15 = 1111_2$ using a 5-bit binary and an SD code. Their representations are as follows:

1) $15_{10} = 16_{10} - 1_{10} = 1000\bar{1}_{SD}$

2) $15_{10} = 16_{10} - 2_{10} + 1_{10} = 100\bar{1}1_{SD}$

3) $15_{10} = 16_{10} - 4_{10} + 3_{10} = 10\bar{1}11_{SD}$

4) etc.

$\boxed{2.1}$

The SD representation, unlike a 2C code, is nonunique. We call a *canonic signed digit* system (CSD) the system with the minimum number of non-zero elements. The following algorithm can be used to produce a classical CSD code.

Algorithm 2.2:	**Classical CSD Coding**

Starting with the LSB substitute all 1 sequences equal or larger than two, with $10\ldots0\bar{1}$.

This CSD coding is the basis for the C utility program **csd3e.exe**[1] on the CD-ROM. This classical CSD code is also unique and an additional property is that the resulting representation has at least one zero between two digits, which may have values 1, $\bar{1}$, or 0.

Example 2.3:   Classical CSD Code

Consider again coding the decimal number 15 using a 5-bit binary and a CSD code. Their representations are: $1111_2 = 1000\bar{1}_{CSD}$. We notice from a comparison with the SD coding from Example 2.1 that only the first representation is a CSD code.

As another example consider the coding of

$$27_{10} = 11011_2 = 1110\bar{1}_{SD} = 100\bar{1}0\bar{1}_{CSD}. \qquad (2.7)$$

We note that, although the first substitution of $011 \rightarrow 10\bar{1}$ does not reduce the complexity, it produces a length-three strike, and the complexity reduces from three additions to two subtractions.

$\boxed{2.3}$

On the other hand, the classical CSD coding does not always produce the optimal CSD coding in terms of hardware complexity, because in Algorithm 2.2 additions are also substituted by subtractions, when there should be no such substitution. For instance 011_2 is coded as $10\bar{1}_{CSD}$, and if this coding is used to produce a constant multiplier the subtraction will need a full-adder

---

[1] You need to copy the program to your harddrive first because the program writes out the results in a file **csd.dat**; you can not start it from the CD directly.

instead of a half-adder for the LSB. The CSD coding given in the following will produce a CSD coding with the minimum number of nonzero terms, but also with the minimum number of subtractions.

Algorithm 2.4:   Optimal CSD Coding

1) Starting with the LSB substitute all 1 sequences larger than two with $10\ldots0\bar{1}$. Also substitute 1011 with $110\bar{1}$.
2) Starting with the MSB, substitute $10\bar{1}$ with 011.

Fractional (CSD) Coding

Many DSP algorithms require the implementation of fractional numbers. Think for instance of trigonometric coefficient like sine or cosine coefficients. Implementation via integer numbers only would result in a large quantization error. The question then is, can we also use the CSD coding to reduce the implementation effort of a fractional constant coefficient? The answer is yes, but we need to be a little careful with the ordering of the operands. In VHDL the analysis of an expression is usually done from left to right, which means an expression like $y = 7 \times x/8$ is implemented as $y = (7 \times x)/8$, and equivalently the expression $y = x/8 \times 7$ is implemented as $y = (x/8) \times 7$. The latter term unfortunately will produce a large quantization error, since the evaluation of $x/8$ is in fact synthesized by the tool[2] as a right shift by three bits, so we will lose the lower three bits of our input x in the computation that follows. Let us demonstrate this with a small HDL design example.

Example 2.5: Fractional CSD Coding

Consider coding the fractional decimal number $0.875 = 7/8$ using a fractional 4-bit binary and CSD code. The 7 can be implemented more efficiently in CSD as $7 = 8 - 1$ and we want to determine the quantization error of the following four mathematically equivalent representations, which give different synthesis results:

$$y0 = 7 \times x/8 = (7 \times x)/8$$
$$y1 = x/8 \times 7 = (x/8) \times 7$$
$$y2 = x/2 + x/4 + x/8 = ((x/2) + (x/4)) + (x/8)$$
$$y3 = x - x/8 = x - (x/8)$$

Using parenthesis in the above equations it is shown how the HDL tool will group the expressions. Multiply and divide have a higher priority than add and subtract and the evaluation is from left to right. The VHDL code[3] of the constant coefficient fractional multiplier is shown next.

```
ENTITY cmul7p8 IS                        ------> Interface
  PORT (    x    : IN  INTEGER RANGE -2**4 TO 2**4-1;
```

---

[2] Most HDL tools only support dividing by power-of-2 values, which can be designed using a shifter, see Sect. 2.5, p. 91.

[3] The equivalent Verilog code cmul7p8.v for this example can be found in Appendix A on page 665. Synthesis results are shown in Appendix B on page 731.

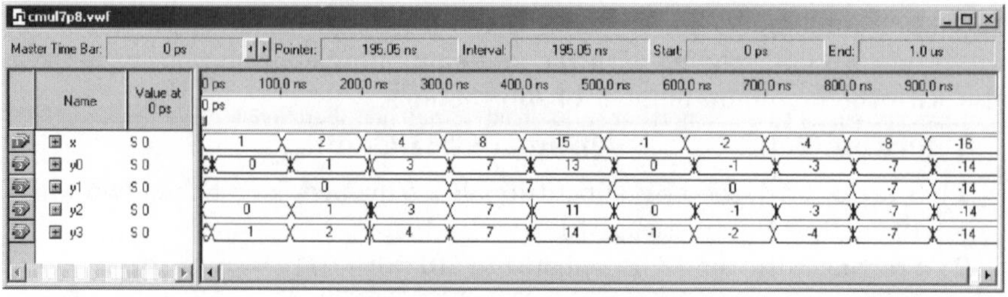

Fig. 2.2. Simulation results for fraction CSD coding.

```
                  y0, y1, y2, y3 : OUT INTEGER RANGE -2**4 TO 2**4-1);
    END;

    ARCHITECTURE fpga OF cmul7p8 IS
    BEGIN

        y0 <= 7 * x / 8;
        y1 <= x / 8 * 7;
        y2 <= x/2 + x/4 + x/8;
        y3 <= x - x/8;

    END fpga;
```

The design uses 48 LEs and no embedded multiplier. A **Registered Performance** can not be measured since there is no register-to-register path. The simulated results of the fractional constant coefficient multiplier is shown in Fig. 2.2. Note the large quantization error for y1. Looking at the results for the input value $x = 4$, we can also see that the CSD coding y3 shows rounding to the next largest integer, while y0 and y2 show rounding to the next smallest integer. For negative value (e.g., -4) we see that the CSD coding y3 shows rounding to the next smallest (i.e., -4) integer, while y0 and y2 show rounding to the next largest (i.e., -3) integer. $\boxed{2.5}$

Carry-Free Adder

The SD number representation can also be used to implement a carry-free adder. Tagaki et al. [30] introduced the scheme presented in Table 2.2. Here, u_k is the interim sum and c_k is the carry of the kth bit (i.e., to be added to u_{k+1}).

Example 2.6: Carry-Free Addition

The addition of 29 to -9 in the SD system is performed below.

Table 2.2. Adding carry-free binaries using the SD representation.

$x_k y_k$	00	01	01	$0\bar{1}$	$0\bar{1}$	11	$\overline{11}$
$x_{k-1} y_{k-1}$	–	neither is $\bar{1}$	at least one is $\bar{1}$	neither is $\bar{1}$	at least one is $\bar{1}$	–	–
c_k	0	1	0	0	$\bar{1}$	1	$\bar{1}$
u_k	0	$\bar{1}$	1	$\bar{1}$	1	0	0

$$
\begin{array}{r}
1\,0\,0\,\bar{1}\,0\,1 \; x_k \\
+ \quad 0\,\bar{1}\,1\,\bar{1}\,1\,1 \; y_k \\
\hline
0\,0\,0\,\bar{1}\,1\,1 \quad c_k \\
1\,\bar{1}\,1\,0\,\bar{1}\,0 \; u_k \\
\hline
1\,\bar{1}\,0\,1\,0\,0 \; s_k
\end{array}
$$

`2.6`

However, due to the ternary logic burden, implementing Table 2.2 with FPGAs requires four-input operands for the c_k and u_k. This translates into a $2^8 \times 4$-bit LUT when implementing Table 2.2.

Multiplier Adder Graph (MAG)

We have seen that the cost of multiplication is a direct function of the number of nonzero elements a_k in A. The CSD system minimizes this cost. The CSD is also the basis for the Booth multiplier [25] discussed in Exercise 2.2 (p. 154).

It can, however, sometimes be more efficient first to factor the coefficient into several factors, and realize the individual factors in an optimal CSD sense [31, 32, 33, 34]. Figure 2.3 illustrates this option for the coefficient 93. The direct binary and CSD codes are given by $93_{10} = 1011101_2 = 110 0\bar{1}01_{\mathrm{CSD}}$,

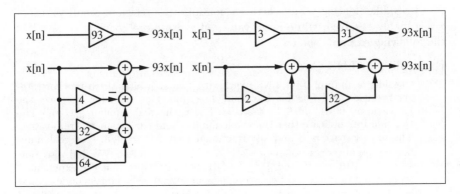

Fig. 2.3. Two realizations for the constant factor 93.

with the 2C requiring four adders, and the CSD requiring three adders. The coefficient 93 can also be represented as $93 = 3 \times 31$, which requires one adder for each factor (see Fig. 2.3). The complexity for the factor number is reduced to two. There are several ways to combine these different factors. The number of adders required is often referred to as the cost of the constant coefficient multiplier. Figure 2.4, suggested by Dempster et al. [33], shows all possible configurations for one to four adders. Using this graph, all coefficients with a cost ranging from one to four can be synthesized with $k_i \in \mathbb{N}_0$, according to:

$$
\begin{aligned}
\textbf{Cost 1:} \quad 1) \quad & A = 2^{k_0}(2^{k_1} \pm 2^{k_2}) \\
\textbf{Cost 2:} \quad 1) \quad & A = 2^{k_0}(2^{k_1} \pm 2^{k_2} \pm 2^{k_3}) \\
2) \quad & A = 2^{k_0}(2^{k_1} \pm 2^{k_2})(2^{k_3} \pm 2^{k_4}) \\
\textbf{Cost 3:} \quad 1) \quad & A = 2^{k_0}(2^{k_1} \pm 2^{k_2} \pm 2^{k_3} \pm 2^{k_4})
\end{aligned}
$$

$$\vdots$$

Using this technique, Table 2.3 shows the optimal coding for all 8-bit integers having a cost between zero and three [5].

Logarithmic Number System (LNS)

The logarithmic number system (LNS) [35, 36] is analogous to the floating-point system with a fixed mantissa and a fractional exponent. In the LNS, a number x is represented as:

$$X = \pm r^{\pm e_x}, \tag{2.8}$$

where r is the system's radix, and e_x is the LNS exponent. The LNS format consists of a sign-bit for the number and exponent, and an exponent assigned I integer bits and F fractional bits of precision. The format in graphical form is shown below:

Sign S_X	Exponent sign S_e	Exponent integer bits I	Exponent fractional bits F

The LNS, like floating-point, carries a nonuniform precision. Small values of x are highly resolved, while large values of x are more coarsely resolved as the following example shows.

Example 2.7: LNS Coding

Consider a radix-2 9-bit LNS word with two sign-bits, three bits for integer precision and four-bit fractional precision. How can, for instance, the LNS coding $00\,011.0010$ be translated into the real number system? The two sign bits indicate that the whole number and the exponent are positive. The integer part is 3 and the fractional part $2^{-3} = 1/8$. The real number representation is therefore $2^{3+1/8} = 2^{3.125} = 8.724$. We find also that $-2^{3.125} = 10\,011.0010$ and $2^{-3.125} = 01\,100.1110$. Note that the exponent is represented in fractional two's complement format. The largest number that can be represented with this 9-bit LNS format is $2^{8-1/16} \approx 2^8 = 256$ and

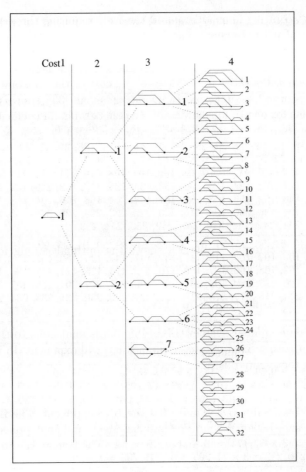

Fig. 2.4. Possible cost one to four graphs. Each node is either an adder or subtractor and each edge is associated with a power-of-two factor (©1995 IEEE [33]).

the smallest is $2^{-8} = 0.0039$, as graphically interpreted in Fig. 2.5a. In contrast, an 8-bit plus sign fixed-point number has a maximal positive value of $2^8 - 1 = 255$, and the smallest nonzero positive value is one. A comparison of the two 9-bit systems is shown in Fig. 2.5b.

2.7

The historical attraction of the LNS lies in its ability to efficiently implement multiplication, division, square-rooting, or squaring. For example, the product $C = A \times B$, where A, B, and C are LNS words, is given by:

$$C = r^{e_a} \times r^{e_b} = r^{e_a + e_b} = r^{e_c}. \tag{2.9}$$

That is, the exponent of the LNS product is simply the sum of the two exponents. Division and high-order operations immediately follow. Unfortunately,

Table 2.3. Cost C (i.e., number of adders) for all 8-bit numbers using the multiplier adder graph (MAG) technique.

C	Coefficient
0	1, 2, 4, 8, 16, 32, 64, 128, 256
1	3, 5, 6, 7, 9, 10, 12, 14, 15, 17, 18, 20, 24, 28, 30, 31, 33, 34, 36, 40, 48, 56, 60, 62, 63, 65, 66, 68, 72, 80, 96, 112, 120, 124, 126, 127, 129, 130, 132, 136, 144, 160, 192, 224, 240, 248, 252, 254, 255
2	11, 13, 19, 21, 22, 23, 25, 26, 27, 29, 35, 37, 38, 39, 41, 42, 44, 46, 47, 49, 50, 52, 54, 55, 57, 58, 59, 61, 67, 69, 70, 71, 73, 74, 76, 78, 79, 81, 82, 84, 88, 92, 94, 95, 97, 98, 100, 104, 108, 110, 111, 113, 114, 116, 118, 119, 121, 122, 123, 125, 131, 133, 134, 135, 137, 138, 140, 142, 143, 145, 146, 148, 152, 156, 158, 159, 161, 162, 164, 168, 176, 184, 188, 190, 191, 193, 194, 196, 200, 208, 216, 220, 222, 223, 225, 226, 228, 232, 236, 238, 239, 241, 242, 244, 246, 247, 249, 250, 251, 253
3	43, 45, 51, 53, 75, 77, 83, 85, 86, 87, 89, 90, 91, 93, 99, 101, 102, 103, 105, 106, 107, 109, 115, 117, 139, 141, 147, 149, 150, 151, 153, 154, 155, 157, 163, 165, 166, 167, 169, 170, 172, 174, 175, 177, 178, 180, 182, 183, 185, 186, 187, 189, 195, 197, 198, 199, 201, 202, 204, 206, 207, 209, 210, 212, 214, 215, 217, 218, 219, 221, 227, 229, 230, 231, 233, 234, 235, 237, 243, 245
4	171, 173, 179, 181, 203, 205, 211, 213
Minimum costs through factorization	
2	$45 = 5 \times 9, 51 = 3 \times 17, 75 = 5 \times 15, 85 = 5 \times 17, 90 = 2 \times 9 \times 5, 93 = 3 \times 31, 99 = 3 \times 33, 102 = 2 \times 3 \times 17, 105 = 7 \times 15, 150 = 2 \times 5 \times 15, 153 = 9 \times 17, 155 = 5 \times 31, 165 = 5 \times 33, 170 = 2 \times 5 \times 17, 180 = 4 \times 5 \times 9, 186 = 2 \times 3 \times 31, 189 = 7 \times 9, 195 = 3 \times 65, 198 = 2 \times 3 \times 33, 204 = 4 \times 3 \times 17, 210 = 2 \times 7 \times 15, 217 = 7 \times 31, 231 = 7 \times 33$
3	$171 = 3 \times 57, 173 = 8 + 165, 179 = 51 + 128, 181 = 1 + 180, 211 = 1 + 210, 213 = 3 \times 71, 205 = 5 \times 41, 203 = 7 \times 29$

addition or subtraction are by comparison far more complex. Addition and subtraction operations are based on the following procedure, where it is assumed that $A > B$.

$$C = A + B = 2^{e_a} + 2^{e_b} = 2^{e_a} \underbrace{\left(1 + 2^{e_b - e_a}\right)}_{\Phi^+(\Delta)} = 2^{e_c}. \tag{2.10}$$

Solving for the exponent e_c, one obtains $e_c = e_a + \phi^+(\Delta)$ where $\Delta = e_b - e_a$ and $\phi^+(u) = \log_2(\Phi^+(\Delta))$. For subtraction a similar table, $\phi^-(u) = \log_2(\Phi^-(\Delta))$, $\Phi^-(\Delta) = (1 - 2^{e_b - e_a})$, can be used. Such tables have been historically used for rational numbers as described in "Logarithmorm Completus," Jurij Vega (1754–1802), containing tables computed by Zech. As a result, the term $\log_2(1 - 2^u)$ is usually referred to as a Zech logarithm.

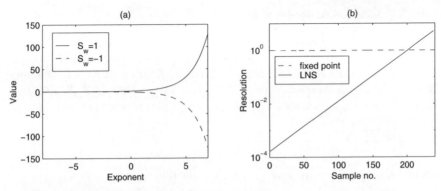

Fig. 2.5. LNS processing. **(a)** Values. **(b)** Resolution.

LNS arithmetic is performed in the following manner [35]. Let $A = 2^{e_a}, B = 2^{e_b}, C = 2^{e_c}$, with S_A, S_B, S_C denoting the sign-bit for each word:

Operation		Action
Multiply	$C = A \times B$	$e_c = e_a + e_b; S_C = S_A \text{ XOR } S_B$
Divide	$C = A/B$	$e_c = e_a - e_b; S_C = S_A \text{ XOR } S_B$
Add	$C = A + B$	$e_c = \begin{cases} e_a + \phi^+(e_b - e_a) \ A \geq B \\ e_b + \phi^+(e_a - e_b) \ B > A \end{cases}$
Subtract	$C = A - B$	$e_c = \begin{cases} e_a + \phi^-(e_b - e_a) \ A \geq B \\ e_b + \phi^-(e_a - e_b) \ B > A \end{cases}$
Square root	$C = \sqrt{A}$	$e_c = e_a/2$
Square	$C = A^2$	$e_c = 2e_a$

Methods have been developed to reduce the necessary table size for the Zech logarithm by using partial tables [35] or using linear interpolation techniques [37]. These techniques are beyond the scope of the discussion presented here.

Residue Number System (RNS)

The RNS is actually an ancient algebraic system whose history can be traced back 2000 years. The RNS is an integer arithmetic system in which the primitive operations of addition, subtraction, and multiplication are defined. The primitive operations are performed concurrently within noncommunicating small-wordlength channels [38, 39]. An RNS system is defined with respect to a positive integer basis set $\{m_1, m_2, \ldots, m_L\}$, where the m_l are all relatively (pairwise) prime. The dynamic range of the resulting system is M, where $M = \prod_{l=1}^{L} m_l$. For signed-number applications, the integer value of X is assumed to be constrained to $X \in [-M/2, M/2)$. RNS arithmetic is defined within a ring isomorphism:

$$\mathbb{Z}_M \cong \mathbb{Z}_{m_1} \times \mathbb{Z}_{m_2} \times \cdots \times \mathbb{Z}_{m_L}, \tag{2.11}$$

where $\mathbb{Z}_M = \mathbb{Z}/(M)$ corresponds to the ring of integers modulo M, called the residue class $\mathrm{mod}\, M$. The mapping of an integer X into an RNS L-tuple $X \leftrightarrow (x_1, x_2, \dots, x_L)$ is defined by $x_l = X \bmod m_l$, for $l = 1, 2, \dots L$. Defining $\Box$ to be the algebraic operations $+, -$ or $*$, it follows that, if $Z, X, Y \in \mathbb{Z}_M$, then:

$$Z = X \Box Y \bmod M \tag{2.12}$$

is isomorphic to $Z \leftrightarrow (z_1, z_2, \dots, z_L),$. Specifically:

$$
\begin{array}{ll}
X & \xrightarrow{(m_1, m_2, \dots, m_L)} \quad (\langle X \rangle_{m_1} \ , \ \langle X \rangle_{m_2} \ , \dots, \ \langle X \rangle_{m_L}) \\
Y & \xleftarrow{(m_1, m_2, \dots, m_L)} \quad (\langle Y \rangle_{m_1} \ , \ \langle Y \rangle_{m_2} \ , \dots, \ \langle Y \rangle_{m_L}) \\
\hline
Z = X \Box Y & \xleftrightarrow{(m_1, m_2, \dots, m_L)} \quad (\langle X \Box Y \rangle_{m_1}, \langle X \Box Y \rangle_{m_2}, \dots, \langle X \Box Y \rangle_{m_L}).
\end{array}
$$

As a result, RNS arithmetic is pairwise defined. The L elements of $Z = (X \Box Y) \bmod M$ are computed concurrently within L small-wordlength mod (m_l) channels whose width is bounded by $w_l = \lceil \log_2(m_l) \rceil$ bits (typical 4 to 8 bits). In practice, most RNS arithmetic systems use small RAM or ROM tables to implement the modular mappings $z_l = x_l \Box y_l \bmod m_l$.

Example 2.8: RNS Arithmetic

Consider an RNS system based on the relatively prime moduli set $\{2, 3, 5\}$ having a dynamic range of $M = 2 \times 3 \times 5 = 30$. Two integers in $\mathbb{Z}_{30}$, say 7_{10} and 4_{10}, have RNS representations $7 = (1, 1, 2)_{\mathrm{RNS}}$ and $4 = (0, 1, 4)_{\mathrm{RNS}}$, respectively. Their sum, difference, and products are 11, 3, and 28, respectively, which are all within $\mathbb{Z}_{30}$. Their computation is shown below.

$$
\begin{array}{lll}
7 \xleftrightarrow{(2,3,5)} (1,1,2) & 7 \xleftrightarrow{(2,3,5)} (1,1,2) & 7 \xleftrightarrow{(2,3,5)} (1,1,2) \\
\underline{+4 \xleftrightarrow{(2,3,5)} +(0,1,4)} & \underline{-4 \xleftrightarrow{(2,3,5)} -(0,1,4)} & \underline{\times 4 \xleftrightarrow{(2,3,5)} \times(0,1,4)} \\
11 \xleftrightarrow{(2,3,5)} (1,2,1) & 3 \xleftrightarrow{(2,3,5)} (1,0,3) & 28 \xleftrightarrow{(2,3,5)} (0,1,3).
\end{array}
$$

2.8

RNS systems have been built as custom VLSI devices [40], GaAs, and LSI [39]. It has been shown that, for small wordlengths, the RNS can provide a significant speed-up using the $2^4 \times 2$-bit tables found in Xilinx FPGAs [41]. For larger moduli, the M2K and M4K tables belonging to the Altera FPGAs are beneficial in designing RNS arithmetic and RNS-to-integer converters. With the ability to support larger moduli, the design of high-precision high-speed FPGA systems becomes a practical reality.

A historical barrier to implementing practical RNS systems, until recently, has been decoding [42]. Implementing RNS-to-integer decoder, division, or magnitude scaling, requires that data first be converted from an RNS format to an integer. The commonly referenced RNS-to-integer conversion methods are called the Chinese remainder theorem (CRT) and the mixed-radix-conversion (MRC) algorithm [38]. The MRC actually produced the digits of a weighted number system representation of an integer while the CRT maps an RNS L-tuple directly to an integer. The CRT is defined below.

$$X \bmod M \equiv \sum_{l=0}^{L-1} \hat{m}_l \langle \hat{m}_l^{-1} x_l \rangle_{m_l} \quad \bmod M, \tag{2.13}$$

where $\hat{m}_l = M/m_l$ is an integer, and $\hat{m}_l^{-1}$ is the multiplicative inverse of $\hat{m}_l \bmod m_l$, i.e., $\hat{m}_l \hat{m}_l^{-1} \equiv 1 \bmod m_l$. Typically, the desired output of an RNS computation is much less than the maximum dynamic range M. In such cases, a highly efficient algorithm, called the $\varepsilon-$CRT [43], can be used to implement a time- and area-efficient RNS to (scaled) integer conversion.

Index Multiplier

There are, in fact, several variations of the RNS. One in common use is based on the use of index arithmetic [38]. It is similar in some respects to logarithmic arithmetic. Computation in the index domain is based on the fact that, if all the moduli are primes, it is known from number theory that there exists a primitive element, a *generator* g, such that:

$$a \equiv g^\alpha \bmod p \tag{2.14}$$

that generates all elements in the field $\mathbb{Z}_p$, excluding zero (denoted $\mathbb{Z}_p/\{0\}$). There is, in fact, a one-to-one correspondence between the integers a in $\mathbb{Z}_p/\{0\}$ and the exponents α in $\mathbb{Z}_{p-1}$. As a point of terminology, the index α, with respect to the generator g and integer a, is denoted $\alpha = \text{ind}_g(a)$.

Example 2.9: Index Coding

Consider a prime moduli $p = 17$; a generator $g = 3$ will generate the elements of $\mathbb{Z}_p/\{0\}$. The encoding table is shown below. For notational purposes, the case $a = 0$ is denoted by $g^{-\infty} = 0$.

a	0	1	2	3	4	5	6	7	8	9	10	11	12	13	14	15	16
$\text{ind}_3(a)$	$-\infty$	0	14	1	12	5	15	11	10	2	3	7	13	4	9	6	8

2.9

Multiplication of RNS numbers can be performed as follows:

1) Map a and b into the index domain, i.e., $a = g^\alpha$ and $b = g^\beta$

2) Add the index values modulo $p - 1$, i.e., $\nu = (\alpha + \beta) \bmod (p - 1)$

3) Map the sum back to the original domain, i.e., $n = g^{\nu}$

If the data being processed is in index form, then only exponent addition $\bmod (p - 1)$ is required. This is illustrated by the following example.

Example 2.10: Index Multiplication

Consider the prime moduli $p = 17$, generator $g = 3$, and the results shown in Example 2.9. The multiplication of $a = 2$ and $b = 4$ proceeds as follows:

$$(\mathrm{ind}_g(2) + \mathrm{ind}_g(4)) \bmod 16 = (14 + 12) \bmod 16 = 10.$$

From the table in Example 2.9 it is seen that $\mathrm{ind}_3(8) = 10$, which corresponds to the integer 8, which is the expected result. $\boxed{2.10}$

Addition in the Index Domain

Most often, DSP algorithms require both multiplication *and* addition. Index arithmetic is well suited to multiplication, but addition is no longer trivial. Technically, addition can be performed by converting index RNS data back into the RNS where addition is simple to implement. Once the sum is computed the result is mapped back into the index domain. Another approach is based on a Zech logarithm. The sum of index-coded numbers a and b is expressed as:

$$d = a + b = g^{\delta} = g^{\alpha} + g^{\beta} = g^{\alpha}\left(1 + g^{\beta-\alpha}\right) = g^{\beta}\left(1 + g^{\alpha-\beta}\right). \qquad (2.15)$$

If we now define the Zech logarithm as

Definition 2.11:   **Zech Logarithm**

$$Z(n) = \mathrm{ind}_g(1 + g^n) \quad\longleftrightarrow\quad g^{Z(n)} = 1 + g^n \qquad (2.16)$$

---

then we can rewrite (2.15) in the following way:

$$g^{\delta} = g^{\beta} \times g^{Z(\alpha-\beta)} \quad\longleftrightarrow\quad \delta = \beta + Z(\alpha - \beta). \qquad (2.17)$$

Adding numbers in the index domain, therefore, requires one addition, one subtraction, and a Zech LUT. The following small example illustrates the principle of adding $2 + 5$ in the index domain.

Example 2.12: Zech Logarithms

A table of Zech logarithms, for a prime moduli 17 and $g = 3$, is shown below.

n	$-\infty$	0	1	2	3	4	5	6	7	8	9	10	11	12	13	14	15
$Z(n)$	0	14	12	3	7	9	15	8	13	$-\infty$	6	2	10	5	4	1	11

The index values for 2 and 5 are defined in the tables found in Example 2.9 (p. 67). It therefore follows that:

$$2 + 5 = 3^{14} + 3^5 = 3^5(1 + 3^9) = 3^{5+Z(9)} = 3^{11} \equiv 7 \mod 17.$$

2.12

The case where $a + b \equiv 0$ needs special attention, corresponding to the case where [44]:

$$-X \equiv Y \mod p \quad \longleftrightarrow \quad g^{\alpha+(p-1)/2} \equiv g^{\beta} \mod p.$$

That is, the sum is zero if, in the index domain, $\beta = \alpha + (p-1)/2 \mod (p-1)$. An example follows.

Example 2.13:   The addition of 5 and 12 in the original domain is given by $5 + 12 = 3^5 + 3^{13} = 3^5(1 + 3^8) = 3^{5+Z(8)} \equiv 3^{-\infty} \equiv 0 \mod 17.$ 2.13

Complex Multiplication using QRNS

Another interesting property of the RNS arises if we process complex data. This special representation, called QRNS, allows very efficient multiplication, which we wish to discuss next.

When the real and imaginary components are coded as RNS digits, the resulting system is called the complex RNS or CRNS. Complex addition in the CRNS requires that two real adds be performed. Complex RNS (CRNS) multiplication is defined in terms of four real products, an addition, and a subtraction. This condition is radically changed when using a variant of the RNS, called the quadratic RNS, or QRNS. The QRNS is based on known properties of Gaussian primes of the form $p = 4k + 1$, where k is a positive integer. The importance of this choice of moduli is found in the factorization of the polynomial $x^2 + 1$ in $\mathbb{Z}_p$. The polynomial has two roots, $\hat{j}$ and $-\hat{j}$, where $\hat{j}$ and $-\hat{j}$ are real integers belonging to the residue class $\mathbb{Z}_p$. This is in sharp contrast with the factoring of $x^2 + 1$ over the complex field. Here, the roots are complex and have the form $x_{1,2} = \alpha \pm j\beta$ where $j = \sqrt{-1}$ is the imaginary operator. Converting a CRNS number into the QRNS is accomplished by the transform $f : \mathbb{Z}_p^2 \to \mathbb{Z}_p^2$, defined as follows:

$$f(a + jb) = ((a + \hat{j}b) \mod p, (a - \hat{j}b) \mod p) = (A, B). \tag{2.18}$$

In the QRNS, addition and multiplication is realized componentwise, and is defined as

$$(a + ja) + (c + jd) \leftrightarrow (A + C, B + D) \quad \mod p \tag{2.19}$$

$$(a + jb)(c + jd) \leftrightarrow (AC, BD) \quad \mod p \tag{2.20}$$

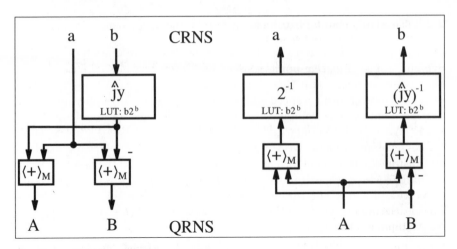

Fig. 2.6. CRNS $\leftrightarrow$ QRNS conversion.

and the square of the absolute value can be computed with

$$|a + jb|^2 \leftrightarrow (A \times B) \mod p. \tag{2.21}$$

The inverse mapping from QRNS digits back to the CRNS is defined by:

$$f^{-1}(A, B) = 2^{-1}(A + B) + j\,(2\hat{\jmath})^{-1}(A - B) \mod p. \tag{2.22}$$

Consider the Gaussian prime $p = 13$ and the complex product of $(a + jb) = (2 + j1)$, $(c + jd) = (3 + j2)$, is $(2 + j1) \times (3 + j2) = (4 + j7) \mod 13$. In this case four real multiplies, a real add, and real subtraction are required to complete the product.

Example 2.14: QRNS Multiplication

The quadratic equation $x^2 \equiv (-1) \mod 13$ has two roots: $\hat{\jmath} = 5$ and $-\hat{\jmath} = -5 \equiv 8 \mod 13$. The QRNS-coded data become:

$$
\begin{aligned}
(a + jb) &= 2 + j \leftrightarrow (2 + 5 \times 1, 2 + 8 \times 1) = (A, B) = (7, 10) \mod 13 \\
(c + jd) &= 3 + j2 \leftrightarrow (3 + 5 \times 2, 3 + 8 \times 2) = (C, D) = (0, 6) \mod 13.
\end{aligned}
$$

Componentwise multiplication yields $(A, B)(C, D) = (7, 10)(0, 6) \equiv (0, 8)$ mod 13, requiring only two real multiplies. The inverse mapping to the CRNS is defined in terms of (2.22), where $2^{-1} \equiv 7$ and $(2\hat{\jmath})^{-1} = 10^{-1} \equiv 4$. Solving the equations for $2x \equiv 1 \mod 13$ and $10x \equiv 1 \mod 13$, produces 7 and 4, respectively. It then follows that

$$f^{-1}(0, 8) = 7(0 + 8) + j\,4(0 - 8) \mod 13 \equiv 4 + j7 \mod 13. \checkmark$$

2.14

Figure 2.6 shows a graphical interpretation of the mapping between CRNS and QRNS.

2.2.3 Floating-Point Numbers

Floating-point systems were developed to provide high resolution over a large dynamic range. Floating-point systems can often provide a solution when fixed-point systems, with their limited dynamic range, fail. Floating-point systems, however, bring a speed and complexity penalty. Most microprocessor floating-point systems comply with the published single- or double-precision IEEE floating-point standard [45, 46], while in FPGA-based systems often employ custom formats. We will therefore discuss in the following standard and custom floating-point formats, and in Sect. 2.6 (p. 104) the design of basic building blocks. Such arithmetic blocks are available from several "intellectual property" providers, or through special request via e-mail to Uwe.Meyer-Baese@ieee.org.

A standard floating-point word consists of a sign-bit s, exponent e, and an unsigned (fractional) normalized mantissa m, arranged as follows:

s	Exponent e	Unsigned mantissa m

Algebraically, a floating-point word is represented by:

$$X = (-1)^S \times 1.m \times 2^{e-\text{bias}}. \tag{2.23}$$

Note that this is a signed magnitude format (see p. 57). The "hidden" one in the mantissa is not present in the binary coding of the floating-point number. If the exponent is represented with E bits then the bias is selected to be

$$\text{bias} = 2^{\text{E}-1} - 1. \tag{2.24}$$

To illustrate, let us determine the decimal value 9.25 in a 12-bit custom floating-point format.

Example 2.15: A (1,6,5) Floating-Point Format

Consider a floating-point representation with a sign bit, E = 6-bit exponent width, and M = 5-bit for the mantissa (not counting the hidden one). Let us now determine the representation of 9.25_{10} in this (1,6,5) floating-point format. Using (2.24) the bias is

$$\text{bias} = 2^{\text{E}-1} - 1 = 31,$$

and the mantissa need to be normalized according the $1.m$ format, i.e.,

$$9.25_{10} = 1001.01_2 = 1.\underbrace{00101}_{m} \times 2^3.$$

The biased exponent is therefore represented with

$$e = 3 + \text{bias} = 34_{10} = 100010_2.$$

Finally, we can represent 9.25_{10} in the (1,6,5) floating-point format with

s	Exponent e	Unsigned mantissa m
0	100010	00101

Besides this fixed-point to floating-point conversion we also need the back conversion from floating-point to integer. So, let us assume the following floating-point number

s	Exponent e	Unsigned mantissa m
1	011111	00000

is given and we wish to find the fixed-point representation of this number. We first notice that the sign bit is one, i.e., it is a negative number. Adding the hidden one to the mantissa and subtracting the bias from the exponent, yields

$$-1.00000_2 \times 2^{31-\text{bias}} = -1.0_2 2^0 = -1.0_{10}.$$

We note that in the floating-point to fixed-point conversion the bias is subtracted from the exponent, while in the fixed-point to floating-point conversion the bias is added to the exponent. | 2.15 |

The IEEE standard 754-1985 for binary floating-point arithmetic [45] also defines some additional useful special numbers to handle, for instance, overflow and underflow. The exponent $e = E_{\max} = 1 \ldots 1_2$ in combination with zero mantissa $m = 0$ is reserved for ∞. Zeros are coded with zero exponent $e = E_{\min} = 0$ and zero mantissa $m = 0$. Note, that due to the signed magnitude representation, plus and minus zero are coded differently. There are two more special numbers defined in the 754 standard, but these additional representations are most often not supported in FPGA floating-point arithmetic. These additional number are *denormals* and $NaN's$ (not a number). With denormalized numbers we can represent numbers smaller than $2^{E_{\min}}$, by allowing the mantissa to represent numbers without the hidden one, i.e., the mantissa can represents numbers smaller than 1.0. The exponent in denormals is code with $e = E_{\min} = 0$, but the mantissa is allowed to be different from zero. NaNs have proven useful in software systems to reduce the number of "exceptions" that are called when an invalid operation is performed. Examples that produce such "quiet" NaNs include:

- Addition or subtraction of two infinities, such as $\infty - \infty$
- Multiplication of zero and infinite, e.g., $0 \times \infty$
- Division of zeros or infinities, e.g., $0/0$ or ∞/∞
- Square root of negative operand

In the IEEE standard 754-1985 for binary floating-point arithmetic NaNs are coded with exponent $e = E_{\max} = 1 \ldots 1_2$ in combination with a nonzero mantissa $m \neq 0$.

We wish now to compare the fixed-point and floating-point representation in terms of precision and dynamic range in the following example.

Example 2.16: 12-Bit Floating- and Fixed-point Representations

Suppose we use again a (1,6,5) floating-point format as in the previous example. The (absolute) largest number we can represent is:

2.2 Number Representation 73

Table 2.4. Example values in (1,6,5) floating-point format.

(1,6,5) format			Decimal	Coding
0	000000	00000	$+0$	$2^{E_{\min}}$
1	000000	00000	-0	$-2^{E_{\min}}$
0	011111	00000	$+1.0$	2^{bias}
1	011111	00000	-1.0	-2^{bias}
0	111111	00000	$+\infty$	$2^{E_{\max}}$
1	111111	00000	$-\infty$	$-2^{E_{\max}}$

$$\pm 1.11111_2 \times 2^{31} \approx \pm 4.23_{10} \times 10^9.$$

The (absolutely measured) smallest number (not including denormals) that can be represented is

$$\pm 1.0_2 \times 2^{1-\text{bias}} = \pm 1.0_2 \times 2^{-30} \approx \pm 9.31_{10} \times 10^{-10}.$$

Note, that $E_{\max} = 1 \ldots 1_2$ and $E_{\min} = 0$ are reserved for zero and infinity in the floating-point format, and must not be used for general number representations. Table 2.4 shows some example coding for the (1,6,5) floating-point format including the special numbers.

For the 12-bit fixed-point format we use one sign bit, 5 integer bits, and 6 fractional bits. The maximum (absolute) values we can represent with this 12-bit fixed-point format are therefore:

$$\pm 11111.111111_2 = \pm (16 + 8 + \cdots \frac{1}{32} + \frac{1}{64})_{10}$$

$$= \pm (32 - \frac{1}{64})_{10} \approx \pm 32.0_{10}.$$

The (absolutely measured) smallest number that this 12-bit fixed-point format represents is

$$\pm 00000.000001_2 = \pm \frac{1}{64}_{10} = \pm 0.015625_{10}.$$

2.16

From this example we notice the larger *dynamic range* of the floating-point representation (4×10^9 compared with 32) but also a higher *precision* of the fixed-point representation. For instance, 1.0 and $1+1/64 = 1.015625$ are code the same in (1,6,5) floating-point format, but can be distinguished in 12-bit fixed-point representation.

Although the IEEE standard 754-1985 for binary floating-point arithmetic [45] is not easy to implement with all its details such as four different rounding modes, denormals, or NaNs, the early introduction in 1985 of the standard helped as it has become the most adopted implementation for microprocessors. The parameters of this IEEE single and double format can be seen from Table 2.5. Due to the fact that already single-precision 754 standard arithmetic designs will require

- a 24×24 bit multiplier, and
- FPGAs allow a more specific dynamic range design (i.e., exponent bit width) and precision (mantissa bit width) design

we find that FPGAs design usually do not adopt the 754 standard and define a special format. Shirazi et al. [47], for instance, have developed a modified format to implement various algorithms on their custom computing machine called SPLASH-2, a multiple-FPGA board based on Xilinx XC4010 devices. They used an 18-bit format so that they can transport two operands over the 36-bit wide system bus of the multiple-FPGA board. The 18-bit format has a 10-bit mantissa, 7-bit exponent and a sign bit, and can represent a range of 3.7×10^{19}.

Table 2.5. IEEE floating-point standard.

	Single	Double
Word length	32	64
Mantissa	23	52
Exponent	8	11
Bias	127	1023
Range	$2^{128} \approx 3.8 \times 10^{38}$	$2^{1024} \approx 1.8 \times 10^{308}$

2.3 Binary Adders

A basic binary N-bit adder/subtractor consists of N full-adders (FA). A full-adder implements the following Boolean equations

$$s_k = x_k \ \text{XOR} \ y_k \ \text{XOR} \ c_k \tag{2.25}$$

$$= x_k \oplus y_k \oplus c_k \tag{2.26}$$

that define the sum-bit. The carry (out) bit is computed with:

$$c_{k+1} = (x_k \ \text{AND} \ y_k) \ \text{OR} \ (x_k \ \text{AND} \ c_k) \ \text{OR} \ (y_k \ \text{AND} \ c_k) \tag{2.27}$$

$$= (x_k \times y_k) + (x_k \times c_k) + (y_k \times c_k) \tag{2.28}$$

In the case of a 2C adder, the LSB can be reduced to a half-adder because the carry input is zero.

The simplest adder structure is called the "ripple carry adder" as shown in Fig. 2.7a in a bit-serial form. If larger tables are available in the FPGA, several bits can be grouped together into one LUT, as shown in Fig. 2.7b. For this "two bit at a time" adder the longest delay comes from the ripple of the carry through all stages. Attempts have been made to reduce the carry delays using techniques such as the carry-skip, carry lookahead, conditional sum,

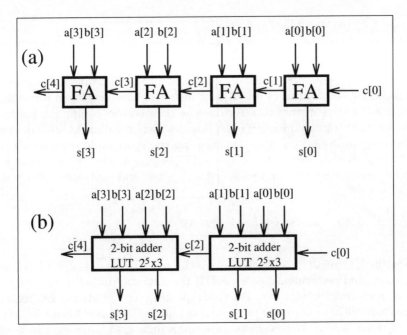

Fig. 2.7. Two's complement adders.

or carry-select adders. These techniques can speed up addition and can be used with older-generation FPGA families (e.g., XC 3000 from Xilinx) since these devices do not provide internal fast carry logic. Modern families, such as the Xilinx Spartan-3 or Altera Cyclone II, possess very fast "ripple carry logic" that is about a magnitude faster than the delay through a regular logic LUT [1]. Altera uses fast tables (see Fig. 1.13, p. 21), while the Xilinx uses hardwired decoders for implementing carry logic based on the multiplexer structure shown in Fig. 2.8, see also Fig. 1.12, p. 19. The presence of the fast-carry logic in modern FPGA families removes the need to develop hardware intensive carry look-ahead schemes.

Figure 2.9 summarizes the size and `Registered Performance` of N-bit binary adders, if implemented with the `lpm_add_sub` megafunction component. Beside the EP2C35F672C6 from the Cyclone II family (that is build currently using a 90-nm process technology), we have also included as a reference the data for mature families. The EP20K200EFC484-2X is from the APEX20KE family and can be found on the Nios development boards, see Chap. 9. The APEX20KE family was introduced in 1999 and used a $0.18\,\mu$m process technology. The EPF10K70RC240-4 is from the FLEX10K family and can be found on the UP2 development boards. The FLEX10K family was introduced in 1995 and used a $0.42\,\mu$m process technology. Although the LE cell structure has not changed much over time we can see from the advance in process technology the improvement in speed. If the operands are

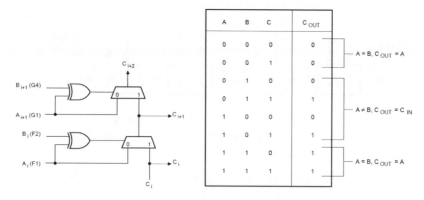

Fig. 2.8. XC4000 fast-carry logic (©1993 Xilinx).

placed in I/O register cells, the delays through the busses of a FPGA are dominant and performance decreases. If the data are routed from local registers, performance improves. For this type of design additional LE register allocation will appear (in the project report file) as increased LE use by a factor of three or four. However, a synchronous registered larger system would not consume any additional resources since the data are registered at the pervious processing stage. A typical design will achieve a speed between these two cases. For Flex10K the adder and register are not merged, and $4 \times N$ LEs are required. LE requirements for the Cyclone II and APEX devices are $3 \times N$ for the speed data shown in Fig. 2.9.

2.3.1 Pipelined Adders

Pipelining is extensively used in DSP solutions due to the intrinsic dataflow regularity of DSP algorithms. Programmable digital signal processor MACs [6, 15, 16] typically carry at least four pipelined stages. The processor:

1) Decodes the command
2) Loads the operands in registers
3) Performs multiplication and stores the product, and
4) Accumulates the products, all concurrently.

The pipelining principle can be applied to FPGA designs as well, at little or no additional cost since each logic element contains a flip-flop, which is otherwise unused, to save routing resources. With pipelining it is possible to break an arithmetic operation into small primitive operations, save the carry and the intermediate values in registers, and continue the calculation in the next clock cycle. Such adders are sometimes called carry save adders[4]

---

[4] The name carry save adder is also used in the context of a Wallace multiplier, see Exercise 2.1, p. 154.

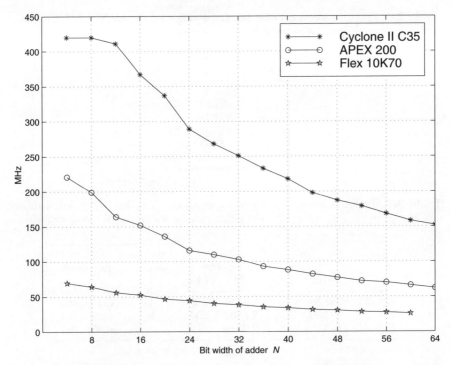

Fig. 2.9. Adder speed and size for Cyclone II, APEX, and Flex10K.

(CSAs) in the literature. Then the question arises: In how many pieces should we divide the adder? Should we use bit level? For Altera's Cyclone II devices a reasonable choice will be always using an LAB with 16 LEs and 16 FFs for one pipeline element. The FLEX10K family has 8 LEs per LAB, while APEX20KE uses 10 LEs per LAB. So we need to consult the datasheet before we make a decision on the size of the pipelining group. In fact, it can be shown that if we try to pipeline (for instance) a 14-bit adder in our Cyclone II devices, the performance does not improve, as reported in Table 2.6, because the pipelined 14-bit adder does not fit in one LAB.

Because the number of flip-flops in one LAB is 16 and we need an extra flip-flop for the carry-out, we should use a maximum block size of 15 bits for maximum `Registered Performance`. Only the blocks with the MSBs can be 16 bits wide, because we do not need the extra flip-flop for the carry. This observation leads to the following conclusions:

1) With one additional pipeline stage we can build adders up to a length $15 + 16 = 31$.
2) With two pipeline stages we can build adders with up to $15+15+16 = 46$-bit length.

Table 2.6. Performance of a 14-bit pipelined adder for the EP2C35F672C6 using synthesis of predefined LPM modules with pipeline option.

Pipeline stages	MHz	LEs
0	395.57	42
1	388.50	56
2	392.31	70
3	395.57	84
4	394.63	98
5	395.57	113

Table 2.7. Performance and resource requirements of adders with and without pipelining. Size and speed are for the maximum bit width, for 31-, 46-, and 61-bit adders.

Bit width	No Pipeline		With pipeline		Pipeline stages	Design file name
	MHz	LEs	MHz	LEs		
$17-31$	253.36	93	316.46	125	1	add1p.vhd
$32-46$	192.90	138	229.04	234	2	add2p.vhd
$47-61$	153.78	183	215.84	372	3	add3p.vhd

3) With three pipeline stages we can build adders with up to $15+15+15+16 = 61$-bit length.

Table 2.7 shows the **Registered Performance** and LE utilization of this kind of pipelined adder. From Table 2.7 it can be concluded that although the bit width increases the **Registered Performance** remains high if we add the appropriate number of pipeline stages.

The following example shows the code of a 31-bit pipelined adder. It turns out that the straight forward implementation of the pipelining would require two registers for the MSBs as shown in Fig. 2.10a. If we instead use adders for the MSBs, we can save a set of LEs, since each LE can implement a full adder, but only one flip-flop. This is graphically interpreted by Fig. 2.10b.

Example 2.17: VHDL Design of 31-bit Pipelined Adder

Consider the VHDL code[5] of a 31-bit pipelined adder that is graphically interpreted in Fig. 2.10. The design runs at 316.46 MHz and uses 125 LEs.

```
LIBRARY ieee;
USE ieee.std_logic_1164.ALL;
USE ieee.std_logic_arith.ALL;
USE ieee.std_logic_unsigned.ALL;
```

---

[5] The equivalent Verilog code **add1p.v** for this example can be found in Appendix A on page 666. Synthesis results are shown in Appendix B on page 731.

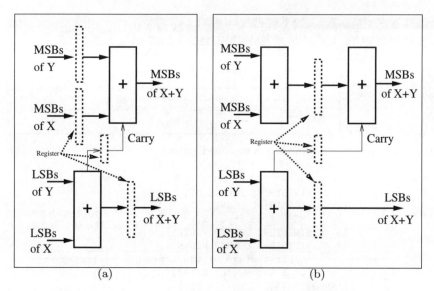

Fig. 2.10. Pipelined adder. **(a)** Direct implementation. **(b)** FPGA optimized approach.

```
ENTITY add1p IS
  GENERIC (WIDTH  : INTEGER := 31; -- Total bit width
           WIDTH1 : INTEGER := 15;  -- Bit width of LSBs
           WIDTH2 : INTEGER := 16);  -- Bit width of MSBs
  PORT (x,y : IN  STD_LOGIC_VECTOR(WIDTH-1 DOWNTO 0);
                                            -- Inputs
        sum : OUT STD_LOGIC_VECTOR(WIDTH-1 DOWNTO 0);
                                            -- Result
        LSBs_Carry : OUT STD_LOGIC;
        clk : IN  STD_LOGIC);
END add1p;

ARCHITECTURE fpga OF add1p IS

  SIGNAL l1, l2, s1                    -- LSBs of inputs
                  : STD_LOGIC_VECTOR(WIDTH1-1 DOWNTO 0);
  SIGNAL r1                            -- LSBs of inputs
                  : STD_LOGIC_VECTOR(WIDTH1 DOWNTO 0);
  SIGNAL l3, l4, r2, s2               -- MSBs of inputs
                  : STD_LOGIC_VECTOR(WIDTH2-1 DOWNTO 0);

BEGIN

  PROCESS -- Split in MSBs and LSBs and store in registers
  BEGIN
   WAIT UNTIL clk = '1';
   -- Split LSBs from input x,y
```

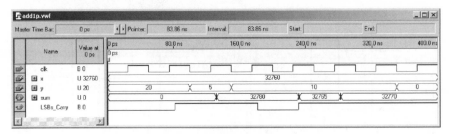

Fig. 2.11. Simulation results for a pipelined adder.

```
      l1 <= x(WIDTH1-1 DOWNTO 0);
      l2 <= y(WIDTH1-1 DOWNTO 0);
   -- Split MSBs from input x,y
      l3 <= x(WIDTH-1 DOWNTO WIDTH1);
      l4 <= y(WIDTH-1 DOWNTO WIDTH1);
-------------- First stage of the adder ------------------
      r1 <= ('0' & l1) + ('0' & l2);
      r2 <= l3 + l4;
------------ Second stage of the adder --------------------
      s1 <= r1(WIDTH1-1 DOWNTO 0);
   -- Add result von MSBs (x+y) and carry from LSBs
      s2 <= r1(WIDTH1) + r2;
   END PROCESS;
   LSBs_Carry <= r1(WIDTH1); -- Add a test signal

   -- Build a single output word of WIDTH = WIDTH1 + WIDHT2
      sum <= s2 & s1 ;     -- Connect s to output pins

   END fpga;
```

The simulated performance of the 15-bit pipelined adder shows Fig. 2.11b. Note that the addition results for 32780 and 32770 produce a carry from the lower 15-bit adder, but there is no carry for $32\,760 + 5 = 32\,765 < 2^{15}$. $\boxed{2.17}$

2.3.2 Modulo Adders

Modulo adders are the most important building blocks in RNS-DSP designs. They are used for both additions and, via index arithmetic, for multiplications. We wish to describe some design options for FPGAs in the following discussion.

A wide variety of *modular* addition designs exists [48]. Using LEs only, the design of Fig. 2.12a is viable for FPGAs. The Altera FLEX devices contain a small number of M2K ROMs or RAMs (EABs) that can be configured as $2^8 \times 8, 2^9 \times 4, 2^{10} \times 2$ or $2^{11} \times 1$ tables and can be used for modulo m_l correction. The next table shows size and **Registered Performance** 6, 7, and 8-bit modulo adder compile for Altera FLEX10K devices [49].

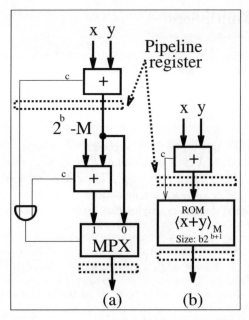

Fig. 2.12. Modular additions. **(a)** MPX-Add and MPX-Add-Pipe. **(b)** ROM-Pipe.

	Pipeline	Bits		
	stages	6	7	8
MPX	0	41.3 MSPS	46.5 MSPS	33.7 MSPS
		27 LE	31 LE	35 LE
MPX	2	76.3 MSPS	62.5 MSPS	60.9 MSPS
		16 LE	18 LE	20 LE
MPX	3	151.5 MSPS	138.9 MSPS	123.5 MSPS
		27 LE	31 LE	35 LE
ROM	3	86.2 MSPS	86.2 MSPS	86.2 MSPS
		7 LE	8 LE	9 LE
		1 EAB	1 EAB	2 EAB

Although the ROM shown in Fig 2.12 provides high speed, the ROM itself produces a four-cycle pipeline delay and the number of ROMs is limited. ROMs, however, are mandatory for the scaling schemes discussed before. The multiplexed-adder (MPX-Add) has a comparatively reduced speed even if a carry chain is added to each column. The pipelined version usually needs the same number of LEs as the unpipelined version but runs about three times as fast. Maximum throughput occurs when the adders are implemented with 3 pipeline stages and 6-bit width channels.

2.4 Binary Multipliers

The product of two N-bit binary numbers, say X and $A = \sum_{k=0}^{N-1} a_k 2^k$, is given by the "pencil and paper" method as:

$$P = A \times X = \sum_{k=0}^{N-1} a_k 2^k X. \qquad (2.29)$$

It can be seen that the input X is successively shifted by k positions and whenever $a_k \neq 0$, then $X2^k$ is accumulated. If $a_k = 0$, then the corresponding shift-add can be ignored (i.e., nop). The following VHDL example uses this "pencil and paper" scheme implemented via FSM to multiply two 8-bit integers. Other FSM design examples can be found in Exercises 2.20, p. 158 and 2.21, p. 159.

Example 2.18: 8-bit Multiplier

The VHDL description[6] of an 8-bit multiplier is developed below. Multiplication is performed in three stages. After **reset**, the 8-bit operands are "loaded" and the product register is set to zero. In the second stage, **s1**, the actual serial-parallel multiplication takes place. In the third step, **s2**, the product is transferred to the output register **y**.

```
PACKAGE eight_bit_int IS            -- User-defined types
  SUBTYPE BYTE IS INTEGER RANGE -128 TO 127;
  SUBTYPE TWOBYTES IS INTEGER RANGE -32768 TO 32767;
END eight_bit_int;

LIBRARY work;
USE work.eight_bit_int.ALL;

LIBRARY ieee;                       -- Using predefined packages
USE ieee.std_logic_1164.ALL;
USE ieee.std_logic_arith.ALL;

ENTITY mul_ser IS                        ------> Interface
  PORT ( clk, reset  : IN   STD_LOGIC;
          x    : IN  BYTE;
          a    : IN  STD_LOGIC_VECTOR(7 DOWNTO 0);
          y    : OUT TWOBYTES);
END mul_ser;

ARCHITECTURE fpga OF mul_ser IS

  TYPE STATE_TYPE IS (s0, s1, s2);
  SIGNAL state    : STATE_TYPE;

BEGIN
  ------> Multiplier in behavioral style
  States: PROCESS(reset, clk)
```

---

[6] The equivalent Verilog code **mul_ser.v** for this example can be found in Appendix A on page 670. Synthesis results are shown in Appendix B on page 731.

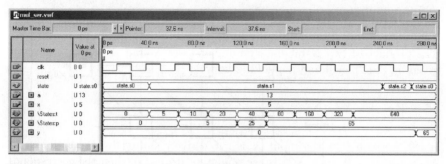

Fig. 2.13. Simulation results for a shift add multiplier.

```
VARIABLE  p, t  : TWOBYTES:=0;          -- Double bit width
VARIABLE count  : INTEGER RANGE 0 TO 7;
BEGIN
  IF reset = '1' THEN
    state <= s0;
  ELSIF rising_edge(clk) THEN
  CASE state IS
    WHEN s0 =>          -- Initialization step
      state <= s1;
      count := 0;
      p := 0;          -- Product register reset
      t := x;          -- Set temporary shift register to x
    WHEN s1 =>              -- Processing step
      IF count = 7 THEN -- Multiplication ready
        state <= s2;
      ELSE
      IF a(count) = '1' THEN
        p := p + t;      -- Add 2^k
      END IF;
      t := t * 2;
      count := count + 1;
      state <= s1;
      END IF;
    WHEN s2 =>          -- Output of result to y and
      y <= p;          -- start next multiplication
      state <= s0;
  END CASE;
  END IF;
END PROCESS States;

  END fpga;
```

Figure 2.13 shows the simulation result of a multiplication of 13 and 5.
The register t shows the partial product sequence of $5, 10, 20, \ldots$. Since
$13_{10} = 00001101_{2C}$, the product register p is updated only three times in
the production of the final result, 65. In state s2 the result 65 is transferred
to the output y of the multiplier. The design uses 121 LEs and and no embedded multiplier. With synthesis style **Speed** its runs with a **Registered
Performance** of 256.15 MHz

2.18

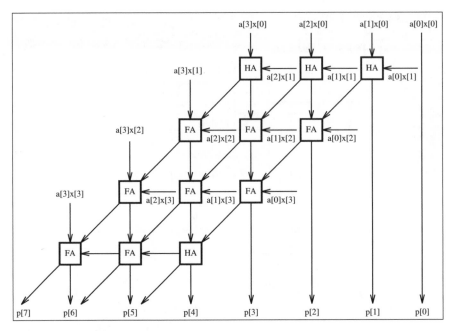

Fig. 2.14. A 4-bit array multiplier.

Because one operand is used in parallel (i.e., X) and the second operand A is used bitwise, the multipliers we just described are called serial/parallel multipliers. If both operands are used serial, the scheme is called a serial/serial multiplier [50], and such a multiplier only needs one full adder, but the latency of serial/serial multipliers is high $\mathcal{O}(N^2)$, because the state machine needs about N^2 cycles.

Another approach, which trades speed for increased complexity, is called an "array," or parallel/parallel multiplier. A 4×4-bit array multiplier is shown in Fig. 2.14. Notice that both operands are presented in parallel to an adder array of N^2 adder cells.

This arrangement is viable if the times required to complete the carry and sum calculations are the same. For a modern FPGA, however, the carry computation is performed faster than the sum calculation and a different architecture is more efficient for FPGAs. The approach for this array multiplier is shown in Fig. 2.15, for an 8×8-bit multiplier. This scheme combines in the first stage two neighboring partial products $a_n X 2^n$ and $a_{n+1} X 2^{n+1}$ and the results are added to arrive at the final output product. This is a direct array form of the "pencil and paper" method and must therefore produce a valid product.

We recognize from Fig. 2.15 that this type of array multiplier gives the opportunity to realize a (parallel) *binary tree* of the multiplier with a total:

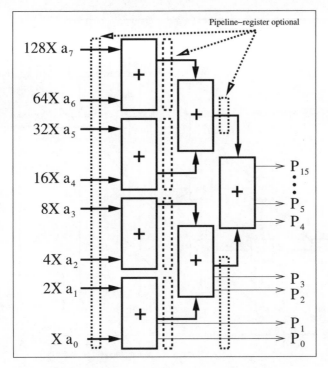

Fig. 2.15. Fast array multiplier for FPGAs.

$$\text{number of stages in the binary tree multiplier} = \log_2(N). \qquad (2.30)$$

This alternative architecture also makes it easier to introduce pipeline stages after each tree level. The necessary number of pipeline stages, according to (2.30), to achieve maximum throughput is:

Bit width	2	$3-4$	$5-8$	$9-16$	$17-32$
Optimal number of pipeline stages	1	2	3	4	5

Since the data are registered at the input and output the number of delays in the simulation would be two larger then the pipeline stage we specified for the `lpm_mul` blocks.

Figure 2.16 reports the **Registered Performance** of pipelined $N \times N$-bit multipliers, using the Quartus II `lpm_mult` function, for 8×8, to 24×24 bits operands. Embedded multiplier are shown with dash lines and up to 16×16-bit the multiplier do not improve with pipelining since they fit in one embedded 18×18-bit array multiplier. The LE-based multiplier are shown with a solid line. Figure 2.17 shows the LEs effort for the multiplier. The pipelined 8×8 bit multiplier outperforms the embedded multiplier if 2 or

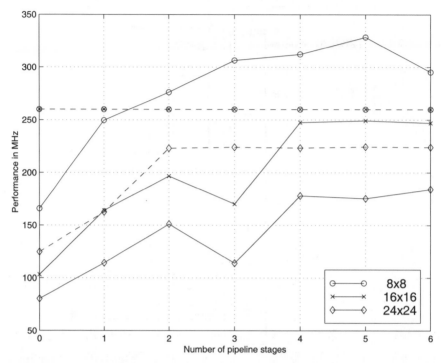

Fig. 2.16. Performance of an array multiplier for FPGAs, LE-based multiplier (**solid line**) and embedded multiplier (**dashed line**).

more pipeline stages are used. We can notice from Fig. 2.16 that, for pipeline delays longer than $\log_2(N)$, there is no essential improvement for LE-based multipliers. The multiplier architecture (embedded or LEs) must be controlled via synthesis options in case we write behavioral code (e.g., p <= a*b). This can be done in the EDA Tools Setting under the Assignments menu. There you find the DSP Block Balancing entry under the Analysis & Synthesis Settings. Select DSP blocks if you like to use the embedded multiplier, Logic Elements to use the LEs only, or Auto, and the synthesis tool will first use the embedded multiplier; if there are not enough then use the LE-based multiplier. If we use the lpm_mul block (see Appendix B, p. 733) we have direct control using the GENERIC MAP parameter DEDICATED_MULTIPLIER_CIRCUITRY => "YES" or "NO".

Other multiplier architectures typically used in the ASIC world include Wallace-tree multipliers and Booth multipliers. They are discussed in Exercises 2.1 (p. 154) and 2.2 (p. 154) but are rarely used in connection with FPGAs.

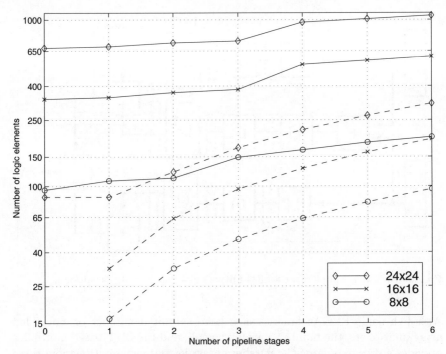

Fig. 2.17. Effort in LEs for array multipliers, LE-based multiplier (**solid line**) and embedded multiplier (**dashed line**).

2.4.1 Multiplier Blocks

A $2N \times 2N$ multiplier can be defined in terms of an $N \times N$ multiplier block [29]. The resulting multiplication is defined as:

$$P = Y \times X = (Y_2 2^N + Y_1)(X_2 2^N + X_1)$$
$$= Y_2 X_2 2^{2N} + (Y_2 X_1 + Y_1 X_2)2^N + Y_1 X_1, \tag{2.31}$$

where the indices 2 and 1 indicate the most significant and least significant N-bit halves, respectively. This partitioning scheme can be used if the capacity of the FPGA is insufficient to implement a multiplier of desired size, or used to implement a multiplier using memory blocks. A 36×36-bit multiplier can be build with four 18×18 bit embedded multipliers and three adders. An 8×8-bit LUT-based multiplier in direct form would require an LUT size of $2^{16} \times 16 = 1\,\text{Mbit}$. The partitioning technique reduces the table size to four $2^8 \times 8$ memory blocks and three adders. A 16×16-bit multiplier requires 16 M4K blocks. The benefit of multiplier implementation via M4Ks versus LE-based is twofold. First, the number of LE is reduced. Secondly, the requirements on the routing resources of the devices are also reduced.

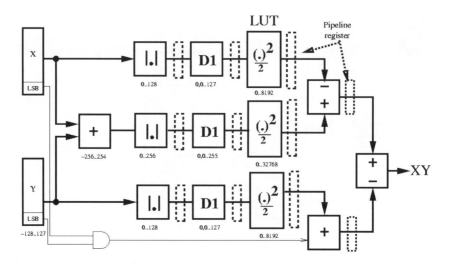

Fig. 2.18. Two's complement 8-bit additive half-square multiplier design.

Although some FPGAs families now have a limited number of embedded array multipliers, the number is usually small, and the LUT-based multiplier provides a way to enlarge the number of fast low-latency multipliers in these devices. In addition, some device families like Cyclone, Flex, or Excalibur do not have embedded multipliers; therefore, the LUT or LE multipliers are the only option.

Half-Square Multiplier

Another way to reduce the memory requirement for LUT-based multipliers is to decrease the bits in the input domain. One bit decrease in the input domain decreases the number of LUT words by a factor of two. An LUT of a square operation of an N-bit word only requires an LUT size of $2^N \times 2^N$. The additive half-square (AHSM) multiplier

$$
Y \times X = \frac{(X+Y)^2 - X^2 - Y^2}{2} =
$$
$$
= \left\lfloor \frac{(X+Y)^2}{2} \right\rfloor - \left\lfloor \frac{X^2}{2} \right\rfloor - \left\lfloor \frac{Y^2}{2} \right\rfloor - \begin{cases} 1 & X, Y \text{ odd} \\ 0 & \text{others} \end{cases} \qquad (2.32)
$$

was introduced by Logan [51]. If the division by 2 is included in the LUT, this requires a correction of -1 in the event that X and Y are odd. A differential half-square multiplier (DHSM) can then be implemented as:

$$
Y \times X = \frac{(X+Y)^2 - X^2 - Y^2}{2}
$$

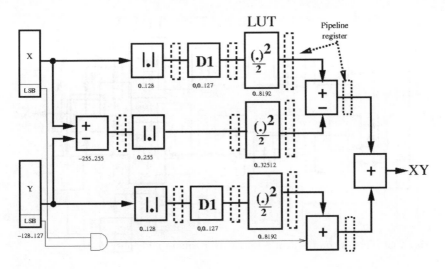

Fig. 2.19. Two's complement 8-bit differential half-square multiplier design.

$$= \left\lfloor \frac{X^2}{2} \right\rfloor + \left\lfloor \frac{Y^2}{2} \right\rfloor - \left\lfloor \frac{(X-Y)^2}{2} \right\rfloor + \begin{cases} 1 & X, Y \text{ odd} \\ 0 & \text{others} \end{cases}. \qquad (2.33)$$

A correction of 1 is required in the event that X and Y are odd. If the numbers are signed, an additional saving is possible by using the diminished-by-one (D1) encoding, see Sect. 2.2.1, p. 56. In D1 coding all numbers are diminished by 1, and the zero gets special encoding [52]. Figure 2.18 shows for 8-bit data the AHSM multiplier, the required LUTs, and the data range of 8-bit input operands. The absolute operation almost allows a reduction by a factor of 2 in LUT words, while the D1 encoding enables a reduction to the next power-of-2 table size that is beneficial for the FPGA design. Since LUT inputs 0 and 1 both have the same square, LUT entry $\lfloor A^2/2 \rfloor$, we share this value and do not need to use special encoding for zero. Without the division by 2, a 17-bit output word would be required. However, the division by two in the squaring table requires an increment (decrement) of the output result for the AHSM (DHSM) in case both input operands are odd values. Figure 2.19 shows a DHSM multiplier that only requires two D1 encoding compared with the AHSM design.

Quarter-Square Multiplier

A further reduction in arithmetic requirements and the number of LUTs can be achieved by using the quarter-square multiplication (QSM) principle that is also well studied in analog designs [53, 54]. The QSM is based on the following equation:

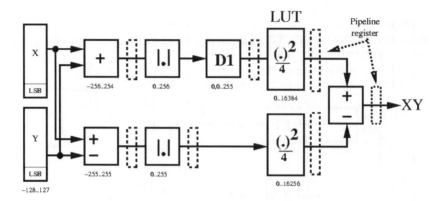

Fig. 2.20. Two's complement 8-bit quarter-square multiplier design.

$$Y \times X = \left\lfloor \frac{(X+Y)^2}{4} \right\rfloor - \left\lfloor \frac{(X-Y)^2}{4} \right\rfloor.$$

It is interesting to note that the division by 4 in (2.34) does not require any correction for operation as in the HSM case. This can be checked as follows. If both operands are even (odd), then the sum and the difference are both even, and the squaring followed by a division of 4 produces no error (i.e., $4|(2u*2v)$). If one operand is odd (even) and the other operand is even (odd), then the sum and the difference after squaring and a division by 4 produce a 0.25 error that is annihilated in both cases. No correction operation is necessary. The direct implementation of (2.34) would require LUTs of $(N+1)$-bit inputs to represent the correct result of $X \pm Y$ as used in [55], which will require four $2^N \times 2^N$ LUTs. Signed arithmetic along with D1 coding will reduce the table to the next power-of-2 value, allowing the design to use only two $2^N \times 2^N$ LUTs compared with the four in [55]. Figure 2.20 shows the D1 QSM circuit.

LUT-Based Multiplier Comparison

For each of the multiplier circuits HDL code can be developed (see Exercises 2.23-2.25, p. 161) and short C programs or MATLAB scripts are necessary to generate the memory initialization files for two's complement, unsigned, and D1 data. The Verilog code from [55] and the half and quarter square designs are then synthesized using the Altera Quartus II software for the popular Cyclone II device from Altera development board. Table 2.8 quantifies the resources required and reports the performance data for the LUT-based multipliers. The table shows the required LUTs for an 8×8-bit signed multiplier, the number of logic elements (LEs), the maximum frequency, and the number of M4K blocks used. Results reveal that the D1 multiplier uses 50% less

LUT resources than proposed in [55] for Cyclone II devices with a moderate increase in LE usage. The D1 QSM doubles the number of fast M4K-based multipliers in the FPGA. Throughput is restricted by the synchronous M4K blocks to 260 MHz in Cyclone II devices.

Comparing the data of Table 2.8 with the data from Figs. 2.16 (p. 86) and 2.17 (p. 87), it can be seen that the LUT-based multiplier reduces the number of LEs but does not improve the Registered Performance.

Table 2.8. Resource and performance data for 8×8-bit signed LUT-based multipliers.

Design	LUT size	LEs	M4K	Reg. Perf. in MHz	Eq. or Ref
Partitioning	$4 \times 2^8 \times 8$	40	4	260.0	(2.31)
Altera's QSM	$2 \times 2^9 \times 16$	34	4	180.9	[55]
D1 AHSM	$2 \times 2^7 \times 16, 2^8 \times 16$	118	3	260.0	(2.32)
D1 DHSM	$2 \times 2^7 \times 16, 2^8 \times 16$	106	3	260.0	(2.33)
D1 QSM	$2 \times 2^8 \times 16$	66	2	260.0	(2.34)

2.5 Binary Dividers

Of all four basic arithmetic operations division is the most complex. Consequently, it is the most time-consuming operation and also the operation with the largest number of different algorithms to be implemented. For a given dividend (or numerator) N and divisor (or denominator) D the division produces (unlike the other basic arithmetic operations) two results: the quotient Q and the remainder R, i.e.,

$$\frac{N}{D} = Q \quad \text{and} \quad R \quad \text{with } |R| < D. \tag{2.34}$$

However, we may think of division as the inverse process of multiplication, as demonstrated through the following equation,

$$N = D \times Q + R, \tag{2.35}$$

it differs from multiplication in many aspects. Most importantly, in multiplication all partial products can be produced parallel, while in division each quotient bit is determined in a sequential "trail-and-error" procedure.

Because most microprocessors handle division as the inverse process to multiplications, referring to (2.35), the numerator is assumed to be the result of a multiplication and has therefore twice the bit width of denominator and quotient. As a consequence, the quotient has to be checked in an awkward procedure to be in the valid range, i.e., that there is no overflow in the quotient. We wish to use a more general approach in which we assume that

$$Q \leq N \quad \text{and} \quad |R| \leq D,$$

i.e., quotient and numerator as well as denominator and remainder are assumed to be of the same bit width. With this bit width assumptions no range check (except $N = 0$) for a valid quotient is necessary.

Another consideration when implementing division comes when we deal with signed numbers. Obviously, the easiest way to handle signed numbers is first to convert both to unsigned numbers and compute the sign of the result as an XOR or modulo 2 add operation of the sign bits of the two operands. But some algorithms, (like the nonrestoring division discussed below), can directly process signed numbers. Then the question arises, how are the sign of quotient and remainder related. In most hardware or software systems (but not for all, such as in the PASCAL programming language), it is assumed that the remainder and the quotient have the same sign. That is, although

$$\frac{234}{50} = 5 \quad \text{and} \quad R = -16 \tag{2.36}$$

meets the requirements from (2.35), we, in general, would prefer the following results

$$\frac{234}{50} = 4 \quad \text{and} \quad R = 34. \tag{2.37}$$

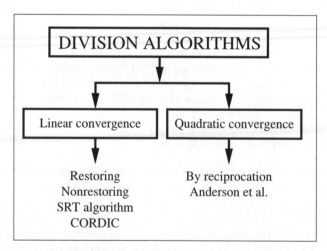

Fig. 2.21. Survey of division algorithms.

Let us now start with a brief overview of the most commonly used division algorithms. Figure 2.21 shows the most popular linear and quadratic convergence schemes. A basic categorization of the linear division algorithms can be done according to the permissible values of each quotient digit generated. In the binary *restoring, nonperforming* or CORDIC algorithms the digits are selected from the set

$$\{0, 1\}.$$

In the binary nonrestoring algorithms a signed-digit set is used, i.e.,

$$\{-1, 1\} = \{\bar{1}, 1\}.$$

In the binary SRT algorithm, named after Sweeney, Robertson, and Tocher [29] who discovered the algorithms at about the same time, the digits from the ternary set

$$\{-1, 0, 1\} = \{\bar{1}, 0, 1\}$$

are used. All of the above algorithms can be extended to higher radix algorithms. The generalized SRT division algorithms of radix r, for instance, uses the digit set

$$\{-2^r - 1, \ldots, -1, 0, 1, \ldots, 2^r - 1\}.$$

We find two algorithms with quadratic convergence to be popular. The first algorithm is the division by reciprocation of the denominator, where we compute the reciprocal with the Newton algorithm for finding zeros. The second quadratic convergence algorithms was developed for the IBM 360/91 in the 1960s by Anderson et al. [56]. This algorithm multiplies numerator and denominator with the same factors and converges $N \rightarrow 1$, which results in $D \rightarrow Q$. Note, that the division algorithms with quadratic convergence produce no remainder.

Although the number of iterations in the quadratic convergence algorithms are in the order of $\log_2(b)$ for b bit operands, we must take into account that each iteration step is more complicated (i.e., uses two multiplications) than the linear convergence algorithms, and speed and size performance comparisons have to be done carefully.

2.5.1 Linear Convergence Division Algorithms

The most obvious sequential algorithms is our "pencil-and-paper" method (which we have used many times before) translated into binary arithmetic. We align first the denominator and load the numerator in the remainder register. We then subtract the aligned denominator from the remainder and store the result in the remainder register. If the new remainder is positive we set the quotient's LSB to 1, otherwise the quotient's LSB is set to zero and we need to restore the previous remainder value by adding the denominator. Finally, we have to realign the quotient and denominator for the next step. The recalculation of the previous remainder is why we call such an algorithm "restoring division." The following example demonstrates a FSM implementation of the algorithm.

Example 2.19: 8-bit Restoring Divider

The VHDL description[7] of an 8-bit divider is developed below. Division is performed in four stages. After **reset**, the 8-bit numerator is "loaded" in the remainder register, the 6-bit denominator is loaded and aligned (by 2^{N-1} for a N bit numerator), and the quotient register is set to zero. In the second and third stages, s1 and s2, the actual serial division takes place. In the fourth step, s3, quotient and remainder are transferred to the output registers. Nominator and quotient are assumed to be 8 bits wide, while denominator and remainder are 6-bit values.

```
-- Restoring Division
LIBRARY ieee;                     -- Using predefined packages
USE ieee.std_logic_1164.ALL;
USE ieee.std_logic_arith.ALL;
USE ieee.std_logic_unsigned.ALL;

ENTITY div_res IS                       ------> Interface
  GENERIC(WN : INTEGER := 8;
          WD : INTEGER := 6;
          PO2WND : INTEGER := 8192; -- 2**(WN+WD)
          PO2WN1 : INTEGER := 128;  -- 2**(WN-1)
          PO2WN : INTEGER := 255);  -- 2**WN-1
  PORT ( clk, reset     : IN  STD_LOGIC;
         n_in  : IN  STD_LOGIC_VECTOR(WN-1 DOWNTO 0);
         d_in  : IN  STD_LOGIC_VECTOR(WD-1 DOWNTO 0);
         r_out : OUT STD_LOGIC_VECTOR(WD-1 DOWNTO 0);
         q_out : OUT STD_LOGIC_VECTOR(WN-1 DOWNTO 0));
END div_res;

ARCHITECTURE flex OF div_res IS

  SUBTYPE TWOWORDS IS INTEGER RANGE -1 TO PO2WND-1;
  SUBTYPE WORD IS INTEGER RANGE 0 TO PO2WN;

  TYPE STATE_TYPE IS (s0, s1, s2, s3);
  SIGNAL s : STATE_TYPE;

BEGIN
-- Bit width:  WN          WD          WN          WD
--           Numerator / Denominator = Quotient and Remainder
-- OR:       Numerator = Quotient * Denominator + Remainder

  States: PROCESS(reset, clk)-- Divider in behavioral style
    VARIABLE  r, d : TWOWORDS :=0;   -- N+D bit width
    VARIABLE  q : WORD;
    VARIABLE count  : INTEGER RANGE 0 TO WN;
  BEGIN
    IF reset = '1' THEN                 -- asynchronous reset
      s <= s0;
    ELSIF rising_edge(clk) THEN
    CASE s IS
```

_____

[7] The equivalent Verilog code div_res.v for this example can be found in Appendix A on page 671. Synthesis results are shown in Appendix B on page 731.

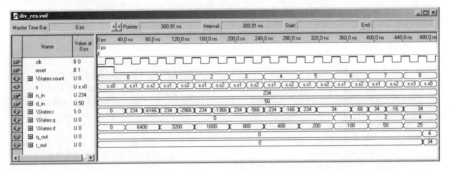

Fig. 2.22. Simulation results for a restoring divider.

```
WHEN s0 =>              -- Initialization step
  s <= s1;
  count := 0;
  q := 0;               -- Reset quotient register
  d := PO2WN1 * CONV_INTEGER(d_in); -- Load denom.
  r := CONV_INTEGER(n_in); -- Remainder = numerator
WHEN s1 =>              -- Processing step
    r := r - d;        -- Subtract denominator
    s <= s2;
WHEN s2 =>             -- Restoring step
  IF r < 0 THEN
    r := r + d;        -- Restore previous remainder
    q := q * 2;        -- LSB = 0 and SLL
  ELSE
    q := 2 * q + 1;   -- LSB = 1 and SLL
  END IF;
  count := count + 1;
  d := d / 2;
  IF count = WN THEN   -- Division ready ?
    s <= s3;
  ELSE
    s <= s1;
  END IF;
WHEN s3 =>                      -- Output of result
  q_out <= CONV_STD_LOGIC_VECTOR(q, WN);
  r_out <= CONV_STD_LOGIC_VECTOR(r, WD);
  s <= s0;                     -- Start next division
END CASE;
END IF;
END PROCESS States;

END flex;
```

Figure 2.22 shows the simulation result of a division of 234 by 50. The register d shows the aligned denominator values $50 \times 2^7 = 6400, 50 \times 2^6 = 3200, \ldots$. Every time the remainder r calculated in step s1 is negative, the previous remainder is restored in step s2. In state s3 the quotient 4 and the remainder 34 are transferred to the output registers of the divider. The design uses

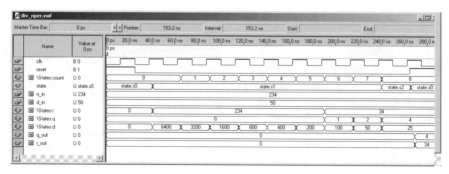

Fig. 2.23. Simulation results for a nonperforming divider.

127 LEs, no embedded multiplier, and runs with a `Registered Performance` of 265.32 MHz.

2.19

The main disadvantage of the restoring division is that we need two steps to determine one quotient bit. We can combine the two steps using a *non-performing* divider algorithm, i.e., each time the denominator is larger than the remainder, we do *not* perform the subtraction. In VHDL we would write the new step as:

```
t := r - d;          -- temporary remainder value
IF t >= 0 THEN       -- Nonperforming test
  r := t;            -- Use new denominator
  q := q * 2 + 1;    -- LSB = 1 and SLL
ELSE
  q := q * 2;        -- LSB = 0 and SLL
END IF;
```

The number of steps is reduced by a factor of 2 (not counting initialization and transfers of results), as can be seen from the simulation in Fig. 2.23. Note also from the simulation shown in Fig. 2.23 that the remainder r is never negative in the nonperforming division algorithms. On the downside the worst case delay path is increased when compared with the restoring division and the maximum `Registered Performance` is expected to be reduced, see Exercise 2.17 (p. 157). The nonperforming divider has two arithmetic operations and the if condition in the worst case path, while the restoring divider has (see step s2) only the if condition and one arithmetic operation in the worst case path.

A similar approach to the nonperforming algorithm, but that does *not* increase the critical path, is the so-called *nonrestoring* division. The idea behind the nonrestoring division is that if we have computed in the restoring division a negative remainder, i.e., $r_{k+1} = r_k - d_k$, then in the next step we will restore r_k by adding d_k and then perform a subtraction of the next aligned denominator $d_{k+1} = d_k/2$. So, instead of adding d_k followed by subtracting $d_k/2$, we can just skip the restoring step and proceed with adding $d_k/2$, when the remainder has (temporarily) a negative value. As a result, we have now quotient bits that can be positive or negative, i.e., $q_k = \pm 1$, but not zero. We can change this signed-digit representation later to a two's complement representation. In conclusion, the nonrestoring algorithms works as follows: every time the remainder after the iteration is positive we store a 1 and subtract the aligned denominator, while for negative remainder, we store a $-1 = \bar{1}$ in the quotient register and add the aligned denominator. To use only one bit in the quotient register we will use a zero in the quotient register to code the -1. To convert this signed-digit quotient back to a two's complement word, the straightforward way is to put all 1s in one word and the zeros, which are actually the coded $-1 = \bar{1}$ in the second word as a one. Then we need just to subtract the two words to compute the two's complement. On the other hand this subtraction of the -1s is nothing other than the complement of the quotient augmented by 1. In conclusion, if q holds the signed-digit representation, we can compute the two's complement via

$$q_{2C} = 2 \times q_{SD} + 1. \tag{2.38}$$

Both quotient and remainder are now in the two's complement representation and have a valid result according to (2.35). If we wish to constrain our results in a way that both have the same sign, we need to correct the negative remainder, i.e., for $r < 0$ we correct this via

$$r := r + D \quad \text{and} \quad q := q - 1.$$

Such a nonrestoring divider will now run faster than the nonperforming divider, with about the same **Registered Performance** as the restoring divider, see Exercise 2.18 (p. 157). Figure 2.24 shows a simulation of the nonrestoring divider. We notice from the simulation that register values of the remainder are allowed now again to be negative. Note also that the abovementioned correction for negative remainder is necessary for this value. The not corrected result is $q = 5$ and $r = -16$ The equal sign correction results in $q = 5 - 1 = 4$ and $r = -16 + 50 = 34$, as shown in Fig. 2.24.

To shorten further the number of clock cycles needed for the division higher radix (array) divider can be built using, for instance, the SRT and radix 4 coding. This is popular in ASIC designs when combined with the carry-save-adder principle as used in the floating-point accelerators of the Pentium microprocessors. For FPGAs with a limited LUT size this higher-order schemes seem to be less attractive.

A totally different approach to improve the latency are the division algorithms with quadratic convergence, which use fast array multiplier. The two most popular versions of this quadratic convergence schemes are discussed in the next section.

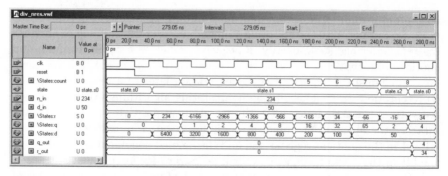

Fig. 2.24. Simulation results for a nonrestoring divider.

2.5.2 Fast Divider Design

The first fast divider algorithm we wish to discuss is the division through multiplication with the reciprocal of the denominator D. The reciprocal can, for instance, be computed via a look-up table for small bit width. The general technique for constructing iterative algorithms, however, makes use of the Newton method for finding a zero. According to this method, we define a function

$$f(x) = \frac{1}{x} - D \quad \rightarrow \quad 0. \tag{2.39}$$

If we define an algorithm such that $f(x_\infty) = 0$ then it follows that

$$\frac{1}{x_\infty} - D = 0 \quad \text{or} \quad x_\infty = \frac{1}{D}. \tag{2.40}$$

Using the tangent the estimation for the next x_{k+1} is calculated using

$$x_{k+1} = x_k - \frac{f(x_k)}{f'(x_k)}, \tag{2.41}$$

with $f(x) = 1/x - D$ we have $f'(x) = 1/x^2$ and the iteration equation becomes

$$\boxed{x_{k+1} = x_k - \frac{\frac{1}{x_k} - D}{\frac{-1}{x_k^2}} = x_k(2 - D \times x_k).} \tag{2.42}$$

Although the algorithm will converge for any initial D, it converges much faster if we start with a normalized value close to 1.0, i.e., we normalized D in such a way that $0.5 \leq D < 1$ or $1 \leq D < 2$ as used for floating-point mantissa, see Sect. 2.6 (p. 104). We can then use an initial value $x_0 = 1$ to get fast convergence. Let us illustrate the Newton algorithm with a short example.

Example 2.20: Newton Algorithm

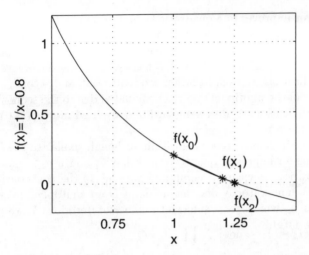

Fig. 2.25. Newton's zero-finding algorithms for $x_\infty = 1/0.8 = 1.25$.

Let us try to compute the Newton algorithm for $1/D = 1/0.8 = 1.25$. The following table shows in the first column the number of the iteration, in the second column the approximation to $1/D$, in the third column the error $x_k - x_\infty$, and in the last column the equivalent bit precision of our approximation.

k	x_k	$x_k - x_\infty$	Eff. bits
0	1.0	−0.25	2
1	1.2	−0.05	4.3
2	1.248	−0.002	8.9
3	1.25	-3.2×10^{-6}	18.2
4	1.25	-8.2×10^{-12}	36.8

Figure 2.25 shows a graphical interpretation of the Newton zero-finding algorithm. The $f(x_k)$ converges rapidly to zero. $\boxed{2.20}$

Because the first iterations in the Newton algorithm only produce a few bits of precision, it may be useful to use a small look-up table to skip the first iterations. A table to skip the first two iterations can, for instance, be found in [29, p. 260].

We note also from the above example the overall rapid convergence of the algorithm. Only 5 steps are necessary to have over 32-bit precision. Many more steps would be required to reach the same precision with the linear convergence algorithms. This quadratic convergence applies for all values not only for our special example. This can be shown as follows:

$$e_{k+1} = x_{k+1} - x_\infty = x_k(2 - D \times x_k) - \frac{1}{D}$$

$$= -D\left(x_k - \frac{1}{D}\right)^2 = -De_k^2,$$

i.e., the error improves in a quadratic fashion from one iteration to the next. With each iteration we double the effective number of bit precision.

Although the Newton algorithm has been successfully used in microprocessor design (e.g., IBM RISC 6000), it has two main disadvantages: First, the two multiplications in each iteration are sequential, and second, the quantization error of the multiplication is accumulated due to the sequential nature of the multiplication. Additional guard bits are used in general to avoid this quantization error.

The following convergence algorithm, although similar to the Newton algorithm, has an improved quantization behavior and uses 2 multiplications in each iteration that can be computed parallel. In the *convergence division* scheme both numerator N and denominator D are multiplied by approximation factors f_k, which, for a sufficient number of iterations k, we find

$$D \prod f_k \to 1 \quad \text{and} \quad N \prod f_k \to Q. \tag{2.43}$$

This algorithm, originally developed for the IBM 360/91, is credited to Anderson et al. [56], and the algorithm works as follows:

Algorithm 2.21:   **Division by Convergence**

1) Normalize N and D such that D is close to 1. Use a normalization interval such as $0.5 \le D < 1$ or $1 \le D < 2$ as used for floating-point mantissa.
2) Initialize $x_0 = N$ and $t_0 = D$.
3) Repeat the following loop until x_k shows the desired precision.
$$f_k = 2 - t_k$$
$$x_{k+1} = x_k \times f_k$$
$$t_{k+1} = t_k \times f_k$$

It is important to note that the algorithm is self-correcting. Any quantization error in the factors does not really matter because numerator and denominator are multiplied with the same factor f_k. This fact has been used in the IBM 360/91 design to reduce the required resources. The multiplier used for the first iteration has only a few significant bits, while in later iteration more multiplier bits are allocated as the factor f_k gets closer to 1.

Let us demonstrate the multiply by convergence algorithm with the following example.

Example 2.22: Anderson–Earle–Goldschmidt–Powers Algorithm

Let us try to compute the division-by-convergence algorithm for $N = 1.5$ and $D = 1.2$, i.e., $Q = N/D = 1.25$ The following table shows in the first column the number of the iteration, in the second column the scaling factor f_k, in the third column the approximation to N/D, in the fourth column the error $x_k - x_\infty$, and in the last column the equivalent bit precision of our approximation.

k	f_k	x_k	$x_k - x_\infty$	Eff. bits
0	$0.8 \approx \frac{205}{256}$	$1.5 \approx \frac{384}{256}$	0.25	2
1	$1.04 \approx \frac{267}{256}$	$1.2 \approx \frac{307}{256}$	-0.05	4.3
2	$1.0016 \approx \frac{257}{256}$	$1.248 \approx \frac{320}{256}$	0.002	8.9
3	$1.0 + 2.56 \times 10^{-6}$	1.25	-3.2×10^{-6}	18.2
4	$1.0 + 6.55 \times 10^{-12}$	1.25	-8.2×10^{-12}	36.8

We note the same quadratic convergence as in the Newton algorithm, see Example 2.20 (p. 99).

The VHDL description[8] of an 8-bit fast divider is developed below. We assume that denominator and numerator are normalized as, for instance, typical for floating-point mantissa values, to the interval $1 \leq N, D < 2$. This normalization step may require essential addition resources (leading-zero detection and two barrelshifters) when the denominator and numerator are not normalized. Nominator, denominator, and quotient are all assumed to be 9 bits wide. The decimal values 1.5, 1.2, and 1.25 are represented in a 1.8-bit format (1 integer and 8 fractional bits) as $1.5 \times 256 = 384, 1.2 \times 256 = 307$, and $1.25 \times 256 = 320$, respectively. Division is performed in three stages. First, the 1.8-formatted denominator and numerator are loaded into the registers. In the second state, s1, the actual convergence division takes place. In the third step, s2, the quotient is transferred to the output register.

```
-- Convergence division after Anderson, Earle, Goldschmidt,
LIBRARY ieee;                              -- and Powers
USE ieee.std_logic_1164.ALL;
USE ieee.std_logic_arith.ALL;
USE ieee.std_logic_unsigned.ALL;


ENTITY div_aegp IS                  ------> Interface
  GENERIC(WN : INTEGER := 9; -- 8 bit plus one integer bit
          WD : INTEGER := 9;
          STEPS : INTEGER := 2;
          TWO : INTEGER := 512; -- 2**(WN+1)
          PO2WN  : INTEGER := 256;  -- 2**(WN-1)
          PO2WN2 : INTEGER := 1023); -- 2**(WN+1)-1
  PORT ( clk, reset : IN  STD_LOGIC;
         n_in       : IN  STD_LOGIC_VECTOR(WN-1 DOWNTO 0);
         d_in       : IN  STD_LOGIC_VECTOR(WD-1 DOWNTO 0);
         q_out      : OUT STD_LOGIC_VECTOR(WD-1 DOWNTO 0));
END div_aegp;

ARCHITECTURE fpga OF div_aegp IS

  SUBTYPE WORD IS INTEGER RANGE 0 TO PO2WN2;

  TYPE STATE_TYPE IS (s0, s1, s2);
  SIGNAL state    : STATE_TYPE;

BEGIN
-- Bit width:  WN          WD          WN          WD
--          Numerator / Denominator = Quotient and Remainder
```

---

[8] The equivalent Verilog code div_aegp.v for this example can be found in Appendix A on page 673. Synthesis results are shown in Appendix B on page 731.

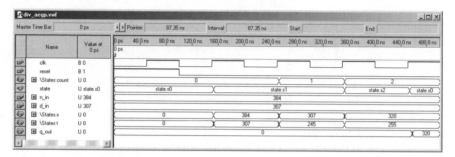

Fig. 2.26. Simulation results for a convergence divider.

```
-- OR:          Numerator = Quotient * Denominator + Remainder

States: PROCESS(reset, clk)-- Divider in behavioral style
   VARIABLE  x, t, f : WORD:=0; -- WN+1 bits
   VARIABLE count  : INTEGER RANGE 0 TO STEPS;
BEGIN
   IF reset = '1' THEN                 -- asynchronous reset
      state <= s0;
   ELSIF rising_edge(clk) THEN
   CASE state IS
      WHEN s0 =>                    -- Initialization step
         state <= s1;
         count := 0;
         t := CONV_INTEGER(d_in); -- Load denominator
         x := CONV_INTEGER(n_in); -- Load numerator
      WHEN s1 =>               -- Processing step
         f := TWO - t;
         x := x * f / PO2WN;
         t := t * f / PO2WN;
         count := count + 1;
         IF count = STEPS THEN -- Division ready ?
            state <= s2;
         ELSE
            state <= s1;
         END IF;
      WHEN s2 =>                       -- Output of results
         q_out <= CONV_STD_LOGIC_VECTOR(x, WN);
         state <= s0;               -- start next division
   END CASE;
   END IF;
END PROCESS States;
```

```
END fpga;
```
Figure 2.26 shows the simulation result of the division 1.5/1.2. The variable f
(which becomes an internal net and is not shown in the simulation) holds the
three scaling factors 205, 267, and 257, sufficient for 8-bit precision results.
The x and t values are multiplied by the scaling factor f and scaled down to
the 1.8 format. x converges to the quotient $1.25 = 320/256$, while t converges
to $1.0 = 255/256$, as expected. In state s3 the quotient $1.25 = 320/256$

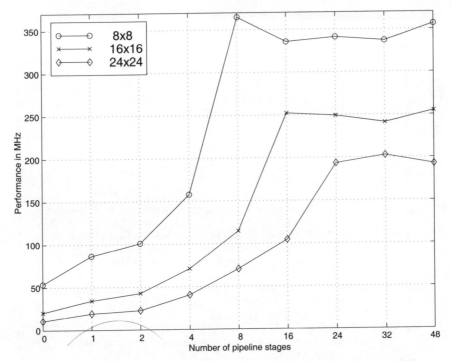

Fig. 2.27. Performance of array divider using the `lpm_divide` macro block.

is transferred to the output registers of the divider. Note that the divider produces no remainder. The design uses 64 LEs, 4 embedded multipliers and runs with a `Registered Performance` of 134.63 MHz. $\boxed{2.22}$

Although the `Registered Performance` of the nonrestoring divider (see Fig. 2.24) is about twice as high, the total latency, however, in the convergence divider is reduced, because the number of processing steps are reduced from 8 to $\lceil\sqrt{8}\rceil = 3$ (not counting initialization in both algorithms). The convergence divider uses less LEs as the nonrestoring divider but also 4 embedded multipliers.

2.5.3 Array Divider

Obviously, as with multipliers, all division algorithms can be implemented in a sequential, FSM-like, way or in the array form. If the array form and pipelining is desired, a good option will then be to use the `lpm_divide` block, which implements an array divider with the option of pipelining, see Appendix B, (p. 749) for a detailed description of the `lpm_divide` block.

Figure 2.27 shows the `Registered Performance` and Fig. 2.28 the LEs necessary for $8 \times 8-$, $16 \times 16-$, and $24 \times 24-$bit array dividers. Note the

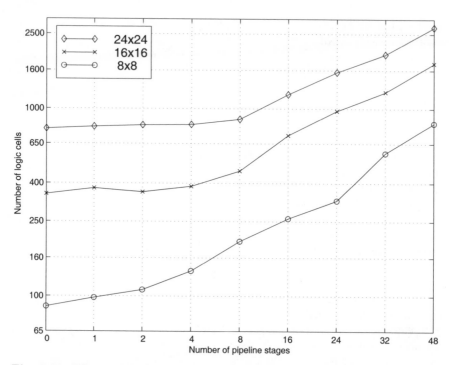

Fig. 2.28. Effort in LEs for array divider using the `lpm_divide` macro block.

logarithmic like scaling for the number of pipeline stages. We conclude from the performance measurement, that the optimal number of pipeline stages is the same as the number of bits in the denominator.

2.6 Floating-Point Arithmetic Implementation

Due to the large gate count capacity of current FPGAs the design of floating-point arithmetic has become a viable option. In addition, the introduction of the embedded 18×18 bit array multiplier in Altera Stratix or Cyclone and Xilinx Virtex II or Spartan III FPGA device families allows an efficient design of custom floating-point arithmetic. We will therefore discuss the design of basic building blocks such as a floating-point adder, subtractor, multiplier, reciprocal and divider, and the necessary conversion blocks to and from fixed-point data format. Such blocks are available from several IP providers, or through special request via e-mail to `Uwe.Meyer-Baese@ieee.org`.

Most of the commercially available floating-point blocks use (typically 3) pipeline stages to increase the throughput. To keep the presentation simple we will not use pipelining. The custom floating-point format we will use is the $(1,6,5)$ floating-point format introduced in Sect. 2.2.3, (p. 71). This format uses 1 sign bit, 6 bits for the exponent and 5 bits for the mantissa. We support special coding for zero and infinities, but we do not support NaNs or denormals. Rounding is done via truncation. The fixed-point format used in the examples has 6 integer bits (including a sign bit) and 6 fractional bits.

2.6.1 Fixed-point to Floating-Point Format Conversion

As shown in Sect. 2.2.3, (p. 71), floating-point numbers use a signed-magnitude format and the first step is therefore to convert the two's complement number to signed-magnitude form. If the sign of the fixed-point number is one, we need to compute the complement of the fixed-point number, which becomes the unnormalized mantissa. In the next step we normalize the mantissa and compute the exponent. For the normalization we first determine the number of leading zeros. This can be done with a LOOP statement within a sequential PROCESS in VHDL. Using this number of leading zeros, we shift the mantissa left, until the first 1 "leaves" the mantissa registers, i.e., the hidden one is also removed. This shift operation is actually the task of a barrelshifter, which can be inferred in VHDL via the SLL instruction. Unfortunately we can not use the SLL with Altera's Quartus II because it is only defined for BIT_VECTOR data type, but not for the STD_LOGIC_VECTOR data type we need for other arithmetic operations. But we can design a barrelshifter in many different ways as Exercise 2.19 (p. 157) shows. Another alternative would be to design a function overloading for the STD_LOGIC_VECTOR that allows a shift operation, see Exercise 1.20, p. 50.

The exponent of our floating-point number is computed as the sum of the bias and the number of integer bits in our fixed-point format minus the leading zeros in the not normalized mantissa.

Finally, we concatenate the sign, exponent, and the normalized mantissa to a single floating-point word if the fixed-point number is not zero, otherwise we set the floating-point word also to zero.

We have assumed that the range of the floating-point number is larger than the range of the fixed-point number, i.e., the special number ∞ will never be used in the conversion.

Figure 2.29 shows the conversion from 12-bit fixed-point data to the (1,6,5) floating-point data for five values ±1, absolute maximum, absolute minimum, and the smallest value. Rows 1 to 3 show the 12-bit fixed-point number and the integer and fractional parts. Rows 4 to 7 show the complete floating-point number, followed by the three parts, sign, exponent, and mantissa. The last row shows the decimal values.

2.6.2 Floating-Point to Fixed-Point Format Conversion

The floating-point to fixed-point conversion is, in general, more complicated than the conversion in the other direction. Depending if the exponent is

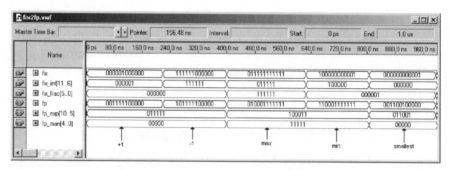

Fig. 2.29. Simulation results for a (1,5,6) fixed-point format to (1,6,5) floating-point conversion.

larger or smaller than the bias we need to implement a left or right shift of the mantissa. In addition, extra consideration is necessary for the special values ±∞ and ±0.

To keep the discussion as simple as possible, we assume in the following that the floating-point number has a larger dynamic range than the fixed-point number, but the fixed-point number has a higher precision, i.e., the number of fractional bits of the fixed-point number is larger than the bits used for the mantissa in the floating-point number.

The first step in the conversion is the correction of the bias in the exponent. We then place the hidden 1 to the left and the (fractional) mantissa to the right of the decimal point of the fixed-point word. We then check if the exponent is too large to be represented with the fixed-point number and set the fixed-point number then to the maximum value. Also, if the exponent is too small, we set the output value to zero. If the exponent is in the valid range that the floating-point number can be represented with the fixed-point format, we shift left the 1.m mantissa value (format see (2.23), p. 71) for positive exponents, and shift right for negative exponent values. This, in general, can be coded with the SLL and SRL in VHDL, respectively, but these BIT_VECTOR operations are not supported in Altera's Quartus II for STD_LOGIC_VECTOR, see Exercise 1.20, p. 50. In the final step we convert the signed magnitude representation to the two's complement format by evaluating the sign bit of the floating-point number.

Figure 2.30 shows the conversion from (1,6,5) floating-point format to (1,5,6) fixed-point data for the five values ±1, absolute maximum, absolute minimum, and the smallest value. The last row shows the decimal values, rows 1 to 4 the 12-bit floating-point number and the three parts, sign, exponent, and mantissa. The rows 5 to 7 show the complete fixed-point number, followed by the integer and fractional parts. Note that the conversion is without any quantization error for ±1 and the smallest value. For the absolute maximum and minimum values, however, the smaller precision in the floating-point numbers gives the imperfect conversion values compared with Fig. 2.29.

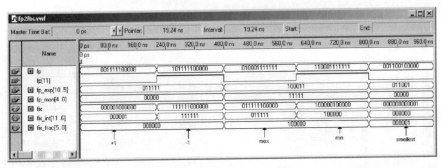

Fig. 2.30. Simulation results for (1,6,5) floating-point format to (1,5,6) fixed-point format conversion.

2.6.3 Floating-Point Multiplication

In contrast to fixed-point operations, multiplication in floating-point is the simplest of all arithmetic operations and we will discuss this first. In general, the multiplication of two numbers in scientific format is accomplished by multiplication of the mantissas and adding of the exponents, i.e.,

$$f_1 \times f_2 = (a_1 2^{e_1}) \times (a_2 2^{e_2}) = (a_1 \times a_2) 2^{e_1 + e_2}.$$

For our floating-point format with an implicit one and a biased exponent this becomes

$$f_1 \times f_2 = (-1)^{s_1} \left(1.m_1 2^{e_1 - \text{bias}}\right) \times (-1)^{s_2} \left(1.m_2 2^{e_2 - \text{bias}}\right)$$

$$= (-1)^{s_1 + s_2 \bmod 2} \underbrace{(1.m_1 \times 1.m_2)}_{m_3} 2^{\overbrace{e_1 + e_2 - \text{bias}}^{e_3} - \text{bias}}$$

$$= (-1)^{s_3} 1.m_3 2^{e_3 - \text{bias}}.$$

We note that the exponent sum needs to be adjusted by the bias, since the bias is included twice in both exponents. The sign of the product is the XOR or modulo-2 sum of the two sign bits of the two operands. We need also to take care of the special values. If one factor is ∞ the product should be ∞ too. Next, we check if one factor is zero and set the product to zero if true. Because we do not support NaNs, this implies that $0 \times \infty$ is set to ∞. Special values may also be produced from original nonspecial operands. If we detect an overflow, i.e.,

$$e_1 + e_2 - \text{bias} \geq E_{\max},$$

we set the product to ∞. Likewise, if we detect an underflow, i.e.,

$$e_1 + e_2 - \text{bias} \leq E_{\min},$$

we set the product to zero. It can be seen that the internal representation of the exponent e_3 of the product, must have two more bits than the two factors,

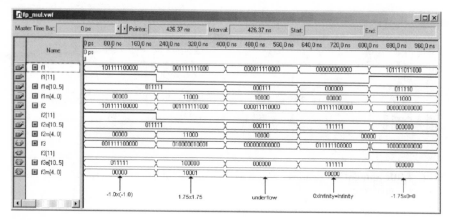

Fig. 2.31. Simulation results for multiplications with floating-point numbers in the (1,6,5) format.

because we need a sign and a guard bit. Fortunately, the normalization of the product $1.m_3$ is relatively simple, because both operands are in the range $1.0 \leq 1.m_{1,2} < 2.0$, the mantissa product is therefore in the range $1.0 \leq 1.m_3 < 4.0$, i.e., a shift by one bit (and exponent adjustment by 1) is sufficient to normalize the product.

Finally, we build the new floating-point number by concatenation of the sign, exponent, and magnitude.

Figure 2.31 shows the multiplication in the (1,6,5) floating-point format of the following values (see also last row in Fig. 2.31):

1) $(-1) \times (-1) = 1.0_{10} = 1.00000_2 \times 2^{31-\text{bias}}$
2) $1.75 \times 1.75 = 3.0625_{10} = 11.0001_2 \times 2^{31-\text{bias}} = 1.10001_2 \times 2^{32-\text{bias}}$
3) exponent: $7 + 7 - \text{bias} = -17 < E_{\min} \rightarrow$ underflow in multiplication
4) $0 \times \infty = \infty$ per definition (NaNs are not supported).
5) $-1.75 \times 0 = -0$

The rows 1 to 4 show the first floating-point number f1 and the three parts: sign, exponent, and mantissa. Rows 5 to 8 show the same for the second operand f2, and rows 9 to 12 the product f3 and the decomposition of the three parts.

2.6.4 Floating-Point Addition

Floating-point addition is more complex than multiplication. Two numbers is scientific format

$$f_3 = f_1 + f_2 = (a_1 2^{e_1}) \pm (a_2 2^{e_2})$$

can only be added if the exponents are the same, i.e., $e_1 = e_2$. Without loss of generality we assume in the following that the second number has the (absolute) smaller value. If this is not true, we just exchange the first and the second number. The next step is now to "denormalize" the smaller number by using the following identity:

$$a_2 2^{e_2} = a_2 / 2^d 2^{e_2 + d}.$$

If we select the normalization factor such as $e_2 + d = e_1$, i.e., $d = e_1 - e_2$, we get

$$a_2 / 2^d 2^{e_2 + d} = a_2 / 2^{e_1 - e_2} 2^{e_1}.$$

Now both numbers have the same exponent and we can, depending on the signs, add or subtract the first mantissa and the aligned second, according to

$$a_3 = a_1 \pm a_2 / 2^{e_1 - e_2}.$$

We need also to check if the second operand is zero. This is the case if $e_2 = 0$ or $d > M$, i.e., the shift operation reduces the second mantissa to zero. If the second operand is zero the first (larger) operand is forwarded to the result f_3.

The two aligned mantissas are added if the two floating-point operands have the same sign, otherwise subtracted. The new mantissa needs to be normalized to have the $1.m_3$ format, and the exponent, initially set to $e_3 = e_1$, needs to be adjusted accordingly to the normalization of the mantissa. We need to determine the number of leading zeros including the first one and perform a shift logic left (SLL). We also need to take into account if one of the operands is a special number, or if over- or underflow occurs. If the first operand is ∞ or the new computed exponent is larger than E_{max} the output is set to ∞. This implies that $\infty - \infty = \infty$ since NaNs are not supported. If the new computed exponent is smaller than E_{min}, underflow has occurred and the output is set to zero. Finally, we concatenate the sign, exponent, and mantissa to the new floating-point number.

Figure 2.32 shows the addition in the (1,6,5) floating-point format of the following values (see also last row in Fig. 2.32):

1) $9.25 + (-10.5) = -1.25_{10} = 1.01000_2 \times 2^{31 - \text{bias}}$
2) $1.0 + (-1.0) = 0$
3) $1.00111_2 \times 2^{2 - \text{bias}} + (-1.00100_2 \times 2^{2 - \text{bias}}) = 0.00011_2 \times 2^{2 - \text{bias}} = 1.1_2 \times 2^{-2 - \text{bias}} \to -2 < E_{min} \to$ underflow
4) $1.01111_2 \times 2^{62 - \text{bias}} + 1.11110_2 \times 2^{62 - \text{bias}} = 11.01101_2 2^{62 - \text{bias}} = 1.12^{63 - \text{bias}} \to 63 \geq E_{max} \to$ overflow
5) $-\infty + 1 = -\infty$

The rows 1 to 4 show the first floating-point number f1 and the three parts: sign, exponent, and mantissa. Rows 5 to 8 show the same for the second operand f2, and rows 9 to 12 show the sum f3 and the decomposition in the three parts, sign, exponent, and mantissa.

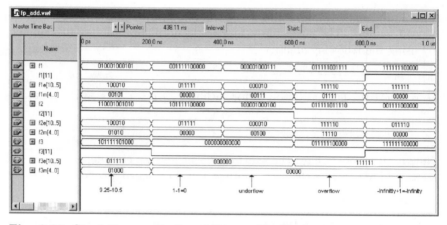

Fig. 2.32. Simulation results for additions with floating-point numbers in the (1,6,5) format.

2.6.5 Floating-Point Division

In general, the division of two numbers in scientific format is accomplished by division of the mantissas and subtraction of the exponents, i.e.,

$$f_1/f_2 = (a_1 2^{e_1})/(a_2 2^{e_2}) = (a_1/a_2)2^{e_1-e_2}.$$

For our floating-point format with an implicit one and a biased exponent this becomes

$$f_1/f_2 = (-1)^{s_1} \left(1.m_1 2^{e_1-\text{bias}}\right)/(-1)^{s_2}\left(1.m_2 2^{e_2-\text{bias}}\right)$$

$$= (-1)^{s_1+s_2 \bmod 2}\underbrace{(1.m_1/1.m_2)}_{m_3}2^{\overbrace{e_1 - e_2 - \text{bias}+\text{bias}}^{e_3}}$$

$$= (-1)^{s_3}1.m_3 2^{e_3+\text{bias}}.$$

We note that the exponent sum needs to be adjusted by the bias, since the bias is no longer present after the subtraction of the exponents. The sign of the division is the XOR or modulo-2 sum of the two sign bits of the two operands. The division of the mantissas can be implemented with any algorithm discussed in Sect. 2.5 (p. 91) or we can use the lpm_divide component. Because the denominator and quotient has to be at least M+1 bits wide, but numerator and quotient have the same bit width in the lpm_divide component, we need to use numerator and quotient with $2 \times (M + 1)$ bits. Because the numerator and denominator are both in the range $1 \leq 1.m_{1,2} < 2$, we conclude that the quotient will be in the range $0.5 \leq 1.m_3 < 2$. It follows that a normalization of only one bit (including the exponent adjustment by 1) is required.

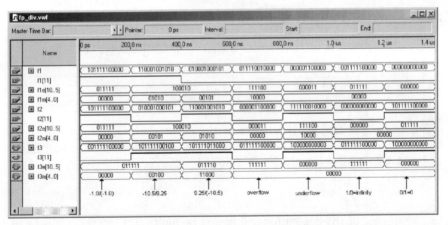

Fig. 2.33. Simulation results for division with floating-point numbers in the (1,6,5) format.

We need also to take care of the special values. The result is ∞ if the numerator is ∞, the denominator is zero, or we detect an overflow, i.e.,

$$e_1 - e_2 + \text{bias} = e_3 \geq E_{\max}.$$

Then we check for a zero quotient. The quotient is set to zero if the numerator is zero, denominator is ∞, or we detect an underflow, i.e.,

$$e_1 - e_2 + \text{bias} = e_3 \leq E_{\min}.$$

In all other cases the result is in the valid range that produces no special result.

Finally, we build the new floating-point number by concatenation of the sign, exponent, and magnitude.

Figure 2.33 shows the division in the (1,6,5) floating-point format of the following values (see also last row in Fig. 2.33):

1) $(-1)/(-1) = 1.0_{10} = 1.00000_2 \times 2^{31-\text{bias}}$
2) $-10.5/9.25_{10} = 1.\overline{135}_{10} \approx 1.001_2 \times 2^{31-\text{bias}}$
3) $9.25/(-10.5)_{10} = 0.880952_{10} \approx 1.11_2 \times 2^{30-\text{bias}}$
4) exponent: $60 - 3 + \text{bias} = 88 > E_{\max} \rightarrow$ overflow in division
5) exponent: $3 - 60 + \text{bias} = -26 < E_{\min} \rightarrow$ underflow in division
6) $1.0/0 = \infty$
7) $0/(-1.0) = -0.0$

Rows 1 to 4 show the first floating-point number and the three parts: sign, exponent, and mantissa. Rows 5 to 8 show the same for the second operand, and rows 9 to 12 show the quotient and the decomposition in the three parts.

2.6.6 Floating-Point Reciprocal

Although the reciprocal function of a floating-point number, i.e.,

$$1.0/f = \frac{1.0}{(-1)^s 1.m 2^e}$$

$$= (-1)^s 2^{-e}/1.m$$

seems to be less frequently used than the other arithmetic functions, it is nonetheless useful since it can also be used in combination with the multiplier to build a floating-point divider, because

$$f_1/f_2 = \frac{1.0}{f_2} \times f_1,$$

i.e., reciprocal of the denominator followed by multiplication is equivalent to the division.

If the bit width of the mantissa is not too large, we may implement the reciprocal of the mantissa, via a look-up table implemented with a `case` statement or with a M4K memory block. Because the mantissa is in the range $1 \leq 1.m < 2$, the reciprocal must be in the range $0.5 < \frac{1}{1.m} \leq 1$. The mantissa normalization is therefore a one-bit shift for all values except $f = 1.0$.

The following include file `fptab5.mif` was generated with the program `fpinv3e.exe` [9] (included on the CD-ROM under `book3e/util`) and shows the first few values for a 5-bit reciprocal look-up table. The file has the following contents:

```
--   This is the floating-point 1/x table for 5 bit data
--   automatically generated with fpinv3e.exe -- DO NOT EDIT!
depth = 32;
width = 5;
address_radix = uns;
data_radix = uns;
content
begin
0 : 0;
1 : 30; --   30.060606
2 : 28; --   28.235294
3 : 27; --   26.514286
4 : 25; --   24.888889
5 : 23; --   23.351351
6 : 22; --   21.894737
7 : 21; --   20.512821
8 : 19; --   19.200000
```

---

[9] You need to copy the program to your harddrive first; you can not start it from the CD directly.

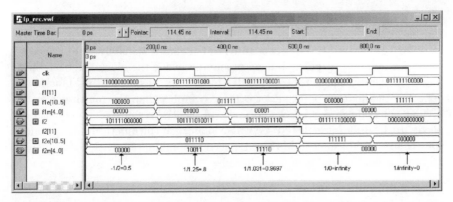

Fig. 2.34. Simulation results for reciprocal with floating-point numbers in the (1,6,5) format.

. . . .

END;

We also need to take care of the special values. The reciprocal of ∞ is 0, and the reciprocal of 0 is ∞. For all other values the new exponent e_2 is computed with

$$e_2 = -(e_1 - \text{bias}) + \text{bias} = 2 \times \text{bias} - e_1.$$

Finally, we build the reciprocal floating-point number by the concatenation of the sign, exponent, and magnitude.

Figure 2.34 shows the reciprocal in the (1,6,5) floating-point format of the following values (see also last row in Fig. 2.34):

1) $-1/2 = -0.5_{10} = -1.0_2 \times 2^{30-\text{bias}}$
2) $1/1.25_{10} = 0.8_{10} \approx (32 + 19)/64 = 1.10011_2 \times 2^{30-\text{bias}}$
3) $1/1.031 = 0.9697_{10} \approx (32 + 30)/64 = 1.11110_2 \times 2^{30-\text{bias}}$
4) $1.0/0 = \infty$
5) $1/\infty = 0.0$

For the first three values the entries (without leading 1) corresponds to the MIF file from above for the address line 0, 8, and 1, respectively. Rows 1 to 4 show the input floating-point number f1 and the three parts: sign, exponent, and mantissa. Rows 5 to 8 show the reciprocal f2 and the decomposition in the three parts. Notice that for the simulation we have to us a clock signal, since for Cyclone II device we can *not* use the M4K blocks without I/O register. If we use a FLEX10K device it would be possible to use the memory block also as asynchronous table only without additional I/O registers, see [57]. In order to align the I/O values in the same time slot without an one clock cycle delay we use a 10 ns offset.

2.6.7 Floating-Point Synthesis Results

In order to measure the `Registered Performance`, registers were added to the input and output ports, but no pipelining inside the block has been used. Table 2.9 shows the synthesis results for all six basic building blocks. As expected the floating-point adder is more complex than the multiplier or the divider. The conversion blocks also use substantial resources. The reciprocal block uses besides the listed LEs also one M4K memory block, or, more specifically, 160 bits of an M4K.

Table 2.9. Synthesis results for floating-point design using the (1,6,5) data format.

Block	MHz	LEs	9 × 9-bit embedded multiplier	M4K memory blocks
fix2fp	97.68	163	–	–
fp2fix	164.8	114	–	–
fp_mul	168.24	63	1	–
fp_add	57.9	181	–	–
fp_div	66.13	153	–	–
fp_rec	331.13	26	–	1

These blocks are available from several "intellectual property" providers, or through special request via e-mail to `Uwe.Meyer-Baese@ieee.org`.

2.7 Multiply-Accumulator (MAC) and Sum of Product (SOP)

DSP algorithms are known to be multiply-accumulate (MAC) intensive. To illustrate, consider the linear convolution sum given by

$$y[n] = f[n] * x[n] = \sum_{k=0}^{L-1} f[k]x[n-k] \tag{2.44}$$

requiring L consecutive multiplications and $L-1$ addition operations per sample $y[n]$ to compute the sum of products (SOPs). This suggests that a $N \times N$-bit multiplier need to be fused together with an accumulator, see Fig. 2.35a. A full-precision $N \times N$-bit product is $2N$ bits wide. If both operands are (symmetric) signed numbers, the product will only have $2N - 1$ significant bits, i.e., two sign bits. The accumulator, in order to maintain sufficient dynamic range, is often designed to be an extra K bits in width, as demonstrated in the following example.

Example 2.23: The Analog Devices PDSP family ADSP21xx contains a 16×16 array multiplier and an accumulator with an extra 8 bits (for a total accumulator width of $32 + 8 = 40$ bits). With this eight extra bits, at least 2^8 accumulations are possible without sacrificing the output. If both operands are symmetric signed, 2^9 accumulation can be performed. In order to produce the desired output format, such modern PDSPs include also a barrelshifter, which allows the desired adjustment within one clock cycle. $\boxed{\text{2.23}}$

This overflow consideration in fixed-point PDSP is important to mainstream digital signal processing, which requires that DSP objects be computed in real time without unexpected interruptions. Recall that checking and servicing accumulator overflow interrupts the data flow and carries a significant temporal liability. By choosing the number of guard bits correctly, the liability can be eliminated.

An alternative approach to the MAC of a conventional PDSP for computing a sum of product will be discussed in the next section.

2.7.1 Distributed Arithmetic Fundamentals

Distributed arithmetic (DA) is an important FPGA technology. It is extensively used in computing the sum of products

$$y = \langle\, \boldsymbol{c}, \boldsymbol{x}\, \rangle = \sum_{n=0}^{N-1} c[n] \times x[n]. \tag{2.45}$$

Besides convolution, correlation, DFT computation and the RNS inverse mapping discussed earlier can also be formulated as such a "sum of products" (SOPs). Completing a filter cycle, when using a conventional arithmetic unit, would take approximately N MAC cycles. This amount can be shortened with pipelining but can, nevertheless, be prohibitively long. This is a fundamental problem when general-purpose multipliers are used.

In many DSP applications, a general-purpose multiplication is technically not required. If the filter coefficients $c[n]$ are known a priori, then technically the partial product term $c[n]x[n]$ becomes a multiplication with a constant (i.e., scaling). This is an important difference and is a prerequisite for a DA design.

The first discussion of DA can be traced to a 1973 paper by Croisier [58] and DA was popularized by Peled and Liu [59]. Yiu [60] extended DA to signed numbers, and Kammeyer [61] and Taylor [62] studied quantization effects in DA systems. DA tutorials are available from White [63] and Kammeyer [64]. DA also is addressed in textbooks [65, 66]. To understand the DA design paradigm, consider the "sum of products" inner product shown below:

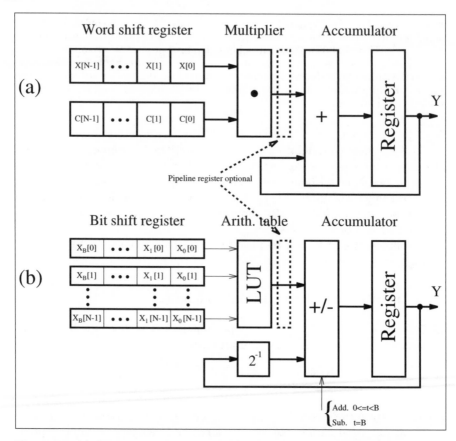

Fig. 2.35. (a) Conventional PDSP and (b) Shift-Adder DA Architecture.

$$y = \langle\, \boldsymbol{c}, \boldsymbol{x}\, \rangle = \sum_{n=0}^{N-1} c[n] \times x[n]$$
$$= c[0]x[0] + c[1]x[1] + \ldots + c[N-1]x[N-1]. \tag{2.46}$$

Assume further that the coefficients $c[n]$ are known constants and $x[n]$ is a variable. An unsigned DA system assumes that the variable $x[n]$ is represented by:

$$x[n] = \sum_{b=0}^{B-1} x_b[n] \times 2^b \quad \text{with } x_b[n] \in [0, 1], \tag{2.47}$$

where $x_b[n]$ denotes the bth bit of $x[n]$, i.e., the nth sample of $\boldsymbol{x}$. The inner product y can, therefore, be represented as:

$$y = \sum_{n=0}^{N-1} c[n] \times \sum_{b=0}^{B-1} x_b[n] \times 2^b. \tag{2.48}$$

Redistributing the order of summation (thus the name "distributed arithmetic") results in:

$$
\begin{aligned}
y = \ & c[0] \left(x_{B-1}[0]2^{B-1} + x_{B-2}[0]2^{B-2} + \ldots + x_0[0]2^0 \right) \\
& + c[1] \left(x_{B-1}[1]2^{B-1} + x_{B-2}[1]2^{B-2} + \ldots + x_0[1]2^0 \right) \\
& \ \ \vdots \\
& + c[N-1] \left(x_{B-1}[N-1]2^{B-1} + \ldots + x_0[N-1]2^0 \right) \\
= \ & \left(c[0]x_{B-1}[0] + c[1]x_{B-1}[1] + \ldots + c[N-1]x_{B-1}[N-1] \right) 2^{B-1} \\
& + \left(c[0]x_{B-2}[0] + c[1]x_{B-2}[1] + \ldots + c[N-1]x_{B-2}[N-1] \right) 2^{B-2} \\
& \ \ \vdots \\
& + \left(c[0]x_0[0] + c[1]x_0[1] + \ldots + c[N-1]x_0[N-1] \right) 2^0,
\end{aligned}
$$

or in more compact form

$$y = \sum_{b=0}^{B-1} 2^b \times \underbrace{\sum_{n=0}^{N-1} c[n] \times x_b[n]}_{f(c[n],x_b[n])} = \sum_{b=0}^{B-1} 2^b \times \sum_{n=0}^{N-1} f\left(c[n], x_b[n] \right). \tag{2.49}$$

Implementation of the function $f(c[n], x_b[n])$ requires special attention. The preferred implementation method is to realize the mapping $f(c[n], x_b[n])$ using one LUT. That is, a 2^N-word LUT is preprogrammed to accept an N-bit input vector $x_b = [x_b[0], x_b[1], \cdots, x_b[N-1]]$, and output $f(c[n], x_b[n])$. The individual mappings $f(c[n], x_b[n])$ are weighted by the appropriate power-of-two factor and accumulated. The accumulation can be efficiently implemented using a shift-adder as shown in Fig. 2.35b. After N look-up cycles, the inner product y is computed.

Example 2.24: Unsigned DA Convolution

A third-order inner product is defined by the inner product equation $y = \langle c, x \rangle = \sum_{n=0}^{2} c[n]x[n]$. Assume that the 3-bit coefficients have the values $c[0] = 2$, $c[1] = 3$, and $c[2] = 1$. The resulting LUT, which implements $f(c[n], x_b[n])$, is defined below:

$x_b[2]$ $x_b[1]$ $x_b[0]$	$f(c[n], x_b[n])$
0 0 0	$1 \times 0 + 3 \times 0 + 2 \times 0 = 0_{10} = 000_2$
0 0 1	$1 \times 0 + 3 \times 0 + 2 \times 1 = 2_{10} = 010_2$
0 1 0	$1 \times 0 + 3 \times 1 + 2 \times 0 = 3_{10} = 011_2$
0 1 1	$1 \times 0 + 3 \times 1 + 2 \times 1 = 5_{10} = 101_2$
1 0 0	$1 \times 1 + 3 \times 0 + 2 \times 0 = 1_{10} = 001_2$
1 0 1	$1 \times 1 + 3 \times 0 + 2 \times 1 = 3_{10} = 011_2$
1 1 0	$1 \times 1 + 3 \times 1 + 2 \times 0 = 4_{10} = 100_2$
1 1 1	$1 \times 1 + 3 \times 1 + 2 \times 1 = 6_{10} = 110_2$

The inner product, with respect to $x[n] = \{x[0] = 1_{10} = 001_2, x[1] = 3_{10} = 011_2, x[2] = 7_{10} = 111_2\}$, is obtained as follows:

Step t	$x_t[2]$ $x_t[1]$ $x_t[0]$	$f[t]$	$+ACC[t-1]$	$=ACC[t]$
0	1 1 1	$6 \times 2^0 +$	0	= 6
1	1 1 0	$4 \times 2^1 +$	6	= 14
2	1 0 0	$1 \times 2^2 +$	14	= 18

As a numerical check, note that

$$y = \langle c, x \rangle = c[0]x[0] + c[1]x[1] + c[2]x[2]$$
$$= 2 \times 1 + 3 \times 3 + 1 \times 7 = 18. \checkmark$$

$\boxed{2.24}$

For a hardware implementation, instead of shifting each intermediate value by b (which will demand an expensive barrelshifter) it is more appropriate to shift the accumulator content itself in each iteration one bit to the right. It is easy to verify that this will give the same results.

The bandwidth of an Nth-order B-bit linear convolution, using general-purpose MACs and DA hardware, can be compared. Figure 2.35 shows the architectures of a conventional PDSP and the same realization using distributed arithmetic.

Assume that a LUT and a general-purpose multiplier have the same delay $\tau = \tau(\text{LUT}) = \tau(\text{MUL})$. The computational latencies are then $B\tau(\text{LUT})$ for DA and $N\tau(\text{MUL})$ for the PDSP. In the case of small bit width B, the speed of the DA design can therefore be significantly faster than a MAC-based design. In Chap. 3, comparisons will be made for specific filter design examples.

2.7.2 Signed DA Systems

In the following, we wish to discuss how (2.46) should be modified, in order to process a signed two's complement number. In two's complement, the MSB is used to distinguish between positive and negative numbers. For instance, from Table 2.1 (p. 57) we see that decimal -3 is coded as $101_2 = -4 + 0 + 1 = -3_{10}$. We use, therefore, the following $(B+1)$-bit representation

$$x[n] = -2^B \times x_B[n] + \sum_{b=0}^{B-1} x_b[n] \times 2^b. \tag{2.50}$$

Combining this with (2.48), the outcome y is defined by:

$$y = -2^B \times f(c[n], x_B[n]) + \sum_{b=0}^{B-1} 2^b \times \sum_{n=0}^{N-1} f(c[n], x_b[n]). \qquad (2.51)$$

To achieve the signed DA system we therefore have two choices to modify the unsigned DA system. They are

- An accumulator with add/subtract control
- Using a ROM with one additional input

Most often the switchable accumulator is preferred, because the additional input bit in the table requires a table with twice as many words. The following example demonstrates the processing steps for the add/sub switch design.

Example 2.25: Signed DA Inner Product

Consider again a third-order inner product defined by the convolution sum $y = \langle c, x \rangle = \sum_{n=0}^{2} c[n]x[n]$. Assume that the data $x[n]$ is given in 4-bit two's complement encoding and that the coefficients are $c[0] = -2$, $c[1] = 3$, and $c[2] = 1$. The corresponding LUT table is given below:

$x_b[2]$ $x_b[1]$ $x_b[0]$			$f(c[k], x_b[n])$
0	0	0	$1 \times 0 + 3 \times 0 - 2 \times 0 = 0_{10}$
0	0	1	$1 \times 0 + 3 \times 0 - 2 \times 1 = -2_{10}$
0	1	0	$1 \times 0 + 3 \times 1 - 2 \times 0 = 3_{10}$
0	1	1	$1 \times 0 + 3 \times 1 - 2 \times 1 = 1_{10}$
1	0	0	$1 \times 1 + 3 \times 0 - 2 \times 0 = 1_{10}$
1	0	1	$1 \times 1 + 3 \times 0 - 2 \times 1 = -1_{10}$
1	1	0	$1 \times 1 + 3 \times 1 - 2 \times 0 = 4_{10}$
1	1	1	$1 \times 1 + 3 \times 1 - 2 \times 1 = 2_{10}$

The values of $x[k]$ are $x[0] = 1_{10} = 0001_{2C}$, $x[1] = -3_{10} = 1101_{2C}$, and $x[2] = 7_{10} = 0111_{2C}$. The output at sample index k, namely y, is defined as follows:

Step t	$x_t[2]$ $x_t[1]$ $x_t[0]$			$f[t] \times 2^t$	$+Y[t-1]=Y[t]$	
0	1	1	1	2×2^0	$+$ 0	$= 2$
1	1	0	0	1×2^1	$+$ 2	$= 4$
2	1	1	0	4×2^2	$+$ 4	$= 20$

	$x_t[2]$ $x_t[1]$ $x_t[0]$			$f[t] \times (-2^t) + Y[t-1]=Y[t]$		
3	0	1	0	$3 \times (-2^3) +$	20	$= -4$

A numerical check results in $c[0]x[0] + c[1]x[1] + c[2]x[2] = -2 \times 1 + 3 \times (-3) + 1 \times 7 = -4$ ✓

2.25

2.7.3 Modified DA Solutions

In the following we wish to discuss two interesting modifications to the basic DA concept, where the first variation reduces the size, and the second increases the speed.

If the number of coefficients N is too large to implement the full word with a single LUT (recall that input LUT bit width = number of coefficients), then we can use partial tables and add the results. If we also add pipeline registers, this modification will not reduce the speed, but can dramatically reduce the size of the design, because the size of a LUT grows exponentially with the address space, i.e., the number of input coefficients N. Suppose the length LN inner product

$$y = \langle c, x \rangle = \sum_{n=0}^{LN-1} c[n]x[n] \tag{2.52}$$

is to be implemented using a DA architecture. The sum can be partitioned into L independent N^{th} parallel DA LUTs resulting in

$$y = \langle c, x \rangle = \sum_{l=0}^{L-1} \sum_{n=0}^{N-1} c[Ll+n]x[Ll+n]. \tag{2.53}$$

This is shown in Fig. 2.36 for a realization of a $4N$ DA design requiring three postadditional adders. The size of the table is reduced from one $2^{4N} \times B$ LUT to four $2^{N} \times B$ tables.

Another variation of the DA architecture increases speed at the expense of additional LUTs, registers, and adders. A basic DA architecture, for a length N^{th} sum-of-product computation, accepts one bit from each of N words. If two bits per word are accepted, then the computational speed can be essentially doubled. The maximum speed can be achieved with the fully pipelined word-parallel architecture shown in Fig. 2.37. Here, a new result of a length four sum-of-product is computed for 4-bit signed coefficients at each LUT cycle. For maximum speed, we have to provide a separate ROM (with identical content) for each bit vector $x_b[n]$. But the maximum speed can become expensive: If we double the input bit width, we need twice as many LUTs, adders and registers. If the number of coefficients N is limited to four or eight this modification gives attractive performance, essentially outperforming all commercially available programmable signal processors, as we will see in Chap. 3.

2.8 Computation of Special Functions Using CORDIC

If a digital signal processing algorithm is implemented with FPGAs and the algorithm uses a nontrivial (transcendental) algebraic function, like $\sqrt{x}$ or

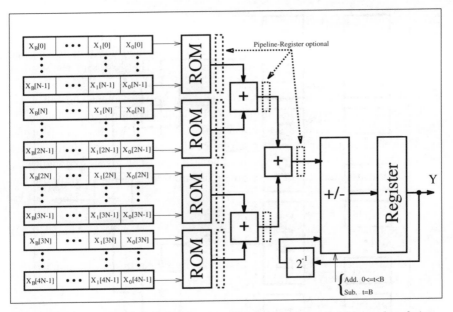

Fig. 2.36. Distributed arithmetic with table partitioning to yield a reduced size.

$\arctan y/x$, we can always use the Taylor series to approximate this function, i.e.,

$$f(x) = \sum_{k=0}^{K} \frac{f^k(x_0)}{k!}(x - x_0)^k, \tag{2.54}$$

where $f^k(x)$ is the kth derivative of $f(x)$ and $k! = k \times (k-1) \ldots \times 1$. The problem is then reduced to a sequence of multiply and add operations. A more efficient, alternative approach, based on the *Coordinate Rotation Digital Computer* (CORDIC) algorithm can also be considered. The CORDIC algorithm is found in numerous applications, such as pocket calculators [67], and in mainstream DSP objects, such as adaptive filters, FFTs, DCTs [68], demodulators [69], and neural networks [40]. The basic CORDIC algorithm can be found in two classic papers by Volder [70] and Walther [71]. Some theoretical extensions have been made, such as the extension of range in the hyperbolic mode, or the quantization error analysis by Hu et al. [72], and Meyer-Bäse et al. [69]. VLSI implementations have been discussed in Ph.D. theses, such as those by Timmermann [73] and Hahn [74]. The first FPGA implementations were investigated by Meyer-Bäse et al. [4, 69]. The realization of the CORDIC algorithm in distributed arithmetic was investigated by Ma [75]. A very detailed overview including details of several applications, was provided by Hu [68] in a 1992 IEEE Signal Processing Magazine review paper.

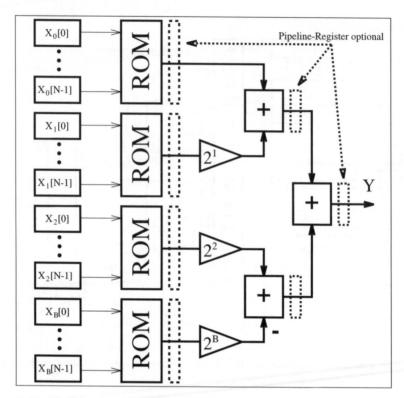

Fig. 2.37. Higher-order distributed arithmetic optimized for speed.

The original CORDIC algorithm by Volder [70] computes a multiplier-free coordinate conversion between rectangular (x, y) and polar (R, θ) coordinates. Walther [71] generalized the CORDIC algorithm to include circular $(m = 1)$, linear $(m = 0)$, and hyperbolic $(m = -1)$ transforms. For each mode, two rotation directions are identified. For *vectoring*, a vector with starting coordinates (X_0, Y_0) is rotated in such a way that the vector finally lies on the abscissa (i.e., x axis) by iteratively converging Y_K to zero. For *rotation*, a vector with a starting coordinate (X_0, Y_0) is rotated by an angle θ_0 in such a way that the final value of the angle register, denoted Z, converges to zero. The angle θ_k is chosen so that each iteration can be performed with an addition and a binary shift. Table 2.10 shows, in the second column, the choice for the rotation angle for the three modes $m = 1, 0$, and -1.

Now we can formally define the CORDIC algorithm as follows:

Table 2.10. CORDIC algorithm modes.

Mode	Angle θ_k	Shift sequence	Radius factor
circular $m = 1$	$\tan^{-1}(2^{-k})$	$0, 1, 2, \ldots$	$K_1 = 1.65$
linear $m = 0$	2^{-k}	$1, 2, \ldots$	$K_0 = 1.0$
hyperbolic $m = -1$	$\tanh^{-1}(2^{-k})$	$1, 2, 3, 4, 4, \ldots$	$K_{-1} = 0.80$

Algorithm 2.26:   **CORDIC Algorithm**

At each iteration, the CORDIC algorithm implements the mapping:
$$\begin{bmatrix} X_{k+1} \\ Y_{k+1} \end{bmatrix} = \begin{bmatrix} 1 & m\delta_k 2^{-k} \\ \delta_k 2^{-k} & 1 \end{bmatrix} \begin{bmatrix} X_k \\ Y_k \end{bmatrix} \tag{2.55}$$
$$Z_{k+1} = Z_k + \delta_k \theta_k,$$
where the angle θ_k is given in Table 2.10, $\delta_k = \pm 1$, and the two rotation directions are $Z_K \to 0$ and $Y_K \to 0$.

This means that six operational modes exist, and they are summarized in Table 2.11. A consequence is that nearly all transcendental functions can be computed with the CORDIC algorithm. With a proper choice of the initial values, the function $X \times Y, Y/X, \sin(Z), \cos(Z), \tan^{-1}(Y), \sinh(Z), \cosh(Z)$, and $\tanh(Z)$ can directly be computed. Additional functions may be generated by choosing appropriate initialization, sometimes combined with multiple modes of operation, as shown in the following listing:

$$\tan(Z)=\sin(Z)/\cos(Z) \qquad \text{Modes: } m=1, 0$$
$$\tanh(Z)=\sinh(Z)/\cosh(Z) \qquad \text{Modes: } m=-1, 0$$
$$\exp(Z)=\sinh(Z)+\cosh(Z) \qquad \text{Modes: } m=-1; \quad x=y=1$$
$$\log_e(W)=2\tanh^{-1}(Y/X) \qquad \text{Modes: } m=-1$$
$$\text{with } X=W+1, Y=W-1$$
$$\sqrt{W}=\sqrt{X^2-Y^2} \qquad \text{Modes: } m=1$$
$$\text{with } X=W+\tfrac{1}{4}, Y=W-\tfrac{1}{4}.$$

Table 2.11. Modes m of operation for the CORDIC algorithm.

m	$Z_K \to 0$	$Y_K \to 0$
1	$X_K = K_1(X_0\cos(Z_0) - Y_0\sin(Z_0))$ $Y_K = K_1(X_0\cos(Z_0) + Y_0\sin(Z_0))$	$X_K = K_1\sqrt{X_0^2+Y_0^2}$ $Z_K = Z_0 + \arctan(Y_0/X_0)$
0	$X_K = X_0$ $Y_K = Y_0 + X_0 \times Z_0$	$X_K = X_0$ $Z_K = Z_0 + Y_0/X_0$
-1	$X_K = K_{-1}(X_0\cosh(Z_0) - Y_0\sinh(Z_0))$ $Y_K = K_{-1}(X_0\cosh(Z_0) + Y_0\sinh(Z_0))$	$X_K = K_{-1}\sqrt{X_0^2+Y_0^2}$ $Z_K = Z_0 + \tanh^{-1}(Y_0/X_0)$

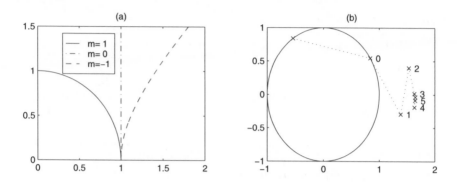

Fig. 2.38. CORDIC. **(a)** Modes. **(b)** Example of circular vectoring.

A careful analysis of (2.55) reveals that the iteration vectors only approach the curves shown in Fig. 2.38a. The length of the vectors changes with each iteration, as shown in Fig. 2.38b. This change in length does *not* depend on the starting angle and after K iterations the same change (called radius factor) always occurs. In the last column of Table 2.10 these radius factors are shown. To ensure that the CORDIC algorithm converges, the sum of all remaining rotation angles must be larger than the actual rotation angle. This is the case for linear and circular transforms. For the hyperbolic mode, all iterations of the form $n_{k+1} = 3n_k + 1$ have to be repeated. These are the iterations $4, 13, 40, 121 \ldots$.

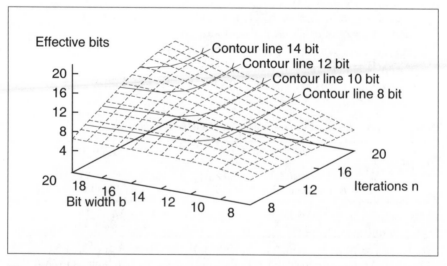

Fig. 2.39. Effective bits in circular mode.

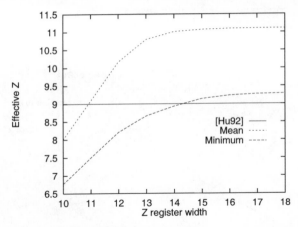

Fig. 2.40. Resolution of phase for circular mode.

Output precision can be estimated using a procedure developed by Hu [76] and illustrated in Fig. 2.39. The graph shows the effective bit precision for the circular mode, depending on the X, Y path width, and the number of iterations. If b bits is the desired output precision, the "rule of thumb" suggests that the X, Y path should have $\log_2(b)$ additional guard bits. From Fig. 2.40, it can also be seen that the bit width of the Z path should have the same precision as that for X and Y.

In contrast to the circular CORDIC algorithm, the effective resolution of a hyperbolic CORDIC cannot be computed analytically because the precision depends on the angular values of $z(k)$ at iteration k. Hyperbolic precision can, however, be estimated using simulation. Figure 2.41 shows the minimum accuracy estimate computed over 1000 test values for each bit-width/number combination of the possible iterations. The 3D representation shows the number of iterations, the bit width of the X/Y path, and the resulting minimum precision of the result in terms of effective bits. The contour lines allow an exchange between the number of iterations and the bit width. For example, to achieve 10-bit precision, one can use a 21-bit X/Y path and 18 iterations, or 14 iterations at 24 bits.

2.8.1 CORDIC Architectures

Two basic structures are used to implement a CORDIC architecture: the more compact state machine or the high-speed, fully pipelined processor.

If computation time is not critical, then a state machine as shown in Fig. 2.42 is applicable. In each cycle, exactly one iteration of (2.55) will be computed. The most complex part of this design is the two barrelshifters. The two barrelshifters can be replaced by a single barrelshifter, using a multiplexer as shown in Fig. 2.43, or a serial (right, or right/left) shifter. Table 2.12

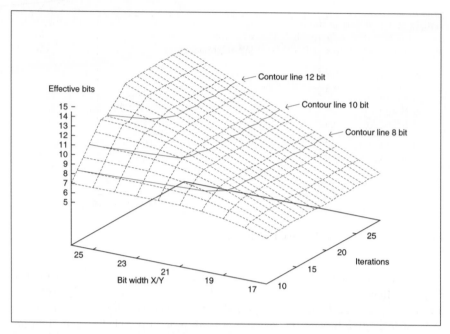

Fig. 2.41. Effective bits in hyperbolic mode.

compares different design options for a 13-bit implementation using Xilinx XC3K FPGAs.

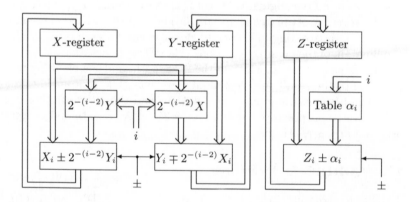

Fig. 2.42. CORDIC state machine.

If high speed is needed, a fully pipelined version of the design shown in Fig. 2.44 can be used. Figure 2.44 shows eight iterations of a circular CORDIC. After an initial delay of K cycles, a new output value becomes

Table 2.12. Effort estimation (Xilinx XC3K) for a CORDIC a machine with 13-bits plus sign for X/Y path. (Abbreviations: Ac=accumulator; BS=barrelshifter; RS=serial right shifter; LRS=serial left/right shifter)

Structure	Registers	Multiplexer	Adder	Shifter	$\sum$LE	Cycle
2BS+2Ac	2×7	0	2×14	2×19.5	81	12
2RS+2Ac	2×7	0	2×14	2×6.5	55	46
2LRS+2Ac	2×7	0	2×14	2×8	58	39
1BS+2Ac	7	3×7	2×14	19.5	75.5	20
1RS+2Ac	7	3×7	2×14	6.5	62.5	56
1LRS+2Ac	7	3×7	2×14	8	64	74
1BS+1Ac	3×7	2×7	14	19.5	68.5	20
1RS+1Ac	3×7	2×7	14	6.5	55.5	92
1LRS+1Ac	3×7	2×7	14	8	57	74

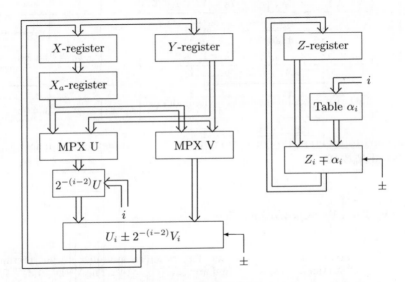

Fig. 2.43. CORDIC machine with reduced complexity.

available after each cycle. As with array multipliers, CORDIC implementations have a quadratic growth in LE complexity as the bit width increases (see Fig. 2.44).

The following example shows the first four steps of a circular-vectoring fully pipelined design.

Example 2.27: Circular CORDIC in Vectoring Mode

The first iteration rotates the vectors from the second or third quadrant to the first or fourth, respectively. The shift sequence is 0,0,1, and 2. The

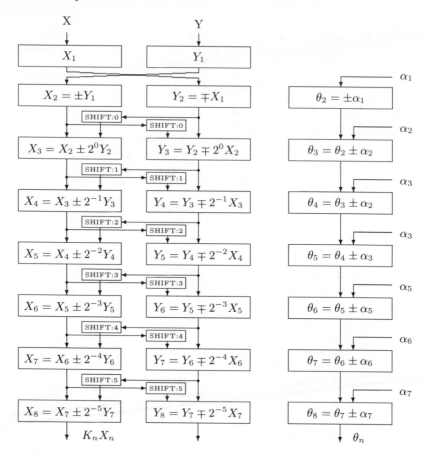

Fig. 2.44. Fast CORDIC pipeline.

rotation angle of the first four steps becomes: $\arctan(\infty) = 90°$, $\arctan(2^0) = 45°$, $\arctan(2^{-1}) = 26.5°$, and $\arctan(2^{-2}) = 14°$. The VHDL code[10] for 8-bit data can be implemented as follows:

```
PACKAGE eight_bit_int IS    -- User-defined types
  SUBTYPE BYTE IS INTEGER RANGE -128 TO 127;
  TYPE ARRAY_BYTE IS ARRAY (0 TO 3) OF BYTE;
END eight_bit_int;

LIBRARY work;
USE work.eight_bit_int.ALL;

LIBRARY ieee;
USE ieee.std_logic_1164.ALL;
USE ieee.std_logic_arith.ALL;
```

---

[10] The equivalent Verilog code `cordic.v` for this example can be found in Appendix A on page 674. Synthesis results are shown in Appendix B on page 731.

```
ENTITY cordic IS                          ------> Interface
     PORT (clk          : IN  STD_LOGIC;
           x_in , y_in : IN  BYTE;
           r, phi, eps : OUT BYTE);
END cordic;

ARCHITECTURE fpga OF cordic IS
  SIGNAL  x, y, z : ARRAY_BYTE:= (0,0,0,0);
BEGIN                                         -- Array of Bytes

  PROCESS                     ------> Behavioral Style
  BEGIN
    WAIT UNTIL clk = '1'; -- Compute last value first in
    r <= x(3);            -- sequential VHDL statements !!
    phi <= z(3);
    eps <= y(3);

    IF y(2) >= 0 THEN            -- Rotate 14 degrees
      x(3) <= x(2) + y(2) /4;
      y(3) <= y(2) - x(2) /4;
      z(3) <= z(2) + 14;
    ELSE
      x(3) <= x(2) - y(2) /4;
      y(3) <= y(2) + x(2) /4;
      z(3) <= z(2) - 14;
    END IF;

    IF y(1) >= 0 THEN            -- Rotate 26 degrees
      x(2) <= x(1) + y(1) /2;
      y(2) <= y(1) - x(1) /2;
      z(2) <= z(1) + 26;
    ELSE
      x(2) <= x(1) - y(1) /2;
      y(2) <= y(1) + x(1) /2;
      z(2) <= z(1) - 26;
    END IF;

    IF y(0) >= 0 THEN            -- Rotate  45 degrees
      x(1) <= x(0) + y(0);
      y(1) <= y(0) - x(0);
      z(1) <= z(0) + 45;
    ELSE
      x(1) <= x(0) - y(0);
      y(1) <= y(0) + x(0);
      z(1) <= z(0) - 45;
    END IF;

-- Test for x_in < 0 rotate 0,+90, or -90 degrees
    IF x_in >= 0 THEN
      x(0) <= x_in;        -- Input in register 0
      y(0) <= y_in;
      z(0) <= 0;
```

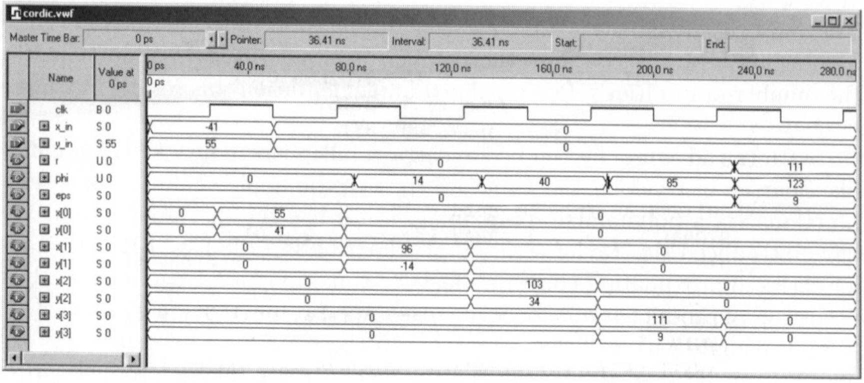

Fig. 2.45. CORDIC simulation results.

```
        ELSIF y_in >= 0 THEN
            x(0) <= y_in;
            y(0) <= - x_in;
            z(0) <= 90;
        ELSE
            x(0) <= - y_in;
            y(0) <= x_in;
            z(0) <= -90;
        END IF;
    END PROCESS;

END fpga;
```

Figure 2.45 shows the simulation of the conversion of $X_0 = -41$, and $Y_0 = 55$. Note that the radius is enlarged to $R = X_K = 111 = 1.618\sqrt{X_0^2 + Y_0^2}$ and the accumulated angle in degrees is $\arctan(Y_0/X_0) = 123°$. The design requires 235 LEs and runs with a `Speed` synthesis optimization at 222.67 MHz using no embedded multiplier.

2.27

The actual LE count in the previous example is larger than that expected for a four-stage 8-bit pipeline design that is $5 \times 8 \times 3 = 120$ LEs. The increase by a factor of two comes from the fact that a FPGA uses an N-bit switchable `LPM_ADD_SUB` megafunction that needs $2N$ LEs. It needs $2N$ LEs because the LE has only three inputs in the fast arithmetic mode, and the switch mode needs four input LUTs. A Xilinx XC4K type LE, see Fig. 1.12, p. 19, would be needed, with four inputs per LE, to reduce the count by a factor of two.

2.9 Computation of Special Functions using MAC Calls

The CORDIC algorithm introduced in the previous section allows one to implement a wide variety of functions at a moderate implementation cost.

The only disadvantage is that some high-precision functions need a large number of iterations, because the number of bits is linearly proportional to the number of iterations. In a pipelined implementation this results in a large latency.

With the advent of fast embedded array multipliers in new FPGA families like Spartan or Cyclone, see Table 1.4 (p. 11), the implementation of special functions via a polynomial approximation has becomes a viable option. We have introduced the Taylor series approximation in (2.54), p. 121. The Taylor series approximation converges fast for some functions, e.g., $\exp(x)$, but needs many product terms for some other special functions, e.g., $\arctan(x)$, to approximate with sufficient precision. In these cases a Chebyshev approximation can be used to shorten the number of iterations or product terms required.

2.9.1 Chebyshev Approximations

The Chebyshev approximation is based on the Chebyshev polynomial

$$T_k(x) = \cos(k \times \arccos(x)) \tag{2.56}$$

defined for the range $-1 \leq x \leq 1$. The $T_k(x)$ may look like trigonometric functions, but using some algebraic identities and manipulations allow us to write (2.56) as a true polynomial. The first few polynomials look like

$$
\begin{aligned}
T_0(x) &= 1 \\
T_1(x) &= x \\
T_2(x) &= 2x^2 - 1 \\
T_3(x) &= 4x^3 - 3x \\
T_4(x) &= 8x^4 - 8x^2 + 1 \\
T_5(x) &= 16x^5 - 20x^3 + 5x \\
T_6(x) &= 32x^6 - 48x^4 + 18x^2 - 1
\end{aligned}
\tag{2.57}
$$

$$\vdots$$

In [77] we find a list of the first 12 polynomials. The first six polynomials are graphical interpreted in Fig.2.46. In general, Chebyshev polynomials obey the following iterative rule

$$T_k(x) = 2xT_{k-1}(x) - T_{k-2}(x) \quad \forall\, k \geq 2. \tag{2.58}$$

A function approximation can now be written as

$$f(x) = \sum_{k=0}^{N-1} c(k)T_k(x). \tag{2.59}$$

Because all discrete Chebyshev polynomials are orthogonal to each other it follows that forward and inverse transform are unique, i.e., bijective [78,

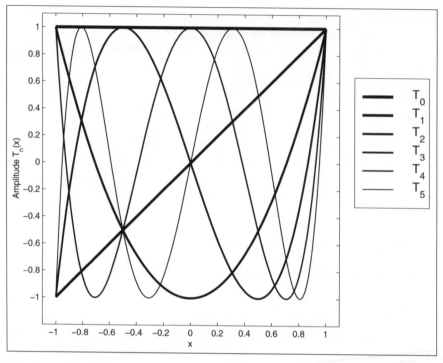

Fig. 2.46. The first 6 Chebyshev polynomials .

p. 191]. The question now is why (2.59) is so much better than, for instance,
a polynomial using the Taylor approximation (2.54)

$$f(x) = \sum_{k=0}^{N-1} \frac{f^k(x_0)}{k!}(x - x_0)^k = \sum_{k=0}^{N-1} p(k)(x - x_0)^k, \qquad (2.60)$$

There are mainly three reasons. First (2.59) is a very close (but not exact)
approximation to the very complicated problem of finding the function approximation with a minimum of the maximum error, i.e., an optimization
of the l_∞ norm $\max(f(x) - \hat{f}(x)) \to \min$. The second reason we prefer
(2.59) is the fact, that a pruned polynomial with $M << N$ still gives a min-
imum/maximum approximation, i.e., a shorter sum still gives a Chebyshev
approximation as if we had started the computation with M as the target
from the very start. Last but not least we gain from the fact that (2.59)
can be computed (for all functions of relevance) with much fewer coefficients
than would be required for a Taylor approximation of the same precision.
Let us study these special function approximation in the following for popular functions, like trigonometric, exponential, logarithmic, and the square
root functions.

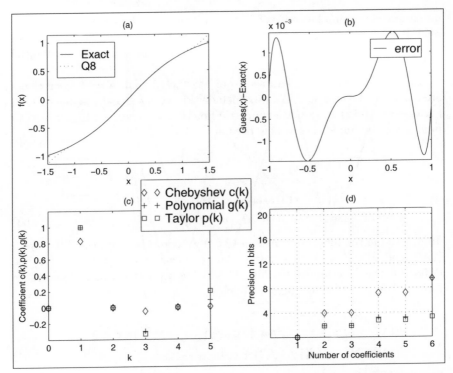

Fig. 2.47. Inverse tangent function approximation. **(a)** Comparison of full-precision and 8-bit quantized approximations. **(b)** Error of quantized approximation for $x \in [-1, 1]$. **(c)** Chebyshev, polynomial from Chebyshev, and Taylor polynomial coefficients. **(d)** Error of the three pruned polynomials.

2.9.2 Trigonometric Function Approximation

As a first example we study the inverse tangent function

$$f(x) = \arctan(x), \qquad (2.61)$$

where x is specified for the range $-1 \leq x \leq 1$. If we need to evaluate function values outside this interval, we can take advantage of the relation

$$\arctan(x) = 0.5 - \arctan(1/x). \qquad (2.62)$$

Embedded multipliers in Altera FPGAs have a basic size of 9×9 bits, i.e., 8 bits plus sign bit data format, or 18×18 bit, i.e., 17 bits plus sign data format. We will therefore in the following always discuss two solutions regarding these two different word sizes.

Fig. 2.47a shows the exact value and approximation for 8-bit quantization, and Fig. 2.47b displays the error, i.e., the difference between the exact function value and the approximation. The error has the typical alternating minimum/maximum behavior of all Chebyshev approximations. The approximation with $N = 6$ already gives an almost perfect approximation. If we use fewer coefficients, e.g., $N = 2$ or $N = 4$, we will have a more-substantial error, see Exercise 2.26 (p. 162).

For 8-bit precision we can see from Fig. 2.47d that $N = 6$ coefficients are sufficient. From Fig. 2.47c we conclude that all even coefficients are zero, because $\arctan(x)$ is an odd symmetric function with respect to $x = 0$. The function to be implemented now becomes

$$f(x) = \sum_{k=0}^{N-1} c(k)T_k(x)$$
$$f(x) = c(1)T_1(x) + c(3)T_3(x) + c(5)T_5(x)$$
$$f(x) = 0.8284T_1(x) - 0.0475T_3(x) + 0.0055T_5(x). \tag{2.63}$$

To determine the function values in (2.63) we can substitute the $T_n(x)$ from (2.57) and solve (2.63). It is however more efficient to use the iterative rule (2.58) for the function evaluation. This is known as Clenshaw's recurrence formula [78, p. 193] and works as follows:

$$d(N) = d(N + 1) = 0$$
$$d(k) = 2xd(k + 1) - d(k + 2) + c(k) \quad k = N - 1, N - 2, \ldots, 1$$
$$f(x) = d(0) = xd(1) - d(2) + c(0) \tag{2.64}$$

For our $N = 6$ system with even coefficients equal to zero we can simplify (2.64) to

$$d(5) = c(5)$$
$$d(4) = 2xc(5)$$
$$d(3) = 2xd(4) - d(5) + c(3)$$
$$d(2) = 2xd(3) - d(4)$$
$$d(1) = 2xd(2) - d(3) + c(1)$$
$$f(x) = xd(1) - d(2). \tag{2.65}$$

We can now start to implement the $\arctan(x)$ function approximation in HDL.

Example 2.28: arctan Function Approximation

If we implement the $\arctan(x)$ using the embedded 9×9 bit multipliers we have to take into account that our values are in the range $-1 \le x < 1$. We therefore use a fractional integer representation in a 1.8 format. In our HDL simulation these fractional numbers are represented as integers and the values are mapped to the range $-256 \le x < 256$. We can use the same number format for our Chebyshev coefficients because they are all less than 1, i.e., we quantize

$$c(1) = 0.8284 = 212/256, \tag{2.66}$$

$$c(3) = -0.0475 = -12/256, \tag{2.67}$$

$$c(5) = 0.0055 = 1/256. \tag{2.68}$$

The following VHDL code[11] shows the arctan(x) approximation using polynomial terms up to $N = 6$.

```
PACKAGE n_bits_int IS             -- User-defined types
    SUBTYPE BITS9 IS INTEGER RANGE -2**8 TO 2**8-1;
    TYPE ARRAY_BITS9_4 IS ARRAY (1 TO 5) of BITS9;
END n_bits_int;

LIBRARY work;
USE work.n_bits_int.ALL;

LIBRARY ieee;
USE ieee.std_logic_1164.ALL;
USE ieee.std_logic_arith.ALL;
USE ieee.std_logic_signed.ALL;

ENTITY arctan IS                      ------> Interface
    PORT (clk      : IN  STD_LOGIC;
          x_in     : IN  BITS9;
          d_o      : OUT ARRAY_BITS9_4;
          f_out    : OUT BITS9);
END arctan;

ARCHITECTURE fpga OF arctan IS

    SIGNAL x,f,d1,d2,d3,d4,d5 : BITS9; -- Auxilary signals
    SIGNAL d : ARRAY_BITS9_4 := (0,0,0,0,0);-- Auxilary array
    -- Chebychev coefficients for 8-bit precision:
    CONSTANT c1 : BITS9 := 212;
    CONSTANT c3 : BITS9 := -12;
    CONSTANT c5 : BITS9 := 1;

BEGIN

    STORE: PROCESS    ------> I/O store in register
    BEGIN
        WAIT UNTIL clk = '1';
        x <= x_in;
        f_out <= f;
    END PROCESS;

    --> Compute sum-of-products:
    SOP: PROCESS (x,d)
    BEGIN
    -- Clenshaw's recurrence formula
        d(5) <= c5;
        d(4) <= x * d(5) / 128;
        d(3) <= x * d(4) / 128 - d(5) + c3;
```

----

[11] The equivalent Verilog code arctan.v for this example can be found in Appendix A on page 676. Synthesis results are shown in Appendix B on page 731.

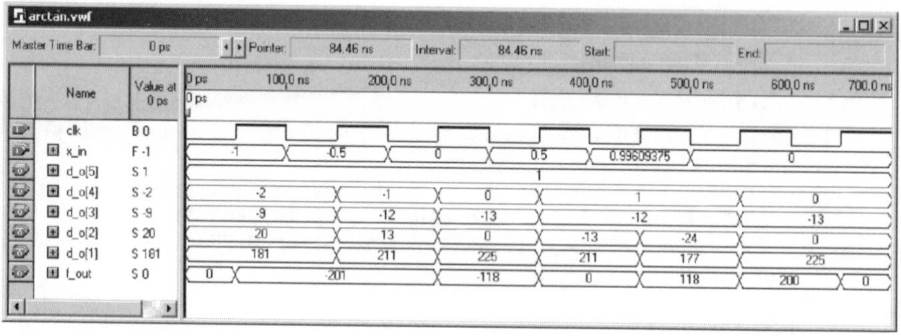

Fig. 2.48. VHDL simulation of the $\arctan(x)$ function approximation for the values $x = -1 = -256/256, x = -0.5 = -128/256, x = 0, x = 0.5 = 128/256, x = 1 \approx 255/256$.

```
d(2) <= x * d(3) / 128 - d(4);
d(1) <= x * d(2) / 128 - d(3) + c1;
f   <= x * d(1) / 256 - d(2); -- last step is different
END PROCESS SOP;

d_o <= d;      -- Provide some test signals as outputs

END fpga;
```

The first PROCESS is used to infer registers for the input and output data. The next PROCESS blocks SOP include the computation of the Chebyshev approximation using Clenshaw's recurrence formula. The iteration variables $d(k)$ are also connected to the output ports so we can monitor them. The design uses 100 LEs, 4 embedded multipliers and has a 32.09 MHz Registered Performance. Comparing FLEX and Cyclone synthesis data we can conclude that the use of embedded multipliers saves many LEs.

A simulation of the arctan function approximation is shown in Fig. 2.48. The simulation shows the result for five different input values:

x	$f(x) = \arctan(x)$	$\hat{f}(x)$	$|error|$	Eff. bits
-1.0	-0.7854	$-201/256 = -0.7852$	0.0053	7.6
-0.5	-0.4636	$-118/256 = -0.4609$	0.0027	7.4
0	0.0	0	0	–
0.5	0.4636	$118/256 = 0.4609$	0.0027	7.4
1.0	0.7854	$200/256 = 0.7812$	0.0053	7.6

Note that, due to the I/O registers, the output values appear with a delay of one clock cycle. 2.28

If the precision in the previous example is not sufficient we can use more coefficients. The odd Chebyshev coefficients for 16-bit precision, for instance, would be

$$c(2k + 1) = (0.82842712, -0.04737854, 0.00487733,$$
$$-0.00059776, 0.00008001, -0.00001282). \qquad (2.69)$$

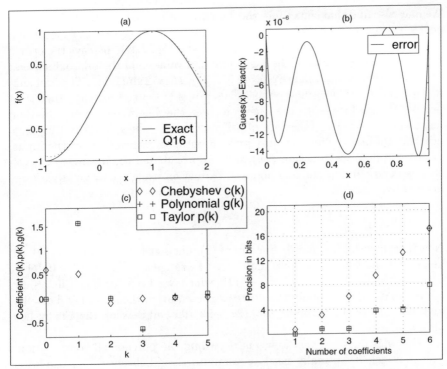

Fig. 2.49. Sine function approximation. **(a)** Comparison of full-precision and 8-bit quantized approximations. **(b)** Error of quantized approximation for $x \in [0, 1]$. **(c)** Chebyshev, polynomial from Chebyshev, and Taylor polynomial coefficients. **(d)** Error of the three pruned polynomials.

If we compare this with the Taylor series coefficient

$$\arctan(x) = x - \frac{x^3}{3} + \frac{x^5}{5} + \ldots (-1)^k \frac{x^{2k+1}}{2k+1} \tag{2.70}$$

$$p(2k+1) = (1, -0.\overline{3}, 0.2, -0.14285714, 0.\overline{1}, -0.\overline{09})$$

we see that the Taylor coefficients converge very slowly compared with the Chebyshev approximation.

There are two more common trigonometric functions. On is the $\sin(x)$ and the other is the $\cos(x)$ function. There is however a small problem with these functions. The argument is usually defined only for the first quadrant, i.e., $0 \le x \le \pi/2$, and the other quadrants values are computed via

$$\sin(x) = -\sin(-x) \qquad \sin(x) = \sin(\pi/2 - x) \tag{2.71}$$

or equivalent for the $\cos(x)$ we use

$$\cos(x) = \cos(-x) \qquad \cos(x) = -\cos(\pi/2 - x). \tag{2.72}$$

We may also find that sometimes the data are normalized $f(x) = \sin(x\pi/2)$ or degree values are used, i.e., $0° \leq x \leq 90°$. Figure 2.49a shows the exact value and approximation for 16-bit quantization, and Fig. 2.49b displays the error, i.e., the difference between the exact function values and the approximation. In Fig. 2.50 the same data are plotted for the $\cos(x\pi/2)$ function. The problem now is that our Chebyshev polynomials are only defined for the range $x \in [-1, 1]$. Which brings up the question, how the Chebyshev approximation has to be modified to take care of different range values? Luckily this does not take too much effort, we just make a linear transformation of the input values. Suppose the function $f(y)$ to be approximated has a range $y \in [a, b]$ then we can develop our function approximation using a change of variable defined by

$$y = \frac{2x - b - a}{b - a}. \tag{2.73}$$

Now if we have for instance in our $\sin(x\pi/2)$ function x in the range $x = [0, 1]$, i.e., $a = 0$ and $b = 1$, it follows that y has the range $y = [(2 \times 0 - 1 - 0)/(1 - 0), (2 \times 1 - 1 - 0)/(1 - 0)] = [-1, 1]$, which we need for our Chebyshev approximation. If we prefer the degree representation then $a = 0$ and $b = 90$, and we will use the mapping $y = (2x - 90)/90$ and develop the Chebyshev approximation in y.

The final question we discuss is regarding the polynomial computation. You may ask if we really need to compute the Chebyshev approximation via the Clenshaw's recurrence formula (2.64) or if we can use instead the direct polynomial approximation, which requires one fewer add operation per iteration:

$$f(x) = \sum_{k=0}^{N-1} p(k)x^k \tag{2.74}$$

or even better use the Horner scheme

$$s(N - 1) = p(N - 1)$$
$$s(k) = s(k + 1) \times x + p(k) \qquad k = N - 2, \ldots 0.$$
$$f = s(0). \tag{2.75}$$

We can of course substitute the Chebyshev functions (2.57) in the approximation formula (2.59), because the $T_n(x)$ do not have terms of higher order than x^n. However there is one important disadvantage to this approach. We will lose the pruning property of the Chebyshev approximation, i.e., if we use in the polynomial approximation (2.74) fewer than N terms, the pruned polynomial will no longer be an l_∞ optimized polynomial. Figure 2.47d (p. 133) shows this property. If we use all 6 terms the Chebyshev and the associated polynomial approximation will have the same precision. If we now prune the polynomial, the Chebyshev function approximation (2.59) using the $T_n(x)$ has more precision than the pruned polynomial using (2.74). The resulting

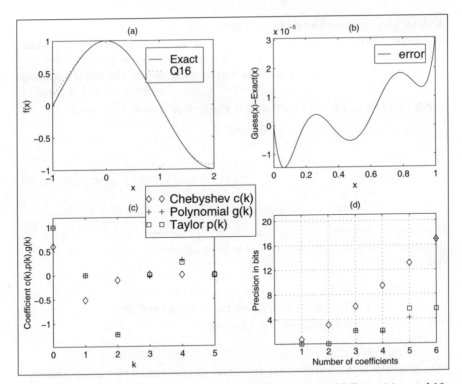

Fig. 2.50. Cosine function approximation. (**a**) Comparison of full-precision and 16-bit quantized approximations. (**b**) Error of quantized approximation for $x \in [0,1]$. (**c**) Chebyshev, polynomial from Chebyshev, and Taylor polynomial coefficients. (**d**) Error of the three pruned polynomials.

precision is much lower than the equivalent pruned Chebyshev function approximation of the same length. In fact it is not much better than the Taylor approximation. So the solution to this problem is not complicated: if we want to shorten the length $M < N$ of our polynomial approximation (2.74) we need to develop first a Chebyshev approximation for length M and then compute the polynomial coefficient $g(k)$ from this pruned Chebyshev approximation. Let us demonstrate this with a comparison of 8- and 16-bit $\arctan(x)$ coefficients. The substitution of the Chebyshev functions (2.57) into the coefficient (2.69) gives the following odd coefficients:

$$g(2k + 1) = (0.99999483, -0.33295711, 0.19534659,$$
$$-0.12044859, 0.05658999, -0.01313038). \tag{2.76}$$

If we now use the length $N = 6$ approximation from (2.66) the odd coefficient will be

$$g(2k + 1) = (0.9982, -0.2993, 0.0876). \tag{2.77}$$

Although the pruned Chebyshev coefficients are the same, we see from a comparison of (2.76) and (2.77) that the polynomial coefficient differ essentially. The coefficient $g(5)$ for instance has a factor of 2 difference.

We can summarize the Chebyshev approximation in the following procedure.

Algorithm 2.29:   Chebyshev Function Approximation

1) Define the number of coefficients N.
2) Transform the variable from x to y using (2.73)
3) Determine the Chebyshev approximation in y.
4) Determine the direct polynomial coefficients $g(k)$ using Clenshaw's recurrence formula.
5) Build the inverse of the mapping y.

If we apply these five steps to our $\sin(x\pi/2)$ function for $x \in [0,1]$ with four nonzero coefficients, we get the following polynomials sufficient for a 16-bit quantization

$$
\begin{aligned}
f(x) &= \sin(x\pi/2) \\
&= 1.57035062x + 0.00508719x^2 - 0.66666099x^3 \\
&\quad + 0.03610310x^4 + 0.05512166x^5 \\
&= (51457x + 167x^2 - 21845x^3 + 1183x^4 + 1806x^5)/32768.
\end{aligned}
$$

Note that the first coefficient is larger than 1 and we need to scale appropriate. This is quite different from the Taylor approximation given by

$$
\sin\left(\frac{x\pi}{2}\right) = \frac{x\pi}{2} - \frac{1}{3!}\left(\frac{x\pi}{2}\right)^3 + \frac{1}{5!}\left(\frac{x\pi}{2}\right)^5
$$
$$
+ \ldots + \frac{(-1)^k}{(2k+1)!}\left(\frac{x\pi}{2}\right)^{2k+1}.
$$

Figure 2.49c shows a graphical illustration. For an 8-bit quantization we would use

$$
\begin{aligned}
f(x) = \sin(x\pi/2) &= 1.5647x + 0.0493x^2 - 0.7890x^3 + 0.1748x^4 \\
&= (200x + 6x^2 - 101x^3 + 22x^4)/128.
\end{aligned}
\tag{2.78}
$$

Although we would expect that, for an odd symmetric function, all even coefficients are zero, this is not the case in this approximation, because we only used the interval $x \in [0,1]$ for the approximation. The $\cos(x)$ function can be derived via the relation

$$
\cos\left(\frac{x\pi}{2}\right) = \sin\left((x+1)\frac{\pi}{2}\right)
\tag{2.79}
$$

or we may also develop a direct Chebyshev approximation. For $x \in [0,1]$ with four nonzero coefficients and get the following polynomial for a 16-bit quantization

$$f(x) = \cos\left(\frac{x\pi}{2}\right)$$
$$= 1.00000780 - 0.00056273x - 1.22706059x^2$$
$$-0.02896799x^3 + 0.31171138x^4 - 0.05512166x^5$$
$$= (32768 - 18x - 40208x^2 - 949x^3 + 10214x^4 - 1806x^5)/32768.$$

For an 8-bit quantization we would use

$$f(x) = \cos\left(\frac{x\pi}{2}\right)$$
$$= (0.9999 + 0.0046x - 1.2690x^2 + 0.0898x^3 + 0.1748x^4 \qquad (2.80)$$
$$= (128 + x - 162x^2 + 11x^3 + 22x^4)/128. \qquad (2.81)$$

Again the Taylor approximation has quite different coefficients:

$$\cos\left(\frac{x\pi}{2}\right) = 1 - \frac{1}{2!}\left(\frac{x\pi}{2}\right)^2 + \frac{1}{4!}\left(\frac{x\pi}{2}\right)^4 + \ldots + \frac{(-1)^k}{(2k)!}\left(\frac{x\pi}{2}\right)^{2k}.$$

Figure 2.49c shows a graphical illustration of the coefficients. We notice from Fig. 2.49d that with the same number (i.e., six) of terms x^k the Taylor approximation only provides about 6 bit accuracy, while the Chebyshev approximation has 16-bit precision.

2.9.3 Exponential and Logarithmic Function Approximation

The exponential function is one of the few functions who's Taylor approximation converges relatively fast. The Taylor approximation is given by

$$f(x) = e^x = 1 + \frac{x}{1!} + \frac{x^2}{2!} + \ldots + \frac{x^k}{k!} \qquad (2.82)$$

with $0 \le x \le 1$. For 16-bit polynomial quantization computed using the Chebyshev coefficients we would use:

$$f(x) = e^x$$
$$= 1.00002494 + 0.99875705x + 0.50977984x^2$$
$$+0.14027504x^3 + 0.06941551x^4$$
$$= (32769 + 32727x + 16704x^2 + 4597x^3 + 2275x^4)/32768.$$

Only terms up to order x^4 are required to reach 16-bit precision. We notice also from Fig. 2.51c that the Taylor and polynomial coefficient computed from the Chebyshev approximation are quite similar. If 8 bits plus sign precision are sufficient, we use

$$f(x) = e^x = 1.0077 + 0.8634x + 0.8373x^2$$
$$= (129 + 111x + 107x^2)/128. \qquad (2.83)$$

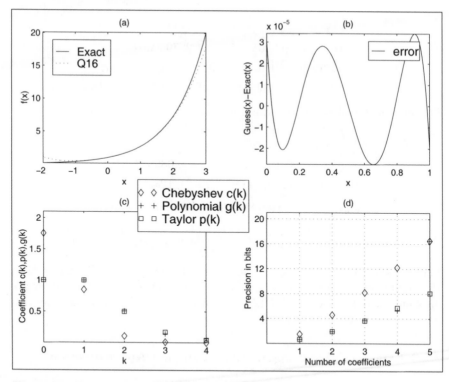

Fig. 2.51. Exponential $f(x) = \exp(x)$ function approximation. **(a)** Comparison of full-precision and 16-bit quantized approximations. **(b)** Error of quantized approximation for $x \in [0, 1]$. **(c)** Chebyshev, polynomial from Chebyshev, and Taylor polynomial coefficients. **(d)** Error of the three pruned polynomials.

Based on the fact that one coefficient is larger than $c(0) > 1.0$ we need to select a scaling factor of 128.

The input needs to be scaled in such a way that $0 \le x \le 1$. If x is outside this range we can use the identity

$$e^{sx} = (e^x)^s \tag{2.84}$$

Because $s = 2^k$ is a power-of-two value this implies that a series of squaring operations need to follow the exponential computation. For a negative exponent we can use the relation

$$e^{-x} = \frac{1}{e^x}, \tag{2.85}$$

or develop a separate approximation. If we like to build a direct function approximation to $f(x) = e^{-x}$ we have to alternate the sign of each second term in (2.82). For a Chebyshev polynomial approximation we get additional minor changes in the coefficients. For a 16-bit Chebyshev polynomial approximation we use

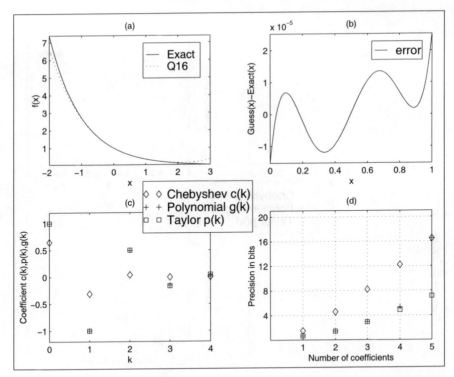

Fig. 2.52. Negative exponential $f(x) = \exp(-x)$ function approximation. **(a)** Comparison of full-precision and 16-bit quantized approximations. **(b)** Error of quantized approximation for $x \in [0,1]$. **(c)** Chebyshev, polynomial from Chebyshev, and Taylor polynomial coefficients. **(d)** Error of the three pruned polynomials.

$$f(x) = e^{-x}$$
$$= 0.99998916 - 0.99945630x + 0.49556967x^2$$
$$-0.15375046x^3 + 0.02553654x^4$$
$$= (65535 - 65500x + 32478x^2 - 10076x^3 + 1674x^4)/65536.$$

where x is defined for the range $x \in [0,1]$. Note that, based on the fact that all coefficients are less than 1, we can select a scaling by a factor of 2 larger than in (2.83). From Fig. 2.52d we conclude that three or five coefficients are required for 8- and 16-bit precision, respectively. For 8-bit quantization we would use the coefficients

$$f(x) = e^{-x} = 0.9964 - 0.9337x + 0.3080x^2$$
$$= (255 - 239x + 79x^2)/256. \tag{2.86}$$

The inverse to the exponential function is the logarithm function, which is typically approximated for the argument in the range $[1,2]$. As notation this is typically written as $f(x) = \ln(1+x)$ now with $0 \le x \le 1$. Figure

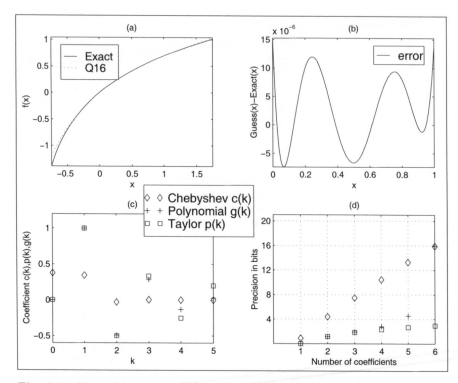

Fig. 2.53. Natural logarithm $f(x) = \ln(1 + x)$ function approximation. **(a)** Comparison of full-precision and 16-bit quantized approximations. **(b)** Error of quantized approximation for $x \in [0, 1]$. **(c)** Chebyshev, polynomial from Chebyshev, and Taylor polynomial coefficients. **(d)** Error of the three pruned polynomials.

2.53a shows the exact and 16-bit quantized approximation for this range. The approximation with $N = 6$ gives an almost perfect approximation. If we use fewer coefficients, e.g., $N = 2$ or $N = 3$, we will have a more substantial error, see Exercise 2.29 (p. 163).

The Taylor series approximation in no longer fast converging as for the exponential function

$$f(x) = \ln\left(1 + x\right) = x - \frac{x^2}{2} + \frac{x^3}{3} + \ldots + \frac{(-1)^{k+1} x^k}{k}$$

as can be seen from the linear factor in the denominator. A 16-bit Chebyshev approximation converges much faster, as can be seen from Fig. 2.53d. Only six coefficients are required for 16-bit precision. With six Taylor coefficients we get less than 4-bit precision. For 16-bit polynomial quantization computed using the Chebyshev coefficients we would use

$$f(x) = \ln(1 + x)$$

$$= 0.00001145 + 0.99916640x - 0.48969909x^2$$
$$+0.28382318x^3 - 0.12995720x^4 + 0.02980877x^5$$
$$= (1 + 65481x - 32093x^2 + 18601x^3 - 8517x^4 + 1954x^5)/65536.$$

Only terms up to order x^5 are required to get 16-bit precision. We also notice from Fig. 2.53c that the Taylor and polynomial coefficient computed from the Chebyshev approximation are similar only for the first three coefficients.

We can now start to implement the $\ln(1 + x)$ function approximation in HDL.

Example 2.30: ln(1+x) Function Approximation

If we implement the $\ln(1 + x)$ using embedded 18×18 bit multipliers we have to take into account that our values x are in the range $0 \leq x < 1$. We therefore use a fractional integer representation with a 2.16 format. We use an additional guard bit that guarantees no problem with any overflow and that $x = 1$ can be exactly represented as 2^{16}. We use the same number format for our Chebyshev coefficients because they are all less than 1.

The following VHDL code[12] shows the $\ln(1 + x)$ approximation using six coefficients.

```
PACKAGE n_bits_int IS              -- User-defined types
   SUBTYPE BITS9 IS INTEGER RANGE -2**8 TO 2**8-1;
   SUBTYPE BITS18 IS INTEGER RANGE -2**17 TO 2**17-1;
   TYPE ARRAY_BITS18_6 IS ARRAY (0 TO 5) of BITS18;
END n_bits_int;

LIBRARY work;
USE work.n_bits_int.ALL;

LIBRARY ieee;
USE ieee.std_logic_1164.ALL;
USE ieee.std_logic_arith.ALL;
USE ieee.std_logic_signed.ALL;

ENTITY ln IS                        ------> Interface
   GENERIC (N : INTEGER := 5);-- Number of coeffcients-1
   PORT (clk      : IN   STD_LOGIC;
         x_in     : IN   BITS18;
         f_out    : OUT  BITS18);
END ln;

ARCHITECTURE fpga OF ln IS

   SIGNAL x, f : BITS18:= 0; -- Auxilary wire
-- Polynomial coefficients for 16-bit precision:
-- f(x) = (1   + 65481 x -32093 x^2 + 18601 x^3
--                          -8517 x^4 + 1954 x^5)/65536
   CONSTANT p : ARRAY_BITS18_6 :=
           (1,65481,-32093,18601,-8517,1954);
   SIGNAL s : ARRAY_BITS18_6 ;
```

---

[12] The equivalent Verilog code ln.v for this example can be found in Appendix A on page 677. Synthesis results are shown in Appendix B on page 731.

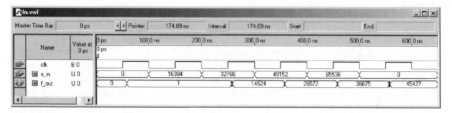

Fig. 2.54. VHDL simulation of the $\ln(1 + x)$ function approximation for the values $x = 0, x = 0.25 = 16384/65536, x = 0.5 = 32768/65536, x = 0.75 = 49152/65536, x = 1.0 = 65536/65536.$

```
BEGIN

    STORE: PROCESS     ------> I/O store in register
    BEGIN
      WAIT UNTIL clk = '1';
      x <= x_in;
      f_out <= f;
    END PROCESS;

    --> Compute sum-of-products:
    SOP: PROCESS (x,s)
    VARIABLE slv : STD_LOGIC_VECTOR(35 DOWNTO 0);
    BEGIN
-- Polynomial Approximation from Chebyshev coeffiecients
    s(N) <= p(N);
    FOR K IN N-1 DOWNTO 0 LOOP
      slv := CONV_STD_LOGIC_VECTOR(x,18)
                  * CONV_STD_LOGIC_VECTOR(s(K+1),18);
      s(K) <= CONV_INTEGER(slv(33 downto 16)) + p(K);
    END LOOP;    -- x*s/65536 problem 32 bits
    f  <= s(0);                -- make visiable outside
    END PROCESS SOP;

  END fpga;
```

The first PROCESS is used to infer the registers for the input and output data. The next PROCESS blocks SOP includes the computation of the Chebyshev approximation using a sum of product computations. The multiply and scale arithmetic is implemented with standard logic vectors data types because the 36-bit products are larger than the valid 32-bit range allowed for integers. The design uses 88 LEs, 10 embedded 9×9-bit multipliers (or half of that for 18×18-bit multipliers) and has a 32.76 MHz Registered Performance.

A simulation of the function approximation is shown in Fig. 2.54. The simulation shows the result for five different input values:

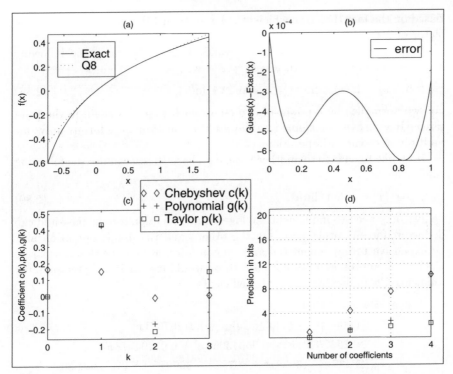

Fig. 2.55. Base 10 logarithm $f(x) = \log_{10}(x)$ function approximation. **(a)** Comparison of full-precision and 8-bit quantized approximations. **(b)** Error of quantized approximation for $x \in [0, 1]$. **(c)** Chebyshev, polynomial from Chebyshev, and Taylor polynomial coefficients. **(d)** Error of the three pruned polynomials.

x	$f(x) = \ln(x)$	$\hat{f}(x)$	$|error|$	Eff. bits
0	0	1	1.52×10^{-5}	16
0.25	$14623.9/2^{16}$	$14624/2^{16}$	4.39×10^6	17.8
0.5	$26572.6/2^{16}$	$26572/2^{16}$	2.11×10^5	15.3
0.75	$36675.0/2^{16}$	$36675/2^{16}$	5.38×10^7	20.8
1.0	$45426.1/2^{16}$	$45427/2^{16}$	1.99×10^5	15.6

Note that, due to the I/O registers, the output values appear with a delay of one clock cycle.

2.30

If we compare the polynomial code of the ln function with Clenshaw's recurrence formula from Example 2.28 (p. 134), we notice the reduction by one adder in the design.

If 8 bit plus sign precision is sufficient, we use

$$f(x) = \ln(1 + x) = 0.0006 + 0.9813x - 0.3942x^2 + 0.1058x^3$$
$$= (251x - 101x^2 + 27x^3)/256. \qquad (2.87)$$

Based on the fact that no coefficient is larger than 1.0 we can select a scaling factor of 256.

If the argument x is not in the valid range $[0, 1]$, using the following algebraic manipulation with $y = sx = 2^k x$ we get

$$\ln(sx) = \ln(s) + \ln(x) = k \times \ln(2) + \ln(x), \tag{2.88}$$

i.e., we normalize by a power-of-two factor such that x is again in the valid range. If we have determined s, the addition arithmetic effort is only one multiply and one add operation.

If we like to change to another base, e.g., base 10, we can use the following rule

$$\log_a(x) = \ln(x) / \ln(a), \tag{2.89}$$

i.e., we only need to implement the logarithmic function for one base and can deduce it for any other base. On the other hand the divide operation may be expensive to implement too and we can alternatively develop a separate Chebyshev approximation. For base 10 we would use, in 16-bit precision, the following Chebyshev polynomial coefficients

$$
\begin{aligned}
f(x) &= \log_{10}(1 + x) \\
&= 0.00000497 + 0.43393245x - 0.21267361x^2 \\
&\quad + 0.12326284x^3 - 0.05643969x^4 + 0.01294578x^5 \\
&= (28438x - 13938x^2 + 8078x^3 - 3699x^4 + 848x^5)/65536
\end{aligned}
$$

for $x \in [0, 1]$. Figure 2.55a shows the exact and 8-bit quantized function of $\log_{10}(1 + x)$. For an 8-bit quantization we would use the following approximation

$$
\begin{aligned}
f(x) &= \log_{10}(1 + x) \\
&= 0.0002 + 0.4262x - 0.1712x^2 + 0.0460x^3 \tag{2.90} \\
&= (109x - 44x^2 + 12x^3)/256, \tag{2.91}
\end{aligned}
$$

which uses only three nonzero coefficients, as shown in Fig. 2.55d.

2.9.4 Square Root Function Approximation

The development of a Taylor function approximation for the square root can not be computed around $x_0 = 0$ because then all derivatives $f^n(x_0)$ would be zero or even worse $1/0$. However, we can compute a Taylor series around $x_0 = 1$ for instance. The Taylor approximation would then be

$$
\begin{aligned}
f(x) &= \sqrt{x} \\
&= \frac{(x-1)^0}{0!} + 0.5\frac{(x-1)^1}{1!} - \frac{0.5^2}{2!}(x-1)^2 + \frac{0.5^2 1.5}{3!}(x-1)^3 - \cdots \\
&= 1 + \frac{x-1}{2} - \frac{(x-1)^2}{8} + \frac{(x-1)^3}{16} - \frac{5}{128}(x-1)^4 + \cdots
\end{aligned}
$$

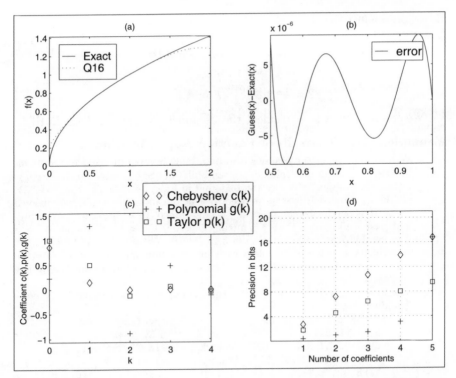

Fig. 2.56. Square root $f(x) = \sqrt{x}$ function approximation. **(a)** Comparison of full-precision and 16-bit quantized approximations. **(b)** Error of quantized approximation for $x \in [0.5, 1)$. **(c)** Chebyshev, polynomial from Chebyshev, and Taylor polynomial coefficients. **(d)** Error of the three pruned polynomials.

The coefficient and the equivalent Chebyshev coefficient are graphically interpreted in Fig.2.56c. For 16-bit polynomial quantization computed using the Chebyshev coefficient we would use

$$
\begin{aligned}
f(x) &= \sqrt{x} \\
&= 0.23080201 + 1.29086721x - 0.88893983x^2 \\
&\quad + 0.48257525x^3 - 0.11530993x^4 \\
&= (7563 + 42299x - 29129x^2 + 15813x^3 - 3778x^4)/32768.
\end{aligned}
$$

The valid argument range is $x \in [0.5, 1)$. Only terms up to order x^4 are required to get 16-bit precision. We also notice from Fig. 2.56c that the Taylor and polynomial coefficients computed from the Chebyshev approximation are not similar. The approximation with $N = 5$ shown in Fig. 2.56a is almost a perfect approximation. If we use fewer coefficients, e.g., $N = 2$ or $N = 3$, we will have a more-substantial error, see Exercise 2.30 (p. 163).

The only thing left to discuss is the question of how to handle argument values outside the range $0.5 \le x < 1$. For the square root operation this can be done by splitting the argument $y = sx$ into a power-of-two scaling factor $s = 2^k$ and the remaining argument with a valid range of $0.5 \le x < 1$. The square root for the scaling factor is accomplished by

$$\sqrt{s} = \sqrt{2^k} = \begin{cases} 2^{k/2} & k \text{ even} \\ \sqrt{2} \times 2^{(k-1)/2} & k \text{ odd} \end{cases} \quad (2.92)$$

We can now start to implement the $\sqrt{x}$ function approximation in HDL.

Example 2.31: Square Root Function Approximation

We can implement the function approximation in a parallel way using N embedded 18×18 bit multiplier or we can build an FSM to solve this iteratively. Other FSM design examples can be found in Exercises 2.20 , p. 158 and 2.21, p. 159. In a first design step we need to scale our data and coefficients in such a way that overflow-free processing is guaranteed. In addition we need a pre- and post-scaling such that x is in the range $0.5 \le x < 1$. We therefore use a fractional integer representation in 3.15 format. We use two additional guard bits that guarantee no problem with any overflow and that $x = 1$ can be exact represented as 2^{15}. We use the same number format for our Chebyshev coefficients because they are all less than 2.
The following VHDL code[13] shows the $\sqrt{x}$ approximation using $N = 5$ coefficients

```
PACKAGE n_bits_int IS          -- User-defined types
  SUBTYPE BITS9 IS INTEGER RANGE -2**8 TO 2**8-1;
  SUBTYPE BITS17 IS INTEGER RANGE -2**16 TO 2**16-1;
  TYPE ARRAY_BITS17_5 IS ARRAY (0 TO 4) of BITS9;
  TYPE STATE_TYPE IS (start,leftshift,sop,rightshift,done);
  TYPE OP_TYPE IS (load, mac, scale, denorm, nop);
END n_bits_int;

LIBRARY work;
USE work.n_bits_int.ALL;

LIBRARY ieee;
USE ieee.std_logic_1164.ALL;
USE ieee.std_logic_arith.ALL;
USE ieee.std_logic_signed.ALL;

ENTITY sqrt IS                     ------> Interface
  PORT (clk, reset : IN  STD_LOGIC;
        x_in       : IN  BITS17;
        a_o, imm_o, f_o    : OUT BITS17;
        ind_o  : OUT INTEGER RANGE 0 TO 4;
        count_o : OUT INTEGER RANGE 0 TO 3;
        x_o,pre_o,post_o : OUT BITS17;
        f_out      : OUT BITS17);
END sqrt;

ARCHITECTURE fpga OF sqrt IS
```

---

[13] The equivalent Verilog code sqrt.v for this example can be found in Appendix A on page 678. Synthesis results are shown in Appendix B on page 731.

```vhdl
SIGNAL s      : STATE_TYPE;
SIGNAL op    : OP_TYPE;

SIGNAL x : BITS17:= 0; -- Auxilary
SIGNAL a,b,f,imm : BITS17:= 0; -- ALU data
-- Chebychev poly coefficients for 16-bit precision:
CONSTANT p : ARRAY_BITS17_5 :=
       (7563,42299,-29129,15813,-3778);
SIGNAL pre, post : BITS17;

BEGIN

  States: PROCESS(clk)     ------> SQRT in behavioral style
   VARIABLE ind  : INTEGER RANGE -1 TO 4:=0;
   VARIABLE count : INTEGER RANGE 0 TO 3;
  BEGIN
    IF reset = '1' THEN           -- Asynchronous reset
      s <= start;
    ELSIF rising_edge(clk) THEN
      CASE s IS                -- Next State assignments
      WHEN start =>            -- Initialization step
        s <= leftshift;
        ind := 4;
        imm <= x_in;    -- Load argument in ALU
        op <= load;
        count := 0;
      WHEN leftshift =>        -- Normalize to 0.5 .. 1.0
        count := count + 1;
        a <= pre;
        op <= scale;
        imm <= p(4);
        IF count = 3 THEN -- Normalize ready ?
          s <= sop;
          op<=load;
          x <= f;
        END IF;
      WHEN sop =>          -- Processing step
        ind := ind - 1;
        a <= x;
        IF ind =-1  THEN -- SOP ready ?
          s <= rightshift;
          op<=denorm;
          a <= post;
        ELSE
          imm <= p(ind);
          op<=mac;
        END IF;
      WHEN rightshift =>   -- Denormalize to original range
        s <= done;
        op<=nop;
      WHEN done =>                 -- Output of results
        f_out <= f;    ------> I/O store in register
        op<=nop;
```

```
        s <= start;                    -- start next cycle
      END CASE;
    END IF;
    ind_o <= ind;
    count_o <= count;
  END PROCESS States;

  ALU: PROCESS
  BEGIN
    WAIT UNTIL clk = '1';
    CASE OP IS
      WHEN load   =>  f  <= imm;
      WHEN mac    =>  f  <= a * f /32768 + imm;
      WHEN scale  =>  f  <= a * f;
      WHEN denorm =>  f  <= a * f /32768;
      WHEN nop    =>  f  <= f;
      WHEN others =>  f  <= f;
    END CASE;
  END PROCESS ALU;

  EXP: PROCESS(x_in)
  VARIABLE slv : STD_LOGIC_VECTOR(16 DOWNTO 0);
  VARIABLE po, pr : BITS17;
  BEGIN
    slv := CONV_STD_LOGIC_VECTOR(x_in, 17);
    pr := 2**14;     -- Compute pre- and post scaling
    FOR K IN 0 TO 15 LOOP
      IF slv(K) = '1' THEN
        pre <= pr;
      END IF;
      pr := pr / 2;
    END LOOP;
    po := 1;       -- Compute pre- and post scaling
    FOR K IN 0 TO 7 LOOP
      IF slv(2*K) = '1' THEN -- even 2^k get 2^k/2
        po := 256*2**K;
      END IF;
--  sqrt(2): CSD Error = 0.0000208 = 15.55 effective bits
-- +1 +0. -1 +0 -1 +0 +1 +0 +1 +0 +0 +0 +0 +0 +1
--  9       7     5     3     1              -5
      IF slv(2*K+1) = '1' THEN -- odd k has sqrt(2) factor
        po := 2**(K+9)-2**(K+7)-2**(K+5)+2**(K+3)
                            +2**(K+1)+2**K/32;
      END IF;
      post <= po;
    END LOOP;

  END PROCESS EXP;

  a_o<=a;   -- Provide some test signals as outputs
  imm_o<=imm;
  f_o <= f;
  pre_o<=pre;
```

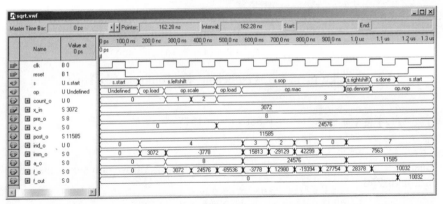

Fig. 2.57. VHDL simulation of the $\sqrt{x}$ function approximation for the value $x = 0.75/8 = 3072/32768$.

```
post_o<=post;
x_o<=x;

END fpga;
```

The code consists of three major PROCESS blocks. The control part is placed in the FSM block, while the arithmetic parts can be found in the ALU and EXP blocks. The first FSM PROCESS is used to control the machine and place the data in the correct registers for the ALU and EXP blocks. In the start state the data are initialized and the input data are loaded into the ALU. In the leftshift state the input data are normalized such that the input x is in the range $x \in [0.5, 1)$. The sop state is the major processing step where the polynomial evaluation takes place using multiply-accumulate operations performed by the ALU. At the end data are loaded for the denormalization step, i.e., rightshift state, that reverses the normalization done before. In the final step the result is transferred to the output register and the FSM is ready for the next square root computation. The ALU PROCESS block performs a $f = a \times f + imm$ operation as used in the Horner scheme (2.75), p. 139 to compute the polynomial function and will be synthesized to a single 18 × 18 embedded multiplier (or two 9 × 9-bit multiplier blocks as reported by Quartus) and some additional add and normalization logic. The block has the form of an ALU, i.e., the signal op is used to determine the current operation. The accumulator register f can be preloaded with an imm operand. The last PROCESS block EXP hosts the computation of the pre- and post-normalization factors according to (2.92). The $\sqrt{2}$ factor for the odd k values of 2^k has been implemented with CSD code computed with the csd3e.exe program. The design uses 336 LEs, 2 embedded 9 × 9-bit multipliers (or half of that for 18 × 18-bit multipliers) and has a 82.16 MHz Registered Performance.

A simulation of the function approximation is shown in Fig. 2.57. The simulation shows the result for the input value $x = 0.75/8 = 0.0938 = 3072/2^{15}$. In the shift phase the input $x = 3072$ is normalized by a pre factor of 8. The normalized result 24 576 is in the range $x \in [0.5, 1) \approx [16\,384, 32\,768)$. Then several MAC operations are computed to arrive at $f = \sqrt{0.75} \times 2^{15} = 28\,378$. Finally a denormalization with a post factor of $\sqrt{2} \times 2^{13} = 11\,585$ takes place and the final result $f = \sqrt{0.75/8} \times 2^{15} = 10\,032$ is transferred to the output register.

| 2.31 |

If 8 bit plus sign precision is sufficient, we would build a square root via

$$
\begin{aligned}
f(x) = \sqrt{x} &= 0.3171 + 0.8801x - 0.1977x^2 \\
&= (81 + 225x - 51x^2)/256.
\end{aligned} \tag{2.93}
$$

Based on the fact that no coefficient is larger than 1.0 we can select a scaling factor of 256.

Exercises

Note: If you have no prior experience with the Quartus II software, refer to the case study found in Sect. 1.4.3, p. 29. If not otherwise noted use the EP2C35F672C6 from the Cyclone II family for the Quartus II synthesis evaluations.

2.1: Wallace has introduced an alternative scheme for a fast multiplier. The basic building block of this type of multiplier is a carry-save adder (CSA). A CSA takes three n-bit operands and produces two n-bit outputs. Because there is no propagation of the carry, this type of adder is sometimes called a 3:2 compress or counter. For an $n \times n$-bit multiplier we need a total of $n - 2$ CSAs to reduce the output to two operands. These operands then have to be added by a (fast) $2n$-bit ripple-carry adder to compute the final result of the multiplier.
(a) The CSA computation can be done in parallel. Determine the minimum number of levels for an $n \times n$-bit multiplier with $n \in [0, 16]$.
(b) Explain why, for FPGAs with fast two's complement adders, these multipliers are not more attractive than the usual array multiplier.
(c) Explain how a pipelined adder in the final adder stage can be used to implement a faster multiplier. Use the data from Table 2.7 (p. 78) to estimate the necessary LE usage and possible speed for:
(c1) an 8×8-bit multiplier
(c2) a 12×12-bit multiplier

2.2: The Booth multiplier used the classical CSD code to reduce the number of necessary add/subtract operations. Starting with the LSB, typically two or three bits (called radix-4 and radix-8 algorithms) are processed in one step. The following table demonstrates possible radix-4 patterns and actions:

x_{k+1}	x_k	x_{k-1}	Accumulator activity	Comment
0	0	0	ACC→ACC +R* (0)	within a string of "0s"
0	0	1	ACC→ACC +R* (X)	end of a string of "1s"
0	1	0	ACC→ACC +R* (X)	
0	1	1	ACC→ACC +R* ($2X$)	end of a string of "1s"
1	0	0	ACC→ACC +R* ($-2X$)	beginning of a string of "1s"
1	0	1	ACC→ACC +R* ($-X$)	
1	1	0	ACC→ACC +R* (-X)	beginning of a string of "1s"
1	1	1	ACC→ACC +R* (0)	within a string of "1s"

The hardware requirements for a state machine implementation are an accumulator and a two's complement shifter.

(a) Let X be a signed 6-bit two's complement representation of $-10 = 110110_{2C}$. Complete the following table for the Booth product $P = XY = -10Y$ and indicate the accumulator activity in each step.

Step	x_5	x_4	x_3	x_2	x_1	x_0	x_{-1}	ACC	ACC + Booth rule
Start	1	1	0	1	1	0	0		
0									
1									
2									

$$(2.94)$$

(b) Compare the latency of the Booth multiplier, with the serial/parallel multiplier from Example 2.18 (p. 82), for the radix-4 and radix-8 algorithms.

2.3: (a) Compile the HDL file **add_2p** with the **QuartusII** compiler with optimization for speed and area. How many LEs are needed? Explain the results.
(b) Conduct a simulation with $15 + 102$.

2.4: Explain how to modify the HDL design **add1p** for subtraction.
(a) Modify the design and simulate as an example:
(b) $3 - 2$
(c) $2 - 3$
(d) Add an asynchronous set to the carry flip-flop to avoid initial wrong sum values. Simulate again $3 - 2$.

2.5: (a) Compile the HDL file **mul_ser** with the Quartus II compiler.
(b) Determine the **Registered Performance** and the used resources of the 8-bit design. What is the total multiplication latency?

2.6: Modify the HDL design file **mul_ser** to multiply 12×12-bit numbers.
(a) Simulate the new design with the values 1000×2000.
(b) Measure the **Registered Performance** and the resources (LEs, multipliers, and M2Ks/M4Ks).
(c) What is the total multiplication latency of the 12×12-bit multiplier?

2.7: (a) Design a state machine in Quartus II to implement the Booth multiplier (see Exercise 2.2) for 6×6 bit signed inputs.
(b) Simulate the four data $\pm 5 \times (\pm 9)$.
(c) Determine the **Registered Performance**.
(d) Determine LE utilization for maximum speed.

2.8: (a) Design a **generic** CSA that is used to build a Wallace-tree multiplier for an 8×8-bit multiplier.

(b) Implement the 8×8 Wallace tree using Quartus II.

(c) Use a final adder to compute the product, and test your multiplier with a multiplication of 100×63.

(d) Pipeline the Wallace tree. What is the maximum throughput of the pipelined design?

2.9: (a) Use the principle of component instantiation, using the predefined macros LPM_ADD_SUB and LPM_MULT, to write the VHDL code for a pipelined complex 8-bit multiplier, (i.e., $(a + jb)(c + jd) = ac - bd + j(ad + bc)$), with all operands $a, b, c,$ and d in 8-bit.

(b) Determine the **Registered Performance**.

(c) Determine LE and embedded multipliers used for maximum speed synthesis.

(d) How many pipeline stages does the optimal single LPM_MULT multiplier have?

(e) How many pipeline stages does the optimal complex multiplier have in total if you use: **(e1)** LE-based multipliers?

(e2) Embedded array multipliers?

2.10: An alternative algorithm for a complex multiplier is:

$$
\begin{array}{lll}
s[1] = a - b & s[2] = c - d & s[3] = c + d \\
m[1] = s[1]d & m[2] = s[2]a & m[3] = s[3]b \\
s[4] = m[1] + m[2] & s[5] = m[1] + m[3] & \\
& (a + jb)(c + jd) = s[4] + js[5] &
\end{array}
\tag{2.95}
$$

which, in general, needs five adders and three multipliers. Verify that if one coefficient, say $c + jd$ is known, then $s[2], s[3],$ and d can be prestored and the algorithm reduces to three adds and three multiplications. Also

(a) Design a pipelined 5/3 complex multiplier using the above algorithm for 8-bit signed inputs. Use the predefined macros LPM_ADD_SUB and LPM_MULT.

(b) Measure the **Registered Performance** and the used resources (LEs, multipliers, and M2Ks/M4Ks) for maximum speed synthesis.

(c) How many pipeline stages does the single LPM_MULT multiplier have?

(d) How many pipeline stages does the complex multiplier have in total is you use

(d1) LE-based multipliers?

(d2) Embedded array multipliers?

2.11: Compile the HDL file cordic with the Quartus II compiler, and

(a) Conduct a simulation (using the waveform file cordic.vwf) with x_in=±30 and y_in=±55. Determine the radius factor for all four simulations.

(b) Determine the maximum errors for radius and phase, compared with an unquantized computation.

2.12: Modify the HDL design cordic to implement stages 4 and 5 of the CORDIC pipeline.

(a) Compute the rotation angle, and compile the VHDL code.

(b) Conduct a simulation with values x_in=±30 and y_in=±55.

(c) What are the maximum errors for radius and phase, compared with the unquantized computation?

2.13: Consider a floating-point representation with a sign bit, E = 7-bit exponent width, and M = 10 bits for the mantissa (not counting the hidden one).

(a) Compute the bias using (2.24) p. 71.

(b) Determine the (absolute) largest number that can be represented.

(c) Determine the (absolutely measured) smallest number (not including denormals) that can be represented.

2.14: Using the result from Exercise 2.13
(a) Determine the representation of $f_1 = 9.25_{10}$ in this $(1,7,10)$ floating-point format.
(b) Determine the representation of $f_2 = -10.5_{10}$ in this $(1,7,10)$ floating-point format.
(c) Compute $f_1 + f_2$ using floating-point arithmetic.
(d) Compute $f_1 * f_2$ using floating-point arithmetic.
(e) Compute f_1/f_2 using floating-point arithmetic.

2.15: For the IEEE single-precision format (see Table 2.5, p. 74) determine the 32-bit representation of:
(a) $f_1 = -0.$
(b) $f_2 = \infty$.
(c) $f_3 = 9.25_{10}$.
(d) $f_4 = -10.5_{10}$.
(e) $f_5 = 0.1_{10}$.
(f) $f_6 = \pi = 3.141593_{10}$.
(g) $f_7 = \sqrt{3}/2 = 0.8660254_{10}$.

2.16: Compile the HDL file `div_res` from Example 2.19 (p. 94) to divide two numbers.
(a) Simulate the design with the values 234/3.
(b) Simulate the design with the values 234/1.
(c) Simulate the design with the values 234/0. Explain the result.

2.17: Design a nonperforming divider based on the HDL file `div_res` from Example 2.19 (p. 94).
(a) Simulate the design with the values 234/50 as shown in Fig. 2.23, p. 96.
(b) Measure the `Registered Performance`, the used resources (LEs, multipliers, and M2Ks/M4Ks) and latency for maximum speed synthesis.

2.18: Design a nonrestoring divider based on the HDL file `div_res` from Example 2.19 (p. 94).
(a) Simulate the design with the values 234/50 as shown in Fig. 2.24, p. 98.
(b) Measure the `Registered Performance`, the used resources (LEs, multipliers, and M2Ks/M4Ks) and latency for maximum speed synthesis.

2.19: Shift operations are usually implemented with a barrelshifter, which can be inferred in VHDL via the SLL instruction. Unfortunately, the SLL is not supported for STD_LOGIC, but we can design a barrelshifter in many different ways to achieve the same function. We wish to design 12-bit barrelshifters, that have the following entity:

```
ENTITY lshift IS                    ------> Interface
   GENERIC (W1 : INTEGER := 12; -- data bit width
            W2 : integer := 4); -- ceil(log2(W1));
   PORT (clk      : IN STD_LOGIC;
         distance : IN STD_LOGIC_VECTOR (W2-1 DOWNTO 0);
         data     : IN STD_LOGIC_VECTOR (W1-1 DOWNTO 0);
         result   : OUT STD_LOGIC_VECTOR (W1-1 DOWNTO 0));
   END;
```

that should be verified via the simulation shown in Fig. 2.58. Use input and output registers for `data` and `result`, no register for the `distance`. Select one of the following devices:

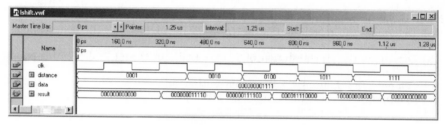

Fig. 2.58. Testbench for the barrel shifter from Exercise 2.19.

(I) EP2C35F672C6 from the Cyclone II family
(II) EPF10K70RC240-4 from the Flex 10K family
(III) EPM7128LC84-7 from the MAX7000S family
(a1) Use a PROCESS and (sequentially) convert each bit of the distance vector in an equivalent power-of-two constant multiplication. Use lshift as the entity name.
(a2) Measure the Registered Performance and the resources (LEs, multipliers, and M2Ks/M4Ks).
(b1) Use a PROCESS and shift (in a loop) the input data always 1 bit only, until the loop counter and distance show the same value. Then transfer the shifted data to the output register. Use lshiftloop as the entity name.
(b2) Measure the Registered Performance and the resources (LEs, multipliers, and M2Ks/M4Ks).
(c1) Use a PROCESS environment and "demux" with a loop statement the distance vector in an equivalent multiplication factor. Then use a single (array) multiplier to perform the multiplication. Use lshiftdemux as the entity name.
(c2) Measure the Registered Performance and the resources (LEs, multipliers, and M2Ks/M4Ks).
(d1) Use a PROCESS environment and convert with a case statement the distance vector to an equivalent multiplication factor. Then use a single (array) multiplier to perform the multiplication. Use lshiftmul as the entity name.
(d2) Measure the Registered Performance and the resources (LEs, multipliers, and M2Ks/M4Ks).
(e1) Use the lpm_clshift megafunction to implement the 12-bit barrelshifter. Use lshiftlpm as the entity name.
(e2) Measure the Registered Performance and the resources (LEs, multipliers, and M2Ks/M4Ks).
(d) Compare all five barrelshifter designs in terms of Registered Performance, resources (LEs, multipliers, and M2Ks/M4Ks), and design reuse, i.e., effort to change data width and the use of software other than Quartus II.

2.20: (a) Design the PREP benchmark 3 shown in Fig. 2.59a with the Quartus II software. The design is a small FSM with eight states, eight data input bits i, clk, rst, and an 8-bit data-out signal o. The next state and output logic is controlled by a positive-edge triggered clk and an asynchronous reset rst, see the simulation in Fig. 2.59c for the function test. The following table shows next state and output assignments,

Current state	Next state	i (Hex)	o (Hex)
start	start	(3c)'	00
start	sa	3c	82
sa	sc	2a	40
sa	sb	1f	20
sa	sa	(2a)'(1f)'	04
sb	se	aa	11
sb	sf	(aa)'	30
sc	sd	–	08
sd	sg	–	80
se	start	–	40
sf	sg	–	02
sg	start	–	01

where x' is the condition not x.

(b) Determine the `Registered Performance` and the used resources (LEs, multipliers, and M2Ks/M4Ks) for a single copy. Compile the HDL file with the synthesis `Optimization Technique` set to `Speed`, `Balanced` or `Area`; this can be found in the `Analysis & Synthesis Settings` section under `EDA Tool Settings` in the `Assignments` menu. Which synthesis options are optimal in terms of LE count and `Registered Performance`?

Select one of the following devices:

(b1) EP2C35F672C6 from the Cyclone II family

(b2) EPF10K70RC240-4 from the Flex 10K family

(b3) EPM7128LC84-7 from the MAX7000S family

(c) Design the multiple instantiation for benchmark 3 as shown in Fig. 2.59b.

(d) Determine the `Registered Performance` and the used resources (LEs, multipliers, and M2Ks/M4Ks) for the design with the maximum number of instantiations of PREP benchmark 3. Use the optimal synthesis option you found in (b) for the following devices:

(d1) EP2C35F672C6 from the Cyclone II family

(d2) EPF10K70RC240-4 from the Flex 10K family

(d3) EPM7128LC84-7 from the MAX7000S family

2.21: (a) Design the PREP benchmark 4 shown in Fig. 2.60a with the Quartus II software. The design is a large FSM with sixteen states, 40 transitions, eight data input bits i[0..7], clk, rst and 8-bit data-out signal o[0..7]. The next state is controlled by a positive-edge-triggered clk and an asynchronous reset rst, see the simulation in Fig. 2.60c for a partial function test. The following shows the output decoder table

Current state	o[7..0]	Current state	o[7..0]
s0	0 0 0 0 0 0 0 0	s1	0 0 0 0 0 1 1 0
s2	0 0 0 1 1 0 0 0	s3	0 1 1 0 0 0 0 0
s4	1 x x x x x x 0	s5	x 1 x x x x 0 x
s6	0 0 0 1 1 1 1 1	s7	0 0 1 1 1 1 1 1
s8	0 1 1 1 1 1 1 1	s9	1 1 1 1 1 1 1 1
s10	x 1 x 1 x 1 x 1	s11	1 x 1 x 1 x 1 x
s12	1 1 1 1 1 1 0 1	s13	1 1 1 1 0 1 1 1
s14	1 1 0 1 1 1 1 1	s15	0 1 1 1 1 1 1 1

where X is the unknown value. Note that the output values does not have an additional output register as in the PREP 3 benchmark. The next state table is:

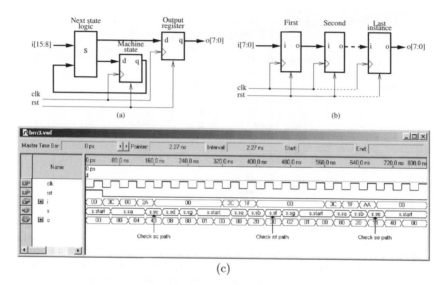

Fig. 2.59. PREP benchmark 3. **(a)** Single design. **(b)** Multiple instantiation. **(c)** Testbench to check function.

Current state	Next state	Condition	Current state	Next state	Condition
s0	s0	$i = 0$	s0	s1	$1 \leq i \leq 3$
s0	s2	$4 \leq i \leq 31$	s0	s3	$32 \leq i \leq 63$
s0	s4	$i > 63$	s1	s0	$i0 \times i1$
s1	s3	$(i0 \times i1)'$	s2	s3	$-$
s3	s5	$-$	s4	s5	$i0 + i2 + i4$
s4	s6	$(i0 + i2 + i4)'$	s5	s5	$i0'$
s5	s7	$i0$	s6	s1	$i6 \times i7$
s6	s6	$(i6 + i7)'$	s6	s8	$i6 \times i7'$
s6	s9	$i6' \times i7$	s7	s3	$i6' \times i7'$
s7	s4	$i6 \times i7$	s7	s7	$i6 \oplus i7$
s8	s1	$(i4 \odot i5)i7$	s8	s8	$(i4 \odot i5)i7'$
s8	s11	$i4 \oplus i5$	s9	s9	$i0'$
s9	s11	$i0$	s10	s1	$-$
s11	s8	$i \neq 64$	s11	s15	$i = 64$
s12	s0	$i = 255$	s12	s12	$i \neq 255$
s13	s12	$i1 \oplus i3 \oplus i5$	s13	s14	$(i1 \oplus i3 \oplus i5)'$
s14	s10	$i > 63$	s14	s12	$1 \leq i \leq 63$
s14	s14	$i = 0$	s15	s0	$i7 \times i1 \times i0$
s15	s10	$i7 \times i1' \times i0$	s15	s13	$i7 \times i1 \times i0'$
s15	s14	$i7 \times i1' \times i0'$	s15	s15	$i7'$

where ik is bit k of input i, the symbol $'$ is the not operation, $\times$ is the Boolean AND operation, $+$ is the Boolean OR operation, $\odot$ is the Boolean equivalence operation, and $\oplus$ is the XOR operation.

(b) Determine the **Registered Performance** and the used resources (LEs, multipliers, and M2Ks/M4Ks) for a single copy. Compile the HDL file with the synthesis **Optimization Technique** set to **Speed**, **Balanced** or **Area**; this can be found

in the Analysis & Synthesis Settings section under EDA Tool Settings in the Assignments menu. Which synthesis options are optimal in terms of LE count and Registered Performance?
Select one of the following devices:

(b1) EP2C35F672C6 from the Cyclone II family

(b2) EPF10K70RC240-4 from the Flex 10K family

(b3) EPM7128LC84-7 from the MAX7000S family

(c) Design the multiple instantiation for benchmark 4 as shown in Fig. 2.60b.

(d) Determine the Registered Performance and the used resources (LEs, multipliers, and M2Ks/M4Ks) for the design with the maximum number of instantiations of PREP benchmark 4. Use the optimal synthesis option you found in (b) for the following devices:

(d1) EP2C35F672C6

(d2) EPF10K70RC240-4

(d3) EPM7128LC84-7

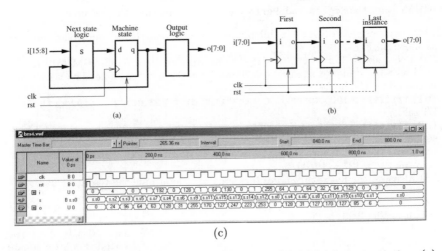

(a)

(b)

(c)

Fig. 2.60. PREP benchmark 4. **(a)** Single design. **(b)** Multiple instantiation. **(c)** Testbench to check function.

2.22: **(a)** Design an 8×8-bit signed multiplier smul8x8 using MK4s memory blocks and the partitioning technique discussed in (2.31), p. 87.

(b) Use a short C or MATLAB script to produce the three required MIF files. You need signed/signed, signed/unsigned, and unsigned/unsigned tables. The last entry in the table should be:

(b1) 11111111 : 11100001;   --> 15 * 15 = 225   for unsigned/unsigned.

(b2) 11111111 : 11110001;   --> -1 * 15 = -15   for signed/unsigned.

(b3) 11111111 : 00000001;   --> -1 * (-1) = 1   for signed/signed.

(c) Verify the design with the three data pairs $-128 \times (-128) = 16384$; $-128 \times 127 = -16256$; $127 \times 127 = 16129$.

(d) Measure the Registered Performance and determine the resources used.

2.23: **(a)** Design an 8×8-bit additive half-square (AHSM) multiplier ahsm8x8 as shown in Fig. 2.18, p. 88.

(b) Use a short C or MATLAB script to produce the two required MIF files. You need a 7- and 8-bit D1 encoded square tables. The first entries in the 7-bit table should be:

```
depth= 128; width = 14;
address_radix = bin; data_radix = bin;
content begin
  0000000 : 00000000000000;   --> (1_d1 * 1_d1)/2 = 0
  0000001 : 00000000000010;   --> (2_d1 * 2_d1)/2 = 2
  0000010 : 00000000000100;   --> (3_d1 * 3_d1)/2 = 4
  0000011 : 00000000001000;   --> (4_d1 * 4_d1)/2 = 8
  0000100 : 00000000001100;   --> (5_d1 * 5_d1)/2 = 12
  ...
```

(c) Verify the design with the three data pairs $-128 \times (-128) = 16384$; $-128 \times 127 = -16256$; $127 \times 127 = 16129$.

(d) Measure the Registered Performance and determine the resources used.

2.24: (a) Design an 8×8-bit differential half-square (DHSM) multiplier dhsm8x8 as shown in Fig. 2.19, p. 89.

(b) Use a short C or MATLAB script to produce the two required MIF files. You need an 8-bit standard square table and a 7-bit D1 encoded table. The last entries in the tables should be:

(b1) 1111111 : 10000000000000; --> (128_d1 * 128_d1)/2 = 8192 for the 7-bit D1 table.

(b2) 11111111 : 111111100000000; --> (255*255)/2 = 32512 for the 8-bit half-square table.

(c) Verify the design with the three data pairs $-128 \times (-128) = 16384$; $-128 \times 127 = -16256$; $127 \times 127 = 16129$.

(d) Measure the Registered Performance and determine the resources used.

2.25: (a) Design an 8×8-bit quarter-square multiplication multiplier qsm8x8 as shown in Fig. 2.20, p. 90.

(b) Use a short C or MATLAB script to produce the two required MIF files. You need an 8-bit standard quarter square table and an 8-bit D1 encoded quarter square table. The last entries in the tables should be:

(b1) 11111111 : 11111110000000; --> (255*255)/4 = 16256 for the 8-bit quarter square table.

(b2) 11111111 : 100000000000000; --> (256_d1 * 256_d1)/4 = 16384 for the D1 8-bit quarter-square table.

(c) Verify the design with the three data pairs $-128 \times (-128) = 16384$; $-128 \times 127 = -16256$; $127 \times 127 = 16129$.

(d) Measure the Registered Performance and determine the resources used.

2.26: Plot the function approximation and the error function as shown in Fig. 2.47a and b (p. 133) for the arctan function for $x \in [-1, 1]$ using the following coefficients:

(a) For $N = 2$ use $f(x) = 0.0000 + 0.8704x = (0 + 223x)/256$.

(b) For $N = 4$ use $f(x) = 0.0000 + 0.9857x + 0.0000x^2 - 0.2090x^3 = (0 + 252x + 0x^2 - 53x^3)/256$.

2.27: Plot the function approximation and the error function as shown, for instance, in Fig. 2.47a and b (p. 133) for the arctan function using the 8-bit precision coefficients, but with increased convergence range and determine the maximum error:

(a) For the arctan(x) approximation the using coefficients from (2.63), p. 134 with $x \in [-2, 2]$

(b) For the sin(x) approximation using the coefficients from (2.78), p. 140 with $x \in [0, 2]$

(c) For the cos(x) approximation using the coefficients from (2.81), p. 141 with $x \in [0, 2]$

(d) For the $\sqrt{1+x}$ approximation using the coefficients from (2.93), p. 153 with $x \in [0, 2]$

2.28: Plot the function approximation and the error function as shown, for instance, in Fig. 2.51a and b (p. 142) for the e^x function using the 8-bit precision coefficients, but with increased convergence range and determine the maximum error:

(a) For the e^x approximation using the coefficients from (2.83), p. 141 with $x \in [-1, 2]$

(b) For the e^{-x} approximation using the coefficients from (2.86), p. 143 with $x \in [-1, 2]$

(c) For the ln($1 + x$) approximation using the coefficients from (2.87), p. 147 with $x \in [0, 2]$

(d) For the $\log_{10}(1 + x)$ approximation using the coefficients from (2.91), p. 148 with $x \in [0, 2]$

2.29: Plot the function approximation and the error function as shown in Fig. 2.53a and b (p. 144) for the ln($1+x$) function for $x \in [0, 1]$ using the following coefficients:

(a) For $N = 2$ use $f(x) = 0.0372 + 0.6794x = (10 + 174x)/256$.

(b) For $N = 3$ use $f(x) = 0.0044 + 0.9182x - 0.2320x^2 = (1 + 235x - 59x^2)/256$.

2.30: Plot the function approximation and the error function as shown in Fig. 2.56a and b (p. 149) for the $\sqrt{x}$ function for $x \in [0.5, 1]$ using the following coefficients:

(a) For $N = 2$ use $f(x) = 0.4238 + 0.5815x = (108 + 149x)/256$.

(b) For $N = 3$ use $f(x) = 0.3171 + 0.8801x - 0.1977x^2 = (81 + 225x - 51x^2)/256$

3. Finite Impulse Response (FIR) Digital Filters

3.1 Digital Filters

Digital filters are typically used to modify or alter the attributes of a signal in the time or frequency domain. The most common digital filter is the linear time-invariant (LTI) filter. An LTI interacts with its input signal through a process called linear convolution, denoted by $y = f * x$ where f is the filter's impulse response, x is the input signal, and y is the convolved output. The linear convolution process is formally defined by:

$$y[n] = x[n] * f[n] = \sum_k x[k]f[n-k] = \sum_k f[k]x[n-k]. \qquad (3.1)$$

LTI digital filters are generally classified as being *finite impulse response* (i.e., FIR), or *infinite impulse response* (i.e., IIR). As the name implies, an FIR filter consists of a finite number of sample values, reducing the above convolution sum to a finite sum per output sample instant. An IIR filter, however, requires that an infinite sum be performed. An FIR design and implementation methodology is discussed in this chapter, while IIR filter issues are addressed in Chap. 4.

The motivation for studying digital filters is found in their growing popularity as a primary DSP operation. Digital filters are rapidly replacing classic analog filters, which were implemented using RLC components and operational amplifiers. Analog filters were mathematically modeled using ordinary differential equations of Laplace transforms. They were analyzed in the time or s (also known as Laplace) domain. Analog prototypes are now only used in IIR design, while FIR are typically designed using direct computer specifications and algorithms.

In this chapter it is assumed that a digital filter, an FIR in particular, has been designed and selected for implementation. The FIR design process will be briefly reviewed, followed by a discussion of FPGA implementation variations.

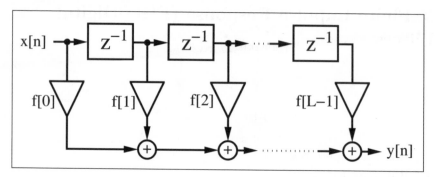

Fig. 3.1. Direct form FIR filter.

3.2 FIR Theory

An FIR with constant coefficients is an LTI digital filter. The output of an FIR of order or length L, to an input time-series $x[n]$, is given by a *finite* version of the convolution sum given in (3.1), namely:

$$y[n] = x[n] * f[n] = \sum_{k=0}^{L-1} f[k]x[n-k], \qquad (3.2)$$

where $f[0] \neq 0$ through $f[L-1] \neq 0$ are the filter's L coefficients. They also correspond to the FIR's impulse response. For LTI systems it is sometimes more convenient to express (3.2) in the z-domain with

$$Y(z) = F(z)X(z), \qquad (3.3)$$

where $F(z)$ is the FIR's *transfer function* defined in the z-domain by

$$F(z) = \sum_{k=0}^{L-1} f[k]z^{-k}. \qquad (3.4)$$

The Lth-order LTI FIR filter is graphically interpreted in Fig. 3.1. It can be seen to consist of a collection of a "tapped delay line," adders, and multipliers. One of the operands presented to each multiplier is an FIR coefficient, often referred to as a "tap weight" for obvious reasons. Historically, the FIR filter is also known by the name "transversal filter," suggesting its "tapped delay line" structure.

The *roots* of polynomial $F(z)$ in (3.4) define the zeros of the filter. The presence of only zeros is the reason that FIRs are sometimes called *all zero filters*. In Chap. 5 we will discuss an important class of FIR filters (called CIC filters) that are *recursive* but also FIR. This is possible because the poles produced by the recursive part are canceled by the nonrecursive part of the filter. The effective pole/zero plot also then has *only* zeros, i.e., is an *all-zero filter* or FIR. We note that nonrecursive filters are always FIR, but recursive filters can be either FIR or IIR. Figure 3.2 illustrates this dependence.

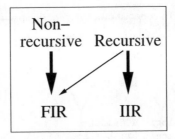

Fig. 3.2. Relation between structure and impulse length.

3.2.1 FIR Filter with Transposed Structure

A variation of the direct FIR model is called the *transposed FIR filter*. It can be constructed from the FIR filter in Fig. 3.1 by:

- Exchanging the input and output
- Inverting the direction of signal flow
- Substituting an adder by a fork, and vice versa

A transposed FIR filter is shown in Fig. 3.3 and is, in general, the preferred implementation of an FIR filter. The benefit of this filter is that we do not need an extra shift register for $x[n]$, and there is no need for an extra pipeline stage for the adder (tree) of the products to achieve high throughput.

The following examples show a direct implementation of the transposed filter.

Example 3.1: Programmable FIR Filter

We recall from the discussion of sum-of-product (SOP) computations using a PDSP (see Sect. 2.7, p. 114) that, for B_x data/coefficient bit width and filter length L, additional $\log_2(L)$ bits for unsigned SOP and $\log_2(L) - 1$ guard bits for signed arithmetic must be provided. For a 9-bit signed data/coefficient and $L = 4$, the adder width must be $9 + 9 + \log_2(4) - 1 = 19$.

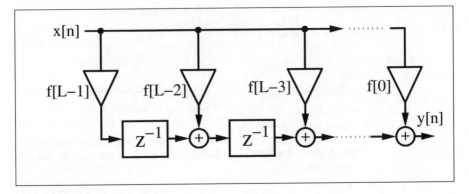

Fig. 3.3. FIR filter in the transposed structure.

The following VHDL code[2] shows the generic specification for an implementation for a length-4 filter.

```
-- This is a generic FIR filter generator
-- It uses W1 bit data/coefficients bits
LIBRARY lpm;                        -- Using predefined packages
USE lpm.lpm_components.ALL;

LIBRARY ieee;
USE ieee.std_logic_1164.ALL;
USE ieee.std_logic_arith.ALL;
USE ieee.std_logic_unsigned.ALL;

ENTITY fir_gen IS                         ------> Interface
  GENERIC (W1 : INTEGER := 9; -- Input bit width
           W2 : INTEGER := 18;-- Multiplier bit width 2*W1
           W3 : INTEGER := 19;-- Adder width = W2+log2(L)-1
           W4 : INTEGER := 11;-- Output bit width
           L  : INTEGER := 4; -- Filter length
        Mpipe : INTEGER := 3-- Pipeline steps of multiplier
           );
  PORT ( clk    : IN STD_LOGIC;
         Load_x : IN  STD_LOGIC;
         x_in   : IN  STD_LOGIC_VECTOR(W1-1 DOWNTO 0);
         c_in   : IN  STD_LOGIC_VECTOR(W1-1 DOWNTO 0);
         y_out  : OUT STD_LOGIC_VECTOR(W4-1 DOWNTO 0));
END fir_gen;

ARCHITECTURE fpga OF fir_gen IS

    SUBTYPE N1BIT IS STD_LOGIC_VECTOR(W1-1 DOWNTO 0);
    SUBTYPE N2BIT IS STD_LOGIC_VECTOR(W2-1 DOWNTO 0);
    SUBTYPE N3BIT IS STD_LOGIC_VECTOR(W3-1 DOWNTO 0);
    TYPE ARRAY_N1BIT IS ARRAY (0 TO L-1) OF N1BIT;
    TYPE ARRAY_N2BIT IS ARRAY (0 TO L-1) OF N2BIT;
    TYPE ARRAY_N3BIT IS ARRAY (0 TO L-1) OF N3BIT;

    SIGNAL  x : N1BIT;
    SIGNAL  y : N3BIT;
    SIGNAL  c : ARRAY_N1BIT; -- Coefficient array
    SIGNAL  p : ARRAY_N2BIT; -- Product array
    SIGNAL  a : ARRAY_N3BIT; -- Adder array

BEGIN

    Load: PROCESS              ------> Load data or coefficient
    BEGIN
      WAIT UNTIL clk = '1';
      IF (Load_x = '0') THEN
        c(L-1) <= c_in;        -- Store coefficient in register
        FOR I IN L-2 DOWNTO 0 LOOP  -- Coefficients shift one
          c(I) <= c(I+1);
```

---

[2] The equivalent Verilog code fir_gen.v for this example can be found in Appendix A on page 680. Synthesis results are shown in Appendix B on page 731.

```
      END LOOP;
   ELSE
      x <= x_in;              -- Get one data sample at a time
   END IF;
END PROCESS Load;


SOP: PROCESS (clk)          ------> Compute sum-of-products
BEGIN
   IF clk'event and (clk = '1') THEN
   FOR I IN 0 TO L-2  LOOP        -- Compute the transposed
      a(I) <= (p(I)(W2-1) & p(I)) + a(I+1); -- filter adds
   END LOOP;
   a(L-1) <= p(L-1)(W2-1) & p(L-1);    -- First TAP has
   END IF;                             -- only a register
   y <= a(0);
END PROCESS SOP;


-- Instantiate L pipelined multiplier
MulGen: FOR I IN 0 TO L-1 GENERATE
Muls: lpm_mult                  -- Multiply p(i) = c(i) * x;
      GENERIC MAP ( LPM_WIDTHA => W1, LPM_WIDTHB => W1,
                    LPM_PIPELINE => Mpipe,
                    LPM_REPRESENTATION => "SIGNED",
                    LPM_WIDTHP => W2,
                    LPM_WIDTHS => W2)
      PORT MAP ( clock => clk, dataa => x,
                 datab => c(I), result => p(I));
      END GENERATE;

   y_out <= y(W3-1 DOWNTO W3-W4);

END fpga;
```

The first process, Load, is used to load the coefficient in a tapped delay line if Load_x=0. Otherwise, a data word is loaded into the x register. The second process, called SOP, implements the sum-of-products computation. The products p(I) are sign-extended by one bit and added to the previous partial SOP. Note also that all multipliers are instantiated by a **generate** statement, which allows the assignment of extra pipeline stages. Finally, the output y_out is assigned the value of the SOP divided by 256, because the coefficients are all assumed to be fractional (i.e., $|f[k]| \leq 1.0$). The design uses 184 LEs, 4 embedded multipliers, and has a 329.06 MHz **Registered Performance**.

To simulate this length-4 filter consider a Daubechies DB4 filter coefficient with

$$G(z) = \frac{(1+\sqrt{3}) + (3+\sqrt{3})z^{-1} + (3-\sqrt{3})z^{-2} + (1-\sqrt{3})z^{-3}}{4\sqrt{2}},$$

$$G(z) = 0.48301 + 0.8365z^{-1} + 0.2241z^{-2} - 0.1294z^{-3}.$$

Quantizing the coefficients to eight bits (plus a sign bit) of precision results in the following model:

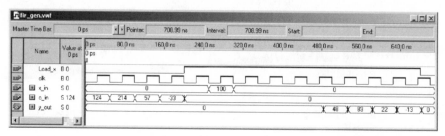

Fig. 3.4. Simulation of the 4-tap programmable FIR filter with Daubechies filter coefficient loaded.

$$G(z) = \left(124 + 214z^{-1} + 57z^{-2} - 33z^{-3}\right)/256$$
$$= \frac{124}{256} + \frac{214}{256}z^{-1} + \frac{57}{256}z^{-2} - \frac{33}{256}z^{-3}.$$

As can be seen from Fig. 3.4, in the first four steps we load the coefficients $\{124, 214, 57, -33\}$ into the tapped delay line. Note that Quartus II can also display signed numbers. As unsigned data the value -33 will be displayed as $512 - 33 = 479$. Then we check the impulse response of the filter by loading 100 into the **x** register. The first valid output is then available after 450 ns.

$$\boxed{3.1}$$

3.2.2 Symmetry in FIR Filters

The center of an FIR's impulse response is an important point of symmetry. It is sometimes convenient to define this point as the 0th sample instant. Such filter descriptions are *a-causal* (centered notation). For an odd-length FIR, the a-causal filter model is given by:

$$F(z) = \sum_{k=-(L-1)/2}^{(L-1)/2} f[k]z^{-k}. \tag{3.5}$$

The FIR's frequency response can be computed by evaluating the filter's transfer function about the periphery of the unity circle, by setting $z = e^{j\omega T}$. It then follows that:

$$F(\omega) = F(e^{j\omega T}) = \sum_k f[k]e^{-j\omega kT}. \tag{3.6}$$

We then denote with $|F(\omega)|$ the filter's *magnitude frequency response* and $\phi(\omega)$ denotes the *phase response*, and satisfies:

$$\phi(\omega) = \arctan\left(\frac{\Im(F(\omega))}{\Re(F(\omega))}\right). \tag{3.7}$$

Digital filters are more often characterized by phase and magnitude than by the z-domain transfer function or the complex frequency transform.

Table 3.1. Four possible linear-phase FIR filters $F(z) = \sum_k f[k]z^{-k}$.

Symmetry L	$f[n] = f[-n]$ odd	$f[n] = f[-n]$ even	$f[n] = -f[-n]$ odd	$f[n] = -f[-n]$ even
Example				
Zeros at	$\pm 120°$	$\pm 90°, 180°$	$0°, 180°$	$0°, 2 \times 180°$

3.2.3 Linear-phase FIR Filters

Maintaining phase integrity across a range of frequencies is a desired system attribute in many applications such as communications and image processing. As a result, designing filters that establish linear-phase versus frequency is often mandatory. The standard measure of the phase linearity of a system is the "group delay" defined by:

$$\tau(\omega) = -\frac{d\phi(\omega)}{d\omega}. \tag{3.8}$$

A perfectly linear-phase filter has a group delay that is constant over a range of frequencies. It can be shown that linear-phase is achieved if the filter is symmetric or antisymmetric, and it is therefore preferable to use the a-causal framework of (3.5). From (3.7) it can be seen that a constant group delay can only be achieved if the frequency response $F(\omega)$ is a purely real or imaginary function. This implies that the filter's impulse response possesses even or odd symmetry. That is:

$$f[n] = f[-n] \quad \text{or} \quad f[n] = -f[-n]. \tag{3.9}$$

An odd-order even-symmetry FIR filter would, for example, have a frequency response given by:

$$F(\omega) = f[0] + \sum_{k>0} f[k]e^{-jk\omega T} + f[-k]e^{jk\omega T} \tag{3.10}$$

$$= f[0] + 2\sum_{k>0} f[k]\cos(k\omega T), \tag{3.11}$$

which is seen to be a purely real function of frequency. Table 3.1 summarizes the four possible choices of symmetry, antisymmetry, even order and odd order. In addition, Table 3.1 graphically displays an example of each class of linear-phase FIR.

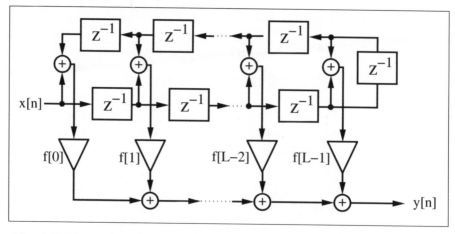

Fig. 3.5. Linear-phase filter with reduced number of multipliers.

The symmetry properties intrinsic to a linear-phase FIR can also be used to reduce the necessary number of multipliers L, as shown in Fig. 3.1. Consider the linear-phase FIR shown in Fig. 3.5 (even symmetry assumed), which fully exploits coefficient symmetry. Observe that the "symmetric" architecture has a multiplier budget per filter cycle exactly half of that found in the direct architecture shown in Fig. 3.1 (L versus $L/2$) while the number of adders remains constant at $L - 1$.

3.3 Designing FIR Filters

Modern digital FIR filters are designed using computer-aided engineering (CAE) tools. The filters used in this chapter are designed using the MATLAB Signal Processing toolbox. The toolbox includes an "Interactive Lowpass Filter Design" demo example that covers many typical digital filter designs, including:

- Equiripple (also known as minimax) FIR design, which uses the Parks–McClellan and Remez exchange methods for designing a linear-phase (symmetric) equiripple FIR. This equiripple design may also be used to design a differentiator or Hilbert transformer.
- Kaiser window design using the inverse DFT method weighted by a Kaiser window.
- Least square FIR method. This filter design also has ripple in the passband and stopband, but the mean least square error is minimized.
- Four IIR filter design methods (Butterworth, Chebyshev I and II, and elliptic) which will be discussed in Chap. 4.

The FIR methods are individually developed in this section. Most often we already know the transfer function (i.e., magnitude of the frequency response) of the desired filter. Such a lowpass specification typically consists of the passband $[0 \ldots \omega_p]$, the transition band $[\omega_p \ldots \omega_s]$, and the stopband $[\omega_s \ldots \pi]$ specification, where the sampling frequency is assumed to be 2π. To compute the filter coefficients we may therefore apply the *direct frequency* method discussed next.

3.3.1 Direct Window Design Method

The discrete Fourier transform (DFT) establishes a direct connection between the frequency and time domains. Since the frequency domain is the domain of filter definition, the DFT can be used to calculate a set of FIR filter coefficients that produce a filter that approximates the frequency response of the target filter. A filter designed in this manner is called a *direct FIR filter*. A direct FIR filter is defined by:

$$f[n] = \mathrm{IDFT}(F[k]) = \sum_k F[k] e^{j2\pi kn/L}. \tag{3.12}$$

From basic signals and systems theory, it is known that the spectrum of a real signal is Hermitian. That is, the real spectrum has even symmetry and the imaginary spectrum has odd symmetry. If the synthesized filter should have only real coefficients, the target DFT design spectrum must therefore be Hermitian or $F[k] = F^*[-k]$, where the $*$ denotes conjugate complex.

Consider a length-16 direct FIR filter design with a rectangular window, shown in Fig. 3.6a, with the passband ripple shown in Fig. 3.6b. Note that the filter provides a reasonable approximation to the ideal lowpass filter with the greatest mismatch occurring at the edges of the transition band. The observed "ringing" is due to the Gibbs phenomenon, which relates to the inability of a finite Fourier spectrum to reproduce sharp edges. The Gibbs ringing is implicit in the direct inverse DFT method and can be expected to be about ±7% over a wide range of filter orders. To illustrate this, consider the example filter with length 128, shown in Fig. 3.6c, with the passband ripple shown in Fig. 3.6d. Although the filter length is essentially increased (from 16 to 128) the ringing at the edge still has about the same quantity. The effects of ringing can only be suppressed with the use of a data "window" that tapers smoothly to zero on both sides. Data windows overlay the FIR's impulse response, resulting in a "smoother" magnitude frequency response with an attendant widening of the transition band. If, for instance, a Kaiser window is applied to the FIR, the Gibbs ringing can be reduced as shown in Fig. 3.7(upper). The deleterious effect on the transition band can also be seen. Other classic window functions are summarized in Table 3.2. They differ in terms of their ability to make tradeoffs between "ringing" and transition bandwidth extension. The number of recognized and published window functions is large. The most common windows, denoted $w[n]$, are:

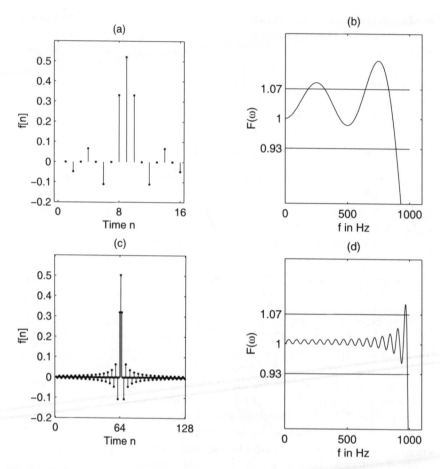

Fig. 3.6. Gibbs phenomenon. **(a)** Impulse response of FIR lowpass with $L = 16$. **(b)** Passband of transfer function $L = 16$. **(c)** Impulse response of FIR lowpass with $L = 128$. **(d)** Passband of transfer function $L = 128$.

- Rectangular: $w[n] = 1$
- Bartlett (triangular) : $w[n] = 2n/N$
- Hanning: $w[n] = 0.5\,(1 - \cos(2\pi n/L)$
- Hamming: $w[n] = 0.54 - 0.46\cos(2\pi n/L)$
- Blackman: $w[n] = 0.42 - 0.5\cos(2\pi n/L) + 0.08\cos(4\pi n/L)$
- Kaiser: $w[n] = I_0\left(\beta\sqrt{1 - (n - L/2)^2/(L/2)^2}\right)$

Table 3.2 shows the most important parameters of these windows.

The 3-dB bandwidth shown in Table 3.2 is the bandwidth where the transfer function is decreased from DC by 3 dB or $\approx 1/\sqrt{2}$. Data windows also generate sidelobes, to various degrees, away from the 0th harmonic. De-

Table 3.2. Parameters of commonly used window functions.

Name	3-dB band-width	First zero	Maximum sidelobe	Sidelobe decrease per octave	Equivalent Kaiser β
Rectangular	$0.89/T$	$1/T$	-13 dB	-6 dB	0
Bartlett	$1.28/T$	$2/T$	-27 dB	-12 dB	1.33
Hanning	$1.44/T$	$2/T$	-32 dB	-18 dB	3.86
Hamming	$1.33/T$	$2/T$	-42 dB	-6 dB	4.86
Blackman	$1.79/T$	$3/T$	-74 dB	-6 dB	7.04
Kaiser	$1.44/T$	$2/T$	-38 dB	-18 dB	3

pending on the smoothness of the window, the third column in Table 3.2 shows that some windows do not have a zero at the first or second zero DFT frequency $1/T$. The maximum sidelobe gain is measured relative to the 0th harmonic value. The fifth column describes the asymptotic decrease of the window per octave. Finally, the last column describes the value β for a Kaiser window that emulates the corresponding window properties. The Kaiser window, based on the first-order Bessel function I_0, is special in two respects. It is nearly optimal in terms of the relationship between "ringing" suppression and transition width, and second, it can be tuned by β, which determines the ringing of the filter. This can be seen from the following equation credited to Kaiser.

$$\beta = \begin{cases} 0.1102(A - 8.7) & A > 50, \\ 0.5842(A - 21)^{0.4} + 0.07886(A - 21) & 21 \leq A \leq 50, \\ 0 & A < 21, \end{cases} \qquad (3.13)$$

where $A = 20\log_{10}\varepsilon_r$ is both stopband attenuation and the passband ripple in dB. The Kaiser window length to achieve a desired level of suppression can be estimated:

$$L = \frac{A - 8}{2.285(\omega_{\text{s}} - \omega_{\text{p}})} + 1. \qquad (3.14)$$

The length is generally correct within an error of ± 2 taps.

3.3.2 Equiripple Design Method

A typical filter specification not only includes the specification of passband ω_{p} and stopband ω_{s} frequencies and ideal gains, but also the allowed deviation (or ripple) from the desired transfer function. The transition band is most often assumed to be arbitrary in terms of ripples. A special class of FIR filter that is particularly effective in meeting such specifications is called the equiripple FIR. An equiripple design protocol minimizes the maximal deviations (ripple error) from the ideal transfer function. The equiripple algorithm applies to a number of FIR design instances. The most popular are:

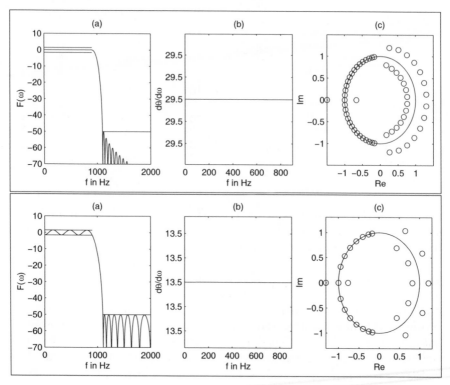

Fig. 3.7. (upper) Kaiser window design with $L = 59$. **(lower)** Parks-McClellan design with $L = 27$.
(a) Transfer function. **(b)** Group delay of passband. **(c)** Zero plot.

- Lowpass filter design (in MATLAB[3] use `firpm(L,F,A,W)`), with tolerance scheme as shown in Fig. 3.8a
- Hilbert filter, i.e., a unit magnitude filter that produces a 90° phase shift for all frequencies in the passband (in MATLAB use `firpm(L, F, A, 'Hilbert')`
- Differentiator filter that has a linear increasing frequency magnitude proportional to ω (in MATLAB use `firpm(L,F,A,'differentiator')`)

The equiripple or minimum-maximum algorithm is normally implemented using the *Parks–McClellan iterative method.* The Parks–McClellan method is used to produce a equiripple or minimax data fit in the frequency domain. It is based on the "alternation theorem" that says that there is exactly one polynomial, a Chebyshev polynomial with minimum length, that fits into a given tolerance scheme. Such a tolerance scheme is shown in Fig. 3.8a, and Fig. 3.8b shows a polynomial that fulfills this tolerance scheme. The length

---

[3] In previous MATLAB versions the function `remez` had to be used.

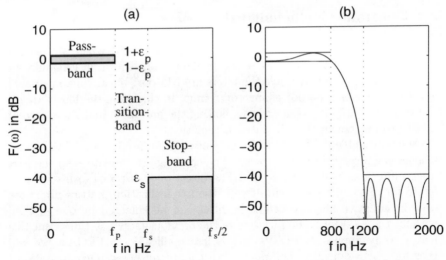

Fig. 3.8. Parameters for the filter design. **(a)** Tolerance scheme **(b)** Example function, which fulfills the scheme.

of the polynomial, and therefore the filter, can be estimated for a lowpass with

$$L = \frac{-10\log_{10}(\varepsilon_p\varepsilon_s) - 13}{2.324(\omega_s - \omega_p)} + 1, \qquad (3.15)$$

where ε_p is the passband and ε_s the stopband ripple.

The algorithm iteratively finds the location of locally maximum errors that deviate from a nominal value, reducing the size of the maximal error per iteration, until all deviation errors have the same value. Most often, the *Remez method* is used to select the new frequencies by selecting the frequency set with the largest peaks of the error curve between two iterations, see [79, p. 478]. This is why the MATLAB equiripple function was called remez in the past (now renamed to firpm for Parks-McClellan).

Compared to the *direct frequency method*, with or without data windows, the advantage of the equiripple design method is that passband and stopband deviations can be specified differently. This may, for instance, be useful in audio applications where the ripple in the passband may be specified to be higher, because the ear only perceives differences larger than 3 dB.

We note from Fig. 3.7(lower) that the equiripple design having the same tolerance requirements as the Kaiser window design enjoys a considerably reduced filter order, i.e., 27 compared with 59.

3.4 Constant Coefficient FIR Design

There are only a few applications (e.g., adaptive filters) where we need a general programmable filter architecture like the one shown in Example 3.1 (p. 167). In many applications, the filters are LTI (i.e., linear time invariant) and the coefficients do not change over time. In this case, the hardware effort can essentially be reduced by exploiting the multiplier and adder (trees) needed to implement the FIR filter arithmetic.

With available digital filter design software the production of FIR coefficients is a straightforward process. The challenge remains to map the FIR design into a suitable architecture. The direct or transposed forms are preferred for maximum speed and lowest resource utilization. Lattice filters are used in adaptive filters because the filter can be enlarged by one section, without the need for recomputation of the previous lattice sections. But this feature only applies to PDSPs and is less applicable to FPGAs. We will therefore focus our attention on the direct and transposed implementations. We will start with possible improvements to the direct form and will then move on to the transposed form. At the end of the section we will discuss an alternative design approach using distributed arithmetic.

3.4.1 Direct FIR Design

The direct FIR filter shown in Fig. 3.1 (p. 166) can be implemented in VHDL using (sequential) PROCESS statements or by "component instantiations" of the adders and multipliers. A PROCESS design provides more freedom to the synthesizer, while component instantiation gives full control to the designer. To illustrate this, a length-4 FIR will be presented as a PROCESS design. Although a length-4 FIR is far too short for most practical applications, it is easily extended to higher orders and has the advantage of a short compiling time. The linear-phase (therefore symmetric) FIR's impulse response is assumed to be given by

$$f[k] = \{-1.0, 3.75, 3.75, -1.0\}. \tag{3.16}$$

These coefficients can be directly encoded into a 5-bit fractional number. For example, 3.75_{10} would have a 5-bit binary representation 011.11_2 where "." denotes the location of the binary point. Note that it is, in general, more efficient to implement only *positive* CSD coefficients, because positive CSD coefficients have fewer nonzero terms and we can take the sign of the coefficient into account when the summation of the products is computed. See also the first step in the RAG algorithm 3.4 discussed later, p. 183.

In a practical situation, the FIR coefficients are obtained from a computer design tool and presented to the designer as floating-point numbers. The performance of a fixed-point FIR, based on floating-point coefficients, needs to be verified using simulation or algebraic analysis to ensure that design specifications remain satisfied. In the above example, the floating-point numbers are 3.75 and 1.0, which can be represented exactly with fixed-point numbers, and the check can be skipped.

Another issue that must be addressed when working with fixed-point designs is protecting the system from *dynamic range overflow*. Fortunately, the *worst-case* dynamic range growth G of an Lth-order FIR is easy to compute and it is:

$$G \leq \log_2 \left(\sum_{k=0}^{L-1} |f[k]| \right). \tag{3.17}$$

The total bit width is then the sum of the input bit width and the bit growth G. For the above filter for (3.16) we have $G = \log_2(9.5) < 4$, which states that the system's internal data registers need to have at least four more integer bits than the input data to insure no overflow. If 8-bit internal arithmetic is used the input data should be bounded by $\pm 128/9.5 = \pm 13$.

Example 3.2: Four-tap Direct FIR Filter

The VHDL design[4] for a filter with coefficients $\{-1, 3.75, 3.75, -1\}$ is shown in the following listing.

```
PACKAGE eight_bit_int IS     -- User-defined types
  SUBTYPE BYTE IS INTEGER RANGE -128 TO 127;
  TYPE ARRAY_BYTE IS ARRAY (0 TO 3) OF BYTE;
END eight_bit_int;

LIBRARY work;
USE work.eight_bit_int.ALL;

LIBRARY ieee;
USE ieee.std_logic_1164.ALL;
USE ieee.std_logic_arith.ALL;

ENTITY fir_srg IS                        ------> Interface
  PORT (clk  :  IN  STD_LOGIC;
        x    :  IN  BYTE;
        y    :  OUT BYTE);
END fir_srg;

ARCHITECTURE flex OF fir_srg IS

  SIGNAL tap : ARRAY_BYTE := (0,0,0,0);
                            -- Tapped delay line of bytes
BEGIN

  p1: PROCESS            ------> Behavioral style
  BEGIN
    WAIT UNTIL clk = '1';
-- Compute output y with the filter coefficients weight.
-- The coefficients are [-1   3.75   3.75   -1].
```

---

[4] The equivalent Verilog code fir_srg.v for this example can be found in Appendix A on page 682. Synthesis results are shown in Appendix B on page 731.

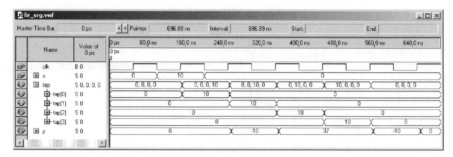

Fig. 3.9. VHDL simulation results of the FIR filter with impulse input 10.

```
-- Division for Altera VHDL is only allowed for
-- powers-of-two values!
   y <= 2 * tap(1) + tap(1) + tap(1) / 2 + tap(1) / 4
        + 2 * tap(2) + tap(2) + tap(2) / 2 + tap(2) / 4
        - tap(3) - tap(0);
   FOR I IN 3 DOWNTO 1 LOOP
      tap(I) <= tap(I-1); -- Tapped delay line: shift one
   END LOOP;
   tap(0) <= x;                    -- Input in register 0
END PROCESS;

END flex;
```

The design is a literal interpretation of the direct FIR architecture found in
Fig. 3.1 (p. 166). The design is applicable to both symmetric and asymmetric
filters. The output of each tap of the tapped delay line is multiplied by the
appropriately weighted binary value and the results are added. The impulse
response y of the filter to an impulse 10 is shown in Fig. 3.9.   3.2

There are three obvious actions that can improve this design:

1) Realize each filter coefficient with an optimized CSD code (see Chap. 2,
Example 2.1, p. 58).

2) Increase effective multiplier speed by pipelining. The output adder
should be arranged in a pipelined balance tree. If the coefficients are coded
as "powers-of-two," the pipelined multiplier and the adder tree can be
merged. Pipelining has low overhead due to the fact that the LE registers
are otherwise often unused. A few additional pipeline registers may be necessary if the number of terms in the tree to be added is not a power of
two.

3) For symmetric coefficients, the multiplication complexity can be reduced
as shown in Fig. 3.5 (p. 172).

The first two actions are applicable to all FIR filters, while the third applies
only to linear-phase (symmetric) filters. These ideas will be illustrated by
example designs.

Table 3.3. Improved FIR filter.

Symmetry	no	yes	no	no	yes	yes
CSD	no	no	yes	no	yes	yes
Tree	no	no	no	yes	no	yes
Speed/MHz	99.17	178.83	123.59	270.20	161.79	277.24
Size/LEs	114	99	65	139	57	81

Example 3.3: Improved Four-tap Direct FIR Filter

The design from the previous example can be improved using a CSD code for the coefficients $3.75 = 2^2 - 2^{-2}$. In addition, symmetry and pipelining can also be employed to enhance the filter's performance. Table 3.3 shows the maximum throughput that can be expected for each different design. CSD coding and symmetry result in smaller, more compact designs. Improvements in `Registered Performance` are obtained by pipelining the multiplier and providing an adder tree for the output accumulation. Two additional pipeline registers (i.e., 16 LEs) are necessary, however. The most compact design is expected using symmetry and CSD coding without the use of an adder tree. The partial VHDL code for producing the filter output y is shown below.

```
t1 <= tap(1) + tap(2); -- Using symmetry
t2 <= tap(0) + tap(3);
IF rising_edge(clk) THEN
  y <= 4 * t1 - t1 / 4 - t2; Apply CSD code and add
  ...
```

The fastest design is obtained when all three enhancements are used. The partial VHDL code, in this case, becomes:

```
WAIT UNTIL clk = '1';   -- Pipelined all operations
t1 <= tap(1) + tap(2);  -- Use symmetry of coefficients
t2 <= tap(0) + tap(3);  -- and pipeline adder
t3 <= 4 * t1 - t1 / 4;  -- Pipelined CSD multiplier
t4 <= -t2;              -- Build a binary tree and add delay
y <=  t3 + t4;
...
```

3.3

Exercise 3.7 (p. 210) discusses the implementation of the filter in more detail.

Direct Form Pipelined FIR Filter

Sometimes a single coefficient has more pipeline delay than all the other coefficients. We can model this delay by $f[n]z^{-d}$. If we now add a positive delay with

$$f[n] = z^d f[n] z^{-d} \tag{3.18}$$

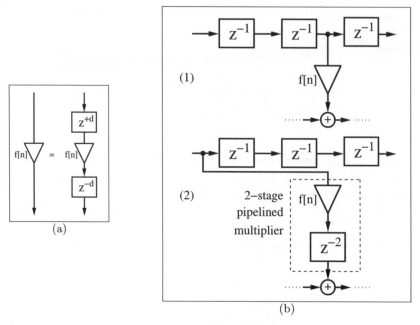

Fig. 3.10. Rephasing FIR filter. **(a)** Principle. **(b)** Rephasing a multiplier. **(1)** Without pipelining. **(2)** With two-stage pipelining.

the two delays are eliminated. Translating this into hardware means that for the direct form FIR filter we have to use the output of the d position previous register.

This principle is shown in Fig. 3.10a. Figure 3.10b shows an example of rephasing a pipelined multiplier that has two delays.

3.4.2 FIR Filter with Transposed Structure

A variation of the direct FIR filter is called the *transposed filter* and has been discussed in Sect. 3.2.1 (p. 167). The transposed filter enjoys, in the case of a constant coefficient filter, the following two additional improvements compared with the direct FIR:

- Multiple use of the repeated coefficients using the reduced adder graph (RAG) algorithm [31, 32, 33, 34]
- Pipeline adders using a carry-save adder

The pipeline adder increases the speed, at additional adder and register costs, while the RAG principle will reduce the size (i.e., number of LEs) of the filter and sometimes also increase the speed. The pipeline adder principle has been discussed in Chap. 2 and here we will focus on the RAG algorithm.

In Chap. 2 it was noted that it can sometimes be advantageous to implement the factors of a constant coefficient, rather than implement the CSD code directly. For example, the CSD code realization of the constant multiplier coefficient 93 requires three adders, while the factors 3×31 only requires two adders, see Fig. 2.3 (p. 61). For a transposed FIR filter, the probability is high that all the coefficients will have several factors in common. For instance, the coefficients 9 and 11 can be built using $8 + 1 = 9$ for the first and $11 = 9 + 2$ for the second. This reduces the total effort by one adder. In general, however, finding the optimal reduced adder graph (RAG) is an *NP*-hard problem. As a result, heuristics must be used. The RAG algorithm first suggested by Dempster and Macleod is described next [33].

Algorithm 3.4:   **Reduced Adder Graph**

1) Reduce all coefficients in the input set to positive odd fundamentals (OF).

2) Evaluate the single-coefficient adder cost of each coefficient using the MAG Table 2.3, p. 64.

3) Remove from the input set all power-of-two values and repeated fundamentals.

4) Create a graph set of all coefficients that can be built with one adder. Remove these coefficients from the input set.

5) Check if a pair of fundamentals in the graph set can be used to generate a coefficient in the input set by using a single adder.

6) Repeat step 5 until no further coefficients are added to the graph set.

This completes the optimal part of the algorithm. Next follows the heuristic part of the algorithm:

7) Add the smallest coefficient requiring two adders (if found) from the input set and its smallest NOF. The OF and one NOF (i.e., auxiliary coefficient) requires two adders using the fundamentals in the graph set.

8) Go to step 5 since the two new fundamentals from step 7 can be used to build other coefficients from the input set.

9) Add the smallest adder cost-3 or higher OF to the graph set and use the minimum NOF sum for this coefficient.

10) Go to step 5 until all coefficients are synthesized.

Steps 1–6 are straightforward, but steps 7–10 are potentially complex since the number of theoretical graphs increases exponentially. To simplify the process it is helpful to use the MAG coding data shown in Table 2.3 (p. 64). Let us briefly review some of the RAG steps that are not so obvious at first glance.

In step 1 all coefficients are reduced to positive odd fundamentals (i.e., power-of-two factors are removed from each coefficient), since this maximizes the number of partial sums, and the negative signs of the coefficients are implemented in the output adder TAPs of the filter. The two coefficient -7 and $28 = 4 \times 7$ would be merged. This works fine except for the unlikely case when all coefficients are negative. Then a sign complement operation has to be added to the filter output.

In step 5 all sums of two extended fundamentals are considered. It may happen that a final division is also required, i.e., $g = (2^u f_1 \pm 2^v f_2)/2^w$. Note that multiplication or division by two can be implemented by left and right shift, respectively, i.e., they do not require hardware resources. For instance the coefficient set $\{7,105,53\}$ MAG coding required one, two, and three adders, respectively. In RAG the set is synthesized as $7 = 8 - 1; 105 = 7 \times 15; 53 = (105 + 1)/2$, requiring only three adders but also a divide/right shift operation.

In step 7 an adder cost-2 coefficient is added and the algorithm selects the auxiliary coefficient, called the non-output fundamental (NOF), with the smallest values. This is motivated by the fact that an additional small NOF will generate more additional coefficients than a larger NOF. For instance, let us assume that the coefficient 45 needs to be added and we must decide which NOF value has to be used. The NOF LUTs lists possible NOFs as 3, 5, 9, or 15. It can now be argued that, if 3 is selected, more coefficients are generated than if any other NOF is used, since $3, 6, 12, 24, 48, \ldots$ can be generated without additional effort from NOF 3. If 15 is used, for instance, as the NOF the coefficients $15, 30, 45, \ldots$, are generated, which produces significantly fewer coefficients than NOF 3.

To illustrate the RAG algorithm, consider coding the coefficients defining the F6 half-band FIR filter of Goodman and Carey [80].

Example 3.5: Reduced Adder Graph for an F6 Half-band Filter

The half-band filter F6 has four nonzero coefficients, namely $f[0], f[1], f[3]$, and $f[5]$, which are $346, 208, -44$, and 9. For a first cost estimation we convert the decimal values (index 10) into binary representations (index 2) and look up the cost for the coefficients in Table 2.3 (p. 64):

$$
\begin{array}{rcccl}
 & & f[k] & & \text{Cost} \\
f[0] = 346_{10} &=& 2 \times 173 &=& 101011010_2 \quad\quad 4 \\
f[1] = 208_{10} &=& 2^4 \times 13 &=& 11010000_2 \quad\quad 2 \\
f[3] = -44_{10} &=& -2^2 \times 11 &=& -101100_2 \quad\quad 2 \\
f[5] = 9_{10} &=& 3^2 &=& 1001_2 \quad\quad 1 \\
 & & \text{Total} & & \quad\quad 9
\end{array}
$$

For the direct CSD code realization, nine adders are required. The RAG algorithms proceeds as follows:

Step	To be realized	Already realized	Action
0)	$\{346, 208, -44, 9\}$	$\{ - \}$	Initialization
1a)	$\{346, 208, 44, 9\}$	$\{ - \}$	No negative coefficients
1b)	$\{173, 13, 11, 9\}$	$\{ - \}$	Remove 2^k factors
2)	$\{173, 13, 11, 9\}$	$\{ - \}$	Look-up coefficients costs: $\{3,2,2,1\}$
3)	$\{173, 13, 11, 9\}$	$\{ - \}$	Remove cost-0 coefficients from set
4)	$\{173, 13, 11\}$	$\{ 9 \}$	Realize cost-1 coefficients: $9 = 8 + 1$
5)	$\{173, 13, 11\}$	$\{9, 11, 13\}$	Build $11 = 9 + 2$ and $13 = 9 + 4$

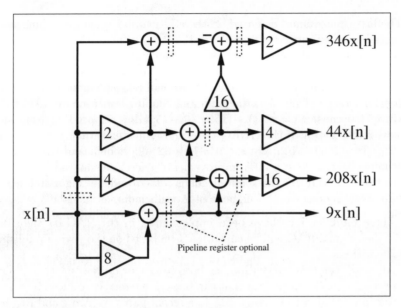

Fig. 3.11. Realization of F6 using RAG algorithm.

Apply the heuristic to the remaining coefficients, starting with the coefficient with the lowest cost and smallest value. It follows that:

Step	Realize	Already realized	Action Find representation
7)	$\{-\}$	$\{9,11,13,173\}$	Add NOF 3: $173 = 11 \times 16 - 3$

Figure 3.11 shows the resulting reduced adder graph. The number of adders is reduced from 9 to 5. The adder path delay is also reduced from 4 to 3. $\boxed{3.5}$

A program `ragopt.exe` that implements the optimal part of the algorithms can be found in the book CD under `book3e/util`. Compared with the original algorithm only some minor improvements have been reported over the years [81].

- The MAG LUT table used has been extended to 14 bits (Gustafsson et al. [82] have actually extended the cost table to 19 bits but do not keep the fundamental table) and all 32 MAG adder cost-4 graph are now considered when computing the minimum NOF sum. Within 14 bits only two coefficients (i.e., 14 709, 15 573) are of cost 5 and, as long as these coefficients are not used, the computed minimum NOF sum list will be optimal in the RAG-95 sense.
- In step 7 all adder cost-2 graph are now considered. There are three such graphs, i.e., a single fundamental followed by an adder cost-2 factor, a sum of two fundamentals, and an adder cost-1 factor or a sum of three fundamentals.

- The last improvement is based on the adder cost-2 selection, which sometimes produced suboptimal results in the RAG-95 algorithm when multiple adder cost-2 coefficients have to be implemented. This can be explained as follows. While the selection of the smallest NOF is motivated by the statistical observation this may lead to suboptional results. For instance, for the coefficient set $\{13, 59, 479\}$ the minimum NOFs values used by RAG-95 are $\{3, 5, 7\}$ because $13 = 4 \times 3 + 1; 59 = 64 - 5; 479 = 59 \times 8 + 7$, resulting in a six-adder graph. If the NOF $\{15\}$ is chosen instead, then all coefficients $(13 = 15 - 2; 59 = 15 \times 4 - 1; 479 = 15 \times 32 - 1)$ benefit and RAG-05 only requires four adders, a 30% improvement. Therefore, instead of selecting the smallest NOF for the smallest adder cost-2 coefficient, a search for the best NOF is done over all adder cost-2 coefficients.

These modifications have been implemented in the RAG-05 algorithm, while the original RAG will be called RAG-95 based on the year of publishing the algorithms.

Although the RAG algorithm has been in use for quite some time, a large set of reliable benchmark data that can be verified and reproduced by anyone was not produced until recently [81]. In a recent paper by Wang and Roy [83], for instance, 60% of the comparison RAG data were declared "unknown." A benchmark should cover filters used in practical applications that are widely published or can easily be computed – a generation of random number filter coefficients that: (a) cannot be verified by a third party, and (b) are of no practical relevance (although used in many publications) are less useful. The problem with a RAG benchmark is that the heuristic part may give different results depending on the exact software implementation or the NOF table used. In addition, since some filters are rather long, a benchmark that lists the whole RAG is not practical in most cases. It is therefore suggested to use a benchmark based on the following equivalence transformation (remembering that the number of output fundamentals is equivalent to the number of adders required):

Theorem 3.6:   RAG Equivalent Transformation

Let $\mathbb{S}_1$ be a coefficient set that can be synthesized by the RAG algorithm with a set of $\mathbb{F}_1$ output fundamentals and $\mathbb{N}_1$ non-output fundamentals, (i.e., internal auxiliary coefficients). A congruent RAG is synthesized if a coefficient set $\mathbb{S}_2$ is used that contains as fundamentals both output and non-output fundamentals from the first set $\mathbb{S}_2 = \mathbb{F}_1 \cup \mathbb{N}_1$.

Proof: Assume that $\mathbb{S}_2$ is synthesized via the RAG algorithm. Now all fundamentals can be synthesized with exactly one adder, since all fundamentals are synthesized in the optimal part of the algorithm. As a result a minimum number $C_2 = \#\mathbb{F}_1 + \#\mathbb{N}_1$ of adders for this fundamental set is used. If now set $\mathbb{S}_1$ is synthesized and generates the same fundamentals (output and non-output) as set $\mathbb{S}_2$, the resulting RAG also uses the minimum number of adders. Since both use the minimum number of adders they must be congruent. q.e.d.

A corollary of Theorem 3.6 is that graphs can now be classified as (guaranteed) optimal and heuristic graphs. An optimal graph has no more than one NOF, while a heuristic graph has more than one NOF. It is only required to provide a list of the NOFs to describe a unique OF graph. If this NOF is added to the coefficient set, all OFs are synthesized via the optimal part of the algorithm, which can easily be programmed. The program `ragopt.exe` that implements the optimal part of the algorithms is in fact available on the book CD under `book3e/util`. Some example benchmarks are given in Table 3.4. The first column shows the filter name, followed by the filter length L, and the bitwidth of the largest coefficient B. Then the reference adder data for CSD coding and CSE coding follows. The idea of the CSE coding is studied in Exercises 3.4 and 3.5 (p. 209) Note that the number of CSD adders given already takes advantage of coefficient symmetry, i.e., $f(k) = f(L - k)$. Common subexpression (CSE) required adder data are used from [83]. For the RAG algorithm the output fundamental (OF) and non-output fundamental (NOF) for RAG-2005 are listed. Note that the number of OFs is already much smaller than the filter length L. We then list in column 8 the adders required in the improved RAG-2005 version. Finally in the last column we list the NOF values that are required to synthesize the RAG filter via the optimal part of the RAG algorithms that is the basis for the program `ragopt.exe`[5] on the book CD under `book3e/util`. `ragopt.exe` uses a MAG LUT `mag14.dat` to determine the MAG costs, and produces two output files: `firXX.dat` that contains the filter data, and a file `ragopt.pro` that has the RAG-n coefficient equations. A `grep` command for lines that start with `Build` yields the equations necessary to construct the RAG-n graph.

It can be seen that the examples from Samueli [84] and Lim and Parker [85] all produce optimal RAG results, i.e., have a maximum of one NOF. Notice particularly for long filters the improvement of RAG compared to CSD and CSE adders. Filters F5-F9 are from the Goodman and Carey set of half-band filters (see Table 5.3, p. 274) and give better results using RAG-05 than RAG-95. The benchmark data from Samueli, and Lim andParker work very well for RAG since the filters are lowpass and therefore taper smoothly to zero at both sides, improving the likelihood of cost-1 output fundamentals. A more-challenging RAG design for DFT coefficients will be discussed in Chap. 6.

Pipelined RAG FIR Filter

Due to the logic delay in the RAG running through several adders, the resulting register performance of the design is not very high even for a small graph. To improve the register performance one can take advantage of the register embedded in each LE that would not otherwise be used. A single register

---

[5] You need to copy the program to your hard drive first; you can not start it from the CD directly.

Table 3.4. Required number of adders for the CSD, CSE, and RAG algorithms for lowpass filters. Prototype filters are from Goodman and Carey [80], Samueli [84], and Lim and Parker [85].

Filter name	L	B	CSD adder	CSE adder	#OF	#NOF	RAG-05 adder	NOF values
F5	11	8	6	-	3	0	3	-
F6	11	9	9	-	4	1	5	3
F7	11	9	7	-	3	1	4	23
F8	15	10	10	-	5	2	7	11, 17
F9	19	13	14	-	5	2	7	13, 1261
S1	25	9	11	6	6	0	6	-
S2	60	14	57	29	26	0	26	-
L1	121	17	145	57	51	1	52	49
L2	63	13	49	23	22	0	22	-
L3	36	11	16	5	5	0	5	-

placed at the output of an adder does therefore not require any additional logic resource. However, power-of-two coefficients that are implemented by shifting the register input word require an additional register not included in the zero-pipeline design. This design with one pipeline stage already enjoys a speed improvement of 50% compared with the non-pipelined design, see Table 3.5(Pipeline stages=1). For the fully pipelined design we need to have the same delay for each incoming path of the adders. For the F6 design one needs to build:

$$x9 <= 8 \times x + x; \quad \text{has delay 1}$$
$$x11 <= x9 + 2 \times x \times z^{-1}; \quad \text{has delay 2}$$
$$x13 <= x9 + 4 \times x \times z^{-1}; \quad \text{has delay 2}$$
$$x3 <= xz^{-1} + 2 \times x \times z^{-1}; \quad \text{has delay 2}$$
$$x173 <= 16 \times x11 - x3; \quad \text{has delay 3}$$

i.e., one extra pipeline register is used for input x, and a maximum delay of three pipeline stage is needed. The pipelined graph is shown in Fig. 3.11 with the dashed register active. Now the coefficients in the RAG are all fully pipelined. Now we need to take care of the different delays of the coefficients. We basically have two options: we can add to the output of *all* coefficients an additional delay, that we achieve the same delay for all coefficients (three in the case of the F6 filter) and then do not need to change the output tap delay line structure; alternative we can use pipeline retiming, i.e., the multiplier outputs need to be aligned in the tap delay line according to their pipeline stages. This is a similar approach to that used in the direct FIR (see Fig. 3.10, p. 182) by aligning the coefficient adder location according to the

Table 3.5. F6 pipeline options for the RAG algorithm.

Pipeline stages	LEs	Fmax (MHz)	Cost LEs/Fmax
0	225	165.95	1.36
1	234	223.61	1.05
max	252	353.86	0.71
Gain% 0/max	-11	114	92

delay, and is shown in Fig. 3.12. Note in order to build only two input adder, we had to use an additional register to delay the x13 coefficient.

For this half-band filter design the pipeline retiming synthesis results shown in Table 3.5 reveal that the design now runs about twice as fast with a moderate (11%) increase in LEs when compared with the unpipelined design. Since the overall cost measured by LEs/Fmax is improved, fully pipelined designs should be preferred.

3.4.3 FIR Filters Using Distributed Arithmetic

A completely different FIR architecture is based on the distributed arithmetic (DA) concept introduced in Sect. 2.7.1 (p. 115). In contrast to a conventional sum-of-products architecture, in distributed arithmetic we always compute the sum of products of a specific bit b over *all* coefficients in one step. This is computed using a small table and an accumulator with a shifter. To illustrate, consider the three-coefficient FIR with coefficients $\{2, 3, 1\}$ found in Example 2.24 (p. 117).

Example 3.7: Distributed Arithmetic Filter as State Machine

A distributed arithmetic filter can be built in VHDL code[6] using the following state machine description:

---

[6] The equivalent Verilog code dafsm.v for this example can be found in Appendix A on page 683. Synthesis results are shown in Appendix B on page 731.

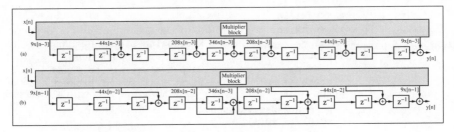

Fig. 3.12. F6 RAG filter with pipeline retiming.

```
LIBRARY ieee;                    -- Using predefined packages
USE ieee.std_logic_1164.ALL;
USE ieee.std_logic_arith.ALL;

ENTITY dafsm IS                          ------> Interface
      PORT (clk, reset : IN STD_LOGIC;
             x0_in, x1_in, x2_in :
                          IN  STD_LOGIC_VECTOR(2 DOWNTO 0);
             lut  : OUT INTEGER RANGE 0 TO 7;
             y    : OUT INTEGER RANGE 0 TO 63);
END dafsm;

ARCHITECTURE fpga OF dafsm IS

  COMPONENT case3  -- User-defined component
    PORT ( table_in   : IN   STD_LOGIC_VECTOR(2 DOWNTO 0);
           table_out  : OUT  INTEGER RANGE 0 TO 6);
  END COMPONENT;

  TYPE STATE_TYPE IS (s0, s1);
  SIGNAL state     : STATE_TYPE;
  SIGNAL x0, x1, x2, table_in
                          : STD_LOGIC_VECTOR(2 DOWNTO 0);
  SIGNAL table_out : INTEGER RANGE 0 TO 7;
BEGIN

  table_in(0) <= x0(0);
  table_in(1) <= x1(0);
  table_in(2) <= x2(0);

  PROCESS (reset, clk)        ------> DA in behavioral style
    VARIABLE p    : INTEGER RANGE 0 TO 63;-- temp. register
    VARIABLE count : INTEGER RANGE 0 TO 3; -- counts shifts
  BEGIN
    IF reset = '1' THEN                -- asynchronous reset
      state <= s0;
    ELSIF rising_edge(clk) THEN
    CASE state IS
      WHEN s0 =>         -- Initialization step
        state <= s1;
        count := 0;
        p := 0;
        x0 <= x0_in;
        x1 <= x1_in;
        x2 <= x2_in;
      WHEN s1 =>                 -- Processing step
        IF count = 3 THEN    -- Is sum of product done ?
          y <= p;            -- Output of result to y and
          state <= s0;       -- start next sum of product
        ELSE
          p := p / 2 + table_out * 4;
          x0(0) <= x0(1);
          x0(1) <= x0(2);
```

```
           x1(0) <= x1(1);
           x1(1) <= x1(2);
           x2(0) <= x2(1);
           x2(1) <= x2(2);
           count := count + 1;
           state <= s1;
        END IF;
    END CASE;
    END IF;
END PROCESS;

LC_Table0: case3
   PORT MAP(table_in => table_in, table_out => table_out);
lut <= table_out; -- Extra test signal

END fpga;
```

The LE table[7] defined as CASE components was generated with the utility program dagen3e.exe. The output is show below.

```
LIBRARY ieee;
USE ieee.std_logic_1164.ALL;
USE ieee.std_logic_arith.ALL;

ENTITY case3 IS
       PORT ( table_in  : IN  STD_LOGIC_VECTOR(2 DOWNTO 0);
              table_out : OUT INTEGER RANGE 0 TO 6);
END case3;

ARCHITECTURE LEs OF case3 IS
BEGIN

-- This is the DA CASE table for
-- the 3 coefficients: 2, 3, 1
-- automatically generated with dagen.exe -- DO NOT EDIT!

  PROCESS (table_in)
  BEGIN
    CASE table_in IS
        WHEN  "000" =>     table_out <=  0;
        WHEN  "001" =>     table_out <=  2;
        WHEN  "010" =>     table_out <=  3;
        WHEN  "011" =>     table_out <=  5;
        WHEN  "100" =>     table_out <=  1;
        WHEN  "101" =>     table_out <=  3;
        WHEN  "110" =>     table_out <=  4;
        WHEN  "111" =>     table_out <=  6;
        WHEN  OTHERS  =>    table_out <=  0;
    END CASE;
  END PROCESS;
END LEs;
```

---

[7] The equivalent Verilog code case3.v for this example can be found in Appendix A on page 684. Synthesis results are shown in Appendix B on page 731.

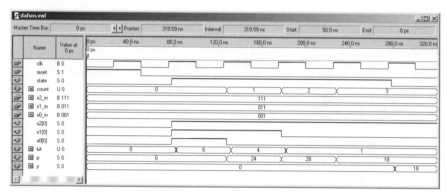

Fig. 3.13. Simulation of the 3-tap FIR filter with input $\{1, 3, 7\}$.

As suggested in Chap. 2, a shift/accumulator is used, which shifts only one position to the right for each step, instead of shifting k positions to the left. The simulation results, shown in Fig. 3.13, report the correct result ($y = 18$) for an input sequence $\{1, 3, 7\}$. The simulation shows the clk, reset, state, and count signals followed by the three input signals. Next the three bits selected from the input word to address the prestored DA LUT are shown. The LUT output values $\{6, 4, 1\}$ are then weighted and accumulated to generate the final output value $y = 18 = 6 + 4 \times 2 + 1 \times 4$. The design uses 32 LEs, no embedded multiplier, no M4K block, and has a 420.17 MHz Registered Performance. | 3.7 |

By defining the distributed arithmetic table with a CASE statement, the synthesizer will use logic cells to implement the LUT. This will result in a fast and efficient design only if the tables are small. For large tables, alternative means must be found. In this case, we may use the 4-kbit embedded memory blocks (M4Ks), which (as discussed in Chap. 1) can be configured as $2^9 \times 9, 2^{10} \times 4, 2^{11} \times 2$ or $2^{12} \times 1$ tables. These design paths are discussed in more detail in the following.

Distributed Arithmetic Using Logic Cells

The DA implementation of an FIR filter is particularly attractive for low-order cases due to LUT address space limitations (e.g., $L \leq 4$). It should be remembered, however, that FIR filters are linear filters. This implies that the outputs of a collection of low-order filters can be added together to define the output of a high-order FIR, as shown in Fig. 2.37 (p. 122). Based on the LEs found in a Cyclone II device, namely $2^4 \times 1$-bit tables, a DA table for four coefficients can be implemented. The number of necessary LEs increases exponentially with order. Typically, the number of LEs is much higher than the number of M4Ks. For example, an EP2C35 contains 35K LEs but only 105 M4Ks. Also, M4Ks can be used to efficiently implement RAMs and FIFOs

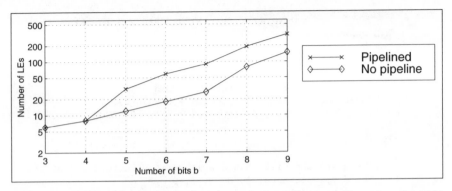

Fig. 3.14. Size comparison of synthesis results for different coding using the CASE statement with b input and outputs.

and other high-valued functions. It is therefore sometimes desirable to use M4Ks economically. On the other side if the design is implemented using larger tables with a $2^b \times b$ CASE statement, inefficient designs can result. The pipelined $2^9 \times 9$ table implemented with one VHDL CASE statement only, for example, required over 100 LEs. Figure 3.14 shows the number of LEs necessary for tables having three to nine bits inputs and outputs using the CASE statement generated with utility program dagen3e.exe.

Another alternative is the design using 4-input LUT only via a CASE statements, and implementing table with more than 4 inputs with an additional (binary tree) multiplexer using $2 \rightarrow 1$ multiplexer only. In this model it is straightforward to add additional pipeline registers to the modular design. For maximum speed, a register must be introduced behind each LUT and $2 \rightarrow 1$ multiplexer. This will, most likely, yield a higher LE count[8] compared to the minimization of the one large LUT. The following example illustrates the structure of a 5-input table.

Example 3.8: Five-input DA Table

The utility program dagen3e.exe accepts filter length and coefficients, and returns the necessary PROCESS statements for the 4-input CASE table followed by a multiplexer. The VHDL output for an arbitrary set of coefficients, namely $\{1, 3, 5, 7, 9\}$, is given[9] in the following listing:

```
LIBRARY ieee;
USE ieee.std_logic_1164.ALL;
USE ieee.std_logic_arith.ALL;

ENTITY case5p IS
        PORT ( clk        : IN   STD_LOGIC;
                table_in  : IN   STD_LOGIC_VECTOR(4 DOWNTO 0);
```

---

[8] A 16:1 multiplexer and is reported with 11 LEs while we need 15 LEs or 2:1 MUX in a tree structure, see Cyclone II Device Handbook p. 5-15 [21].

[9] The equivalent Verilog code case5p.v for this example can be found in Appendix A on page 685. Synthesis results are shown in Appendix B on page 731.

```
                        table_out : OUT INTEGER RANGE 0 TO 25);
      END case5p;

      ARCHITECTURE LEs OF case5p IS

        SIGNAL lsbs : STD_LOGIC_VECTOR(3 DOWNTO 0);
        SIGNAL msbs0 : STD_LOGIC_VECTOR(1 DOWNTO 0);
        SIGNAL table0out00, table0out01 : INTEGER RANGE 0 TO 25;

      BEGIN

      -- These are the distributed arithmetic CASE tables for
      -- the 5 coefficients: 1, 3, 5, 7, 9
      -- automatically generated with dagen.exe -- DO NOT EDIT!

        PROCESS
        BEGIN
          WAIT UNTIL clk = '1';
          lsbs(0) <= table_in(0);
          lsbs(1) <= table_in(1);
          lsbs(2) <= table_in(2);
          lsbs(3) <= table_in(3);
          msbs0(0) <= table_in(4);
          msbs0(1) <= msbs0(0);
        END PROCESS;

        PROCESS        -- This is the final DA MPX stage.
        BEGIN          -- Automatically generated with dagen.exe
          WAIT UNTIL clk = '1';
          CASE msbs0(1) IS
            WHEN  '0' =>    table_out <= table0out00;
            WHEN  '1' =>    table_out <= table0out01;
            WHEN  OTHERS  =>    table_out <= 0;
          END CASE;
        END PROCESS;

        PROCESS        -- This is the DA CASE table 00 out of 1.
        BEGIN          -- Automatically generated with dagen.exe
          WAIT UNTIL clk = '1';
          CASE lsbs IS
            WHEN "0000" =>    table0out00 <= 0;
            WHEN "0001" =>    table0out00 <= 1;
            WHEN "0010" =>    table0out00 <= 3;
            WHEN "0011" =>    table0out00 <= 4;
            WHEN "0100" =>    table0out00 <= 5;
            WHEN "0101" =>    table0out00 <= 6;
            WHEN "0110" =>    table0out00 <= 8;
            WHEN "0111" =>    table0out00 <= 9;
            WHEN "1000" =>    table0out00 <= 7;
            WHEN "1001" =>    table0out00 <= 8;
            WHEN "1010" =>    table0out00 <= 10;
            WHEN "1011" =>    table0out00 <= 11;
            WHEN "1100" =>    table0out00 <= 12;
```

```
        WHEN  "1101" =>   table0out00 <=  13;
        WHEN  "1110" =>   table0out00 <=  15;
        WHEN  "1111" =>   table0out00 <=  16;
        WHEN   OTHERS  =>    table0out00 <=  0;
      END CASE;
    END PROCESS;

    PROCESS      -- This is the DA CASE table 01 out of 1.
    BEGIN        -- Automatically generated with dagen.exe
      WAIT UNTIL clk = '1';
      CASE lsbs IS
        WHEN  "0000" =>   table0out01 <=  9;
        WHEN  "0001" =>   table0out01 <=  10;
        WHEN  "0010" =>   table0out01 <=  12;
        WHEN  "0011" =>   table0out01 <=  13;
        WHEN  "0100" =>   table0out01 <=  14;
        WHEN  "0101" =>   table0out01 <=  15;
        WHEN  "0110" =>   table0out01 <=  17;
        WHEN  "0111" =>   table0out01 <=  18;
        WHEN  "1000" =>   table0out01 <=  16;
        WHEN  "1001" =>   table0out01 <=  17;
        WHEN  "1010" =>   table0out01 <=  19;
        WHEN  "1011" =>   table0out01 <=  20;
        WHEN  "1100" =>   table0out01 <=  21;
        WHEN  "1101" =>   table0out01 <=  22;
        WHEN  "1110" =>   table0out01 <=  24;
        WHEN  "1111" =>   table0out01 <=  25;
        WHEN   OTHERS  =>    table0out01 <=  0;
      END CASE;
    END PROCESS;
  END LEs;
```

The five inputs produce two CASE tables and a $2 \rightarrow 1$ bus multiplexer. The multiplexer may also be realized with a component instantiation using the LPM function busmux. The program dagen3e.exe writes a VHDL file with the name caseX.vhd, where X is the filter length that is also the input bit width. The file caseXp.vhd is the same table, except with additional pipeline registers. The component can be used directly in a state machine design or in an unrolled filter structure.

☐ 3.8

Referring to Fig. 3.14, it can be seen that the structured VHDL code improves on the number of required LEs. Figure 3.15 compares the different design methods in terms of speed. We notice that the busmux generated VHDL code allows to run all pipelined designs with the maximum speed of 464 MHz outperforming the M4Ks by nearly a factor two. Without pipeline stages the synthesis tools is capable to reduce the LE count essentially, but Registered Performance is also reduced. Note that still a busmux design is used. The synthesis tool is not able to optimize one (large) case statement in the same way. Although we get a high Registered Performance using eight pipeline stages for a $2^9 \times 9$ table with 464 MHz the design may now be too large for some applications. We may also consider the partitioning technique

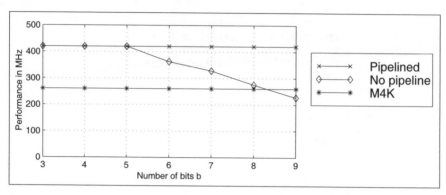

Fig. 3.15. Speed comparison for different coding styles using the CASE statement.

(Exercise 3.6, p. 210), shown in Fig. 2.36 (p. 121), or implementation with an M4K, discussed next.

DA Using Embedded Array Blocks

As mentioned in the last section, it is not economical to use the 4-kbit M4Ks for a short FIR filter, mainly because the number of available M4Ks is limited. Also, the maximum registered speed of an M4K is 260 MHz, and an LE table implementation may be faster. The following example shows the DA implementation using a component instantiation of the M4K.

Example 3.9: Distributed Arithmetic Filter using M4Ks

The CASE table from the last example can be replaced by a M4K ROM. The ROM table is defined by file darom3.mif. The default input and output configuration of the M4K is given by "REGISTERED." If it is not desirable to have a registered configuration, set LPM_ADDRESS_CONTROL => "UNREGISTERED" or LPM_OUTDATA => "UNREGISTERED." Note that in Cyclone II at least one input must be registered. With Flex devices we can also build asynchronous, i.e., non registered M2K ROMs. The VHDL code[10] for the DA state machine design is shown below:

```
LIBRARY lpm;
USE lpm.lpm_components.ALL;

LIBRARY ieee;                 -- Using predefined packages
USE ieee.std_logic_1164.ALL;
USE ieee.std_logic_arith.ALL;
USE ieee.std_logic_unsigned.ALL;  -- Contains conversion
                                  -- VECTOR -> INTEGER
ENTITY darom IS                   ------> Interface
  PORT (clk, reset  : IN STD_LOGIC;
        x_in0, x_in1, x_in2
```

---

[10] The equivalent Verilog code darom.v for this example can be found in Appendix A on page 687. Synthesis results are shown in Appendix B on page 731.

```
                              : IN STD_LOGIC_VECTOR(2 DOWNTO 0);
      lut  : OUT INTEGER RANGE 0 TO 7;
      y    : OUT INTEGER RANGE 0 TO 63);
END darom;

ARCHITECTURE fpga OF darom IS
  TYPE STATE_TYPE IS (s0, s1);
  SIGNAL state                      : STATE_TYPE;
  SIGNAL x0, x1, x2, table_in, mem
                          : STD_LOGIC_VECTOR(2 DOWNTO 0);
  SIGNAL table_out                  : INTEGER RANGE 0 TO 7;
BEGIN

  table_in(0) <= x0(0);
  table_in(1) <= x1(0);
  table_in(2) <= x2(0);

  PROCESS (reset, clk)        ------> DA in behavioral style
    VARIABLE  p   : INTEGER RANGE 0 TO 63; --Temp. register
    VARIABLE count : INTEGER RANGE 0 TO 3;
  BEGIN                                 -- Counts the shifts
    IF reset = '1' THEN             -- Asynchronous reset
      state <= s0;
    ELSIF rising_edge(clk) THEN
    CASE state IS
      WHEN s0 =>              -- Initialization step
        state <= s1;
        count := 0;
        p := 0;
        x0 <= x_in0;
        x1 <= x_in1;
        x2 <= x_in2;
      WHEN s1 =>             -- Processing step
        IF count = 3 THEN    -- Is sum of product done ?
          y <= p / 2 + table_out * 4; -- Output of result
          state <= s0;               -- to y andstart next
        ELSE                         -- sum of product
          p := p / 2 + table_out * 4;
          x0(0) <= x0(1);
          x0(1) <= x0(2);
          x1(0) <= x1(1);
          x1(1) <= x1(2);
          x2(0) <= x2(1);
          x2(1) <= x2(2);
          count := count + 1;
          state <= s1;
        END IF;
    END CASE;
    END IF;
  END PROCESS;

  rom_1: lpm_rom
    GENERIC MAP ( LPM_WIDTH => 3,
```

```
                    LPM_WIDTHAD => 3,
                    LPM_OUTDATA => "REGISTERED",
                    LPM_ADDRESS_CONTROL => "UNREGISTERED",
                    LPM_FILE => "darom3.mif")
        PORT MAP(outclock => clk,address => table_in,q => mem);

    table_out <= CONV_INTEGER(mem);
    lut <= table_out;

  END fpga;
```

Compared with Example 3.7 (p. 189), we now have a component instantiation of the LPM_ROM. Because there is a need to convert between the STD_LOGIC_VECTOR output of the ROM and the integer, we have used the package std_logic_unsigned from the library ieee. The latter contains the CONV_INTEGER function for unsigned STD_LOGIC_VECTOR.

The include file darom3.mif was generated with the program dagen3e.exe. The file has the following contents:

```
--  This is the DA MIF table for the 3 coefficients: 2, 3, 1
--  automatically generated with dagen3e.exe
-- DO NOT EDIT!
WIDTH = 3; DEPTH = 8; ADDRESS_RADIX = uns; DATA_RADIX = uns;
CONTENT BEGIN
        0    :    0;
        1    :    2;
        2    :    3;
        3    :    5;
        4    :    1;
        5    :    3;
        6    :    4;
        7    :    6;
END;
```

The design runs at 218.29 MHz and uses 27 LEs, and one M4K memory block (more precisely, 24 bits of an M4K).

The simulation results, shown in Fig. 3.16, are very similar to the dafsm simulation shown in Fig. 3.13 (p, 3.13). Due to the mandatory 1 clock cycle delay of the synchronous M4K memory block we notice a delay by one clock cycle in the lut output signal; the result ($y = 18$) for the input sequence $\{1, 3, 7\}$, however, is still correct. The simulation shows the clk, reset, state, and count signals followed by the three input signals. Next the three bits selected from the input word to address the prestored DA LUT are shown. The LUT output values $\{6, 4, 1\}$ are then weighted and accumulated to generate the final output value $y = 18 = 6 + 4 \times 2 + 1 \times 4$.

3.9

But M4Ks have only a single address decoder and if we implement a $2^3 \times 3$ table, a complete M4K would be consumed unnecessarily, and it can not be used elsewhere. For longer filters, however, the use of M4Ks is attractive because:

- M4Ks have registered throughput at a constant 260 MHz, and
- Routing effort is reduced

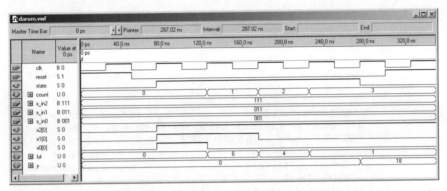

Fig. 3.16. Simulation of the 3-tap FIR M4K-based DA filter with input $\{1, 3, 7\}$.

Signed DA FIR Filter

A signed DA filter will require a signed accumulator. The following example shows the VHDL code for the previously studied three-coefficient example, 2.25 from Chap. 2 (p. 119).

Example 3.10: Signed DA FIR Filter

For the signed DA filter, an additional state is required. See the variable count[11] to process the sign bit.

```
LIBRARY ieee;                    -- Using predefined packages
USE ieee.std_logic_1164.ALL;
USE ieee.std_logic_arith.ALL;

ENTITY dasign IS                        ------> Interface
       PORT (clk, reset : IN STD_LOGIC;
               x_in0, x_in1, x_in2
                         : IN  STD_LOGIC_VECTOR(3 DOWNTO 0);
               lut  : out INTEGER RANGE -2 TO 4;
               y    : OUT INTEGER RANGE -64 TO 63);
END dasign;

ARCHITECTURE fpga OF dasign IS

  COMPONENT case3s     -- User-defined components
    PORT ( table_in : IN  STD_LOGIC_VECTOR(2 DOWNTO 0);
           table_out : OUT INTEGER RANGE -2 TO 4);
  END COMPONENT;

  TYPE STATE_TYPE IS (s0, s1);
  SIGNAL state       : STATE_TYPE;
  SIGNAL table_in    : STD_LOGIC_VECTOR(2 DOWNTO 0);
  SIGNAL x0, x1, x2  : STD_LOGIC_VECTOR(3 DOWNTO 0);
  SIGNAL table_out   : INTEGER RANGE -2 TO 4;
```

---

[11] The equivalent Verilog code case3s.v for this example can be found in Appendix A on page 688. Synthesis results are shown in Appendix B on page 731.

```
BEGIN

  table_in(0) <= x0(0);
  table_in(1) <= x1(0);
  table_in(2) <= x2(0);

  PROCESS (reset, clk)          ------> DA in behavioral style
    VARIABLE  p : INTEGER RANGE -64 TO 63:= 0; -- Temp. reg.
    VARIABLE count : INTEGER RANGE 0 TO 4; -- Counts the
  BEGIN                                        -- shifts
    IF reset = '1' THEN                 -- asynchronous reset
      state <= s0;
    ELSIF rising_edge(clk) THEN
    CASE state IS
      WHEN s0 =>          -- Initialization step
        state <= s1;
        count := 0;
        p := 0;
        x0 <= x_in0;
        x1 <= x_in1;
        x2 <= x_in2;
      WHEN s1 =>             -- Processing step
        IF count = 4 THEN -- Is sum of product done?
          y <= p;         -- Output of result to y and
          state <= s0; -- start next sum of product
        ELSE
          IF count = 3 THEN            -- Subtract for last
          p := p / 2 - table_out * 8; -- accumulator step
          ELSE
          p := p / 2 + table_out * 8;  -- Accumulation for
          END IF;                      -- all other steps
            FOR k IN 0 TO 2 LOOP   -- Shift bits
              x0(k) <= x0(k+1);
              x1(k) <= x1(k+1);
              x2(k) <= x2(k+1);
            END LOOP;
          count := count + 1;
          state <= s1;
        END IF;
    END CASE;
    END IF;
  END PROCESS;

  LC_Table0: case3s
    PORT MAP(table_in => table_in, table_out => table_out);
  lut <= table_out; -- Extra test signal

END fpga;
```
The LE table (component case3s.vhd) was generated using the program
dagen3e.exe. The VHDL code[12] is shown below:

---

[12] The equivalent Verilog code case3s.v for this example can be found in Appendix A on page 690. Synthesis results are shown in Appendix B on page 731.

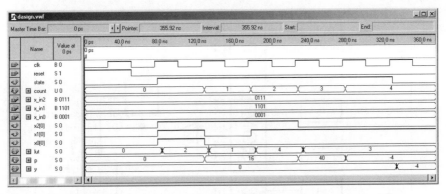

Fig. 3.17. Simulation of the 3-tap signed FIR filter with input $\{1, -3, 7\}$.

```
LIBRARY ieee;
USE ieee.std_logic_1164.ALL;
USE ieee.std_logic_arith.ALL;

ENTITY case3s IS
  PORT ( table_in  : IN   STD_LOGIC_VECTOR(2 DOWNTO 0);
         table_out : OUT  INTEGER RANGE -2 TO 4);
END case3s;

ARCHITECTURE LEs OF case3s IS
BEGIN

-- This is the DA CASE table for
-- the 3 coefficients: -2, 3, 1
-- automatically generated with dagen.exe -- DO NOT EDIT!

  PROCESS (table_in)
  BEGIN
    CASE table_in IS
      WHEN  "000" =>    table_out <=  0;
      WHEN  "001" =>    table_out <=  -2;
      WHEN  "010" =>    table_out <=  3;
      WHEN  "011" =>    table_out <=  1;
      WHEN  "100" =>    table_out <=  1;
      WHEN  "101" =>    table_out <=  -1;
      WHEN  "110" =>    table_out <=  4;
      WHEN  "111" =>    table_out <=  2;
      WHEN  OTHERS  =>    table_out <=  0;
    END CASE;
  END PROCESS;
END LEs;
```

Figure 3.17 shows the simulation for the input sequence $\{1, -3, 7\}$. The simulation shows the clk, reset, state, and count signals followed by the four input signals. Next the three bits selected from the input word to address the prestored DA LUT are shown. The LUT output values $\{2, 1, 4, 3\}$ are then weighted and accumulated to generate the final output value $y =$

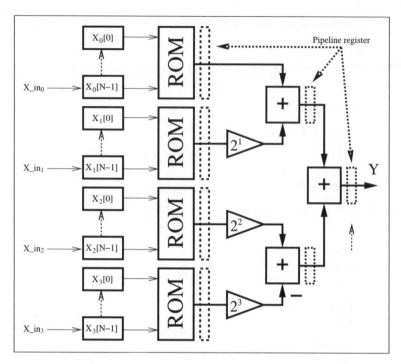

Fig. 3.18. Parallel implementation of a distributed arithmetic FIR filter.

$2+1\times2+4\times4-3\times8 = -4$. The design uses 56 LEs, no embedded multiplier, and has a 236.91 MHz `Registered Performance`. $\boxed{3.10}$

To accelerate a DA filter, unrolled loops can be used. The input is applied sample by sample (one word at a time), in a bit-parallel form. In this case, for each bit of input a separate table is required. While the table size varies (input bit width equals number of filter taps), the contents of the tables are the same. The obvious advantage is a reduction of VHDL code size, if we use a component definition for the LE tables, as previously presented. To demonstrate, the unrolling of the 3-coefficients, 4-bit input example, previously considered, is developed below.

Example 3.11: Loop Unrolling for DA FIR Filter

In a typical FIR application, the input values are processed in word parallel form (i.e., see Fig. 3.18). The following VHDL code[3] illustrates the unrolled DA code, according to Fig. 3.18.

```
LIBRARY ieee;                    -- Using predefined packages
USE ieee.std_logic_1164.ALL;
```

---

[3] The equivalent Verilog code `dapara.v` for this example can be found in Appendix A on page 691. Synthesis results are shown in Appendix B on page 731.

```
USE ieee.std_logic_arith.ALL;

ENTITY dapara IS                        ------> Interface
      PORT (clk  : IN STD_LOGIC;
            x_in : IN  STD_LOGIC_VECTOR(3 DOWNTO 0);
            y    : OUT INTEGER RANGE -46 TO 44);
END dapara;

ARCHITECTURE fpga OF dapara IS
  TYPE ARRAY4x3 IS ARRAY (0 TO 3)
                        OF STD_LOGIC_VECTOR(2 DOWNTO 0);
  SIGNAL x : ARRAY4x3;
  TYPE IARRAY IS ARRAY (0 TO 3) OF INTEGER RANGE -2 TO 4;
  SIGNAL h : IARRAY;
  SIGNAL s0 : INTEGER RANGE -6 TO 12;
  SIGNAL s1 : INTEGER RANGE -10 TO 8;
  SIGNAL t0, t1, t2, t3 : INTEGER RANGE -2 TO 4;
  COMPONENT case3s
    PORT ( table_in  : IN  STD_LOGIC_VECTOR(2 DOWNTO 0);
           table_out : OUT INTEGER RANGE -2 TO 4);
  END COMPONENT;

BEGIN

  PROCESS                        ------> DA in behavioral style
  BEGIN
    WAIT UNTIL clk = '1';
    FOR l IN 0 TO 3 LOOP  -- For all four vectors
      FOR k IN 0 TO 1 LOOP  -- shift all bits
        x(l)(k) <= x(l)(k+1);
      END LOOP;
    END LOOP;
    FOR k IN 0 TO 3 LOOP  -- Load x_in in the
      x(k)(2) <= x_in(k); -- MSBs of the registers
    END LOOP;
    y <= h(0) + 2 * h(1) + 4 * h(2) - 8 * h(3);
-- Pipeline register and adder tree
--   t0 <= h(0); t1 <= h(1); t2 <= h(2); t3 <= h(3);
--   s0 <= t0 + 2 * t1; s1 <= t2 - 2 * t3;
--   y <= s0 + 4 * s1;
  END PROCESS;

  LC_Tables: FOR k IN 0 TO 3 GENERATE -- One table for each
  LC_Table: case3s                    -- bit in x_in
            PORT MAP(table_in => x(k), table_out => h(k));
  END GENERATE;

END fpga;
```

The design uses four tables of size $2^3 \times 4$ and all tables have the *same* content as the table in Example 3.10 (p. 199). Figure 3.19 shows the simulation for the input sequence $\{1, -3, 7\}$. Because the input is applied serially (and bit-parallel) the expected result $-4_{10} = 11111000_{2C}$ is computed at the 400-ns interval.

3.11

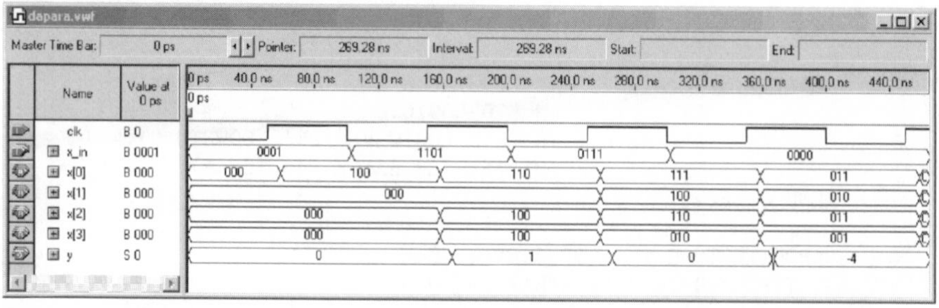

Fig. 3.19. Simulation results for the parallel distributed arithmetic FIR filter.

The previous design requires no embedded multiplier, 33 LEs, no M4K memory block, and runs at 214.96 MHz. An important advantage of the DA concept, compared with the general-purpose MAC design, is that pipelining is easy achieved. We can add additional pipeline registers to the table output and at the adder-tree output with no cost. To compute y, we replace the line

```
y <= h(0) + 2 * h(1) + 4 * h(2) - 8 * h(3);
```

In a first step we only pipeline the adders. We use the signals s0 and s1 for the pipelined adder within the PROCESS statement, i.e.,

```
s0 <= h(0) + 2 * h(1); s1 <= h(2) - 2 * h(3);
y <= s0 + 4 * s1;
```

and the Registered Performance increase to 368.60 MHz, and about the same number of LEs are used. For a fully pipeline version we also need to store the case LUT output in registers; the partial VHDL code then becomes:

```
t0 <= h(0); t1 <= h(1); t2 <= h(2); t3 <= h(3);
s0 <= t0 + 2 * t1; s1 <= t2 - 2 * t3;
y <= s0 + 4 * s1;
```

The size of the design increases to 47 LEs, because the registers of the LE that hold the case tables can no longer be used for the x input shift register. But the Registered Performance increases from 214.96 MHz to 420 MHz.

3.4.4 IP Core FIR Filter Design

Altera and Xilinx usually also offer with the full subscription an FIR filter generator, since this is one of the most often used intellectual property (IP) blocks. For an introduction to IP blocks see Sect. 1.4.4, p. 35.

FPGA vendors in general prefer distributed arithmetic (DA)-based FIR filter generators since these designs are characterized by:

- fully pipelined architecture
- short compile time

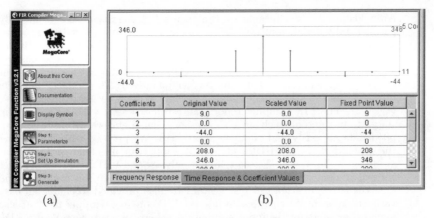

Fig. 3.20. IP design of FIR **(a)** IP toolbench. **(b)** Coefficient specification.

- good resource estimation
- area results independent from the coefficient values, in contrast to the RAG algorithm

DA-based filters do not require any coefficient optimization or the computation of a RAG graph, which may be time consuming when the coefficient set is large. DA-based code generation including all VHDL code and testbenches is done in a few seconds using the vendor's FIR compilers [86].

Let us have a look at the FIR filter generation of an F6 filter from Goodman and Carey [80] that we had discussed before, see Example 3.5, p. 184. But this time we use the Altera FIR compiler [86] to build the filter. The Altera FIR compiler MegaCore function generates FIR filters optimized for Altera devices. Stratix and Cyclone II devices are supported but no mature devices from the APEX or Flex family. You can use the IP toolbench MegaWizard design environment to specify a variety of filter architectures, including fixed-coefficient, multicycle variable, and multirate filters. The FIR compiler includes a coefficient generator, but can also load and use predefined (for instance computed via MATLAB) coefficients from a file.

Example 3.12: F6 Half-band Filter IP Generation

To start the Altera FIR compiler we select the `MegaWizard Plug-In Manager` under the `Tools` menu and the library selection window (see Fig. 1.23, p. 39) will pop up. The FIR compiler can be found under `DSP→Filters`. You need to specify a design name for the core and then proceed to the `ToolBench`. We first parameterize the filter and, since we want to use the F6 coefficients, we select `Edit Coefficient Set` and load the coefficient filter by selecting `Imported Coefficient Set`. The coefficient file is a simple text file with each line listing a single coefficient, starting with the first coefficient in the first line. The coefficients can be integer or floating-point numbers, which will then be quantized by the tool since only integer-coefficient filters can be generated with the FIR compiler. The coefficients are shown in the impulse response window as shown in Fig. 3.20b and can be modified if needed.

After loading the coefficients we can then select the Structure to be fully parallel, fully serial, multi-bit serial, or multicycle. We select Distributed Arithmetic: Fully Parallel Filter. We set the input coefficient width to 8 bit and let the tool compute the output bitwidth based on the method Actual Coefficients. We select Coefficient Scaling as None since our integer coefficients should not be further quantized. The transfer function in integer and floating-point should therefore be seen as matching lines, see Fig. 3.21. The FIR compiler reports an estimated size of 312 LEs. We skip step 2 from the toolbench since the design is small and we will use the compiled data to verify and simulate the design. We proceed with step 3 and the generation of the VHDL code and all supporting files follows. These files are listed in Table 3.6. We see that not only are the VHDL and Verilog files generated along with their component files, but MATLAB (bit accurate) and Quartus II (cycle accurate) test vectors are also provided to enable an easy verification path. We then instantiate the FIR core in a wrapper file that also includes registers for the input and output values. We then compile the HDL code of the filter to enable a timing simulation and provide precise resource data. The impulse response simulation of the F6 filter is shown in Figure 3.22. We see that two additional control signals rdy_to_ld and done have been synthesized, although we did not ask for them. | 3.12 |

The design from the Example 3.12 requires 426 LEs and runs at 362.84 MHz. Without the wrapper file the LE count (404 LEs) is still slightly higher than the estimation of 312 LEs. The overall cost metric measured as the quotient LEs/Fmax is 1.17 and is better than RAG without pipelining, since the DA is fully pipelined, as you can see from the large initial delay of the impulse response. For an appropriate comparison we should compare the DA-based design with the fully pipelined RAG design. The cost of the DA design is higher than the fully pipelined RAG design, see Table 3.5, p. 189. But the Registered Performance of the DA-based IP core is slightly higher than the fully pipelined RAG design.

3.4.5 Comparison of DA- and RAG-Based FIR Filters

In the last section we followed a detailed case study of the F6 half-band RAG- and DA-based FIR filter designs. The question now would be whether the results were just one single (atypical) example or if the results in terms of speed/size/cost are typical. In order to answer this question a set of larger filters has been designed using VHDL for fully pipelined RAG (see Exercises 3.13-3.29, p. 212) and should be compared with the synthesis data using the FIR core compiler from Altera that implements a fully parallel DA filter [86]. Table 3.7 shows the results for three half-band filters (F6, F8, and F9) from Goodman and Carey [80], two from Samueli [84], and two from Lim and Parker [85]. The first column shows the filter name, followed by the pipeline stages used. No-pipeline and fully pipelined data are reported, but no one-pipeline design data, as in Table 3.5 (p. 189). The third column shows the

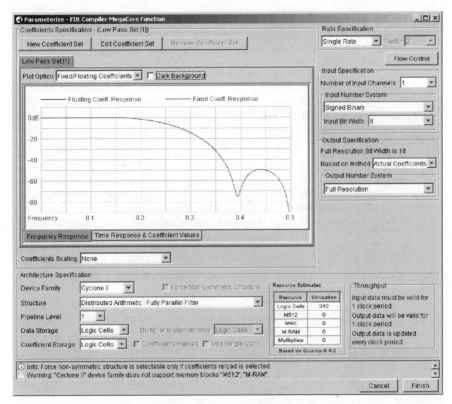

Fig. 3.21. IP parametrization of FIR core according to the F6 Example 3.5, p. 184.

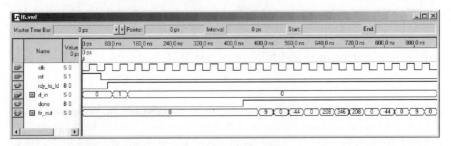

Fig. 3.22. FIR core timing simulation result.

filter length L. The next three columns show the synthesis data for the RAG-based filters, namely LEs, `Registered Performance` and the cost (area $\times$ delay) measured as the quotient LEs/Fmax. Columns 7-9 show the same three values for the DA-based designs generated with Altera's FIR compiler. For each filter two rows are used to show the data for zero/no and fully pipelined designs. Finally in the last rows the average value for zero- and fully pipelined designs are given and, at the very end, a comparison of the gain/loss of RAG

Table 3.6. IP files generation for FIR core.

File	Description
f6_core.vhd	A MegaCore function variation file, which defines a top-level VHDL description of the custom MegaCore function
f6_core_inst.vhd	VHDL sample instantiation file
f6_core.cmp	A VHDL component declaration for the MegaCore function variation
f6_core.inc	An AHDL include declaration file for the MegaCore function variation function
f6_core_bb.v	Verilog HDL black-box file for the MegaCore function variation
f6_core.bsf	Quartus II symbol file to be used in the Quartus II block diagram editor
f6_core_st.v	Generated FIR filter netlist
f6_core _constraints.tcl	This file contains the necessary constraints to achieve FIR filter size and speed
f6_core_mlab.m	This file provides a MATLAB simulation model for the customized FIR filter
f6_core_tb.m	This file provides a MATLAB testbench for the customized FIR filter
f6_core.vec	This file provides simulation test vectors to be used simulating the customized FIR filter with the Quartus II software
f6_core.html	The MegaCore function report file

zero- and fully pipelined and fully pipelined RAG compared with DA-based designs are given.

It can be seen from Table 3.7, that

- Fully pipelined RAG filters enjoy size reductions averaging 71% compared with DA-based designs.
- The fully pipelined RAG filter requires on average only 6% more LEs than the RAG design without pipelining.
- The **Register Performance** of the DA-based FIR filters is on average 8% higher than fully pipelined RAG designs.
- The overall cost, measured as LEs/Fmax, is on average 56% better for RAG-based compared with DA-based designs when a fully pipeline approach is used.

It can also be seen from Table 3.7 that, without pipelining (pipe=0), the DA-based approach gives better results. With a 6% increase in area, the cost for RAG pipelining is quite reasonable.

Table 3.7. Size, speed and cost comparison of DA and RAG algorithm.

Filter name	Pipe stages	L	LEs	RAG Fmax (MHz)	Cost $\frac{LEs}{Fmax}$	LEs	DA Fmax (MHz)	Cost $\frac{LEs}{Fmax}$
F6	0	11	225	165.95	1.36			
	max		234	353.86	0.71	396	332.34	1.19
F8	0	15	326	135.85	2.40			
	max		360	323.42	1.11	570	340.72	1.67
F9	0	19	461	97.26	4.74			
	max		534	304.04	1.76	717	326.16	2.20
S1	0	25	460	130.63	3.52			
	max		492	296.65	1.66	985	356.51	2.76
L3	0	36	651	205.3	3.17			
	max		671	310.37	2.16	1406	321.3	4.38
S2	0	60	1672	129.97	12.86			
	max		1745	252.91	6.90	2834	289.02	9.81
L2	0	63	1446	134.95	10.72			
	max		1531	265.53	5.77	2590	282.41	9.17
Mean	0		745	140.34	5.53			
Mean	max		793	296.60	2.86	1357	321.21	4.45
Gain%			RAG-0/RAG-max			RAG-max/DA		
			−6	111	93	71	−8	56

Exercises

Note: If you have no prior experience with the Quartus II software, refer to the case study found in Sect. 1.4.3, p. 29. If not otherwise noted use the EP2C35F672C6 from the Cyclone II family for the Quartus II synthesis evaluations.

3.1: A filter has the following specification: sampling frequency 2 kHz; passband 0–0.4 kHz, stopband 0.5–1 kHz; passband ripple, 3 dB, and stopband ripple, 48 dB. Use the MATLAB software and the "Interactive Lowpass Filter Design" demo from the Signal Processing Toolbox for the filter design.
(a1) Design a direct filter with a Kaiser window.
(a2) Determine the filter length and the absolute ripple in the passband.
(b1) Design an equiripple filter (use the functions remex or firpm).
(b2) Determine the filter length and the absolute ripple in the passband.

3.2: **(a)** Compute the RAG for a length-11 half-band filter F5 that has the nonzero coefficients $f[0] = 256, f[\pm 1] = 150, f[\pm 3] = -25, f[\pm 5] = 3$.
(b) What is the minimum output bit width of the filter, if the input bit width is 8 bits?
(c1) Write and compile (with the Quartus II compiler) the HDL code for the filter.
(c2) Simulate the filter with impulse and step responses.
(d) Write the VHDL code for the filter in distributed arithmetic, using the state machine approach with the table realized as LPM_ROM.

3.3: (a) Compute the RAG for length-11 half-band filter F7 that has the nonzero coefficients $f[0] = 512, f[\pm1] = 302, f[\pm3] = -53, f[\pm5] = 7$.
(b) What is the minimum output bit width of the filter, if the input bit width is 8 bits?
(c1) Write and compile (with the Quartus II compiler) the VHDL code for the filter.
(c2) Simulate the filter with impulse and step response.

3.4: Hartley [87] has introduced a concept to implement constant coefficient filters, by exploiting common subexpressions across coefficients. For instance, the filter

$$y[n] = \sum_{k=0}^{L-1} a[k]x[n-k], \tag{3.19}$$

with three coefficients $a[k] = \{480, -302, 31\}$. The CSD code of these three coefficients is given by

	512	256	128	64	32	16	8	4	2	1
480 :	1	0	0	0	-1	0	0	0	0	0
-302 :	0	-1	0	-1	0	1	0	0	1	0
31 :	0	0	0	0	1	0	0	0	0	-1

From the table we note that the pattern $\begin{vmatrix} 1 & 0 \\ 0 & -1 \end{vmatrix}$ can be found four times. If we therefore build the temporary variable $h[n] = 2x[n] - x[n-1]$, we can compute the filter output with

$$y[n] = 256h[n] - 16h[n] - 32h[n-1] + h[n-1]. \tag{3.20}$$

(a) Verify (3.20) by substituting $h[n] = 2x[n] - x[n-1]$.
(b) How many adders are required to yield the direct CSD implementation of (3.19) and the implementation with subexpression sharing?
(c1) Implement the filter with subexpression sharing with Quartus II for 8-bit inputs.
(c2) Simulate the impulse response of the filter.
(c3) Determine the **Registered Performance** and the used resources (LEs, multipliers, and M4Ks).

3.5: Use the subexpression method from Exercise 3.4 to implement a 4-tap filter with the coefficients $a[k] = \{-1406, -1109, -894, 2072\}$.
(a) Find the CSD code and the subexpression representation for the most frequent pattern.
(b) Substitute for the subexpression a 2 or -2, respectively. Apply the subexpression sharing one more time to the reduced set.
(c) Determine the temporary equations and check by substitution back into (3.19).
(d) How many adders are required to yield the direct CSD implementation of (3.19) and the implementation with subexpression sharing?
(e1) Implement the filter with subexpression sharing with Quartus II for 8-bit inputs.
(e2) Simulate the impulse response of the filter.
(e3) Determine the **Registered Performance** and the used resources (LEs, multipliers, and M4Ks).

3.6: (a1) Use the program dagen3e.exe to compile a DA table for the coefficients $\{20, 24, 21, 100, 13, 11, 19, 7\}$ using multiple CASE statements.

Synthesize the design for maximum speed and determine the resources (LEs, multipliers, and M4Ks) and **Registered Performance**.
(a2) Simulate the design using power-of-two $2^k; 0 \leq k \leq 7$ input values.
(b) Use the partitioning technique to implement the same table using two sets, namely $\{20, 24, 21, 100\}$ and $\{13, 11, 19, 7\}$, and an additional adder. Synthesize the design for maximum speed and determine the size and **Registered Performance**.
(b2) Simulate the design using power-of-two $2^k; 0 \leq k \leq 7$ input values.
(c) Compare the designs from (a) and (b).

3.7: Implement 8-bit input/output improved 4-tap $\{-1, 3.75, 3.75, -1\}$ filter designs according to the listing in Table 3.3, p. 181. For each filter write the HDL code and determine the resources (LEs, multipliers, and M4Ks) and **Registered Performance**.
(a) Synthesize `fir_sym.vhd` as the filter using symmetry.
(b) Synthesize `fir_csd.vhd` as the filter using CSD coding.
(c) Synthesize `fir_tree.vhd` as the filter using an adder tree.
(d) Synthesize `fir_csd_sym.vhd` as the filter using CSD coding and symmetry.
(e) Synthesize `fir_csd_sym_tree.vhd` as the filter using all three improvements.

3.8: (a) Write a short MATLAB program that plots the
(a1) impulse response,
(a2) frequency response, and
(a2) the pole/zero plot for the half-band filter F3, see Table 5.3, p. 274.
Hint: Use the MATLAB functions: `filter, stem, freqz, zplane`.
(b) What is the bit growth of the F3 filter? What is the total required output bit width for an 8-bit input?
(c) Use the `csd3e.exe` program from the CD to determine the CSD code for the coefficients.
(c) Use the `ragopt.exe` program from the CD to determine the reduced adder graph (RAG) of the filter coefficients.

3.9: Repeat Exercise 3.8 for the CFIR filter of the GC4114 communication IC. Try the WWW to download a datasheet if possible. The 31 filter coefficients are:
$-23, -3, 103, 137, -21, -230, -387, -235, 802, 1851, 81, -4372, -4774, 5134, 20\,605,$
$28\,216, 20\,605, 5134, -4774, -4372, 81, 1851, 802, -235, -387, -230, -21, 137, 103,$
$-3, -23.$

3.10: Download the datasheet for the GC4114 from the WWW. Use the results from Exercise 3.9.
(a) Design the 31-tap symmetric CFIR compensation filter as CSD FIR filter in transposed form (see Fig. 3.3, p. 167) for 8-bit input and an asynchronous reset. Try to match the simulation shown in Fig. 3.23.
(b) For the device EP2C35F672C6 from the Cyclone II family determine the resources (LEs, multipliers, and M4Ks) and the **Registered Performance**.

3.11: Download the datasheet for the GC4114 from the WWW. Use the results from Exercise 3.9.
(a) Design the 31-tap symmetric CFIR compensation filter using distributed arithmetic. Use the `dagen3e.exe` program from the CD to generate the HDL code for the coefficients. Note you should use always groups of four coefficients each and add the results in an adder tree.
(b) Design the DA FIR filter in the full parallel form (see Fig. 3.18, p. 202) for 8-bit input and an asynchronous reset. Take advantage of the coefficient symmetry. Try to match the simulation shown in Fig. 3.24.

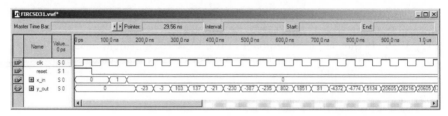

Fig. 3.23. Testbench for the CSD FIR filter in Exercise 3.10.

(c) For the device EP2C35F672C6 from the Cyclone II family determine the resources (LEs, multipliers, and M4Ks) and the `Registered Performance`.

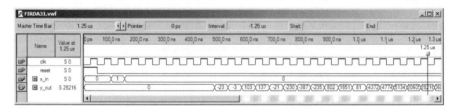

Fig. 3.24. Testbench for the DA-based FIR filter in Exercise 3.11.

3.12: Repeat Exercise 3.8 for the half-band filter F4, see Table 5.3, p. 274.

3.13: Repeat Exercise 3.8 for the half-band filter F5, see Table 5.3, p. 274.

3.14: Use the results from Exercise 3.13 and report the HDL code, resources (LEs, multipliers, and M4Ks) and `Registered Performance` for the HDL design of the F5 half-band HDL FIR filter as:
(a) an RAG filter without pipelining
(b) a fully pipelined RAG filter
(c) a DA fully pipelined filter using an FIR core generator

3.15: Repeat Exercise 3.8 for the half-band filter F6, see Table 5.3, p. 274.

3.16: Use the results from Exercise 3.15 and report the HDL code, resources (LEs, multipliers, and M4Ks) and `Registered Performance` for the HDL design of the F6 half-band HDL FIR filter as:
(a) an RAG filter without pipelining
(b) a fully pipelined RAG filter
(c) a DA fully pipelined filter using an FIR core generator

3.17: Repeat Exercise 3.8 for the half-band filter F7, see Table 5.3, p. 274.

3.18: Repeat Exercise 3.8 for the half-band filter F8, see Table 5.3, p. 274.

3.19: Use the results from Exercise 3.18 and report the HDL code, resources (LEs, multipliers, and M4Ks) and `Registered Performance` for the HDL design of the F8 half-band HDL FIR filter as:

(a) an RAG filter without pipelining
(b) a fully pipelined RAG filter
(c) a DA fully pipelined filter using an FIR core generator

3.20: FIR features design. In this problem we want to compare the influence of additional features like reset and enable for different device families. Use the results from Exercise 3.18 for the CSD code. For all following HDL F8 CSD designs with 8-bit input determine the resources (LEs, multipliers, and M2Ks/M4Ks) and Registered Performance. As the device use the EP2C35F672C6 from the Cyclone II family and the EPF10K70RC240-4 from the Flex 10K family.
(a) Design the F8 CSD FIR filter in direct form (see Fig. 3.1, p. 166).
(b) Design the F8 CSD FIR filter in transposed form (see Fig. 3.3, p. 167).
(c) Add a synchronous reset to the transposed FIR from (b).
(d) Add an asynchronous reset to the transposed FIR from (b).
(e) Add a synchronous reset and enable to the transposed FIR from (b).
(f) Add an asynchronous reset and enable to the transposed FIR from (b).
(g) Tabulate your resources (LEs, multipliers, and M2Ks/M4Ks) and Registered Performance results from (a)-(g). What conclusions can be drawn for Flex and Cyclone II devices from the measurements?

3.21: Repeat Exercise 3.8 for the half-band filter F9, see Table 5.3, p. 274.

3.22: Use the results from Exercise 3.21 and report the HDL code, resources (LEs, multipliers, and M4Ks) and Registered Performance for the HDL design of the F9 half-band HDL FIR filter as:
(a) an RAG filter without pipelining
(b) a fully pipelined RAG filter
(c) a DA fully pipelined filter using an FIR core generator

3.23: Repeat Exercise 3.8 for the Samueli filter S1 [84]. The 25 filter coefficients are: $1, 3, 1, 8, 7, 10, 20, 1, 40, 34, 56, 184, 246, 184, 56, 34, 40, 1, 20, 10, 7, 8, 1, 3, 1$.

3.24: Use the results from Exercise 3.23 and report the HDL code, resources (LEs, multipliers, and M4Ks) and Registered Performance for the HDL design of the Samueli filter S1 FIR filter as:
(a) an RAG filter without pipelining
(b) a fully pipelined RAG filter
(c) a DA fully pipelined filter using an FIR core generator

3.25: Repeat Exercise 3.8 for the Samueli filter S2 [84]. The 60 filter coefficients are: $31, 28, 29, 22, 8, -17, -59, -116, -188, -268, -352, -432, -500, -532, -529, -464,$ $-336, -129, 158, 526, 964, 1472, 2008, 2576, 3136, 3648, 4110, 4478, 4737, 4868,$ $4868, 4737, 4478, 4110, 3648, 3136, 2576, 2008, 1472, 964, 526, 158, -129, -336,$ $-464, -529, -532, -500, -432, -352, -268, -188, -116, -59, -17, 8, 22, 29, 28,$ 31.

3.26: Use the results from Exercise 3.25 and report the HDL code, resources (LEs, multipliers, and M2Ks/M4Ks) and Registered Performance for the HDL design of the Samueli filter S2 FIR filter as:
(a) an RAG filter without pipelining.
(b) a fully pipelined RAG filter.
(c) a DA fully pipelined filter using an FIR core generator.

3.27: Repeat Exercise 3.8 for the Lim and Parker L2 filter [85]. The 63 filter coefficients are: $3, 6, 8, 7, 1, -9, -19, -24, -20, -5, 15, 31, 33, 16, -15, -46, -59, -42, 4,$

61, 99, 92, 29, $-71, -164, -195, -119$, 74, 351, 642, 862, 944, 862, 642, 351, 74, $-119, -195, -164, -71$, 29, 92, 99, 61, 4, $-42, -59, -46, -15$, 16, 33, 31, 15, -5, $-20, -24, -19, -9$, 1, 7, 8, 6, 3.

3.28: Use the results from Exercise 3.27 and report the HDL code, resources (LEs, multipliers, and M4Ks) and `Registered Performance` for the HDL design of the Lim and Parker L2 filter FIR filter as:
(a) an RAG filter without pipelining
(b) a fully pipelined RAG filter
(c) a DA fully pipelined filter using an FIR core generator

3.29: Repeat Exercise 3.8 for the Lim and Parker L3 filter [85]. The 36 filter coefficients are: $10, 1, -8, -14, -14, -3$, 10, 20, 24, $9, -18, -40, -48, -20, 36$, 120, 192, 240, 240, 192, 120, 36, $-20, -48, -40, -18$, 9, 24, 20, 10, $-3, -14, -14, -8$, 1, 10.

3.30: Use the results from Exercise 3.29 and report the HDL code, resources (LEs, multipliers, and M4Ks) and `Registered Performance` for the HDL design of the Lim and Parker filter L3 FIR filter as:
(a) an RAG filter without pipelining
(b) a fully pipelined RAG filter
(c) a DA fully pipelined filter using an FIR core generator

4. Infinite Impulse Response (IIR) Digital Filters

Introduction

In Chap. 3 we introduced the FIR filter. The most important properties that make the FIR attractive (+) or unattractive (−) for selective applications include:

+ FIR linear-phase performance is easily achieved.
+ Multiband filters are possible.
+ The Kaiser window method allows iterative-free design.
+ FIRs have a simple structure for decimators and interpolators (see Chap. 5).
+ Nonrecursive filters are always stable and have no limit cycles.
+ It is easy to get high-speed, pipelined designs.
+ FIRs typically have low coefficient and arithmetic roundoff error budgets, and well-defined quantization noise.
− Recursive FIR filters may be unstable because of imperfect pole/zero annihilation.
− The sophisticated Parks–McClellan algorithms must be available for minimax filter design.
− High filter length requires high implementation effort.

Compared to an FIR filter, an IIR filter can often be much more efficient in terms of attaining certain performance characteristics with a given filter order. This is because the IIR filter incorporates feedback and is capable of realizing both zeros and poles of a system transfer function, whereas the FIR filter is an all-zero filter. In this chapter, the fundamentals of IIR filter design will be developed. The traditional approach to the design of IIR filters involves the transformation of an analog filter, with defined feedback specifications, into the digital domain. This is a reasonable approach, mainly because the art of designing analog filters is highly advanced, and many standard tables are available, i.e., [88]. We will review the four most important classes of these analog prototype filters in this chapter, namely Butterworth, Chebyshev I and II, and elliptic filters.

The IIR will be shown to overcome many of the deficiencies of the FIR, but to have some less desirable properties as well. The general desired (+) and undesired (−) properties of an IIR filter are:

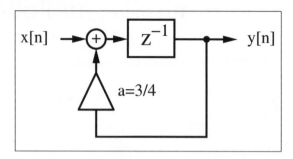

Fig. 4.1. First-order IIR filter used as lossy integrator.

+ Standard design using an analog prototype filter is well understood.

+ Highly selective filters can be realized with low-order designs that can run at high speeds.

+ Design using tables and a pocket calculator is possible.

+ For the same tolerance scheme, filters are short, compared with FIR filters.

+ Closed-loop design algorithms can be used.

− Nonlinear-phase response is typical, i.e., it is difficult to get linear-phase response. (Using an allpass filter for phase compensation results in twice the complexity.)

− Limit cycles may occur for integer implementation.

− Multiband design is difficult; only low, high, or bandpass filters are designed.

− Feedback can introduce instabilities. (Most often, the mirror pole to the unit circle can be used to produce the same magnitude response, and the filter will be stable.)

− It is more difficult to get high-speed, pipelined designs

To demonstrate the possible benefits of using IIR filters, we will discuss a first-order IIR filter example.

Example 4.1: Lossy Integrator I

One of the basic tasks of a filter may be to smooth a noisy signal. Assume that a signal $x[n]$ is received in the presence of wideband zero-mean random noise. Mathematically, an integrator could be used to suppress the effects of the noise. If the average value of the input signal is to be preserved over a finite time interval, a *lossy* integrator is often used to process the signal with additive noise. Figure 4.1 displays a simple first-order lossy integrator that satisfies the discrete-time difference equation:

$$y[n+1] = \frac{3}{4}y[n] + x[n]. \tag{4.1}$$

As we can see from the impulse response in Fig. 4.2a, the same functionality of the first-order lossy integrator can be achieved with a 15-tap FIR filter. The step response to the lossy integrator is shown in Fig. 4.2b.

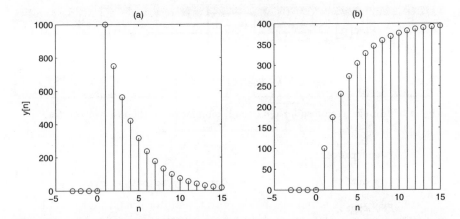

Fig. 4.2. Simulation of lossy integrator with $a = 3/4$. **(a)** Impulse response for $x[n] = 1000\delta[n]$. **(b)** Step response for $x[n] = 100\sigma[n]$.

The following VHDL code[1] shows a possible implementation of this IIR filter.

```
PACKAGE n_bit_int IS              -- User-defined type
   SUBTYPE BITS15 IS INTEGER RANGE -2**14 TO 2**14-1;
END n_bit_int;

LIBRARY work;
USE work.n_bit_int.ALL;

LIBRARY ieee;
USE ieee.std_logic_1164.ALL;
USE ieee.std_logic_arith.ALL;

ENTITY iir IS
   PORT (x_in  : IN  BITS15;      -- Input
         y_out : OUT BITS15;      -- Result
         clk   : IN  STD_LOGIC);
END iir;

ARCHITECTURE fpga OF iir IS

   SIGNAL x, y : BITS15 := 0;

BEGIN

   PROCESS      -- Use FF for input and recursive part
   BEGIN
      WAIT UNTIL clk = '1';
      x   <= x_in;
      y   <= x + y / 4 + y / 2;
```

---

[1] The equivalent Verilog code iir.v for this example can be found in Appendix A on page 692. Synthesis results are shown in Appendix B on page 731.

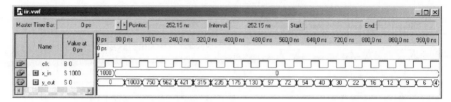

Fig. 4.3. Impulse response for Quartus II simulation of the lossy integrator.

```
    end process;

    y_out <= y;               -- Connect y to output pins

  END fpga;
```
Registers have been implemented using a WAIT statement inside a PROCESS
block, while the multiplication and addition is implemented using CSD code.
The design uses 62 LEs, no embedded multiplier, and has a 160.69 MHz
Registered Performance, if synthesized with the Speed option. The response
of the filter to an impulse of amplitude 1000, shown in Fig. 4.3, agrees with
the MATLAB simulated results presented in Fig. 4.2a.   4.1

An alternative design approach using a "standard logic vector" data type
and LPM_ADD_SUB megafunctions is discussed in Exercise 4.6 (p. 241). This
second approach will produce longer VHDL code but will have the benefit of
direct control, at the bit level, over the sign extension and multiplier.

4.1 IIR Theory

A nonrecursive filter incorporates, as the name implies, no feedback. The
impulse response of such a filter is finite, i.e., it is an FIR filter. A recursive
filter, on the other hand has feedback, and is expected, in general, to have
an infinite impulse response, i.e., to be an IIR filter. Figure 4.4a shows filters
with separate recursive and nonrecursive parts. A *canonical* filter is produced
if these recursive and nonrecursive parts are merged together, as shown in
Fig. 4.4b. The transfer function of the filter from Fig. 4.4 can be written as:

$$F(z) = \frac{\sum_{l=0}^{L-1} b[l]z^{-l}}{1 - \sum_{l=1}^{L-1} a[l]z^{-l}}. \tag{4.2}$$

The difference equation for such a system yields:

$$y[n] = \sum_{l=0}^{L-1} b[l]x[n-l] + \sum_{l=1}^{L-1} a[l]y[n-l]. \tag{4.3}$$

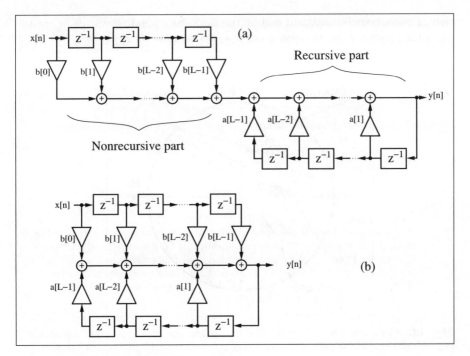

Fig. 4.4. Filter with feedback.

Comparing this with the difference equation for the FIR filter (3.2) on p. 166, we find that the difference equation for recursive systems depends not only on the L previous values of the input sequence $x[n]$, but also on the $L - 1$ previous values of $y[n]$.

If we compute poles and zeros of $F(z)$, we see that the nonrecursive part, i.e., the numerator of $F(z)$, produces the *zeros* p_{0l}, while the denominator of $F(z)$ produces the *poles* $p_{\infty l}$.

For the transfer function, the *pole/zero plot* can be used to look up the most important properties of the filter. If we substitute $z = e^{j\omega T}$ in the z-domain transfer function, we can construct the Fourier transfer function

$$F(\omega) = |F(\omega)|e^{j\theta(\omega)} = \frac{\displaystyle\prod_{l=0}^{L-2} p_{0l} - e^{j\omega T}}{\displaystyle\prod_{l=0}^{L-2} p_{\infty l} - e^{j\omega T})} = \frac{\exp(j\sum_l \beta_l)\displaystyle\prod_{l=0}^{L-2} v_l}{\exp(j\sum_l \alpha_l)\displaystyle\prod_{l=0}^{L-2} u_l} \quad (4.4)$$

by graphical means. This is shown in Fig. 4.5, for a specific amplitude (i.e., gain) and phase value. The gain at a specific frequency ω_0 is the quotient of the zero vectors v_l and the pole vectors u_l. These vectors start at a specific zero or pole, respectively, and end at the frequency point, $e^{j\omega_0 T}$, of interest. The phase gain for the example from Fig. 4.5 becomes $\theta(\omega_0) = \beta_0 + \beta_1 - \alpha_0$.

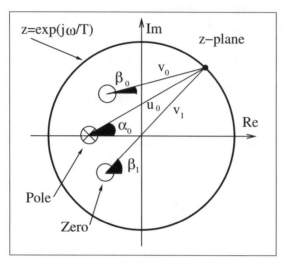

Fig. 4.5. Computation of transfer function using the pole/zero plot. Amplitude gain $= u_0 u_1 / v_0$, phase gain $= \beta_0 + \beta_1 - \alpha_0$.

Using the connection between the transfer function in the Fourier domain and the pole/zero plot, we can already deduce several properties:

1) A *zero on the unit circle* $p_0 = e^{j\omega_0 T}$ (with no annihilating pole) produces a zero in the transfer function in the Fourier domain at the frequency ω_0.
2) A *pole on the unit circle* $p_\infty = e^{j\omega_0 T}$ (and no annihilating zero) produces an infinite gain in the transfer function in the Fourier domain at the frequency ω_0.
3) A *stable* filter with all poles inside the unit circle can have any type of input signal.
4) A *real filter* has single poles and zeros on the real axis, while complex poles and zeros appear always in pairs, i.e., if $a_0 + ja_1$ is a pole or zero, $a_0 - ja_1$ must also be a pole or zero.
5) A *linear-phase* (i.e., constant group delay) filter has all poles and zeros symmetric to the unit circle or at $z = 0$.

If we combine observations 3 and 5, we find that, for a stable linear-phase system, all zeros must be symmetric to the unit circle and only poles at $z = 0$ are permitted.

An IIR filter (with poles $z \neq 0$) can therefore be only approximately linear-phase. To achieve this approximation a well-known principle from analog filter design is used: an allpass has a unit gain, and introduces a nonzero phase gain, which is used to achieve linearization in the frequency range of interest, i.e., the passband.

4.2 IIR Coefficient Computation

In classical IIR design, a digital filter is designed that approximates an ideal filter. The ideal digital filter model specifications are mathematically converted into a set of specifications from an analog filter model using the *bilinear z-transform* given by:

$$s = \frac{z-1}{z+1}. \tag{4.5}$$

A classic analog Butterworth, Chebyshev, or elliptic model can be synthesized from these specifications, and is then mapped into a digital IIR using this bilinear z-transform.

An analog *Butterworth* filter has a magnitude-squared frequency response given by:

$$|F(\omega)|^2 = \frac{1}{1 + \left(\frac{\omega}{\omega_s}\right)^{2N}}. \tag{4.6}$$

The poles of $|F(\omega)|^2$ are distributed along a circular arc at locations separated by π/N radians. More specifically, the transfer function is N times differentiable at $\omega = 0$. This results in a locally smooth transfer function around 0 Hz. An example of a Butterworth filter model is shown in Fig. 4.6(upper). Note that the tolerance scheme for this design is the same as for the Kaiser window and equiripple design shown in Fig. 3.7 (p. 176).

An analog *Chebyshev* filter of Type I or II is defined in terms of a Chebyshev polynomial $V_N(\omega) = \cos(N \cos(\omega))$, which forces the filter poles to reside on an ellipse. The magnitude-squared frequency response of a Type I filter is represented by:

$$|F(\omega)|^2 = \frac{1}{1 + \varepsilon^2 V_N^2 \left(\frac{\omega}{\omega_s}\right)}. \tag{4.7}$$

An example of a typical Type I magnitude frequency and impulse response is shown in Fig. 4.7(upper). Note the ripple in the passband, and smooth stopband behavior.

The Type II magnitude-squared frequency response is modeled as:

$$|F(\omega)|^2 = \frac{1}{1 + \left(\varepsilon^2 V_N^2 \left(\frac{\omega}{\omega_s}\right)^{-1}\right)}. \tag{4.8}$$

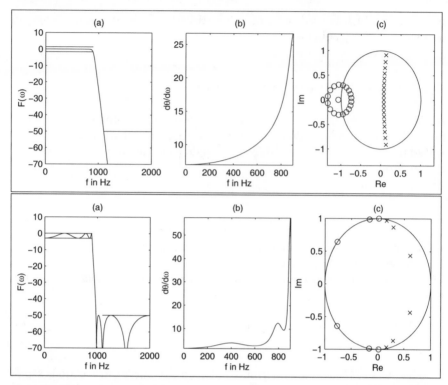

Fig. 4.6. Filter design with MATLAB toolbox. **(upper)** Butterworth filter and **(lower)** elliptic Filter.
(a) Transfer function. **(b)** Group delay of passband. **(c)** Pole/zero plot. ($\times$ = pole; $\circ$ = zero).

An example of a typical Type II magnitude frequency and impulse response is shown in Fig. 4.7(lower). Note that in this case a smooth passband results, and the stopband now exhibits ripple behavior.

An analog *elliptic* prototype filter is defined in terms of the solution to the Jacobian elliptic function, $U_N(\omega)$. The magnitude-squared frequency response is modeled as:

$$|F(\omega)|^2 = \frac{1}{1 + \varepsilon^2 U_N^2 \left(\frac{\omega}{\omega_s}\right)^{-1}}. \tag{4.9}$$

The magnitude-squared and impulse response of a typical elliptic filter is shown in Fig. 4.6(lower). Observe that the elliptic filter exhibits ripple in both the passband and stopband.

If we compare the four different IIR filter implementations, we find that a Butterworth filter has order 19, a Chebyshev has order 8, while the elliptic design has order 6, for the same tolerance scheme shown in Fig. 3.8 (p. 177).

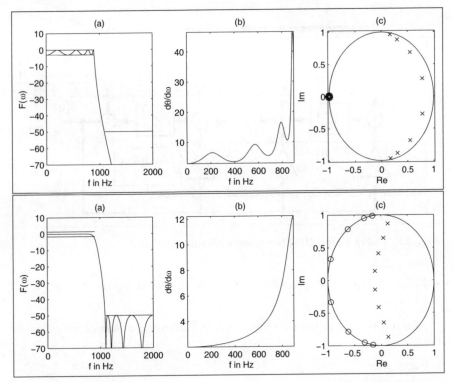

Fig. 4.7. Chebyshev filter design with MATLAB toolbox. Chebyshev I **(upper)** and Chebyshev II **(lower)**.
(a) Transfer function. **(b)** Group delay of passband. **(c)** Pole/zero plot ($\times$ = pole; o = zero).

If we compare Figs. 4.6 and 4.7, we find that for the filter with shorter order the ripple increases, and the group delay becomes highly nonlinear. A good compromise is most often the Chebyshev Type II filter with medium order, a flat passband, and tolerable group delay.

4.2.1 Summary of Important IIR Design Attributes

In the previous section, classic IIR types were presented. Each model provides the designer with tradeoff choices. The attributes of classic IIR types are summarized as follows:

- **Butterworth:** Maximally flat passband, flat stopband, wide transition band
- **Chebyshev I:** Equiripple passband, flat stopband, moderate transition band
- **Chebyshev II:** Flat passband, equiripple stopband, moderate transition band

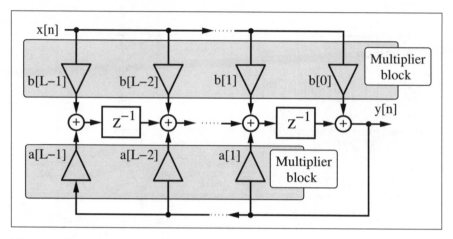

Fig. 4.8. Direct I form IIR filter using multiplier blocks.

- **Elliptic:** Equiripple passband, equiripple stopband, narrow transition band

For a given set of filter requirement, the following observations generally hold:

- Filter order
  - Lowest: Elliptic
  - Medium: Chebyshev I or II
  - Highest: Butterworth
- Passband characteristics
  - Equiripple: Elliptic, Chebyshev I
  - Flat: Butterworth, Chebyshev II
- Stopband characteristics
  - Equiripple: Elliptic, Chebyshev II
  - Flat: Butterworth, Chebyshev I
- Transition band characteristics
  - Narrowest: Elliptic
  - Medium: Chebyshev I+II
  - Widest: Butterworth

4.3 IIR Filter Implementation

Obtaining an IIR transfer function is generally considered to be a straightforward exercise, especially if design software like MATLAB is used. IIR filters can be developed in the context of many architectures. The most important structures are summarized as follows:

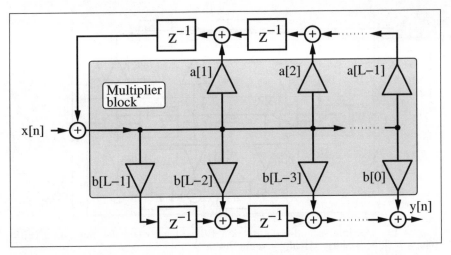

Fig. 4.9. Direct II form IIR filter using multiplier blocks.

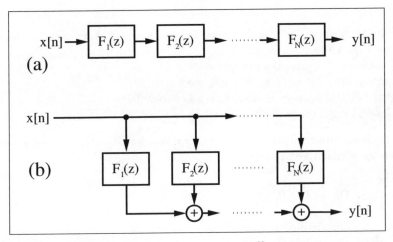

Fig. 4.10. (a) Cascade implementation $F(z) = \prod_{k=1}^{N} F_k(z)$. (b) Parallel implementation $F(z) = \sum_{k=1}^{N} F_k(z)$.

- Direct I form (see Fig. 4.8)
- Direct II form (see Fig. 4.9)
- Cascade of first- or second-order systems (see Fig. 4.10a)
- Parallel implementation of first- or second-order systems (see Fig. 4.10b).
- BiQuad implementation of a typical second-order section found in basic cascade or parallel designs (see Fig. 4.11)
- Normal [89], i.e., cascade of first- or second-order state variable systems (see Fig. 4.10a)

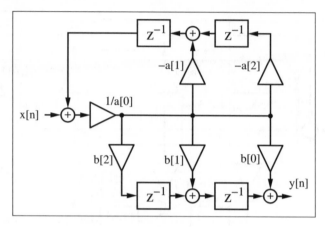

Fig. 4.11. Possible second-order section BiQuad with transfer function $F(z) = (b[0] + b[1]z^{-1} + b[2]z^{-2})/(a[0] + a[1]z^{-1} + a[2]z^{-2})$.

- Parallel normal, i.e., parallel first- or second-order state variable systems (see Fig. 4.10b)
- Continued fraction structures
- Lattice filter (after Gray–Markel, see Fig. 4.12)
- Wave digital implementation (after Fettweis [90])
- General state space filter

Each architecture serves a unique purpose. Some of the general selection rules are summarized below:

- Speed
  - High: Direct I & II
  - Low: Wave
- Fixed-point arithmetic roundoff error sensitivity
  - High: Direct I & II
  - Low: Normal, Lattice
- Fixed-point coefficient roundoff error sensitivity
  - High: Direct I & II
  - Low: Parallel, Wave
- Special properties
  - Orthogonal weight outputs: Lattice
  - Optimized second-order sections: Normal
  - Arbitrary IIR specification: State variable

With the help of software tools like MATLAB, the coefficients can easily be converted from one architecture to another, as demonstrated by the following example.

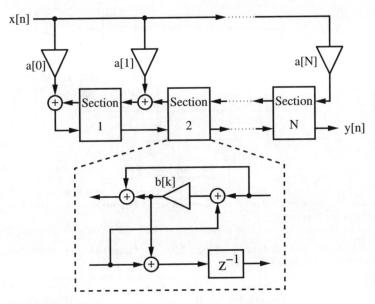

Fig. 4.12. Lattice filter.

Example 4.2: Butterworth Second-order System

Assume we wish to design a Butterworth filter (order $N = 10$, passband $Fp = 0.3$ Fs) realized by second-order systems. We can use the following MATLAB code to generate the coefficients:

```
N=10;Fp=0.3;
[B,A]=butter(N,Fp)
[sos, gain]=tf2sos(B,A)
```

i.e., we first compute the Butterworth coefficient using the function `butter()`, and then convert this filter coefficient using the "transfer function to second-order section" function `tf2sos` to compute the BiQuad coefficients. We will get the following results using MATLAB for the second-order sections:

$b[0,i]$	$b[1,i]$	$b[2,i]$	$a[0,i]$	$a[1,i]$	$a[2,i]$
1.0000	2.1181	1.1220	1.0000	−0.6534	0.1117
1.0000	2.0703	1.0741	1.0000	−0.6831	0.1622
1.0000	1.9967	1.0004	1.0000	−0.7478	0.2722
1.0000	1.9277	0.9312	1.0000	−0.8598	0.4628
1.0000	1.8872	0.8907	1.0000	−1.0435	0.7753

and the gain is $4.9614 10^{-5}$.

Figure 4.13 shows the transfer function, group delay, and the pole/zero plot of the filter. Note that all zeros are near $z_{0i} = -1$, which can also be seen from the numerator coefficients of the second-order systems. Note also the rounding error in $b[1,i] = 2$ and $b[0,i] = b[2,i] = 1$.   4.2

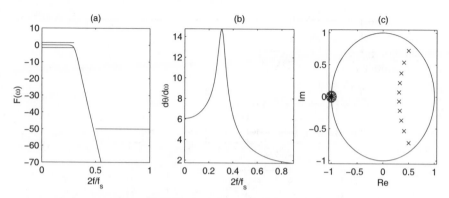

Fig. 4.13. Tenth-order Butterworth filter showing **(a)** magnitude, **(b)** phase, and **(c)** group delay response.

Table 4.1. Data for eighth-order elliptic filter by Crochiere and Oppenheim [91] sorted according the costs $M \times W$.

Type	Word-length W	Mults M	Adds	Delays	Cost $M \times W$
Wave	11.35	12	31	10	136
Cascade	11.33	13	16	8	147
Parallel	10.12	18	16	8	182
Lattice	13.97	17	32	8	238
Direct I	20.86	16	16	16	334
Direct II	20.86	16	16	8	334
Cont.-frac	22.61	18	16	8	408

4.3.1 Finite Wordlength Effects

Crochiere and Oppenheim [91] have shown that the coefficient wordlength required for a digital filter is closely related to the coefficient sensitivities. Implementation of the same IIR filter can therefore lead to a wide range of required wordlengths. To illustrate some of the dynamics of this problem, consider an eighth-order elliptic filter analyzed by Crochiere and Oppenheim [91]. The resulting eighth-order transfer function was implemented with a Wave, Cascade, Parallel, Lattice, Direct I and II, and Continuous Fraction architecture. The estimated coefficient wordlength to meet a specific maximal passband error criterion was conservatively estimated as shown in the second column of Table 4.1. As a result, it can be seen that the Direct form needs more wordlength than the Wave or Parallel structure. This has led to the conclusion that a Wave structure gives the best complexity (MW) in terms of the bit-width (W) multiplier product (M), as can be seen from column six of Table 4.1.

In the context of FIR filters (see Chap. 3), the reduced adder graph (RAG) technique was introduced in order to simplify the design of a block of several multipliers [92, 93]. Dempster and Macleod have evaluated the eighth-order elliptic filter from above, in the context of RAG multiplier implementation strategies. A comparison is presented in Table 4.2. The second column displays the multiplier block size. For a Direct II architecture, two multiplier blocks, of size 9 and 7, are required. For a Wave architecture, no two coefficients have the same input, and, as a result, no multiplier blocks can be developed. Instead, eleven individual multipliers must be implemented. The third column displays the number of adders/subtractors B for a canonical signed digit (CSD) design required to implement the multiplier blocks. Column four shows the same result for single-optimized multiplier adder graphs (MAG) [94]. Column five shows the result for the reduced adder graph. Column six shows the overall adder/wordwidth product for a RAG design. Table 4.2 shows that Cascade and Parallel forms give comparable or better results, compared with Wave digital filters, because the multiplier block size is an essential criterion when using the RAG algorithms. Delays have not been considered for the FPGA design, because all the logic cells have an associated flip-flop.

4.3.2 Optimization of the Filter Gain Factor

In general, we derive the IIR integer coefficients from floating-point filter coefficients by first normalizing to the maximum coefficient, and then multiplying with the desired gain factor, i.e., bit-width $2^{\mathrm{round}(W)}$. However, most often it is more efficient to select the gain factor within a range, $2^{\lfloor W \rfloor} \ldots 2^{\lceil W \rceil}$. There will be essentially no change in the transfer function, because the coefficients must be rounded anyway, after multiplying by the gain factor. If we apply, for instance, this search in the range $2^{\lfloor W \rfloor} \ldots 2^{\lceil W \rceil}$ for the cascade filter in the Crochiere and Oppenheim design example from above (gain used in Table 4.2 was $2^{\lfloor 11.33 \rfloor - 1} = 1024$), we get the data reported in Table 4.3.

Table 4.2. Data for eighth-order elliptic filter implemented using CSD, MAG, and RAG strategies [92].

Type	Block size	CSD B	MAG B	RAG B	RAG $W(B+A)$
Cascade	$4 \times 3, 2 \times 1$	26	26	24	453
Parallel	$11 \times 9, 4 \times 2, 1 \times 1$	31	30	29	455
Wave	11×1	58	63	22	602
Lattice	$1 \times 9, 8 \times 1$	33	31	29	852
Direct I	1×16	103	83	36	1085
Direct II	$1 \times 9, 1 \times 7$	103	83	41	1189
Cont.-frac	18×1	118	117	88	2351

Table 4.3. Variation of the gain factor to minimize filter complexity of the cascade filter.

	CSD	MAG	RAG
Optimal gain	1122	1121	1121
# adders for optimal gain	23	21	18
# adders for gain = 1024	26	26	24
Improvement	12%	19%	25%

We note, from the comparison shown in Table 4.3 a substantial improvement in the number of adders required to implement the multiplier. Although the optimal gain factor for MAG and RAG in this case is the same, it can be different.

4.4 Fast IIR Filter

In Chap. 3, FIR filter `Registered Performance` was improved using pipelining (see Table 3.3, p. 181). In the case of FIR filters, pipelining can be achieved at essentially no cost. Pipelining IIR filters, however, is more sophisticated and is certainly not free. Simply introducing pipeline registers for all adders will, especially in the feedback path, very likely change the pole locations and therefore the transfer function of the IIR filter. However strategies that do not change the transfer function and still allow a higher throughput have been reported in the literature. The reported methods that look promising to improve IIR filter throughput are:

- Look-ahead interleaving in the time domain [95]
- Clustered look-ahead pole/zero assignment [96, 97]
- Scattered look-ahead pole/zero assignment [95, 98]
- IIR decimation filter design [99]
- Parallel processing [100]
- RNS implementation [39, Sect. 4.2][49]

The first five methods are based on filter architecture or signal flow techniques, and the last is based on computer arithmetic (see Chap. 2). These techniques will be demonstrated with examples. To simplify the VHDL representation of each case, only a first-order IIR filter will be considered, but the same ideas can be applied to higher-order IIR filters and can be found in the literature references.

4.4.1 Time-domain Interleaving

Consider the differential equation of a first-order IIR system, namely

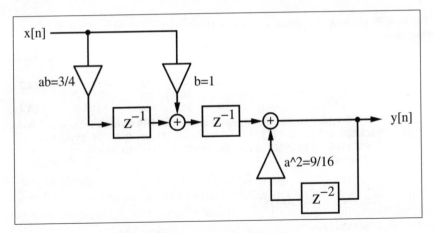

Fig. 4.14. Lossy integrator with look-ahead arithmetic.

$$y[n+1] = ay[n] + bx[n].$$ (4.10)

The output of the first-order system, namely $y[n+1]$, can be computed using a look-ahead methodology by substituting $y[n+1]$ into the differential equation for $y[n+2]$. That is

$$y[n+2] = ay[n+1] + bx[n+1] = a^2y[n] + abx[n] + bx[n+1].$$ (4.11)

The equivalent system is shown in Fig. 4.14.

This concept can be generalized by applying the look-ahead transform for $(S-1)$ steps, resulting in:

$$y[n+S] = a^S y[n] + \underbrace{\sum_{k=0}^{S-1} a^k bx[n+S-1-k]}_{(\eta)}.$$ (4.12)

It can be seen that the term (η) defines an FIR filter having coefficients $\{b, ab, a^2b, \ldots, a^{S-1}b\}$, that can be pipelined using the pipelining techniques presented in Chap. 3 (i.e., pipelined multiplier and pipelined adder trees). The recursive part of (4.12) can now also be implemented with an S-stage pipelined multiplier for the coefficient a^S. We will demonstrate the look-ahead design with the following example.

Example 4.3: Lossy Integrator II

Consider again the lossy integrator from Example 4.1 (p. 216), but now with look-ahead. Figure 4.14 shows the look-ahead lossy integrator, which is a combination of a nonrecursive part (i.e., FIR filter for x), and a recursive part with delay 2 and coefficient 9/16.

$$y[n+2] = \frac{3}{4}y[n+1] + x[n+1] = \frac{3}{4}\left(\frac{3}{4}y[n] + x[n]\right) + x[n+1]$$

$$= \frac{9}{16}y[n] + \frac{3}{4}x[n] + x[n+1]. \tag{4.13}$$

$$y[n] = \frac{9}{16}y[n-2] + \frac{3}{4}x[n-2] + x[n-1] \tag{4.14}$$

$$\tag{4.15}$$

The VHDL code[2] shown below, implements the IIR filter in look-ahead form.

```
PACKAGE n_bit_int IS              -- User-defined type
  SUBTYPE BITS15 IS INTEGER RANGE -2**14 TO 2**14-1;
END n_bit_int;

LIBRARY work;
USE work.n_bit_int.ALL;

LIBRARY ieee;
USE ieee.std_logic_1164.ALL;
USE ieee.std_logic_arith.ALL;

ENTITY iir_pipe IS
  PORT ( x_in  : IN   BITS15;    -- Input
         y_out : OUT  BITS15;    -- Result
         clk   : IN   STD_LOGIC);
END iir_pipe;

ARCHITECTURE fpga OF iir_pipe IS

  SIGNAL  x, x3, sx, y, y9 : BITS15 := 0;

BEGIN

  PROCESS   -- Use FFs for input, output and pipeline stages
  BEGIN
    WAIT UNTIL clk = '1';
    x   <= x_in;
    x3  <= x / 2 + x / 4;   -- Compute x*3/4
    sx  <= x + x3; -- Sum of x element i.e. output FIR part
    y9  <= y / 2 + y / 16;  -- Compute y*9/16
    y   <= sx + y9;         -- Compute output
  END PROCESS;

  y_out <= y ;     -- Connect register y to output pins

END fpga;
```

The pipelined adder and multiplier in this example are implemented in two steps. In the first stage, $\frac{9}{16}y[n]$ is computed. In the second stage, $x[n+1] + \frac{3}{4}x[n]$ and $\frac{9}{16}y[n]$ are added. The design uses 124 LEs, no embedded multiplier and has a 207.08 MHz Registered Performance. The response of the filter to an impulse of amplitude 1000 is shown in Fig. 4.15.

4.3

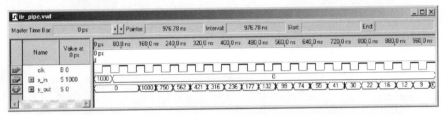

Fig. 4.15. VHDL simulation of impulse response of the look-ahead lossy integrator.

Comparing the look-ahead scheme with the 62 LEs and 160.69 MHz solution reported in Example 4.1 (p. 216), we find that look-ahead pipelining requires many more resources, but attains a speed-up of about 30%. The comparison of the two filter's response to the impulse with amplitude 1000 shown in Fig. 4.3 (p. 218) and Fig. 4.15 reveals that the look-ahead scheme has an additional overall delay, and that the quantization effect differs by a ± 2 amount between the two methodologies.

An alternative design approach, using a standard logic vector data type and LPM_ADD_SUB megafunctions, is discussed in Exercise 4.7 (p. 241). The second approach will produce longer VHDL code, but will have the benefit of direct control at the bit level of the sign extension and multiplier.

4.4.2 Clustered and Scattered Look-Ahead Pipelining

Clustered and scattered look-ahead pipelining schemes add self-canceling poles and zeros to the design to facilitate pipelining of the recursive portion of the filter. In the *clustered* method, additional pole/zeros are introduced in such a way that in the denominator of the transfer function the coefficients for $z^{-1}, z^{-2}, \ldots, z^{-(S-1)}$ become zero. The following example shows clustering for a second-order filter.

Example 4.4: Clustering Method

A second-order transfer function is assumed to have a pole at $1/2$ and $3/4$ and a transfer function given by:

$$F(z) = \frac{1}{1 - 1.25z^{-1} + 0.375z^{-2}} = \frac{1}{(1 - 0.5z^{-1})(1 - 0.75z^{-1})}. \quad (4.16)$$

Adding a canceling pole/zero at $z = -1.25$ results in a new transfer function

$$F(z) = \frac{1 + 1.25z^{-1}}{1 - 1.1875z^{-2} + 0.4688z^{-3}}. \quad (4.17)$$

The recursive part of the filter can now be implemented with an additional pipeline stage.

4.4

The problem with clustering is that the cancelled pole/zero pair may lie outside the unit circle, as is the case in the previous example (i.e., $z_\infty =$

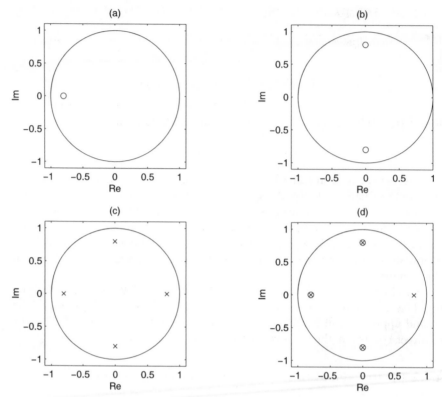

Fig. 4.16. Pole/zero plot for scattered look-ahead first-order IIR filter.
(a) $F_1(z) = (1 + az^{-1})$. **(b)** $F_2(z) = 1 + a^2z^{-2}$. **(c)** $F_3(z) = 1/(1 - a^4z^{-4})$.
(d) $F(z) = \prod_k F_k(z) = (1 + az^{-1})(1 + a^2z^{-2})/(1 - a^4z^{-4}) = 1/(1 - az^{-1})$.

-1.25). This introduces instability into the design if the pole/zero annihilating is not perfect. In general, a second-order system with poles at r_1, r_2 and with one extra canceling pair, has a pole location at $-(r_1 + r_2)$, which lies outside the unit circle for $|r_1 + r_2| > 1$. Soderstrand et al. [97], have described a stable clustering method, which in general introduces more than one canceling pole/zero pair.

The *scattered look-ahead* approach does not introduce stability problems. It introduces $(S - 1)$ canceling pole/zero pairs located at $z_k = pe^{j\pi k/S}$, for an original filter with a pole located at p. The denominator of the transfer function has, as a result, only zero coefficients associated with the terms z^0, z^S, z^{-2S}, etc.

Example 4.5: Scattered Look-Ahead Method

Consider implementing a second-order system having poles located at $z_{\infty 1} = 0.5$ and $z_{\infty 2} = 0.75$ with two additional pipeline stages. A second-order transfer function of a filter with poles at $1/2$ and $3/4$ has the transfer function

$$F(z) = \frac{1}{1 - 1.25z^{-1} + 0.375z^{-2}} = \frac{1}{(1 - 0.5z^{-1})(1 - 0.75z^{-1})}. \quad (4.18)$$

Note that in general a pole/zero pair at p and p^* results in a transfer function of

$$(1 - pz^{-1})(1 - p^*z^{-1}) = 1 - (p + p^*)z^{-1} + rr^*z^{-2}$$

and in particular with $p = r \times \exp(j2\pi/3)$ it follows that

$$(1 - pz^{-1})(1 - p^*z^{-1}) = 1 - 2r\cos(2\pi/3)z^{-1} + r^2z^{-2}$$
$$= 1 + rz^{-1} + r^2z^{-2}.$$

The scattered look-ahead introduces two additional pipeline stages by adding pole/zero pairs at $0.5e^{\pm j2\pi/3}$ and $0.75e^{\pm j2\pi/3}$. Adding a canceling pole/zero at this location results in

$$F(z) = \frac{1}{1 - 1.25z^{-1} + 0.375z^{-2}}$$
$$\times \frac{(1 + 0.5z^{-1} + 0.25z^{-2})(1 + .75z^{-1} + 0.5625z^{-2})}{(1 + 0.5z^{-1} + 0.25z^{-2})(1 + .75z^{-1} + 0.5625z^{-2})}$$
$$= \frac{1 + 1.25z^{-1} + 1.1875z^{-2} + 0.4687z^{-3} + 0.1406z^{-4}}{1 - 0.5469z^{-3} + 0.0527z^{-6}}$$
$$= \frac{512 + 640z^{-1} + 608z^{-2} + 240z^{-3} + 72z^{-4}}{512 - 280z^{-3} + 27z^{-6}},$$

and the recursive part can be implemented with two additional pipeline stages.

4.5

It is interesting to note that for a first-order IIR system, clustered and scattered look-ahead methods result in the same pole/zero canceling pair lying on a circle around the origin with angle differences $2\pi/S$. The nonrecursive part can be realized with a "power-of-two decomposition" according to

$$(1 + az^{-1})(1 + a^2z^{-2})(1 + a^4z^{-4})\cdots. \quad (4.19)$$

Figure 4.16 shows such a pole/zero representation for a first-order section, which enables an implementation with four pipeline stages in the recursive part.

4.4.3 IIR Decimator Design

Martinez and Parks [99] have introduced, in the context of decimation filters (see Chap. 5, p. 254), a filter design algorithm based on the minimax method. The resulting transfer function satisfies

$$F(z) = \frac{\sum\limits_{l=0}^{L} b[l]z^{-l}}{1 - \sum\limits_{n=0}^{N/S} a[n]z^{-nS}}. \quad (4.20)$$

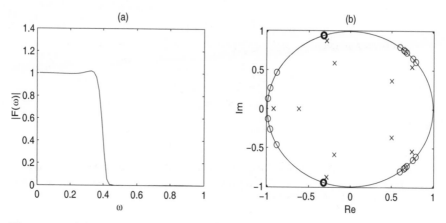

Fig. 4.17. (a) Transfer function, and (b) pole/zero distribution of a 37-order Martinez–Parks IIR filter with $S = 5$.

That is, only every other S coefficient in the denominator is nonzero. In this case, the recursive part (i.e., the denominator) can be pipelined with S stages. It has been found that in the resulting pole/zero distribution, all zeros are on the unit circle, as is usual for an elliptic filter, while the poles lie on circles, whose main axes have a difference in angle of $2\pi/S$, as shown in Fig. 4.17b.

4.4.4 Parallel Processing

In a parallel-processing filter implementation [100], P parallel IIR paths are formed, each running at a $1/P$ input sampling rate. They are combined at the output using a multiplexer, as shown in Fig. 4.18. Because a multiplexer, in general, will be faster than a multiplier and/or adder, the parallel approach will be faster. Furthermore, each path P has a factor of P more time to compute its assigned output.

To illustrate, consider again a first-order system and $P = 2$. The look-ahead scheme, as in (4.11)

$$y[n + 2] = ay[n + 1] + x[n + 1] = a^2y[n] + ax[n] + x[n + 1] \qquad (4.21)$$

is now split into even $n = 2k$ and odd $n = 2k - 1$ output sequences, obtaining

$$y[n + 2] = \begin{cases} y[2k + 2]=a^2y[2k] + ax[2k] + x[2k + 1] \\ y[2k + 1]=a^2y[2k - 1] + ax[2k - 1] + x[2k] \end{cases}, \qquad (4.22)$$

where $n, k \in \mathbb{Z}$. The two equations are the basis for the following parallel IIR filter FPGA implementation.

Example 4.6: Lossy Integrator III

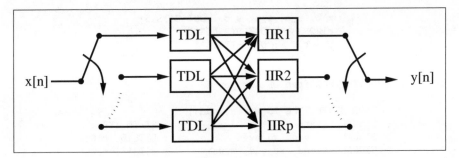

Fig. 4.18. Parallel IIR implementation. The tapped delay lines (TDL) run with a $1/p$ input sampling rate.

Consider implementing a parallel lossy integrator, with $a = 3/4$, as an extension to the methods presented in Examples 4.1 (p. 216) and 4.3 (p. 231). A two-channel parallel lossy integrator, which is a combination of two non-recursive parts (i.e., an FIR filter for x), and two recursive parts with delay 2 and coefficient 9/16, is shown in Fig. 4.19. The VHDL code[3] shown below implements the design.

```
    PACKAGE n_bit_int IS              -- User-defined type
       SUBTYPE BITS15 IS INTEGER RANGE -2**14 TO 2**14-1;
    END n_bit_int;

    LIBRARY work;
    USE work.n_bit_int.ALL;

    LIBRARY ieee;
    USE ieee.std_logic_1164.ALL;
    USE ieee.std_logic_arith.ALL;

    ENTITY iir_par IS                       ------> Interface
       PORT ( clk, reset : IN  STD_LOGIC;
              x_in       : IN  BITS15;
```

---

[3] The equivalent Verilog code iir_par.v for this example can be found in Appendix A on page 693. Synthesis results are shown in Appendix B on page 731.

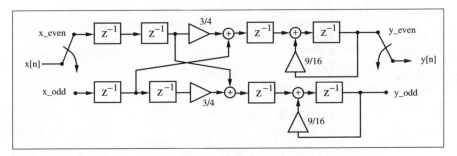

Fig. 4.19. Two-path parallel IIR filter implementation.

```
        x_e, x_o, y_e, y_o : OUT BITS15;
        clk2      : OUT STD_LOGIC;
        y_out     : OUT BITS15);
END iir_par;

ARCHITECTURE fpga OF iir_par IS

  TYPE STATE_TYPE IS (even, odd);
  SIGNAL  state                    : STATE_TYPE;
  SIGNAL  x_even, xd_even          : BITS15 := 0;
  SIGNAL  x_odd, xd_odd, x_wait    : BITS15 := 0;
  SIGNAL  y_even, y_odd, y_wait, y : BITS15 := 0;
  SIGNAL  sum_x_even, sum_x_odd    : BITS15 := 0;
  SIGNAL  clk_div2                 : STD_LOGIC;

BEGIN

  Multiplex: PROCESS (reset, clk) --> Split x into even and
  BEGIN              -- odd samples; recombine y at clk rate
    IF reset = '1' THEN            -- asynchronous reset
      state <= even;
    ELSIF rising_edge(clk) THEN
    CASE state IS
      WHEN even =>
        x_even <= x_in;
        x_odd <= x_wait;
        clk_div2 <= '1';
        y <= y_wait;
        state <= odd;
      WHEN odd =>
        x_wait <= x_in;
        y <= y_odd;
        y_wait <= y_even;
        clk_div2 <= '0';
        state <= even;
      END CASE;
      END IF;
  END PROCESS Multiplex;

  y_out <= y;
  clk2  <= clk_div2;
  x_e <= x_even; -- Monitor some extra test signals
  x_o <= x_odd;
  y_e <= y_even;
  y_o <= y_odd;

  Arithmetic: PROCESS
  BEGIN
    WAIT UNTIL clk_div2 = '0';
    xd_even <= x_even;
    sum_x_even <= (xd_even * 2 + xd_even) /4 + x_odd;
    y_even <= (y_even * 8 + y_even )/16 + sum_x_even;
    xd_odd <= x_odd;
```

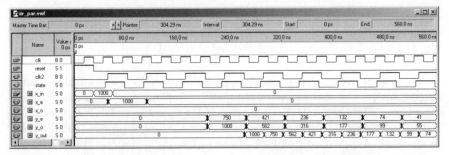

Fig. 4.20. VHDL simulation of the response of the parallel IIR filter to an impulse 1000.

```
sum_x_odd <= (xd_odd * 2 + xd_odd) /4 + xd_even;
y_odd   <= (y_odd * 8 + y_odd) / 16 + sum_x_odd;
END PROCESS Arithmetic;

END fpga;
```

The design is realized with two PROCESS statements. In the first, PROCESS Multiplex, x is split into even and odd indexed parts, and the output y is recombined at the clk rate. In addition, the first PROCESS statement generates the second clock, running at clk/2. The second block implements the filter's arithmetic according to (4.22). The design uses 268 LEs, no embedded multiplier, and has a 168.12 MHz Registered Performance. The simulation is shown in Fig. 4.20.

4.6

The disadvantage of the parallel implementation, compared with the other methods presented, is the relatively high implementation cost of 268 LEs.

4.4.5 IIR Design Using RNS

Because the residue number system (RNS) uses an intrinsically short wordlength, it is an excellent candidate to implement fast (recursive) IIR filters. In

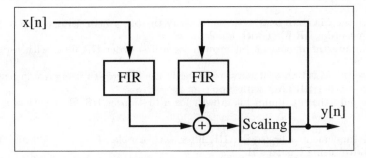

Fig. 4.21. RNS implementation of IIR filters using two FIR sections and scaling.

a typical IIR-RNS design, a system is implemented as a collection of recursive and nonrecursive systems, each defined in terms of an FIR structure (see Fig. 4.21). Each FIR may be implemented in RNS-DA, using a quarter-square multiplier, or in the index domain, as developed in Chap. 2 (p. 67).

For a stable filter, the recursive part should be scaled to control dynamic range growth. The scaling operation may be implemented with mixed radix conversion, Chinese remainder theorem (CRT), or the $\epsilon-$CRT method. For high-speed designs, it is preferable to add an additional pipeline delay based on the clustered or scattered look-ahead pipelining technique [39, Sect. 4-2]. An RNS recursive filter design will be developed in detail in Sect. 5.3. It will be seen that RNS design will improve speed from 50 MHz to more than 70 MHz.

Exercises

Note: If you have no prior experience with the Quartus II software, refer to the case study found in Sect. 1.4.3, p. 29. If not otherwise noted use the EP2C35F672C6 from the Cyclone II family for the Quartus II synthesis evaluations.

4.1: A filter has the following specification: sampling frequency 2 kHz; passband 0–0.4 kHz, stopband 0.5–1 kHz; passband ripple, 3 dB, and stopband ripple, 48 dB. Use the MATLAB software and the "Interactive Lowpass Filter Design" demo from the Signal Processing Toolbox for the filter design.
(a1) Design a Butterworth filter (called BUTTER).
(a2) Determine the filter length and the absolute ripple in the passband.
(b1) Design a Chebyshev type I filter (called CHEBY1).
(b2) Determine the filter length and the absolute ripple in the passband.
(c1) Design a Chebyshev type II filter (called CHEBY2).
(c2) Determine the filter length and the absolute ripple in the passband.
(d1) Design an elliptic filter (called ELLIP).
(d2) Determine the filter length and the absolute ripple in the passband.

4.2: (a) Compute the maximum bit growth for a first-order IIR filter with a pole at $z_\infty = 3/4$.
(a2) Use the MATLAB or C software to verify the bit growth using a step response of the first-order IIR filter with a pole at $z_\infty = 3/4$.
(b) Compute the maximum bit growth for a first-order IIR filter with a pole at $z_\infty = 3/8$.
(b2) Use the MATLAB or C software to verify the bit growth using a step response of the first-order IIR filter with a pole at $z_\infty = 3/8$.
(c) Compute the maximum bit growth for a first-order IIR filter with a pole at $z_\infty = p$.

4.3: (a) Implement a first-order IIR filter with a pole at $z_{\infty 0} = 3/8$ and 12-bit input width, using Quartus II.
(b) Determine the the `Registered Performance` and the used resources (LEs, multipliers, and M4Ks).

(c) Simulate the design with an input impulse of amplitude 100.
(d) Compute the maximum bit growth for the filter.
(e) Verify the result from (d) with a simulation of the step response with amplitude 100.

4.4: (a) Implement a first-order IIR filter with a pole at $z_{\infty 0} = 3/8$, 12-bit input width, and a look-ahead of one step, using Quartus II.
(b) Determine the `Registered Performance` and the used resources (LEs, multipliers, and M4Ks).
(c) Simulate the design with an input impulse of amplitude 100.

4.5: (a) Implement a first-order IIR filter with a pole at $z_{\infty 0} = 3/8$, 12-bit input width, and a parallel design with two paths, using Quartus II.
(b) Determine the `Registered Performance` and the used resources (LEs, multipliers, and M4Ks).
(c) Simulate the design with an input impulse of amplitude 100.

4.6: (a) Implement a first-order IIR filter as in Example 4.1 (p. 216), using a 15-bit `std_logic_vector`, and implement the adder with two `lpm_add_sub` megafunctions, using Quartus II.
(b) Determine the `Registered Performance` and the used resources (LEs, multipliers, and M4Ks).
(c) Simulate the design with an input impulse of amplitude 1000, and compare the results to Fig. 4.3 (p. 218).

4.7: (a) Implement a first-order pipelined IIR filter from Example 4.3 (p. 231) using a 15-bit `std_logic_vector`, and implement the adder with four `lpm_add_sub` megafunctions, using Quartus II.
(b) Determine the `Registered Performance` and the used resources (LEs, multipliers, and M4Ks).
(c) Simulate the design with an input impulse of amplitude 1000, and compare the results to Fig. 4.15 (p. 233).

4.8: Shajaan and Sorensen have shown that an IIR Butterworth filter can be efficiently designed by implementing the coefficients as signed-power-of-two (SPT) values [101]. The transfer function of a cascade filter with N sections

$$F(z) = \prod_{l=1}^{N} S[l] \frac{b[l,0] + b[l,1]z^{-1} + b[l,2]z^{-2}}{a[l,0] + a[l,1]z^{-1} + a[l,2]z^{-2}} \tag{4.23}$$

should be implemented using the second-order sections shown in Fig. 4.11 (p. 226). A tenth-order filter, as discussed in Example 4.2 (p. 227), can be realized with the following SPT filter coefficients [101]:

l	$S[l]$	$1/a[l,0]$	$a[l,1]$	$a[l,2]$
1	2^{-1}	1	$-1 - 2^{-4}$	$1 - 2^{-2}$
2	2^{-1}	2^{-1}	$-1 - 2^{-1}$	$1 - 2^{-5}$
3	2^{-1}	2^{-1}	$-1 - 2^{-1}$	$2^{-1} + 2^{-5}$
4	1	2^{-1}	$-1 - 2^{-2}$	$2^{-2} + 2^{-5}$
5	2^{-1}	2^{-1}	$-1 - 2^{-1}$	$2^{-2} + 2^{-4}$

We choose $b[0] = b[2] = 0.5$ and $b[1] = 1$ because the zeros of the Butterworth filter are all at $z = -1$.
(a) Compute and plot the transfer function of the first BiQuad and the complete

filter.

(b) Implement and simulate the first BiQuad for 8-bit inputs.

(c) Build and simulate the five-stage filter with Quartus II.

(d) Determine the **Registered Performance** and the used resources (LEs, multipliers, and M4Ks) of the filter.

4.9: (a) Design a tenth-order order lowpass Butterworth filter using MATLAB with a cut-off frequency of 0.3 the sampling frequency, i.e, [b,a]=butter(10,0.3).

(b) Plot the transfer function using freqz() for ∞ bits (i.e., real coefficients) and 12-bit fractional bits. What is the stopband suppression in dB at 0.5 of the Nyquist frequency?

Hint: round(a*2^B)/2^B has B fractional bits.

(c) Generate a pole/zero plot using zplane() for ∞ bits and 12 fractional bits.

(d) Plot the impulse response for an impulse of amplitude 100 using filter() and stem() for coefficients with 12-bit fractional bits. Also plot the response to impulse of amplitude 100 of the recursive part only, i.e., set the FIR part to b=[1];

(e) For the 12 fractional bit filter determine using the csd3e.exe program from the CD the CSD representation for all coefficients a and b from part (a).

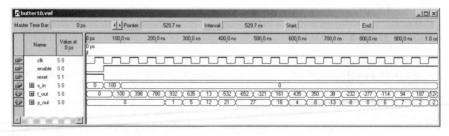

Fig. 4.22. IIR Butterworth testbench for Exercise 4.10.

4.10: (a) Using the results from Exercise 4.9 develop the VHDL code for the tenth-order Butterworth filter for 8-bit inputs. As internal data format use a 14.12 bit format, i.e., 14 integer and 12 fractional bits. You need to scale the input and output by 2^{12} and use an internal 26-bit format. Use the direct form II, i.e., Fig. 4.9 (p. 225) for your design. Make sure that the transfer function of Fig. 4.9 and the MATLAB representation of the transfer functions match.

Recommendation: you can start with the recursive part first and try to match the simulation from Exercise 4.9(d). Then add the nonrecursive part.

(b) Add an active-high enable and active-high asynchronous reset to the design.

(c) Try to match the simulation from Exercise 4.9(d) shown in simulation Fig.4.22, where t_out is the output of the recursive part.

(d) For the device EP2C35F672C6 from the Cyclone II family determine the resources (LEs, multipliers, and M4Ks) and **Registered Performance**.

4.11: (a) Design the PREP benchmark 5 shown in Fig. 4.23a with the Quartus II software. The design has a 4×4 unsigned array multiplier followed by an 8-bit accumulator. If mac = '1' accumulation is performed otherwise the adder output s shows the multiplier output without adding q. rst is an asynchronous reset and the 8-bit register is positive-edge triggered via clk, see the simulation in Fig. 4.23c

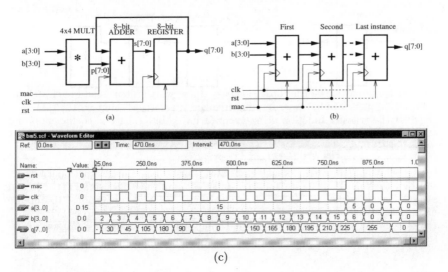

Fig. 4.23. PREP benchmark 5. (**a**) Single design. (**b**) Multiple instantiation. (**c**) Testbench to check function.

for the function test.

(**b**) Determine the `Registered Performance` and the used resources (LEs, multipliers, and M4Ks) for a single copy. Compile the HDL file with the synthesis `Optimization Technique` set to `Speed`, `Balanced` or `Area`; this can be found in the `Analysis & Synthesis Settings` section under `EDA Tool Settings` in the `Assignments` menu. Which synthesis options are optimal for size or `Registered Performance`?

Select one of the following devices:

(**b1**) EP2C35F672C6 from the Cyclone II family

(**b2**) EPF10K70RC240-4 from the Flex 10K family

(**b3**) EPM7128LC84-7 from the MAX7000S family

(**c**) Design the multiple instantiation for benchmark 5 as shown in Fig. 4.23b.

(**d**) Determine the `Registered Performance` and the used resources (LEs, multipliers, and M2Ks/M4Ks) for the design with the maximum number of instantiations of PREP benchmark 5. Use the optimal synthesis option you found in (b) for the following devices:

(**d1**) EP2C35F672C6 from the Cyclone II family

(**d2**) EPF10K70RC240-4 from the Flex 10K family

(**d3**) EPM7128LC84-7 from the MAX7000S family

4.12: (**a**) Design the PREP benchmark 6 shown in Fig. 4.24a with the Quartus II software. The design is positive-edge triggered via `clk` and includes a 16-bit accumulator with an asynchronous reset `rst`; see the simulation in Fig. 4.24c for the function test.

(**b**) Determine the `Registered Performance` and the used resources (LEs, multipliers, and M2Ks/M4Ks) for a single copy. Compile the HDL file with the synthesis `Optimization Technique` set to `Speed`, `Balanced` or `Area`; this can be found in the `Analysis & Synthesis Settings` section under `EDA Tool Settings` in the `Assignments` menu. Which synthesis options are optimal for size or `Registered`

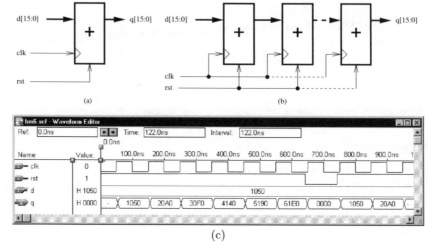

Fig. 4.24. PREP benchmark 6. **(a)** Single design. **(b)** Multiple instantiation. **(c)** Testbench to check function.

Performance?
Select one of the following devices:
(b1) EP2C35F672C6 from the Cyclone II family
(b2) EPF10K70RC240-4 from the Flex 10K family
(b3) EPM7128LC84-7 from the MAX7000S family
(c) Design the multiple instantiation for benchmark 6 as shown in Fig. 4.24b.
(d) Determine the **Registered Performance** and the used resources (LEs, multipliers, and M2Ks/M4Ks) for the design with the maximum number of instantiations of PREP benchmark 6. Use the optimal synthesis option you found in (b) for the following devices:
(d1) EP2C35F672C6 from the Cyclone II family
(d2) EPF10K70RC240-4 from the Flex 10K family
(d3) EPM7128LC84-7 from the MAX7000S family

5. Multirate Signal Processing

Introduction

A frequent task in digital signal processing is to adjust the sampling rate according to the signal of interest. Systems with different sampling rates are referred to as *multirate* systems. In this chapter, two typical examples will illustrate decimation and interpolation in multirate DSP systems. We will then introduce polyphase notation, and will discuss some efficient decimator designs. At the end of the chapter we will discuss filter banks and a quite new, highly celebrated addition to the DSP toolbox: wavelet analysis.

5.1 Decimation and Interpolation

If, after A/D conversion, the signal of interest can be found in a small frequency band (typically, lowpass or bandpass), then it is reasonable to filter with a lowpass or bandpass filter and to reduce the sampling rate. A narrow filter followed by a downsampler is usually referred to as a *decimator* [79]. [1] The filtering, downsampling, and the effect on the spectrum is illustrated in Fig. 5.1.

We can reduce the sampling rate up to the limit called the "Nyquist rate," which says that the sampling rate must be higher than the bandwidth of the signal, in order to avoid aliasing. Aliasing is demonstrated in Fig. 5.2 for a lowpass signal. Aliasing is irreparable, and should be avoided at all cost.

For a bandpass signal, the frequency band of interest must fall within an *integer band*. If f_s is the sampling rate, and R is the desired downsampling factor, then the band of interest must fall between

$$k\frac{f_\mathrm{s}}{2R} < f < (k+1)\frac{f_\mathrm{s}}{2R} \qquad k \in \mathbb{N}. \tag{5.1}$$

If it does not, there may be aliasing due to "copies" from the negative frequency bands, although the sampling rate may still be higher than the Nyquist rate, as shown in Fig. 5.3.

Increasing the sampling rate can be useful, in the D/A conversion process, for example. Typically, D/A converters use a sample-and-hold of first-order

---

[1] Some authors refer to a downsampler as a decimator.

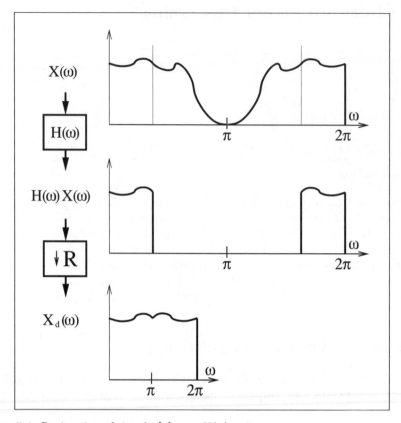

Fig. 5.1. Decimation of signal $x[n]$ $\circ\!\!-\!\!\bullet$ $X(\omega)$.

at the output, which produces a step-like output function. This can be compensated for with an analog $1/\text{sinc}(x)$ compensation filter, but most often a digital solution is more efficient. We can use, in the digital domain, an *expander* and an additional filter to get the desired frequency band. We note, from Fig. 5.4, that the introduced zeros produce an extra copy of the baseband spectrum that must first be removed before the signal can be processed with the D/A converter. The much smoother output signal of such an *interpolation*[2] can be seen in Fig. 5.5.

5.1.1 Noble Identities

When manipulating signal flow graphs of multirate systems it is sometimes useful to rearrange the filter and downsampler/expander, as shown in Fig. 5.6. These are the so-called "Noble" relations [102]. For the decimator, it follows

$$(\downarrow R)\, F(z) = F(z^R)\, (\downarrow R), \tag{5.2}$$

---

[2] Some authors refer to the expander as an interpolator.

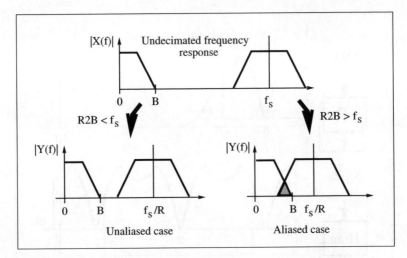

Fig. 5.2. Unaliased and aliased decimation cases.

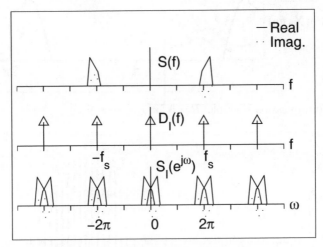

Fig. 5.3. Integer band violation (©1995 VDI Press [4]).

i.e., if the downsampling is done first, we can reduce the filter length $F(z^R)$ by a factor of R.

For the interpolator, the Noble relation is defined as

$$F(z)\,(\uparrow R) = (\uparrow R)\,F(z^R), \tag{5.3}$$

i.e., in an interpolation putting the filter before the expander results in an R-times shorter filter.

These two identities will become very useful when we discuss polyphase implementation in Sect. 5.2 (p. 249).

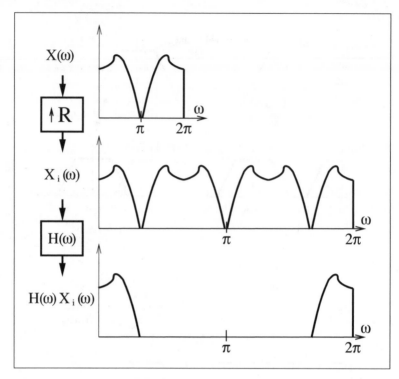

Fig. 5.4. Interpolation example. $R = 3$ for $x[n] \circ\!\!-\!\!\bullet X(\omega)$.

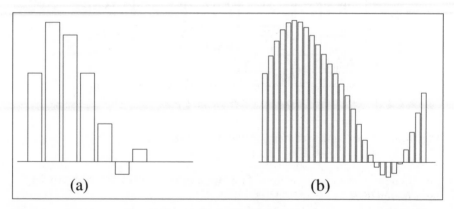

Fig. 5.5. D/A conversion. **(a)** Low oversampling, high degradation. **(b)** High oversampling, low degradation.

5.1.2 Sampling Rate Conversion by Rational Factor

If the input and output rate of a multirate system is *not* an integer factor, then a rational change factor R_1/R_2 in the sampling rate can be used. More

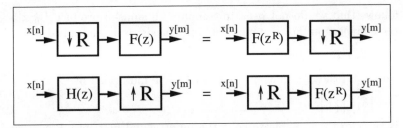

Fig. 5.6. Equivalent multirate systems (Noble relation).

precisely, we first use an interpolator to increase the sampling rate by R_1, and then use a decimator to downsample by R_2. Since the filters used for interpolation and decimation are both lowpass filters, it follows, from the upper configuration in Fig. 5.7, that we only need to implement the lowpass filter with the smaller passband frequency, i.e.,

$$f_p = \min\left(\frac{\pi}{R_1}, \frac{\pi}{R_2}\right). \tag{5.4}$$

This is graphically interpreted in the lower configuration of Fig. 5.7. We will discuss later in Sect. 5.6, p. 280 different design options of this system.

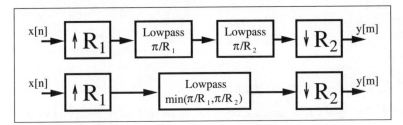

Fig. 5.7. Noninteger decimation system. (*upper*) Cascade of an interpolator and a decimator. (*lower*) Result combining the lowpass filters.

5.2 Polyphase Decomposition

Polyphase decomposition is very useful when implementing decimation or interpolation in IIR or FIR filter and filter banks. To illustrate this, consider the polyphase decomposition of an FIR decimation filter. If we add downsampling by a factor of R to the FIR filter structure shown in Fig. 3.1 (p. 166), we find that we only need to compute the outputs $y[n]$ at time instances

$$y[0], y[R], y[2R], \dots . \tag{5.5}$$

It follows that we do not need to compute all sums-of-product $f[k]x[n-k]$ of the convolution. For instance, $x[0]$ only needs to be multiplied by

$$f[0], f[R], f[2R], \ldots .$$ (5.6)

Besides $x[0]$, these coefficients only need to be multiplied by

$$x[R], x[2R], \ldots .$$ (5.7)

It is therefore reasonable to split the input signal first into R separate sequences according to

$$x[n] = \sum_{r=0}^{R-1} x_r[n]$$
$$x_0[n] = \{x[0], x[R], \ldots\}$$
$$x_1[n] = \{x[1], x[R+1], \ldots\}$$
$$\vdots$$
$$x_{R-1}[n] = \{x[R-1], x[2R-1], \ldots\}$$

and also to split the filter $f[n]$ into R sequences

$$f[n] = \sum_{r=0}^{R-1} f_r[n]$$
$$f_0[n] = \{f[0], f[R], \ldots\}$$
$$f_1[n] = \{f[1], f[R+1], \ldots\}$$
$$\vdots$$
$$f_{R-1}[n] = \{f[R-1], f[2R-1], \ldots\}.$$

Figure 5.8 shows a decimator filter implemented using polyphase decomposition. Such a decimator can run R times faster than the usual FIR filter followed by a downsampler. The filters $f_r[n]$ are called polyphase filters, because they all have the same magnitude transfer function, but they are separated by a sample delay, which introduces a phase offset.

A final example illustrates the polyphase decomposition.

Example 5.1: Polyphase Decimator Filter

Consider a Daubechies length-4 filter with $G(z)$ and $R = 2$.

$$G(z) = \left((1 + \sqrt{3}) + (3 + \sqrt{3})z^{-1} + (3 - \sqrt{3})z^{-2} + (1 - \sqrt{3})z^{-3} \right) \frac{1}{4\sqrt{2}}$$

$$G(z) = 0.48301 + 0.8365z^{-1} + 0.2241z^{-2} - 0.1294z^{-3}.$$

Quantizing the filter to 8 bits of precision results in the following model:

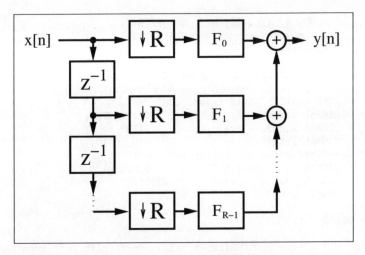

Fig. 5.8. Polyphase realization of a decimation filter.

$$G(z) = \left(124 + 214z^{-1} + 57z^{-2} - 33z^{-3}\right)/256$$
$$G(z) = G_0(z^2) + z^{-1}G_1(z^2)$$
$$= \underbrace{\left(\frac{124}{256} + \frac{57}{256}z^{-2}\right)}_{G_0(z^2)} + z^{-1}\underbrace{\left(\frac{214}{256} - \frac{33}{256}z^{-2}\right)}_{G_1(z^2)},$$

and it follows that
$$G_0(z) = \frac{124}{256} + \frac{57}{256}z^{-1} \qquad G_1(z) = \frac{214}{256} - \frac{33}{256}z^{-1}. \tag{5.8}$$

The following VHDL code[3] shows the polyphase implementation for DB4.

```
PACKAGE n_bits_int IS            -- User-defined types
  SUBTYPE BITS8 IS INTEGER RANGE -128 TO 127;
  SUBTYPE BITS9 IS INTEGER RANGE -2**8 TO 2**8-1;
  SUBTYPE BITS17 IS INTEGER RANGE -2**16 TO 2**16-1;
  TYPE ARRAY_BITS17_4 IS ARRAY (0 TO 3) of BITS17;
END n_bits_int;

LIBRARY work;
USE work.n_bits_int.ALL;

LIBRARY ieee;
USE ieee.std_logic_1164.ALL;
USE ieee.std_logic_arith.ALL;
USE ieee.std_logic_signed.ALL;

ENTITY db4poly IS                      ------> Interface
  PORT (clk, reset      : IN  STD_LOGIC;
        x_in            : IN  BITS8;
        clk2            : OUT STD_LOGIC;
```

---

[3] The equivalent Verilog code db4poly.v for this example can be found in Appendix A on page 697. Synthesis results are shown in Appendix B on page 731.

```
          x_e, x_o, g0, g1 : OUT BITS17;
          y_out             : OUT BITS9);
END db4poly;

ARCHITECTURE fpga OF db4poly IS

  TYPE STATE_TYPE IS (even, odd);
  SIGNAL state                 : STATE_TYPE;
  SIGNAL x_odd, x_even, x_wait : BITS8 := 0;
  SIGNAL clk_div2              : STD_LOGIC;
  -- Arrays for multiplier and taps:
  SIGNAL r  : ARRAY_BITS17_4 := (0,0,0,0);
  SIGNAL x33, x99, x107      : BITS17;
  SIGNAL y      : BITS17 := 0;

BEGIN

  Multiplex: PROCESS(reset, clk) ----> Split into even and
  BEGIN                          -- odd samples at clk rate
    IF reset = '1' THEN          -- Asynchronous reset
      state <= even;
    ELSIF rising_edge(clk) THEN
      CASE state IS
        WHEN even =>
          x_even <= x_in;
          x_odd  <= x_wait;
          clk_div2 <= '1';
          state <= odd;
        WHEN odd =>
          x_wait <= x_in;
          clk_div2 <= '0';
          state <= even;
      END CASE;
    END IF;
  END PROCESS Multiplex;


  AddPolyphase: PROCESS (clk_div2,x_odd,x_even,x33,x99,x107)
  VARIABLE m  : ARRAY_BITS17_4 ;
  BEGIN
-- Compute auxiliary multiplications of the filter
    x33  <= x_odd * 32 + x_odd;
    x99  <= x33 * 2 + x33;
    x107 <= x99 + 8 * x_odd;
-- Compute all coefficients for the transposed filter
    m(0) := 4 * (32 * x_even - x_even);        -- m[0] = 127
    m(1) := 2 * x107;                          -- m[1] = 214
    m(2) := 8 * (8 * x_even - x_even) + x_even;-- m[2] = 57
    m(3) := x33;                               -- m[3] = -33
------> Compute the filters and infer registers
    IF clk_div2'event and (clk_div2 = '0') THEN
------------ Compute filter G0
      r(0) <=  r(2) + m(0);    -- g[0] = 127
      r(2) <=  m(2);           -- g[2] = 57
```

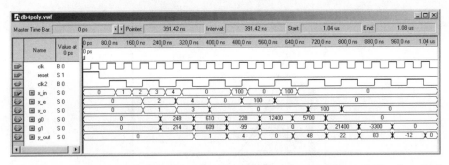

Fig. 5.9. VHDL simulation of the polyphase implementation of the length-4 Daubechies filter.

```
------------ Compute filter G1
    r(1) <=  -r(3) + m(1);    -- g[1] = 214
    r(3) <=  m(3);            -- g[3] = -33
------------ Add the polyphase components
    y <= r(0) + r(1);
  END IF;
END PROCESS AddPolyphase;


x_e <= x_even; -- Provide some test signal as outputs
x_o <= x_odd;
clk2 <= clk_div2;
g0 <= r(0);
g1 <= r(1);


y_out <= y / 256; -- Connect to output

END fpga;
```

The first PROCESS is the FSM, which includes the control flow and the splitting of the input stream at the sampling rate into even and odd samples. The second PROCESS includes the reduced adder graph (RAG) multiplier, and the last PROCESS hosts the two filters in a transposed structure. Although the output is scaled, there is potential growth by the amount $\sum |g_k| = 1.673 < 2^1$. Therefore the output y_out was chosen to have an additional guard bit. The design uses 173 LEs, no embedded multiplier, and has a 136.65 MHz **Registered Performance**.

A simulation of the filter is shown in Fig. 5.9. The first four input samples are a triangle function to demonstrate the splitting into even and odd samples. Impulses with an amplitude of 100 are used to verify the coefficients of the two polyphase filters. Note that the filter is *not* shift invariant. $\boxed{\text{5.1}}$

From the VHDL simulation shown in Fig. 5.9, it can be seen that such a decimator is no longer shift invariant, resulting in a technically nonlinear system. This can be validated by applying a single impulse. Initializing at an even-indexed sample, the response is $G_0(z)$, while for an odd-indexed sample, the response is $G_1(z)$.

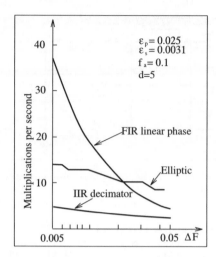

Fig. 5.10. Comparison of computational effort for decimators $\Delta F = f_\mathrm{p} - f_\mathrm{s}$.

5.2.1 Recursive IIR Decimator

It is also possible to apply polyphase decomposition to recursive filters and to get the speed benefit, if we follow the idea from Martinez and Parks [99], in the transfer function

$$F(z) = \frac{\sum\limits_{l=0}^{L-1} a[l] z^{-l}}{1 - \sum\limits_{l=1}^{K-1} b[l] z^{-lR}}. \tag{5.9}$$

i.e., the recursive part has only each Rth coefficient. We have already discussed such a design in the context of IIR filters (Fig. 4.17, p. 236). Figure 5.10 shows that, depending on the transition width ΔF of the filter, an IIR decimator offers substantial savings compared with an FIR decimator.

5.2.2 Fast-running FIR Filter

An interesting application of polyphase decomposition is the so-called *fast-running* FIR filter. The basic idea of this filter is the following: If we decompose the input signal $x[n]$ into R polyphase components, we can use Winograd's short convolution algorithms to implement a fast filter. Let us demonstrate this with an example for $R = 2$.

Example 5.2: Fast-Running FIR filter

We decompose the input signal $X(z)$ and filter $F(z)$ into even and odd polyphase components, i.e.,

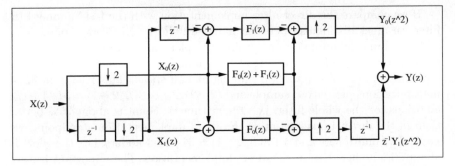

Fig. 5.11. Fast-running FIR filter with $R = 2$.

$$X(z) = \sum_n x[n]z^{-n} = X_0(z^2) + z^{-1}X_1(z^2) \tag{5.10}$$

$$F(z) = \sum_n f[n]z^{-n} = F_0(z^2) + z^{-1}F_1(z^2). \tag{5.11}$$

The convolution in the time domain of $x[n]$ and $f[n]$ yields a polynomial multiply in the z-domain. It follows for the output signal $Y(z)$ that

$$Y(z) = Y_0(z^2) + z^{-1}Y_1(z^2) \tag{5.12}$$
$$= (X_0(z^2) + z^{-1}X_1(z^2))(F_0(z^2) + z^{-1}F_1(z^2)). \tag{5.13}$$

If we split (5.13) into the polyphase components $Y_0(z)$ and $Y_1(z)$ we get

$$Y_0(z) = X_0(z)F_0(z) + z^{-1}X_1(z)F_1(z) \tag{5.14}$$
$$Y_1(z) = X_1(z)F_0(z) + X_0(z)F_1(z). \tag{5.15}$$

If we now compare (5.13) with a 2×2 linear convolution

$$A(z) \times B(z) = (a[0] + z^{-1}a[1])(b[0] + z^{-1}b[1]) \tag{5.16}$$
$$= a[0]b[0] + z^{-1}(a[0]b[1] + a[1]b[0]) + a[1]b[1]z^{-2}, \tag{5.17}$$

we notice that the factors for z^{-1} are the same, but for $Y_0(z)$ we must compute an extra delay to get the right phase relation. Winograd [103] has compiled a list of short convolution algorithms, and a linear 2×2 convolution can be computed using three multiplications and six adds with

$$
\begin{array}{lll}
a[0] = x[0] - x[1] & a[1] = x[0] & a[2] = x[1] - x[0] \\
b[0] = f[0] - f[1] & b[1] = f[0] & b[2] = f[1] - f[0] \\
c[k] = a[k]b[k] & k = 0, 1, 2 & \\
y[0] = c[1] + c[2] & y[1] = c[1] - c[0]. &
\end{array} \tag{5.18}
$$

With the help of this short convolution algorithm, we can now define the fast-running filter as follows:

$$
\begin{bmatrix} Y_0 \\ Y_1 \end{bmatrix} =
\begin{bmatrix} 0 & 1 & -1 \\ -1 & 1 & 0 \end{bmatrix}
\begin{bmatrix} F_0 & 0 & 0 \\ 0 & F_0 + F_1 & 0 \\ 0 & 0 & F_1 \end{bmatrix}
\begin{bmatrix} 1 & -1 \\ 1 & 0 \\ 1 & -z^{-1} \end{bmatrix}
\begin{bmatrix} X_0 \\ X_1 \end{bmatrix}. \tag{5.19}
$$

Figure 5.11 shows the graphical interpretation.

$\boxed{5.2}$

If we compare the direct filter implementation with the fast-running FIR filter we must distinguish between hardware effort and average number of adder and multiplier operations. A direct implementation would have L multipliers and $L-1$ adders running at full speed. For the fast-running filter we have three filters of length $L/2$ running at half speed. This results in $3L/4$ multiplications per output sample and $(2+2)/2+3/2(L/2-1) = 3L/4+1/2$ additions for the whole filter, i.e., the arithmetic count is about 25% better than in the direct implementation. From an implementation standpoint, we need $3L/2$ multipliers and $4+3(L/2-1) = 3L/2+1$ adders, i.e., the effort is about 50% higher than in the direct implementation. The important feature in Fig. 5.11 is that the fast-running filter basically runs at twice the speed of the direct implementation. Using a higher number R of decomposition may further increase the maximum throughput. The general methology for R polyphase signals with f_a as input rate is now as follows:

Algorithm 5.3:   **Fast-Running FIR Filter**

1) Decompose the input signal into R polyphase signals, using A_e adders to form R sequences at a rate of f_a/R.

2) Filter the R sequences with R filters of length L/R.

3) Use A_a additions to compute the polyphase representation of the output $Y_k(z)$. Use a final output multiplexer to generate the output signal $Y(z)$.

Note that the computed partial filter of length L/R may again be decomposed, using Algorithm 5.3. Then the question arises: When should we stop the iterative decomposition? Mou and Duhamel [104] have compiled a table with the goal of minimizing the average arithmetic count. Table 5.1 shows the optimal decomposition. The criterion used was a minimum total number of multiplications and additions, which is typical for a MAC-based design. In Table 5.1, all partial filters that should be implemented based on Algorithm 5.3 are underlined.

For a larger length than 60, a fast convolution using the FFT is more efficient, and will be discussed in Chap. 6.

5.3 Hogenauer CIC Filters

A very efficient architecture for a high decimation-rate filter is the "cascade integrator comb" (CIC) filter introduced by Hogenauer [106]. The CIC (also known as the Hogenauer filter), has proven to be an effective element in high-decimation or interpolation systems. One application is in wireless communications, where signals, sampled at RF or IF rates, need to be reduced to baseband. For narrowband applications (e.g., cellular radio), decimation rates in excess of 1000 are routinely required. Such systems are sometimes referred to as channelizers [107]. Another application area is in $\Sigma\Delta$ data converters [108].

Table 5.1. Computational effort for the recursive FIR decomposition [104, 105].

L	Factors	$M + A$	$\frac{M+A}{L}$	L	Factors	$M + A$	$\frac{M+A}{L}$
2	direct	6	3	22	$\underline{11} \times 2$	668	30.4
3	direct	15	5	24	$2^2 \times \underline{3} \times 2$	624	26
4	$\underline{2} \times 2$	26	6.5	25	$\underline{5} \times 5$	740	29.6
5	direct	45	9	26	$\underline{13} \times 2$	750	28.9
6	$\underline{3} \times 2$	56	9.33	27	$3^2 \times 3$	810	30
8	$2^2 \times 2$	94	11.75	30	$\underline{5} \times \underline{3} \times 2$	912	30.4
9	$\underline{3} \times 3$	120	13.33	32	$2^4 \times 2$	1006	31.44
10	$\underline{5} \times 2$	152	15.2	33	$\underline{11} \times 3$	1248	37.8
12	$\underline{2} \times \underline{3} \times 2$	192	16	35	$\underline{7} \times 5$	1405	40.1
14	$\underline{7} \times 2$	310	22.1	36	$2^2 \times \underline{3} \times 3$	1260	35
15	$\underline{5} \times 3$	300	20	39	$\underline{13} \times 3$	1419	36.4
16	$2^3 \times 2$	314	19.63	55	$\underline{11} \times 5$	2900	52.7
18	$\underline{2} \times \underline{3} \times 3$	396	22	60	$\underline{5} \times \underline{2} \times \underline{3} \times 2$	2784	46.4
20	$\underline{5} \times \underline{2} \times 2$	472	23.6	65	$\underline{13} \times 5$	3345	51.46
21	$\underline{7} \times 3$	591	28.1				

CIC filters are based on the fact that perfect pole/zero canceling can be achieved. This is only possible with exact integer arithmetic. Both two's complement and the residue number system have the ability to support error-free arithmetic. In the case of two's complement, arithmetic is performed modulo 2^b, and, in the case of the RNS, modulo M.

An introductory case study will be used to demonstrate.

5.3.1 Single-Stage CIC Case Study

Figure 5.12 shows a first-order CIC filter without decimation in 4-bit arithmetic. The filter consists of a (recursive) integrator (I-section), followed by a 4-bit differentiator or comb (C-section). The filter is realized with 4-bit values, which are implemented in two's complement arithmetic, and the values are bounded by $-8_{10} = 1000_{2C}$ and $7_{10} = 0111_{2C}$.

Figure 5.13 shows the impulse response of the filter. Although the filter is recursive, the impulse response is finite, i.e., it is a *recursive* FIR filter. This

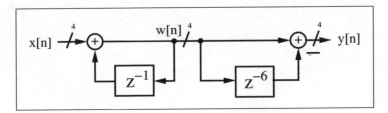

Fig. 5.12. Moving average in 4-bit arithmetic.

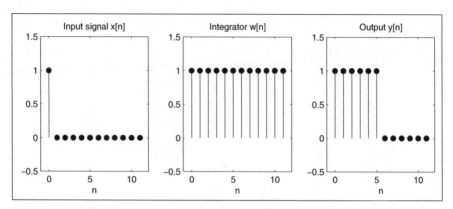

Fig. 5.13. Impulse response of the filter from Fig. 5.12.

is unusual because we generally expect a recursive filter to be an IIR filter. The impulse response shows that the filter computes the sum

$$y[n] = \sum_{k=0}^{D-1} x[n-k], \qquad (5.20)$$

where D is the delay found in the comb section. The filter's response is a *moving average* defined over D contiguous sample values. Such a moving average is a very simple form of a lowpass filter. The same moving-average filter implemented as a nonrecursive FIR filter, would require $D-1 = 5$ adders, compared with one adder and one subtractor for the CIC design.

A recursive filter having a known pole location has its largest steady-state sinusoidal output when the input is an "eigenfrequency" signal, one whose pole directly coincides with a pole of the recursive filter. For the CIC section, the eigenfrequency corresponds to the frequency $\omega = 0$, i.e., a step input. The step response of the first-order moving average given by (5.20) is a ramp for the first D samples, and a constant $y[n] = D = 6$ thereafter, as shown in Fig. 5.14. Note that although the integrator $w[n]$ shows frequent overflows, the output is still correct. This is because the comb subtraction also uses two's complement arithmetic, e.g., at the time of the first wrap-around, the actual integrator signal is $w[n] = -8_{10} = 1000_{2C}$, and the delay signal is $w[n-6] = 2_{10} = 0010_{2C}$. This results in $y[n] = -8_{10} - 2_{10} = 1000_{2C} - 0010_{2C} = 0110_{2C} = 6_{10}$, as expected. The accumulator would continue to count upward until $w[n] = -8_{10} = 1000_{2C}$ is again reached. This pattern would continue as long as the step input is present. In fact, as long as the output $y[n]$ is a valid 4-bit two's complement number in the range $[-8, 7]$, the exact arithmetic of the two's complement system will automatically compensate for the integrator overflows.

In general, a 4-bit filter width is usually much too small for a typical application. The Harris IC HSP43220, for instance, has five stages and uses a

Table 5.2. RNS mapping for the set $(2, 3, 5)$.

$a =$	0	1	2	3	4	5	6	7	8	9	10	11	12	13	14	15
$a \bmod 2$	0	1	0	1	0	1	0	1	0	1	0	1	0	1	0	1
$a \bmod 3$	0	1	2	0	1	2	0	1	2	0	1	2	0	1	2	0
$a \bmod 5$	0	1	2	3	4	0	1	2	3	4	0	1	2	3	4	0

$a =$	16	17	18	19	20	21	22	23	24	25	26	27	28	29	30
$a \bmod 2$	0	1	0	1	0	1	0	1	0	1	0	1	0	1	0
$a \bmod 3$	1	2	0	1	2	0	1	2	0	1	2	0	1	2	0
$a \bmod 5$	1	2	3	4	0	1	2	3	4	0	1	2	3	4	0

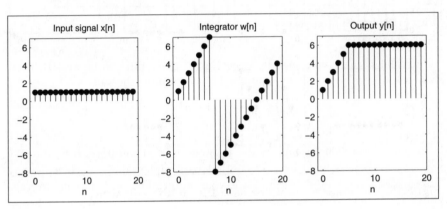

Fig. 5.14. Step response (eigenfrequency test) of the filter from Fig. 5.12.

66-bit integrator width. To reduce the adder latency, it is therefore reasonable to use a multibase RNS system. If we use, for instance, the set $\mathbb{Z}_{30} = \{2, 3, 5\}$, it can be seen from Table 5.2 that a total of $2 \times 3 \times 5 = 30$ unique values can be represented. The mapping is unique (bijective) and is proven by the Chinese remainder theorem.

Figure 5.15 displays the step response of the illustrated RNS implementation. The filter's output, $y[n]$, has been reconstructed using data from Table 5.2. The output response is identical with the sample value obtained in the two's complement case (see Fig. 5.14). A mapping that preserves the structure is called a *homomorphism*. A bijective homomorphism is called an *isomorphism* (notation $\cong$), which can be expressed as:

$$\mathbb{Z}_{30} \cong \mathbb{Z}_2 \times \mathbb{Z}_3 \times \mathbb{Z}_5. \tag{5.21}$$

5.3.2 Multistage CIC Filter Theory

The transfer function of a general CIC system consisting of S stages is given by:

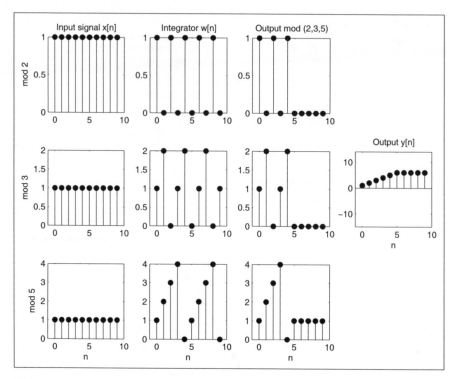

Fig. 5.15. Step response of the first-order CIC in RNS arithmetic.

$$F(z) = \left(\frac{1 - z^{-RD}}{1 - z^{-1}} \right)^S , \qquad (5.22)$$

where D is the number of delays in the comb section, and R the downsampling (decimation) factor.

It can be seen from (5.22) that $F(z)$ is defined with respect to RDS zeros and S poles. The RD zeros generated by the numerator term $(1 - z^{-RD})$ are located on $2\pi/(RD)$-radian centers beginning at $z = 1$. Each distinct zero appears with multiplicity S. The S poles of $F(z)$ are located at $z = 1$, i.e., at the zero frequency (DC) location. It can immediately be seen that they are annihilated by S zeros of the CIC filter. The result is an S-stage moving average filter. The maximum dynamic range growth occurs at the DC frequency (i.e., $z = 1$). The maximum dynamic range growth is

$$B_{\text{grow}} = (RD)^S \quad \text{or} \quad b_{\text{grow}} = \log_2 (B_{\text{grow}}) \text{ bits.} \qquad (5.23)$$

Knowledge of this value is important when designing a CIC filter, since the need for exact arithmetic as shown in the single-state CIC example. In practice, the worst-case gain can be substantial, as evidenced by a 66-bit dynamic range built into commercial CIC filters (e.g., the Harris HSP43220 [107] channelizer), typically designed using two's complement arithmetic.

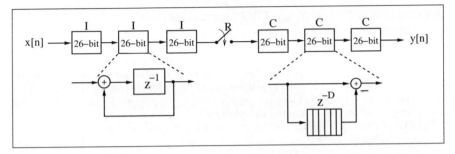

Fig. 5.16. CIC filter. Each stage 26-bit.

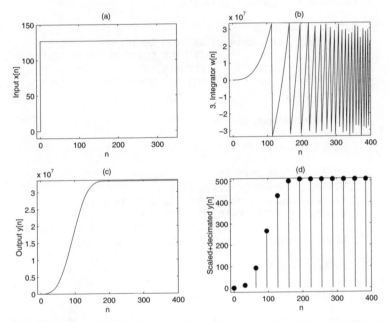

Fig. 5.17. MATLAB simulation of the three-stage CIC filter shown in Fig. 5.16.

Figure 5.16 shows a three-stage CIC filter that consists of a three-stage integrator, a sampling rate reduction by R, and a three-stage comb. Note that all integrators are implemented first, then the decimator, and finally the comb sections. The rearrangement saves a factor R of delay elements in the comb sections. The number of delays D for a high-decimation rate filter is typically one or two.

A three-stage CIC filter with an input wordwidth of eight bits, along with $D = 2$, $R = 32$, or $DR = 2 \times 32 = 64$, would require an internal wordwidth of $W = 8 + 3\log_2(64) = 26$ bits to ensure that run-time overflow would not occur. The output wordwidth would normally be a value significantly less than W, such as 10 bits.

Example 5.4: Three-Stage CIC Decimator I

The worst-case gain condition can be forced by supplying a step (DC) signal to the CIC filter. Fig. 5.17a shows a step input signal with amplitude 127. Figure 5.17b displays the output found at the third integrator section. Observe that run-time overflows occur at a regular rate. The CIC output shown in Fig. 5.17c is interpolated (smoothed) for display at the input sampling rate. The output shown in Fig. 5.17d is scaled to 10-bit precision and displayed at the decimated sample rate.

The following VHDL code[5] shows the CIC example design.

```
LIBRARY ieee;
USE ieee.std_logic_1164.ALL;
USE ieee.std_logic_arith.ALL;
USE ieee.std_logic_signed.ALL;

ENTITY cic3r32 IS
    PORT ( clk, reset : IN  STD_LOGIC;
               x_in       : IN  STD_LOGIC_VECTOR(7 DOWNTO 0);
               clk2       : OUT STD_LOGIC;
               y_out      : OUT STD_LOGIC_VECTOR(9 DOWNTO 0));
END cic3r32;

ARCHITECTURE fpga OF cic3r32 IS

  SUBTYPE word26 IS STD_LOGIC_VECTOR(25 DOWNTO 0);

  TYPE    STATE_TYPE IS (hold, sample);
  SIGNAL  state   : STATE_TYPE ;
  SIGNAL  count   : INTEGER RANGE 0 TO 31;
  SIGNAL  x : STD_LOGIC_VECTOR(7 DOWNTO 0) :=
                      (OTHERS => '0');  -- Registered input
  SIGNAL  sxtx : STD_LOGIC_VECTOR(25 DOWNTO 0);
                                       -- Sign extended input
  SIGNAL  i0, i1 , i2 : word26 := (OTHERS=>'0');
                                -- I section  0, 1, and 2
  SIGNAL  i2d1, i2d2, c1, c0 : word26 := (OTHERS=>'0');
                                -- I and COMB section 0
  SIGNAL  c1d1, c1d2, c2 : word26 := (OTHERS=>'0');-- COMB1
  SIGNAL  c2d1, c2d2, c3 : word26 := (OTHERS=>'0');-- COMB2

BEGIN

  FSM: PROCESS (reset, clk)
  BEGIN
    IF reset = '1' THEN                 -- Asynchronous reset
      state <= hold;
      count <= 0;
      clk2  <= '0';
    ELSIF rising_edge(clk) THEN
      IF count = 31 THEN
        count <= 0;
```

---

[5] The equivalent Verilog code cic3r32.v for this example can be found in Appendix A on page 694. Synthesis results are shown in Appendix B on page 731.

```
         state <= sample;
         clk2  <= '1';
      ELSE
         count <= count + 1;
         state <= hold;
         clk2  <= '0';
      END IF;
   END IF;
END PROCESS FSM;

sxt: PROCESS (x)
BEGIN
  sxtx(7 DOWNTO 0) <= x;
  FOR k IN 25 DOWNTO 8 LOOP
    sxtx(k) <= x(x'high);
  END LOOP;
END PROCESS sxt;

Int: PROCESS -- 3 integrator sections
BEGIN
  WAIT UNTIL clk = '1';
     x    <= x_in;
     i0   <= i0 + sxtx;
     i1   <= i1 + i0 ;
     i2   <= i2 + i1 ;
END PROCESS Int;

Comb: PROCESS -- 3 comb sections
BEGIN
   WAIT UNTIL clk = '1';
   IF state = sample THEN
      c0   <= i2;
      i2d1 <= c0;
      i2d2 <= i2d1;
      c1   <= c0 - i2d2;
      c1d1 <= c1;
      c1d2 <= c1d1;
      c2   <= c1  - c1d2;
      c2d1 <= c2;
      c2d2 <= c2d1;
      c3   <= c2  - c2d2;
   END IF;
END PROCESS Comb;

   y_out <= c3(25 DOWNTO 16);   -- i.e., c3 / 2**16
```

  END fpga;
The designed filter includes a finite state machine (FSM), a sign extension,
sxt: PROCESS, and two arithmetic PROCESS blocks. The FSM: PROCESS contains the clock divider for the comb section. The Int: PROCESS realizes the
three integrators. The Comb: PROCESS includes the three comb filters, each
having a delay of two samples. The filter uses 337 LEs, no embedded multiplier and has a 282.17 MHz Registered Performance. Note that the filter

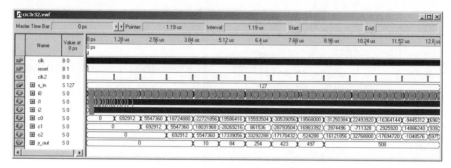

Fig. 5.18. VHDL simulation of the three-stage CIC filter shown in Fig. 5.16.

would require many more LEs without the early downsampling. The early downsampling saves $3 \times 32 \times 26 = 2496$ registers or LEs.

If we compare the filter outputs (Fig. 5.18 shows the VHDL output y_out, and the response $y[n]$ from the MATLAB simulation shown in Fig. 5.17d we see that the filter behaves as expected.   5.4

Hogenauer [106] noted, based on a careful analysis, that some of the lower significant bits from early stages can be eliminated without sacrificing system integrity. Figure 5.19 displays the system's magnitude frequency response for a design using full (worst-case) wordwidth in all stages, and using the wordlength "pruning" policy suggested by Hogenauer.

5.3.3 Amplitude and Aliasing Distortion

The transfer function of an S-stage CIC filter was reported to be

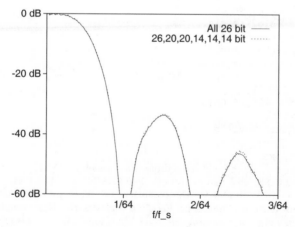

Fig. 5.19. CIC transfer function (f_s is sampling frequency at the input).

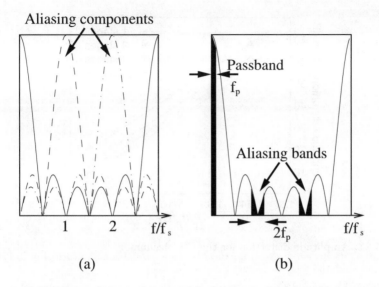

Fig. 5.20. Transfer function of a three-stage CIC decimator. Note that f_s is the sampling frequency at the lower rate.

$$F(z) = \left(\frac{1 - z^{-RD}}{1 - z^{-1}}\right)^S.$$

(5.24)

The amplitude distortion and the maximum aliasing component can be computed in the frequency domain by evaluating $F(z)$ along the arc $z = e^{j2\pi fT}$. The magnitude response becomes

$$|F(f)| = \left(\frac{\sin(2\pi fTRD/2)}{\sin(2\pi fT/2)}\right)^S,$$

(5.25)

which can be used to directly compute the *amplitude distortion* at the passband edge ω_p. Figure 5.20 shows $|F(f - k\frac{1}{2R})|$ for a three-stage CIC filter with $R = 3$, $D = 2$, and $RD = 6$. Observe that several copies of the CIC filter's low-frequency response are aliased in the baseband.

It can be seen that the maximum aliasing component can be computed from $|F(f)|$ at the frequency

$$f|_{\text{Aliasing has maximum}} = 1/(2R) - f_p.$$

(5.26)

Most often, only the first aliasing component is taken into consideration, because the second component is smaller. Figure 5.21 shows the amplitude distortion at f_p for different ratios of $f_p/(Df_s)$.

Figure 5.22 shows, for different values of S, R, and D, the maximum aliasing component for a special ratio of passband frequency and sampling frequency, f_p/f_s.

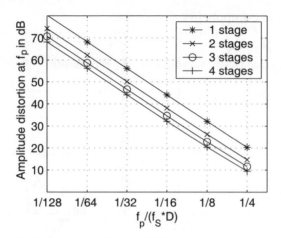

Fig. 5.21. Amplitude distortion for the CIC decimator.

It may be argued that the amplitude distortion can be corrected with a cascaded FIR compensation filter, which has a transfer function $1/|F(z)|$ in the passband, but the aliasing distortion can *not* be repaired. Therefore, the acceptable aliasing distortion is most often the dominant design parameter.

5.3.4 Hogenauer Pruning Theory

The total internal wordwidth is defined as the sum of the input wordwidth and the maximum dynamic growth requirement (5.23), or algebraically:

$$B_{\text{intern}} = B_{\text{input}} + B_{\text{growth}}. \tag{5.27}$$

If the CIC filter is designed to perform exact arithmetic with this wordwidth at all levels, no run-time overflow will occur at the output. In general, input and output bit width of a CIC filter are in the same range. We find then that quantization introduced through pruning in the output is, in general, larger than quantization introduced by also pruning some LSBs at previous stages. If $\sigma_{T,2S+1}^2$ is the quantization noise introduced through pruning in the output, Hogenauer suggested to set it equal to the sum of the noise σ_k^2 introduced by all previous sections. For a CIC filter with S integrator and S comb sections, it follows that:

$$\sum_{k=1}^{2S} \sigma_{T,k}^2 = \sum_{k=1}^{2S} \sigma_k^2 P_k^2 \leq \sigma_{T,2S+1}^2 \tag{5.28}$$

$$\sigma_{T,k}^2 = \frac{1}{2S}\sigma_{T,2S+1}^2 \tag{5.29}$$

$$P_k^2 = \sum_n (h_k[n])^2 \quad k = 1, 2, \ldots, 2S, \tag{5.30}$$

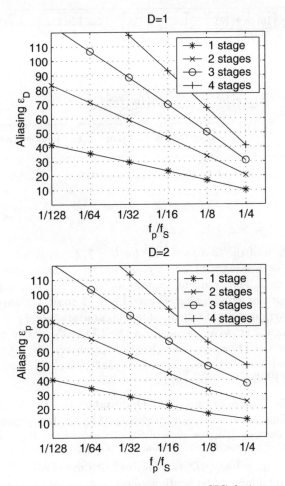

Fig. 5.22. Maximum aliasing for one- to four stage CIC decimator.

where P_k^2 is the power gain from stage k to the output. Compute next the number of bits B_k, which should be pruned by

$$B_k = \left\lfloor 0.5 \log_2 \left(P_k^{-2} \times \frac{6}{N} \times \sigma_{T,2S+1}^2 \right) \right\rfloor \tag{5.31}$$

$$\sigma_{T,k}^2 \big|_{k=2S+1} = \frac{1}{12} 2^{2B_k} = \frac{1}{12} 2^{2(B_{\text{in}} - B_{\text{out}} + B_{\text{growth}})}. \tag{5.32}$$

The power gain $P_k^2, k = S+1, \ldots, 2S$ for the comb sections can be computed using the binomial coefficient

$$H_k(z) = \sum_{n=0}^{2S+1-k} (-1)^n \binom{2S+1-k}{n} z^{-kRD}$$

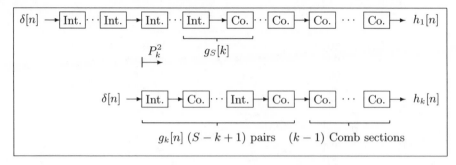

Fig. 5.23. Rearrangement to simplify the computation of P_k^2 (©1995 VDI Press [4]).

$$k = S, S + 1, \ldots, 2S. \qquad (5.33)$$

For computation of the first factor P_k^2 for $k = 1, 2, \ldots, S$, it is useful to keep in mind that each integrator/comb pair produces a finite (moving average) impulse response. The resulting system for stage k is therefore a series of $S - k + 1$ integrator/comb pairs followed by $k - 1$ comb sections. Figure 5.23 shows this rearrangement for a simplified computation of P_k^2.

The program `cic.exe` (included on the CD-ROM under `book3e/util`) computes this CIC pruning. The program produces the impulse response `cicXX.imp` and a configuration file `cicXX.dat`, where `XX` must be specified. The following design example explains the results.

Example 5.5: Three-Stages CIC Decimator II

Let us design the same overall CIC filter as in Example 4 (p. 262) but this time with bit pruning. The row data of the decimator were: $B_{\text{input}} = 8, B_{\text{output}} = 10$, Bit $R = 32$, and $D = 2$. Obviously, the bit growth is

$$B_{\text{growth}} = \lceil \log_2(RD^S) \rceil = \log_2(64^3) \lceil 3 \times 6 \rceil = 18, \qquad (5.34)$$

and the total internal bit width becomes

$$B_{\text{intern}} = B_{\text{input}} + B_{\text{growth}} = 8 + 18 = 26. \qquad (5.35)$$

The program `cic.exe` shows the following results:

```
-- -----------------------------------------------------------
-- Program for the design of a CIC decimator.
-- -----------------------------------------------------------
--      Input bit width      Bin  =      8
--      Output bit width     Bout =     10
--      Number of stages     S    =      3
--      Decimation factor    R    =     32
--      COMB delay           D    =      2
--      Frequency resolution DR   =     64
--      Passband freq. ratio P    =      8
-- -----------------------------------------------------------
-- ---------------- Results of the Design ----------------
-- -----------------------------------------------------------
-- -------- Computed bit width:
-- -------- Maximum bit growth over all stages        =     18
```

```
-- -------- Maximum bit width including sign Bmax+1 =    26
-- Stage   1 INTEGRATOR. Bit width :  26
-- Stage   2 INTEGRATOR. Bit width :  21
-- Stage   3 INTEGRATOR. Bit width :  16
-- Stage   1 COMB.       Bit width :  14
-- Stage   2 COMB.       Bit width :  13
-- Stage   3 COMB.       Bit width :  12
-- ------- Maximum aliasing component : 0.002135 = 53.41 dB
-- ------- Amplitude distortion       : 0.729769 =  2.74 dB
```

$$\boxed{5.5}$$

The design charts shown in Figs. 5.21 and 5.22 may also be used to compute the maximum aliasing component and the amplitude distortion. If we compare this data with the tables provided by Hogenauer then the aliasing suppression is 53.4 dB (for Delay = 2 [106, Table II]), and the passband attenuation is 2.74 dB [106, Table I]. Note that the Table I provided by Hogenauer are normalized with the comb delay, while the program cic.exe does not normalize with the comb delay.

The following design example demonstrates the detailed bit-width design, using Quartus II.

Example 5.6: Three-Stage CIC Decimator III

The data for the design should be the same as for Example 5.4 (p. 262), but we now consider the pruning as computed in Example 5.5 (p. 268).

The following VHDL code[6] shows the CIC example design with pruning.

```vhdl
LIBRARY ieee;
USE ieee.std_logic_1164.ALL;
USE ieee.std_logic_arith.ALL;
USE ieee.std_logic_signed.ALL;

ENTITY cic3s32 IS
  PORT ( clk, reset  : IN STD_LOGIC;
           x_in       : IN STD_LOGIC_VECTOR(7 DOWNTO 0);
           clk2       : OUT STD_LOGIC;
           y_out      : OUT STD_LOGIC_VECTOR(9 DOWNTO 0));
END cic3s32;

ARCHITECTURE fpga OF cic3s32 IS

  SUBTYPE word26 IS STD_LOGIC_VECTOR(25 DOWNTO 0);
  SUBTYPE word21 IS STD_LOGIC_VECTOR(20 DOWNTO 0);
  SUBTYPE word16 IS STD_LOGIC_VECTOR(15 DOWNTO 0);
  SUBTYPE word14 IS STD_LOGIC_VECTOR(13 DOWNTO 0);
  SUBTYPE word13 IS STD_LOGIC_VECTOR(12 DOWNTO 0);
  SUBTYPE word12 IS STD_LOGIC_VECTOR(11 DOWNTO 0);

  TYPE    STATE_TYPE IS (hold, sample);
```

---

[6] The equivalent Verilog code cic3s32.v for this example can be found in Appendix A on page 696. Synthesis results are shown in Appendix B on page 731.

```
      SIGNAL  state       : STATE_TYPE ;
      SIGNAL  count       : INTEGER RANGE 0 TO 31;
      SIGNAL  x           : STD_LOGIC_VECTOR(7 DOWNTO 0)
                          := (OTHERS => '0'); -- Registered input
      SIGNAL  sxtx  : STD_LOGIC_VECTOR(25 DOWNTO 0);
                                        -- Sign extended input
      SIGNAL  i0  : word26 := (OTHERS => '0');   -- I section 0
      SIGNAL  i1  : word21 := (OTHERS => '0');   -- I section 1
      SIGNAL  i2  : word16 := (OTHERS => '0');   -- I section 2
      SIGNAL  i2d1, i2d2, c1, c0 : word14 := (OTHERS => '0');
                                        -- I and COMB section 0
      SIGNAL  c1d1, c1d2, c2 : word13 := (OTHERS=>'0');--COMB 1
      SIGNAL  c2d1, c2d2, c3 : word12 := (OTHERS=>'0');--COMB 2

  BEGIN

    FSM: PROCESS (reset, clk)
    BEGIN
      IF reset = '1' THEN                -- Asynchronous reset
        state <= hold;
        count <= 0;
        clk2  <= '0';
      ELSIF rising_edge(clk) THEN
        IF count = 31 THEN
          count <= 0;
          state <= sample;
          clk2  <= '1';
        ELSE
          count <= count + 1;
          state <= hold;
          clk2  <= '0';
        END IF;
      END IF;
    END PROCESS FSM;

    Sxt : PROCESS (x)
    BEGIN
      sxtx(7 DOWNTO 0) <= x;
      FOR k IN 25 DOWNTO 8 LOOP
        sxtx(k) <= x(x'high);
      END LOOP;
    END PROCESS Sxt;

    Int: PROCESS
    BEGIN
    WAIT
      UNTIL clk = '1';
        x   <= x_in;
        i0  <= i0 + x;
        i1  <= i1 + i0(25 DOWNTO 5);  -- i.e., i0/32
        i2  <= i2 + i1(20 DOWNTO 5);  -- i.e., i1/32
    END PROCESS Int;
```

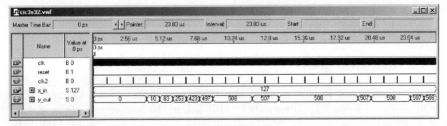

Fig. 5.24. VHDL simulation of the three-stage CIC filter, implemented with bit pruning.

```
Comb: PROCESS
BEGIN
  WAIT UNTIL clk = '1';
  IF state = sample THEN
    c0   <= i2(15 DOWNTO 2);    -- i.e., i2/4
    i2d1 <= c0;
    i2d2 <= i2d1;
    c1   <= c0 - i2d2;
    c1d1 <= c1(13 DOWNTO 1);    -- i.e., c1/2
    c1d2 <= c1d1;
    c2   <= c1(13 DOWNTO 1) - c1d2;
    c2d1 <= c2(12 DOWNTO 1);    -- i.e., c2/2
    c2d2 <= c2d1;
    c3   <= c2(12 DOWNTO 1) - c2d2;
  END IF;
END PROCESS Comb;

  y_out <= c3(11 DOWNTO 2);    -- i.e., c3/4

END fpga;
```

The design has the same architecture as the unscaled CIC shown in Example 5.4 (p. 262). The design consists of a finite state machine (FSM), a sign extension `sxt: PROCESS`, and two arithmetic `PROCESS` blocks. The `FSM: PROCESS` contains the clock divider for the comb sections. The `Int: PROCESS` realizes the three integrators. The `Comb: PROCESS` includes the three comb sections, each having a delay of two. But now, all integrator and comb sections are designed with the bit width suggested by Hogenauer's pruning technique. This reduces the size of the design to 205 LEs and the design now runs at 284.58 MHz.   5.6

This design does not improve the speed (282.17 versus 284.58 MHz), but saves a substantial number of LEs (about 30%), compared with the design considered in Example 5.4 (p. 262). Comparing the filter output of the VHDL simulations, shown in Figs. 5.24 and 5.18 (p. 264), different LSB quantization behavior can be noted (see Exercise 5.11, p. 338). In the pruned design, "noise" possesses the asymptotic behavior of the LSB ($507 \leftrightarrow 508$).

The design of a CIC *interpolator* and its pruning technique is discussed in Exercise 5.24, p. 340.

5.3.5 CIC RNS Design

The design of a CIC filter using the RNS was proposed by Garcia et al. [49]. A three-stage CIC filter, with 8-bit input, 10-bit output, $D = 2$, and $R = 32$ was implemented. The maximum wordwidth was 26 bits. For the RNS implementation, the 4-moduli set $(256, 63, 61, 59)$, i.e., one 8-bit two's complement and three 6-bit moduli, covers this range (see Fig. 5.25). The output was scaled using an ε-CRT requiring eight tables and three two's complement adders [43, Fig. 1], or (as shown in Fig. 5.26) using a base removal scaling (BRS) algorithm based on two 6-bit moduli (after [42]), and an ε-CRT for the remaining two moduli, for a total of five modulo adders and nine ROM tables, or seven tables (if multiplicative inverse ROM and the ε-CRT are combined). The following table shows the speed in MSPS and the number of LEs and EABs used for the three scaling schemes for a FLEX10K device.

Type	ε-CRT	BRS ε-CRT (Speed data for BRS m_4 only)	BRS ε-CRT combined ROM
MSPS	58.8	70.4	58.8
#LE	34	87	87
#Table (EAB)	8	9	7

The decrease in speed to 58.8 MSPS, for the scaling schemes 1 and 3, is the result of the need for a 10-bit ε-CRT. It should be noted that this does not reduce the *system* speed, since scaling is applied at the lower (output) sampling rate. For the BRS ε-CRT, it is assumed that only the BRS m_4 part (see Fig. 5.26) must run at the input sampling rate, while BRS m_3 and ε-CRT run at the output sampling rate.

Some resources can be saved if a scaling scheme, similar to Example 5.5 (p. 268), and illustrated in Fig. 5.25, is used. With this scheme, the BRS ε-CRT scheme must be applied to reduce the bit width in the earlier sections of the filter. The early use of ROMs decreases the possible throughput from 76.3 to 70.4 MSPS, which is the maximum speed of the BRS with m_4. At the output, the efficient ε-CRT scheme was applied.

The following table summarizes the three implemented filter designs on a FLEX10K device, without including the scaling data.

Type	2C 26-bit	RNS $8, 6, 6, 6$-bit	Detailed bit width RNS design
MSPS	49.3	76.3	70.4
#LEs	343	559	355

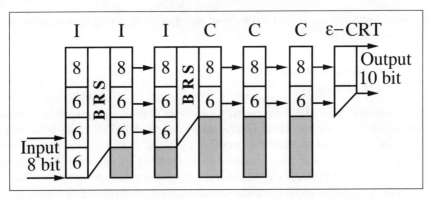

Fig. 5.25. CIC filter. Detail of design with base removal scaling (BRS).

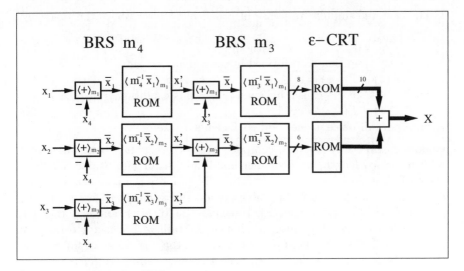

Fig. 5.26. BRS and ε-CRT conversion steps.

5.4 Multistage Decimator

If the decimation rate R is large it can be shown that a multistage design can be realized with less effort than a single-stage converter. In particular, S stages, each having a decimation capability of R_k, are designed to have an overall downsampling rate of $R = R_1 R_2 \cdots R_S$. Unfortunately, passband imperfections, such as ripple deviation, accumulate from stage to stage. As a result, a passband deviation target of ε_p must normally be tightened on the order of $\varepsilon_p' = \varepsilon_p/S$ to meet overall system specifications. This is obviously a worst-case assumption, in which all short filters have the maximum ripple at the same frequencies, which is, in general, too pessimistic. It is often more reasonable to try an initial value near the given passband specification ε_p, and then selectively reduce it if necessary.

5.4.1 Multistage Decimator Design Using Goodman–Carey Half-band Filters

Goodman and Carey [80] proposed to develop multistage systems based on the use of CIC and half-band filters. As the name implies, a *half-band filter* has a passband and stopband located at $\omega_s = \omega_p = \pi/2$, or midway in the baseband. A half-band filter can therefore be used to change the sampling rate by a factor of two. If the half-band filter has *point symmetry* relative to $\omega = \pi/2$, then all even coefficients (except the center tap) become zero.

Definition 5.7:   **Half-band Filter**

The centered impulse response of a half-band filter obeys the following rule

$$f[k] = 0 \qquad k = \text{even without } k = 0. \tag{5.36}$$

The same condition transformed in the z-domain reads

$$F(z) + F(-z) = c, \tag{5.37}$$

where $c \in \mathbb{C}$. For a causal half-band filter this condition translates to

$$F(z) - F(-z) = cz^{-d}, \tag{5.38}$$

since now all (except one) odd coefficients are zero.

Goodman and Carey [80] have compiled a list of integer half-band filters that, with increased length, have smaller amplitude distortions. Table 5.3 shows the coefficients of these half-band filters. To simplify the representation, the coefficients were noted with a center tap located at $d = 0$. F1 is the moving-average filter of length L, i.e., it is Hogenauer's CIC filter, and may therefore be used in the first stage also, to change the rate with a factor other than two. Figure 5.27 shows the transfer function of the nine different filters. Note that in the logarithmic plot of Fig. 5.27, the point symmetry (as is usual for half-band filters) cannot be observed.

Table 5.3. Centered coefficients of the half-band filter F1 to F9 from Goodman and Carey [80].

Name	L	Ripple	$f[0]$	$f[1]$	$f[3]$	$f[5]$	$f[7]$	$f[9]$
F1	3	–	1	1				
F2	3	–	2	1				
F3	7	–	16	9	−1			
F4	7	36 dB	32	19	−3			
F5	11	–	256	150	−25	3		
F6	11	49 dB	346	208	−44	9		
F7	11	77 dB	512	302	−53	7		
F8	15	65 dB	802	490	−116	33	−6	
F9	19	78 dB	8192	5042	−1277	429	−116	18

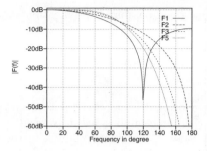 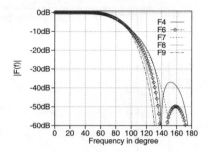

Fig. 5.27. Transfer function of the half-band filter F1 to F9.

The basic idea of the Goodman and Carey multistage decimator design is that, in the first stages, filters with larger ripple and less complexity can be applied, because the passband-to-sampling frequency ratio is relatively small. As the passband-to-sampling frequency ratio increases, we must use filters with less distortion. The algorithm stops at $R = 2$. For the final decimation ($R = 2$ to $R = 1$), a longer half-band filter must be designed.

Goodman and Carey have provided the design chart shown in Fig. 5.28. Initially, the input oversampling factor R and the necessary attenuation in the passband and stopband $A = A_p = A_s$ must be computed. From this starting point, the necessary filters for $R, R/2, R/4, \ldots$ can be drawn as a horizontal line (at the same stopband attenuation). The filters F4 and F6–F9 have ripple in the passband (see Exercise 5.8, p. 337), and if several such filters are used it may be necessary to adjust ε_p. We may, therefore, consider the following adjustment

$$A = -20 \log_{10} \varepsilon_p \quad \text{for} \quad \text{F1–F3, F5} \tag{5.39}$$

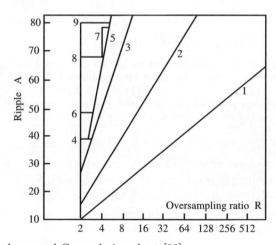

Fig. 5.28. Goodman and Carey design chart [80].

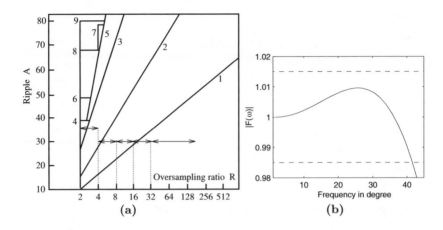

Fig. 5.29. Design example for Goodman and Carey half-band filter. (a) Design chart. (b) Transfer function $|F(\omega)|$.

$$A = -20 \log_{10} \min \left(\frac{\varepsilon_p}{S'}, \varepsilon_s \right) \quad \text{for} \quad \text{F4, F6–F9,} \tag{5.40}$$

where S' is the number of stages with ripple.

We will demonstrate the multistage design with the following example.

Example 5.8: Multistage Half-band Filter Decimator

We wish to develop a decimator with $R = 160, \varepsilon_p = 0.015$, and $\varepsilon_s = 0.031 = 30$ dB, using the Goodman and Carey design approach.

At first glance, we can conclude that we need a total of five filters and mark the starting point at $R = 160$ and 30 dB in Fig. 5.29a. From 160 to 32, we use a CIC filter of length $L = 5$. This CIC filter is followed by two F2 filter and one F3 filter to reach $R = 8$. Now we need a filter with ripple. It follows that

$$A = -20 \log_{10} \min \left(\frac{0.015}{1}, 0.031 \right) = 36.48 \text{ dB.} \tag{5.41}$$

From Fig. 5.28, we conclude that for 36 dB the filter F4 is appropriate. We may now compute the whole filter transfer function $|F(\omega)|$ by using the Noble relation (see Fig. 5.6, p. 249) $F(z) = \text{F1}(z)\text{F2}(z^5)\text{F2}(z^{10})\text{F3}(z^{20})\text{F4}(z^{40})$, who's passband is shown in Fig. 5.29b. Figure 5.29a shows the design algorithm, using the design chart from Fig. 5.28.

5.8

Example 5.8 shows that considering only the filter with ripple in (5.40) was sufficient. Using a more pessimistic approach, with $S = 6$, we would have obtained $A = -20 \log(0.015/6) = 52$ dB, and we would have needed filter F8, with essentially higher effort. It is therefore better to start with an optimistic assumption and possibly correct this later.

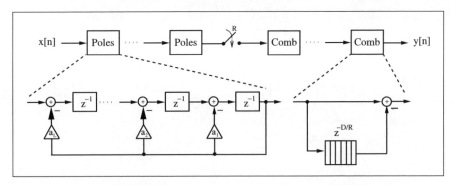

Fig. 5.30. Cascading of frequency-sampling filters to save a factor of R delays for multirate signal processing [4, Sect. 3.4].

5.5 Frequency-Sampling Filters as Bandpass Decimators

The CIC filters discussed in Sect. 5.3 (p. 256) belong to a larger class of systems called *frequency-sampling filters* (FSFs). Frequency-sampling filters can be used, as channelizer or decimating filter, to decompose the information spectrum into a set of discrete subbands, such as those found in multiuser communication systems. A classic FSF consists of a comb filter cascaded with a bank of frequency-selective resonators [4, 65]. The resonators independently produce a collection of poles that selectively annihilate the zeros produced by the comb prefilter. Gain adjustments are applied to the output of the resonators to shape the resulting magnitude frequency response of the overall filter. An FSF can also be created by cascading all-pole filter sections with all-zero filter (comb) sections, as suggested in Fig. 5.30. The delay of the comb section, $1 \pm z^{-D}$, is chosen so that its zeros cancel the poles of the all-pole prefilter as shown in Fig. 5.31. Wherever there is a complex pole, there is also an annihilating complex zero that results in an all-zero FIR, with the usual linear-phase and constant group-delay properties.

Frequency-sampling filters are of interest to designers of multirate filter banks due, in part, to their intrinsic low complexity and linear-phase behavior. FSF designs rely on exact pole-zero annihilation and are often found in embedded applications. Exact FSF pole-zero annihilation, can be guaranteed by using polynomial filters defined over an *integer ring* using the two's complement or the residue number system (RNS). The poles of an FSF filter developed in this manner can reside on the periphery of the unit circle. This conditionally unstable location is acceptable, due to the guarantee of exact pole-zero cancellation. Without this guarantee, the designer would have to locate the poles of the resonators within the unit circle, with a loss in performance. In addition, by allowing the FSF poles and zeros to reside on the unit circle, a multiplier-less FSF can be created, with an attendant reduction in complexity and an increase in data bandwidth.

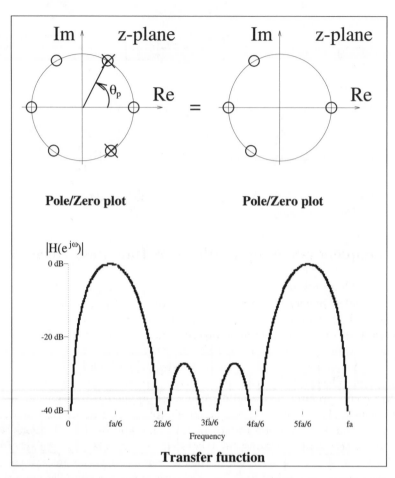

Fig. 5.31. Example of pole/zero-compensation for a pole angle of 60° and comb delay $D = 6$.

Consider the filter shown in Fig. 5.30. It can be shown that first-order filter sections (with integer coefficients) produce poles at angles of 0° and 180°. Second-order sections, with integer coefficients, can produce poles at angles of 60°, 90°, and 120°, according to the relationship $2\cos(2\pi K/D)=1$, 0, and -1. The frequency selectivity of higher-order sections is shown in Table 5.4. The angular frequencies for all polynomials having integer coefficients with roots on the unit circle, up to order six, are reported. The building blocks listed in Table 5.4 can be used to efficiently design and implement such FSF filters. For example, a two's complement (i.e., RNS single modulus) filter bank was developed for use as a constant-Q speech processing filter bank. It covers a frequency range from 900 to 8000 Hz [109, 110], using 16 kHz sampling frequency. An integer coefficient half-band filter HB6 [80] anti-aliasing

Table 5.4. Filters with integer coefficients producing unique angular pole locations up to order six. Shown are the filter coefficients and nonredundant angular locations of the roots on the unit circle.

$C_k(z)$	Order	a_0	a_1	a_2	a_3	a_4	a_5	a_6	θ_1	θ_2	θ_3
$-C_1(z)$	1	1	-1						$0°$		
$C_2(z)$	1	1	1						$180°$		
$C_6(z)$	2	1	-1	1					$60°$		
$C_4(z)$	2	1	0	1					$90°$		
$C_3(z)$	2	1	1	1					$120°$		
$C_{12}(z)$	4	1	0	-1	0	1			$30°$	$150°$	
$C_{10}(z)$	4	1	-1	1	-1	1			$36°$	$108°$	
$C_8(z)$	4	1	0	0	0	1			$45°$	$135°$	
$C_5(z)$	4	1	1	1	1	1			$72°$	$144°$	
$C_{16}(z)$	6	1	0	0	-1	0	0	1	$20.00°$	$100.00°$	$140.00°$
$C_{14}(z)$	6	1	-1	1	-1	1	-1	1	$25.71°$	$77.14°$	$128.57°$
$C_7(z)$	6	1	1	1	1	1	1	1	$51.42°$	$102.86°$	$154.29°$
$C_9(z)$	6	1	0	0	1	0	0	1	$40.00°$	$80.00°$	$160.00°$

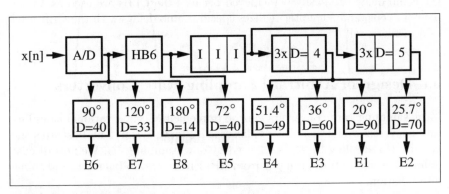

Fig. 5.32. Design of a filter bank consisting of a half-band and CIC prefilter and FSF comb-resonator sections.

filter and a third-order multiplier-free CIC filter (also known as Hogenauer filter [106] see Sect. 5.3, p. 256), was then added to the design to suppress unwanted frequency components, as shown in Fig. 5.32. The bandwidth of each resonator can be independently tuned by the number of stages and delays in the comb section. The number of stages and delays is optimized to meet the desired bandwidth requirements. All frequency-selective filters have two stages and delays.

The filter bank was prototyped using a Xilinx XC4000 FPGA with the complexity reported in Table 5.5. Using high-level design tools (XBLOCKS from Xilinx), the number of used CLBs was typically 20% higher than the

Table 5.5. Number of CLBs used in Xilinx XC4000 FPGAs (notation: F20D90 means filter pole angle $20.00°$, delay comb $D = 90$). Total: actual 1572 CLBs, nonrecursive FIR: 11292 CLBs

	F20D90	F25D70	F36D60	F51D49	F72D40	F90D40
Theory	122	184	128	164	124	65
Practice	160	271	190	240	190	93
Nonre. FIR	2256	1836	1924	1140	1039	1287

	F120D33	F180D14	HB6	III	D4	D5
Theory	86	35	122	31	24	24
Practice	120	53	153	36	33	33
Nonre. FIR	1260	550				

theoretical prediction obtained by counting adders, flip-flops, ROMs, and RAMs.

The design of an FSF can be manipulated by changing the comb delay, channel amplitude, or the number of sections. For example, adaptation of the comb delay can easily be achieved because the CLBs are used as 32×1 memory cells, and a counter realizes specific comb delays with the CLB used as a memory cell.

5.6 Design of Arbitrary Sampling Rate Converters

Most sampling rate converters can be implemented via a rational sampling rate converter system as already discussed. Figure 5.7, p. 249 illustrates the system. Upsampling by R_1 is followed by downsampling R_2. We now discuss different design options, ranging from IIR, FIR filters, to Lagrange and spline interpolation.

To illustrate, let us look at the design procedure for a rational factor change in the sampling rate with interpolation by $R_1 = 3$ followed by a decimation by $R_2 = 4$, i.e., a rate change by $R = R_1/R_2 = 3/4 = 0.75$.

Example 5.9: $R = 0.75$ Rate Changer I

An interpolation by 3 followed by a decimation by 4 with a centered lowpass filter has to be designed. As shown in Fig. 5.7, p. 249 we only need to implement one lowpass with a cut-off frequency of $\min(\frac{\pi}{3}, \frac{\pi}{4}) = \frac{\pi}{4}$. In MATLAB the frequencies in the filter design procedures are normalized to $f_2/2 = \pi$ and the design of a tenth order Chebyshev II filter with a 50 dB stopband attenuation is accomplished by

```
[B, A] = cheby2(10,50,0.25)
```

A Chebyshev II filter was chosen because it has a flat passband, ripple in the stopband, and moderate filter length and is therefore a good choice in many applications. If we want to reduce the filter coefficient sensitivity to

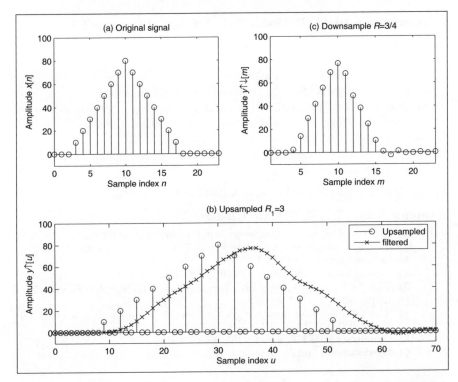

Fig. 5.33. IIR filter simulation of rational rate change. **(a)** Original signal **(b)** upsampled and filtered version of the original signal. **(c)** Signal after downsampling.

quantization, we can use an implementation form with biquad sections rather than the direct forms, see Example 4.2 (p. 227). In MATLAB we use

```
[SOS, gain]=tf2sos(B,A)
```

Using this IIR filter we can now simulated the rational rate change. Figure 5.33 shows a simulation for a triangular test signal. Figure 5.33a shows the original input sequence. (b) the signal after upsampling by 3 and after filtering with the IIR filter, and (c) the signal after downsampling. | 5.9 |

Although the rational interpolation via an IIR filter is not perfect we notice that the triangular shape is well preserved, but ringing of the filter next to the triangle when the signal should be zero can be observed. We may ask if the interpolation can be improved if we use an exact lowpass that we can build in the frequency domain. Instead of using the IIR filter in the time domain, we may try to interpolate by means of a DFT/FFT-based frequency method [111]. In order to keep the frame processing simple we choose a DFT or FFT whose length N is a multiple of the rate change factors, i.e., $N = k \times R_1 \times R_2$ with $k \in \mathbb{N}$. The necessary processing steps can be summarized as follows:

Algorithm 5.10:   Rational Rate Change using an FFT

The algorithm to compute the rate change by $R = R_1/R_2$ via an FFT is as follows:

1) Select a block of $k \times R_2$ samples.
2) Interpolate with $(R_1 - 1)$ zeros between each sample.
3) Compute the FFT of size $N = k \times R_1 \times R_2$.
4) Apply the lowpass filter operation in the frequency domain.
5) Compute the IFFT of size $N = k \times R_1 \times R_2$.
6) Compute finally the output sequence by downsampling by R_1, i.e., keep $k \times R_2$ samples.

Let us illustrate this algorithm with a small numerical example.

Example 5.11: $R = 0.75$ **Rate Changer II**

Let us assume we have a triangular input sequence x to interpolate by $R = R_1/R_2 = 3/4 = 0.75$. and we select $k = 1$. The steps are as follows:

1) Original block $x = (1, 2, 3, 4)$.
2) Interpolation 3 gives $x_i = (1, 0, 0, 2, 0, 0, 3, 0, 0, 4, 0, 0)$.
3) The FFT gives $X_i = (10, -2 + \mathrm{j}2, -2, 2 - \mathrm{j}2, 10, -2 + \mathrm{j}2, -2, -2 - \mathrm{j}2, 10, -2 + \mathrm{j}2, -2, -2 - \mathrm{j}2)$.
4) The Lowpass filter operation in the frequency domain. $X_{\mathrm{lp}} = (10, -2 + \mathrm{j}2, -2, 0, 0, 0, 0, 0, 0, 0, -2, -2 - \mathrm{j}2)$.
5) Compute the IFFT, $y = 3 \times \mathrm{ifft}(X_{\mathrm{lp}})$.
6) Downsampling finally gives $y = (0.5000, 2.6340, 4.3660)$.

5.11

Let us now apply this block processing to the triangular sequence shown in Fig. 5.34a. From the results shown in Fig. 5.34b we notice the border effects between the blocks when compared with the FFT interpolation results with full-length input data as shown in Fig. 5.34d. This is due to the fact that the underlying assumption of the DFT is that the signals in time and frequency are periodic. We may try to improve the quality and reduce the border disturbance by applying a window function that tapers smoothly to zero at the borders. This will however also reduce the number of useful samples in our output sequence and we need to implement an overlapping block processing. This can be improved by using longer (i.e., $k > 1$) FFTs and removing the leading and tailing samples. Figure 5.34c shows the result for $k = 2$ with removal of 50% of the lead and tail samples. We may therefore ask, why not using a very long FFT, which produces the best results, as can be seem from full length FFT simulation result as shown in Fig. 5.34d? The reason we prefer the short FFT comes from the computational perspective: the longer FFT will require more effort per output sample. Although with $k > 1$ we have more output values per FFT available and overall need fewer FFTs, the computational effort per sample of a radix-2 FFT is $\mathrm{ld}(N)/2$ complex multiplications, because the FFT requires $\mathrm{ld}(N)N/2$ complex multiplications for

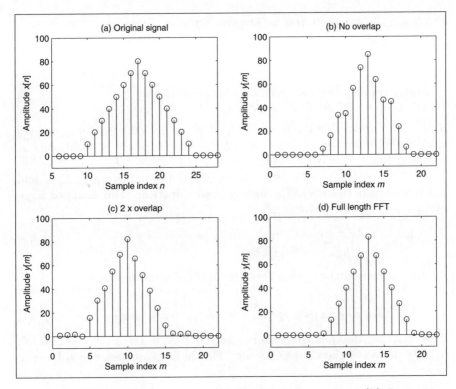

Fig. 5.34. FFT based rational rate change. **(a)** Original signal **(b)** Decimation without overlap. **(c)** 50% overlap. **(c)** Full length FFT.

the N-point radix-2 FFT. A short FFT reduces therefore the computational effort.

Given the contradiction that longer FFTs are used to produce a better approximation while the computational effort requires shorter FFTs, we now want to discuss briefly what computational simplifications can be made to simplify the computation of the two long FFTs in Algorithm 5.10. Two major savings are basically possible.

The first is in the forward transform: the interpolated sequence has many zeros, and if we use a Cooley–Tuckey decimation-in-time-based algorithm, we can group all nonzero values in one DFT block and the remaining $k(R_1 \times R_2 - R_2)$ in the other $R_1 - 1$ groups. Then basically only one FFT of size $k \times R_2$ needs to be computed compared with the full-length $N = k \times R_1 \times R_2$ FFT.

The second simplification can be implemented by computing the down-sampling in the frequency domain. For downsampling by two we compute

$$F_{\downarrow 2}(k) = F(k) + F(k + N/2) \tag{5.42}$$

for all $k \leq N/2$. The reason is that the downsampling in the time domain leads to a Nyquist frequency repetition of the base band scaled by a factor of 2. In Algorithm 5.10 we need downsampling by a factor R_2, which we compute as follows

$$F_{\downarrow R_2}(k) = \sum_n F(k + nN/R_2). \tag{5.43}$$

If we now consider that due to the lowpass filtering in the frequency domain many samples are set to zero, the summations necessary in (5.43) can be further reduced. The IFFT required is only of length $k \times R_1$.

   To illustrate the saving let us assume that the implemented FFT and IFFT both require $\mathrm{ld}(N)N/2$ complex multiplications. The modified algorithm is improved by a factor of

$$F = \frac{2\frac{kR_1R_2}{2}\mathrm{ld}(kR_1R_2)}{\frac{kR_1}{2}\mathrm{ld}(kR_1) + \frac{kR_2}{2}\mathrm{ld}(kR_2)}. \tag{5.44}$$

For the simulation above with $R = 3/4$ and 50% overlap we get

$$F = \frac{\mathrm{ld}(24)24}{\mathrm{ld}(6)6/2 + \mathrm{ld}(8)8/2} = 5.57. \tag{5.45}$$

If we can use the Winograd short-term DFT algorithms (see Sect. 6.1.6, p. 359) instead of the Cooley–Tuckey FFT the improvement would be even larger.

5.6.1 Fractional Delay Rate Change

In some applications the input and output sampling rate quotients are close to 1, as in the previous example with $R = 3/4$. As another examples consider a change from the CD rate (44.1 kHz) to DAT player rate (48 kHz), which requires a rational factor change factor of $R = 147/160$. In this case the direct approach using a sampling rate interpolation followed by a sampling rate reduction would required a very high sampling rate for the lowpass interpolation filter. In case of the CD→DAT change for instance the filter must run with 147 times the input sampling rate.

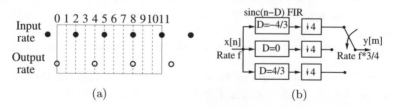

(a)   (b)

Fig. 5.35. (a) Input and output sampling grid for fractional delay rate change. (b) Filter configuration for $R = 3/4$ sinc filter system.

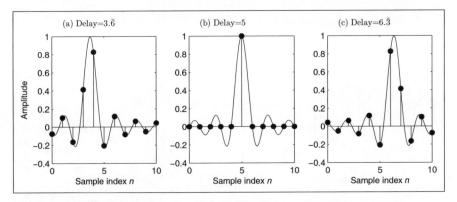

Fig. 5.36. Fractional delay filter with delays $D = 5 - 4/3, 5$, and $5 + 4/3$.

In these large-R_1 cases we may consider implementation of the rate change with the help of fractional delays. We will briefly review the idea and then discuss the HDL code for two different versions. The idea will be illustrated by a rate change of $R = 3/4$. Figure 5.35a shows the input and output sampling grid for a system. For each block of four input values, the system computes three interpolated output samples. From a filter perspective we need three filters with unit transfer functions: one filter with a zero delay and two filters that implement the delays of $D = \pm 4/3$. A filter with unit frequency is a sinc or $\sin(t)/t = \mathrm{sinc}(t)$ filter in the time domain. We must allow an initial delay to make the filter realizable, i.e., causal. Figure 5.35b shows the filter configuration and Fig. 5.35 shows the three length-11 filter impulse responses for the delays $5 - 4/3 = 3.\bar{6}, 5$, and $5 + 4/3 = 6.\bar{3}$.

We can now apply these three filters to our triangular input signal and for each block of four input samples we compute three output samples. Figure 5.37c shows the simulation result for the length-11 sinc filter. Figure 5.37b shows that length-5 filters produce too much ripple to be useful. The length-11 sinc filters produce a much smoother triangular output, but some ripple due to the Gibbs phenomenon can be observed next to the triangular function when the output should be zero.

Let us now have a look at the HDL implementation of the fractional delay rate changer using sinc filters.

Example 5.12: R= 0.75 Rate Changer III

The following VHDL code[7] shows the sinc filter design for an $R = 3/4$ rate change.

```
PACKAGE n_bits_int IS           -- User-defined types
  SUBTYPE BITS8  IS INTEGER RANGE -128 TO 127;
  SUBTYPE BITS9  IS INTEGER RANGE -2**8 TO 2**8-1;
  SUBTYPE BITS17 IS INTEGER RANGE -2**16 TO 2**16-1;
```

---

[7] The equivalent Verilog code `rc_sinc.v` for this example can be found in Appendix A on page 697. Synthesis results are shown in Appendix B on page 731.

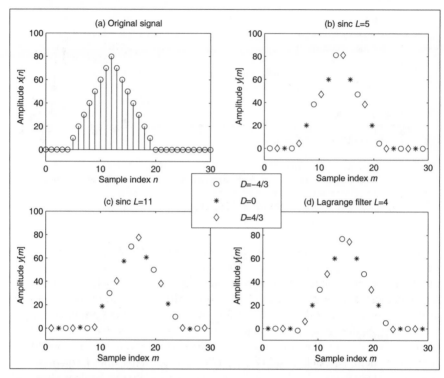

Fig. 5.37. Fraction delay interpolation **(a)** Original signal **(b)** Filtering with sinc filter of length 5. **(c)** Filtering with sinc filter of length 11. **(d)** Interpolation using Lagrange interpolation.

```
TYPE ARRAY_BITS8_11 IS ARRAY (0 TO 10) of BITS8;
TYPE ARRAY_BITS9_11 IS ARRAY (0 TO 10) of BITS9;
TYPE ARRAY_BITS8_3  IS ARRAY (0 TO 2) of BITS8;
TYPE ARRAY_BITS8_4  IS ARRAY (0 TO 3) of BITS8;
TYPE ARRAY_BITS17_11 IS ARRAY (0 TO 10) of BITS17;
END n_bits_int;

LIBRARY work;
USE work.n_bits_int.ALL;

LIBRARY ieee;
USE ieee.std_logic_1164.ALL;
USE ieee.std_logic_arith.ALL;
USE ieee.std_logic_signed.ALL;

ENTITY rc_sinc IS                        ------> Interface
  GENERIC (OL : INTEGER := 2; -- Output buffer length -1
           IL : INTEGER := 3; -- Input buffer length -1
           L  : INTEGER := 10 -- Filter length -1
           );
```

```
  PORT (clk                        : IN  STD_LOGIC;
        x_in                       : IN  BITS8;
        reset                      : IN  STD_LOGIC;
        count_o                    : OUT INTEGER RANGE 0 TO 12;
        ena_in_o, ena_out_o,ena_io_o : OUT BOOLEAN;
        f0_o, f1_o, f2_o           : OUT BITS9;
        y_out                      : OUT BITS9);
END rc_sinc;

ARCHITECTURE fpga OF rc_sinc IS

  SIGNAL count  : INTEGER RANGE 0 TO 12; -- Cycle R_1*R_2
  SIGNAL ena_in, ena_out, ena_io : BOOLEAN; -- FSM enables
  -- Constant arrays for multiplier and taps:
  CONSTANT c0  : ARRAY_BITS9_11
               := (-19,26,-42,106,212,-53,29,-21,16,-13,11);
  CONSTANT c2  : ARRAY_BITS9_11
               := (11,-13,16,-21,29,-53,212,106,-42,26,-19);
  SIGNAL x : ARRAY_BITS8_11 := (0,0,0,0,0,0,0,0,0,0,0);
                            -- TAP registers for 3 filters
  SIGNAL ibuf : ARRAY_BITS8_4 := (0,0,0,0); -- in registers
  SIGNAL obuf : ARRAY_BITS8_3 := (0,0,0);   -- out registers
  SIGNAL f0, f1, f2      : BITS9 := 0; -- Filter outputs

BEGIN

  FSM: PROCESS (reset, clk)     ------> Control the system
  BEGIN                         -- sample at clk rate
    IF reset = '1' THEN         -- Asynchronous reset
      count <= 0;
    ELSIF rising_edge(clk) THEN
      IF count = 11 THEN
        count <= 0;
      ELSE
        count <= count + 1;
      END IF;
      CASE count IS
        WHEN 2 | 5 | 8 | 11 =>
          ena_in <= TRUE;
        WHEN others =>
          ena_in <= FALSE;
      END CASE;
      CASE count IS
        WHEN 4 | 8 =>
          ena_out <= TRUE;
        WHEN others =>
          ena_out <= FALSE;
      END CASE;
      IF COUNT = 0 THEN
        ena_io <= TRUE;
      ELSE
        ena_io <= FALSE;
      END IF;
```

```
      END IF;
    END PROCESS FSM;

    INPUTMUX: PROCESS              ------> One tapped delay line
    BEGIN
      WAIT UNTIL clk = '1';
      IF ENA_IN THEN
        FOR I IN IL DOWNTO 1 LOOP
          ibuf(I) <= ibuf(I-1);        -- shift one
        END LOOP;
        ibuf(0) <= x_in;                -- Input in register 0
      END IF;
    END PROCESS;

    OUPUTMUX: PROCESS              ------> One tapped delay line
    BEGIN
      WAIT UNTIL clk = '1';
      IF ENA_IO THEN  -- store 3 samples in output buffer
        obuf(0) <= f0 ;
        obuf(1) <= f1;
        obuf(2) <= f2 ;
      ELSIF ENA_OUT THEN
        FOR I IN OL DOWNTO 1 LOOP
          obuf(I) <= obuf(I-1);        -- shift one
        END LOOP;
      END IF;
    END PROCESS;

    TAP: PROCESS                   ------> One tapped delay line
    BEGIN                          -- get 4 samples at one time
      WAIT UNTIL clk = '1';
      IF ENA_IO THEN
        FOR I IN 0 TO 3 LOOP -- take over input buffer
          x(I) <= ibuf(I);
        END LOOP;
        FOR I IN 4 TO 10 LOOP -- 0->4; 4->8 etc.
          x(I) <= x(I-4);          -- shift 4 taps
        END LOOP;
      END IF;
    END PROCESS;

    SOP0: PROCESS (clk, x) --> Compute sum-of-products for f0
    VARIABLE sum : BITS17;
    VARIABLE p : ARRAY_BITS17_11;
    BEGIN
      FOR I IN 0 TO L LOOP -- Infer L+1  multiplier
        p(I) := c0(I) * x(I);
      END LOOP;
      sum := p(0);
      FOR I IN 1 TO L  LOOP       -- Compute the direct
        sum := sum + p(I);          -- filter adds
      END LOOP;
      IF clk'event and clk = '1' THEN
```

```
      f0 <= sum /256;
    END IF;
END PROCESS SOP0;

SOP1: PROCESS (clk, x) --> Compute sum-of-products for f1
BEGIN
    IF clk'event and clk = '1' THEN
      f1 <= x(5);  -- No scaling, i.e. unit inpulse
    END IF;
END PROCESS SOP1;

SOP2: PROCESS (clk, x) --> Compute sum-of-products for f2
VARIABLE sum : BITS17;
VARIABLE p : ARRAY_BITS17_11;
BEGIN
    FOR I IN 0 TO L LOOP -- Infer L+1  multiplier
      p(I) := c2(I) * x(I);
    END LOOP;
    sum := p(0);
    FOR I IN 1 TO L  LOOP        -- Compute the direct
      sum := sum + p(I);         -- filter adds
    END LOOP;
    IF clk'event and clk = '1' THEN
      f2 <= sum /256;
    END IF;
END PROCESS SOP2;

f0_o <= f0;        -- Provide some test signal as outputs
f1_o <= f1;
f2_o <= f2;
count_o <= count;
ena_in_o <= ena_in;
ena_out_o <= ena_out;
ena_io_o <= ena_io;

y_out <= obuf(OL); -- Connect to output

END fpga;
```

The first **PROCESS** is the FSM, which includes the control flow and generation of the enable signals for the input and output buffers, and the enable signal **ena_io** for the three filters. The full round takes 12 clock cycles. The next three **PROCESS** blocks include the input buffer, output buffer, and the TAP delay line. Note that only one tapped delay line is used for all three filters. The final three **PROCESS** blocks include the sinc filter. The output **y_out** was chosen to have an additional guard bit. The design uses 448 LEs, 19 embedded multiplier and has a 61.93 MHz **Registered Performance**.

A simulation of the filter is shown in Fig. 5.38. The simulation first shows the control and enable signals of the FSM. A triangular input **x_in** is used. The three filter outputs only update once every 12 clock cycles. The filter output values (**f0,f1,f2**) are arranged in the correct order to generate the output **y_out**. Note that the filter values 20 and 60 from **f1** appear unchanged in the output sequence, while the other values are interpolated.

5.12

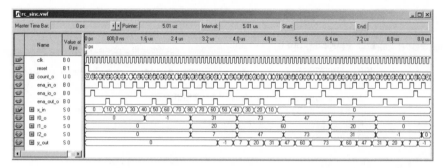

Fig. 5.38. VHDL simulation of the $R = 3/4$ rate change using three sinc filter.

Notice that in this particular case the filters are implemented in the direct rather than in the transposed form, because now we only need one tapped delay line for all three filters. Due to the complexity of the design the coding style for this example was based more on clarity than efficiency. The filters can be improved if MAG coding is used for the filter coefficients and a pipelined adder tree is applied at the filter summations, see Exercise 5.15 (p. 338).

5.6.2 Polynomial Fractional Delay Design

The implementation of the fractional delay via a set of lowpass filters is attractive as long as the number of delays (i.e., nominator R_1 in the rate change factor $R = R_1/R_2$ to be implemented) is small. For large R_1 however, as for examples required for the CD→DAT conversion with rate change factor $R = 147/160$, this implies a large implementation effort, because 147 different lowpass filters need to be implemented. It is then more attractive to compute the fractional delay via a polynomial approximation using Lagrange or spline polynomials [112, 113]. An N-point so-called Lagrange polynomial approximation will be of the type:

$$p(t) = c_0 + c_1 t + c_2 t^2 + \ldots + c_{N-1} t^{N-1}, \tag{5.46}$$

where typically 3-4 points should be enough, although some high-quality audio application use up to 10 terms [114]. The Lagrange polynomial approximation has the tendency to oscillate at the ends, and the interval to be estimated should be at the center of the polynomial. Figure 5.39 illustrates this fact for a signal with just two nonzero values. It can also be observed that a length-4 polynomial already gives a good approximation for the center interval, i.e., $0 \leq n \leq 1$, and that the improvement from a length-4 to a length-8 polynomial is not significant. A bad choice of the approximation interval would be the first or last interval, e.g., the range $3 \leq n \leq 4$ for a length-8 polynomial. This choice would show large oscillations and errors. The input sample set in use should therefore be placed symmetric around

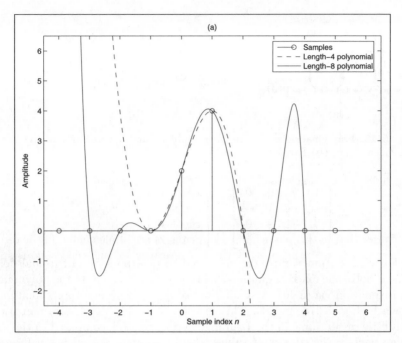

Fig. 5.39. Polynomial approximation using a short and long polynomial.

the interval for which the fractional delay should be approximated such that $0 \leq d \leq 1$. We use input samples at times $-N/2-1, \ldots, N/2$. For four points for example we will use input samples at $-1, 0, 1, 2$. In order to fit the polynomial $p(t)$ through the sample points we substitute the sample times and $x(t)$ values into (5.46) and solve this equation for the coefficients c_k. This matrix equation $\boldsymbol{V}\boldsymbol{c} = \boldsymbol{x}$ leads to a so-called Lagrange polynomial [78, 112] and for $N = 4$, for instance, we get:

$$
\begin{bmatrix} 1 & t_{-1} & t_{-1}^2 & t_{-1}^3 \\ 1 & t_0 & t_0^2 & t_0^3 \\ 1 & t_1 & t_1^2 & t_1^3 \\ 1 & t_2 & t_2^2 & t_2^3 \end{bmatrix} \times \begin{bmatrix} c_0 \\ c_1 \\ c_2 \\ c_3 \end{bmatrix} = \begin{bmatrix} x(n-1) \\ x(n) \\ x(n+1) \\ x(n+2) \end{bmatrix} \tag{5.47}
$$

with $t_k = k$; we need to solve this equation for the unknown coefficients c_n. We also notice that the matrix for the t_k is a Vandermonde matrix a popular matrix type we also use for the DFT. Each line in the Vandermonde matrix is constructed by building the power series of a basic element, i.e., $t_k^l = 1, t_k, t_k^2, \ldots$. Substitution of the t_k and matrix inversion leads to

$$
\begin{bmatrix} c_0 \\ c_1 \\ c_2 \\ c_3 \end{bmatrix} = \begin{bmatrix} 1 & -1 & 1 & -1 \\ 1 & 0 & 0 & 0 \\ 1 & 1 & 1 & 1 \\ 1 & 2 & 4 & 8 \end{bmatrix}^{-1} \times \begin{bmatrix} x(n-1) \\ x(n) \\ x(n+1) \\ x(n+2) \end{bmatrix}
$$

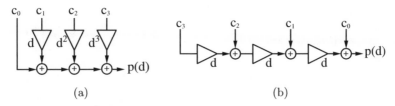

Fig. 5.40. Fractional delay via an $N = 4$ polynomial approximation. **(a)** Direct implementation. **(b)** Farrow structure.

$$= \begin{bmatrix} 0 & 1 & 0 & 0 \\ -\frac{1}{3} & -\frac{1}{2} & 1 & -\frac{1}{6} \\ \frac{1}{2} & -1 & \frac{1}{2} & 0 \\ -\frac{1}{6} & \frac{1}{2} & -\frac{1}{2} & \frac{1}{6} \end{bmatrix} \times \begin{bmatrix} x(n-1) \\ x(n) \\ x(n+1) \\ x(n+2) \end{bmatrix} \tag{5.48}$$

For each output sample we now need to determine the fractional delay value d, solve the matrix equation (5.48), and finally compute the polynomial approximation via (5.46). A simulation using the Lagrange approximation is shown in Fig. 5.37d, p. 286. This give a reasonably exact approximation, with little or no ripple in the Lagrange approximation, compared to the sinc design next to the triangular values where the input and output values are supposed to be zero.

The polynomial evaluation (5.46) can be more efficiently computed if we use the Horner scheme instead of the direct evaluation of (5.46). Instead of

$$p(d) = c_0 + c_1 d + c_2 d^2 + c_3 d^3 \tag{5.49}$$

we use for $N = 4$ the Horner scheme

$$p(d) = c_0 + d(c_1 + d(c_2 + c_3 d)). \tag{5.50}$$

The advantage is that we do not need to evaluate the power of d^k values. This was first suggested by Farrow [115] and is therefore called in the literature the Farrow structure [116, 117, 118, 119]. The Farrow structure for four coefficients is shown in Fig. 5.40b.

Let use now look at an implementation example of the polynomial fractional delay design.

Example 5.13: $R = 0.75$ Rate Changer IV

The following VHDL code[8] shows the Farrow design using a Lagrange polynomial of order 3 for a $R = 3/4$ rate change.

```
PACKAGE n_bits_int IS              -- User-defined types
  SUBTYPE BITS8 IS INTEGER RANGE -128 TO 127;
  SUBTYPE BITS9 IS INTEGER RANGE -2**8 TO 2**8-1;
  SUBTYPE BITS17 IS INTEGER RANGE -2**16 TO 2**16-1;
  TYPE ARRAY_BITS8_4 IS ARRAY (0 TO 3) of BITS8;
```

---

[8] The equivalent Verilog code **farrow.v** for this example can be found in Appendix A on page 697. Synthesis results are shown in Appendix B on page 731.

```
END n_bits_int;

LIBRARY work;
USE work.n_bits_int.ALL;

LIBRARY ieee;
USE ieee.std_logic_1164.ALL;
USE ieee.std_logic_arith.ALL;
USE ieee.std_logic_signed.ALL;

ENTITY farrow IS                            ------> Interface
  GENERIC (IL : INTEGER := 3); -- Input puffer length -1
  PORT (clk                    : IN   STD_LOGIC;
        x_in                   : IN   BITS8;
        reset                  : IN   STD_LOGIC;
        count_o                : OUT INTEGER RANGE 0 TO 12;
        ena_in_o, ena_out_o    : OUT BOOLEAN;
        c0_o, c1_o, c2_o, c3_o : OUT BITS9;
        d_out, y_out           : OUT BITS9);
END farrow;

ARCHITECTURE fpga OF farrow IS

  SIGNAL count  : INTEGER RANGE 0 TO 12; -- Cycle R_1*R_2
  CONSTANT delta : INTEGER := 85; -- Increment d
  SIGNAL ena_in, ena_out : BOOLEAN; -- FSM enables
  SIGNAL x, ibuf : ARRAY_BITS8_4 := (0,0,0,0); -- TAP reg.
  SIGNAL  d : BITS9 := 0; -- Fractional Delay scaled to 8 bits
  -- Lagrange matrix outputs:
  SIGNAL c0, c1, c2, c3      : BITS9 := 0;

BEGIN

  FSM: PROCESS (reset, clk)    ------> Control the system
  VARIABLE dnew : BITS9 := 0;
  BEGIN                        -- sample at clk rate
    IF reset = '1' THEN             -- Asynchronous reset
      count <= 0;
      d <= delta;
    ELSIF rising_edge(clk) THEN
      IF count = 11 THEN
        count <= 0;
      ELSE
        count <= count + 1;
      END IF;
      CASE count IS
        WHEN 2 | 5 | 8 | 11 =>
          ena_in <= TRUE;
        WHEN others =>
          ena_in <= FALSE;
      END CASE;
      CASE count IS
        WHEN 3 | 7 | 11 =>
```

```
              ena_out <= TRUE;
           WHEN others =>
              ena_out <= FALSE;
        END CASE;
  -- Compute phase delay
        IF ENA_OUT THEN
           dnew := d + delta;
           IF dnew >= 255 THEN
            d <= 0;
           ELSE
            d <= dnew;
           END IF;
        END IF;
     END IF;
END PROCESS FSM;


TAP: PROCESS                  ------> One tapped delay line
BEGIN
  WAIT UNTIL clk = '1';
  IF ENA_IN THEN
    FOR I IN 1 TO IL LOOP
      ibuf(I-1) <= ibuf(I);       -- Shift one
    END LOOP;
    ibuf(IL) <= x_in;             -- Input in register IL
  END IF;
END PROCESS;


GET: PROCESS     ------> Get 4 samples at one time
BEGIN
  WAIT UNTIL clk = '1';
  IF ENA_OUT THEN
    FOR I IN 0 TO IL LOOP -- take over input buffer
      x(I) <= ibuf(I);
    END LOOP;
  END IF;
END PROCESS;


--> Compute sum-of-products:
SOP: PROCESS (clk, x, d, c0, c1, c2, c3, ENA_OUT)
VARIABLE y : BITS9;
BEGIN
-- Matrix multiplier iV=inv(Vandermonde) c=iV*x(n-1:n+2)'
--      x(0)     x(1)          x(2)       x(3)
-- iV=   0      1.0000          0          0
--    -0.3333   -0.5000      1.0000     -0.1667
--     0.5000   -1.0000      0.5000        0
--    -0.1667    0.5000     -0.5000      0.1667
  IF ENA_OUT THEN
    IF clk'event AND clk = '1' THEN
      c0 <= x(1);
      c1 <= -85 * x(0)/256 - x(1)/2 + x(2) - 43 * x(3)/256;
      c2 <= (x(0) + x(2)) /2 - x(1) ;
      c3 <= (x(1) - x(2))/2 + 43 * (x(3) - x(0))/256;
```

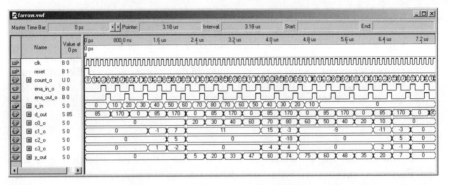

Fig. 5.41. VHDL simulation of the $R = 3/4$ rate change using Lagrange polynomials and a Farrow combiner.

```
        END IF;

    -- Farrow structure = Lagrange with Horner schema
    -- for u=0:3, y=y+f(u)*d^u; end;
      y := c2 + (c3 * d) / 256; -- d is scale by 256
      y := (y * d) / 256 + c1;
      y := (y * d) / 256 + c0;

        IF clk'event AND clk = '1' THEN
          y_out <= y; -- Connect to output + store in register
        END IF;
      END IF;
    END PROCESS SOP;

      c0_o <= c0;      -- Provide some test signals as outputs
      c1_o <= c1;
      c2_o <= c2;
      c3_o <= c3;
      count_o <= count;
      ena_in_o <= ena_in;
      ena_out_o <= ena_out;
      d_out <= d;

    END fpga;
```

The HDL code for the control is similar to the rc_sinc design discussed in Example 5.9 (p. 280). The first PROCESS is the FSM, which includes the control flow and generation of the enable signals for input, output buffer, and the computation of the delay D. The full round takes 12 clock cycles. The next two PROCESS blocks include the input buffer and the TAP delay line. Note that only one tapped delay line is used for all four polynomial coefficients c_k. The SOP PROCESS blocks includes the Lagrange matrix computation and the Farrow combiner. The output y_out was chosen to have an additional guard bit. The design uses 279 LEs, 6 embedded multipliers and has a 43.91 MHz Registered Performance.

A simulation of the filter is shown in Fig. 5.41. The simulation shows first the control and enable signals of the FSM. A triangular input x_in is used.

The three filter outputs only update once every four clock cycles, i.e., three times in an overall cycle. The filter output values are weighted using the Farrow structure to generate the output y_out. Note that only the first and second Lagrange polynomial coefficients are nonzero, due to the fact that a triangular input signal does not have higher polynomial coefficient. Notice also that the filter values 20 and 60 from c0 appear unchanged in the output sequence (because $D = 0$ at these points in time), while the other values are interpolated.

$\boxed{5.13}$

Although the implementation data for the Lagrange interpolation with the Farrow combiner and the sinc filter design do not differ much for our example design with $R = R_1/R_2 = 3/4$, larger differences occur when we try to implement rate changes with large values of R_1. The discussed Farrow design only needs to be changed in the enable signal generation. The effort for the Lagrange interpolation and Farrow combiner remain the same, while for a sinc filter the design effort will be proportional to the number of filters to be implemented, i.e., R_1, see Exercise 5.16 (p. 338). The only disadvantage of the Farrow combiner is the long latency due to the sequential organization of the multiplications, but this can be improved by adding pipeline stages for the multipliers and coefficient data, see Exercise 5.17 (p. 339).

5.6.3 B-Spline-Based Fractional Rate Changer

Polynomial approximation using Lagrange polynomials is smooth in the center but has the tendency to have large ripples at the end of the polynomials, see Fig. 5.38, p. 290. Much smoother behavior is promised when B-spline approximation functions are used. In contrast to Lagrange polynomials B-splines are of finite length and a B-spline of degree N must be N-times differentiable, hence the smooth behavior. Depending on the border definitions, several versions of B-splines can be defined [78, p.113-116], but the most popular are those defined via the integration of the box function, as shown in Fig. 5.42. A B-spline of degree zero is integrated to give a triangular B-spline, degree-one B-spline integrated yields a quadratic function, etc.

An analytic description of a B-spline is possible [120, 121] using the following representation of the ramp function:

$$(t - \tau)_+ = \begin{cases} t - \tau & \forall\, t > \tau \\ 0 & \text{otherwise} \end{cases}. \tag{5.51}$$

This allows us to represent the Nth-degree symmetric B-spline as

$$\beta^N(t) = \frac{1}{N!} \sum_{k=0}^{N+1} (-1)^k \binom{N+1}{k} \left(t - k + \frac{N+1}{2}\right)_+^N. \tag{5.52}$$

All segments of the B-splines use polynomials of degree N and are therefore N-times differentiable, resulting in a smooth behavior also at the end of the

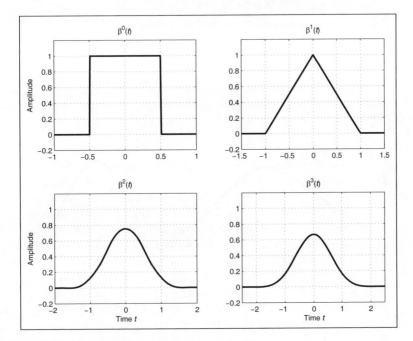

Fig. 5.42. B-spline functions of degree zero to three.

B-splines. Zero- and first-degree B-splines give box and triangular representations, respectively; quadratic and cubic B-splines are next. Cubic B-splines are the most popular type used in DSP although for very high-quality speech processing degree six has been used [114]. For a cubic B-spline (5.52) for instance we get

$$
\beta^3(t) = \frac{1}{6} \sum_{k=0}^{4} (-1)^k \binom{4}{k} (t - k + 2)_+^3
$$

$$
= \frac{1}{6}(t+2)_+^3 - \frac{2}{3}(t+1)_+^3 + t_+^3 - \frac{2}{3}(t-1)_+^3 + \frac{1}{6}(t-2)_+^3
$$

$$
= \underbrace{\frac{1}{6}(t+2)^3}_{t>-2} - \underbrace{\frac{2}{3}(t+1)^3}_{t>-1} + \underbrace{t^3}_{t>0} - \underbrace{\frac{2}{3}(t-1)^3}_{t>1} + \underbrace{\frac{1}{6}(t-2)^3}_{t>2}. \quad (5.53)
$$

We can now use this cubic B-spline for the reconstruction of the spline, by summation of the weighted sequence of the B-splines, i.e.,

$$
\hat{y}(t) = \sum_k x(k)\, \beta^3(t - k). \tag{5.54}
$$

This weighted sum is shown in Fig. 5.43 as a bold line. Although $\hat{y}(t)$ is quite smooth, we can also observe that the spline $\hat{y}(t)$ does not go exactly through the sample points, i.e., $\hat{y}(k) \neq x(k)$. Such a B-spline reconstruction is called

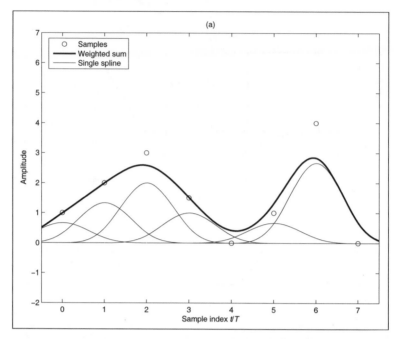

Fig. 5.43. Spline approximation using cubic B-splines.

in the literature a B-spline *approximation*. From a B-spline *interpolation*, however, we expect that the weighted sum goes exactly through our sample points [122]. For cubic B-splines for instance it turns out [117, 123] that the cubic B-spline applies a filter weight whose z-transform is given by

$$H(z) = \frac{z + 4 + z^{-1}}{6} \tag{5.55}$$

to the sample points. To achieve a perfect interpolation we therefore need to apply an inverse cubic B-spline filter, i.e.,

$$F(z) = 1/H(z) = \frac{6}{z + 4 + z^{-1}} \tag{5.56}$$

to our input samples. Unfortunately the pole/zero plot of this filter reveals that this IIR filter in not stable and, if we apply this filter to our input sequence, we may produce an increasing signal for the impulse response, see Exercise 5.18 (p. 339). Unser et al. [124] suggest spliting the filter into a stable, causal part and a stable, a-causal filter part and applying the a-causal filter starting with the last value of the output of the first causal filter. While this works well in image processing with a finite number of samples in each image line, in a continuous signal processing scheme this is not practical, especially when the filters are implemented with finite arithmetic.

However another approach that can be used for continuous signal processing is to approximate the filter $F(z) = 1/H(z)$ by an FIR filter. It turns

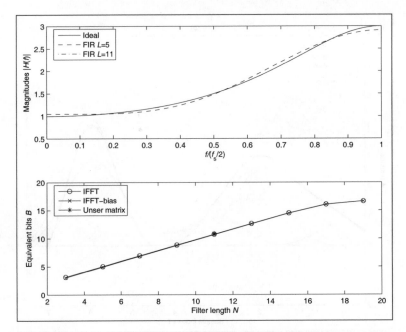

Fig. 5.44. FIR compensation filter design for cubic B-spline interpolation.

out that even very few FIR coefficients give a good approximation, because
the transfer function does not have any sharp edges, see Fig. 5.44. We just
need to compute the transfer function of the IIR filter and then take the
IFFT of the transfer function to determine the FIR time values. We may also
apply a bias correction if a DC shift is critical in the application. Unser et
al. [125] suggested an algorithm to optimize the filter coefficient set, but due
to the nature of the finite coefficient precision and finite coefficient set, the
gain compared with the direct IFFT method is not significant, see Fig. 5.44
and Exercise 5.19, p. 339.

Now we can apply this FIR filter first to our input samples and then use a
cubic B-spline reconstruction. As can be seen from Fig. 5.45 we have in fact
an interpolation, i.e., the reconstructed function goes through our original
sampling points, i.e., $\hat{y}(k) = x(k)$.

The only thing left to do is to develop a fractional delay B-spline interpolation and to determine the Farrow filter structure. We want to use the
popular cubic B-spline set and only consider fractional delays in the range
$0 \le d \le 1$. For an interpolation with four points we use the samples at time
instances $t = -1, 0, 1, 2$ of the input signal [116, p. 780]. With the B-spline
representation (5.53) and the weighted sum (5.54) we also find that four
B-spline segments have to be considered and we arrive at

$$y(d) = x(n+2)\beta^3(d-2) + x(n+1)\beta^3(d-1)$$

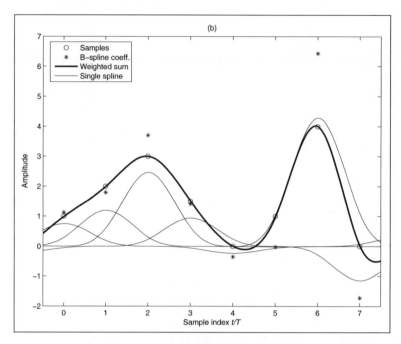

Fig. 5.45. Interpolation using cubic B-splines and FIR compensation filter.

$$+x(n)\beta^3(d) + x(n-1)\beta^3(d-1) \tag{5.57}$$

$$= x(n+2)\frac{d^3}{6} + x(n+1)\left[\frac{1}{6}(d+1)^3 - \frac{2}{3}d^3\right] + x(n) \times$$

$$\left[d^3 - \frac{2}{3}(d+1)^3 + \frac{1}{6}(d+2)^3\right] + x(n-1)\left[-\frac{1}{6}(d-1)^3\right] \tag{5.58}$$

$$= x(n+2)\frac{d^3}{6} + x(n+1)\left[-\frac{d^3}{2} + \frac{d^2}{2} + \frac{d}{2} + \frac{1}{6}\right]$$

$$+x(n)\left[\frac{d^3}{2} - d^2 + \frac{2}{3}\right] + x(n-1)\left[-\frac{d^3}{6} + \frac{d^2}{2} - \frac{d}{2} + \frac{1}{6}\right]. \tag{5.59}$$

In order to realize this in a Farrow structure we need to summarize according to the factors d^k, which yields the following four equations:

$$\begin{array}{llllll}
d^0: & 0 & +x(n+1)/6 & +2x(n)/3 & +x(n-1)/6 & = c_0 \\
d^1: & 0 & +x(n+1)/2 & +0 & -x(n-1)/2 & = c_1 \\
d^2: & 0 & +x(n+1)/2 & -x(n) & +x(n-1)/2 & = c_2 \\
d^3: & x(n+2)/6 & -x(n+1)/2 & x(n)/2 & -x(n-1)/6 & = c_3
\end{array} \tag{5.60}$$

This Farrow structure can be translated directly into a B-spline rate changer as discussed in Exercise 5.23, p. 340.

5.6.4 MOMS Fractional Rate Changer

One aspect[9] often overlooked in traditional design of interpolation kernels $\phi(t)$ is the *order* of the approximation, which is an essential parameter in the quality of the interpolation result. Here the order is defined by the rate of decrease of the square error (i.e., L^2 norm) between the original function and the reconstructed function when the sampling step vanishes. In terms of implementation effort the *support* or length of the interpolation function is a critical design parameter. It is now important to notice that the B-splines used in the last section is both maximum order and minimum support (MOMS) [126]. The question then is whether or not the B-spline is the only kernel that has degree $L - 1$, support of length L, and order L. It turns out that there is a whole class of functions that obey this MOMS behavior. This class of interpolating polynomials can be described as

$$\phi(t) = \beta^N(t) + \sum_{k=1}^{N} p(k)\frac{\mathrm{d}^k \, \beta^N(t)}{\mathrm{d}t^k}. \tag{5.61}$$

Since B-splines are built via successive convolution with the box function the differentiation can be computed via

$$\frac{\mathrm{d}\beta^{k+1}(t)}{\mathrm{d}t} = \beta^k\left(t + \frac{1}{2}\right) - \beta^k\left(t - \frac{1}{2}\right). \tag{5.62}$$

From (5.61) it can be seen that we have a set of design parameters $p(k)$ at hand that can be chosen to meet certain design goals. In many designs symmetry of the interpolation kernel is desired, forcing all odd coefficients $p(k)$ to zero. A popular choice is $N = 3$, i.e., the cubic spline type, and it follows then that

$$\begin{aligned}
\phi(t) &= \beta^3(t) + p(2)\frac{\mathrm{d}^2\beta^3(t)}{\mathrm{d}t^2} \\
&= \beta^3(t) + p(2)\left(\beta^1(t+1) - 2\beta^1(t) + \beta^1(t-1)\right)
\end{aligned} \tag{5.63}$$

and only the design parameter $p(2)$ needs to be determined. We may, for instance, try to design a direct interpolating function that requires no compensation filter at all. Those I-MOMS occur for $p(2) = -1/6$, and are identical to the Lagrange interpolation (see Exercise 5.20, p. 339) and therefore give suboptimal interpolation results. Figure 5.46(b) shows the I-MOMS interpolation of degree three. Another design goal may be to minimize the interpolation error in the L^2 norm sense. These O-MOMS require $p(2) = 1/42$, and the approximation error is a magnitude smaller than for I-MOMS [127]. Figure 5.46(c) shows the O-MOMS interpolation kernel for degree three. We may also use an iterative method to maximize a specific application the S/N of the interpolation. For a specific set of five images, for instance, $p(2) = 1/28$ has been found to perform 1 dB better than O-MOMS [128].

---

[9] This section was suggested by P. Thévenaz from EPFL.

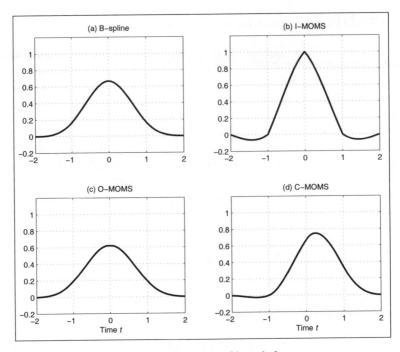

Fig. 5.46. Possible MOMS kernel functions of length four.

Unfortunately the compensation filter required for O-MOMS has (as for B-splines) an instable pole location and an FIR approximation has to be used in a continuous signal processing scheme. The promised gain via the small L^2 error of the O-MOMS will therefore most likely not result in much overall gain if the FIR has to be built in finite-precision arithmetic. If we give up the symmetry requirements of the kernel then we can design MOMS functions in such a way that the interpolation function sampled at integer points $\phi(k)$ is a causal function, i.e., $\phi(-1) = 0$, as can be seen from Fig. 5.46(d). This C-MOMS function is fairly smooth since $p(2) = p(3) = 0$. The C-MOMS requirement demands $p(1) = -1/3$ and we get the asymmetric interpolation function, but with the major advantage that a simple one-pole stable IIR compensation filter with $F(z) = 1.5/(1 + 0.5z^{-1})$ can be used. No FIR approximation is necessary as for B-splines or O-MOMS [127]. It is now interesting to observe that the C-MOMS maxima and sample points no longer have the same time location as in the symmetric kernel, e.g., B-spline case. To see this compare Fig. 5.45 with Fig. 5.47. However, the weighted sum of the C-MOMS goes thorough the sample point as we expect for a spline interpolation. Experiments with C-MOMS splines shows that in terms of interpolation C-MOMS performs better than B-splines and a little worse than O-MOMS, when O-MOMS is implemented at full precision.

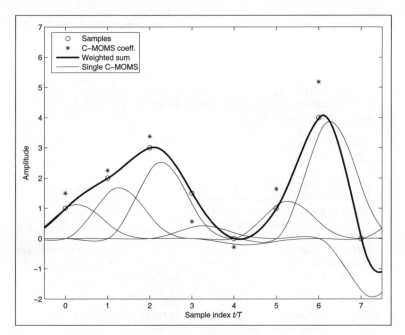

Fig. 5.47. C-MOMS interpolation using IIR compensation filter.

The only thing left to do is to compute the equation for the Farrow re-sampler. We may use (5.53) to compute the interpolation function $\phi(t) = \beta^3(t) - 1/3 \mathrm{d}\beta^3(t)/\mathrm{d}t$ and then the Farrow structure is sorted according to the delays d^k. Alternatively we can use the precomputed equation (5.60) for the B-splines and apply the differentiation directly to the Farrow matrix, i.e., we compute $c_k^{\mathrm{new}} = c_k - kc_{k+1}/3$ for $k = 1, 2,$ and 3, since $p(1) = -1/3$ and $p(2) = p(3) = 0$ for cubic C-MOMS. The same principle can also be applied to compute the Farrow equations for O-MOMS and I-MOMS, see Exercise 5.20, p. 339. For C-MOMS this yields the following four equations:

$$
\begin{aligned}
d^0 &: & 0 & & & +2x(n)/3 & +1x(n-1)/3 &= c_0 \\
d^1 &: & 0 & & +x(n+1)/6 & +2x(n)/3 & -5x(n-1)/6 &= c_1 \\
d^2 &: & -x(n+2)/6 & +x(n+1) & -3x(n)/2 & +2x(n-1)/3 &= c_2 \\
d^3 &: & x(n+2)/6 & -x(n+1)/2 & +x(n)/2 & -x(n-1)/6 &= c_3
\end{aligned}
\tag{5.64}
$$

We now develop the VHDL code for the cubic C-MOMS fractional rate changer.

Example 5.14: $R= 0.75$ Rate Changer V

The following VHDL code[10] shows an $R = 3/4$ rate change using a C-MOMS spline polynomial of degree three.

---

[10] The equivalent Verilog code `cmomc.v` for this example can be found in Appendix A on page 697. Synthesis results are shown in Appendix B on page 731.

```
PACKAGE n_bits_int IS            -- User-defined types
  SUBTYPE BITS8 IS INTEGER RANGE -128 TO 127;
  SUBTYPE BITS9 IS INTEGER RANGE -2**8 TO 2**8-1;
  SUBTYPE BITS17 IS INTEGER RANGE -2**16 TO 2**16-1;
  TYPE ARRAY_BITS8_4 IS ARRAY (0 TO 3) of BITS8;
  TYPE ARRAY_BITS9_3 IS ARRAY (0 TO 2) of BITS9;
  TYPE ARRAY_BITS17_5 IS ARRAY (0 TO 4) of BITS17;
END n_bits_int;

LIBRARY work;
USE work.n_bits_int.ALL;

LIBRARY ieee;
USE ieee.std_logic_1164.ALL;
USE ieee.std_logic_arith.ALL;
USE ieee.std_logic_signed.ALL;

ENTITY cmoms IS                         ------> Interface
  GENERIC (IL : INTEGER := 3);-- Input puffer length -1
  PORT (clk              : IN  STD_LOGIC;
        x_in             : IN  BITS8;
        reset            : IN  STD_LOGIC;
        count_o          : OUT INTEGER RANGE 0 TO 12;
        ena_in_o, ena_out_o : OUT BOOLEAN;
        t_out               : out INTEGER RANGE 0 TO 2;
        d1_out           : out BITS9;
        c0_o, c1_o, c2_o, c3_o : OUT BITS9;
        xiir_o, y_out    : OUT BITS9);
END cmoms;

ARCHITECTURE fpga OF cmoms IS

  SIGNAL count  : INTEGER RANGE 0 TO 12; -- Cycle R_1*R_2
  SIGNAL t      : INTEGER RANGE 0 TO 2;
  SIGNAL ena_in, ena_out : BOOLEAN; -- FSM enables
  SIGNAL x, ibuf : ARRAY_BITS8_4 := (0,0,0,0); -- TAP regs.
  SIGNAL xiir : BITS9 := 0; -- iir filter output
  -- Precomputed value for d**k
  CONSTANT d1 : ARRAY_BITS9_3 := (0,85,171);
  CONSTANT d2 : ARRAY_BITS9_3 := (0,28,114);
  CONSTANT d3 : ARRAY_BITS9_3 := (0,9,76);
  -- Spline matrix output:
  SIGNAL c0, c1, c2, c3      : BITS9 := 0;

BEGIN
  t_out <= t;
  d1_out <= d1(t);
  FSM: PROCESS (reset, clk)    ------> Control the system
  BEGIN                              -- sample at clk rate
    IF reset = '1' THEN              -- Asynchronous reset
      count <= 0;
      t <= 1;
    ELSIF rising_edge(clk) THEN
```

```
          IF count = 11 THEN
            count <= 0;
          ELSE
            count <= count + 1;
          END IF;
          CASE count IS
            WHEN 2 | 5 | 8 | 11 =>
               ena_in <= TRUE;
             WHEN others =>
               ena_in <= FALSE;
          END CASE;
          CASE count IS
            WHEN 3 | 7 | 11 =>
               ena_out <= TRUE;
             WHEN others =>
               ena_out <= FALSE;
          END CASE;
  -- Compute phase delay
          IF ENA_OUT THEN
            IF t >= 2 THEN
              t <= 0;
            ELSE
              t <= t + 1;
            END IF;
          END IF;
       END IF;
     END PROCESS FSM;

  --  Coeffs: H(z)=1.5/(1+0.5z^-1)
    IIR: PROCESS (clk)               ------> Behavioral Style
      VARIABLE x1 : BITS9 := 0;
    BEGIN    -- Compute iir coefficients first
      IF rising_edge(clk) THEN   -- iir:
        IF ENA_IN THEN
          xiir <= 3 * x1 / 2 - xiir / 2;
          x1 := x_in;
        END IF;
      END IF;
    END PROCESS;

    TAP: PROCESS                ------> One tapped delay line
    BEGIN
      WAIT UNTIL clk = '1';
      IF ENA_IN THEN
        FOR I IN 1 TO IL LOOP
          ibuf(I-1) <= ibuf(I);       -- Shift one
        END LOOP;
        ibuf(IL) <= xiir;           -- Input in register IL
      END IF;
    END PROCESS;

    GET: PROCESS     ------> Get 4 samples at one time
    BEGIN
```

```
        WAIT UNTIL clk = '1';
        IF ENA_OUT THEN
          FOR I IN 0 TO IL LOOP -- take over input buffer
            x(I) <= ibuf(I);
          END LOOP;
        END IF;
      END PROCESS;

      -- Compute sum-of-products:
      SOP: PROCESS (clk, x, c0, c1, c2, c3, ENA_OUT)
      VARIABLE y, y0, y1, y2, y3, h0, h1 : BITS17;
      BEGIN                            -- pipeline registers
-- Matrix multiplier C-MOMS matrix:
--     x(0)        x(1)        x(2)        x(3)
--     0.3333      0.6667      0           0
--    -0.8333      0.6667      0.1667      0
--     0.6667     -1.5         1.0        -0.1667
--    -0.1667      0.5        -0.5         0.1667
      IF ENA_OUT THEN
        IF clk'event and clk = '1' THEN
          c0 <= (85 * x(0) + 171 * x(1))/256;
          c1 <= (171 * x(1) - 213 * x(0) + 43 * x(2)) / 256;
          c2 <= (171 * x(0) - 43 * x(3))/256 - 3*x(1)/2 + x(2);
          c3 <= 43 * (x(3) - x(0)) / 256 +  (x(1) - x(2))/2;
-- No Farrow structure, parallel LUT for delays
-- for u=0:3, y=y+f(u)*d^u; end;
          y  :=  h0 + h1;
          h0 :=  y0 + y1;
          h1 :=  y2 + y3;
          y0 :=  c0 * 256;
          y1 :=  c1 * d1(t);
          y2 :=  c2 * d2(t);
          y3 :=  c3 * d3(t);
        END IF;
      END IF;
      y_out <= y/256; -- Connect to output
      y_full <= y;

    END PROCESS SOP;
      c0_o <= c0; -- Provide some test signal as outputs
      c1_o <= c1;
      c2_o <= c2;
      c3_o <= c3;
      count_o <= count;
      ena_in_o <= ena_in;
      ena_out_o <= ena_out;
      xiir_o <= xiir;

  END fpga;
```

The HDL code for the control is similar the the rc_sinc design discussed in Example 5.9, p. 280. The first PROCESS is the FSM and includes the control flow and the generation of the enable signals for input and output buffer. The computation of the index for the delay $d1 = d^1$ and its power representation

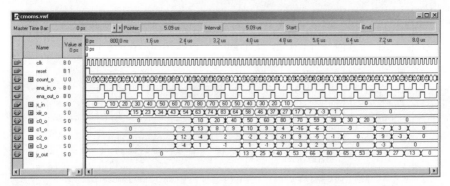

Fig. 5.48. VHDL simulation of the $R = 3/4$ rate change using cubic C-MOMS splines and a one-pole IIR compensation filter.

d2= d^2 and d3= d^3 are precomputed and stored in tables as constant. The full round takes 12 clock cycles. The IIR PROCESS blocks include the IIR compensation filter. The next two PROCESS blocks include the input buffer and the TAP delay line. Note that only one tapped delay line is used for all four polynomial coefficients c_k. The SOP PROCESS block includes the cubic C-MOMS matrix computation and the output combiner. Note that no Farrow structure is used to speed up the computation with a parallel multiplier/adder tree structure. This speeds up the design by a factor of 2. The output y_out was chosen to have an additional guard bit. The design uses 372 LEs, 10 embedded multipliers and has an 85.94 MHz Registered Performance.

A simulation of the filter is shown in Fig. 5.48. The simulation shows first the control and enable signals of the FSM. A rectangular input x_in similar to that in Fig. 5.49 is used. The IIR filter output shows the sharpening of the edges. The C-MOMS matrix output values c_k are weighted by d^k and summed to generate the output y_out.

$$\boxed{5.14}$$

As with the Lagrange interpolation we may also use a Farrow combiner to compute the output y_out. This is particular interesting if array multipliers are available and we have large values of R_1 and therefore large constant table requirements, see Exercise 5.21 (p. 339).

Finally let us demonstrate the limits of our rate change methods. One particularly difficult problem [129] is the rate change for a rectangular input signal, since we know from the Gibbs phenomenon (see Fig. 3.6, p. 174) that any finite filter has the tendency to introduce ringing at a rectangular edge. Within a DAT recorder two frequencies 32 kHz and 48 kHz are in use and a conversion between them is a common task. The rational rate change factor in this case is $R = 3/2$, if we increase the sampling rate from 32 to 48 kHz. This rate change is shown for a rectangular wave in Fig. 5.49 using O-MOMS spline interpolation. The FIR prefilter output shown in Fig. 5.49b emphasizes the edges. Figure 5.49c shows the result of the O-MOMS cubic spline rate changer without FIR prefiltering. Although the signal without the filter seems smoother, a closer look reveals that the edges in the O-MOMS cubic spline

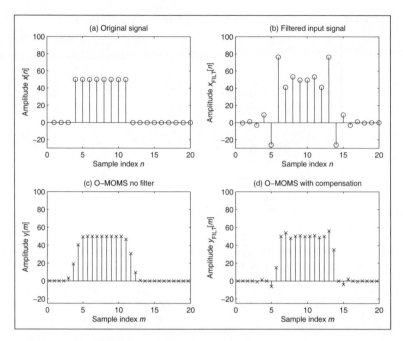

Fig. 5.49. O-MOMS-based fractional $R = 3/2$ rate change. **(a)** Original signal. **(b)** Original signal filter with length-11 FIR compensation filter. **(c)** O-MOMS approximation (no compensation filter). **(d)** O-MOMS rate change using a compensation filter.

interpolation are now better preserved than without prefiltering, as shown in Fig. 5.49d. But it can still be seen that, even with O-MOMS and a length-11 FIR compensation filter at full precision, the Gibbs phenomenon is visible.

5.7 Filter Banks

A *digital filter bank* is a collection of filters having a common input or output, as shown in Fig. 5.50. One common application of the *analysis filter bank* shown in Fig. 5.50a is spectrum analysis, i.e., to split the input signal into R different so-called subband signals. The combination of several signals into a common output signal, as shown in Fig. 5.50b, is called a *synthesis filter bank*. The analysis filter may be nonoverlapping, slightly overlapping, or substantially overlapping. Figure 5.51 shows an example of a slightly overlapping filter bank, which is the most common case.

Another important characteristic that distinguishes different classes of filter banks is the bandwidth and spacing of the center frequencies of the filters. A popular example of a *nonuniform filter bank* is the octave-spaced or *wavelet filter bank*, which will be discussed in Sect. 5.8 (p. 328). In *uniform*

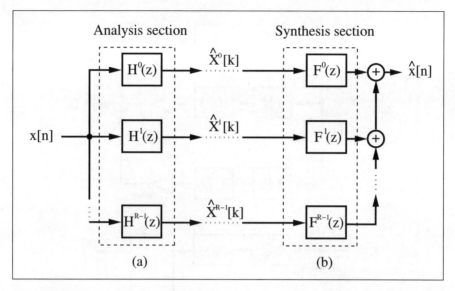

Fig. 5.50. Typical filter bank decomposition system showing **(a)** analysis, and **(b)** synthesis filters.

filter banks, all filters have the same bandwidth and sampling rates. From the implementation standpoint, uniform, maximal decimating filter banks are often preferred, because they can be realized with the help of an FFT algorithm, as shown in the next section.

5.7.1 Uniform DFT Filter Bank

In a maximal decimating, or critically sampled filter bank, the decimation or interpolation R is equal to the number of bands K. We call it a *DFT filter bank* if the rth band filter $h^r[n]$ is computed from the "modulation" of a single prototype filter $h[n]$, according to

$$h^r[n] = h[n]W_R^{rn} = h[n]e^{-j2\pi rn/R}. \tag{5.65}$$

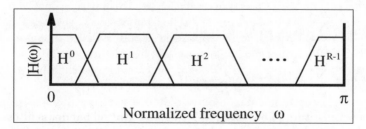

Fig. 5.51. R channel filter bank, with a small amount of overlapping.

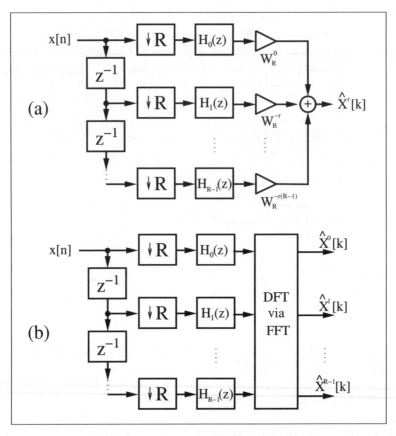

Fig. 5.52. (a) Analysis DFT filter bank for channel k. (b) Complete analysis DFT filter bank.

An efficient implementation of the R channel filter bank can be generated if we use polyphase decomposition (see Sect. 5.2, p. 249) of the filter $h^r[n]$ and the input signal $x[n]$. Because each of these bandpass filters is critically sampled, we use a decomposition with R polyphase signals according to

$$h[n] = \sum_{k=0}^{R-1} h_k[n] \leftrightarrow h_k[m] = h[mR - k] \tag{5.66}$$

$$x[n] = \sum_{k=0}^{R-1} x_k[n] \leftrightarrow x_k[m] = x[mR - k]. \tag{5.67}$$

If we now substitute (5.66) into (5.65), we find that all bandpass filters $h^r[n]$ share the same polyphase filter $h_k[n]$, while the "twiddle factors" for each filter are different. This structure is shown in Fig. 5.52a for the rth filter $h^r[n]$.

It is now obvious that this "twiddle multiplication" for $h^r[n]$ corresponds to the rth DFT component, with an input vector of $\hat{x}_0[n], \hat{x}_1[n], \ldots, \hat{x}_{R-1}[n]$. The computation for the whole analysis band can be reduced to filtering with R polyphase filters, followed by a DFT (or FFT) of these R filtered components, as shown in Fig. 5.52b. This is obviously much more efficient than direct computation using the filter defined in (5.65) (see Exercise 5.6, p. 336).

The polyphase filter bank for the uniform DFT *synthesis bank* can be developed as an inverse operation to the analysis bank, i.e., we can use the R spectral components $\hat{X}^r[k]$ as input for the inverse DFT (or FFT), and reconstruct the output signal using a polyphase interpolator structure, shown in Fig. 5.53. The reconstruction bandpass filter becomes

$$f^r[n] = \frac{1}{R} f[n] W_R^{-rn} = f[n] e^{j2\pi rn/R}. \tag{5.68}$$

If we now combine the analysis and synthesis filter banks, we can see that the DFT and IDFT annihilate each other, and *perfect reconstruction* occurs if the convolution of the included polyphase filter gives a unit sample function, i.e.,

$$h_r[n] * f_r[n] = \begin{cases} 1 & n = d \\ 0 & \text{else.} \end{cases} \tag{5.69}$$

In other words, the two polyphase functions must be inverse filters of each other, i.e.,

$$H_r(z) \times F_r(z) = z^{-d}$$
$$F_r(z) = \frac{z^{-d}}{H_r(z)},$$

where we allow a delay d in order to have causal (realizable) filters. In a practical design, these ideal conditions cannot be met exactly by two FIR filters. We can use approximation for the two FIR filters, or we can combine an FIR and IIR, as shown in the following example.

Example 5.15: DFT Filter Bank

The *lossy integrator* studied in Example 4.3 (p. 231) should be interpreted in the context of a DFT filter bank with $R = 2$. The difference equation was

$$y[n+1] = \frac{3}{4} y[n] + x[n]. \tag{5.70}$$

The impulse response of this filter in the z-domain is

$$F(z) = \frac{z^{-1}}{1 - 0.75z^{-1}}. \tag{5.71}$$

In order to get two polyphase filters, we use a similar scheme as for the "scattered look-ahead" modification (see Example 4.5, p. 234), i.e., we introduce an additional pole/zero pair at the mirror position. Multiplying nominator and denominator by $(1 + 0.75z^{-1})$ yields

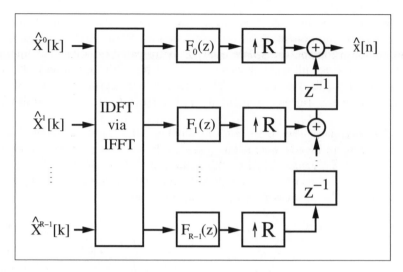

Fig. 5.53. DFT synthesis filter bank.

$$F(z) = \underbrace{\frac{0.75z^{-2}}{1 - 0.75^2 z^{-2}}}_{H_0(z^2)} + z^{-1} \underbrace{\frac{1}{1 - 0.75^2 z^{-2}}}_{H_1(z^2)} \tag{5.72}$$

$$= H_0\left(z^2\right) + z^{-1} H_1\left(z^2\right), \tag{5.73}$$

which gives the two polyphase filters:

$$H_0(z) = \frac{0.75z^{-1}}{1 - 0.75^2 z^{-1}} = 0.75z^{-1} + 0.4219z^{-2} + 0.2373z^{-3} + \dots \tag{5.74}$$

$$H_1(z) = \frac{1}{1 - 0.75^2 z^{-1}} = 1 + 0.5625z^{-1} + 0.3164z^{-2} + \dots . \tag{5.75}$$

We can approximate these impulse responses with a nonrecursive FIR, but to get less than 1% error we must use about 16 coefficients. It is therefore much more efficient if we use the two recursive polyphase IIR filters defined by (5.74) and (5.75). After decomposition with the polyphase filters, we then apply a 2-point DFT, which is given by

$$\boldsymbol{W} = \begin{bmatrix} 1 & 1 \\ 1 & -1 \end{bmatrix}.$$

The whole analysis filter bank can now be constructed as shown in 5.54a. For the synthesis bank, we first compute the inverse DFT using

$$\boldsymbol{W}^{-1} = \frac{1}{2} \begin{bmatrix} 1 & 1 \\ 1 & -1 \end{bmatrix}.$$

In order to get a perfect reconstruction we must find the inverse polyphase filter to $h_0[n]$ and $h_1[n]$. This is not difficult, because the $H_r(z)'s$ are single-pole IIR filters, and $F_r(z) = z^{-d}/H_r(z)$ must therefore be two-tap FIR filters.

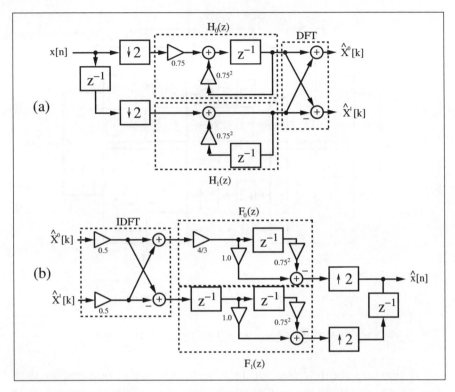

Fig. 5.54. Critically sampled uniform DFT filter bank for $R = 2$. **(a)** Analysis filter bank. **(b)** Synthesis filter bank.

Using (5.74) and (5.75), we find that $d = 1$ is already sufficient to get causal filters, and it is

$$F_0[n] = \frac{4}{3}\left(1 - 0.75^2 z^{-1}\right) \tag{5.76}$$

$$F_1[n] = z^{-1} - 0.75^2 z^{-2}. \tag{5.77}$$

The synthesis bank is graphically interpreted in Fig. 5.54b.   5.15

5.7.2 Two-channel Filter Banks

Two-channel filter banks are an important tool for the design of general filter banks and wavelets. Figure 5.55 shows an example of a two-channel filter bank that splits the input $x[n]$ using lowpass ($G(z)$) and highpass ($H(z)$) "analysis" filters. The resulting signal $\hat{x}[n]$ is reconstructed using lowpass and highpass "synthesis" filters. Between the analysis and synthesis sections

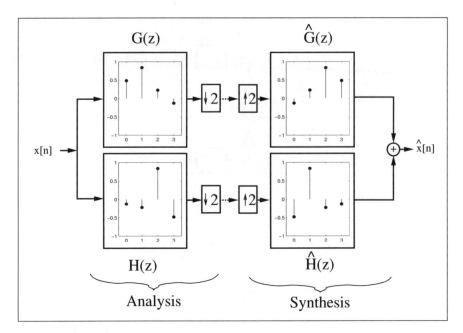

Fig. 5.55. Two-channel filter bank using Daubechies filter of length-4.

are decimation and interpolation by 2 units. The signal between the decimators and interpolators is often quantized, and nonlinearly processed for enhancement, or compressed.

It is common practice to define only the lowpass filter $G(z)$, and to use its definition to specify the highpass filter $H(z)$. The construction rule is normally given by

$$h[n] = (-1)^n g[n] \;\; \circ\!\!-\!\!\bullet \;\; H(z) = G(-z), \qquad (5.78)$$

which defines the filters to be mirrored pairs. Specifically, in the frequency domain, $|H(e^{j\omega})| = |G(e^{j(\omega-\pi)})|$. This is a *quadrature mirror filter (QMF)* bank, because the two filters have mirror symmetry to $\pi/2$.

For the synthesis shown in Fig. 5.55, we first use an expander (a sampling rate increase of 2), and then two separate reconstruction filters, $\hat{G}(z)$ and $\hat{H}(z)$, to reconstruct $\hat{x}[n]$. A challenging question now is, can the input signal be perfectly reconstructed, i.e., can we satisfy

$$\hat{x}[n] = x[n - d]? \qquad (5.79)$$

That is, a perfectly reconstructed signal has the same shape as the original, up to a phase (time) shift. Because $G(z)$ and $H(z)$ are not ideal rectangular filters, achieving perfect reconstruction is not a trivial problem. Both filters produce essential aliasing components after the downsampling by 2, as shown in Fig. 5.55. The simple orthogonal filter bank that satisfies (5.79) is attributed to Alfred Haar (circa 1910) [130].

Example 5.16: Two-Channel Haar Filter Bank I

The filter transfer functions of the two-channel QMF filter bank from Fig. 5.56 are[12]

$$G(z) = 1 + z^{-1} \qquad H(z) = 1 - z^{-1}$$
$$\hat{G}(z) = \frac{1}{2}(1 + z^{-1}) \qquad \hat{H}(z) = \frac{1}{2}(-1 + z^{-1}).$$

Using data found in the table in Fig. 5.56, it can be verified that the filter produces a perfect reconstruction of the input. The input sequence $x[0], x[1], x[2], \ldots$, processed by $G(z)$ and $H(z)$, yields the sum $x[n] + x[n-1]$ and difference $x[n] - x[n-1]$, respectively. The downsampling followed by upsampling forces every second value to zero. After applying the synthesis filter and combining the output we again get the input sequence delayed by one, i.e., $\hat{x}[n] = x[n-1]$, a *perfect* reconstruction with $d = 1$. ⬛5.16

In the following we will discuss the general relationships the four filters must obey to get a perfect reconstruction. It is useful to remember that decimation and interpolation by 2 of a signal $s[k]$ is equivalent to multiplying $S(z)$ by the sequence $\{1, 0, 1, 0, \ldots,\}$. This translates, in the z-domain, to

$$S_{\downarrow\uparrow}(z) = \frac{1}{2}\left(S(z) + S(-z)\right). \tag{5.80}$$

If this signal is applied to the two-channel filter bank, the lowpass path $X_{\downarrow\uparrow G}(z)$ and highpass path $X_{\downarrow\uparrow H}(z)$ become

$$X_{\downarrow\uparrow G}(z) = \frac{1}{2}\left(X(z)G(z) + X(-z)G(-z)\right), \tag{5.81}$$

$$X_{\downarrow\uparrow H}(z) = \frac{1}{2}\left(X(z)H(z) + X(-z)H(-z)\right). \tag{5.82}$$

After multiplication by the synthesis filter $\hat{G}(z)$ and $\hat{H}(z)$, and summation of the results, we get $\hat{X}(z)$ as

$$\begin{aligned}
\hat{X}(z) &= X_{\downarrow\uparrow G}(z)\hat{G}(z) + X_{\downarrow\uparrow H}(z)\hat{H}(z) \\
&= \frac{1}{2}\left(G(z)\hat{G}(z) + H(z)\hat{H}(z)\right)X(z) \\
&\quad + \frac{1}{2}\left(G(-z)\hat{G}(z) + H(-z)\hat{H}(z)\right)X(-z).
\end{aligned} \tag{5.83}$$

The factor of $X(-z)$ shows the aliasing component, while the term at $X(z)$ shows the amplitude distortion. For a *perfect* reconstruction this translates into the following:

---

[12] Sometimes the amplitude factors are chosen in such a way that *orthonormal* filters are obtained, i.e., $\sum_n |h[n]|^2 = 1$. In this case, the filters have an amplitude factor of $1/\sqrt{2}$. This will complicate a hardware design significantly.

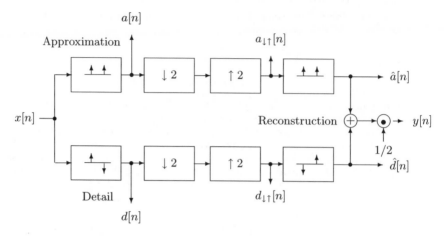

a[n]

Approximation

$a_{\downarrow\uparrow}[n]$

$\hat{a}[n]$

$x[n]$

Reconstruction

$y[n]$

1/2

$\hat{d}[n]$

Detail

$d_{\downarrow\uparrow}[n]$

d[n]

	\multicolumn{5}{c}{Time step n}
	0	1	2	3	4
$x[n]$	$x[0]$	$x[1]$	$x[2]$	$x[3]$	$x[4]$
$a[n]$	$x[0]$	$x[0]+x[1]$	$x[1]+x[2]$	$x[2]+x[3]$	$x[3]+x[4]$
$d[n]$	$x[0]$	$x[1]-x[0]$	$x[2]-x[1]$	$x[3]-x[2]$	$x[4]-x[3]$
$a_{\downarrow\uparrow}[n]$	$x[0]$	0	$x[1]+x[2]$	0	$x[3]+x[4]$
$d_{\downarrow\uparrow}[n]$	$x[0]$	0	$x[2]-x[1]$	0	$x[4]-x[3]$
$\hat{a}[n]$	$x[0]$	$x[0]$	$x[1]+x[2]$	$x[1]+x[2]$	$x[3]+x[4]$
$\hat{d}[n]$	$-x[0]$	$x[0]$	$x[1]-x[2]$	$x[2]-x[1]$	$x[3]-x[4]$
$\hat{x}[n]$	0	$x[0]$	$x[1]$	$x[2]$	$x[3]$

Fig. 5.56. Two-channel Haar-QMF bank.

Theorem 5.17:   **Perfect Reconstruction**

A perfect reconstruction for a two-channel filter bank, as shown in
Fig. 5.55, is achieved if
1) $G(-z)\hat{G}(z) + H(-z)\hat{H}(z) = 0$, i.e., the reconstruction is free of aliasing.
2) $G(z)\hat{G}(z) + H(z)\hat{H}(z) = 2z^{-d}$, i.e., the amplitude distortion has amplitude one.

Let us check this condition for the Haar filter bank.

Example 5.18: Two-Channel Haar Filter bank II

The filters of the two-channel Haar QMF bank were defined by

$$G(z) = 1 + z^{-1} \quad H(z) = 1 - z^{-1}$$
$$\hat{G}(z) = \frac{1}{2}(1 + z^{-1}) \quad \hat{H}(z) = \frac{1}{2}(-1 + z^{-1}).$$

The two conditions from Theorem 5.17 can be proved with:

1) $G(-z)\hat{G}(z) + H(-z)\hat{H}(z)$

$$= \frac{1}{2}(1 - z^{-1})(1 + z^{-1}) + \frac{1}{2}(1 + z^{-1})(-1 + z^{-1})$$

$$= \frac{1}{2}(1 - z^{-2}) + \frac{1}{2}(-1 + z^{-2}) = 0 \quad \checkmark$$

2) $G(z)\hat{G}(z) + H(z)\hat{H}(z)$

$$= \frac{1}{2}(1 + z^{-1})^2 + \frac{1}{2}(1 - z^{-1})(-1 + z^{-1})$$

$$= \frac{1}{2}\left((1 + 2z^{-1} + z^{-2}) + (-1 + 2z^{-1} - z^{-2})\right) = 2z^{-1} \quad \checkmark$$

$$\boxed{5.18}$$

For the proof using Theorem 5.17, it can be noted that the perfect reconstruction condition does not change if we switch the analysis and synthesis filters.

In the following we will discuss some restrictions that can be made in the filter design to fulfill the condition from Theorem 5.17 more easily.

First, we limit the filter choice by using the following:

Theorem 5.19:   Aliasing-Free Two-Channel Filter Bank

A two-channel filter bank is *aliasing-free* if
$$G(-z) = -\hat{H}(z) \qquad and \qquad H(-z) = \hat{G}(z). \tag{5.84}$$

This can be checked if we use (5.84) for the first condition of Theorem 5.17. Using a length-4 filter, these two conditions can be interpreted as follows:

$$g[n] = \{g[0], g[1], g[2], g[3]\} \rightarrow \hat{h}[n] = \{-g[0], g[1], -g[2], g[3]\}$$
$$h[n] = \{h[0], h[1], h[2], h[3]\} \rightarrow \hat{g}[n] = \{h[0], -h[1], h[2], -h[3]\}.$$

With the restriction of the filters as in Theorem 5.19, we can now simplify the second condition in Theorem 5.17. It is useful to define first an auxiliary *product filter* $F(z) = G(z)\hat{G}(z)$. The second condition from Theorem 5.17 becomes

$$G(z)\hat{G}(z) + H(z)\hat{H}(z) = F(z) - \hat{G}(-z)G(-z) \quad = F(z) - F(-z) \tag{5.85}$$

and we finally get

$$\boxed{F(z) - F(-z) = 2z^{-d},} \tag{5.86}$$

i.e., the product filter must be a *half-band filter*.[13] The construction of a perfect reconstruction filter bank uses the following three simple steps:

---

[13] For the definition of a half-band filter, see p. 274.

Algorithm 5.20:   Perfect-Reconstruction Two-Channel Filter Bank

1) Define a normalized causal half-band filter according to (5.86).
2) Factor the filter $F(z)$ in $F(z) = G(z)\hat{G}(z)$.
3) Compute $H(z)$ and $\hat{H}(z)$ using (5.84), i.e., $\hat{H}(z) = -G(-z)$ and $H(z) = \hat{G}(-z)$.

We wish to demonstrate Algorithm 5.20 with the following example. To simplify the notation we will, in the following example, write a combination of a length L filter for $G(z)$, and length N for $\hat{G}(z)$, as an L/N filter.

Example 5.21: Perfect-Reconstructing Filter Bank Using F3

The (normalized) causal half-band filter F3 (Table 5.3, p. 274) of length 7 has the following z-domain transfer function

$$\text{F3}(z) = \frac{1}{16}\left(-1 + 9z^{-2} + 16z^{-3} + 9z^{-4} - z^{-6}\right). \tag{5.87}$$

Using (5.86) we first verify that $\text{F3}(z) - \text{F3}(-z) = 2z^{-3}$. The zeros of the transfer function are at $z_{01-4} = -1$, $z_{05} = 2 + \sqrt{3} = 3.7321$, and $z_{06} = 2 - \sqrt{3} = 0.2679 = 1/z_{05}$. There are different choices for factoring $F(z) = G(z)\hat{G}(z)$. A 5/3 filter is, for instance,
a) $G(z) = (-1 + 2z^{-1} + 6z^{-2} + 2z^{-3} - z^{-4})/8$ and $\hat{G}(z) = (1 + 2z^{-1} + z^{-2})/2$.
We may design a 4/4 filter as:

b) $G(z) = \frac{1}{4}(1 + z^{-1})^3$ and $\hat{G}(z) = \frac{1}{4}(-1 + 3z^{-1} + 3z^{-2} - z^{-3})$.
Another configuration of the 4/4 configuration uses the Daubechies filter configuration, which is often found in wavelet applications and has the form:

c) $G(z) = \frac{1-\sqrt{3}}{4\sqrt{2}}(1 + z^{-1})^2(-z_{05} + z^{-1})$ and $\hat{G}(z) = -\frac{1+\sqrt{3}}{4\sqrt{2}}(1 + z^{-1})^2(-z_{06} + z^{-1})$.

Figure 5.57 shows these three combinations, along with their pole/zero plots.

5.21

For the Daubechies filter, the condition $H(z) = -z^{-N}G(-z^{-1})$ holds in addition, i.e., highpass and lowpass polynomials are mirror versions of each other. This is a typical behavior in *orthogonal* filter banks.

From the pole/zero plots shown in Fig. 5.57, for $F(z) = G(z)\hat{G}(z)$ the following conclusions can be made:

Corollary 5.22:   Factorization of a Half-band Filter

1) To construct a *real* filter, we must always group the conjugate symmetric zeros at (z_0 and z_0^*) in the same filter.
2) For *linear-phase* filters, the pole/zero plot must be symmetrical to the unit circle ($z = 1$). Zero pairs at (z_0 and $1/z_0$) must be assigned to the *same* filter.
3) To have *orthogonal* filters that are mirror polynomials of each other, ($F(z) = U(z)U(z^{-1})$), all pairs z_0 and $1/z_0$ must be assigned to *different* filters.

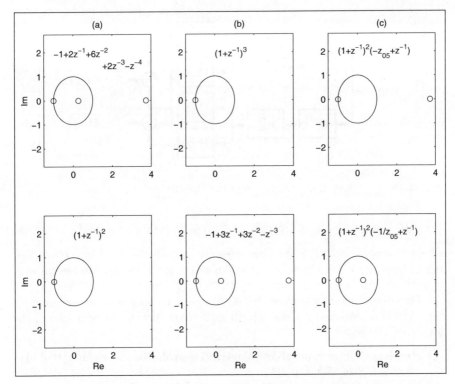

Fig. 5.57. Pole/zero plot for different factorization of the half-band filter F3. Upper row $G(z)$. lower row $\hat{G}(z)$. **(a)** Linear-phase 5/3 filter. **(b)** Linear-phase 4/4 filter. **(c)** 4/4 Daubechies filter.

We note that some of the above conditions can *not* be fulfilled at the same time. In particular, rules 2 and 3 represent a contradiction. Orthogonal, linear-phase filters are, in general, not possible, except when all zeros are on the unit circle, as in the case of the Haar filter bank.

If we classify the filter banks from Example 5.21, we find that configurations (a) and (b) are real linear-phase filters, while (c) is a real orthogonal filter.

Implementing Two-Channel Filter Banks

We will now discuss different options for implementing two-channel filter banks. We will first discuss the general case, and then special simplifications that are possible if the filters are QMF, linear-phase, or orthogonal. We will only discuss the analysis filter bank, as synthesis may be achieved with graph transposition.

Polyphase two-channel filter banks. In the general case, with two filters $G(z)$ and $H(z)$, we can realize each filter as a polyphase filter

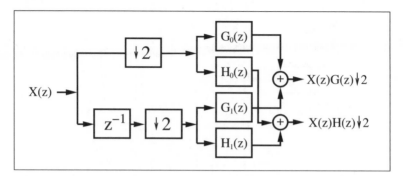

Fig. 5.58. Polyphase implementation of the two-channel filter bank.

$$H(z) = H_0(z^2) + z^{-1}H_1(z^2) \qquad G(z) = G_0(z^2) + z^{-1}G_1(z^2), \qquad (5.88)$$

which is shown in Fig. 5.58. This does not reduce the hardware effort ($2L$ multipliers and $2(L-1)$ adders are still used), but the design can be run with twice the usual sampling frequency, $2f_s$.

These four polyphase filters have only half the length of the original filters. We may implement these length $L/2$ filters directly or with one of the following methods:

1) Run-length filter using short Winograd convolution algorithms [104], discussed in Sect. 5.2.2, p. 254.

2) Fast convolution using FFT (discussed in Chap. 6) or NTTs (discussed in Chap. 7).

3) Using advanced arithmetic concepts discussed in Chap. 3, such as distribute arithmetic, reduced adder graph, or residue number system.

Using the fast convolution FFT/NTT techniques has the additional benefit that the forward transform for each polyphase filter need only be done once, and also, the inverse transform can be applied to the spectral sum of the two components, as shown in Fig. 5.59. But, in general, FFT methods only give improvements for longer filters, typically, larger than 32; however, the typical two-channel filter length is less than 32.

Lifting. Another general approach to constructing fast and efficient two-channel filter banks is the *lifting* scheme introduced recently by Swelden [131] and Herley and Vetterli [132]. The basic idea is the use of cross-terms (called lifting and dual-lifting), as in a lattice filter, to construct a longer filter from a short filter, while preserving the perfect reconstruction conditions. The basic structure is shown in Fig. 5.60.

Designing a lifting scheme typically starts with the "lazy filter bank," with $G(z) = \hat{H}(z) = 1$ and $H(z) = \hat{G}(z) = z^{-1}$. This channel bank fulfills both conditions from Theorem 5.17 (p. 316), i.e., it is a perfect reconstruction filter bank. The following question arises: if we keep one filter fixed, what are

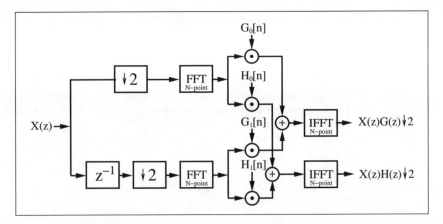

Fig. 5.59. Two-channel filter bank with polyphase decomposition and fast convolution using the FFT (©1999 Springer Press [5]).

filters $S(z)$ and $T(z)$ such that the filter bank is still a perfect reconstruction? The answer is important, and not trivial:

$$\text{Lifting: } G'(z) = G(z) + \hat{G}(-z)S(z^2) \quad \text{for any } S(z^2). \qquad (5.89)$$
$$\text{Dual-Lifting: } \hat{G}'(z) = \hat{G}(z) + G(-z)T(z^2) \quad \text{for any } T(z^2). \qquad (5.90)$$

To check, if we substitute the lifting equation into the perfect reconstruction condition from Theorem 5.17 (p. 316), and we see that both conditions are fulfilled if $\hat{G}(z)$ and $\hat{H}(z)$ still meet the conditions of Theorem 5.19 (p. 317) for the aliasing free filter bank (Exercise 5.9, p. 337).

The conversion of the Daubechies length-4 filter bank into lifting steps demonstrates the design.

Example 5.23: Lifting Implementation of the DB4 Filter

One filter configuration in Example 5.21 (p. 318) was the Daubechies length-4 filter [133, p. 195]. The filter coefficients were

$$G(z) =$$

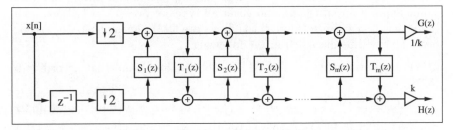

Fig. 5.60. Two-channel filter implementation using lifting and dual-lifting steps.

$$\left((1 + \sqrt{3}) + (3 + \sqrt{3})z^{-1} + (3 - \sqrt{3})z^{-2} + (1 - \sqrt{3})z^{-3}\right)\frac{1}{4\sqrt{2}}$$

$$H(z) =$$

$$\left(-(1 - \sqrt{3}) + (3 - \sqrt{3})z^{-1} - (3 + \sqrt{3})z^{-2} + (1 + \sqrt{3})z^{-3}\right)\frac{1}{4\sqrt{2}}.$$

A possible implementation uses two lifting steps and one dual-lifting step. The differential equations that produce a two-channel filter bank based on the above equation are

$$h_1[n] = x[2n + 1] - \sqrt{3}x[2n]$$

$$g_1[n] = x[2n] + \frac{\sqrt{3}}{4}h_1[n] + \frac{\sqrt{3} - 2}{4}h_1[n - 1]$$

$$h_2[n] = h_1[n] + g_1[n + 1]$$

$$g[n] = \frac{\sqrt{3} + 1}{\sqrt{2}}g_1[n]$$

$$h[n] = \frac{\sqrt{3} - 1}{\sqrt{2}}h_2[n].$$

Note that the early decimation and splitting of the input into even $x[2n]$ and odd $x[2n - 1]$ sequences allows the filter to run with $2f_s$. This structure can be directly translated into hardware and can be implemented using Quartus II (Exercise 5.10, p. 337). The implementation will use five multiplications and four adders. The reconstruction filter bank can be constructed based on graph transposition, which is, in the case of the differential equations, a reversing of the operations and flipping of the signs. | 5.23 |

Daubechies and Sweldens [134], have shown that *any* (bi)orthogonal wavelet filter bank can be converted into a sequence of lifting and dual-lifting steps. The number of multipliers and adders required then depends on the number of lifting steps (more steps gives less complexity) and can reach up to 50% compared with the direct polyphase implementation. This approach seems especially promising if the bit width of the multiplier is small [135]. On the other hand, the lattice-like structure does not allow use of reduced adder graph (RAG) techniques, and for longer filters the direct polyphase approach will often be more efficient.

Although the techniques (polyphase decomposition and lifting) discussed so far improve speed or size and cover all types of two-channel filters, additional savings can be achieved if the filters are QMF, linear-phase, or orthogonal. This will be discussed in the following.

QMF implementation. For QMF [136] we have found that according to (5.78),

$$h[n] = (-1)^n g[n] \quad \circ\!\!-\!\!\bullet \quad H(z) = G(-z). \tag{5.91}$$

But this implies that the polyphase filters are the same (except the sign), i.e.,

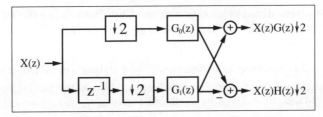

Fig. 5.61. Polyphase realization of the two-channel QMF bank (©1999 Springer Press [5]).

$$G_0(z) = H_0(z) \quad G_1(z) = -H_1(z). \tag{5.92}$$

Instead of the four filters from Fig. 5.58, for QMF we only need two filters and an additional "Butterfly," as shown in Fig. 5.61. This saves about 50%. For the QMF filter we need:

$$L \text{ real adders} \quad L \text{ real multipliers,} \tag{5.93}$$

and the filter can run with twice the usual input-sampling rate.

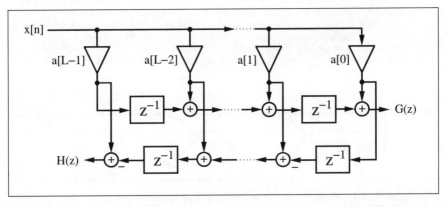

Fig. 5.62. Orthogonal two-channel filter bank using the transposed FIR structure.

Orthogonal filter banks. An orthogonal filter pair[14] obeys the conjugate mirror filter (CQF) [137] condition, defined by

$$H(z) = z^{-N} G(-z^{-1}). \tag{5.94}$$

If we use the transposed FIR filter shown in Fig. 5.62, we need only half the number of multipliers. The disadvantage is that we can *not* benefit from polyphase decomposition to double the speed.

---

[14] The orthogonal filter name comes from the fact that the scalar product of the filters, for a shift by two (i.e., $\sum g[k]h[k - 2l] = 0, k, l \in \mathbb{Z}$), is zero.

Another alternative is realization of the CQF bank using the lattice filter shown in Fig. 5.63. The following example demonstrates the conversion of the direct FIR filter into a lattice filter.

Example 5.24: Lattice Daubechies $L = 4$ Filter Implementation

One filter configuration in Example 5.21 (p. 318) was the Daubechies length-4 filter [133, p. 195]. The filter coefficients were

$$G(z) = \frac{(1 + \sqrt{3}) + (3 + \sqrt{3})z^{-1} + (3 - \sqrt{3})z^{-2} + (1 - \sqrt{3})z^{-3}}{4\sqrt{2}}$$

$$= 0.48301 + 0.8365z^{-1} + 0.2241z^{-2} - 0.1294z^{-3} \tag{5.95}$$

$$H(z) = \frac{-(1 - \sqrt{3}) + (3 - \sqrt{3})z^{-1} - (3 + \sqrt{3})z^{-2} + (1 + \sqrt{3})z^{-3}}{4\sqrt{2}}$$

$$= 0.1294 + 0.2241z^{-1} - 0.8365z^{-2} + 0.48301z^{-3}. \tag{5.96}$$

The transfer function for a two-channel lattice with two stages is

$$G(z) = \left(1 + a[0]z^{-1} - a[0]a[1]z^{-2} + a[1]z^{-3}\right)s \tag{5.97}$$

$$H(z) = \left(-a[1] - a[0]a[1]z^{-1} - a[0]z^{-2} + z^{-3}\right)s. \tag{5.98}$$

If we now compare (5.95) with (5.97) we find

$$s = \frac{1 + \sqrt{3}}{4\sqrt{2}} \qquad a[0] = \frac{3 + \sqrt{3}}{4\sqrt{2}s} \qquad a[1] = \frac{1 - \sqrt{3}}{4\sqrt{2}s}. \tag{5.99}$$

We can now translate this structure direct into hardware and implement the filter bank with Quartus II as shown in the following VHDL[15] code.

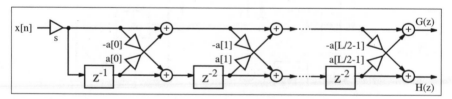

Fig. 5.63. Lattice realization for the orthogonal two-channel filter bank (©1999 Springer Press [5]).

```
PACKAGE n_bits_int IS                 -- User-defined types
   SUBTYPE BITS8 IS INTEGER RANGE -128 TO 127;
   SUBTYPE BITS9 IS INTEGER RANGE -2**8 TO 2**8-1;
   SUBTYPE BITS17 IS INTEGER RANGE -2**16 TO 2**16-1;
   TYPE ARRAY_BITS17_4 IS ARRAY (0 TO 3) OF BITS17;
END n_bits_int;

LIBRARY work;
USE work.n_bits_int.ALL;
```

---

[15] The equivalent Verilog code `db4latti.v` for this example can be found in Appendix A on page 709. Synthesis results are shown in Appendix B on page 731.

```
LIBRARY ieee;
USE ieee.std_logic_1164.ALL;
USE ieee.std_logic_arith.ALL;
USE ieee.std_logic_unsigned.ALL;

ENTITY db4latti IS                        ------> Interface
  PORT (clk, reset : IN  STD_LOGIC;
        clk2       : OUT STD_LOGIC;
        x_in       : IN  BITS8;
        x_e, x_o   : OUT BITS17;
        g, h       : OUT BITS9);
END db4latti;

ARCHITECTURE fpga OF db4latti IS

  TYPE STATE_TYPE IS (even, odd);
  SIGNAL state                    : STATE_TYPE;
  SIGNAL sx_up, sx_low, x_wait    : BITS17 := 0;
  SIGNAL clk_div2                 : STD_LOGIC;
  SIGNAL sxa0_up, sxa0_low        : BITS17 := 0;
  SIGNAL up0, up1, low0, low1     : BITS17 := 0;

BEGIN

  Multiplex: PROCESS (reset, clk) ----> Split into even and
  BEGIN                            -- odd samples at clk rate
    IF reset = '1' THEN            -- Asynchronous reset
      state <= even;
    ELSIF rising_edge(clk) THEN
      CASE state IS
        WHEN even =>
        -- Multiply with 256*s=124
          sx_up   <= 4 * (32 *   x_in - x_in);
          sx_low  <= 4 * (32 * x_wait - x_wait);
          clk_div2 <= '1';
          state <= odd;
        WHEN odd =>
          x_wait <= x_in;
          clk_div2 <= '0';
          state <= even;
      END CASE;
    END IF;
  END PROCESS;

---------- Multipy a[0] = 1.7321
  sxa0_up  <= (2*sx_up  - sx_up /4)
                            - (sx_up /64 + sx_up/256);
  sxa0_low <= (2*sx_low - sx_low/4)
                            - (sx_low/64 + sx_low/256);
---------- First stage -- FF in lower tree
  up0  <= sxa0_low + sx_up;
  LowerTreeFF: PROCESS
```

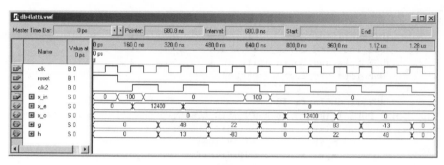

Fig. 5.64. VHDL simulation of the Daubechies length-4 lattice filter bank.

```
BEGIN
  WAIT UNTIL clk = '1';
  IF clk_div2 = '1' THEN
    low0 <= sx_low - sxa0_up;
  END IF;
END PROCESS;

---------- Second stage   a[1]=0.2679
up1  <= (up0 - low0/4) - (low0/64 + low0/256);
low1 <= (low0 + up0/4) + (up0/64  + up0/256);

x_e  <= sx_up;    -- Provide some extra test signals
x_o  <= sx_low;
clk2 <= clk_div2;

OutputScale: PROCESS
BEGIN
  WAIT UNTIL clk = '1';
    IF clk_div2 = '1' THEN
      g <=  up1 / 256;
      h <= low1 / 256;
    END IF;
END PROCESS;

END fpga;
```

This VHDL code is a direct translation of the lattice shown in Fig. 5.63. The incoming stream is multiplied by $s = 0.48 \approx 124/256$. Next, the cross-term product multiplications, with $a[0] = 1.73 \approx (2 - 2^{-2} - 2^{-6} - 2^{-8})$, of the first stage are computed. It follows that the stage 1 additions and the lower tree signal must be delayed by one sample. In the second stage, the cross multiplication by $a[1] = 0.27 \approx (2^{-2} + 2^{-6} + 2^{-8})$ and the final output addition are implemented. The design uses 418 LEs, no embedded multiplier, and has a 58.81 MHz **Registered Performance**.

The VHDL simulation is shown in Fig. 5.64. The simulation shows the response to an impulse with amplitude 100 at even and odd positions for the filters $G(z)$ and $H(z)$, respectively.

5.24

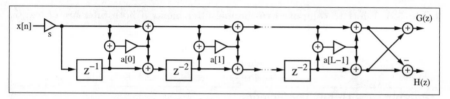

Fig. 5.65. Lattice filter to realize linear-phase two-channel filter bank (©1999 Springer Press [5]).

If we compare the size of the lattice with the direct polyphase implementation of $G(z)$ shown in Example 5.1 on p. 250 (LEs multiplied by two), we note that both designs have about the same size ($208 \times 2 = 416$ LEs, versus 331 LEs). Although the lattice implementation needs only five multipliers, compared with eight multipliers for the polyphase implementation, we note that in the polyphase implementation we can use the RAG technique to implement the coefficients of the transposed filter, while in the lattice we must implement single multipliers, which, in general, are less efficient.

Linear-phase two-channel filter bank. We have already seen in Chap. 3 that if a linear filter has even or odd symmetry, 50% of multiplier resources can be saved. The same symmetry also applies for polyphase decomposition of the filters if the filters, have even length. In addition, these filters may run at twice the speed.

If $G(z)$ and $H(z)$ have the same length, another implementation using lattice filters can further decrease the implementation effort, as shown in Fig. 5.65. Notice that the lattice is different from the lattice used for the orthogonal filter bank shown in Fig. 5.63.

The following example demonstrates how to convert a direct architecture into a lattice filter.

Example 5.25: Lattice for $L = 4$ Linear-Phase Filter

One filter configuration in Example 5.21 (p. 318) was a linear-phase filter pair, with both filters of length 4. The filters are

$$G(z) = \frac{1}{4}\left(1 + 3z^{-1} + 3z^{-2} + 1z^{-3}\right) \tag{5.100}$$

and

$$H(z) = \frac{1}{4}\left(-1 + 3z^{-1} + 3z^{-2} - 1z^{-3}\right). \tag{5.101}$$

The transfer functions for the two-channel length-4 linear-phase lattice filters are:

$$G(z) = \left((1 + a[0]) + a[0]z^{-1} + a[0]z^{-2} + (1 + a[0])z^{-3}\right)s \tag{5.102}$$

$$H(z) = \left(-(1 + a[0]) + a[0]z^{-1} + a[0]z^{-2} - (1 + a[0])z^{-3}\right)s. \tag{5.103}$$

Comparing (5.100) with (5.102), we find

$$s = -1/2 \qquad a[0] = -1.5. \tag{5.104}$$

5.25

Table 5.6. Effort to compute two-channel filter banks if both filter are of length L.

Type	Number of real multipliers	Number of real adders	see Fig.	Speed	Can use RAG ?
Polyphase with any coefficients					
Direct FIR filtering	$2L$	$2L - 2$	5.58	$2f_s$	✓
Lifting	$\approx L$	$\approx L$	5.60	$2f_s$	−
Quadrature mirror filter (QMF)					
Identical polyphase filter	L	L	5.61	$2f_s$	✓
Orthogonal filter					
Transposed FIR filter	L	$2L - 2$	5.62	f_s	✓
Lattice	$L + 1$	$3L/4$	5.63	$2f_s$	−
Linear-phase filter					
Symmetric filter	L	$2L - 2$	3.5	$2f_s$	✓
Lattice	$L/2$	$3L/2 - 1$	5.65	$2f_s$	−

Note that, compared with the direct implementation, only about one quarter of the multipliers are required.

The disadvantage of the linear-phase lattice is that not all linear-phase filters can be implemented. Specifically, $G(z)$ must be even symmetric, $H(z)$ must be odd symmetric, and both filters must be of the same length, with an even number of samples.

Comparison of implementation options. Finally, Table 5.6 compares the different implementation options, which include the general case and special types like QMF, linear-phase and orthogonal.

Table 5.6 shows the required number of multipliers and adders, the reference figure, the maximum input rate, and the structurally important question of whether the coefficients can be implemented using reduced adder graph technique, or occur as single-multiplier coefficients. For shorter filters, the lattice structure seems to be attractive, while for longer filters, RAG will most often produce smaller and faster designs. Note that the number of multipliers and adders in Table 5.6 are an estimate of the hardware effort required for the filter, and *not* the typical number found in the literature for the computational effort per input sample in a PDSP/μP solution [104, 138].

Excellent additional literature about two-channel filter banks is available (see [102, 135, 138, 139]).

5.8 Wavelets

A time-frequency representation of signals processed through transform methods has proven beneficial for audio and image processing [135, 140, 141]. Many signals subject to analysis are known to have statistically constant properties for only short time frames (e.g., speech or audio signals). It is therefore reasonable to analyze such signals in a short window, compute the signal parameter, and slide the window forward to analyze the next frame. If this analysis is based on Fourier transforms, it is called a *short-term Fourier transform* (STFT).

A short-term Fourier transform (STFT) is formally defined by

$$X(\tau, f) = \int_{-\infty}^{\infty} x(t)\, w(t - \tau)\, e^{-j2\pi f t}\, dt, \qquad (5.105)$$

i.e., it slides a window function $w(t - \tau)$ over the signal $x(t)$, and produces a continuous time–frequency map. The window should taper smoothly to zero, both in frequency and time, to ensure localization in frequency Δ_f and time Δ_t of the mapping. One weight function, the Gaussian function $(g(t) = e^{-t^2})$, is optimal in this sense, and provides the minimum (Heisenberg principle) product $\Delta_f \Delta_t$ (i.e., best localization), as proposed by Gabor in 1949 [142]. The discretization of the Gabor transform leads to the discrete Gabor transform (DGT). The Gabor transform uses identical resolution windows throughout the time and frequency plane (see Fig. 5.67a). Every rectangle in Fig. 5.67a has exactly the same shape, but often a constant Q (i.e., the quotient of bandwidth to center frequency) is desirable, especially in audio and image processing. That is, for high frequencies we wish to have broadband filters and short sampling intervals, while for low frequencies, the bandwidth should be small and the intervals larger. This can be accomplished with the continuous wavelet transform (CWT), introduced by Grossmann and Morlet [143],

$$\mathrm{CWT}(\tau, f) = \int_{-\infty}^{\infty} x(t)\, h\left(\frac{t - \tau}{s}\right) dt, \qquad (5.106)$$

where $h(t)$, known from the Heugens principle in physics, is called a small wave or *wavelet*. Some typical wavelets are displayed in Fig. 5.68.

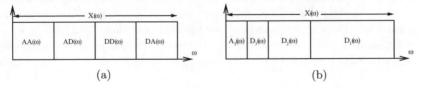

(a)   (b)

Fig. 5.66. Frequency distribution for **(a)** Fourier (constant bandwidth) and **(b)** constant Q.

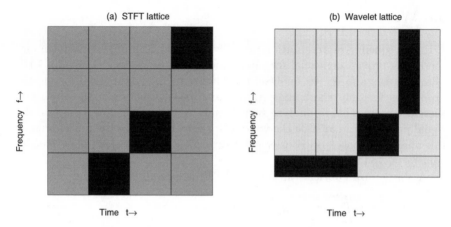

Fig. 5.67. Time frequency grids for a chirp signal. **(a)** Short-term Fourier transform. **(b)** Wavelet transform.

If we use now as a wavelet

$$h(t) = \left(e^{j2\pi kt} - e^{-k^2/2} \right) e^{-t^2/2} \tag{5.107}$$

we still enjoy the "optimal" properties of the Gaussian window, but now with different scales in time and frequency. This so-called Morlet transform is also subject to quantization, and is then called the discrete Morlet transformation (DMT) [144]. In the discrete case the lattice points in time and frequency are shown in Fig. 5.67b. The exponential term $e^{-k^2/2}$ in (5.107) was introduced such that the wavelet is DC free. The following examples show the excellent performance of the Gaussian window.

Example 5.26: Analysis of a Chirp Signal

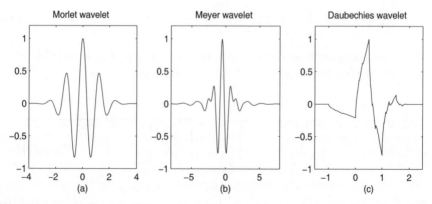

Fig. 5.68. Some typical wavelets from Morlet, Meyer, and Daubechies.

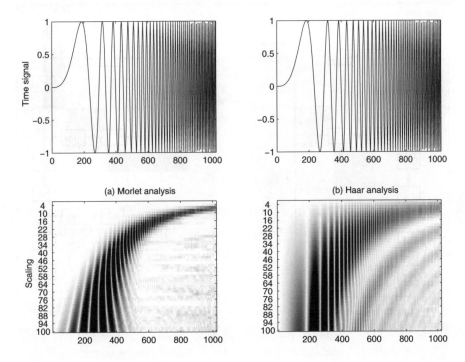

Fig. 5.69. Analysis of a chirp signal with **(a)** Discrete Morlet transform. **(b)** Haar transform.

Figure 5.69 shows the analysis of a constant amplitude signal with increasing frequency. Such signals are called chirp signals. If we applied the Fourier transform we would get a uniform spectrum, because all frequencies are present. The Fourier spectrum does not preserve time-related information. If we use instead an STFT with a Gaussian window, i.e., the Morlet transform, as shown in Fig. 5.69a, we can clearly see the increasing frequency. But the Gaussian window shows the best localization of all windows. On the other hand, with a Haar window we would have less computational effort, but, as can be seen from Fig. 5.69b, the Haar window will achieve less precise time-frequency localization of the signal.
$\boxed{5.26}$

Both DGT and DMT provide good localization by using a Gaussian window, but both are computationally intensive. An efficient multiplier-free implementation is based on two ideas [144]. First, the Gaussian window can be sufficiently approximated by a convolution of ($\geq$ 3) rectangular functions, and second, single-passband frequency-sampling filters (FSF) can be efficiently implemented by defining algebraic integers over polynomial rings, as introduced in [144].

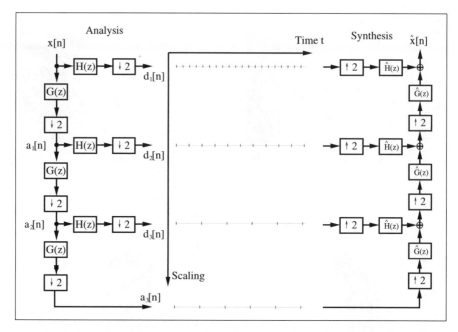

Fig. 5.70. Wavelets tree decomposition in three octaves (©1999 Springer Press [5]).

In the following, we wish to focus our attention on a newly popular analysis method called the *discrete wavelet transform*, which better exploits the auditory and visual human perception mode (i.e., constant Q), and also can often be more efficiently computed, using $\mathcal{O}(n)$ complexity algorithms.

5.8.1 The Discrete Wavelet Transformation

A discrete-time version of the analog model leads to the *discrete wavelet transform* (DWT). In practical applications, the DWT is restricted to the discrete time *dyadic DWT* with $a = 2$, and will be considered in the following. The DWT achieves the constant Q bandwidth distribution shown in Fig. 5.66b and Fig. 5.67b by always applying the two-channel filter bank in a filter tree to the lowpass signal, as shown in Fig. 5.70.

We now wish to focus on what conditions for the CWT wavelet allow it to be realized with a two-channel DWT filter bank. We may argue that if we sample a continuous wavelet at an appropriate rate (above the Nyquist rate), we may call the sampled version a DWT. But, in general, only those continuous wavelet transforms that can be realized with a two-channel filter bank are called DWT.

Closely related to whether a continuous wavelet $\psi(t)$ can be realized with a two-channel DWT, is the question of whether the *scaling equation*

$$\phi(t) = \sum_n g[n]\,\phi(2t - n) \tag{5.108}$$

exists, where the actual wavelet is computed with

$$\psi(t) = \sum_n h[n]\phi(2t - n), \tag{5.109}$$

where $g[n]$ is a lowpass, and $h[n]$ a highpass filter. Note that $\phi(t)$ and $\psi(t)$ are continuous functions, while $g[n]$ and $h[n]$ are sample sequences (but still may also be IIR filters). Note that (5.108) is similar to the *self-similarity* ($\phi(t) = \phi(at)$) exhibited by fractals. In fact, the scaling equation may iterate to a fractal, but that is, in general, not the desired case, because most often a smooth wavelet is desired. The smoothness can be improved if we use a filter with maximal numbers of zeros at π.

We consider now backwards reconstruction: we start with the filter $g[n]$, and construct the corresponding wavelet. This is the most common case, especially if we use the half-band design from Algorithm 5.20 (p. 318) to generate perfect reconstruction filter pairs of the desired length and property.

To get a graphical interpretation of the wavelet, we start with a rectangular function (box function) and build, according to (5.108), the following graphical iteration:

$$\phi^{(k+1)}(t) = \sum_n g[n]\,\phi^{(k)}(2t - n). \tag{5.110}$$

If this converges to a stable $\phi(t)$, the (new) wavelet is found. This iteration obviously converges for the Haar filter $\{1, 1\}$ immediately after the first iteration, because the sum of two box functions scaled and added is again a box function, i.e.,

Let us now graphically construct the wavelet that belongs to the filter $g[n] = \{1, 1, 1, 1\}$, which we will call Hutlet4 [145].

Example 5.27: Hutlet of Length-4

We start with four box functions weighted by $g[n] = \{1, 1, 1, 1\}$. The sum shown in Fig. 5.71a is the starting $\phi^{(1)}(t)$. This function is scaled by two, and the sum gives a two-step function. After 10 iterations we already get a very smooth trapezoid function. If we now use the QMF relation, from (5.78) (p. 314), to construct the actual wavelet, we get the Hutlet4, which has two triangles as shown in Fig. 5.72.

5.27

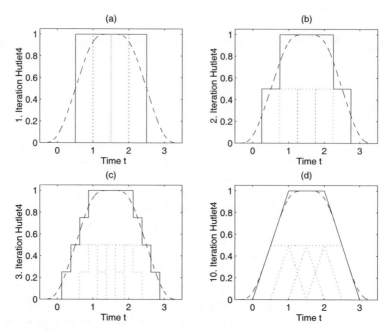

Fig. 5.71. Iteration steps $1, 2, 3$, and 10 for Hutlet4. (**solid line:** $\phi^{(k+1)}(t)$; **dotted line:** $\phi^{(k)}(2t - n)$; and ideal Hut-function: **dashed**)

We note that $g[n]$ is the impulse response of the moving-average filter, and can be implemented as an one-stage CIC filter [146]. Figure 5.72 shows all scaling functions and wavelets for this type of wavelet with even length coefficients.

As noted before, the iteration defined by (5.110) may also converge to a fractal. Such an example is shown in Fig. 5.73, which is the wavelet for the length-5 "moving average filter." This indicates the challenge of the filter selection $g[n]$: it may converge to a smooth or, totally chaotic function, depending only on an apparently insignificant property like the length of the filter!

We still have not explained why the two-scale equation (5.108) is so important for the DWT. This can be better understood if we rearrange the downsampler (compressor) and filter in the analysis part of the DWT, using the "Noble" relation

$$(\downarrow M)\, H(z) = H(z^M)\, (\downarrow M), \tag{5.111}$$

which was introduced in Sect. 5.1.1, p. 246. The results for a three-level filter bank are shown in Fig. 5.74. If we compute the impulse response of the cascade sequences, i.e.,

$$H(z) \leftrightarrow d_1[k/2]$$

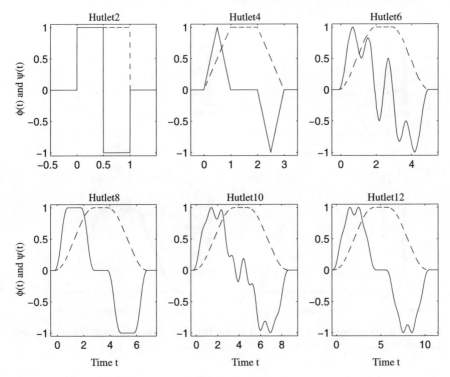

Fig. 5.72. The Hutlet wavelet family **(solid line)** and scaling function **(dashed line)** after 10 iterations (©1999 Springer Press [5]).

$$G(z)H(z^2) \leftrightarrow d_2[k/4]$$
$$G(z)G(z^2)H(z^4) \leftrightarrow d_3[k/8]$$
$$G(z)G(z^2)G(z^4) \leftrightarrow a_3[k/8],$$

we find that a_3 is an approximation to the scaling function, while d_3 gives an approximation to the mother wavelet, if we compare the graphs with the continuous wavelet shown in Fig. 5.68 (p. 330).

This is not always possible. For instance, for the Morlet wavelet shown in Fig. 5.68 (p. 330), no scaling function can be found, and a realization using the DWT is *not* possible.

Two-channel DWT design examples for the Daubechies length-4 filter have already been discussed, in combination with polyphase representation (Example 5.1, p. 250), and the lattice implementation of orthogonal filters in Example 5.24 (p. 324).

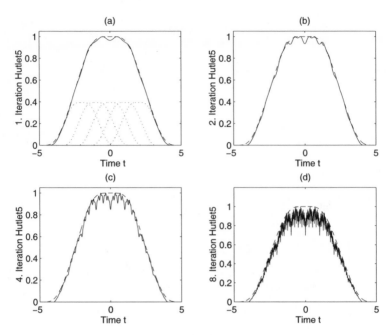

Fig. 5.73. Iteration step 1,2,4, and 8 for Hutlet5. The sequence converges to a fractal!

Exercises

Note: If you have no prior experience with the Quartus II software, refer to the case study found in Sect. 1.4.3, p. 29. If not otherwise noted use the EP2C35F672C6 from the Cyclone II family for the Quartus II synthesis evaluations.

5.1: Let $F(z) = 1 + z^{-d}$. For which d do we have a half-band filter according to Definition 5.7 (p. 274)?

5.2: Let $F(z) = 1 + z^{-5}$ be a half-band filter.
(a) Draw $|F(\omega)|$. What kind of symmetry does this filter have?
(b) Use Algorithm 5.20 (p. 318) to compute a perfectly reconstructing real filter bank. What is the total delay of the filter bank?

5.3: Use the half-band filter F3 from Example 5.21 (p. 318) to build a perfect-reconstruction filter bank, using Algorithm 5.20 (p. 318), of length
(a) 1/7.
(b) 2/6.

5.4: How many different filter pairs can be built, using F3 from Example 5.21 (p. 318), if both filters are
(a) Complex.
(b) Real.

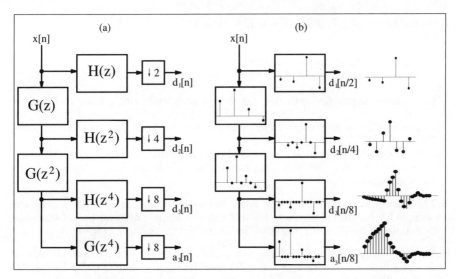

Fig. 5.74. DWT filter bank rearrange using Noble relations. (**a**) Transfer function in the z-domain. (**b**) Impulse response for the length-4 Daubechies filters.

(**c**) Linear-phase.
(**d**) Orthogonal filter bank.

5.5: Use the half-band filter $F2(z) = 1 + 2z^{-1} + z^{-2}$ to compute, based on Algorithm 5.20 (p. 318), all possible perfect-reconstructing filter banks.

5.6: (**a**) Compute the number of real additions and multiplications for a direct implementation of the critically sampled uniform DFT filter bank shown in Fig. 5.50 (p. 309). Assume the length L analysis and synthesis filters have complex coefficients, and the inputs are real valued.
(**b**) Assume an FFT algorithm is used that needs $(15N \log_2(N))$ real additions and multiplications. Compute the total effort for a uniform DFT filter bank, using the polyphase representation from Figs. 5.52 (p. 310) and 5.53 (p. 312), for R of length L complex filters.
(**c**) Using the results from (a) and (b) compute the effort for a critically sampled DFT filter bank with $L = 64$ and $R = 16$.

5.7: Use the lossy integrator from Example 5.15 (p. 311) to implement an $R = 4$ uniform DFT filter bank.
(**a**) Compute the analysis polyphase filter $H_k(z)$.
(**b**) Determine the synthesis filter $F_k(z)$ for perfect reconstruction.
(**c**) Determine the 4×4 DFT matrix. How many real additions and multiplications are used to compute the DFT?
(**d**) Compute the total computational effort of the whole filter bank, in terms of real additions and multiplications per input sample.

5.8: Analyze the frequency response of each Goodman and Carey half-band filter from Table 5.3 (p. 274). Zoom in on the passband to estimate the ripple of the filter.

5.9: Prove the perfect reconstruction for the *lifting* and *dual-lifting* scheme from (5.89) and (5.90) on p. 321.

5.10: (a) Implement the Daubechies length-4 filter using the lifting scheme from Example 5.23 (p. 321), with 8-bit input and coefficient, and 10-bit output quantization.
(b) Simulate the design with two impulses of amplitude 100, similar to Fig. 5.64 (p. 326).
(c) Determine the `Registered Performance` and the used resources (LEs, multipliers, and M4Ks).
(d) Compare the lifting design with the direct polyphase implementation (Example 5.1, p. 250) and with the lattice implementation (Example 5.24, p. 324), in terms of size and speed.

5.11: Use component instantiation of the two designs from Example 5.4 (p. 262) and Example 5.6 (p. 269) to compute the difference of the two filter outputs. Determine the maximum positive and negative deviation.

5.12: (a) Use the reduced adder graph design from Fig. 3.11 (p. 185) to build a half-band filter F6 (see Table 5.3, p. 274) for 8-bit inputs using Quartus II. Use the transposed FIR structure (Fig. 3.3, p. 167) as the filter architecture.
(b) Verify the function via a simulation of the impulse response.
(c) Determine the `Registered Performance` and the used resources (LEs, multipliers, and M4Ks), of the F6 design.

5.13: (a) Compute the polyphase representation for F6 from Table 5.3, p. 274.
(b) Implement the polyphase filter F6 with decimation $R = 2$ for 8-bit inputs with Quartus II.
(c) Verify the function via a simulation of the impulse (one at even and one at odd) response.
(d) Determine the `Registered Performance` and the used resources (LEs, multipliers, and M4Ks) of the polyphase design.
(e) What are the advantages and disadvantages of the polyphase design, when compared with the direct implementation from Exercise 5.12 (p. 338), in terms of size and speed.

5.14: (a) Compute the 8-bit quantized DB4 filters $G(z)$ by multiplication of (5.95) with 256 and taking the integer part. Use the programm `csd3e.exe` from the CD-ROM or the data from Table 2.3, p. 64.
(b1) Design the filter $G(z)$ only from Fig. 5.62, p. 323 for 9-bit inputs with Quartus II. Assume that input and coefficient are signed, i.e., only one additional guard bit is required for a filter of length 4.
(b2) Determine the `Registered Performance` and the used resources (LEs, multipliers, and M4Ks) of the filter $G(z)$.
(b3) What are the advantages and disadvantages of the CSD design, when compared with the programmable FIR filter from Example 3.1 (p. 167), in terms of size and speed.
(c1) Design the filter bank with $H(z)$ and $G(z)$ from Fig. 5.62, p. 323.
(c2) Determine the `Registered Performance` and the used resources (LEs, multipliers, and M4Ks) of the filter bank.
(c3) What are the advantages and disadvantages of the CSD filter bank design, when compared with the lattice design from Example 5.24, (p. 324), in terms of size and speed.

5.15: **(a)** Use the MAG coding from Table 2.3 (p. 64) to build the sinc filter from Example 5.12, (p. 285).
(a1) Determine the `Registered Performance` and the used resources (LEs, multipliers, and M4Ks) of the MAG `rc_sinc` design.
(b) Implement a pipelined adder tree to improve the throughput of the filter `rc_sinc` design.
(b1) Determine the `Registered Performance` and the used resources (LEs, multipliers, and M4Ks) for the improved design.

5.16: **(a)** Use the sinc filter data from Example 5.12, (p. 285) to estimate the implementation effort for an $R = 147/160$ rate changer. Assume that the two filters each account for 50% of the resources.
(b) Use the Farrow filter data from Example 5.13, (p. 292) to estimate the implementation effort for an $R = 147/160$ rate changer.
(c) Compare the two design options from (a) and (b) for small and large values of R_1 in terms of required LEs and `Registered Performance`.

5.17: The Farrow combiner from Example 5.13, (p. 292) uses several multiplications in series. Pipeline register for the data and multiplier can be added to perform a maximum-speed design.
(a) How many pipeline stages (total) are required for a maximum-speed design if we use:
(a1) an embedded array multiplier?
(a2) an LE-based multiplier?
(b) Design the HDL code for a maximum-speed design using:
(b1) an embedded array multiplier
(b2) an LE-based multiplier
(c) Determine the `Registered Performance` and the used resources (LEs, multipliers, and M4Ks) for the improved designs.

5.18: **(a)** Compute and plot using MATLAB or C the impulse response of the IIR filter given in (5.56), p. 298. Is this filter stable?
(b) Simulate the IIR filter using the following input signal: $x = [0, 0, 0, 1, 2, 3, 0, 0, 0]$; using the filter F= `[1 4 1]/6` determine
(b1) x filtered by $F(z)$ followed by $1/(F(z))$ and plot the results.
(b2) x filtered by $1/F(z)$ followed by $F(z)$ and plot the results.
(c) Split the filter $1/F(z)$ in a stable causal and a-causal part and apply the causal first-to-last sample while the a-causal is applied last-to-first sample. Repeat the simulation in (b).

5.19: **(a)** Plot the filter transfer function of the IIR filter given in (5.56), p. 298.
(b) Build the length-11 IFFT of the filter and apply a DC correction, i.e., $\sum_k h(k) = 0$.
(c) Unser et al. [125] determined the following length-11 FIR approximation for the IIR: $[-0.0019876, 0.00883099, -0.0332243, 0.124384, -0.46405, 1.73209, -0.46405, 0.124384, -0.0332243, 0.00883099, -0.0019876]$. Plot the impulse response and transfer function of this filter.
(d) Determine the error of the two solutions in (b) and (c) by computing the convolution with the filter (5.55) and building the square sum of the elements (excluding 1).

5.20: Use the Farrow equation (5.60) (p. 300) for B-splines to determine the Farrow matrix for the c_k for
(a) I-MOMS with $\phi(t) = \beta^3(t) - \frac{1}{6}\frac{d^2\beta^3(t)}{dt^2}$
(b) O-MOMS with $\phi(t) = \beta^3(t) + \frac{1}{42}\frac{d^2\beta^3(t)}{dt^2}$

5.21: Study the FSM part of the cubic B-spline interpolator from Example 5.14, (p. 303) for an $R = 147/160$ rate changer.
(a) Assume that the delays d^k are stored in LUTs or M4Ks tables. What is the required table size if d^k are quantized to
(a1) 8 bit unsigned?
(a2) 16 bit unsigned?
(b) Determine the first five phase values for (a1) and (a2). Is 8 bit sufficient precision?
(c) Assume that the Farrow structure is used, i.e., the delays d^k are computed successively using the Horner scheme. What are the FSM hardware requirements for this solution?

5.22: Use the results from Exercise 5.20 and the fractional rate change design from Example 5.14, (p. 303) to design an $R = 3/4$ rate changer using O-MOMS.
(a) Determine the RAG-n for the FIR compensation filter with the following coefficients: $(-0.0094, 0.0292, -0.0831, 0.2432, -0.7048, 2.0498, -0.7048, 0.2432, \ldots)$
$= (-1, 4, -11, 31, -90, 262, -90, 31, -11, 4, -1)/128$.
(b) Replace the IIR filter with the FIR filter from (a) and adjust the Farrow matrix coefficients as determined in Exercise 5.20(b).
(c) Verify the functionality with a triangular test function as in Fig. 5.48, p. 307.
(d) Determine the **Registered Performance** and the used resources (LEs, embedded multipliers, and M4Ks) for the O-MOMS design.

5.23: Use the Farrow matrix (5.60) (p. 300) and the fractional rate change design from Example 5.14, (p. 303) to design an $R = 3/4$ rate changer using B-splines.
(a) Determine the RAG-n for the FIR compensation filter with the following coefficients: $(0.0085, -0.0337, 0.1239, -0.4645, 1.7316, -0.4645, 0.1239, -0.0337, 0.0085)$
$= (1, -4, 16, -59, 222, -59, 16, -4, 1)/128$
(b) Replace the IIR filter with the FIR filter from (a).
(c) Verify the functionality with a triangular test function as in Fig. 5.48, p. 307.
(d) Determine the **Registered Performance** and the used resources (LEs, embedded multipliers, and M4Ks) for the B-spline design.

5.24: (a) The GC4114 has a four-stage CIC interpolator with a variable sampling change factor R. Try to download and study the datasheet for the GC4114 from the WWW.
(b) Write a short C or MATLAB program that computes the bit growth $B_k = \log_2(G_k)$ for the CIC interpolator using Hogenauers [106] equation:

$$G_k = \begin{cases} 2^k & k = 1, 2, \ldots, S \\ \frac{2^{2S-k}(RD)^{k-S}}{R} & k = S+1, S+2, \ldots, 2S \end{cases} \qquad (5.112)$$

where D is the delay of the comb, S is number of stages, and R is the interpolation factor. Determine for the GC4114 ($S = 4$ stages, delay comb $D = 1$) the output bit growth for R=8, R=32, and R=16 384.
(c) Write a MATLAB program to simulate a four-stage CIC interpolator with delay 1 in the comb and R=32 upsampling. Try to match the simulation shown in Fig. 5.75.
(d) Measure the bit growth for each stage using the program from (c) for a step input and compare the results to (b).

5.25: Using the results from Exercise 5.24
(a) Design a four-stage CIC interpolator with delay $D = 1$ in the comb and $R = 32$ rate change for 16-bit input. Use for the internal bit width the output bit width you determined in Exercise 5.24(b) for R=32. Try to match the MATLAB simulation

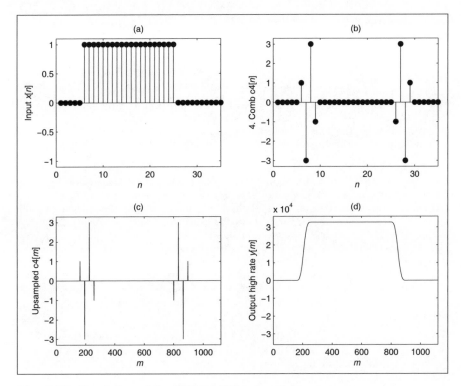

Fig. 5.75. Simulation of the GC4114 CIC interpolator.

shown in Fig. 5.75 with the HDL simulation for the input and output.

(b) Design the four-stage CIC interpolator now with the detailed bit width determined in Exercise 5.24(b). Try to match the MATLAB simulation with the HDL simulation for the input and output.

(c) Determine the `Registered Performance` and the used resources (LEs, multipliers, and M2Ks/M4Ks) for the two designs from (a) and (b) using:

(c1) the device EPF10K70RC240-4 from the UP2 board

(c2) the device EP2C35F672C6 from the Cyclone II family

6. Fourier Transforms

The discrete Fourier transform (DFT) and its fast implementation, the fast Fourier transform (FFT), have played a central role in digital signal processing.

DFT and FFT algorithms have been invented (and reinvented) in many variations. As Heideman et al. [147] pointed out, we know that Gauss used an FFT-type algorithm that today we call the Cooley–Tukey FFT. In this chapter we will discuss the most important algorithms summarized in Fig. 6.1.

We will follow the terminology introduced by Burrus [148], who classified FFT algorithms simply by the (multidimensional) index maps of their input and output sequences. We will therefore call all algorithms that do *not* use a multidimensional index map, DFT algorithms, although some of them, such as the Winograd DFT algorithms, enjoy an essentially reduced computational effort. DFT and FFT algorithms do not "stand alone": the most

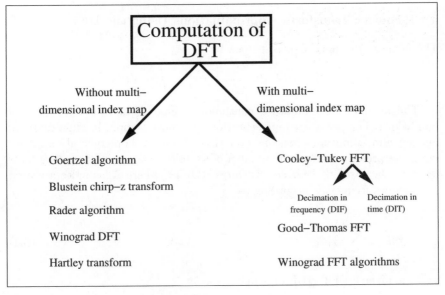

Fig. 6.1. Classifications of DFT and FFT algorithms.

efficient implementations often result in a combination of DFT and FFT algorithms. For instance, the combination of the Rader prime algorithm and the Good–Thomas FFT results in excellent VLSI implementations. The literature provides many FFT design examples. We find implementations with PDSPs and ASICs [149, 150, 151, 152, 153, 154]. FFTs have also been developed using FPGAs for 1-D [155, 156, 157] and 2-D transforms [47, 158].

We will discuss in this chapter the four most important DFT algorithms and the three most often used FFT algorithms, in terms of computational effort, and will compare the different implementation issues. At the end of the chapter, we will discuss Fourier-related transforms, such as the DCT, which is an important tool in image compression (e.g., JPEG, MPEG). We start with a short review of definitions and the most important properties of the DFT.

For more detailed study, students should be aware that DFT algorithms are covered in basic DSP books [5, 79, 159, 160], and a wide variety of FFT books are also available [66, 161, 162, 163, 164, 165].

6.1 The Discrete Fourier Transform Algorithms

We will start with a review of the most important DFT properties and will then review basic DFT algorithms introduced by Bluestein, Goertzel, Rader, and Winograd.

6.1.1 Fourier Transform Approximations Using the DFT

The Fourier transform pair is defined by

$$X(f) = \int_{-\infty}^{\infty} x(t)e^{-j2\pi ft}\, dt \longleftrightarrow x(t) = \int_{-\infty}^{\infty} X(f)e^{j2\pi ft}\, df. \qquad (6.1)$$

The formulation assumes a continuous signal of infinite duration and bandwidth. For practical representation, we must sample in time and frequency, and amplitudes must be quantized. From an implementation standpoint, we prefer to use a finite number of samples in time and frequency. This leads to the *discrete Fourier transform* (DFT), where N samples are used in time and frequency, according to

$$X[k] = \sum_{n=0}^{N-1} x[n]e^{-j2\pi kn/N} = \sum_{n=0}^{N-1} x[n]W_N^{kn}, \qquad (6.2)$$

and the inverse DFT (IDFT) is defined as

$$x[n] = \frac{1}{N}\sum_{k=0}^{N-1} X[k]e^{j2\pi kn/N} = \frac{1}{N}\sum_{k=0}^{N-1} X[k]W_N^{-kn}, \qquad (6.3)$$

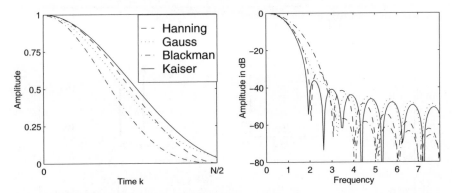

Fig. 6.2. Window functions in time and frequency.

or, in vector/matrix notation

$$X = Wx \leftrightarrow x = \frac{1}{N}W^*X. \tag{6.4}$$

If we use the DFT to approximate the Fourier spectrum, we must remember the effect of sampling in time and frequency, namely:

- By sampling in *time*, we get a periodic spectrum with the sampling frequency f_S. The approximation of a Fourier transform by a DFT is reasonable only if the frequency components of $x(t)$ are concentrated on a smaller range than the Nyquist frequency $f_S/2$, as stated in the "Shannon sampling theorem."
- By sampling in the *frequency* domain, the time function becomes periodic, i.e., the DFT assumes the time series to be periodic. If an N-sample DFT is applied to a signal that does not complete an integer number of cycles within an N-sample window, a phenomenon called *leakage* occurs. Therefore, if possible, we should choose the sampling frequency and the analysis window in such a way that it covers an integer number of periods of $x(t)$, if $x(t)$ is periodic.

A more practical alternative for decreasing leakage is the use of a window function that tapers smoothly to zero on both sides. Such window functions were already discussed in the context of FIR filter design in Chap. 3 (see Table 3.2, p. 175). Figure 6.2 shows the time and frequency behavior of some typical windows [107, 166].

An example illustrates the use of a window function.

Example 6.1: Windowing

Figure 6.3a shows a sinusoidal signal that does not complete an integer number of periods in its sample window. The Fourier transform of the signal should ideally include only the two Dirac functions at $\pm\omega_0$, as displayed in Fig. 6.3b. Figures 6.3c and d show the DFT analysis with different windows. We note that the analysis with the box function has somewhat more ripple

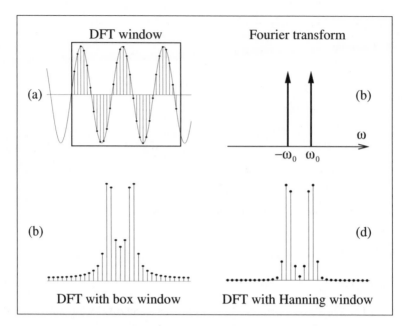

Fig. 6.3. Analysis of periodic function through the DFT, using window functions.

than the analysis with the Hanning window. An exact analysis would also show that the main lope width with Hanning analysis is larger than the width achieved with the box function, i.e., no window.
$$\boxed{6.1}$$

6.1.2 Properties of the DFT

The most important properties of the DFT are summarized in Table 6.1. Many properties are identical with the Fourier transform, e.g., the transform is unique (bijective), the superposition applies, and real and imaginary parts are related through the Hilbert transform.

The similarity of the forward and inverse transform leads to an alternative inversion algorithm. Using the vector/matrix notation (6.4) of the DFT

$$X = Wx \leftrightarrow x = \frac{1}{N}W^*X, \tag{6.5}$$

we can conclude

$$x^* = \frac{1}{N}\left(W^*X\right)^* = \frac{1}{N}WX^*, \tag{6.6}$$

i.e., we can use the DFT of X^* scaled by $1/N$ to compute the inverse DFT.

Table 6.1. Theorems of the DFT

Theorem	$x[n]$	$X[k]$
Transform	$x[n]$	$\sum_{n=0}^{N-1} x[n]\mathrm{e}^{-\mathrm{j}2\pi nk/N}$
Inverse Transform	$\frac{1}{N}\sum_{k=0}^{N-1} X[k]\mathrm{e}^{\mathrm{j}2\pi nk/N}$	$X[k]$
Superposition	$s_1 x_1[n] + s_2 x_2[n]$	$s_1 X_1[k] + s_2 X_2[k]$
Time reversal	$x[-n]$	$X[-k]$
Conjugate complex	$x^*[n]$	$X^*[-k]$
Split		
Real part	$\Re(x[n])$	$(X[k] + X^*[-k])/2$
Imaginary part	$\Im(x[n])$	$(X[k] + X^*[-k])/(2\mathrm{j})$
Real even part	$x_e[n] = (x[n] + x[-n])/2$	$\Re(X[k])$
Real odd part	$x_o[n] = (x[n] - x[-n])/2$	$\mathrm{j}\Im(X[k])$
Symmetry	$X[n]$	$Nx[-k]$
Cyclic convolution	$x[n] \circledast f[n]$	$X[k]F[k]$
Multiplication	$x[n] \times f[n]$	$\frac{1}{N}X[k] \circledast F[k]$
Periodic shift	$x[n-d \bmod N]$	$X[k]\mathrm{e}^{-\mathrm{j}2\pi dk/N}$
Parseval theorem	$\sum_{n=0}^{N-1} \|x[n]\|^2$	$\frac{1}{N}\sum_{k=0}^{N-1} \|X[k]\|^2$

DFT of a Real Sequence

We now turn to some additional computational savings for DFT (and FFT) computations, when the input sequence is real. In this case, we have two options: we can compute with one N-point DFT the DFT of two N-point sequences, or we can compute with an N-point DFT a length $2N$ DFT of a real sequence.

If we use the Hilbert property from Table 6.1, i.e., a real sequence has an even-symmetric real spectrum and an odd imaginary spectrum, the following algorithms can be synthesized [161].

Algorithm 6.2:   Length $2N$ Transform with N-point DFT

The algorithm to compute the $2N$-point DFT $X[k] = X_r[k] + jX_i[k]$ from the time sequence $x[n]$ is as follows:

1) Build an N-point sequence $y[n] = x[2n] + jx[2n+1]$ with $n = 0, 1, \ldots N - 1$.

2) Compute $y[n] \circ\!\!-\!\!\bullet\, Y[k] = Y_r[k] + jY_i[k]$. where $\Re(Y[k]) = Y_r[k]$ is the real and $\Im(Y[k]) = Y_i[k]$ is the imaginary part of $Y[k]$, respectively.

3) Compute

$$X_r[k] = \frac{Y_r[k] + Y_r[-k]}{2} + \cos(\pi k/N)\,\frac{Y_i[k] + Y_i[-k]}{2}$$

$$-\sin(\pi k/N)\,\frac{Y_r[k] - Y_r[-k]}{2}$$

$$X_i[k] = \frac{Y_i[k] - Y_i[-k]}{2} - \sin(\pi k/N)\,\frac{Y_i[k] + Y_i[-k]}{2}$$

$$-\cos(\pi k/N)\,\frac{Y_r[k] - Y_r[-k]}{2}$$

with $k = 0, 1, \ldots N - 1$.

The computational effort, therefore, besides an N-point DFT (or FFT), is 4 N real additions and multiplications, from the twiddle factors $\pm \exp(j\pi k/N)$.

To transform two length-N sequences with a length-N DFT, we use the fact (see Table 6.1) that a real sequence has an even spectrum, while the spectrum of a purely imaginary sequence is odd. This is the basis for the following algorithm.

Algorithm 6.3:   Two Length N Transforms with one N-point DFT

The algorithm to compute the N-point DFT $g[n] \circ\!\!-\!\!\bullet\, G[k]$ and $h[n] \circ\!\!-\!\!\bullet\, H[k]$ is as follows:

1) Build an N-point sequence $y[n] = h[n] + jg[n]$ with $n = 0, 1, \ldots N - 1$.

2) Compute $y[n] \circ\!\!-\!\!\bullet\, Y[k] = Y_r[k] + jY_i[k]$, where $\Re(Y[k]) = Y_r[k]$ is the real and $\Im(Y[k]) = Y_i[k]$ is the imaginary part of $Y[k]$, respectively.

3) Compute, finally

$$H[k] = \frac{Y_r[k] + Y_r[-k]}{2} + j\frac{Y_i[k] - Y_i[-k]}{2}$$

$$G[k] = \frac{Y_i[k] + Y_i[-k]}{2} - j\frac{Y_r[k] - Y_r[-k]}{2},$$

with $k = 0, 1, \ldots N - 1$.

The computational effort, therefore, besides an N-point DFT (or FFT), is 2 N real additions, to form the correct two N-point DFTs.

Fast Convolution Using DFT

One of the most frequent applications of the DFT (or FFT) is the computation of convolutions. As with the Fourier transform, the convolution in time is

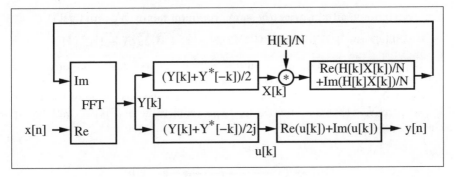

Fig. 6.4. Real convolution using a complex FFT [66].

done by multiplying the two transformed sequences: the two time sequences are transformed in the frequency domain, we compute a (scalar) pointwise product, and we transform the product back into the time domain. The main difference, compared with the Fourier transform, is that now the DFT computes a *cyclic*, and not a linear, convolution. This must be considered when implementing fast convolution with the FFT. This leads to two methods called "overlap save" and "overlap add." In the overlap save method, we basically discharge the samples at the border that are corrupted by the cyclic convolution. In the overlap add method, we zero-pad the filter and signal in such a way that we can directly add the partial sequences to a common product stream.

Most often the input sequences for the fast convolution are real. An efficient convolution may therefore be accomplished with a real transform, such as the Hartley transform discussed in Exercise 6.15, p. 393. We may also construct an FFT-like algorithm for the Hartley transform, and can get about twice the performance compared with a complex transform [167].

If we wish to utilize an available FFT program, we may use one of the previously discussed Algorithms, 6.2 or 6.3, for real sequences. An alternative approach is shown in Fig. 6.4. It shows a similar approach to Algorithm 6.3, where we implemented two N-point transforms with one N-point DFT, but in this case we use the "real" part for a DFT, and the imaginary part for the IDFT, which is needed for the back transformation, according to the convolution theorem.

It is assumed that the DFT of the real-valued filter (i.e., $F[k] = F[-k]^*$) has been computed offline and, in addition, in the frequency domain we need only $N/2$ multiplications to compute $X[k]F[k]$.

6.1.3 The Goertzel Algorithm

A single spectral component $X[k]$ in the DFT computation is given by

$$X[k] = x[0] + x[1]W_N^k + x[2]W_N^{2k} + \ldots + x[N-1]W_N^{(N-1)k}.$$

We can combine all x[n] with the same common factor W_N^k, and get

$$X[k] = x[0] + W_N^k \left(x[1] + W_N^k \left(x[2] + \ldots + W_N^k x[N-1]\right) \ldots\right).$$

It can be noted that this results in a possible recursive computation of $X[k]$. This is called the Goertzel algorithm, and is graphically interpreted by Fig. 6.5. The computation of $y[n]$ starts with the last value of the input sequence $x[N-1]$. After step three, a spectrum value of $X[k]$ is available at the output.

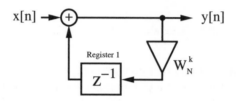

Step	$x[n]$	Register 1	$y[n]$
0	$x[3]$	0	$x[3]$
1	$x[2]$	$W_4^k x[3]$	$x[2] + W_4^k x[3]$
2	$x[1]$	$W_4^k x[2] + W_4^{2k} x[3]$	$x[1] + W_4^k x[2] + W_4^{2k} x[3]$
3	$x[0]$	$W_4^k x[1]$ $+W_4^{2k} x[2] + W_4^{3k} x[3]$	$x[0] + W_4^k x[1]$ $+W_4^{2k} x[2] + W_4^{3k} x[3]$

Fig. 6.5. The length-4 Goertzel algorithm.

If we have to compute several spectral components, we can reduce the complexity if we combine factors of the type $e^{\pm j 2\pi n/N}$. This will result in second-order systems having a denominator according to

$$z^2 - 2z \cos\left(\frac{2\pi n}{N}\right) + 1.$$

All complex multiplications are then reduced to real multiplications.

In general, the Goertzel algorithm can be attractive if only a few spectral components have to be computed. For the whole DFT, the effort is of order N^2, and therefore yields no advantage compared with the direct DFT computation.

6.1.4 The Bluestein Chirp-z Transform

In the Bluestein chirp-z transform (CZT) algorithm, the DFT exponent nk is quadratic expanded to

$$nk = -(k-n)^2/2 + n^2/2 + k^2/2. \tag{6.7}$$

The DFT therefore becomes

$$X[k] = W_N^{k^2/2} \sum_{n=0}^{N-1} \left(x[n] W_N^{n^2/2} \right) W_N^{-(k-n)^2/2}. \tag{6.8}$$

This algorithm is graphically interpreted in Fig. 6.6. This results in the following

---

Algorithm 6.4:   **Bluestein Chirp-z Algorithm**

The computation of the DFT is done in three steps, namely

1) N multiplication of $x[n]$ with $W_N^{n^2/2}$
2) Linear convolution of $x[n] W_N^{n^2/2} * W_N^{n^2/2}$
3) N multiplications with $W_N^{k^2/2}$

For a complete transform, we therefore need a length-N convolution and $2N$ complex multiplications. The advantage, compared with the Rader algorithms, is that there is no restriction to primes in the transform length N. The CZT can be defined for every length.

Narasimha et al. [168] and others have noticed that in the CZT algorithm many coefficients of the FIR filter part are trivial or identical. For instance, the length-8 CZT has an FIR filter of length 14, but there are only four different complex coefficients, as graphically interpreted in Fig. 6.7. These four coefficients are $1, j$, and $\pm e^{22.5°}$, i.e., we only have two nontrivial real coefficients to implement.

It may be of general interest what the *maximum* DFT length for a fixed number C_N of (complex) coefficients is. This is shown in the following table.

DFT length	8	12	16	24	40	48	72	80	120	144	168	180	240	360	504
C_N	4	6	7	8	12	14	16	21	24	28	32	36	42	48	64

As mentioned before, the number of different complex coefficients does not directly correspond to the implementation effort, because some coefficients

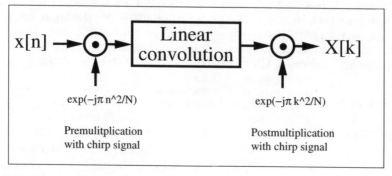

Fig. 6.6. The Bluestein chirp-z algorithm.

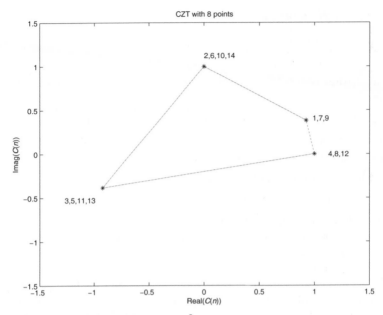

Fig. 6.7. CZT coefficients $C(n) = e^{j2\pi \frac{n^2/2 \bmod 8}{8}}$; $n = 1, 2, \dots, 14$.

may be trivial (i.e., ± 1 or $\pm j$) or may show symmetry. In particular, the power-of-two length transform enjoys many symmetries, as can be seen from Fig. 6.8. If we compute the maximum DFT length for a specific number of nontrivial real coefficients, we find as maximum-length transforms:

DFT length	10	16	20	32	40	48	50	80	96	160	192
sin/cos	2	3	5	6	8	9	10	11	14	20	25

Length 16 and 32 are therefore the maximum length DFTs with only three and fix real multipliers, respectively.

In general, power-of-two lengths are popular FFT building blocks, and the following table therefore shows, for length $N = 2^n$, the effort when implementing the CZT filter in transposed form.

N	C_N	sin/cos	CSD adder	RAG adder	NOFs for 14-bit coefficients
8	4	2	23	7	3,5,33,49,59
16	7	3	91	8	3,25,59,63,387
32	12	6	183	13	3,25,49,73,121,375
64	23	11	431	18	5,25,27,93,181,251,7393
128	44	22	879	31	5,15,25,175,199,319,403,499,1567
256	87	42	1911	49	5,25,765,1443,1737,2837,4637

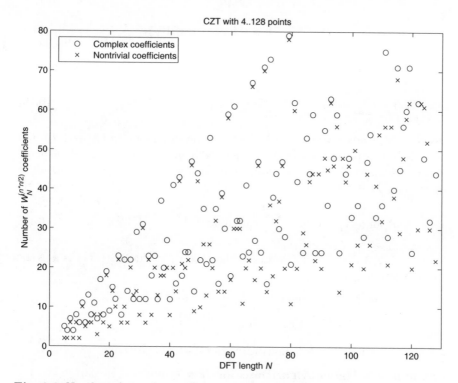

Fig. 6.8. Number of complex coefficients and nontrivial real multiplications for the CZT.

The first column shows the DFT length N. The second column shows the total number of complex exponentials C_N. The worst-case effort for C_N complex coefficients is that $2C_N$ real, nontrivial coefficients must be implemented. The actual number of different nontrivial real coefficients is shown in column three. We note when comparing columns two and three that for power-of-two lengths the symmetry and trivial coefficients reduce the number of nontrivial coefficients. The next columns show, for CZT DFTs up to length 256, the effort (i.e., number of adders) for a 15-bit (14-bit unsigned plus sign bit) coefficient precision implementation, using the CSD and RAG algorithms (discussed in Chap. 2), respectively. For CSD coding no coefficient symmetry is consider and the adder count is quite high. We note that the RAG algorithm when compared with CSD can essentially reduce the effort for DFT lengths larger than 16.

6.1.5 The Rader Algorithm

The Rader algorithm [169, 170] to compute a DFT,

$$X[k] = \sum_{n=0}^{N-1} x[n] W_N^{nk} \qquad k, n \in \mathbb{Z}_N; \quad \text{ord}(W_N) = N \tag{6.9}$$

is defined only for prime length N. We first compute the DC component with

$$X[0] = \sum_{n=0}^{N-1} x[n]. \tag{6.10}$$

Because $N = p$ is a prime, we know from the discussion in Chap. 2 (p. 67) that there is a primitive element, a *generator* g, that generates all elements of n and k in the field $\mathbb{Z}_p$, excluding zero, i.e., $g^k \in \mathbb{Z}_p/\{0\}$. We substitute n by $g^n \bmod N$ and k with $g^k \bmod N$, and get the following index transform:

$$X[g^k \bmod N] - x[0] = \sum_{n=0}^{N-2} x[g^n \bmod N] W_N^{g^{n+k} \bmod N} \tag{6.11}$$

for $k \in \{1, 2, 3, \ldots, N-1\}$. We note that the right side of (6.11) is a cyclic convolution, i.e.,

$$\left[x[g^0 \bmod N], x[g^1 \bmod N], \ldots, x[g^{N-2} \bmod N] \right]$$
$$\circledast \left[W_N, W_N^g, \ldots, W_N^{g^{N-2} \bmod N} \right]. \tag{6.12}$$

An example with $N = 7$ demonstrates the Rader algorithms.

Example 6.5: Rader Algorithms for $N = 7$

For $N = 7$, we know that $g = 3$ is a primitive element (see, for instance, [5], Table B.7), and the index transform is

$$[3^0, 3^1, 3^2, 3^3, 3^4, 3^5] \bmod 7 \equiv [1, 3, 2, 6, 4, 5]. \tag{6.13}$$

We first compute the DC component

$$X[0] = \sum_{n=0}^{6} x[n] = x[0] + x[1] + x[2] + x[3] + x[4] + x[5] + x[6],$$

and in the second step, the cyclic convolution of $X[k] - x[0]$

$$[x[1], x[3], x[2], x[6], x[4], x[5]] \circledast [W_7, W_7^3, W_7^2, W_7^6, W_7^4, W_7^5],$$

or in matrix notation

$$
\begin{bmatrix} X[1] \\ X[3] \\ X[2] \\ X[6] \\ X[4] \\ X[5] \end{bmatrix}
=
\begin{bmatrix}
W_7^1 & W_7^3 & W_7^2 & W_7^6 & W_7^4 & W_7^5 \\
W_7^3 & W_7^2 & W_7^6 & W_7^4 & W_7^5 & W_7^1 \\
W_7^2 & W_7^6 & W_7^4 & W_7^5 & W_7^1 & W_7^3 \\
W_7^6 & W_7^4 & W_7^5 & W_7^1 & W_7^3 & W_7^2 \\
W_7^4 & W_7^5 & W_7^1 & W_7^3 & W_7^2 & W_7^6 \\
W_7^5 & W_7^1 & W_7^3 & W_7^2 & W_7^6 & W_7^4
\end{bmatrix}
\begin{bmatrix} x[1] \\ x[3] \\ x[2] \\ x[6] \\ x[4] \\ x[5] \end{bmatrix}
+
\begin{bmatrix} x[0] \\ x[0] \\ x[0] \\ x[0] \\ x[0] \\ x[0] \end{bmatrix}. \tag{6.14}
$$

This is graphically interpreted using an FIR filter in Fig. 6.9.

We now verify the $p = 7$ Rader DFT formula, using a test triangular signal $x[n] = 10\lambda[n]$ (i.e., a triangle with step size 10). Directly interpreting (6.14), one obtains

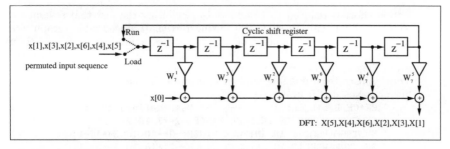

Fig. 6.9. Length $p = 7$ Rader prime-factor DFT implementation.

$$
\begin{bmatrix} X[1] \\ X[3] \\ X[2] \\ X[6] \\ X[4] \\ X[5] \end{bmatrix} = \begin{bmatrix} W_7^1 & W_7^3 & W_7^2 & W_7^6 & W_7^4 & W_7^5 \\ W_7^3 & W_7^2 & W_7^6 & W_7^4 & W_7^5 & W_7^1 \\ W_7^2 & W_7^6 & W_7^4 & W_7^5 & W_7^1 & W_7^3 \\ W_7^6 & W_7^4 & W_7^5 & W_7^1 & W_7^3 & W_7^2 \\ W_7^4 & W_7^5 & W_7^1 & W_7^3 & W_7^2 & W_7^6 \\ W_7^5 & W_7^1 & W_7^3 & W_7^2 & W_7^6 & W_7^4 \end{bmatrix} \begin{bmatrix} 20 \\ 40 \\ 30 \\ 70 \\ 50 \\ 60 \end{bmatrix} + \begin{bmatrix} 10 \\ 10 \\ 10 \\ 10 \\ 10 \\ 10 \end{bmatrix}
$$

$$
= \begin{bmatrix} -35 + j72 \\ -35 + j8 \\ -35 + j28 \\ -35 - j72 \\ -35 - j8 \\ -35 - 28 \end{bmatrix}.
$$

The value of $X[0]$ is the sum of the time series, which is $10+20+\cdots+70 = 280$.

6.5

In addition, in the Rader algorithms we may use the symmetries of the complex pairs $e^{\pm j2k\pi/N}, k \in [0, N/2]$, to build more-efficient FIR realizations (Exercise 6.5, p. 391). Implementing a Rader prime-factor DFT is equivalent to implementing an FIR filter, which we discussed in Chap. 3. In order to implement a fast FIR filter, a fully pipelined DA or the transposed filter structure using the RAG algorithm is attractive. The RAG FPGA implementation is illustrated in the following example.

Example 6.6: Rader FPGA Implementation

An RAG implementation of the length-7 Rader algorithm is accomplished as follows. The first step is quantizing the coefficients. Assuming that the input values and coefficients are to be represented as a signed 8-bit word, the quantized coefficients are:

k	0	1	2	3	4	5	6
$\mathrm{Re}\{256 \times W_7^k\}$	256	160	-57	-231	-231	-57	160
$\mathrm{Im}\{256 \times W_7^k\}$	0	-200	-250	-111	111	250	200

A direct-form implementation of all the individual coefficients would (consulting Table 2.3, p. 64) consume 24 adders for the constant coefficient multipliers. Using the transposed structure, the individual coefficient implementation effort is reduced to 11 adders by exploiting the fact that several co-efficients differ only in sign. Optimizing further (reduced adder graph, see Fig. 2.4, p. 63), the number of adders reaches a minimum of seven (see Factor: PROCESS and Coeffs: PROCESS below). This is more than a three times improvement over the direct FIR architecture. The following VHDL code[1] illustrates a possible implementation of the length-7 Rader DFT, using transposed FIR filters.

```
PACKAGE B_bit_int IS      ------> User-defined types
   SUBTYPE WORD8 IS INTEGER RANGE -2**7 TO 2**7-1;
   SUBTYPE WORD11 IS INTEGER RANGE -2**10 TO 2**10-1;
   SUBTYPE WORD19 IS INTEGER RANGE -2**18 TO 2**18-1;
   TYPE ARRAY_WORD IS ARRAY (0 to 5) OF WORD19;
END B_bit_int;

LIBRARY work;
USE work.B_bit_int.ALL;

LIBRARY ieee;
USE ieee.std_logic_1164.ALL;
USE ieee.std_logic_arith.ALL;
USE ieee.std_logic_unsigned.ALL;

ENTITY rader7 IS                       ------> Interface
   PORT ( clk, reset    : IN  STD_LOGIC;
          x_in          : IN  WORD8;
          y_real, y_imag : OUT WORD11);
END rader7;

ARCHITECTURE fpga OF rader7 IS

   SIGNAL  count    : INTEGER RANGE 0 TO 15;
   TYPE    STATE_TYPE IS (Start, Load, Run);
   SIGNAL  state    : STATE_TYPE ;
   SIGNAL  accu     : WORD11 := 0;         -- Signal for X[0]
   SIGNAL  real, imag : ARRAY_WORD := (0,0,0,0,0,0);
                                  -- Tapped delay line array
   SIGNAL  x57, x111, x160, x200, x231, x250 : WORD19 := 0;
                        -- The (unsigned) filter coefficients
   SIGNAL  x5, x25, x110, x125, x256  : WORD19 ;
                        -- Auxiliary filter coefficients
   SIGNAL  x, x_0 : WORD8;  -- Signals for x[0]

BEGIN

   States: PROCESS (reset, clk)-----> FSM for RADER filter
   BEGIN
      IF reset = '1' THEN               -- Asynchronous reset
         state <= Start;
      ELSIF rising_edge(clk) THEN
         CASE state IS
            WHEN Start =>               -- Initialization step
```

---

[1] The equivalent Verilog code rader7.v for this example can be found in Appendix A on page 710. Synthesis results are shown in Appendix B on page 731.

```
            state <= Load;
            count <= 1;
            x_0 <= x_in;          -- Save x[0]
            accu <= 0 ;           -- Reset accumulator for X[0]
            y_real  <= 0;
            y_imag  <= 0;
        WHEN Load => -- Apply x[5],x[4],x[6],x[2],x[3],x[1]
          IF count = 8 THEN       -- Load phase done ?
            state <= Run;
          ELSE
            state <= Load;
            accu  <= accu + x ;
          END IF;
          count <= count + 1;
        WHEN Run => -- Apply again x[5],x[4],x[6],x[2],x[3]
          IF count = 15 THEN      -- Run phase done ?
            y_real <= accu;    -- X[0]
            y_imag <= 0;  -- Only re inputs i.e. Im(X[0])=0
            state  <= Start;      -- Output of result
          ELSE                    -- and start again
            y_real <= real(0) / 256 + x_0;
            y_imag <= imag(0) / 256;
            state <= Run;
          END IF;
          count <= count + 1;
      END CASE;
    END IF;
END PROCESS States;

Structure: PROCESS      -- Structure of the two FIR
BEGIN                   -- filters in transposed form
  WAIT UNTIL clk = '1';
  x <= x_in;
  -- Real part of FIR filter in transposed form
  real(0) <= real(1) + x160  ;    -- W^1
  real(1) <= real(2) - x231  ;    -- W^3
  real(2) <= real(3) - x57   ;    -- W^2
  real(3) <= real(4) + x160  ;    -- W^6
  real(4) <= real(5) - x231  ;    -- W^4
  real(5) <= -x57   ;             -- W^5

  -- Imaginary part of FIR filter in transposed form
  imag(0) <= imag(1) - x200  ;    -- W^1
  imag(1) <= imag(2) - x111  ;    -- W^3
  imag(2) <= imag(3) - x250  ;    -- W^2
  imag(3) <= imag(4) + x200  ;    -- W^6
  imag(4) <= imag(5) + x111  ;    -- W^4
  imag(5) <= x250;                -- W^5
END PROCESS Structure;

Coeffs: PROCESS  -- Note that all signals
BEGIN            -- are globally defined
  WAIT UNTIL clk = '1';
```

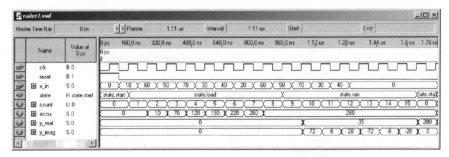

Fig. 6.10. VHDL simulation of a seven-point Rader algorithm.

```
-- Compute the filter coefficients and use FFs
  x160    <= x5 * 32;
  x200    <= x25 * 8;
  x250    <= x125 * 2;
  x57     <= x25 + x * 32;
  x111    <= x110 + x;
  x231    <= x256 - x25;
END PROCESS Coeffs;

Factors: PROCESS (x, x5, x25)    -- Note that all signals
BEGIN                            -- are globally defined
-- Compute the auxiliary factor for RAG without an FF
  x5      <= x * 4 + x;
  x25     <= x5 * 4 + x5;
  x110    <= x25 * 4 + x5 * 2;
  x125    <= x25 * 4 + x25;
  x256    <= x * 256;
END PROCESS Factors;

  END fpga;
```

The design consists of four blocks of statements within the four PROCESS statements. The first – Stages: PROCESS – is the state machine, which distinguishes the three processing phases, Start, Load, and Run. The second – Structure: PROCESS – defines the two FIR filter paths, real and imaginary. The third item implements the multiplier block using the reduced adder graph. The forth block – Factor: PROCESS – implements the unregistered factors of the RAG algorithm. It can be seen that all coefficients are realized by using six adders and one subtractor. The design uses 443 LEs, no embedded multiplier, and has a 137.06 MHz Registered Performance. Figure 6.10 displays simulation results using Quartus II for a triangular input sequence $x[n] = \{10, 20, 30, 40, 50, 60, 70\}$. Note that the input and output sequences, starting at 1 μs, occur in permuted order, and negative results appear signed if we use the signed data type in the simulator. Finally, at 1.7 μs, $X[0] = 280$ is forwarded to the output and rader7 is ready to process the next input frame.

6.6

Because the Rader algorithm is restricted to prime lengths there is less symmetry in the coefficients, compared with the CZT. The following table shows, for primes length $2^n \pm 1$, the implementation effort of the circular filter in transposed form.

DFT length	sin/cos	CSD adder	RAG adder	NOFs for 14-bit coefficients
7	6	52	13	7,11,31,59,101,177,319
17	16	138	23	3,35,103,415,1153,1249,8051
31	30	244	38	3,9,133,797,877,975,1179,3235
61	60	496	66	5,39,51,205,265,3211
127	124	1060	126	5

The first column shows the cyclic convolution length N, which is also the number of complex coefficients. Comparing column two and the worst case with $2N$ real sin/cos coefficients, we see that symmetry and trivial coefficients reduce the number of nontrivial coefficients by a factor of 2. The next two columns show the effort for a 14-bit (plus sign) coefficient precision implementation using CSD or RAG algorithms, respectively. The last column shows the auxiliary coefficient, i.e., NOFs used by RAG. Note the advantage of RAG for longer filters. It can be seen from this table that the effort for CSD-type filters can be estimated by $BN/2$, where B is the coefficient bit width (14 in this table) and N is the filter length. For RAG, the effort (i.e., the number of adders) is only N, i.e., a factor $B/2$ improvement over CSD for longer filters (for $B = 14$, a factor $\approx 14/2 = 7$ of improvement). For longer filters, RAG needs only one additional adder for each additional coefficient, because the already-synthesized coefficient produces a dense grid of small coefficients.

6.1.6 The Winograd DFT Algorithm

The first algorithm with a reduced number of multiplications necessary we want to discuss is the *Winograd DFT* algorithm. The Winograd algorithm is a combination of the Rader algorithm (which translates the DFT into a cyclic convolution), and Winograd's [103] short convolution algorithm, which we have already used to implement fast-running FIR filters (see Sect. 5.2.2, p. 254).

The length is therefore restricted to primes or powers of primes. Table 6.2 gives an overview of the number of arithmetic operations necessary.

The following example for $N = 5$ demonstrates the steps to build a Winograd DFT algorithm.

Example 6.7: $N = 5$ **Winograd DFT Algorithm**

An alternative representation of the Rader algorithm, using $X[0]$ instead of $x[0]$, is given by [5]

Table 6.2. Effort for the Winograd DFT with real inputs. Trivial multiplications are those by ± 1 or $\pm j$. For complex inputs, the number of operations is twice as large.

Block length	Total number of real multiplications	Total number nontrivial multiplications	Total number of real additions
2	2	0	2
3	3	2	6
4	4	0	8
5	6	5	17
7	9	8	36
8	8	2	26
9	11	10	44
11	21	20	84
13	21	20	94
16	18	10	74
17	36	35	157
19	39	38	186

$$X[0] = \sum_{n=0}^{4} x[n] = x[0] + x[1] + x[2] + x[3] + x[4]$$

$$X[k] - X[0]$$
$$= [x[1], x[2], x[4], x[3]] \circledast [W_5 - 1, W_5^2 - 1, W_5^4 - 1, W_5^3 - 1]$$
$$k = 1, 2, 3, 4.$$

If we implement the cyclic convolution of length 4 with a Winograd algorithm that costs only five nontrivial multiplications, we get the following algorithm:

$$X[k] = \sum_{n=0}^{4} x[n] e^{-j2\pi kn/5} \qquad k = 0, 1, \ldots, 4$$

$$\begin{bmatrix} X[0] \\ X[4] \\ X[3] \\ X[2] \\ X[1] \end{bmatrix} = \begin{bmatrix} 1 & 0 & 0 & 0 & 0 \\ 1 & 1 & 1 & 1 & 0 & -1 \\ 1 & 1 & -1 & 1 & 1 & 0 \\ 1 & 1 & -1 & -1 & -1 & 0 \\ 1 & 1 & 1 & -1 & 0 & 1 \end{bmatrix}$$

$$\times \mathrm{diag}(1, \frac{1}{2}(\cos(2\pi/5) + \cos(4\pi/5)) - 1,$$

$$\frac{1}{2}(\cos(2\pi/5) - \cos(4\pi/5)), j\sin(2\pi/5),$$

$$j(-\sin(2\pi/5) + \sin(4\pi/5)), j(\sin(2\pi/5) + \sin(4\pi/5)))$$

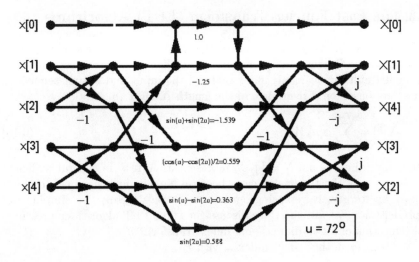

Fig. 6.11. Winograd 5-point DFT signal flow graph.

$$\times \begin{bmatrix} 1 & 1 & 1 & 1 & 1 \\ 0 & 1 & 1 & 1 & 1 \\ 0 & 1 & -1 & -1 & 1 \\ 0 & 1 & -1 & 1 & -1 \\ 0 & 1 & 0 & 0 & -1 \\ 0 & 0 & -1 & 1 & 0 \end{bmatrix} \begin{bmatrix} x[0] \\ x[1] \\ x[2] \\ x[3] \\ x[4] \end{bmatrix}.$$

The total computational effort is therefore only 5 or 10 real nontrivial multiplications for real or imaginary input sequences $x[n]$, respectively. The signal flow graph shown in Fig. 6.11 shows also how to implement the additions in an efficient fashion.

6.7

It is quite convenient to use a matrix notation for the Winograd DFT algorithm, and so we get

$$\boldsymbol{W}_{N_l} = \boldsymbol{C}_l \times \boldsymbol{B}_l \times \boldsymbol{A}_l, \tag{6.15}$$

where $\boldsymbol{A}_l$ incorporates the input addition, $\boldsymbol{B}_l$ is the diagonal matrix with the Fourier coefficients, and $\boldsymbol{C}_l$ includes the output additions. The only disadvantage is that now it is not as easy to define the exact steps of the short convolution algorithms, because the sequence in which input and output additions are computed is lost with this matrix representation.

This combination of Rader algorithms and a short Winograd convolution, known as the Winograd DFT algorithm, will be used later, together with index mapping to introduce the Winograd FFT algorithm. This is the FFT algorithm with the least number of real multiplications among all known FFT algorithms.

6.2 The Fast Fourier Transform (FFT) Algorithms

As mentioned in the introduction of this chapter, we use the terminology introduced by Burrus [148], who classified all FFT algorithms simply by different (multidimensional) index maps of the input and output sequences. These are based on a transform of the length N DFT (6.2)

$$X[k] = \sum_{n=0}^{N-1} x[n]\, W_N^{nk} \tag{6.16}$$

into a multidimensional $N = \prod_l N_l$ representation. It is, in general, sufficient to discuss only the two-factor case, because higher dimensions can be built simply by iteratively replacing again one of these factors. To simplify our representation we will therefore discuss the three FFT algorithms presented only in terms of a two-dimensional index transform.

We transform the (time) index n with

$$n = An_1 + Bn_2 \bmod N \qquad \begin{cases} 0 \leq n_1 \leq N_1 - 1 \\ 0 \leq n_2 \leq N_2 - 1, \end{cases} \tag{6.17}$$

where $N = N_1 N_2$, and $A, B \in \mathbb{Z}$ are constants that must be defined later. Using this index transform, a two-dimensional mapping $f : \mathbb{C}^N \to \mathbb{C}^{N_1 \times N_2}$ of the data is built, according to

$$
\begin{aligned}
& [x[0]\ x[1]\ x[2] \cdots x[N-1]] \\
& = \begin{bmatrix}
x[0,0] & x[0,1] & \cdots & x[0, N_2 - 1] \\
x[1,0] & x[1,1] & \cdots & x[1, N_2 - 1] \\
\vdots & \vdots & \ddots & \vdots \\
x[N_1 - 1, 0] & x[N_1 - 1, 1] & \cdots & x[N_1 - 1, N_2 - 1]
\end{bmatrix}.
\end{aligned} \tag{6.18}
$$

Applying another index mapping k to the output (frequency) domain yields

$$k = Ck_1 + Dk_2 \bmod N \qquad \begin{cases} 0 \leq k_1 \leq N_1 - 1 \\ 0 \leq k_2 \leq N_2 - 1, \end{cases} \tag{6.19}$$

where $C, D \in \mathbb{Z}$ are constants that must be defined later. Because the DFT is bijective, we must choose A, B, C, and D in such a way that the transform representation is still unique, i.e., a bijective projection. Burrus [148] has determined the general conditions for how to choose A, B, C, and D for specific N_1 and N_2 such that the mapping is bijective (see Exercises 6.7 and 6.8, p. 392). The transforms given in this chapter are all unique.

An important point in distinguishing different FFT algorithms is the question of whether N_1 and N_2 are allowed to have a common factor, i.e., $\gcd(N_1, N_2) > 1$, or whether the factors must be coprime. Sometimes algorithms with $\gcd(N_1, N_2) > 1$ are referred to as *common-factor algorithms* (CFAs), and algorithms with $\gcd(N_1, N_2) = 1$ are called *prime-factor algorithms* (PFAs). A CFA algorithm discussed in the following is the Cooley–Tukey FFT, while the Good–Thomas and Winograd FFTs are of the PFA type. It should be emphasized that the Cooley–Tukey algorithm may indeed realize FFTs with two factors, $N = N_1 N_2$, which are coprime, and that for a PFA the factors N_1 and N_2 must only be coprime, i.e., they must *not* be primes themselves. A transform of length $N = 12$ factored with $N_1 = 4$ and $N_2 = 3$, for instance, can therefore be used for both CFA FFTs and PFA FFTs!

6.2.1 The Cooley–Tukey FFT Algorithm

The Cooley–Tukey FFT is the most universal of all FFT algorithms, because any factorization of N is possible. The most popular Cooley–Tukey FFTs are those where the transform length N is a power of a basis r, i.e., $N = r^\nu$. These algorithms are often referred to as radix-r algorithms.

The index transform suggested by Cooley and Tukey (and earlier by Gauss) is also the simplest index mapping. Using (6.17) we have $A = N_2$ and $B = 1$, and the following mapping results

$$n = N_2 n_1 + n_2 \qquad \begin{cases} 0 \leq n_1 \leq N_1 - 1 \\ 0 \leq n_2 \leq N_2 - 1. \end{cases} \tag{6.20}$$

From the valid range of n_1 and n_2, we conclude that the modulo reduction given by (6.17) need not be explicitly computed.

For the inverse mapping from (6.19) Cooley and Tukey, choose $C = 1$ and $D = N_1$, and the following mapping results

$$k = k_1 + N_1 k_2 \qquad \begin{cases} 0 \leq k_1 \leq N_1 - 1 \\ 0 \leq k_2 \leq N_2 - 1. \end{cases} \tag{6.21}$$

The modulo computation can also be skipped in this case. If we now substitute n and k in W_N^{nk} according to (6.20) and (6.21), respectively, we find

$$W_N^{nk} = W_N^{N_2 n_1 k_1 + N_1 N_2 n_1 k_2 + n_2 k_1 + N_1 n_2 k_2}. \tag{6.22}$$

Because W is of order $N = N_1 N_2$, it follows that $W_N^{N_1} = W_{N_2}$ and $W_N^{N_2} = W_{N_1}$. This simplifies (6.22) to

$$W_N^{nk} = W_{N_1}^{n_1 k_1} W_N^{n_2 k_1} W_{N_2}^{n_2 k_2}. \tag{6.23}$$

If we now substitute (6.23) in the DFT from (6.16) it follows that

$$X[k_1, k_2] = \sum_{n_2=0}^{N_2-1} W_{N_2}^{n_2 k_2} \left(W_N^{n_2 k_1} \underbrace{\sum_{n_1=0}^{N_1-1} x[n_1, n_2] W_{N_1}^{n_1 k_1}}_{N_1\text{-point transform}} \right) \tag{6.24}$$

$$\underbrace{\phantom{X[k_1, k_2] = \sum_{n_2=0}^{N_2-1} W_{N_2}^{n_2 k_2} \left(W_N^{n_2 k_1} \sum_{n_1=0}^{N_1-1} x[n_1, n_2] W_{N_1}^{n_1 k_1} \right)}}_{\tilde{x}[n_2, k_1]}$$

$$= \sum_{n_2=0}^{N_2-1} W_{N_2}^{n_2 k_2} \, \bar{x}[n_2, k_1]. \tag{6.25}$$

$\underbrace{\qquad\qquad\qquad\qquad}_{}$

N_2-point transform

We now can define the complete Cooley–Tukey algorithm

Algorithm 6.8:　　　**Cooley–Tukey Algorithm**

An $N = N_1 N_2$-point DFT can be done using the following steps:
1) Compute an index transform of the input sequence according to (6.20).
2) Compute the N_2 DFTs of length N_1.
3) Apply the twiddle factors $W_N^{n_2 k_1}$ to the output of the first transform stage.
4) Compute N_1 DFTs of length N_2.
5) Compute an index transform of the output sequence according to (6.21).

The following length-12 transform demonstrates these steps.

Example 6.9: Cooley–Tukey FFT for $N = 12$

Assume $N_1 = 4$ and $N_2 = 3$. It follows then that $n = 3n_1 + n_2$ and $k = k_1 + 4k_2$, and we can compute the following tables for the index mappings:

n_2	n_1				k_2	k_1			
	0	1	2	3		0	1	2	3
0	$x[0]$	$x[3]$	$x[6]$	$x[9]$	0	$X[0]$	$X[1]$	$X[2]$	$X[3]$
1	$x[1]$	$x[4]$	$x[7]$	$x[10]$	1	$X[4]$	$X[5]$	$X[6]$	$X[7]$
2	$x[2]$	$x[5]$	$x[8]$	$x[11]$	2	$X[8]$	$X[9]$	$X[10]$	$X[11]$

With the help of this transform we can construct the signal flow graph shown in Fig. 6.12. It can be seen that first we must compute three DFTs with four points each, followed by the multiplication with the twiddle factors, and finally we compute four DFTs each having length 3.　　　　　　　　　　　$\boxed{6.9}$

For direct computation of the 12-point DFT, a total of $12^2 = 144$ complex multiplications and $11^2 = 121$ complex additions are needed. To compute the Cooley–Tukey FFT with the same length we need a total of 12 complex multiplication for the twiddle factors, of which 8 are trivial (i.e., ± 1 or $\pm j$) multiplications. According to Table 6.2 (p. 360), the length-4 DFTs can be computed using 8 real additions and no multiplications. For the length-3 DFTs, we need 4 multiplications and 6 additions. If we implement the (fixed coefficient) complex multiplications using 3 additions and 3 multiplications (see Algorithm 6.10, p. 367), and consider that $W^0 = 1, W^3 = -j$ and $W^6 = -1$ are trivial, the total effort for the 12-point Cooley–Tukey FFT is given by

$$3 \times 16 + 4 \times 3 + 4 \times 12 = 108 \quad \text{real additions and}$$
$$4 \times 3 + 4 \times 4 = 28 \quad \text{real multiplications.}$$

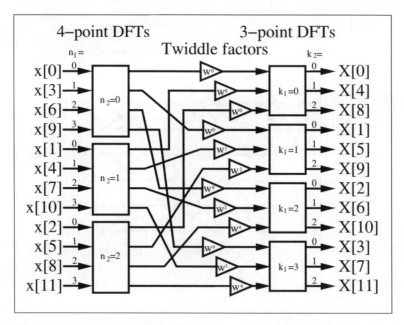

Fig. 6.12. Cooley–Tukey FFT for $N = 12$.

For the direct implementation we would need $2 \times 11^2 + 12^2 \times 3 = 674$ real additions and $12^2 \times 3 = 432$ real multiplications. It is now obvious why the Cooley–Tukey algorithm is called the "fast Fourier transform" (FFT).

Radix-r Cooley–Tukey Algorithm

One important fact that distinguishes the Cooley–Tukey algorithm from other FFT algorithms is that the factors for N can be chosen arbitrarily. It is therefore possible to use a radix-r algorithm in which $N = r^S$. The most popular algorithms are those of basis $r = 2$ or $r = 4$, because the necessary basic DFTs can, according to Table 6.2 (p. 360), be implemented without any multiplications. For $r = 2$ and S stages, for instance, the following index mapping results

$$n = 2^{S-1} n_1 + \cdots + 2 n_{S-1} + n_S \tag{6.26}$$
$$k = k_1 + 2 k_2 + \cdots + 2^{S-1} k_S. \tag{6.27}$$

For $S > 2$ a common practice is that in the signal flow graph a 2-point DFT is represented with a *Butterfly*, as shown in Fig. 6.13 for an 8-point transform. The signal flow graph representation has been simplified by using the fact that all arriving arrows at a node are added, while the constant coefficient multiplications are symbolized through a factor at an arrow. A radix-r algorithm has $\log_r(N)$ *stages*, and for each *group* the same type of twiddle factor occurs.

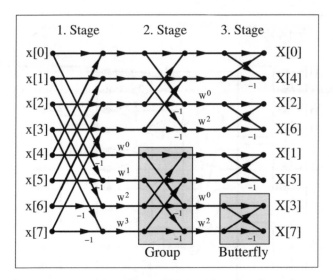

Fig. 6.13. Decimation-in-frequency algorithm of length-8 for radix-2.

It can be seen from the signal flow graph in Fig. 6.13 that the computation can be done *"in-place,"* i.e., the memory location used by a butterfly can be overwritten, because the data are no longer needed in the next computational steps. The total number of twiddle factor multiplications for the radix-2 transform is given by

$$\log_2(N)N/2, \tag{6.28}$$

because only every second arrow has a twiddle factor.

Because the algorithm shown in Fig. 6.13 starts in the frequency domain to split the original DFT into shorter DFTs, this algorithm is called a decimation-in-frequency (DIF) algorithm. The input values typically occur in natural order, while the index of the frequency values is in bit-reversed order. Table 6.3 shows the characteristic values of a DIF radix-2 algorithm.

Table 6.3. Radix-2 FFT with frequency decimation.

	Stage 1	Stage 2	Stage 3	$\cdots$	Stage $\log_2(N)$
Number of groups	1	2	4	...	$N/2$
Butterflies per group	$N/2$	$N/4$	$N/8$	...	1
Increment exponent twiddle factors	1	2	4	...	$N/2$

We may also construct an algorithm with decimation in time (DIT). In this case, we start by splitting the input (time) sequence, and we find that all frequency values will appear in natural order (Exercise 6.10, p. 392).

The necessary index transform for index 41, for an radix-2 and radix-4 algorithm, is shown in Fig. 6.14. For a radix-2 algorithm, a reversing of the bit sequence, a *bitreverse*, is necessary. For a radix-4 algorithm we must first build "digits" of two bits, and then reverse the order of these digits. This operation is called *digitreverse*.

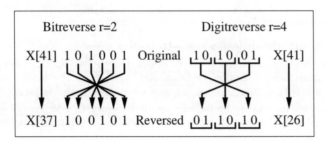

Fig. 6.14. Bitreverse and digitreverse.

Radix-2 Cooley–Tukey Algorithm Implementation

A radix-2 FFT can be efficiently implemented using a butterfly processor which includes, besides the butterfly itself, an additional complex multiplier for the twiddle factors.

A radix-2 butterfly processor consists of a complex adder, a complex subtraction, and a complex multiplier for the twiddle factors. The complex multiplication with the twiddle factor is often implemented with four real multiplications and two add/subtract operations. However, it is also possible to build the complex multiplier with only three real multiplications and three add/subtract operations, because one operand is precomputed. The algorithm works as follows:

Algorithm 6.10:   Efficient Complex Multiplier

The complex twiddle factor multiplication $R + jI = (X + jY) \times (C + jS)$ can be simplified, because C and S are precomputed and stored in a table. It is therefore also possible to store the following three coefficients

$$C, \quad C + S, \quad \text{and} \quad C - S. \tag{6.29}$$

With these three precomputed factors we first compute

$$E = X - Y, \quad \text{and then} \quad Z = C \times E = C \times (X - Y). \tag{6.30}$$

We can then compute the final product using

$$R = (C - S) \times Y + Z \tag{6.31}$$

$$I = (C + S) \times X - Z. \tag{6.32}$$

To check:

$$R = (C - S)Y + C(X - Y)$$
$$= CY - SY + CX - CY = CX - SY \checkmark$$
$$I = (C + S)X - C(X - Y)$$
$$= CX + SX - CX + CY = CY + SX. \checkmark$$

The algorithm uses three multiplications, one addition, and two subtractions, at the cost of an additional, third table.

The following example demonstrates the implementation of this twiddle factor complex multiplier.

Example 6.11: Twiddle Factor Multiplier

Let us first choose some concrete design parameters for the twiddle factor multiplier. Let us assume we have 8-bit input data, the coefficients should have 8 bits (i.e., 7 bits plus sign), and we wish to multiply by $e^{j\pi/9} = e^{j20°}$. Quantized to 8 bits, the twiddle factor becomes $C + jS = 128 \times e^{j\pi/9} = 121 + j39$. If we use an input value of $70 + j50$, then the expected result is

$$(70 + j50)e^{j\pi/9} = (70 + j50)(121 + j39)/128$$
$$= (6520 + j8780)/128 = 50 + j68.$$

If we use Algorithm 6.10 to compute the complex multiplication, the three factors become:

$$C = 121, \quad C + S = 160, \quad \text{and} \quad C - S = 82.$$

We note from the above that, in general, the tables $C + S$ and $C - S$ must have one more bit of precision than the C and S tables.

The following VHDL code[2] implements the twiddle factor multiplier.

```
LIBRARY lpm;
USE lpm.lpm_components.ALL;

LIBRARY ieee;
USE ieee.std_logic_1164.ALL;
USE ieee.std_logic_arith.ALL;

ENTITY ccmul IS
  GENERIC (W2  : INTEGER := 17;    -- Multiplier bit width
           W1  : INTEGER := 9;     -- Bit width c+s sum
           W   : INTEGER := 8);    -- Input bit width
    PORT (clk  : STD_LOGIC;   -- Clock for the output register
          x_in, y_in, c_in                        -- Inputs
             : IN  STD_LOGIC_VECTOR(W-1 DOWNTO 0);
          cps_in, cms_in                          -- Inputs
             : IN  STD_LOGIC_VECTOR(W1-1 DOWNTO 0);
          r_out, i_out                            -- Results
             : OUT STD_LOGIC_VECTOR(W-1 DOWNTO 0));
  END ccmul;
```

---

[2] The equivalent Verilog code ccmul.v for this example can be found in Appendix A on page 713. Synthesis results are shown in Appendix B on page 731.

```
ARCHITECTURE fpga OF ccmul IS

  SIGNAL x, y, c : STD_LOGIC_VECTOR(W-1 DOWNTO 0);
                                    -- Inputs and outputs
  SIGNAL r, i, cmsy, cpsx, xmyc           -- Products
                   : STD_LOGIC_VECTOR(W2-1 DOWNTO 0);
  SIGNAL xmy, cps, cms, sxtx, sxty        -- x-y etc.
                   : STD_LOGIC_VECTOR(W1-1 DOWNTO 0);

BEGIN
    x   <= x_in;   -- x
    y   <= y_in;   -- j * y
    c   <= c_in;   -- cos
    cps <= cps_in; -- cos + sin
    cms <= cms_in; -- cos - sin

  PROCESS
  BEGIN
    WAIT UNTIL clk='1';
    r_out <= r(W2-3 DOWNTO W-1);   -- Scaling and FF
    i_out <= i(W2-3 DOWNTO W-1);   -- for output
  END PROCESS;
---------- ccmul with 3 mul. and 3 add/sub --------------
  sxtx <= x(x'high) & x;        -- Possible growth for
  sxty <= y(y'high) & y;        -- sub_1 -> sign extension

  sub_1: lpm_add_sub                   -- Sub: x - y;
    GENERIC MAP ( LPM_WIDTH => W1, LPM_DIRECTION => "SUB",
                  LPM_REPRESENTATION => "SIGNED")
    PORT MAP (dataa => sxtx, datab => sxty, result => xmy);

  mul_1: lpm_mult           -- Multiply (x-y)*c = xmyc
    GENERIC MAP ( LPM_WIDTHA => W1, LPM_WIDTHB => W,
                  LPM_WIDTHP => W2, LPM_WIDTHS => W2,
                  LPM_REPRESENTATION => "SIGNED")
    PORT MAP ( dataa => xmy, datab => c, result => xmyc);

  mul_2: lpm_mult              -- Multiply (c-s)*y = cmsy
    GENERIC MAP ( LPM_WIDTHA => W1, LPM_WIDTHB => W,
                  LPM_WIDTHP => W2, LPM_WIDTHS => W2,
                  LPM_REPRESENTATION => "SIGNED")
    PORT MAP ( dataa => cms, datab => y, result => cmsy);

  mul_3: lpm_mult              -- Multiply (c+s)*x = cpsx
    GENERIC MAP ( LPM_WIDTHA => W1, LPM_WIDTHB => W,
                  LPM_WIDTHP => W2, LPM_WIDTHS => W2,
                  LPM_REPRESENTATION => "SIGNED")
    PORT MAP ( dataa => cps, datab => x, result => cpsx);

  sub_2: lpm_add_sub      -- Sub: i <= (c-s)*x - (x-y)*c;
    GENERIC MAP ( LPM_WIDTH => W2, LPM_DIRECTION => "SUB",
                  LPM_REPRESENTATION => "SIGNED")
    PORT MAP ( dataa => cpsx, datab => xmyc, result => i);
```

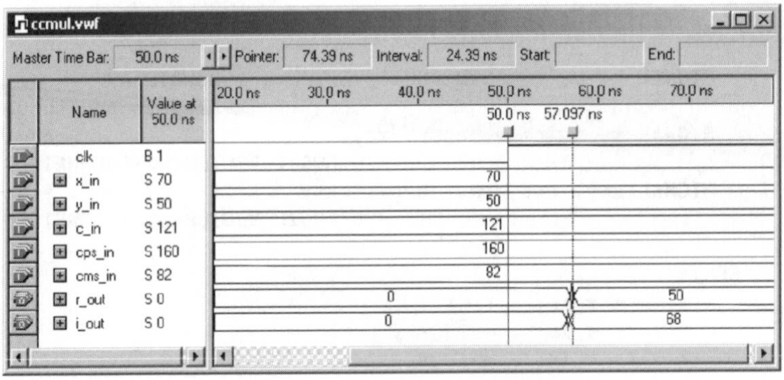

Fig. 6.15. VHDL simulation of a twiddle factor multiplier.

```
add_1: lpm_add_sub          -- Add: r <= (x-y)*c + (c+s)*y;
    GENERIC MAP ( LPM_WIDTH => W2, LPM_DIRECTION => "ADD",
                  LPM_REPRESENTATION => "SIGNED")
    PORT MAP ( dataa => cmsy, datab => xmyc, result => r);

END fpga;
```

The twiddle factor multiplier is implemented using component instantiations of three `lpm_mult` and three `lpm_add_sub` modules. The output is scaled such that it has the same data format as the input. This is reasonable, because multiplication with a complex exponential $e^{j\phi}$ should not change the magnitude of the complex input. To ensure short latency (for an in-place FFT), the complex multiplier only has output registers, with no internal pipeline registers. With only one output register, it is impossible to determine the `Registered Performance` of the design, but from the simulation results in Fig. 6.15, it can be estimated. The design uses 39 LEs, 3 embedded multipliers, and may run faster, if the `lpm_mult` components can be pipelined (see Fig. 2.16, p. 86). | 6.11 |

An in-place implementation, i.e., with only *one* data memory, is now possible, because the butterfly processor is designed without pipeline stages. If we introduce additional pipeline stages (one for the butterfly and three for the multiplier) the size of the design increases insignificantly (see Exercise 6.23, p. 395), however, the speed increases significantly. The price for this pipeline design is the cost for extra data memory for the whole FFT, because data read and write memories must now be separated, i.e., no in-place computation can be done.

Using the twiddle factor multiplier introduced above, it is now possible to design a butterfly processor for a radix-2 Cooley–Tukey FFT.

Example 6.12: Butterfly Processor

To prevent overflow in the arithmetic, the butterfly processor computes the two (scaled) butterfly equations

$$D_{re} + j \times D_{im} = ((A_{re} + j \times A_{im}) + (B_{re} + j \times B_{im}))/2$$
$$E_{re} + j \times E_{im} = ((A_{re} + j \times A_{im}) - (B_{re} + j \times B_{im}))/2$$

Then the temporary result $E_{re} + j \times E_{im}$ must be multiplied by the twiddle factor.

The VHDL code[3] of the whole butterfly processor is shown in the following.

```
LIBRARY lpm;
USE lpm.lpm_components.ALL;

LIBRARY ieee;
USE ieee.std_logic_1164.ALL;
USE ieee.std_logic_arith.ALL;

PACKAGE mul_package IS      -- User-defined components
  COMPONENT ccmul
    GENERIC (W2 : INTEGER := 17;    -- Multiplier bit width
             W1 : INTEGER := 9;     -- Bit width c+s sum
             W  : INTEGER := 8);    -- Input bit width
    PORT
    (clk   : IN STD_LOGIC; -- Clock for the output register
     x_in, y_in, c_in: IN  STD_LOGIC_VECTOR(W-1 DOWNTO 0);
                                                   -- Inputs
     cps_in, cms_in  : IN  STD_LOGIC_VECTOR(W1-1 DOWNTO 0);
                                                   -- Inputs
     r_out, i_out    : OUT STD_LOGIC_VECTOR(W-1 DOWNTO 0));
                                                   -- Results
  END COMPONENT;
END mul_package;

LIBRARY work;
USE work.mul_package.ALL;


LIBRARY ieee;
USE ieee.std_logic_1164.ALL;
USE ieee.std_logic_arith.ALL;

LIBRARY lpm;
USE lpm.lpm_components.ALL;

LIBRARY ieee;
USE ieee.std_logic_1164.ALL;
USE ieee.std_logic_arith.ALL;
USE ieee.std_logic_unsigned.ALL;

ENTITY bfproc IS
  GENERIC (W2 : INTEGER := 17;    -- Multiplier bit width
           W1 : INTEGER := 9;     -- Bit width c+s sum
           W  : INTEGER := 8);    -- Input bit width
  PORT
```

---

[3] The equivalent Verilog code bfproc.v for this example can be found in Appendix A on page 715. Synthesis results are shown in Appendix B on page 731.

```
(clk               : STD_LOGIC;
 Are_in, Aim_in, c_in,                    -- 8 bit inputs
 Bre_in, Bim_in    : IN  STD_LOGIC_VECTOR(W-1 DOWNTO 0);
 cps_in, cms_in    : IN  STD_LOGIC_VECTOR(W1-1 DOWNTO 0);
                                   -- 9 bit coefficients
 Dre_out, Dim_out,                        -- 8 bit results
 Ere_out, Eim_out  : OUT STD_LOGIC_VECTOR(W-1 DOWNTO 0)
                              := (OTHERS => '0'));
END bfproc;

ARCHITECTURE fpga OF bfproc IS

  SIGNAL dif_re, dif_im                      -- Bf out
                      : STD_LOGIC_VECTOR(W-1 DOWNTO 0);
  SIGNAL Are, Aim, Bre, Bim : INTEGER RANGE -128 TO 127:=0;
                              -- Inputs as integers
  SIGNAL c              : STD_LOGIC_VECTOR(W-1 DOWNTO 0)
                              := (OTHERS => '0'); -- Input
  SIGNAL cps, cms      : STD_LOGIC_VECTOR(W1-1 DOWNTO 0)
                              := (OTHERS => '0'); -- Coeff in
BEGIN

  PROCESS    -- Compute the additions of the butterfly using
  BEGIN      -- integers and store inputs in flip-flops
    WAIT UNTIL clk = '1';
    Are      <= CONV_INTEGER(Are_in);
    Aim      <= CONV_INTEGER(Aim_in);
    Bre      <= CONV_INTEGER(Bre_in);
    Bim      <= CONV_INTEGER(Bim_in);
    c        <= c_in;                    -- Load from memory cos
    cps      <= cps_in;         -- Load from memory cos+sin
    cms      <= cms_in;         -- Load from memory cos-sin
    Dre_out <= CONV_STD_LOGIC_VECTOR( (Are + Bre )/2, W);
    Dim_out <= CONV_STD_LOGIC_VECTOR( (Aim + Bim )/2, W);
  END PROCESS;

  -- No FF because butterfly difference "diff" is not an
  PROCESS (Are, Bre, Aim, Bim)          -- output port
  BEGIN
    dif_re <= CONV_STD_LOGIC_VECTOR(Are/2 - Bre/2, 8);
    dif_im <= CONV_STD_LOGIC_VECTOR(Aim/2 - Bim/2, 8);
  END PROCESS;

  ---- Instantiate the complex twiddle factor multiplier ----
  ccmul_1: ccmul                     -- Multiply (x+jy)(c+js)
    GENERIC MAP ( W2 => W2, W1 => W1, W => W)
    PORT MAP  ( clk => clk, x_in => dif_re, y_in => dif_im,
                c_in => c, cps_in => cps, cms_in => cms,
                r_out => Ere_out, i_out => Eim_out);

END fpga;
```

The butterfly processor is implemented using one adder, one subtraction, and the twiddle factor multiplier instantiated as a component. Flip-flops

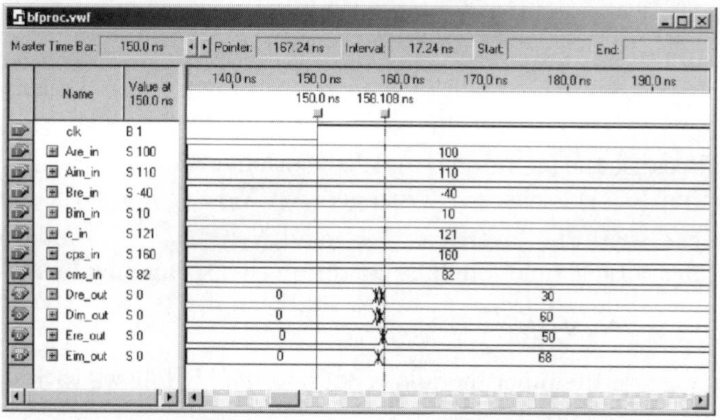

Fig. 6.16. VHDL simulation of a radix-2 butterfly processor.

have been implemented for input A, B, the three table values, and the output port D, in order to have single input/output registered design. The design uses 131 LEs, 3 embedded multipliers, and has a 95.73 MHz **Registered Performance**. Figure 6.16 shows the simulation for the zero-pipeline design, for the inputs $A = 100 + \mathrm{j}110$, $B = -40 + \mathrm{j}10$, and $W = \mathrm{e}^{\mathrm{j}\pi/9}$.   6.12

6.2.2 The Good–Thomas FFT Algorithm

The index transform suggested by Good [171] and Thomas [172] transforms a DFT of length $N = N_1 N_2$ into an "actual" two-dimensional DFT, i.e., there are *no* twiddle factors as in the Cooley–Tukey FFT. The price we pay for the twiddle factor free flow is that the factors must be coprime (i.e., $\gcd(N_k, N_l) = 1$ for $k \neq l$), and we have a somewhat more complicated index mapping, as long as the index computation is done "online" and no precomputed tables are used for the index mapping.

If we try to eliminate the twiddle factors introduced through the index mapping of n and k according to (6.17) and (6.19), respectively, it follows from

$$
\begin{aligned}
W_N^{nk} &= W_N^{(An_1+Bn_2)(Ck_1+Dk_2)} \\
&= W_N^{ACn_1k_1+ADn_1k_2+BCk_1n_2+BDn_2k_2} \\
&= W_N^{N_2n_1k_1} W_N^{N_1k_2n_2} = W_{N_1}^{n_1k_1} W_{N_2}^{k_2n_2},
\end{aligned}
\tag{6.33}
$$

that we must fulfill all the following necessary conditions at the same time:

$$
\langle AD \rangle_N = \langle BC \rangle_N = 0
\tag{6.34}
$$

$$
\langle AC \rangle_N = N_2
\tag{6.35}
$$

$$
\langle BD \rangle_N = N_1.
\tag{6.36}
$$

The mapping suggested by Good [171] and Thomas [172] fulfills this condition and is given by

$$A = N_2 \quad B = N_1 \quad C = N_2\langle N_2^{-1}\rangle_{N_1} \quad D = N_1\langle N_1^{-1}\rangle_{N_2}. \tag{6.37}$$

To check: Because the factors AD and BC both include the factor $N_1 N_2 = N$, it follows that (6.34) is checked. With $\gcd(N_1, N_2) = 1$ and a theorem due to Euler, we can write the inverse as $N_2^{-1} \bmod N_1 = N_2^{\phi(N_1)-1} \bmod N_1$ where ϕ is the Euler totient function. The condition (6.35) can now be rewritten as

$$\langle AC\rangle_N = \langle N_2 N_2 \langle N_2^{\phi(N_1)-1}\rangle_{N_1}\rangle_N. \tag{6.38}$$

We can now solve the inner modulo reduction, and it follows with $\nu \in \mathbb{Z}$ and $\nu N_1 N_2 \bmod N = 0$ finally

$$\langle AC\rangle_N = \langle N_2 N_2 (N_2^{\phi(N_1)-1} + \nu N_1)\rangle_N = N_2. \tag{6.39}$$

The same argument can be applied for the condition (6.36), and we have shown that all three conditions from (6.34)–(6.36) are fulfilled if the Good–Thomas mapping (6.37) is used. $\qquad \square$

In conclusion, we can now define the following theorem

Theorem 6.13:   Good–Thomas Index Mapping

The index mapping suggested by Good and Thomas for n is

$$n = N_2 n_1 + N_1 n_2 \bmod N \quad \begin{cases} 0 \leq n_1 \leq N_1 - 1 \\ 0 \leq n_2 \leq N_2 - 1 \end{cases} \tag{6.40}$$

and as index mapping for k results

$$k = N_2\langle N_2^{-1}\rangle_{N_1} k_1 + N_1\langle N_1^{-1}\rangle_{N_2} k_2 \bmod N \quad \begin{cases} 0 \leq k_1 \leq N_1 - 1 \\ 0 \leq k_2 \leq N_2 - 1 \end{cases}. \tag{6.41}$$

The transform from (6.41) is identical to the Chinese remainder theorem 2.13 (p. 67). It follows, therefore, that k_1 and k_2 can simply be computed via a modulo reduction, i.e., $k_l = k \bmod N_l$.

If we now substitute the Good–Thomas index map in the equation for the DFT matrix (6.16), it follows that

$$X[k_1, k_2] = \sum_{n_2=0}^{N_2-1} W_{N_2}^{n_2 k_2} \underbrace{\left(\underbrace{\sum_{n_1=0}^{N_1-1} x[n_1, n_2] W_{N_1}^{n_1 k_1}}_{N_1\text{-point transform}} \right)}_{\bar{x}[n_2, k_1]} \tag{6.42}$$

$$= \underbrace{\sum_{n_2=0}^{N_2-1} W_{N_2}^{n_2 k_2}\, \bar{x}[n_2, k_1],}_{N_2\text{-point transform}} \tag{6.43}$$

i.e., as claimed at the beginning, it is an "actual" two-dimensional DFT transform without the twiddle factor introduced by the mapping suggested by Cooley and Tukey. It follows that the Good–Thomas algorithm, although similar to the Cooley–Tukey Algorithm 6.8, has a different index mapping and no twiddle factors.

Algorithm 6.14:   Good–Thomas FFT Algorithm

An $N = N_1 N_2$-point DFT can be computed according to the following steps:
1) Index transform of the input sequence, according to (6.40).
2) Computation of N_2 DFTs of length N_1.
3) Computation of N_1 DFTs of length N_2.
4) Index transform of the output sequence, according to (6.41).

An $N = 12$ transform shown in the following example demonstrates the steps.

Example 6.15: Good–Thomas FFT Algorithm for $N = 12$

Suppose we have $N_1 = 4$ and $N_2 = 3$. Then a mapping for the input index according to $n = 3n_1 + 4n_2 \bmod 12$, and $k = 9k_1 + 4k_2 \bmod 12$ for the output index results, and we can compute the following index mapping tables

n_2	n_1				k_2	k_1			
	0	1	2	3		0	1	2	3
0	$x[0]$	$x[3]$	$x[6]$	$x[9]$	0	$X[0]$	$X[9]$	$X[6]$	$X[3]$
1	$x[4]$	$x[7]$	$x[10]$	$x[1]$	1	$X[4]$	$X[1]$	$X[10]$	$X[7]$
2	$x[8]$	$x[11]$	$x[2]$	$x[5]$	2	$X[8]$	$X[5]$	$X[2]$	$X[11]$

Using these index transforms we can construct the signal flow graph shown in Fig. 6.17. We realize that the first stage has three DFTs each having four points and the second stage four DFTs each of length 3. Multiplication by twiddle factors between the stages is not necessary. $\boxed{\text{6.15}}$

6.2.3 The Winograd FFT Algorithm

The Winograd FFT algorithm [103] is based on the observation that the inverse DFT matrix (6.4) (without prefactor N^{-1}) of dimension $N_1 \times N_2$, with $\gcd(N_1, N_2) = 1$, i.e.,

$$x[n] = \sum_{k=0}^{N-1} X[k] W_N^{-nk} \tag{6.44}$$

$$\boldsymbol{x} = \boldsymbol{W}_N^* \, \boldsymbol{X} \tag{6.45}$$

can be rewritten using the *Kronecker product*[4] with two quadratic IDFT matrices each, with dimension N_1 and N_2, respectively. As with the index

---

[4] A Kronecker product is defined by

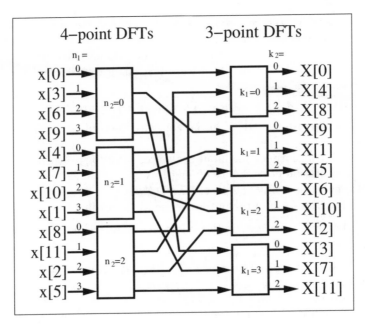

Fig. 6.17. Good–Thomas FFT for $N = 12$.

mapping for the Good–Thomas algorithm, we must write the indices of $X[k]$ and $x[n]$ in a two-dimensional scheme and then read out the indices row by row. The following example for $N = 12$ demonstrates the steps.

Example 6.16: IDFT using Kronecker Product for $N = 12$

Let $N_1 = 4$ and $N_2 = 3$. Then we have the output index transform $k = 9k_1 + 4k_2 \bmod 12$ according to the Good–Thomas index mapping:

$$
\begin{aligned}
\boldsymbol{A} \otimes \boldsymbol{B} &= [a[i,j]]\boldsymbol{B} \\
&= \begin{bmatrix} a[0,0]\boldsymbol{B} & \cdots & a[0,L-1]\boldsymbol{B} \\ \vdots & & \vdots \\ a[K-1,0]\boldsymbol{B} & \cdots & a[K-1,L-1]\boldsymbol{B} \end{bmatrix}
\end{aligned}
$$

where $\boldsymbol{A}$ is a $K \times L$ matrix.

$$
\begin{bmatrix} X[0] \\ X[1] \\ X[2] \\ X[3] \\ X[4] \\ X[5] \\ X[6] \\ X[7] \\ X[8] \\ X[9] \\ X[10] \\ X[11] \end{bmatrix}
\rightarrow
\begin{array}{c|cccc}
k_2 & & & k_1 & \\
 & 0 & 1 & 2 & 3 \\ \hline
0 & X[0] & X[9] & X[6] & X[3] \\
1 & X[4] & X[1] & X[10] & X[7] \\
2 & X[8] & X[5] & X[2] & X[11]
\end{array}
\rightarrow
\begin{bmatrix} X[0] \\ X[9] \\ X[6] \\ X[3] \\ X[4] \\ X[1] \\ X[10] \\ X[7] \\ X[8] \\ X[5] \\ X[2] \\ X[11] \end{bmatrix}.
$$

We can now construct a length-12 IDFT with

$$
\begin{bmatrix} x[0] \\ x[9] \\ x[6] \\ x[3] \\ x[4] \\ x[1] \\ x[10] \\ x[7] \\ x[8] \\ x[5] \\ x[2] \\ x[11] \end{bmatrix}
=
\begin{bmatrix}
W_{12}^0 & W_{12}^0 & W_{12}^0 \\
W_{12}^0 & W_{12}^{-4} & W_{12}^{-8} \\
W_{12}^0 & W_{12}^{-8} & W_{12}^{-4}
\end{bmatrix}
\otimes
\begin{bmatrix}
W_{12}^0 & W_{12}^0 & W_{12}^0 & W_{12}^0 \\
W_{12}^0 & W_{12}^{-3} & W_{12}^{-6} & W_{12}^{-9} \\
W_{12}^0 & W_{12}^{-6} & W_{12}^0 & W_{12}^{-6} \\
W_{12}^0 & W_{12}^{-9} & W_{12}^{-6} & W_{12}^{-3}
\end{bmatrix}
\begin{bmatrix} X[0] \\ X[9] \\ X[6] \\ X[3] \\ X[4] \\ X[1] \\ X[10] \\ X[7] \\ X[8] \\ X[5] \\ X[2] \\ X[11] \end{bmatrix}.
$$

$$\boxed{6.16}$$

So far we have used the Kronecker product to (re)define the IDFT. Using the short-hand notation $\tilde{x}$ for the permuted sequence x, we may use the following matrix/vector notation:

$$\tilde{x} = W_{N_1} \otimes W_{N_2} \, \tilde{X}. \tag{6.46}$$

For these short DFTs we now use the Winograd DFT Algorithm 6.7 (p. 359), i.e.,

$$W_{N_l} = C_l \times B_l \times A_l, \tag{6.47}$$

where A_l incorporate the input additions, B_l is a diagonal matrix with the Fourier coefficients, and C_l includes the output additions. If we now substitute (6.47) into (6.46), and use the fact that we can change the sequence of matrix multiplications and Kronecker product computation (see for instance [5, App. D]), we get

$$
\begin{aligned}
W_{N_1} \otimes W_{N_2} &= (C_1 \times B_1 \times A_1) \otimes (C_2 \times B_2 \times A_2) \\
&= (C_1 \otimes C_2)(B_1 \otimes B_2)(A_1 \otimes A_2).
\end{aligned} \tag{6.48}
$$

Because the matrices A_l and C_l are simple addition matrices, the same applies for its Kronecker products, $A_1 \otimes A_2$ and $C_1 \otimes C_2$. The Kronecker product of two quadratic diagonal matrices of dimension N_1 and N_2, respectively, obviously also gives a diagonal matrix of dimension $N_1 N_2$. The total number of necessary multiplications is therefore identical to the number of diagonal elements of $\boldsymbol{B} = \boldsymbol{B}_1 \otimes \boldsymbol{B}_2$, i.e., $M_1 M_2$, if M_1 and M_2, respectively, are the number of multiplications used to compute the smaller Winograd DFTs according to Table 6.2 (p. 360).

We can now combine the different steps to construct a Winograd FFT.

Theorem 6.17:   Winograd FFT Design

A $N = N_1 N_2$-point transform with coprimes N_1 and N_2 can be constructed as follows:

1) Index transform of the input sequence according to the Good–Thomas mapping (6.40), followed by a row read of the indices.
2) Factorization of the DFT matrix using the Kronecker product.
3) Substitute the length N_1 and N_2 DFT matrices through the Winograd DFT algorithm.
4) Centralize the multiplications.

After successful construction of the Winograd FFT algorithm, we can compute the Winograd FFT using the following three steps:

Theorem 6.18:   Winograd FFT Algorithm

1) Compute the preadditions $\boldsymbol{A}_1$ and $\boldsymbol{A}_2$.
2) Compute $M_1 M_2$ multiplications according to the matrix $\boldsymbol{B}_1 \otimes \boldsymbol{B}_2$.
3) Compute postadditions according to $\boldsymbol{C}_1$ and $\boldsymbol{C}_2$.

Let us now look at a construction of a Winograd FFT of length-12, in detail in the following example.

Example 6.19: Winograd FFT of Length 12

To build a Winograd FFT, we have, according to Theorem 6.17, to compute the necessary matrices used in the transform. For $N_1 = 3$ and $N_2 = 4$ we have the following matrices:

$$\boldsymbol{A}_1 \otimes \boldsymbol{A}_2 = \begin{bmatrix} 1 & 1 & 1 \\ 0 & 1 & 1 \\ 0 & 1 & -1 \end{bmatrix} \otimes \begin{bmatrix} 1 & 1 & 1 & 1 \\ 1 & -1 & 1 & -1 \\ 1 & 0 & -1 & 0 \\ 0 & 1 & 0 & -1 \end{bmatrix} \tag{6.49}$$

$$\boldsymbol{B}_1 \otimes \boldsymbol{B}_2 = \operatorname{diag}(1, -3/2, \sqrt{3}/2) \otimes \operatorname{diag}(1, 1, 1, -i) \tag{6.50}$$

$$\boldsymbol{C}_1 \otimes \boldsymbol{C}_2 = \begin{bmatrix} 1 & 0 & 0 \\ 1 & 1 & i \\ 1 & 1 & -i \end{bmatrix} \otimes \begin{bmatrix} 1 & 0 & 0 & 0 \\ 0 & 1 & 0 & 0 \\ 1 & 0 & -1 & 0 \\ 0 & 0 & 1 & -1 \end{bmatrix} . \tag{6.51}$$

Combining these matrices according to (6.48) results in the Winograd FFT algorithm. Input and output additions can be realized multiplier free, and the total number of real multiplication becomes $2 \times 3 \times 4 = 24$.   6.19

So far we have used the Winograd FFT to compute the IDFT. If we now want to compute the DFT with the help of an IDFT, we can use a technique we used in (6.6) on p. 346 to compute the IDFT with help of the DFT. Using matrix/vector notation we find

$$x^* = (W_N^* \, X)^* \tag{6.52}$$
$$x^* = W_N \, X^*, \tag{6.53}$$

if $W_N = [e^{2\pi j n k/N}]$ with $n, k \in \mathbb{Z}_N$ is a DFT. The DFT can therefore be computed using the IDFT with the following steps: Compute the conjugate complex of the input sequence, transform the sequence with the IDFT algorithm, and compute the conjugate complex of the output sequence.

It is also possible to use the Kronecker product algorithms, i.e., the Winograd FFT, to compute the DFT directly. This leads to a slide-modified output index mapping, as the following example shows.

Example 6.20: A 12-point DFT can be computed using the following Kronecker product formulation:

$$
\begin{bmatrix} X[0] \\ X[3] \\ X[6] \\ X[9] \\ X[4] \\ X[7] \\ X[10] \\ X[1] \\ X[8] \\ X[11] \\ X[2] \\ X[5] \end{bmatrix}
=
\begin{bmatrix} W_{12}^0 & W_{12}^0 & W_{12}^0 \\ W_{12}^0 & W_{12}^4 & W_{12}^8 \\ W_{12}^0 & W_{12}^8 & W_{12}^4 \end{bmatrix}
\otimes
\begin{bmatrix} W_{12}^0 & W_{12}^0 & W_{12}^0 & W_{12}^0 \\ W_{12}^0 & W_{12}^3 & W_{12}^6 & W_{12}^9 \\ W_{12}^0 & W_{12}^6 & W_{12}^0 & W_{12}^6 \\ W_{12}^0 & W_{12}^9 & W_{12}^6 & W_{12}^3 \end{bmatrix}
\begin{bmatrix} x[0] \\ x[9] \\ x[6] \\ x[3] \\ x[4] \\ x[1] \\ x[10] \\ x[7] \\ x[8] \\ x[5] \\ x[2] \\ x[11] \end{bmatrix} . \tag{6.54}
$$

The input sequence $x[n]$ can be considered to be in the order used for Good–Thomas mapping, while in the (frequency) output index mapping for $X[k]$, each first and third element are exchanged, compared with the Good–Thomas mapping. | 6.20 |

6.2.4 Comparison of DFT and FFT Algorithms

It should now be apparent that there are many ways to implement a DFT. The choice begins with the selection of a short DFT algorithm from among those shown in Fig. 6.1 (p. 343). The short DFT can then be used to develop long DFTs, using the indexing schemes provided by Cooley–Tukey, Good–Thomas, or Winograd. A common objective in choosing an implementation is minimum multiplication complexity. This is a viable criterion when the implementation cost of multiplication is much higher compared with other operations, such as additions, data access, or index computation.

Table 6.4. Number of real multiplications for a length-12 complex input FFT algorithm (twiddle factor multiplications by W^0 are not counted). A complex multiplication is assumed to use four real multiplications.

DFT Method	Index mapping		
	Good–Thomas Fig. 6.17 p. 376	Cooley–Tukey Fig. 6.2 p. 363	Winograd Example 6.16 p. 376
Direct	$4 \times 12^2 = 4 \times 144 = 576$		
RPFA	$4(3(4-1)^2$ $+4(3-1)^2) = 172$	$4(43+6) = 196$	$-$
WFTA	$3 \times 0 \times 2$ $+4 \times 2 \times 2 = 16$	$16 + 4 \times 6 = 40$	$2 \times 3 \times 4 = 24$

Figure 6.18 shows the number of multiplications required for various FFT lengths. It can be concluded that the Winograd FFT is most attractive, based purely on a multiply complexity criterion. In this chapter, the design of $N = 4 \times 3 = 12$-point FFTs has been presented in several forms. A comparison of a direct, Rader prime-factor algorithms, and Winograd DFT algorithms

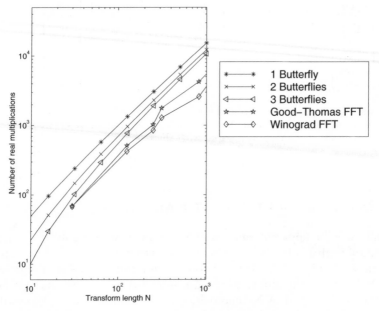

Fig. 6.18. Comparison of different FFT algorithm based on the number of necessary real multiplications.

Table 6.5. Important properties for FFT algorithms of length $N = \prod N_k$.

Property	Cooley–Tukey	Good–Thomas	Winograd
Any transform		no	
Length	yes	$\gcd(N_k, N_l) = 1$	
Maximum			
order of W	N	$\max(N_k)$	
Twiddle			
factors needed	yes	no	no
# Multiplications	bad	fair	best
# Additions	fair	fair	fair
# Index comput-			
ation effort	best	fair	bad
Data in-place	yes	yes	no
Implementation	small	can use RPFA,	small size for
advantages	butterfly	fast, simple	full parallel, medium-
	processor	FIR array	size FFT (< 50)

used for the basic DFT blocks, and the three different index mappings called Good–Thomas, Cooley–Tukey, and Winograd FFT, is presented in Table 6.4.

Besides the number of multiplications, other constraints must be considered, such as possible transform lengths, number of additions, index computation overhead, coefficient or data memory size, and run-time code length. In many cases, the Cooley–Tukey method provides the best overall solution, as suggested by Table 6.5.

Some of the published FPGA realizations are summarized in Table 6.6. The design by Goslin [157] is based on a radix-2 FFT, in which the butterflies have been realized using distributed arithmetic, discussed in Chap. 2. The design by Dandalis et al. [173], is based on an approximation of the DFT using the so-called arithmetic Fourier transform and will be discussed in Sect. 7.1. The ERNS FFT, from Meyer-Bäse et al. [174], uses the Rader algorithm in combination with the number theoretic transform, which will also be discussed in Chap. 7.

With FPGAs reaching complexities of more than 1M gates today, full integration of an FFT on a single FPGA is viable. Because the design of such an FFT block is labor intensive, it most often makes sense to utilize commercially available "intellectual property" (IP) blocks (sometimes also called "virtual components" VCs). See, for instance, the IP partnership programs at www.xilinx.com or www.altera.com. The majority of the commercially available designs are based on radix-2 or radix-4.

Table 6.6. Comparison of some FPGA FFT implementations [5].

Name	Data type	FFT type	N-point FFT time	Clock rate P	Internal RAM/ ROM	Design aim/ source
Xilinx FPGA	8 bit	Radix=2 FFT	$N = 256$ 102.4 µs	70 MHz 4.8 W @3.3 V	No	573 CLBs [157]
Xilinx FPGA	16 bit	AFT	$N = 256$ 82.48 µs	50 MHz 15.6 W @ 3.3 V	No	[173] 2602 CLBs
			42.08 µs	29.5 W		4922 CLBs
Xilinx FPGA ERNS- NTT	12.7 bit	FFT using NTT	$N = 97$ 9.24 µs	26 MHz 3.5 W @3.3 V	No	1178 CLBs [174]

6.2.5 IP Core FFT Design

Altera and Xilinx offer FFT generators, since this is, besides the FIR filter, one of the most often used intellectual property (IP) blocks. For an introduction to IP blocks see Sect. 1.4.4, p. 35. Xilinx has some free fixed-length and bitwidth hard core [175], but usually the FFT parameterized cores have to be purchased from the FPGA vendors for a (reasonable) licence fee.

Let us have a look at the 256-point FFT generation that we discussed before, see Example 6.12, p. 370, for a radix-2 butterfly processor. But this time we use the Altera FFT compiler [176] to build the FFT, including the FSM to control the processing. The Altera FFT MegaCore function is a high performance, highly parameterizable FFT processor optimized for Altera devices. Stratix and Cyclone II devices are supported but no mature devices from the APEX or Flex family. The FFT function implements a radix-2/4 decimation-in-frequency (DIF) FFT algorithm for transform lengths of 2^S, where $6 \leq S \leq 14$. Internally a block-floating-point architecture is used to maximize signal dynamic range in the transform calculation. You can use the IP toolbench MegaWizard design environment to specify a variety of FFT architectures, including $4 \times 2+$ and $3 \times 5+$ butterfly architectures and different parallel architectures. The FFT compiler includes a coefficient generator for the twiddle factors that are stored in M4K blocks.

Example 6.21: Length-256 FFT IP Generation

To start the Altera FFT compiler we select `MegaWizard Plug-In Manager` under the `Tools` menu, and the library selection window (see Fig. 1.23, p. 39) will pop up. The FFT generator can be found under DSP→Transform. You need to specify a design name for the core and we can then proceed to the `ToolBench`, see Fig. 6.19a. We first select `Parameters` and choose as the FFT length 256, and set the data and coefficient precision to 16 bits. We then have

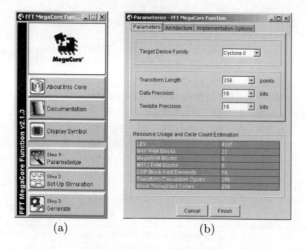

Fig. 6.19. IP design of an FFT (**a**) IP toolbench. (**b**) Coefficient specification.

a choice of different architecture: `Streaming`, `Buffered Burst`, and `Burst`. The different architectures use more or fewer additional buffers to allow block-by-block processing that requires, in the case of the `Streaming` architecture, no additional clock cycles. Every 256 cycles, we submit a new data set to the FFT core and, after some processing, get a new 256-point data set. With the $3*\times5+$ selection in the `Implementation Options` the logic resource estimation for the Cyclone II family will be as follows:

Resource	Streaming	Buffered burst	Burst
LEs	4581	4638	4318
M4K	22	18	10
Mega RAM	0	0	0
M512	0	0	0
DSP block 9-bit	18	18	18
Transform calculation cycles	256	258	262
Block throughput cycles	256	331	775

Step 2 from the toolbench will generate a simulation model that is required for the ModelSim simulation. We proceed with step 3, and the generation of the VHDL code and all supporting files follows. We then instantiate the FFT core in a wrapper file that we can compile with Quartus II. The coefficient files for the twiddle factor, along with MATLAB testbenches and ModelTech simulation files and scripts, are generated within a few seconds. This files are listed in Table 6.7. We see that not only are the VHDL and Verilog files generated along with their component file, but MATLAB (bit accurate) and ModelSim (cycle accurate) test vectors are provided to enable an easy verification path. Unfortunately there is no vector file *.vwf that can be used for the Quartus II simulation and we need to put together our own testbench. As test data we use a short triangular sequence generated in MATLAB as follows:

```
x=[(1:8)*20,zeros(1,248)];
Y=fft(x);
```

with the following instruction we quantize and list the first five samples scaled by 2^{-3} as in the Quartus II simulation:

```
sprintf('%d ',real(round(Y(1:5)*2^-3)))
sprintf('%d ',imag(round(Y(1:5)*2^-3))),
```

and the (expected) test data will be

```
90 89  87  84  79  73  67  59  50  41 ...      (real)
0 -10 -20 -30 -39 -47 -55 -61 -66 -70 ...      (imag)
```

Table 6.7. IP files generation for FFT core.

File	Description
fft256.vhd	A MegaCore function variation file, which defines a top-level VHDL description of the custom MegaCore function
fft256_inst.vhd	A VHDL sample instantiation file
fft256.cmp	A VHDL component declaration for the MegaCore function variation
fft256.inc	An AHDL include declaration file for the MegaCore function variation function.
fft256_bb.v	Verilog HDL black-box file for the MegaCore function variation
fft256.bsf	Quartus II symbol file to be used in the Quartus II block diagram editor
fft256.vho	Functional model used by the ModelSim simulation
fft_tb.vhd	Testbench used by the ModelSim simulation (can not be used with Quartus II simulator)
fft256_vho_msim.tcl	Compile script used by the ModelSim simulation
fft256_model.m	This file provides a MATLAB simulation model for the customized FFT
fft256_wave.do	Waveform scripts used by the ModelSim simulation
*.txt	Two text files with random real and imaginary input data
fft256_tb.m	This file provides a MATLAB testbench for the customized FFT
*.hex	Six sin/cos twiddle coefficient tables
f6_core.vec	This file provides simulation test vectors to be used simulating the customized FFT with the Quartus II software
fft256.html	The MegaCore function report file

The simulation of the FFT block is shown in Figs. 6.20 and 6.21. We see that the processing works in several steps. After **reset** is low we set the

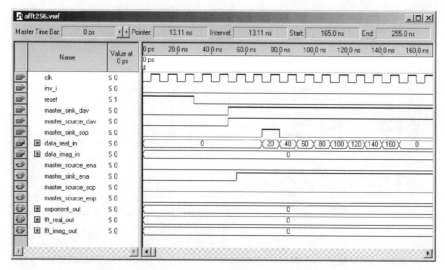

Fig. 6.20. Quartus II simulation of the IP FFT block initialization steps.

data available signal `.._dav` from the sink and source. Then the FFT block response with a high in the enable signal `master_sink_ena`. We then set the signal processing start flag `mast_sink_sop` to high for one clock cycle. At the same time we apply the first input data (i.e., value 20 in our test) to the FFT block followed by the next data in each clock cycles. After 256 clock cycles all input data are processed. Since the FFT uses `Steaming` mode a total latency of one extra block is required and the first data are available after $256 \times 2 \times 10$ ns $\approx 5\,\mu$s, as indicated by the `master_source_sop` signal, see Fig. 6.21a. After 256 clock cycles all output data are transmitted as shown by the pulse in the `master_source_eop` signal (see Fig. 6.21b) and the FFT will output the next block. Notice that the output data shows little quantization, but have a block exponent of -3, i.e., are scaled by $1/8$. This is a result of the block floating-point format used inside the block to minimize the quantization noise in the multistage FFT computation. To unscale use a barrelshifter and shift all real and imaginary data according to this exponent value. $\boxed{6.21}$

The design from the previous example runs at 144.09 MHz and requires 4461 LEs, 18 embedded multipliers of size 9×9 bits (i.e., nine blocks of size 18×18 bits), and 19 M4Ks embedded memory blocks, see the floorplan in Fig. 6.22. If we compare this with the estimation of 22 M4Ks, 18 multipliers, 4581 LEs given by the FFT core toolbench, we observe no error for the multiplier, a 3% error for the LEs, and an 18% error for the M4Ks estimation.

6.3 Fourier-Related Transforms

The *discrete cosine transform* (DCT) and *discrete sine transform* (DST) are not DFTs, but they can be computed using a DFT. However, DCTs and DSTs

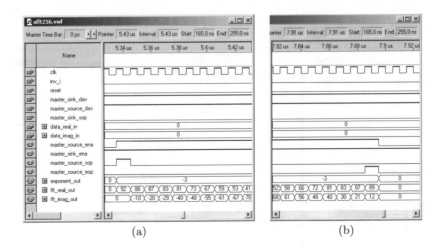

Fig. 6.21. FFT core simulation output results. **(a)** Start of the output frame. **(b)** End of the output frame.

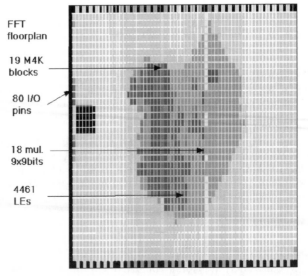

Fig. 6.22. Annotated Quartus II floorplan of the 256-point FFT.

can not be used directly to compute fast convolution, by multiplying the transformed spectra and an inverse transform, i.e., the convolution theorem does *not* hold. The applications for DCTs and DSTs are therefore not as broad as those for FFTs, but in some applications, like image compression, DCTs are (due to their close relationship to the Kahunen–Loevé transform) very popular. However, because DCTs and DSTs are defined by sine and cosine "kernels," they have a close relation to the DFT, and will be presented in this chapter. We will begin with the definition and properties of DCTs and DSTs, and will then present an FFT-like fast computation algorithm to implement the DCT. All DCTs obey the following transform pattern observed by Wang [177]:

$$X[k] = \sum_n x[n] C_N^{n,k} \longleftrightarrow x[n] = \sum_k X[k] C_N^{n,k}. \tag{6.55}$$

The kernel functions $C_N^{n,k}$, for four different DCT instances, are defined by

DCT-I: $C_N^{n,k} = \sqrt{2/N} c[n] c[k] \cos\left(nk\frac{\pi}{N}\right)$ $n, k = 0, 1, \ldots, N$

DCT-II: $C_N^{n,k} = \sqrt{2/N} c[k] \cos\left(k(n+\frac{1}{2})\frac{\pi}{N}\right)$ $n, k = 0, 1, \ldots, N-1$

DCT-III: $C_N^{n,k} = \sqrt{2/N} c[n] \cos\left(n(k+\frac{1}{2})\frac{\pi}{N}\right)$ $n, k = 0, 1, \ldots, N-1$

DCT-IV: $C_N^{n,k} = \sqrt{2/N} \cos\left((k+\frac{1}{2})(n+\frac{1}{2})\frac{\pi}{N}\right)$ $n, k = 0, 1, \ldots, N-1$,

where $c[m] = 1$ except $c[0] = 1/\sqrt{2}$. The DST has the same structure, but the cosine terms are replaced by sine terms. DCTs have the following properties:

1) DCTs implement functions using cosine bases.
2) All transforms are *orthogonal*, i.e., $C \times C^t = k[n]I$.
3) A DCT is a real transform, unlike the DFT.
4) DCT-I is its own inverse.
5) DCT-II is the inverse of DCT-III, and vice versa.
6) DCT-IV is its own inverse. Type IV is symmetric, i.e., $C = C^t$.
7) The convolution property of the DCT is not the same as the convolution multiplication relationship in the DFT.
8) The DCT is an approximation of the Kahunen–Loevé transformation (KLT).

The two-dimensional 8×8 transform of the DCT-II is used most often in image compression, i.e., in the H.261, H.263, and MPEG standards for video and in the JPEG standard for still images. Because the two-dimensional transform is *separable* into two dimensions, we compute the two-dimensional DCT by row transforms followed by column transforms, or vice versa (Exercise 6.17, p. 394). We will therefore focus on the implementation of one-dimensional transforms.

6.3.1 Computing the DCT Using the DFT

Narasimha and Peterson [178] have introduced a scheme describing how to compute the DCT with the help of the DFT [179, p. 50]. The mapping of the DCT to the DFT is attractive because we can then use the wide variety of FFT-type algorithms. Because DCT-II is used most often, we will further develop the relationship of the DFT and DCT-II. To simplify the representation, we will skip the scaling operation, since it can be included at the end of the DFT or FFT computation. Assuming that the transform length is even, we can rewrite the DCT-II transform

$$X[k] = \sum_{n=0}^{N-1} x[n] \cos\left(k\left(n+\frac{1}{2}\right)\frac{\pi}{N}\right),$$ (6.56)

using the following permutation

$$y[n] = x[2n] \quad \text{and} \quad y[N-n-1] = x[2n+1]$$
$$\text{for} \quad n = 0, 1, \dots, N/2 - 1.$$

It follows then that

$$X[k] = \sum_{n=0}^{N/2-1} y[n] \cos\left(k(2n+\frac{1}{2})\frac{\pi}{N}\right)$$
$$+ \sum_{n=0}^{N/2-1} y[N-n-1] \cos\left(k(2n+\frac{3}{2})\frac{\pi}{N}\right)$$
$$X[k] = \sum_{n} y[n] \cos\left(k(2n+\frac{1}{2})\frac{\pi}{N}\right).$$ (6.57)

If we now compute the DFT of $y[n]$ denoted with $Y[k]$, we find that

$$X[k] = \Re\left(W_{4N} Y[k]\right)$$
$$= \cos\left(\frac{\pi k}{2N}\right) \Re(Y[k]) - \sin\left(\frac{\pi k}{2N}\right) \Im(Y[k]).$$ (6.58)

This can be easily transformed in a C or MATLAB program (see Exercise 6.17, p. 394), and can be used to compute the DCT with the help of a DFT or FFT.

6.3.2 Fast Direct DCT Implementation

The symmetry properties of DCTs have been used by Byeong Lee [180] to construct an FFT-like DCT algorithm. Because of its similarities to a radix-2 Cooley–Tukey FFT, the resulting algorithm is sometimes referred to as the *fast DCT* or simply FCT. Alternatively, a fast DCT algorithm can be developed using a matrix structure [181]. A DCT can be obtained by "transposing" an inverse DCT (IDCT) since the DCT is known to be an orthogonal transform. IDCT Type II was introduced in (6.55) and, noting that $\hat{X}[k] = c[k]X[k]$, it follows that

$$x[n] = \sum_{k=0}^{N-1} \hat{X}[k] C_N^{n,k}, \qquad n = 0, 1, \dots, N-1.$$ (6.59)

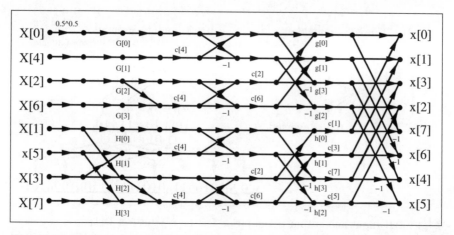

Fig. 6.23. 8-point fast DCT flow graph with the short-hand notation $c[p] = 1/(2\cos(p\pi/16))$.

Decomposing $x[n]$ into even and odd parts it can be shown that $x[n]$ can be reconstructed by two $N/2$ DCTs, namely

$$G[k] = \hat{X}[2k], \tag{6.60}$$

$$H[k] = \hat{X}[2k+1] + \hat{X}[2k-1], \qquad k = 0, 1, \ldots, N/2 - 1. \tag{6.61}$$

In the time domain, we get

$$g[n] = \sum_{k=0}^{N/2-1} G[k] C_{N/2}^{n,k}, \tag{6.62}$$

$$h[n] = \sum_{k=0}^{N/2-1} H[k] C_{N/2}^{n,k}, \qquad k = 0, 1, \ldots, N/2 - 1. \tag{6.63}$$

The reconstruction becomes

$$x[n] = g[n] + 1/(2C_N^{n,k})h[n], \tag{6.64}$$

$$x[N - 1 - n] = g[n] - 1/(2C_N^{n,k})h[n], \tag{6.65}$$

$$n = 0, 1, \ldots, N/2 - 1.$$

By repeating this process, we can decompose the DCT further. Comparing (6.62) with the radix-2 FFT twiddle factor shown in Fig. 6.13 (p. 366) shows that a division seems to be necessary for the FCT. The twiddle factors $1/(2C_N^{n,k})$ should therefore be precomputed and stored in a table. Such a table approach is also appropriate for the Cooley–Tukey FFT, because the "online" computation of the trigonometric function is, in general, too time consuming. We will demonstrate the FCT with the following example.

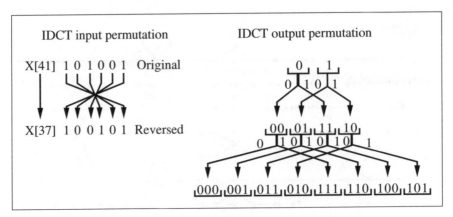

Fig. 6.24. Input and output permutation for the 8-point fast DCT.

Example 6.22: A 8-point FCT

For an 8-point FCT (6.60)–(6.65) become

$$G[k] = \hat{X}[2k], \tag{6.66}$$
$$H[k] = \hat{X}[2k+1] + \hat{X}[2k-1], \qquad k = 0,1,2,3. \tag{6.67}$$

and in the time domain we get

$$g[n] = \sum_{k=0}^{3} G[k]C_4^{n,k}, \tag{6.68}$$

$$h[n] = \sum_{k=0}^{3} H[k]C_4^{n,k}, \qquad n = 0,1,2,3. \tag{6.69}$$

The reconstruction becomes

$$x[n] = g[n] + 1/(2C_8^{n,k})h[n], \tag{6.70}$$
$$x[N-1-n] = g[n] - 1/(2C_8^{n,k})h[n], \qquad n = 0,1,2,3. \tag{6.71}$$

Equations (6.66) and (6.67) form the first stage in the flow graph in Fig. 6.23, and (6.70) and (6.71) build the last stage in the flow graph. $\boxed{6.22}$

In Fig. 6.23, the input sequence $\hat{X}[k]$ is applied in bit-reversed order. The order of the output sequence $x[n]$ is generated in the following manner: starting with the set $(0,1)$ we form the new set by adding a prefix 0 and 1. For the prefix 1, all bits of the previous pattern are inverted. For instance, from the sequence 10 we get the two babies 010 and $1\overline{1}\overline{0} = 101$. This scheme is graphically interpreted in Fig. 6.24.

Exercises

Note: If you have no prior experience with the Quartus II software, refer to the case study found in Sect. 1.4.3, p. 29. If not otherwise noted use the EP2C35F672C6 from the Cyclone II family for the Quartus II synthesis evaluations.

6.1: Compute the 3-dB bandwidth, first zero, maximum sidelobe, and decrease per octave, for a rectangular and triangular window using the Fourier transform.

6.2: (a) Compute the cyclic convolution of $x[n] = \{3, 1, -1\}$ and $f[n] = \{2, 1, 5\}$.
(b) Compute the DFT matrix $\boldsymbol{W}_3$ for $N = 3$.
(c) Compute the DFT of $x[n] = \{3, 1, -1\}$ and $f[n] = \{2, 1, 5\}$.
(d) Now compute $Y[k] = X[k]F[k]$, followed by $\boldsymbol{y} = \boldsymbol{W}_3^{-1}\boldsymbol{Y}$, for the signals from part (c).
Note: use a C compiler or MATLAB for part (c) and (d).

6.3: A single spectral component $X[k]$ in the DFT computation

$$X[k] = x[0] + x[1]W_N^k + x[2]W_N^{2k} + \ldots + x[N-1]W_N^{(N-1)k}$$

can be rearranged by collecting all common factors W_N^k, such that we get

$$X[k] = x[0] + W_N^k(x[1] + W_N^k(x[2] + \ldots + W_N^k x[N-1])\ldots)).$$

This results in a possibly recursive computation of $X[k]$. This is called the Goertzel algorithm and is graphically interpreted by Fig. 6.5 (p. 350). The Goertzel algorithm can be attractive if only a few spectral components must be computed. For the whole DFT, the effort is of order N^2 and there is no advantage compared with the direct DFT computation.
(a) Construct the recursive signal flow graph, including input and output register, to compute a single $X[k]$ for $N = 5$.
For $N = 5$ and $k = 1$, compute all registers contents for the following input sequences:
(b) $\{20, 40, 60, 80, 100\}$.
(c) $\{j20, j40, j60, j80, j100\}$.
(d) $\{20 + j20, 40 + j40, 60 + j60, 80 + j80, 100 + j100\}$.

6.4: The Bluestein chirp-z algorithm was defined in Sect. 6.1.4 (p. 350). This algorithm is graphically interpreted in Fig. 6.6 (p. 351).
(a) Determine the CZT algorithms for $N = 4$.
(b) Using C or MATLAB, determine the CZT for the triangular sequence $x[n] = \{0, 1, 2, 3\}$.
(c) Using C or MATLAB, extend the length to $N = 256$, and check the CZT results with an FFT of the same length. Use a triangular input sequence, $x[n] = n$.

6.5: (a) Design a direct implementation of the nonrecursive filter for the $N = 7$ Rader algorithm.
(b) Determine the coefficients that can be combined.
(c) Compare the realizations from (a) and (b) in terms of realization effort.

6.6: Design a length $N = 3$ Winograd DFT algorithm and draw the signal flow graph.

6.7: (a) Using the two-dimensional index transform $n = 3n_1 + 2n_2 \mod 6$, with $N_1 = 2$ and $N_2 = 3$, determine the mapping (6.18) on p. 362. Is this mapping bijective?

(b) Using the two-dimensional index transform $n = 2n_1 + 2n_2 \mod 6$, with $N_1 = 2$ and $N_2 = 3$, determine the mapping (6.18) on p. 362. Is this mapping bijective?

(c) For $\gcd(N_1, N_2) > 1$, Burrus [148] found the following conditions such that the mapping is bijective:

$$A = aN_2 \text{ and } B \neq bN_1 \text{ and } \gcd(a, N_1) = \gcd(B, N_2) = 1$$

or

$$A \neq aN_2 \text{ and } B = bN_1 \text{ and } \gcd(A, N_1) = \gcd(b, N_2) = 1,$$

with $a, b \in \mathbb{Z}$. Suppose $N_1 = 9$ and $N_2 = 15$. For $A = 15$, compute all possible values for $B \in \mathbb{Z}_{20}$.

6.8: For $\gcd(N_1, N_2) = 1$, Burrus [148] found that in the following conditions the mapping is bijective:

$$A = aN_2 \text{ and/or } B = bN_1 \text{ and } \gcd(A, N_1) = \gcd(B, N_2) = 1, \qquad (6.72)$$

with $a, b \in \mathbb{Z}$. Assume $N_1 = 5$ and $N_2 = 8$. Determine whether the following mappings are possibly bijective index mappings:
(a) $A = 8, B = 5$.
(b) $A = 8, B = 10$.
(c) $A = 24, B = 15$.
(d) For $A = 7$, compute all valid $B \in \mathbb{Z}_{20}$.
(e) For $A = 8$, compute all valid $B \in \mathbb{Z}_{20}$.

6.9: (a) Draw the signal flow graph for a radix-2 DIF algorithm where $N = 16$.
(b) Write a C or MATLAB program for the DIF radix-2 FFT.
(c) Test your FFT program with a triangular input $x[n] = n + jn$ with $n \in [0, N-1]$.

6.10: (a) Draw the signal flow graph for a radix-2 DIT algorithm where $N = 8$.
(b) Write a C or MATLAB program for the DIT radix-2 FFT.
(c) Test your FFT program with a triangular input $x[n] = n + jn$ with $n \in [0, N-1]$.

6.11: For a common-factor FFT the following 2D DFT (6.24; p. 363) is used:

$$X[k_1, k_2] = \sum_{n_2=0}^{N_2-1} W_{N_2}^{n_2 k_2} \left(W_N^{n_2 k_1} \sum_{n_1=0}^{N_1-1} x[n_1, n_2] W_{N_1}^{n_1 k_1} \right) \qquad (6.73)$$

(a) Compile a table for the index map for a $N = 16$ radix-4 FFT with: $n = 4n_1 + n_2$ and $k = k_1 + 4k_2$, and $0 \leq n_1, k_1 \leq N_1$ and $0 \leq n_2, k_2 \leq N_2$.
(b) Complete the signal flow graph (x, X and twiddle factors) for the $N = 16$ radix 4 shown in Fig.6.25.
(c) Compute the 16-point FFT for $x = [0\ 1,\ 0,\ 0,\ 0,\ 2,\ 0,\ 0,\ 0,\ 3,\ 0,\ 0,\ 0,\ 4,\ 0,\ 0]$ using the following steps:
(c1) Map the input data and compute the DFTs of the first stage.
(c2) Multiply the (none-zero DFTs) with the twiddle factors (hint: $w = \exp(-j2\pi/16)$).
(c3) Compute the second level DFT.
(c4) Sort the output sequence X in the right order (use two fractional digits).
Note: Consider using a C compiler or MATLAB for part (c).

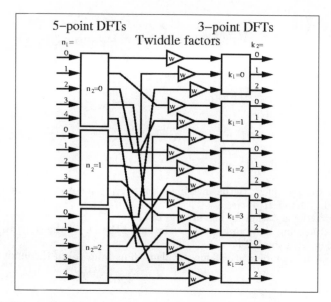

Fig. 6.25. Incomplete 16-point radix-4 FFT signal flow graph.

6.12: Draw the signal flow graph for an $N = 12$ Good–Thomas FFT, such that no crossings occur in the signal flow graph.
(Hint: Use a 3D representation of the row and column DFTs)

6.13: The index transform for FFTs by Burrus and Eschenbacher [182] is given by

$$n = N_2 n_1 + N_1 n_2 \bmod N \qquad \begin{cases} 0 \le n_1 \le N_1 - 1 \\ 0 \le n_2 \le N_2 - 1, \end{cases} \qquad (6.74)$$

and

$$k = N_2 k_1 + N_1 k_2 \bmod N \qquad \begin{cases} 0 \le k_1 \le N_1 - 1 \\ 0 \le k_2 \le N_2 - 1. \end{cases} \qquad (6.75)$$

(a) Compute the mapping for n and k with $N_1 = 3$ and $N_2 = 4$.
(b) Compute W^{nk}.
(c) Substitute W^{nk} from (b) in the DFT matrix.
(d) What type of FFT algorithm is this?
(e) Can the Rader algorithm be used to compute the DFTs of length N_1 or N_2?

6.14: (a) Compute the DFT matrices $\boldsymbol{W}_2$ and $\boldsymbol{W}_3$.
(b) Compute the Kronecker product $\boldsymbol{W}_6' = \boldsymbol{W}_2 \otimes \boldsymbol{W}_3$.
(c) Compute the index for the vectors $\boldsymbol{X}$ and $\boldsymbol{x}$, such that $\boldsymbol{X} = \boldsymbol{W}_6' \boldsymbol{x}$ is a DFT of length 6.
(d) Compute the index mapping for $x[n]$ and $X[k]$, with $\boldsymbol{x} = \boldsymbol{W}_2^* \otimes \boldsymbol{W}_3^* \boldsymbol{X}$ being the IDFT.

6.15: The discrete Hartley transformation (DHT) is a transform for real signals. A length N transform is defined by

$$H[n] = \sum_{k=0}^{N-1} \text{cas}(2\pi nk/N)\, h[k], \tag{6.76}$$

with $\text{cas}(x) = \sin(x) + \cos(x)$. The relation with the DFT ($f[k] \overset{\text{DFT}}{\longleftrightarrow} F[n]$) is

$$H[n] = \Re\{F[n]\} - \Im\{F[n]\} \tag{6.77}$$
$$F[n] = E[n] - jO[n] \tag{6.78}$$
$$E[n] = \frac{1}{2}\left(H[n] + H[-n]\right) \tag{6.79}$$
$$O[n] = \frac{1}{2}\left(H[n] - H[-n]\right), \tag{6.80}$$

where $\Re$ is the real part, $\Im$ the imaginary part, $E[n]$ the even part of $H[n]$, and $O[n]$ the odd part of $H[n]$.

(a) Compute the equation for the inverse DHT.
(b) Compute (using the frequency convolution of the DFT) the steps to compute a convolution with the DHT.
(c) Show possible simplifications for the algorithms from (b), if the input sequence is even.

6.16: The DCT-II form is:

$$X[k] = c[k]\sqrt{\frac{2}{N}} \sum_{n=0}^{N-1} x[n] \cos\left(\frac{2\pi}{4N}(2n+1)k\right) \tag{6.81}$$

$$c[k] = \begin{cases} \sqrt{1/2} & k = 0 \\ 1 & \text{otherwise} \end{cases}. \tag{6.82}$$

(a) Compute the equations for the inverse transform.
(b) Compute the DCT matrix for $N = 4$.
(c) Compute the transform of $x[n] = \{1, 2, 2, 1\}$ and $x[n] = \{1, 1, -1, -1\}$.
(d) What can you say about the DCT of even or odd symmetric sequences?

6.17: The following MATLAB code can be used to compute the DCT-II transform (assuming even length $N = 2^n$), with the help of a radix-2 FFT (see Exercise 6.9).

```
function X = DCTII(x)
  N = length(x);                  % get length
  y = [ x(1:2:N); x(N:-2:2) ];    % re-order elements
  Y = fft(y);                     % Compute the FFT
  w = 2*exp(-i*(0:N-1)'*pi/(2*N))/sqrt(2*N); % get weights
  w(1) = w(1) / sqrt(2);          % make it unitary
  X = real(w .* Y);               % compute pointwise product
```

(a) Compile the program with C or MatLab.
(b) Compute the transform of $x[n] = \{1, 2, 2, 1\}$ and $x[n] = \{1, 1, -1, -1\}$.

6.18: Like the DFT, the DCT is a separable transform and, we can therefore implement a 2D DCT using 1D DCTs. The 2D $N \times N$ transform is given by

$$X[n_1, n_2] =$$
$$\frac{c[n_1]c[n_2]}{4} \sum_{k=0}^{N-1}\sum_{l=0}^{N-1} x[k, l] \cos\left(n_1(k + \frac{1}{2})\frac{\pi}{N}\right) \cos\left(n_2(l + \frac{1}{2})\frac{\pi}{N}\right), \tag{6.83}$$

where $c[0] = 1/\sqrt{2}$ and $c[m] = 1$ for $m \neq 0$.

Use the program introduced in Exercise 6.17 to compute an 8×8 DCT transform by

(a) First row followed by column transforms.

(b) First column followed by row transforms.

(c) Direct implementation of (6.83).

(d) Compare the results from (a) and (b) for the test data $x[k, l] = k + l$ with $k, l \in [0, 7]$

6.19: **(a)** Implement a first-order system according to Exercise 6.3, to compute the Goertzel algorithm for $N = 5$ and $n = 1$, and 8-bit coefficient and input data, using Quartus II.

(b) Determine the `Registered Performance` and the used resources (LEs, multipliers, and M4Ks).

Simulate the design with the three input sequences:

(c) $\{20, 40, 60, 80, 100\}$,

(d) $\{j20, j40, j60, j80, j100\}$, and

(e) $\{20 + j20, 40 + j40, 60 + j60, 80 + j80, 100 + j100\}$.

6.20: **(a)** Design a `Component` to compute the (real input) 4-point Winograd DFT (from Example 6.16, p. 376) using Quartus II. The input and output precision should be 8 bits and 10 bit, respectively.

(b) Determine the `Registered Performance` and the used resources (LEs, multipliers, and M4Ks).

Simulate the design with the three input sequences:

(c) $\{40, 70, 100, 10\}$.

(d) $\{0, 30, 60, 90\}$.

(e) $\{80, 110, 20, 50\}$.

6.21: **(a)** Design a `Component` to compute the (complex input) 3-point Winograd DFT (from Example 6.16, p. 376) using Quartus II. The input and output precision should be 10 bits and 12 bits, respectively.

(b) Determine the `Registered Performance` and the used resources (LEs, multipliers, and M4Ks).

(c) Simulate the design with the input sequences $\{180, 220, 260\}$.

6.22: **(a)** Using the designed 3- and 4-point `Components` from Exercises 6.20 and 6.21, use component instantiation to design a fully parallel 12-point Good–Thomas FFT similar to that shown in Fig. 6.17 (p. 376), using Quartus II. The input and output precision should be 8 bit and 12 bit, respectively.

(b) Determine the `Registered Performance` and the used resources (LEs, multipliers, and M4Ks).

(c) Simulate the design with the input sequences $x[n] = 10n$ with $0 \leq n \leq 11$.

6.23: **(a)** Design a component `ccmulp` similar to the one shown in Example 6.11 (p. 368), to compute the twiddle factor multiplication. Use three pipeline stages for the multiplier and one for the input subtraction $X - Y$, using Quartus II. The input and output precision should again be 8 bits.

(b) Conduct a simulation to ensure that the pipelined multiplier correctly computes $(70 + j50)(121 + j39)$.

(c) Determine the `Registered Performance` and the used resources (LEs, multipliers, and M4Ks) of the twiddle factor multiplier.

(d) Now implement the whole pipelined butterfly processor.

(e) Conduct a simulation, with the data from Example 6.12 (p. 370).

(f) Determine the `Registered Performance` and the used resources (LEs, multipliers, and M4Ks) of the whole pipelined butterfly processor.

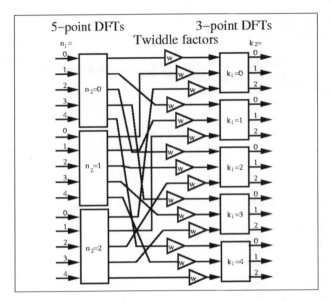

Fig. 6.26. Incomplete 15-point CFA FFT signal flow graph.

6.24: (a) Compute the cyclic convolution of $x[n] = \{1, 2, 3, 4, 5\}$ and $f[n] = \{-1, 0, -2, 0, 4\}$.
(b) Compute the DFT matrix $\boldsymbol{W}_5$ for $N = 5$.
(c) Compute the DFT of $x[n]$ and $f[n]$.
(d) Now compute $Y[k] = X[k]F[k]$, followed by $\boldsymbol{y} = \boldsymbol{W}_5^{-1}\boldsymbol{Y}$, for the signals from part (c).
Note: use a C compiler or MATLAB for part (c) and (d).

6.25: For a common-factor FFT the following 2D DFT (6.24; p. 363) is used:

$$X[k_1, k_2] = \sum_{n_2=0}^{N_2-1} W_{N_2}^{n_2 k_2} \left(W_N^{n_2 k_1} \sum_{n_1=0}^{N_1-1} x[n_1, n_2] W_{N_1}^{n_1 k_1} \right). \tag{6.84}$$

(a) Compile a table for the index map for a $N = 15$, $N_1 = 5$, and $N_2 = 3$ FFT with $n = 3n_1 + n_2$ and $k = k_1 + 5k_2$.
(b) Complete the signal flow graph shown in Fig. 6.26 for the $N = 15$ transform.
(c) Compute the 15-point FFT for $x = [0, 1, 0, 0, 2, 0, 0, 3, 0, 0, 4, 0, 0, 5, 0]$ using the following steps:
(c1) Map the input data and compute the DFTs of the first stage.
(c2) Multiply the (nonzero DFTs) with the twiddle factors, i.e., $w = \exp(-j2\pi/15)$.
(c3) Compute the second-level DFT.
(c4) Sort the output sequence X into the right order (use two fractional digits).
Note: use a C compiler or MATLAB for part (c).

6.26: For a prime-factor FFT the following 2D DFT (6.42); p. 374 is used:

$$X[k_1, k_2] = \sum_{n_2=0}^{N_2-1} W_{N_2}^{n_2 k_2} \left(\sum_{n_1=0}^{N_1-1} x[n_1, n_2] W_{N_1}^{n_1 k_1} \right). \tag{6.85}$$

(a) Compile a table for the index map for a $N = 15, N_1 = 5, N_2 = 3$ FFT with $n = 3n_1 + 5n_2$ mod 15 and $k = 6k_1 + 10k_2$ mod 15, and $0 \leq n_1, k_1 \leq N_1$ and $0 \leq n_2, k_2 \leq N_2$.
(b) Draw the signal flow graph for the $N = 15$ transform.
(c) Compute the 15-point FFT for $x = [0, 0, 5, 0, 0, 1, 0, 0, 2, 0, 0, 3, 0, 0, 4]$ using the following steps:
(c1) Map the input data and compute the DFTs of the first stage.
(c2) Compute the second-level DFTs.
(c3) Sort the output sequence X into the right order (use two fractional digits).
Note: use a C compiler or MATLAB to verify part (c).

6.27: (a) Develop the table for the 5×2 Good–Thomas index map (see Theorem 6.13, p. 374) for $N_1 = 5$ and $N_2 = 2$.
(b) Develop a program in MATLAB or C to compute the (real-input) five-point Winograd DFT using the signal flow graph shown in Fig. 6.11, p. 6.11. Test your code using the two input sequences $\{10 , 30, 50, 70, 90 \}$ and $\{60, 80, 100, 20, 40\}$.
(c) Develop a program in MATLAB or C to compute the (complex input) two-point Winograd DFT. Test your code using the two input sequences $\{250, 300\}$ and $\{-50 + -j67, -j85\}$.
(d) Combine the two programs from (b) and (c) and build a Good–Thomas 5×2 FFT using the mapping from (a). Test your code using the input sequences $x[n] = 10n$ with $1 <= n <= 10$.

6.28: (a) Design a five-point (real-input) DFT in HDL. The input and output precision should be 8 and 11 bits, respectively. Use registers for the input and output. Add a synchronous enable signal to the registers. Quantize the center coefficients using the program csd.exe from the CD and use a CSD coding with at least 8-bit precision.
(b) Simulate the design with the two input sequences $\{10, 30, 50, 70, 90\}$ and $\{60, 80, 100, 20, 40\}$, and match the simulation shown in Fig. 6.27.
(c) Determine the Registered Performance and the used resources (LEs, embedded multipliers, and M4Ks) of the five-point DFT.

6.29: (a) Design a two-point (complex input) Winograd DFT in HDL. Input and output precision should be 11 and 12 bits, respectively. Use registers for the input and output. Add a synchronous enable signal to the registers.
(b) Simulate the design with the two input sequences $\{250, 300\}$ and $\{-50 + j67, -j85\}$, and match the simulation shown in Fig. 6.28.
(c) Determine the Registered Performance and the used resources (LEs, embedded multipliers, and M4Ks) of the two-point DFT.

6.30: (a) Using the designed five- and two-point components from Exercises 6.28 and 6.29 use component instantiation to design a fully parallel 10-point Good–Thomas FFT similar to your software code from Exercise 6.27. The input and output precision should be 8 and 12 bits, respectively. Add an asynchronous reset for the I/O FSM and I/O registers. Use a signal ENA to indicate when a set of I/O values has been transferred.
(b) Simulate the design with the input $x[n] = 10n$ with $1 \leq n \leq 10$. Try to match the simulation shown in Fig. 6.29.
(c) Determine the Registered Performance and the used resources (LEs, embedded multipliers, and M4Ks) of the 10-point Good–Thomas FFT.

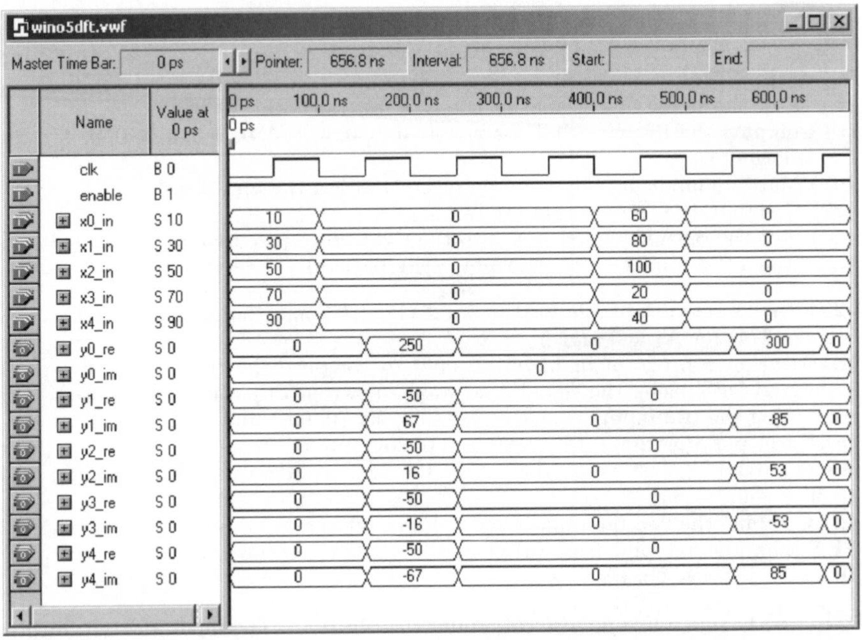

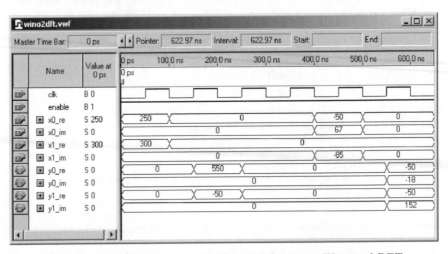

Fig. 6.27. VHDL simulation of a five-point real-input Winograd DFT.

Fig. 6.28. VHDL simulation of a two-point complex-input Winograd DFT.

6.31: Fast IDCT design.
(a) Develop a fast IDCT (MATLAB or C) code for the length-8 transform according to Fig. 6.23 (p. 389). Note that the scaling for $X[0]$ is $\sqrt{1/2}$ and the DCT scaling $\sqrt{2/N}$ according to (6.55) is not shown in Fig. 6.23 (p. 389).
(a) Verify your program with the MATLAB function idct for the sequence $X =$

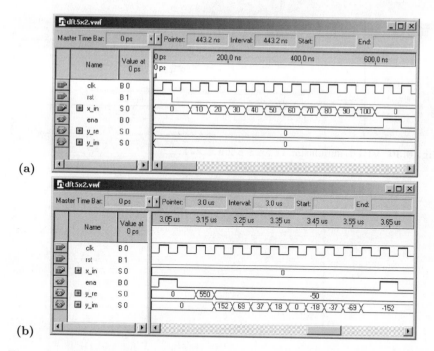

Fig. 6.29. VHDL simulation of a 10-point Good–Thomas FFT. **(a)** Begin of frame. **(b)** End of frame.

$10, 20, \ldots 80$.

(c) Determine the maximum bit growth for each spectral component. Hint: in MATLAB take advantage of the functions `abs`, `max`, `dctmtx`, and `sum`.

(d) Using the program `csd.exe` from the CD determine for each coefficient $c[p] = 0.5/\cos(p/16)$ the CSD presentation for at least 8-bit precision.

(e) For the input sequence $X = 10, 20, \ldots, 80$ compute the output in float and integer format.

(f) Tabulate the intermediate values behind the first-, second-, and third-stage multiplication by $c[p]$. As input use the sequence X from (e) with additional four guard bits, i.e., scaled by $2^4 = 16$.

6.32: (a) Develop the HDL code for the length-8 transform according to Fig. 6.23 (p. 6.23). Include an asynchronous reset and a signal `ena` when the transform is ready. Use serial I/O. Input `x_in` should be 8 bit, as the output `y_out` and internal data format use a 14 integer format with four fractional bits, i.e., scale the input ($\times 16$) and output ($/16$) in order to implement the four fractional bits.

(b) Use the data from Exercise 6.31(f) to debug the HDL code. Match the simulation from Fig. 6.30 for the input and output sequences.

(c) Determine the `Registered Performance` and the used resources (LEs, embedded multipliers, and M4Ks) of the 8-point IDCT.

(d) Determine the maximum output error in percent comparing the HDL and software results from Exercise 6.31.

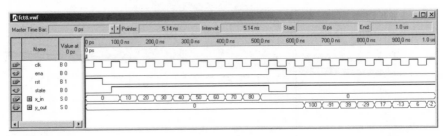

Fig. 6.30. VHDL simulation of an 8-point IDCT.

7. Advanced Topics

Several algorithms exist that enable FPGAs to outperform PDSPs by an order of magnitude, due to the fact that FPGAs can be built with bitwise implementations. Such applications are the focus of this chapter.

For number theoretic transforms (NTTs), the essential advantage of FP-GAs is that it is possible to implement modulo arithmetic in any desired bit width. NTTs are discussed in detail in Sect. 7.1.

For error control and cryptography, two basic building blocks are used: Galois field arithmetic and linear feedback shift registers (LFSR). Both can be efficiently implemented with FPGAs, and are discussed in Sect. 7.2. If, for instance, an N-bit LFSR is used as an M-multistep number generator, this will give an FPGA at least an MN speed advantage over a PDSPs or microprocessor.

Finally, in Sect. 7.3, communication systems designed with FPGAs will demonstrate low system costs, high throughput, and the possibility of fast prototyping. A comprehensive discussion of both coherent and incoherent receivers will close this chapter.

7.1 Rectangular and Number Theoretic Transforms (NTTs)

Fast implementation of convolution, and discrete Fourier transform (DFT) computations, are frequent problems in signal and image processing. In practice these operations are most often implemented using fast Fourier transform (FFT) algorithms. NTTs can, in some instances, outperform FFT-based systems. In addition, it is also possible to use a rectangular transform, like the Walsh–Hadamard or the arithmetic Fourier transform, to get an approximation of the DFT or convolution, as will be discussed at the end of Sect. 7.1.

In 1971, Pollard [183] defined the NTT, over a finite group, as the transform pair

$$x[n] = N^{-1} \sum_{k=0}^{N-1} X[k]\alpha^{-nk} \bmod M \leftrightarrow X[k] = \sum_{n=0}^{N-1} x[k]\alpha^{kn} \bmod M, \quad (7.1)$$

where $N \times N^{-1} \equiv 1$ exists, and $\alpha \in \mathbb{Z}_M$ ($\mathbb{Z}_M = \{0, 1, 2, \ldots, M-1\}$, and $\mathbb{Z}_M \cong \mathbb{Z}/M\mathbb{Z}$) is an element of order N, i.e., $\alpha^N \equiv 1$ and $\alpha^k \not\equiv 1$ for all $k \in \{1, 2, 3, \ldots, N-1\}$ in the finite group $(\mathbb{Z}_M, \times)$ (see Exercise 7.1, p. 472).

It is important to be able to ensure that, for a given tuple (α, M, N), such a transform pair exists. Clearly, α must be of order N modulo M. In order to ensure that the inverse NTT (INTT) exists, other requirements are:

1) The multiplicative inverse $N^{-1} \bmod M$ must exist, i.e., the equation $x \times N \equiv 1 \bmod M$ must have a solution $x \in \mathbb{Z}_M$.

2) The determinant of the transform matrix $|A| = |[\alpha^{kn}]|$ must be nonzero so that the matrix is invertible, i.e., A^{-1} exists.

1) It can only be concluded that a multiplicative inverse exists if α and M do not share a common factor, or in short notation, $\gcd(\alpha, M) = 0$.

2) For the second condition, a well-known fact from algebra is used: The NTT matrix is a special case of the *Vandermonde matrix* (with $a[k] = \alpha_N^k$), and it follows for the determinant

$$\det(V) = \begin{vmatrix} 1 & a[0] & a[0]^2 & \cdots & a[0]^{L-1} \\ 1 & a[1] & a[1]^2 & \cdots & a[1]^{L-1} \\ \vdots & \vdots & & \ddots & \vdots \\ 1 & a[L-1] & a[L-1]^2 & \cdots & a[L-1]^{L-1} \end{vmatrix} = \prod_{k>l}(a[k] - a[l]). \quad (7.2)$$

For $\det(V) \neq 0$, it is required that $a[k] \neq a[l] \; \forall \; k \neq l$. Since the calculations are, in fact, modulo M, a second constraint arises. Specifically, there cannot be a zero multiplier in the determinant (i.e., $\gcd\left(\prod_{k>l} a_k - a_l, M\right) = 1$).

In conclusion, to check the existence of an NTT, it must be verified that:

Theorem 7.1:   **Existence of an NTT over $\mathbb{Z}_M$**

An NTT of length N for α defined over $\mathbb{Z}_M$ exists, if:

1) $\gcd(\alpha, M) = 1$.

2) α is of order N, i.e.,

$$\alpha^n \bmod M \begin{cases} = 1 & n = N \\ \neq 1 & 1 \leq n < N. \end{cases} \quad (7.3)$$

3) The inverse $\det(A)^{-1}$ exist, i.e., $\gcd(\alpha^l - 1, M) = 1$ for $l = 1, 2, \ldots, N-1$.

For $\mathbb{Z}_p, p = $ prime, all the conditions shown above are automatically satisfied. In $\mathbb{Z}_p$ elements up to an order $p-1$ can be found. But transforms length $p-1$ are, in general, of limited practical interest, since in this case "general" multiplications and modulo reductions are necessary, and it is more appropriate to use a "normal" FFT in binary or QRNS arithmetic [184] and [39, paper 5-6].

There are no useful transforms in the ring $M = 2^b$. But it is possible to use the next neighbors, $2^b \pm 1$. If primes are used, then conditions 1 and 3 are automatically satisfied. We therefore need to discuss what kind of primes $2^b \pm 1$ are known.

Mersenne and Fermat Numbers. Primes of the form $2^b - 1$ were first investigated by the French mathematician Marin Mersenne (1588–1648). Using the geometry series

$$\left(1 + 2^q + 2^{2q} + \ldots + 2^{qr-1}\right)\left(2^q - 1\right) = 2^{qr} - 1$$

it can be concluded that exponent b of a Mersenne prime must also be a prime. This is necessary, but not sufficient, as the example $2^{11} - 1 = 23 \times 89$ shows. The first Mersenne primes $2^b - 1$ have exponents

$$b = 2, 3, 5, 7, 13, 17, 31, 61, 89, 107, 127, 521, 607, 1279. \tag{7.4}$$

Primes of the type $2^b + 1$ are known from one of Fermat's old letters. Fermat conjectured that all numbers $2^{(2^t)} + 1$ are primes but, as for Mersenne primes, this is necessary but not sufficient. It is necessary because if b is odd, i.e., $b = q2^t$ then

$$2^{q2^t} = \left(2^{(2^t)} + 1\right)\left(2^{(q-1)2^t} - 2^{(q-2)2^t} + 2^{(q-3)2^t} - \cdots + 1\right)$$

is not prime, as in the case of $(2^4 + 1)|(2^{12} + 1)$, i.e., $17|4097$. There are five known Fermat primes

$$F_0 = 3 \quad F_1 = 5 \quad F_2 = 17 \quad F_3 = 257 \quad F_4 = 65537, \tag{7.5}$$

but Euler (1707-1783) showed that 641 divides $F_5 = 2^{32} + 1$. Up to F_{21} there are no Fermat primes, which reduce the possible prime Fermat primes for NTTs to the first five.

7.1.1 Arithmetic Modulo $2^b \pm 1$

In Chap. 2, the one's complement (1C) and diminished-by-one (D1) coding were reviewed. Consult Table 2.1 (p. 57) for C1 and D1 coding. It was claimed that C1 coding can efficiently represent arithmetic modulo $2^b - 1$. This is used to build Mersenne NTTs, as suggested by Rader [185]. D1 coding efficiently represents arithmetic modulo $2^b + 1$, and is therefore preferred for Fermat NTTs, as suggested by Leibowitz [52].

The following table illustrates again the 1C and D1 arithmetic for computing addition.

1C	D1
$s = a + b + c_N$	if$((a == 0)\&\&(b == 0))s = 0$
	else $s = a + b + \overline{c_N}$

where a and b are the input operands, s is the sum and c_N the carry bit of the intermediate sum $a + b$ without modulo reduction. To implement the 1C addition, first form the intermediate B-bit sum. Then add the carry of the MSB c_N to the LSB. In D1 arithmetic, the carry must first be inverted before adding it to the LSB. The hardware requirement to add modulo $2^B \pm 1$ is therefore a total of two adders. The second adder may be built using half-adders, because one operand, besides the carry in the LSB, is zero.

Example 7.2: As an example, compute $10 + 7 \bmod M$.

Decimal	1C $M = 15$	D1 $M = 17$
7	0111	00110
+10	+1010	+01001
17	10001	01111
Correction	$+1 = 0010$	$+1 = 1.0000$
Check:	$17_{10} \bmod 15 = 2$	$17_{10} \bmod 17 = 0$

7.2

Subtraction is defined in terms of an *additive inverse*. Specifically, $B = -A$ is said to be the additive inverse of A if $A + B = 0$. How the additive inverse is built can easily be seen by consulting Table 2.1 (p. 57). Additive inverse production is

1C	D1
$\bar{a}$	if$(\mathrm{zf}(a)! = 1)\bar{a}$

It can be seen that a bitwise complement must first be computed. That is sufficient in the case of 1C, and for the nonzero elements in D1, coding. But for the zero in D1, the bitwise complement should be inhibited.

Example 7.3: The computation of the inverse of two is as follows

Decimal	1C $M = 15$	D1 $M = 17$
2	0010	0001
-2	1101	1110

which can be verified using the data provided in Table 2.1 (p. 57).

7.3

The simplest α for an NTT is 2. Depending on $M = 2^b \pm 1$, the arithmetic codings (C1 for Mersenne transforms and D1 for Fermat NTTs) is selected first. The only necessary multiplications are then those with $\alpha^k = 2^k$. These multiplications are implemented, as shown in Chap. 2, by a binary (left) rotation by k bit positions. The leftmost outgoing bit, i.e., carry c_N, is copied to the LSB. For the D1 coding (other than where $A = 0$) a complement of the carry bit must be computed, as the following table shows:

1C	D1
$\text{shl}(X, k, c_N)$	if$(X! = 0)$ $\text{shl}(X, k, \overline{c_N})$

The following example illustrates the multiplications by $\alpha^k = 2^k$ used most frequently in NTTs.

Example 7.4: Multiplication by 2^k for 1C and D1 Coding

The following table shows the multiplication of ± 2 by 2, and finally a multiplication of 2 by $8 = 2^3$ to demonstrate the modulo operation for 1C and D1 coding.

Decimal	1C $M = 15$	D1 $M = 17$
2×2^1	0010	0001
$= 4$	0100	0011
-2×2^1	1101	1110
$= -4$	1011	1100
2×2^3	0010	0001
$= 16$	0001	1111

which can be verified using the data found in Table 2.1 (p. 57). [7.4]

7.1.2 Efficient Convolutions Using NTTs

In the last section we saw that with α being a power of two, multiplication was reduced to data shifts that can be built efficiently and fast with FPGAs, if the modulus is $M = 2^b \pm 1$. Obviously this can be extended to complex αs of the kind $2^u \pm j2^v$. Multiplication of complex αs can also be reduced to simple data shifts.

In order to avoid general multiplications and general modulo operations, the following constraints when building NTTs should be taken into account:

> **Theorem 7.5:   Constraints for Practical Useful NTTs**
>
> A NTT is only of practical interest if
> 1) The arithmetic is modulo $M = 2^b \pm 1$.
> 2) All multiplications $x[k]\alpha^{kn}$ can be realized with a maximum of 2 modulo additions.

7.1.3 Fast Convolution Using NTTs

Fast cyclic convolution of two sequences x and h may be performed by multiplying two transformed sequences [66, 170, 185], as described by the following theorem.

Theorem 7.6:   **Convolution by NTT**

Let x and y be sequences of length N defined modulus M, and $z = \langle x \circledast y \rangle_M$ be the circular convolution of x and y. Let $X = \text{NTT}(x)$, and $Y = \text{NTT}(y)$ be the length-N NTTs of x and y computed over M. Then
$$z = \text{NTT}^{-1}(X \odot Y). \tag{7.6}$$

To prove the theorem, it must first be known that the commutative, associative, and distributive laws hold in a ring modulo M. That these properties hold is obvious, since $\mathbb{Z}$ is an integral domain (a commutative ring with unity) [186, 187].

Specifically, the circular convolution outcome, $y[n]$, is given by

$$y[n] = \left\langle N^{-1} \sum_{l=0}^{N-1} \left(\sum_{m=0}^{N-1} x[m] \alpha^{ml} \right) \left(\sum_{k=0}^{N-1} h[k] \alpha^{kl} \right) \alpha^{-ln} \right\rangle_M. \tag{7.7}$$

Applying the properties of commutation, association, and distribution, the sums and products can be rearranged, giving

$$y[n] = \left\langle \sum_{k=0}^{N-1} \sum_{m=0}^{N-1} x[m] h[k] \left(N^{-1} \sum_{l=0}^{N-1} \alpha^{(m+k-n)l} \right) \right\rangle_M. \tag{7.8}$$

Clearly for combinations of m, n, and k such that $\langle m + n - k \rangle \equiv 0 \bmod N$, the sum over l gives N ones and is therefore equal to N. However, for $\langle m + n - k \rangle_N \equiv r \neq 0$, the sum is given by

$$\sum_{l=0}^{N-1} \alpha^{rl} = 1 + \alpha^r + \alpha^{2r} + \ldots + \alpha^{r(N-1)} = \frac{1 - \alpha^{rN}}{1 - \alpha^r} \equiv 0 \tag{7.9}$$

for $\alpha^r \neq 1$. Because α is of order N, and $r < N$, it follows that $\alpha^r \neq 1$. It follows that for the sum over l, (7.8) becomes

$$N^{-1} \sum_{l=0}^{N-1} \alpha^{(m+k-n)l} = \begin{cases} \langle N N^{-1} \equiv 1 \rangle_M & \text{for } m + l - n \equiv 0 \bmod N \\ 0 & \text{for } m + l - n \not\equiv 0 \bmod N \end{cases}.$$

It is now possible to eliminate either the sum over k, using $k \equiv \langle n - m \rangle$, or the sum over m, using $m \equiv \langle n - k \rangle$. The first case gives

$$y[n] = \left\langle \sum_{m=0}^{N-1} x[m] h[\langle n - m \rangle_N] \right\rangle_M, \tag{7.10}$$

while the second case gives

$$y[n] = \left\langle \sum_{k=0}^{N-1} h[k] x[\langle n - k \rangle_N] \right\rangle_M. \qquad \square \tag{7.11}$$

The following example demonstrates the convolution.

Example 7.7: Fermat NTT of Length 4

Compute the cyclic convolution of length-4 time series $x[n] = \{1, 1, 0, 0\}$ and $h[n] = \{1, 0, 0, 1\}$, using a Fermat NTT modulo 257.

Solution: For the NTT of length 4 modulo $M = 257$, the element $\alpha = 16$ has order 4. In addition, using a symmetric range $[-128, \ldots, 128]$, we need $4^{-1} \equiv -64 \bmod 257$ and $16^{-1} \equiv -64 \bmod 257$. The transform and inverse transform matrices are given by

$$T = \begin{bmatrix} 1 & 1 & 1 & 1 \\ 1 & 16 & -1 & -16 \\ 1 & -1 & 1 & -1 \\ 1 & -16 & -1 & 16 \end{bmatrix} \qquad T^{-1} = \begin{bmatrix} 1 & 1 & 1 & 1 \\ 1 & -16 & -1 & 16 \\ 1 & -1 & 1 & -1 \\ 1 & 16 & -1 & -16 \end{bmatrix}. \tag{7.12}$$

The transform of $x[n]$ and $h[n]$ is followed by the multiplication element by element, of the transformed sequence. The result for $y[n]$, using the INTT, is shown in the following

$$
\begin{array}{rcl}
n, k & = & \{0, \quad 1, \quad 2, \quad 3\} \\
x[n] & = & \{1, \quad 1, \quad 0, \quad 0\} \\
X[k] & = & \{2, \quad 17, \quad 0, \quad -15\} \\
h[n] & = & \{1, \quad 0, \quad 0, \quad 1\} \\
H[k] & = & \{2, \quad -15, \quad 0, \quad 17\} \\
X[k] \times H[k] & = & \{4 \quad 2, \quad 0 \quad 2\} \\
y[n] = x[n] \circledast h[n] & = & \{2 \quad 1 \quad 0 \quad 1\}.
\end{array}
$$

7.7

Wordlength limitations for NTT. When using an NTT to perform convolution, remember that all elements of the output sequence $y[n]$ must be bounded by M. This is true (for simplicity, unsigned coding is assumed) if

$$x_{\max} h_{\max} L \leq M. \tag{7.13}$$

If the bit widths $B_x = \log_2(x_{\max})$, $B_h = \log_2(h_{\max})$, $B_L = \log_2(L)$, and $B_M = \log_2(M)$ are used, it follows that for $B_x = B_h$ the maximum bit width of the input is bounded by

$$\boxed{B_x = \frac{B_M - B_L}{2},} \tag{7.14}$$

with the additional constraint that $M = 2^b \pm 1$, and α is a power of two. It follows that very few prime M transforms exist. Table 7.1 displays the most useful choices of αs, and the attendant transform length (i.e., order of αs) of Mersenne and Fermat NTTs.

If complex transforms and nonprime Ms are also considered, then the number and length of the transform becomes larger, and the complexity also increases. In general, for nonprime modul, the conditions from Theorem 7.1 (p. 402) should be checked. It is still possible to utilize Mersenne or Fermat arithmetic, by using the following congruence

$$a \bmod u \equiv (a \bmod (u \times v)) \bmod u, \tag{7.15}$$

Table 7.1. Prime $M = 2^b \pm 1$ NTTs including complex transforms.

Mersenne $M = 2^b - 1$		Fermat $M = 2^b + 1$	
α	$\text{ord}_M(\alpha)$	α	$\text{ord}_M(\alpha)$
2	b	2	b
-2	$2b$	$\sqrt{2}$	$2b$
$\pm 2\mathrm{j}$	$4b$	$1 + \mathrm{j}$	$4b$
$1 \pm \mathrm{j}$	$8b$		

which states that everything is first computed modulo $M = u \times v = 2^b \pm 1$, and only the output sequence need be computed modulo u, which is the valid module regarding Theorem 7.1. Although using $M = u \times v = 2^b \pm 1$ increases the internal bit width, 1C or D1 arithmetic can be used. They have lower complexity than modulo arithmetic modulo u, and this will, in general, reduce the overall effort for the system.

Such nonprime NTTs are called *pseudotransforms*, i.e., *pseudo-Mersenne* transforms or *pseudo-Fermat* transforms. The following example demonstrates the construction for a pseudo-Fermat transform.

Example 7.8: A Fermat NTT of Length 50

Using the MATLAB utility `order.m` (see Exercise 7.1, p. 472), it can be determined that $\alpha = 2$ is of order 50 modulo $2^{25} + 1$. From Theorem 7.1, we know that $\gcd(\alpha^2 - 1, M) = 3$, and a length 50 transform does *not* exist modulo $2^{25} + 1$. It is therefore necessary to identify the "bad" factors in $M = (2^b \pm 1)$, those that do not have order 50, and exclude these factors by using the final modulo operation in (7.15).

<u>Solution:</u> Using the standard MATLAB function `factor(2^25+1)`, the prime-factors of M are:

$$2^{25} + 1 = 3 \times 11 \times 251 \times 4051. \tag{7.16}$$

The order of $\alpha = 2$ for the single factor can be computed with the algorithm given in Exercise 7.1 on p. 472. They are

$$\begin{array}{ll} \text{ord}_3(2) = 2 & \text{ord}_{11}(2) = 10 \\ \text{ord}_{251}(2) = 50 & \text{ord}_{4051}(2) = 50. \end{array} \tag{7.17}$$

In order to have an NTT of length 50, a final modulo reduction with $(2^{25} + 1)/33$ must be computed. ⬛ 7.8

Comparing Fermat and Mersenne NTT implementations, consider that

- A Mersenne NTT of length b, with b primes, can be converted by the chirp-z transform (CZT), or the Rader prime factor theorem (PFT) [169], into a cyclic convolution, as shown in Fig. 7.1a. In addition this allows a simplified bus structure if a multi-FPGA implementation [174] is used.
- Fermat NTTs with $M = 2^{(2^t)} + 1$ have a power-of-two length $N = 2^t$, and can therefore be implemented with the usual Cooley–Tukey radix-2-type FFT algorithm, which we discussed in Chap. 6.

Table 7.2. Data for some Agarwal–Burrus NTTs, to compute cyclic convolution using real Fermat NTTs ($b = 2^t, t = 0$ to 4) or pseudo-Fermat NTTs $t = 5, 6$.

Module	α	1D	2D
$2^b + 1$	2	$2b$	$2b^2$
$2^b + 1$	$\sqrt{2}$	$4b$	$8b^2$

7.1.4 Multidimensional Index Maps for NTTs and the Agarwal–Burrus NTT

For NTTs, in general the transform length N is proportional to the bit width b. This constraint makes it *impossible* to build long (one-dimensional) transforms, because the necessary bit width will be tremendous. It is possible to try the multidimensional index maps, called Good–Thomas and Cooley–Tukey, which we discussed in Chap. 6. If these methods are applied to NTTs, the following problems arise:

- In Cooley–Tukey algorithms of length $N = N_1 N_2$, an element of order N in the twiddle factors is needed. It follows that the transform length is not increased, compared with the one-dimensional case, and will result in large bit width. It is therefore not attractive.
- If Good–Thomas mapping is applied, there is *no* need for an element of length N, for a length $N = N_1 N_2$ transform. However, two coprime length transforms N_1 and N_2 are needed for the same M. That is impossible for NTTs, if the transforms listed in Table 7.1 (p. 408) are used. The only way to make Good–Thomas NTTs work is to use different extension fields, as reported in [174], or to use them in combination with Winograd short-convolution algorithms, but this will also increase the complexity of the implementation.

An alternative method suggested by Agarwal and Burrus [188] seems to be more attractive. In the Agarwal–Burrus algorithm, a one-dimensional array is also first mapped into a two-dimensional array, but in contrast to the Good–Thomas methods, the lengths N_1 and N_2 must *not* be coprime. The Agarwal–Burrus algorithm can be understood as a generalization of the overlap-save method, where periodic extensions of the signals are built. If an α of order $2L$ is used, a convolution of size

$$\boxed{N = 2L^2} \tag{7.18}$$

can be built. From Table 7.2, it can be seen that this two-dimensional method improves the maximum length of the transforms.

To compute the Agarwal–Burrus NTT, the following five steps are used:

Algorithm 7.9:   **Agarwal–Burrus NTT**

The cyclic convolution of $x \circledast h$ of length $N = 2L^2$, with an NTT of length L, is accomplished with the following steps:

1) Index transformation of the one-dimensional sequence into a two-dimensional array according to

$$
x = \begin{bmatrix}
x[0] & x[L] & \cdots & x[N-L] \\
x[1] & x[L+1] & \cdots & x[N-L+1] \\
\vdots & \vdots & \ddots & \vdots \\
x[L-1] & x[2L-1] & \cdots & x[N-1] \\
0 & 0 & \cdots & 0 \\
\vdots & \vdots & \ddots & \vdots \\
0 & 0 & 0 & 0
\end{bmatrix} \tag{7.19}
$$

$$
h = \begin{bmatrix}
h[N-L+1] & h[L] & \cdots & h[N-2L+1] \\
\vdots & \vdots & \ddots & \vdots \\
h[N-1] & h[L-1] & \cdots & h[N-L-1] \\
h[0] & h[L] & \cdots & h[N-L] \\
h[1] & h[L+1] & \cdots & h[N-L+1] \\
\vdots & \vdots & \ddots & \vdots \\
h[L-1] & h[2L-1] & \cdots & h[N-1]
\end{bmatrix}. \tag{7.20}
$$

2) Computation of the row transforms $\left[\begin{smallmatrix} \rightarrow \\ \vdots \end{smallmatrix}\right]$ followed by the column transforms $[\downarrow\downarrow \cdots]$.

3) Computation of the element-by-element matrix multiplication, $Y = H \odot X$.

4) Inverse transforms of the columns $[\downarrow\downarrow \cdots]$ followed by the inverse row transforms $\left[\begin{smallmatrix} \rightarrow \\ \vdots \end{smallmatrix}\right]$.

5) Reconstruction of the output sequence from the lower part of y, according to

$$
y = \begin{bmatrix}
\vdots & \vdots & \ddots & \vdots \\
y[0] & y[L] & \cdots & y[N-L] \\
y[1] & y[L+1] & \cdots & y[N-L+1] \\
\vdots & \vdots & \ddots & \vdots \\
y[L-1] & y[2L-1] & \cdots & y[N-1]
\end{bmatrix}. \tag{7.21}
$$

The Agarwal–Burrus NTT can be demonstrated with the following example:

Example 7.10: Length 8 Agarwal–Burrus NTT

An NTT modulo 257 of length 4 exists for $\alpha = 16$. Compute the convolution of $X(z) = 1 + z^{-1} + z^{-2} + z^{-3}$ with $F(z) = 1 + 2z^{-1} + 3z^{-2} + 4z^{-3}$ using a

Fermat NTT modulo 257.

Solution: First, the index maps and transforms of $x[n]$ and $f[n]$ are computed. It follows that

$$x = \begin{bmatrix} 1\ 1\ 0\ 0 \\ 1\ 1\ 0\ 0 \\ 0\ 0\ 0\ 0 \\ 0\ 0\ 0\ 0 \end{bmatrix} \longleftrightarrow X = \begin{bmatrix} 4 & 34 & 0 & 227 \\ 34 & 32 & 0 & 2 \\ 0 & 0 & 0 & 0 \\ 227 & 2 & 0 & 225 \end{bmatrix} \tag{7.22}$$

$$f = \begin{bmatrix} 0\ 0\ 0\ 0 \\ 0\ 2\ 4\ 0 \\ 1\ 3\ 0\ 0 \\ 2\ 4\ 0\ 0 \end{bmatrix} \longleftrightarrow F = \begin{bmatrix} 16 & 143 & 255 & 112 \\ 253 & 114 & 66 & 206 \\ 249 & 212 & 255 & 51 \\ 253 & 45 & 195 & 145 \end{bmatrix}. \tag{7.23}$$

Now, an element-by-element multiplication is computed, which results in

$$Y = \begin{bmatrix} 64 & 236 & 0 & 238 \\ 121 & 50 & 0 & 155 \\ 0 & 0 & 0 & 0 \\ 120 & 90 & 0 & 243 \end{bmatrix} \longleftrightarrow y = \begin{bmatrix} 2 & 6 & 4 & 0 \\ 0 & 2 & 6 & 4 \\ 1 & 6 & 9 & 4 \\ 3 & 10 & 7 & 0 \end{bmatrix}. \tag{7.24}$$

From the lower half of y, the element of $y[n] = \{1, 3, 6, 10, 9, 7, 4, 0\}$ can be seen.

$\boxed{7.10}$

With the Agarwal–Burrus NTT, a double-size intermediate memory is needed, but much longer transforms can be computed. The two-dimensional principle can easily be extended to three-dimensional index maps, but most often the transform length achieved with the two-dimensional method will be sufficient. For instance, for $\alpha = 2$ and $b = 32$, the transform length is increased from 64 in the one-dimensional case to $2^{11} = 2048$ in the two-dimensional case.

7.1.5 Computing the DFT Matrix with NTTs

Most often DFTs and NTTs are used to compute convolution, and it can be attractive to use NTTs to compute this convolution with FPGAs, because 1C and D1 can be efficiently implemented. But sometimes it is necessary to compute the DFT to estimate the Fourier spectrum. Then a question arises: Is it possible to use the more efficient NTT to compute the DFT? This question has been addressed in detail by Siu and Constantinides [189].

The idea is as follows: For prime p-length DFTs, the Rader algorithm can be used, which converts the task into a length $p - 1$ cyclic convolution. This cyclic convolution is then computed by an NTT of the original sequence and the DFT twiddle factors in the NTT domain, multiplication elementwise, and the back conversion. These processing steps are illustrated in Fig. 7.1b. The principle is demonstrated in the following example.

Example 7.11: Rader Algorithm for $N = 5$

For $N = 5$, a generator is $g = 2$, which gives the following index map, $\{2^0, 2^1, 2^2, 2^3\} \bmod 5 \equiv \{1, 2, 4, 3\}$. First, the DC component is computed with

Table 7.3. Building blocks to compute DFT with Fermat NTT.

DFT length	Number ring	α	Number of real Mul.	Shift-Add.
3	$F_1, F_2, F_3, F_4, F_5, F_6$	$2^2, 2^4, 2^8, 2^{16}, 2^{32}, 2^{64}$	2	6
5	$F_1, F_2, F_3, F_4, F_5, F_6$	$2, 2^2, 2^4, 2^8, 2^{16}, 2^{32}$	4	20
17	F_3, F_4, F_5, F_6	$2, 2^2, 2^4, 2^8$	16	144
257	F_6	$\sqrt{2}$	256	4544
13	$F_1, F_2, F_3, F_4, F_5, F_6$	$2, 2^2, 2^4, 2^8, 2^{16}, 2^{32}$	16	104
97	F_4, F_5, F_6	$2, 2^2, 2^3, 2^4$	128	1408
193	F_5, F_6	$2, 2^2$	256	3200
769	F_6	$\sqrt{2}$	1024	16448

$$X[0] = \sum_{n=0}^{4} x[n] = x[0] + x[1] + x[2] + x[3] + x[4]$$

and in the second step, $X[k] - x[0]$, the cyclic convolution

$$\{x[1], x[2], x[4], x[3]\} \circledast \{W_5^1, W_5^2, W_5^4, W_5^3\}.$$

Now the NTT is applied to the (reordered) sequences $x[n]$ and W_5^k, as shown in Example 7.7 (p. 407). The transformed sequences are then multiplied element by element, in the NTT domain, and finally the INTT is computed.

7.11

For Mersenne NTTs a problem arises, in that the NTT itself is of prime length, and therefore the length increased by one can *not* be of prime length. But for a Fermat NTT, the length is 2^t, since $M = 2^t + 1$, which is a prime. Siu and Constantinides found eight such short-length DFT building blocks to be useful. These basic building blocks are summarized in Table 7.3.

The first part of Table 7.3 shows blocks that do not need an index transform. In the second part are listed the building blocks that have two coprime factors. They are $13 - 1 = 3 \times 4$, $97 - 1 = 3 \times 32$, $193 - 1 = 3 \times 64$, and $769 - 1 = 3 \times 256$. The disadvantage of the two-factor case is that, in a two-dimensional index map, for only one dimension every second transform of the twiddle factor becomes zero.

In the multidimensional map, it is also possible to implement a radix-2 FFT-like algorithm, or to combine Fermat NTTs with other NTT algorithms, such as the (pseudo-) Fermat NTT transform, (pseudo-) Mersenne transform, Lagrange interpolation, Eisenstein NTTs or a short convolution algorithm such as the Winograd algorithm [66, 189].

In the following, the techniques for the two-factor case using a length $13 - 1 = 3 \times 4$ multidimensional index map are reviewed. This is similar to the discussion in Chap. 6 for FFTs.

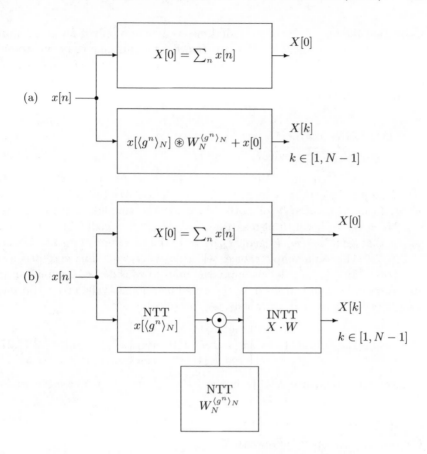

Fig. 7.1. The use of NTTs in Rader's prime-length algorithm for computing the DFT. (**a**) Rader's original algorithm. (**b**) Modification of the Rader prime algorithm using NTTs.

7.1.6 Index Maps for NTTs

To directly realize the NTT matrix is generally too expensive. This problem may be resolved by suitable multidimensional techniques. Burrus [148] gives a systematic overview of different common and prime factor maps, from one dimension to multiple dimensions. The mapping is explained for the two-dimensional case. Higher-order mapping is equivalent. The mapping from the one-dimensional cyclic length-N convolution from (7.1), into a two-dimensional convolution with dimension $N = N_1 \times N_2$, can be written in linear form as follows:

$$n = M_1 n_1 + M_2 n_2 \bmod N, \tag{7.25}$$

where $n_1 \in \{0, 1, 2, \ldots, N_1-1\}$ and $n_2 \in \{0, 1, 2, \ldots, N_2-1\}$. For $\gcd(N_1, N_2) \neq 1$, the well-known Cooley–Tukey FFT algorithm may be used. Burrus [148] shows that the map is cyclic in both dimensions if and only if N_1 and N_2 are relatively prime, i.e., $\gcd(N_1, N_2) = 1$. In order for this map to be one-to-one and onto (i.e., a bijection), the mapping constants M_1 and M_2 must satisfy certain conditions. For the relatively prime case, the conditions to make the mapping bijective are:

$$[M_1 = \beta N_1 \ and/or \ M_2 = \gamma N_1] \ and$$
$$\gcd(M_1, N_1) = \gcd(M_2, M_2) = 1. \tag{7.26}$$

As an example, consider $N_1 = 3$ and $N_2 = 4$, $N = 12$. From condition (7.26) we see that it is necessary to choose M_1 (a multiple of N_2), or M_2 (a multiple of N_1), or both. Make M_1 the simplest multiple of N_2, i.e., $M_1 = N_2 = 4$, which also satisfies $\gcd(M_1, N_1) = \gcd(4, 3) = 1$. Then, noting that $\gcd(M_2, N_2) = \gcd(M_2, 4) = 1$, the possible values for M_2 are $\{1, 3, 5, 7, 9, 11\}$. As a simple choice, select $M_2 = N_1 = 3$. The map becomes $n = \langle 4n_1 + 3n_2 \rangle_{12}$. Now let us apply the map to consider a 12-point convolution example. The transform of the one-dimensional cyclic array $x[n]$ into a 3×4 two-dimensional array $x[n_1, n_2]$, produces

$$[x[0]x[1]x[2]\ldots x[11]] \leftrightarrow \begin{bmatrix} x[0] & x[3] & x[6] & x[9] \\ x[4] & x[7] & x[10] & x[1] \\ x[8] & x[11] & x[2] & x[5] \end{bmatrix}. \tag{7.27}$$

To recover the sequence $X[k]$ from the $X[k_1, k_2]$, use the Chinese remainder theorem, as suggested by Good [171],

$$k = \langle (N_2^{-1} \bmod N_1)N_2k_1 + (N_1^{-1} \bmod N_2)N_1k_2 \rangle_N. \tag{7.28}$$

The α matrix can now be rewritten as

$$X[k_1, k_2] = \sum_{n_1=0}^{N_1-1} \left(\sum_{n_2=0}^{N_2-1} x[n_1, n_2]\alpha_{N_2}^{n_2 k_2} \right) \alpha_{N_1}^{n_1 k_1}, \tag{7.29}$$

where α_{N_i} is an element of order N_i. Having mapped the original sequence $x[n]$ into the two-dimensional array $x[n_1, n_2]$, the desired matrix can be evaluated by the following two steps:

1) Perform an N_2-point NTT on each row of the matrix $x[n_1, n_2]$.
2) Perform an N_1-point NTT on each column of the resultant matrix, to yield $X[k_1, k_2]$.

These processing steps are shown in Fig. 7.2. The input map is given by (7.27), while the output map can be computed with (7.28),

$$k = \langle \langle 4^{-1} \rangle_3 4k_1 + \langle 3^{-1} \rangle_4 3k_2 \rangle_{12} = \langle 4k_1 + 9k_2 \rangle_{12}. \tag{7.30}$$

The array $X[k_1, k_2]$ will therefore have the following arrangement:

$$[X[0]X[1]X[2]\ldots X[11]] \leftrightarrow \begin{bmatrix} X[0] & X[9] & X[6] & X[3] \\ X[4] & X[1] & X[10] & X[7] \\ X[8] & X[5] & X[2] & X[11] \end{bmatrix}. \tag{7.31}$$

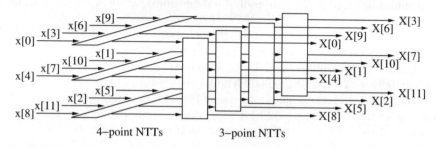

Fig. 7.2. Two-dimensional map. First stage: three 4-point NTTs. Second stage: four 3-point NTTs.

Length 97 DFT case study. In recent years programmable digital signal processors (e.g., TMS320; Motorola 56K; AT&T 32C) have become the dominant vehicle to implement fast convolution via FFT algorithms. These PDSPs provide a fast (real) multiplier with typical cycle times of 10 to 50 ns. There are also some NTT implementations [190], but NTT implementations need modulo arithmetic, which is not supported by general-purpose PDSPs. Dedicated accelerators, such as the FNT from McClellan [190], use 90 standard ECL 10K ICs. In recent years, field-programmable gate arrays (FPGAs) have become dense enough and fast enough to implement typical high-speed DSP applications [4, 158]. It is possible to implement several arithmetic cores with only one FPGA, producing good packaging, speed, and power characteristics. FPGAs, with their fine granularity, can implement modulo arithmetic efficiently, without penalty, as in the PDSP case.

In NTT implementation of Fermat number arithmetic, the previously discussed speed and hardware advantages, compared with conventional FFT implementations, become an even bigger advantage for an FPGA implementation. By implementing the DFT algorithm with the Rader prime convolution strategy, the required I/O performance can be further reduced.

To clarify the NTT design paradigm, a length-97 DFT in the Fermat number system, F_4 and F_5, for real input data, will be shown. A Xilinx XC4K multi-FPGA board has been used to implement this design, as reported in [174].

For modulo Fermat number arithmetic (modulo $2^n + 1$) it is advantageous to use, instead of the usual two's complement arithmetic (2C), the "Diminished one" (D1) number system from Leibowitz [52]. Negative numbers are the same as in 2C, and positive numbers are diminished by one. The zero is encoded as a zero string and the MSB "ZERO-FLAG" is one. Therefore the diminished system consists of a ZERO-FLAG and integer bits x_k. For 2C, the MSB is the sign bit, while for D1 the second MSB is the sign bit. With this encoding the basic operations of 2C↔D1 conversion, negation, addition,

and multiplication by 2^m can easily be determined, as shown in Sect. 7.1.1
(p. 403).

The rough processing steps of the 97-point transform are shown in
Fig. 7.1b. A direct length-96 implementation for a single NTT will cost at
least 96×2 barrel shifters and 96×2 accumulators and, therefore, approximately $96(2 \times 32 + 2 \times 18) = 9600$ Xilinx combinatorial logic blocks (CLBs).
Therefore it seemed reasonable to use a 32×3 index map, as described in
the last section. The length-32 FFT now becomes a simpler length-32 Fermat
NTT, and the length-3 transform has $\alpha^k = 1; \hat{\jmath}$ and $-1 - \hat{\jmath}$ with $\hat{\jmath}^2 = \hat{\jmath} + 1$.
The 32-point FNT can be realized with the usual radix-2 FFT-type algo-
rithm, while the length-3 transform can be implemented by a two-tap FIR
filter. The following table gives CLB utilization estimates for Xilinx XC4000
FPGAs, for F_4:

Length-32 FNT	Length-3 FIR NTT	14 Multipliers 32-bit	Length-3 NTTS^{-1}	Two length-32 FNT^{-1}
104	108	462	288	216

The design consumes a total of 1178 CLBs. To get high throughput, the
buffer memory between the blocks must be doubled. Two real buffers for the
first FNT, and three complex buffers, are required. If the buffers are realized
internally, an additional 748 CLBs are required, which will also minimize the
I/O requirements. If 80% utilization is assumed, then about six XC4010s are
needed for the design, including the buffer memory.

The time-critical path in the design is the length-32 FNT. To maximize
the throughput, a three-stage pipeline is used inside the butterfly. For a 5-ns
FPGA, the butterfly speed is 28 ns for F_4, and 38.5 ns for F_5. For three length-
32 FNTs, five stages, each with 16 butterflies, must be computed. This gives
a total transform time of 7.15 μs for F_4, and 9.24 μs for F_5, for the length-97
DFT. To set this result in perspective, the time for the butterfly computation
gives a fair comparison. A TMS320C50 PDSP with a 50-ns cycle time needs
17 cycles for a butterfly [191], or 850 ns, assuming zero wait-state memory.
Another "conventional" FPGA fixed-point arithmetic design [158] uses four
serial/parallel multipliers (2 MHz), and therefore has a latency of 500 ns for
the butterfly.

7.1.7 Using Rectangular Transforms to Compute the DFT

Rectangular transforms also map an input sequence in an image domain, but
do not necessarily have the DFT structure, i.e., $\boldsymbol{A} = [\alpha^{nk}]$. Examples are Haar
transforms [130], the Walsh–Hadamard transform [71], a ruff-quantized DFT
[5], or the arithmetic Fourier transform [173, 192, 193]. These rectangular
transforms, in general, do *not* support cyclic convolution, but they may be
used to approximate the DFT spectrum [165]. The advantage of rectangular

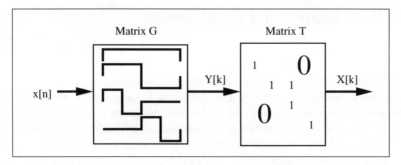

Fig. 7.3. DFT computation using rectangular transform and map matrix T.

transforms is that the coefficients are from the set $\{-1, 0, 1\}$ and they do not need any multiplications.

How to compute the DFT is shown in Fig. 7.3. In order to have a useful system, it is assumed that the rectangular transform can be computed with low effort, and the second transform using the matrix T, which maps the rectangular transform to the DFT vectors, has only a few nonzero elements.

Table 7.4. Comparison of different transforms to approximate the DFT [5].

Transform	Number of base	Algorithmic complexity	Zeros in 16×16 T Matrix
Walsh	N	$N \log_2(N)$	66
Hadamard	N	$N \log_2(N)$	66
Haar	N	$2N$	18
AFT	$N+1$	N^2	82
QDFT	$2N$	$(N/8)^2+3N$	86

Table 7.4 compares different implementations. The algorithmic complexity of the Walsh–Hadamard and Haar transforms is most interesting, but from the number of zeros in the second transform T it can be concluded that the arithmetic Fourier transform and the ruff-quantized DFT are more attractive for approximating the DFT.

7.2 Error Control and Cryptography

Modern communications systems, such as pagers, mobile phones or satellite transmission systems, use algorithms to correct transmission errors, since error-correction coding better utilizes the band-limited channel capacity than special modulation schemes (see Fig. 7.4). In addition, most systems also use cryptography algorithms, not just to protect messages against unauthorized listeners, but also to protect messages against unauthorized changes.

In a typical transmission scheme, such as that shown in Fig. 7.5, the *encoder* (for error correction or cryptography) is placed between the data source and the actual modulation. On the receiver side, the *decoder* is located between demodulation and the data destination (sink). Often an encoder and decoder are combined in one circuit, referred to as a CODEC.

Typical error correction and cryptographic algorithms use finite field arithmetic and are therefore more suitable for FPGAs than they are for PDSPs [195]. Bitwise operations or linear feedback shift registers (LFSR) can be very efficiently realized with FPGAs. Some CODEC schemes use large tables, and one objective when selecting the appropriate algorithms for FPGAs is therefore to find out which algorithms are most suitable. The

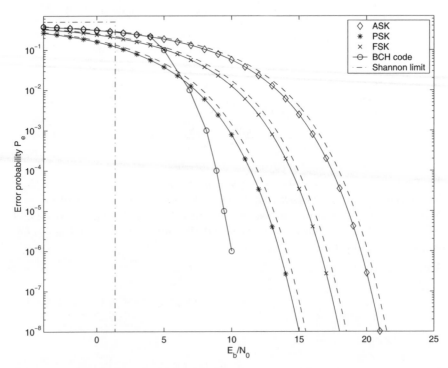

Fig. 7.4. Performance of modulation schemes [194]. **Solid line** coherent demodulation and **dashed line** incoherent demodulation.

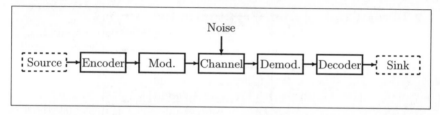

Fig. 7.5. Typical communications system configuration.

algorithms presented in this section are mainly based on previous publications [4] and have been used to develop a paging system for low frequencies [196, 197, 198, 199, 200], and an error-correction scheme for radio-controlled watches [201, 202].

It is impossible in a short section to present the whole theory of error correction and cryptography. We will present the basic ideas and suggest, for further investigation, one of the excellent textbooks in this area [163, 203, 204, 205, 206, 207, 208].

7.2.1 Basic Concepts from Coding Theory

The simplest way to protect a digital transmission against random errors is to repeat the message several times. This is called *repetition code*. For a repetition of 5, for instance, the message is sent five times, i.e.,

$$0 \Leftrightarrow 00000 \tag{7.32}$$

$$1 \Leftrightarrow 11111, \tag{7.33}$$

where the left side shows the k information bits and the right side the n-bit codewords. The minimum distance between two codewords, also called the *Hamming distance d^**, is also n and the repetition code is of the form $(n, k, d^*) = (5, 1, 5)$. With such a code it is possible to correct up to $\lfloor (n-1)/2 \rfloor$ random errors. But from the perspective of channel efficiency, this code is not very attractive. If our system is two-way then it is more efficient to use a technique such as a parity check and an *automatic repeat request* (ARQ) for any detected parity error. Such parity checks are used, for instance, in PC memory.

Error correction using a Hamming code. If a few more parity check bits are added, it is possible to *correct* a word with a parity error.

If the parities $P_{1,0}, P_{1,1}, P_{1,2}$, and $P_{1,3}$ are computed using modulo 2 operations, i.e., XOR, according to

$$
\begin{aligned}
P_{1,0} &= i_{21} \oplus i_{22} \oplus i_{23} \oplus i_{24} \ ; \oplus \ i_{25} \oplus i_{26} \oplus i_{27} \\
P_{1,1} &= i_{21} \qquad \oplus i_{23} \qquad \oplus i_{25} \qquad \oplus i_{27} \\
P_{1,2} &= i_{21} \oplus i_{22} \qquad\qquad \oplus i_{25} \oplus i_{26} \\
P_{1,3} &= i_{21} \oplus i_{22} \oplus i_{23} \oplus i_{24}
\end{aligned}
$$

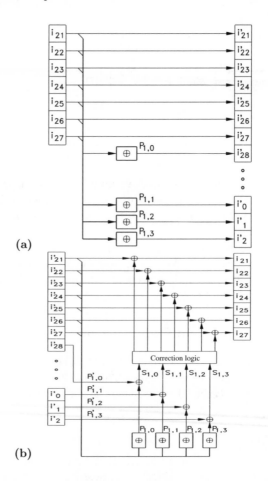

Fig. 7.6. (a) Coder and (b) Decoder for Hamming code.

then the parity detector is $i'_{28}(= P_{1,0})$ and three additional bits are necessary to locate the error position. Figure 7.6a shows the encoder, and Fig. 7.6b the decoder including the correction logic. On the decoder side the incoming parities are XOR'd with the newly computed parities. This forms the so-called *syndrome* $(S_{1,0} \cdots S_{1,3})$. The parities have been chosen in such a way that the syndrome pattern corresponds to the position of the bit in binary code, i.e., a $3 \to 7$ demultiplexer can be used to decode the error location.

For a more compact representation of the decoder, the following parity check matrix H can be used

$$\boldsymbol{H} = \left[\boldsymbol{P}^T{:}\boldsymbol{I}\right] = \begin{bmatrix} 1\,1\,1\,1\,1\,1\,1\,1\,0\,0\,0 \\ 1\,0\,1\,0\,1\,0\,1\,0\,1\,0\,0 \\ 1\,1\,0\,0\,1\,1\,0\,0\,0\,1\,0 \\ 1\,1\,1\,1\,0\,0\,0\,0\,0\,0\,1 \end{bmatrix}. \tag{7.34}$$

Table 7.5. Estimated effort for error correction with Hamming code.

Block Hamming code	CLB effort for		
	Minutes (11,7,3)	Hours (10,6,3)	Date (27,22,3)
Register	6	6	14
Syndrome computation	5	5	16
Correction logic	4	4	22
Output register	4	4	11
Sum	19	19	63
Total		101	

It is possible to describe the encoder using a generator matrix. $G = [I\vdots P]$, i.e., the generator matrix consists of a systematic identity matrix I followed by the parity-bits matrix P. A codeword v is computed by multiplying (modulo 2) the information word i with the generator matrix G:

$$v = i \times G. \tag{7.35}$$

The (de)coders shown in Fig. 7.6 are those for a (11,7,3) Hamming code, and it is possible to detect *and* correct one error. In general, it can be shown that for 4 parity bits, up to 15 information bits, can be used, i.e., a (15,11,3) Hamming code has been *shortened* to a (11,7,3) code.

A Hamming code with distance 3 generally has a $(2^m - 1, 2^m - m, 3)$ structure. The dates in radio-controlled watches, for instance, are coded with 22 bits, and a (31,26,3) Hamming code can be shortened to a (27,22,3) code to achieve a single-error correcting code. The parity check matrix becomes:

$$H = \begin{bmatrix} 1 & 0 & 1 & 0 & 1 & 0 & 1 & 0 & 1 & 0 & 1 & 0 & 1 & 0 & 1 & 0 & 1 & 0 & 1 & 0 & 1 & 0 & 1 & 0 & 1 & 0 & 0 & 0 & 0 \\ 1 & 1 & 0 & 0 & 1 & 1 & 0 & 0 & 1 & 1 & 0 & 0 & 1 & 1 & 0 & 0 & 1 & 1 & 0 & 0 & 1 & 1 & 0 & 1 & 0 & 0 & 0 \\ 1 & 1 & 1 & 1 & 0 & 0 & 0 & 0 & 1 & 1 & 1 & 1 & 0 & 0 & 0 & 0 & 1 & 1 & 1 & 1 & 0 & 0 & 0 & 0 & 1 & 0 & 0 \\ 1 & 1 & 1 & 1 & 1 & 1 & 1 & 1 & 0 & 0 & 0 & 0 & 0 & 0 & 0 & 0 & 1 & 1 & 1 & 1 & 1 & 1 & 0 & 0 & 0 & 1 & 0 \\ 1 & 1 & 1 & 1 & 1 & 1 & 1 & 1 & 1 & 1 & 1 & 1 & 1 & 1 & 1 & 0 & 0 & 0 & 0 & 0 & 0 & 0 & 0 & 0 & 0 & 0 & 1 \end{bmatrix}.$$

Again, the syndromes can be sorted in such a way that the correction logic is a simple $5 \to 22$ demultiplexer.

Table 7.5 shows the estimated effort in CLBs using Xilinx XC3K FP-GAs for an error-correction unit for radio-controlled watches that uses three separate data blocks for minutes, hours, and date.

In conclusion, with an additional 3+3+5=11 bits and the parity bits for the minutes using about 100 CLBs, it is possible to correct one error in each of the three blocks.

Survey of Error Correction Codes

After the introductory case study in the last section, commonly used codes and possible encoder and decoder implementations will be discussed next.

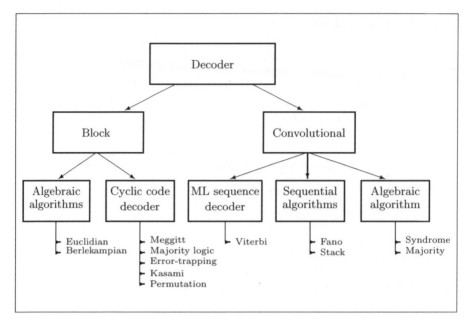

Fig. 7.7. Decoder for error correction.

Most often the effort for the decoder is of greater concern, since many communications systems like pager or radios use one sender and several receivers. Figure 7.7 shows a diagram of possible decoders.

Some nearly optimal decoders use huge tables and are not included in Fig. 7.7. The difference between block and convolutional codes is based on whether "memory" is used in the code generation. Both methods are characterized by the code rate R, which is the quotient of the information bits and the code length, i.e., $R = k/n$. For tree codes with memory, the actual output block, which is n bits long, depends not only on the present k information bits, but also on the previous m symbols, as shown in Fig. 7.8. Characteristics of convolution codes are the memory length $\nu = m \times k$, as well the distance profile, the free distance d_f, and the minimum distance d_m (see, for instance, [163]). Block codes can most often be constructed with algebraic methods using Galois fields, but tree codes are often only found in computer simulations.

Our discussion will be limited to *linear* codes, i.e., codes where the sum of two codewords is again a codeword, because this simplifies the decoder implementation. For linear codes, the Hamming distance can always be computed as the difference between a codeword and the zero word, which simplifies comparisons of the performance of the code. Linear tree codes are often called *convolutional codes*, because the codes can be built using an FIR-like structure. Convolutional codes may be *catastrophic* or *noncatastrophic*. In the case of a catastrophic code, a single error will be propagated forever. It can be

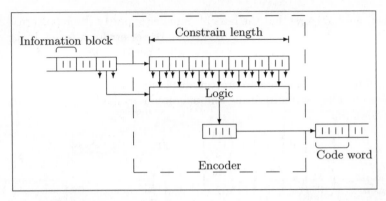

Fig. 7.8. Parameters of the convolutional encoders.

shown that systematic convolutional codes are always noncatastrophic. It is also common to distinguish between *random error* correction codes and *burst error* correction codes. In burst error correction, there may be a long burst of errors (or erasures). In random error correction code, the capability to correct errors is not limited to consecutive bits – the error may have a random position in the received codeword.

Coding bounds. With *coding bounds* we can compare different coding schemes. The bounds show the maximum error correction capability of the code. A decoder can never be better than the upper bound of the code, and sometimes to reduce the complexity of the decoder it is necessary to decode less than the theoretical bound.

A simple but still good, rough estimation is the Singleton bound or the Hamming bound. The *Singleton bound* states that the minimum Hamming distance d^* is upper bounded by the number of parity bits $(n - k)$. It is also known [163, p. 256] that the number of correctable errors t and the number of erasures e for a code is upper bounded by the Hamming distance. This gives the following bounds:

$$e + 2t + 1 \leq d^* \leq n - k + 1. \qquad (7.36)$$

A code with $d^* = n - k + 1$ is called *maximum distance separable*, but besides the repetition code and the parity check code, there are no binary maximum distance separable codes [163, p. 431]. Following the example in the last section from Table 7.5, with 11 parity bits the upper bound can be used to correct up to five errors.

For a t-error-correcting binary code, the following *Hamming bound* provides a good estimation:

$$2^{n-k} \geq \sum_{m=0}^{t} \binom{n}{m}. \qquad (7.37)$$

Equation (7.37) says that the possible number of parity check patterns (2^{n-k}) must be greater than or equal to the number of error patterns. If the equal sign is valid in (7.37), such codes are called *perfect codes*. A perfect code is, for instance, the Hamming code discussed in the last section. If it is desired, for instance, to find a code to protect all 44 bits transmitted in one minute for radio-controlled watches, using the maximum-available 13 parity bits, then it follows that

$$2^{13} > \binom{44}{0} + \binom{44}{1} + \binom{44}{2} \qquad \text{but} \tag{7.38}$$

$$2^{13} < \binom{44}{0} + \binom{44}{1} + \binom{44}{2} + \binom{44}{3}, \tag{7.39}$$

i.e., it should be possible to find a code with the capability to correct two random errors but none with three errors. In the following sections we will review such block encoders and decoders, and then discuss convolutional encoders and decoders.

7.2.2 Block Codes

The linear cyclic binary BCH codes (from Bose, Chaudhuri, and Hocquen-ghem) and the subclass of Reed–Solomon codes, consist of a large class of block codes. BCH codes have various known efficient decoders in the time and frequency domains. In the following, we will illustrate the shortening of a (63,50,6) to a (57,44,6) BCH code. The algorithm is discussed in detail by Blahut [163, pp. 162–6].

The code is based on a transformation of $GF(2^6)$ to $GF(2)$. To describe $GF(2^6)$, a primitive polynomial of degree 6 is needed, such as $P(x) = x^6 + x + 1$. To compute the generator polynomial, the least common multiple of the first $d-1 = 5$ minimal polynomials in $GF(2^6)$ must be computed. If α denotes a primitive element in $GF(2^6)$, it follows then that $\alpha^0 = 1$ and $m_{1(x)} = x - 1$. The minimum polynomials of α, α^2 and α^4 are identical $m_{\alpha(x)} = x^6 + x + 1$, and the minimum polynomial to α^3 is $m_{\alpha^3(x)} = x^6 + x^4 + x^2 + x + 1$. It is now possible to build the generator polynomial, $g(x)$:

$$g(x) = m_{1(x)} \times m_{\alpha(x)} \times m_{\alpha^3(x)} \tag{7.40}$$

$$= x^{13} + x^{12} + x^{11} + x^{10} + x^9 + x^8 + x^6 + x^3 + x + 1. \tag{7.41}$$

Using this generator polynomial (to compute the parity bits), it is now a straight forward procedure to build the encoder and decoder.

Encoder. Since a systematic code is desired, the first codeword bits are identical with the information bits. The parity bits $p(x)$ are computed by modulo reduction of the information bits $i(x)$ shifted in order to get a systematic code according to:

$$p(x) = i(x) \times x^{n-k} \bmod g(x). \tag{7.42}$$

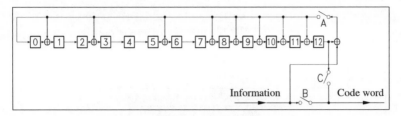

Fig. 7.9. Encoder for (57,44,6) BCH code.

Such a modulo reduction can be achieved with a recursive shift register as shown in Fig. 7.9. The circuit works as follows: In the beginning, switches A and B are closed and C is open. Next, the information bits are applied (MSB first) and directly transferred to the codeword. At the same time, the recursive shift register computes the parity bits. After the information bits are all processed, switches A and B are opened and C is closed. The parity bits are now shifted into the codeword.

Decoder. The decoder is usually more complex than the encoder. A Meggitt decoder can be used for decoding in the time domain, and frequency decoding is also possible, but it needs a detailed understanding of the algebraic properties of BCH codes ([163, pp. 166−200], [203, pp. 81−107], [204, pp. 65−73]). Such frequency decoders for FPGAs are already available as intellectual property (IP) blocks, sometimes also called "virtual components," VC (see [19, 20, 209]).

The Meggitt decoder (shown in Fig. 7.10) is very efficient for codes with only a few errors to be corrected, since the decoder uses the cyclic properties of BCH codes. Only errors in the highest bit position are corrected and then a cyclic shift is computed, so that eventually all corrupted bits pass the MSB position and are corrected.

In order to use a shortened code and to regain the cyclic properties of the codes, a forward incoupling of the received data $a(x)$ must be computed. This condition can be gained for code shortened by b bits using the condition

$$s(x) = a(x)i(x)\bmod g(x) = x^{n-k+b}i(x)\bmod g(x). \qquad (7.43)$$

For the shortened (57,44,6) BCH code this becomes

$$\begin{aligned}
a(x) &= x^{63-50+6}\bmod g(x) = x^{19}\bmod g(x)\\
&= x^{19}\bmod (x^{13}+x^{12}+x^{11}+x^{10}+x^9+x^8+x^6+x^3+x+1)\\
&= x^{10}+x^7+x^6+x^5+x^3+x+1.
\end{aligned}$$

The developed code has the ability to correct two errors. If only the error in the MSB need be corrected, a total of $1 + \binom{56}{1} = 1 + 56 = 57$ different error patterns must be stored, as shown in Table 7.6. The 57 syndrome values can be computed through a simulation and are listed in [202, B.3].

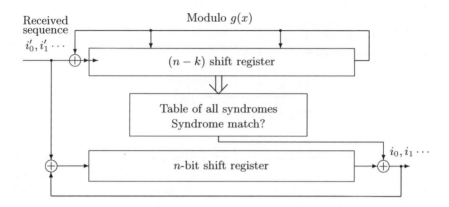

Fig. 7.10. Basic blocks of the Meggitt decoder.

Now all the building blocks are available for constructing the Meggitt decoder for the (57,44,6) BCH code. The decoder is shown in Fig. 7.11.

The Meggitt decoder has two stages. In the initialization phase, the syndrome is computed by processing the received bits modulo the generator polynomial $g(x)$. This takes 57 cycles. In the second phase, the actual error correction takes place. The content of the syndrome register is compared with the values of the syndrome table. If an entry is found, the table delivers a one, otherwise it delivers a zero. This hit bit is then XOR'd with the received bits in the shift register. In this way, the error is removed from the shift register. The hit bit is also wired to the syndrome register, to remove the error pattern from the syndrome register. Once again the syndrome and the shift register are clocked, and the next correction can be done. At the end, the shift register should include the corrected word, while the syndrome register should contain the all-zero word. If the syndrome is not zero, then more than two errors have occurred, and these can not be corrected with this BCH code.

Table 7.6. Table of possible error patterns.

No.	Error pattern						
1	0	0	0	$\cdots$	0	0	1
2	0	0	0	$\cdots$	0	1	1
3	0	0	0	$\cdots$	1	0	1
$\vdots$				$\ldots\ldots\ldots\ldots\ldots\ldots\ldots$			
56	0	1	0	$\cdots$	0	0	1
57	1	0	0	$\cdots$	0	0	1

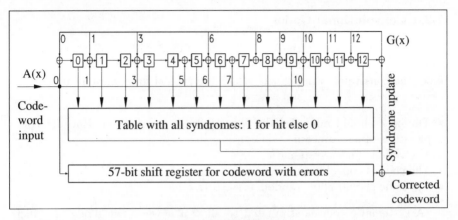

Fig. 7.11. Meggitt decoder for (57,44,6) BCH code.

Our only concern for an FPGA implementation of the Meggitt decoder is the large number (13) of inputs for the syndrome table, because the LUTs of FPGAs typically have 4 to 8 inputs. It is possible to use an external EPROM or (for Altera Flex 10K) four 2-kbit EABs to implement a table of size $2^{13} \times 1$. The syndrome is wired to the address lines, which deliver a hit (one) for the 57 syndromes, and otherwise a zero. It is also possible to use the logic synthesis tool to compute the table with internal logic blocks on the FPGA. The Xilinx XNFOPT (used in [202]) needs 132 LUTs, each with $2^4 \times 2$ bits. If modern binary decision diagrams (BBDs) synthesizer type [210, 211, 212] are used, this number can (at the cost of additional delays) be reduced to 58 LUTs with a size of $2^4 \times 2$ bits [213]. Table 7.7 shows the estimated effort, using Flex 10K, for the Meggitt decoder using the different kinds of syndrome tables.

Table 7.7. Estimated effort for Altera FLEX devices, for the three versions of the Meggitt decoder based on XC3K implementations [4]. (EABs are used as $2^{11} \times 1$ ROMs.)

	Syndrome table		
Function group	Using EABs	Only LEs	BDD [213]
Interface	36 LEs	36 LEs	36 LEs
Syndrome table	2 LEs, 4 EABs	264 LEs	116 LEs
64-bit FIFO	64 LEs	64 LEs	64 LEs
Meggitt decoder	12 LEs	12 LEs	12 LEs
State machine	21 LEs	21 LEs	21 LEs
Total	135 LEs, 4 EABs	397 LEs	249 LEs

7.2.3 Convolutional Codes

We also want to explore the kind of convolutional error-correcting decoders
that are suitable for an FPGA realization. To simplify the discussion, the
following constraints are defined, which are typical for communications systems:

- The code should minimize the complexity of the decoder. Encoder com-
  plexity is of less concern.
- The code is linear systematic.
- The code is convolutional.
- The code should allow random error correction.

A systematic code is stipulated to allow a power-down mode, in which
only incoming bits are received without error correction [201]. A random-
error-correction code is stipulated if the channel is slow fading.

Figure 7.12 shows a diagram of the possible tree codes, while Fig. 7.7
(p. 422) shows possible decoders. Fano and stack decoders are not very suitable for an FPGA implementation because of the complexity of organizing a
stack [200]. A conventional μP/μC realization is much more suitable here. In
the following sections, maximum-likelihood sequence decoders and algebraic
algorithms are compared regarding hardware complexity, measured in CLBs
usage for the Xilinx XC3K FPGA, and achievable error correction.

Viterbi maximum likelihood sequence decoder. The Viterbi decoder
deals with an erroneous sequence by determining the corresponding sender
sequence with the minimum Hamming distance. Put differently, the algorithm
finds the optimal path through the trellis diagram, and is therefore an *optimal*
memoryless noisy-sequence estimator (MLSE).

The advantage of the Viterbi decoder is its constant decoding time and
MLSE optimality. The disadvantage lies in its high memory requirements and
resulting limitation to codes with very short constraint length. Figures 7.13
and 7.14 show an $R = k/n = 1/2$ encoder and the attendant trellis diagram.
The constraint length $\nu = m \times k$ is 2, so the trellis has 2^ν nodes. Each node
has $2^k = 2$ outgoing and at most $2^k = 2$ incoming edges. For a binary trellis
($k = 1$) like this, it is convenient to show a zero as an upward edge and a one
as a downward edge.

For MLSE decoding it is sufficient to store only the 2^ν paths (and their
metrics) passing through the nodes at a given level, because the MLSE path
must pass through one of these nodes. Incoming paths with a smaller metric
than the "survivor" with the highest metric need not be stored, because these
paths will never be part of the MLSE path. Nevertheless, the maximum
metric at any given time may not be part of the MLSE path if it is part
of a short erroneous sequence. Voting down such a local error is analogous
to demodulating a digital FM signal with memory [214]. Simulation results
in [163, p. 381] and [203, pp. 120−3] show that it is sufficient to construct a

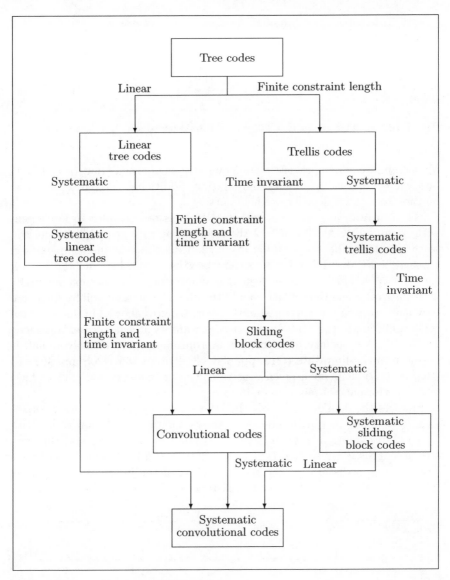

Fig. 7.12. Survey of tree codes [163].

path memory of four to five times the constraint length. Infinite path memory yields no significant improvement.

The Viterbi decoder hardware consists of three main parts: path memory with output decoder (see Fig. 7.15), survivor computation, and maximum detection (see Fig. 7.16). The path memory is $4\nu 2^{\nu}$ bits, consuming $2\nu 2^{\nu}$ CLBs. The output decoder uses $(1 + 2 + \ldots + 2^{\nu-1})$ 2-to-1 multiplexers.

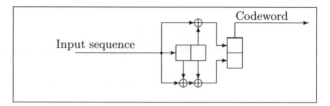

Fig. 7.13. Encoder for an $R = 1/2$ convolutional decoder.

Metric update adders, registers and comparisons are each $(\lceil \log_2(\nu * n)\rceil + 1)$ bits wide. For the maximum computation, additional comparisons, 2-to-1 multiplexers and a decoder are necessary.

The hardware for decoders with $k > 1$ seems too complex to implement with today's FPGAs. For $n > 2$ the information rate $R = 1/n$ is too low, so the most suitable code rate is $R = 1/2$. Table 7.8 lists the complexity in CLBs in a XC3K FPGA for constraint lengths $\nu = 2, 3, 4$, and the general case, for $R = 1/2$. It can be seen that complexity increases exponentially with constraint length ν, which should thus be as short as possible. Although very few errors can be corrected in the short window allowed by such a small constraint length, the MLSE algorithm guarantees acceptable performance.

Next it is necessary to choose an appropriate generating polynomial. It is shown in the literature ([215, pp. 306–8], [216, p. 465], [205, pp. 402–7], [163, p. 367]) that, for a given constraint length, *nonsystematic* codes have better performance than systematic codes, but using a nonsystematic code contradicts the demand for using the information bits without error correction. *Quick look in* (QLI) codes are nonsystematic convolution codes with $R = 1/2$, providing *free distance* values as good as any known code for constraint lengths $\nu = 2$ to 4 [217]. The advantage of QLI codes is that only one XOR gate is necessary for the reconstruction of the information sequence. QLIs with $\nu = 2, 3$, and 4 have a free distance of $d_f = 5, 6$, and 7, respectively

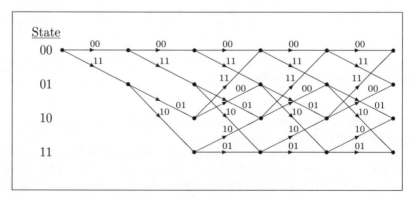

Fig. 7.14. Trellis for $R = 1/2$ convolutional decoder.

Path memory and output decoder

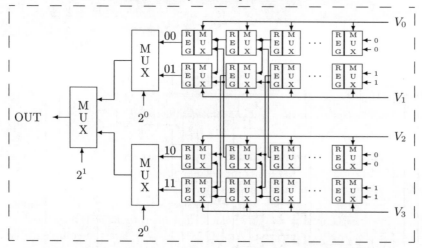

Fig. 7.15. Viterbi decoder with constraint length 4ν and $2^{\nu=2}$ nodes: path memory and output decoder.

[216, p. 465]. This seems to be a good compromise for low power consump-

Table 7.8. Hardware complexity in CLBs for an $R = 1/2$ Viterbi decoder for $\nu = 2, 3, 4$, and the general case.

Function	$\nu = 2$	$\nu = 3$	$\nu = 4$	$\nu \in \mathbb{N}$
Path memory	16	48	128	$4 \times \nu \times 2^{\nu-1}$
Output decoder	1,5	3,5	6,5	$1 + 2 + \ldots + 2^{\nu-2}$
Metric ΔM	4	4	4	4
Metric clear	1	2	4	$\lceil (2 + 4 + \ldots + 2^{\nu-1})/4 \rceil$
Metric adder	24	64	128	$(\lceil \log_2(n\nu) \rceil + 1) \times 2^{\nu+1}$
Survivor-MUX	6	24	48	$(\lceil \log_2(n\nu) \rceil + 1) \times 2^{\nu-1}$
Metric compare	6	24	48	$(\lceil \log_2(n\nu) \rceil + 1) \times 2^{\nu-1}$
Maximum compare	4,5	14	30	$(\lceil \log_2(n\nu) \rceil + 1) \times \frac{1}{2} \times (1 + 2 + \ldots 2^{\nu-1})$
MUX	3	12	28	$(2 + \ldots + 2^{\nu-1}) \times \frac{1}{2} \times (\lceil \log_2(n\nu) \rceil + 1)$
Decoder	1	2	4	$\lceil (2 + \ldots + 2^{\nu-1})/4 \rceil$
State machine	4	4	4	
Sum:	67	197.5	428.5	

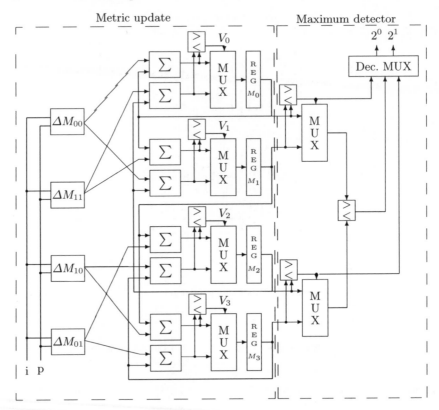

Fig. 7.16. Viterbi decoder with constraint length 4ν and $2^{\nu=2}$ nodes: metric calculation.

tion. The upper part of Table 7.9 shows the generating polynomials in octal notation.

Error-correction performance of the QLI decoder. To compute the error-correction performance of the QLI decoder, it is convenient to use the "union bound" method. Because QLI codes are linear, error sequences can be computed as a difference from the zero sequence. An MLSE decoder will make an incorrect decision if a sequence that starts at the null state, and differs from the null-word at j separate time steps, contains at least $j/2$ ones. The probability of this occurrence is

$$P_j = \begin{cases} \sum_{i=(j+1)/2}^{j} \binom{j}{i} p^i q^{j-i} & \text{for odd j} \\ \frac{1}{2}\binom{j}{j/2} p^{j/2} q^{j/2} + \sum_{i=j/2+1}^{j} \binom{j}{i} p^i q^{j-i} & \text{for even j.} \end{cases} \tag{7.44}$$

Now the only thing necessary for a bit-error probability formula is to compute the number w_j of paths with weight j for the code, which is an easily programmable task [200, C.4]. Because P_j decreases exponentially with

Table 7.9. Union-bound weights for a Viterbi decoder with $\nu = 2$ to 4 using QLI codes.

Code	O1 = 7 O2 = 5	O1 = 74 O2 = 54	O1 = 66 O2 = 46
Constraint length	$\nu = 2$	$\nu = 3$	$\nu = 4$
Distance		Weight w_j	
0-4	0	0	0
5	1	0	0
6	4	2	0
7	12	7	4
8	32	18	12
9	80	49	26
10	192	130	74
11	448	333	205
12	1024	836	530
13	2304	2069	1369
14	5120	5060	3476
15	11 264	12 255	8470
16	24 576	29 444	19 772
17	53 079	64 183	43 062
18	109 396	126 260	83 346
19	103 665	223 980	147 474
20	262 144	351 956	244 458

increasing j, only the first few w_j must be computed. Table 7.9 shows the w_j for $j = 0$ to 20. The total error probability can now be computed with:

$$P_b < \frac{1}{k} \sum_{j=0}^{\infty} w_j P_j. \tag{7.45}$$

Syndrome algebraic decoder. The syndrome decoder (Fig. 7.17) and encoder (Fig. 7.18), like standard block decoders, computes a number of parity bits from the data sequence. The decoder's newly computed parity bits are XOR'd with the received parity bits to create the "syndrome" word, which will be nonzero if an error occurs in transmission. The error position and value are determined from the syndrome value. In contrast to block codes, where only one generator polynomial is used, convolutional codes at data rate $R = k/n$ have $k + 1$ generating polynomials. The complete generator may be written in a compact $n \times k$ generator matrix. For the encoder of Fig. 7.18 the matrix is

$$\boldsymbol{G}(x) = \begin{bmatrix} 1 & x^{21} + x^{20} + x^{19} + x^{17} + x^{16} + x^{13} + x^{11} + 1 \end{bmatrix}. \tag{7.46}$$

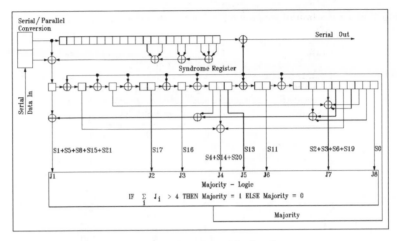

Fig. 7.17. Trial and error majority decoder with $J = 8$.

For a systematic code the matrix has the form $\boldsymbol{G}(x) = [\boldsymbol{I} \vdots \boldsymbol{P}(x)]$. The parity check matrix $\boldsymbol{H}(x) = [-\boldsymbol{P}(x)^T \vdots \boldsymbol{I}]$ is easily computed, given that $\boldsymbol{G} \times \boldsymbol{H}^T = \boldsymbol{0}$. The desired syndrome vector is thus $\boldsymbol{S} = \boldsymbol{v} \times \boldsymbol{H}^T$, where $\boldsymbol{v}$ is the received bit sequence.

The syndrome decoder now looks up the calculated syndrome in a table to find the correct sequence. To keep the table small, only sequences with an error at the first bit position are included. If the decoder needs to correct errors of more than one bit, we cannot clear the syndrome after the correction. Instead, the syndrome value must be subtracted from a syndrome register (see the "Majority" signal in Fig. 7.17).

A 22-bit table would be necessary for the standard convolutional decoder, but it is unfortunately difficult to implement a good FPGA look-up table with more than 4 to 11 bit addresses [201]. Majority codes, a special class of syndrome-decodable codes, offer an advantage here. This type of canonical self-orthogonal code (CSOC) has exclusively ones in the first row of the $\{A_k\}$ parity check matrix (where the J columns are used as an orthogonal set to compute the syndrome) [205, p. 284]. Thus, every error in the first-bit position

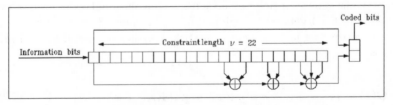

Fig. 7.18. Systematic $(44, 22)$ encoder with rate $R = 1/2$ and constraint length $\nu = 22$.

Table 7.10. Some majority-decodable "trial and error" codes [205, p. 406].

J	t_{MD}	ν	Generating polynomial	Orthogonal equation
2	1	2	$1+x$	s_0, s_1
4	2	6	$1+x^3+x^4+x^5$	s_0, s_3, s_4, s_1+s_5
6	3	12	$1+x^6+x^7+x^9+x^{10}+x^{11}$	$s_0, s_6, s_7, s_9, s_1+s_3+s_{10},$ $s_4+s_8+s_{11}$
8	4	22	$1+x^{11}+x^{13}+x^{16}+x^{17}+x^{19}$ $+x^{20}+x^{21}$	$s_0, s_{11}, s_{13}, s_{16}, s_{17}, s_2+s_3+s_6+$ $s_{19}, s_4+s_{14}+s_{20}, s_1+s_5+s_8+$ $s_{15}+s_{21}$
10	5	36	$1+x^{18}+x^{19}+x^{27}+x^{28}+x^{29}$ $+x^{30}+x^{32}+x^{33}+x^{35}$	$s_0, s_{18}, s_{19}, s_{27}, s_1+s_9+s_{28}, s_{10}+$ $s_{20}+s_{29}, s_{11}+s_{30}+s_{31},$ $s_{13}+s_{21}+s_{23}+s_{32}, s_{14}+$ $s_{33}+s_{34}, s_2+s_3+s_{16}+s_{24}+$ $s_{26}+s_{35}$

will cause at least $\lceil J/2 \rceil$ ones in the syndrome register. The decoding rule is therefore

$$e_0^i = \begin{cases} 1 & \text{for} \quad \sum_{k=1}^{J} A_k > \lceil J/2 \rceil \\ 0 & \text{otherwise} \end{cases} . \tag{7.47}$$

Thus the name "majority code": instead of the expensive syndrome table only a majority vote is needed. Massey [205, p. 289] has designed a class of majority codes, called trial and error codes, which, instead of evaluating the syndrome vector directly, manipulate a combination of syndrome bits to get a vector orthogonal to e_0^i. This small additional hardware cost results in slightly better error correction performance than the conventional CSOC codes. Table 7.10 lists some trial and error codes with data rate $R = 1/2$. Figure 7.17 shows a trial and error decoder with $J = 8$. Table 7.11 shows the complexity in CLBs of decoders with $J = 4$ to 10.

Error-correction capability of the trial and error decoder. To calculate the error-correction performance of trial and error codes, we must first

Table 7.11. Complexity in CLBs of a majority decoder with $J = 4$ to 10.

Function	$J=4$	$J=6$	$J=8$	$J=10$
Register	6	12	22	36
XOR-Gate	2	4	7	11
Majority-circuit	1	5	7	15
Sum	9	22	36	62

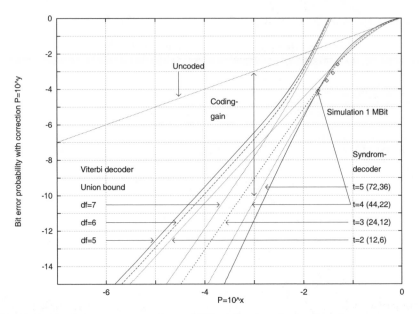

Fig. 7.19. Performance comparison of Viterbi and majority decoders.

note that in a window twice the constraint length, the codes allow up to $\lfloor J/2 \rfloor$-bit errors [163, p. 440]:

$$P(J) = \sum_{k=0}^{\lfloor J/2 \rfloor} \binom{2\nu}{k} p^k (1-p)^{2\nu-k}. \tag{7.48}$$

A computer simulation of 10^6 bits, in Fig. 7.19, reveals good agreement with this equation. The equivalent single-error probability P_B of an (n,k) code can be computed with

$$P(J) = P(0) = (1 - P_B)^k \tag{7.49}$$

$$\rightarrow P_B = 1 - e^{\ln(P(J))/k}. \tag{7.50}$$

Final comparison. Figure 7.19 shows the error-correction performance of Viterbi and majority decoders. For a comparable hardware cost (Viterbi, $\nu = 2$, $d_f = 5$, 67 CLBs and trial and error, $t = 5$, 62 CLBs) the better performance of the majority decoder, due to the greater constraint length permitted, is immediately apparent. The optimal MLSE property of the Viterbi algorithm cannot compensate for its short constraint length.

7.2.4 Cryptography Algorithms for FPGAs

Many communication systems use data-stream ciphers to protect relevant information, as shown in Fig. 7.20. The key sequence K is more or less a

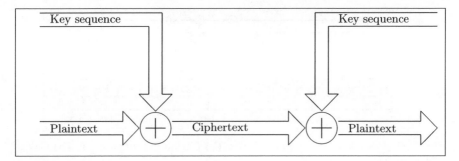

Fig. 7.20. The principle of a synchronous data-stream cipher.

"pseudorandom sequence" (known to the sender and the receiver), and with the modulo 2 property of the XOR function, the plaintext P can be reconstructed at the receiver side, because

$$P \oplus K \oplus K = P \oplus 0 = P. \tag{7.51}$$

In the following, we compare an algorithm based on a linear-feedback shift register (LFSR) and a "data encryption standard" (DES) cryptographic algorithm. Neither algorithm requires large tables and both are suitable for an FPGA implementation.

Linear Feedback Shift Registers Algorithm

LFSRs with maximal sequence length are a good approach for an ideal security key, because they have good statistical properties (see, for instance, [218, 219]). In other words, it is difficult to analyze the sequence in a cryptographic attack, an analysis called *cryptoanalysis*. Because bitwise designs are possible with FPGAs, such LFSRs are more efficiently realized with FPGAs than PDSPs. Two possible realizations of a LFSR of length 8 are shown in Fig. 7.21.

For the XOR LFSR there is always the possibility of the all-zero word, which should never be reached. If the cycle starts with any nonzero word, the cycle length is always $2^l - 1$. Sometimes, if the FPGA wakes up with an all-zero state, it is more convenient to use a "mirrored" or inverted LFSR circuit. If the all-zero word is a valid pattern and produces exactly the inverse sequence, it is necessary to substitute the XOR with a "not XOR" or XNOR gate. Such LFSRs can easily be designed using a PROCESS statement in VHDL, as the following example shows.

Example 7.12: Length 6 LFSR

The following VHDL code[2] implements a LFSR of length 6.

---

[2] The equivalent Verilog code lfsr.v for this example can be found in Appendix A on page 716. Synthesis results are shown in Appendix B on page 731.

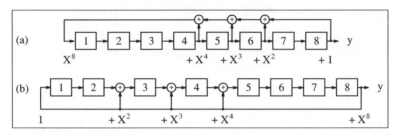

Fig. 7.21. Possible realizations of LFSRs. **(a)** Fibonacci configuration. **(b)** Galois configuration.

```
LIBRARY ieee;
USE ieee.std_logic_1164.ALL;
USE ieee.std_logic_arith.ALL;

ENTITY lfsr IS                          ------> Interface
  PORT ( clk : IN  STD_LOGIC;
         y   : OUT STD_LOGIC_VECTOR(6 DOWNTO 1));
END lfsr;

ARCHITECTURE fpga OF lfsr IS

  SIGNAL ff : STD_LOGIC_VECTOR(6 DOWNTO 1)
                                         := (OTHERS => '0');
BEGIN

    PROCESS            -- Implement length 6 LFSR with xnor
    BEGIN
      WAIT UNTIL clk = '1';
      ff(1) <= NOT (ff(5) XOR ff(6));
      FOR I IN 6 DOWNTO 2 LOOP    -- Tapped delay line:
        ff(I) <= ff(I-1);          -- shift one
      END LOOP;
    END PROCESS ;

    PROCESS (ff)
    BEGIN                -- Connect to I/O cell
      FOR k IN 1 TO 6 LOOP
        y(k) <= ff(k);
      END LOOP;
    END PROCESS;

  END fpga;
```

From the simulation of the design in Fig. 7.22, it can be concluded that the LFSR goes through all possible bit patterns, which results in the maximum sequence length of $2^6 - 1 = 63 \approx 630\,\text{ns}/10\,\text{ns}$. The design uses 6 LEs, no embedded multiplier, and has a 420.17 MHz **Registered Performance.** $\boxed{\text{7.12}}$

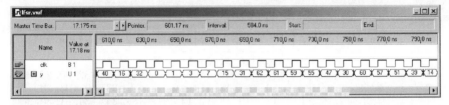

Fig. 7.22. LFSR simulation.

Note that a complete cycle of an LFSR sequence fulfills the three criteria for optimal length $2^l - 1$ pseudorandom sequences defined by Golomb [220, p. 188]:

1) The number of 1s and 0s in a cycle differs by no more than one.
2) Runs of length k (e.g., $111\cdots$ sequence, $000\cdots$ sequence) have a total fractional part of all runs of $1/2^k$.
3) The autocorrelation function $C(\tau)$ is constant for $\tau \in [1, n-1]$.

LFSRs are usually constructed from primitive polynomials in GF(2) using the circuits shown in Fig. 7.21. Stahnke [206] has compiled a list of such primitive polynomials up to order 168. This paper is available online at http://www.jstor.org. With today's available algebraic software packages like MAPLE, MUPAD, or MAGMA such a list can easily be extended. The following is a code example for MAPLE to compute the primitive polynomials of type $x^l + x^a + 1$ with the smallest a.

```
with(numtheory):
for l from 2 by 1 to 45 do
  for a from 1 by 1 to l-1 do
    if (Primitive(x^l+x^a+1) mod 2) then
    print(l,a);
    break;
  fi;
  od;
od;
```

Table 7.12 shows the necessary XOR list of the first 45 maximum length LFSRs according to Fig. 7.21a. For instance, the entry for polynomial fourteen (14, 13, 11, 9) means the primitive polynomial is

$$p_{14}(x) = x^{14} + x^{14-13} + x^{14-11} + x^{14-9} + 1$$
$$= x^{14} + x^5 + x^3 + x + 1.$$

For $l > 2$ these primitive polynomials always have "twins," which are also primitive polynomials [221]. These are the "time" reversed versions $x^l + x^{l-a} + 1$.

Stahnke [19, XAPP52] has computed primitive polynomials of type $x^l + x^a + 1$. There are no primitive polynomials with four elements, i.e. $(x^l +$

Table 7.12. A list of the first 45 LFSR.

l	Exponents	l	Exponents	l	Exponents
1	1	16	16, 14, 13, 11	31	31, 28
2	2, 1	17	17, 14	32	32, 30, 29, 23
3	3, 2	18	18, 11	33	33, 20
4	4, 3	19	19, 18, 17, 14	34	34, 31, 30, 26
5	5, 3	20	20, 17	35	35, 33
6	6, 5	21	21, 19	36	36, 25
7	7, 6	22	22, 21	37	37, 36, 33, 31
8	8, 6, 5, 4	23	23, 18	38	37, 36, 33, 31
9	9, 5	24	24, 23, 21, 20	39	39, 35
10	10, 7	25	25, 22	40	40, 37, 36, 35
11	11, 9	26	26, 25, 24, 30	41	41, 38
12	12, 11, 8, 6	27	27, 26, 25, 22	42	42, 39, 38, 35
13	13, 12, 10, 9	28	28, 25	43	43, 41, 40, 36
14	14, 13, 11, 9	29	29, 27	44	44, 42, 41, 37
15	15, 14	30	30, 29, 26, 24	45	45, 44, 43, 41

$x^b + x^a + 1$) for $l < 45$. But it is possible to find polynomials of the type $x^l + x^{a+b} + x^b + x^a + 1$, which Stahnke used for those l where a polynomial of the type $x^l + x^a + 1$ ($l = 8, 12, 13$, etc.) does not exist.

The LFSRs with four elements in Table 7.12 were computed to have the maximum sum (i.e., $a + b$) for the tap exponents. We will see later that, for multistep LFSR implementations, this usually gives the minimum complexity.

If n random bits are used at once, it is possible to clock our LFSR n times. In general, it is *not* a good idea to use just the lowest n bits of our LFSR, since this will lead to weak random properties, i.e., low cryptographic security. But it is possible to compute the equation for n-bit shifts, so that only one clock cycle is needed to generate n new random bits. The necessary equation can be computed more easily if a "state-space" description of the LFSR is used, as the following example shows.

Example 7.13: Three Steps-at-Once LFSR

Let us assume a primitive polynomial of length 6, e.g., $p = x^6 + x + 1$, is used to compute random sequences. The task now is to compute three "new" bits in one clock cycle. To obtain the required equation, the state-space description of our LFSR must first be computed, i.e., $\boldsymbol{x}(t+1) = \boldsymbol{A}\boldsymbol{x}(t)$

$$
\begin{bmatrix} x_6(t+1) \\ x_5(t+1) \\ x_4(t+1) \\ x_3(t+1) \\ x_2(t+1) \\ x_1(t+1) \end{bmatrix} = \begin{bmatrix} 0\,1\,0\,0\,0\,0 \\ 0\,0\,1\,0\,0\,0 \\ 0\,0\,0\,1\,0\,0 \\ 0\,0\,0\,0\,1\,0 \\ 0\,0\,0\,0\,0\,1 \\ 1\,1\,0\,0\,0\,0 \end{bmatrix} \begin{bmatrix} x_6(t) \\ x_5(t) \\ x_4(t) \\ x_3(t) \\ x_2(t) \\ x_1(t) \end{bmatrix}. \tag{7.52}
$$

With this state-space description, the actual values $\boldsymbol{x}(t)$ and the transition matrix $\boldsymbol{A}$ are used to compute the new values $\boldsymbol{x}(t+1)$. To compute the values for $\boldsymbol{x}(t+2)$, simply compute $\boldsymbol{x}(t+2) = \boldsymbol{A}\boldsymbol{x}(t+1) = \boldsymbol{A}^2\boldsymbol{x}(t)$. The next iteration gives $\boldsymbol{x}(t+3) = \boldsymbol{A}^3\boldsymbol{x}(t)$. The equations for an n-step-at-once LFSR can therefore be computed by evaluating $\boldsymbol{A}^n \bmod 2$. For $n = 3$ it follows that

$$\boldsymbol{A}^3 \bmod 2 = \begin{bmatrix} 0 & 0 & 0 & 1 & 0 & 0 \\ 0 & 0 & 0 & 0 & 1 & 0 \\ 0 & 0 & 0 & 0 & 0 & 1 \\ 1 & 1 & 0 & 0 & 0 & 0 \\ 0 & 1 & 1 & 0 & 1 & 0 \\ 0 & 0 & 1 & 1 & 0 & 1 \end{bmatrix}. \tag{7.53}$$

As expected, for the register x_6 to x_4 there is a shift of three positions, while the other three values x_1 to x_3 are computed using an EXOR operation. The following VHDL code[3] implements this three-step LFSR.

```
LIBRARY ieee;
USE ieee.std_logic_1164.ALL;
USE ieee.std_logic_arith.ALL;


ENTITY lfsr6s3 IS                     ------> Interface
  PORT ( clk : IN  STD_LOGIC;
         y   : OUT STD_LOGIC_VECTOR(6 DOWNTO 1));
END lfsr6s3;

ARCHITECTURE fpga OF lfsr6s3 IS

  SIGNAL ff : STD_LOGIC_VECTOR(6 DOWNTO 1) := (OTHERS => '0');

BEGIN

  PROCESS    -- Implement three step length-6 LFSR with xnor
  BEGIN
    WAIT UNTIL clk = '1';
    ff(6) <= ff(3);
    ff(5) <= ff(2);
    ff(4) <= ff(1);
    ff(3) <= NOT (ff(5) XOR ff(6));
    ff(2) <= NOT (ff(4) XOR ff(5));
    ff(1) <= NOT (ff(3) XOR ff(4));
  END PROCESS ;

  PROCESS (ff)
  BEGIN                   -- Connect to I/O cell
    FOR k IN 1 TO 6 LOOP
      y(k) <= ff(k);
    END LOOP;
  END PROCESS;

END fpga;
```

Figure 7.23 shows a simulation of the three-step LFSR design. Comparing the simulation of this LFSR in Fig. 7.23 with the simulation of the single-step LFSR in Fig. 7.22, it can be concluded that now every third sequence value occurs. The cycle length is reduced from $2^6 - 1$ to $(2^6 - 1)/3 = 21$. The

---

[3] The equivalent Verilog code lfsr6s3.v for this example can be found in Appendix A on page 717. Synthesis results are shown in Appendix B on page 731.

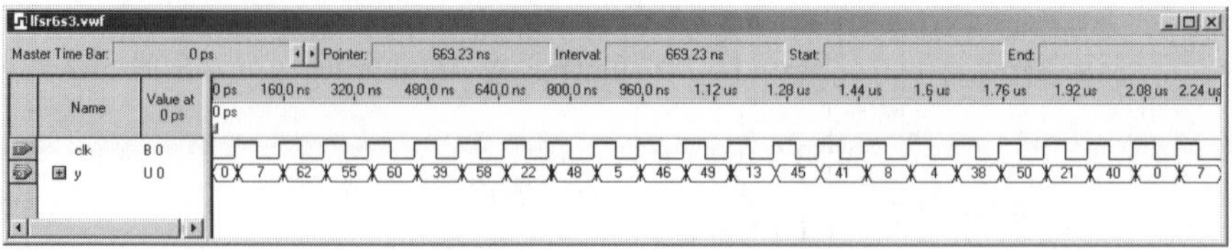

Fig. 7.23. Multistep LFSR simulation.

design uses 6 LEs, no embedded multiplier, and has a 420.17 MHz `Registered`
`Performance`.   7.13

To implement such a multistep LFSR, we want to select the primitive
polynomial that results in the lowest circuit effort, which can be computed
by counting the nonzero entries in the A^k mod 2 matrix, and/or the maximum fan-in for the register, which corresponds to the number of ones in
each row. For a few shifts, the fan-in for the circuit from Fig. 7.21a may be
advantageous. It can also be observed [4] that if the feedback signals are close
in the A matrix, some entries in the A^k matrix may become zero, due to
the modulo 2 operations. As mentioned earlier, the two and four-tap LFSR
data in Table 7.12 were therefore computed to yield the maximum sum of all
taps. For the same sum, the primitive polynomial that has the larger value
for the smallest tap was selected, e.g., (11, 12) is better than (10, 13). This
was chosen because tap l is mandatory for the maximum-length LFSR, and
the other values should be close to this tap.

If, for instance, Stanke's $s_{14,a}(x) = x^{14} + x^{12} + x^{11} + x + 1$ primitive
polynomial is used, this will result in 58 entries for an $n = 8$ multistep LFSR,
while if the LFSR from Table 7.12, $p_{14} = x^{14} + x^5 + x^3 + x^1 + 1$ (i.e., taps
14,13,11,9) is used, the A^8 mod 2 matrix has only 35 entries (Exercise 7.6,
p. 474). Fig. 7.24 shows the total number of ones for the LFSR for the two
polynomials with the two different implementations from Fig. 7.21, while
Fig. 7.25 shows the maximum fan-in (i.e., the maximum needed input bit
width for a LC) for this LFSR. It can be concluded from the two figures that
a careful choice of the polynomial and LFSR structure can provide substantial
savings. For the multistep LFSR synthesis, it can be seen from Fig. 7.25 that
the LFSR of Fig. 7.21b has fewer fan-ins (i.e., smaller LC input bit width), but
for longer multistep k, the effort seems similar for the primitive polynomials
from Table 7.12.

---

[4] It is obviously not applicable to select the LFSR with the smallest implementation effort, because there are $\phi(2^l - 1)/l$ primitive polynomials, where $\phi(x)$ is
the Euler function that computes the number of coprimes to x. For instance, a
16-bit register has $\phi(2^{16} - 1)/16 = 2048$ different primitive polynomials [221]!

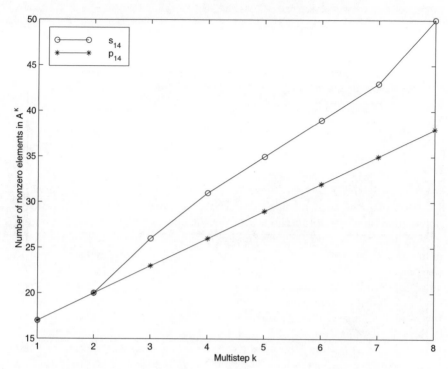

Fig. 7.24. Number of ones in $\boldsymbol{A}_{14}^k$.

Combining LFSR

An additional gain in performance in cryptographic security can be achieved if several LFSR registers are combined into one key generator. Several linear and nonlinear combinations exist [208], [207, pp. 150−173]. Meaningful for implementation effort and security are nonlinear combinations with thresholds. For a combination of three different LFSRs with length L_1, L_2, and L_3 the *linear complexity*, which is the equivalent length of *one* LFSR (which may be synthesized with the Berlekamp−Massey algorithm, for instance, [207, pp. 141−9]), provides

$$L_{\text{ges}} = L_1 \times L_2 + L_2 \times L_3 + L_1 \times L_3. \tag{7.54}$$

Figure 7.26 shows a realization for such a scheme.

Since the key in the selected paging format has 50 bits, a total length of $2 \times 50 = 100$ registers was chosen, and the three feedback polynomials are:

$$p_{33}(x) = x^{33} + x^6 + x^4 + x + 1 \tag{7.55}$$

$$p_{29}(x) = x^{29} + x^2 + 1 \tag{7.56}$$

$$p_{38}(x) = x^{38} + x^6 + x^5 + x + 1. \tag{7.57}$$

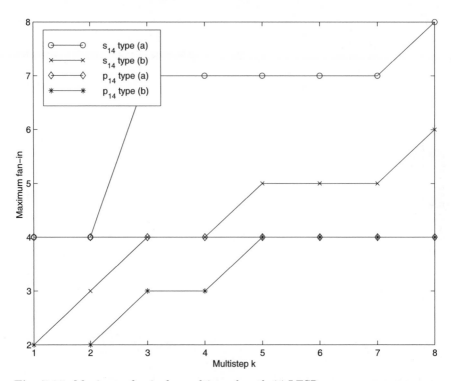

Fig. 7.25. Maximum fan-in for multistep length-14 LFSR.

All the polynomials are *primitive*, which guarantees that the length of all three shift-register sequences gives a maximum. For the linear complexity of the combination it follows that:

$$L_1 = 33; \qquad L_2 = 29; \qquad L_3 = 38$$
$$L_{\text{total}} = 33 \times 29 + 33 \times 38 + 29 \times 38 = 3313.$$

Table 7.13. Cost, measured in CLBs, of a 3K Xilinx FPGA.

Function group	CLBs
50-bit key register	25
100-bit shift register	50
Feedback	3
Threshold	0.5
XOR with message	0.5
Total	79

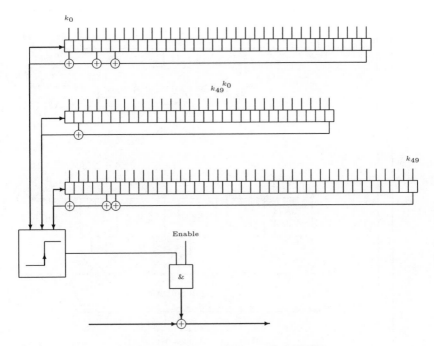

Fig. 7.26. Realization of the data-stream cipher with 3 LFSR.

After each coding the key is lost, and an additional 50 registers are needed to store the key. The 50-bit key is used twice. Table 7.13 shows the hardware resources required with Xilinx FPGAs of the 3K family.

DES-Based Algorithm. The data encryption standard (DES), outlined in Fig. 7.27, is typically used in a block cipher. By selecting the "output feedback mode" (OFB) it is also possible to use the modified DES in a data-stream cipher (see Fig. 7.28). The other modes (ECB, CBC, or CFB) of the DES are, in general, not applicable for communication systems, due to the "avalanche effect": A single-bit error in the transmission will alter approximately 50% of all bits in a block.

We will review the principles of the DES algorithm and then discuss suitable modifications for FPGA implementations.

The DES comprises a finite state machine translating plaintext blocks into ciphertext blocks. First the block to be substituted is loaded into the state register (32 bits). Next it is expanded (to 48 bits), combined with the key (also 48 bits) and substituted in eight $6 \rightarrow 4$ bit-width S-boxes. Finally, permutations of single bits are performed. This cycle may be (if desired, with a changing key) applied several times. In the DES, the key is usually

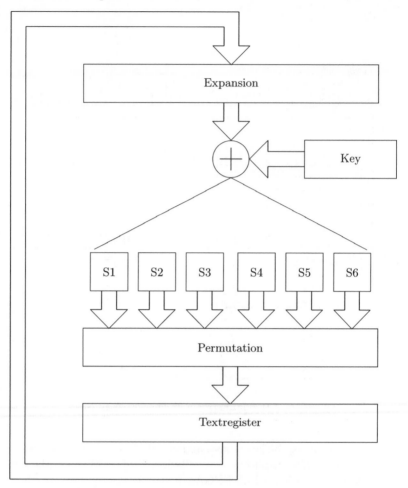

Fig. 7.27. State machine for a block encryption system (DES).

shifted one or two bits so that after 16 rounds the key is back in the original position. Because the DES can therefore be seen as an iterative application of the Feistel cipher (shown in Fig. 7.29), the S-boxes must not be invertible. To simplify an FPGA realization some modifications are useful, such as a reduction of the length of the state register to 25 bits. No expansion is used. Use the final permutations as listed in Table 7.14.

Because most FPGAs only have four to five input look-up tables (LUTs), S-boxes with five inputs have been designed, as displayed in Table 7.15.

Although the intention was to use the OFB mode only, the S-boxes in Table 7.16 were generated in such a manner that they can be inverted. The modified DES may therefore also be used as a normal block cipher (electronic code book).

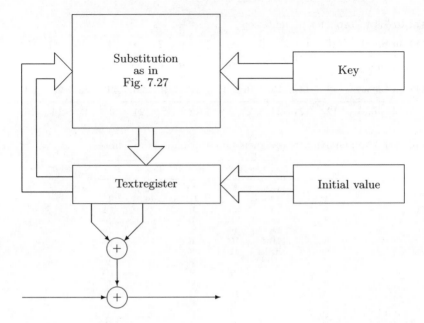

Fig. 7.28. Block cipher in the OFB-mode used as data-stream cipher.

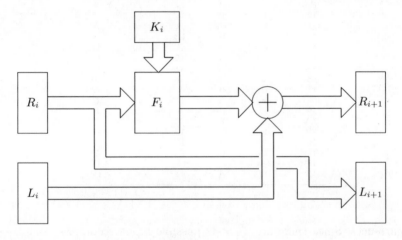

Fig. 7.29. Principle of the Feistel network.

A reasonable test for S-boxes is the dependency matrix. This matrix shows, for every input/output combination, the probability that an output bit changes if an input bit is changed. With the avalanche effect the ideal probability is $1/2$. Table 7.16 shows the dependency matrix for the new five S-boxes. Instead of the probability, the table shows the absolute number of

Table 7.14. Table for permutation.

From bit no.	0	1	2	3	4	5	6	7	8	9	10	11	12
To bit no.	20	4	5	10	15	21	0	6	11	16	22	1	7

From bit no.	13	14	15	16	17	18	19	20	21	22	23	24
To bit no.	12	17	23	2	8	13	18	24	3	9	14	19

Table 7.15. The five new designed substitution boxes (S-boxes).

Input	Box 1	Box 2	Box 3	Box 4	Box 5
0	1E	F	14	19	6
1	13	1	1D	14	E
2	14	13	16	D	1A
3	1	1F	B	4	3
4	1A	19	5	1C	B
5	1B	1C	E	1A	1E
6	E	12	8	1E	0
7	B	11	F	1	2
8	D	8	4	C	1D
9	10	7	C	F	C
A	3	1B	1E	1B	18
B	0	0	13	1D	17
C	4	1A	10	5	1
D	6	C	1	15	15
E	A	1D	18	E	1B
F	17	2	17	13	9
10	19	B	1C	17	19
11	16	1E	A	9	A
12	7	18	1B	3	4
13	1C	D	3	10	14
14	1D	5	19	A	13
15	5	14	D	16	11
16	2	15	0	12	10
17	1F	9	2	1F	12
18	F	3	15	B	5
19	11	10	6	2	F
1A	C	6	7	6	8
1B	18	17	12	18	16
1C	9	4	1F	11	1C
1D	15	16	1A	8	7
1E	8	E	9	7	D
1F	12	A	11	0	1F

occurrences. Since there are $2^5 = 32$ possible input vectors for each S-box, the ideal value is 16. A random generator was used to generate the S-boxes. The reason that some values differ much from the ideal 16 may lie in the desired inversion.

The hardware effort of the DES-based algorithm is summarized in Table 7.17.

Cryptographic performance comparison. We will next discuss the cryptographic performance analysis of the LFSR- and DES-based algorithms. Several security tests have been defined and the following comparison shows the

Table 7.16. Dependency matrix for the five substitution boxes (ideal value is 16).

Box 1					Box 2					Box 3				
20	12	20	20	20	20	16	20	12	20	20	12	16	16	16
12	20	12	16	16	20	20	20	16	16	16	20	16	16	16
12	16	16	12	8	12	20	20	16	8	16	16	20	12	12
16	16	20	12	16	16	24	12	16	12	16	8	12	16	20
20	16	20	12	12	16	20	16	20	20	20	12	12	20	12

Box 4					Box 5				
20	16	20	20	16	12	20	8	12	20
12	16	12	16	20	20	12	16	24	20
20	16	16	20	16	16	12	12	20	16
20	16	16	20	24	16	20	16	20	12
16	12	28	20	16	12	16	16	12	24

Table 7.17. Hardware effort of the modified DES based algorithm.

Function group	CLBs
25-bit key register	12.5
25-bit additions	12.5
25-bit state register	12.5
Five S-boxes 5→5	25
Permutation	0
25-bit initialization vector	12.5
Multiplex: Initialization vector/S-box	12.5
XOR with message	1
Total	87.5

two most interesting (the others do not show clear differences between the two schemes). For both tests, 100 random keys were generated.

1) Using different keys, the generated sequences were analyzed. In each random key, one bit was changed and the number of bit changes in the plaintext was recorded. On average, about 50% of the bits should be inverted (avalanche effect).

2) Similar to Test 1, but this time the number of changes in the output sequence were analyzed, depending on the changed position of the key. Again, 50% of the bits should change in sign.

For both tests, plaintext with 64-bit length were used (see again Fig. 7.20, p. 437). The plaintext is arbitrary. For Test 1, all variations over each individual key position were accumulated. For Test 2, all changes depending on the position in the output sequence were accumulated. This test was performed for 100 random keys. For Test 1, the ideal value is $64 \times 0.5 \times 100 = 3200$ and for Test 2 the optimal value is $50 \times 0.5 \times 100 = 2500$. Figures 7.30 and 7.31

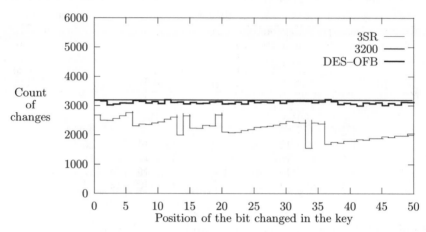

Fig. 7.30. Results of test 1.

display the results. They clearly show that the DES-OFB scheme is much more sensitive to changes in the key than is the scheme with three LFSRs. The conclusion from Test 2 is that the SR scheme needs about 32 steps until a change in the key will affect the output sequence. For the DES-OFB scheme, only the first four samples differ considerably from the ideal value of 2500.

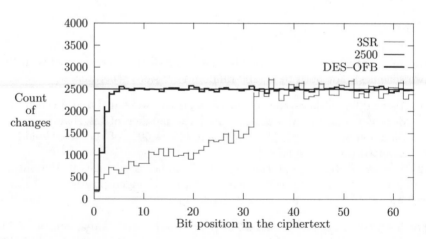

Fig. 7.31. Results of test 2.

Due to the superior test results, the DES-OFB scheme may be preferred over the LFSR scheme.

A final note on encryption security. In general, it is not easy to conclude that an encryption system is secure. Besides the fact that a key may be stolen, the fact that a fast crack algorithm is not *now* known does not *prove* that there are no such fast algorithms. Differential power attack algorithms, for instance, recently showed how to explore weakness in the implementation rather than a weakness in the algorithms itself [222]. There is also the problem of a "brute force attack" using more powerful computers and/or parallel attacks. A good example is the 56-bit key DES algorithm, which was the standard for many years but was finally declared insecure in 1997. The DES was first cracked by a network of volunteer computer owners on the Internet, which cracked the key in 39 days. Later, in July 1997, the Electronic Frontier Foundation (EFF) finished the design of a cracker machine. It has been documented in a book [223], including all schematics and software source code, which can be downloaded from `http://www.eff.org/`. This cracker machine performs an exhaustive key search and can crack any 56-bit key in less than five days. It was built out of custom chips, each of which has 24 cracker units. Each of the 29 boards used consists of 64 "Deep Crack" chips, i.e., a total of 1856 chips, or 44 544 units, are in use. The system cost was $250,000. When DES was introduced in 1977 the system costs were estimated at $20 million, which corresponds to about $40 million today. This shows a good approximation to "Moore's law," which says that every 18 months the size or speed or price of microprocessors improves by a factor 2. From 1977 to 1998 the price of such a machine should drop to $40 \times 10^6/2^{22/1.5} \approx$ fifteen hundred dollars, i.e., it should be affordable to build a DES cracker today (as was proven by the EFF).

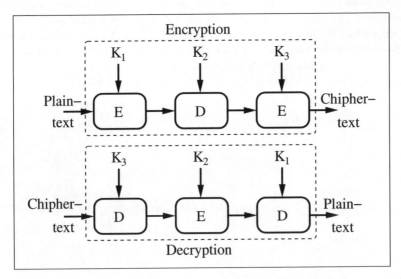

Fig. 7.32. Triple DES (K_l = keys; E=single Encryption; D=single Decryption).

Table 7.18. Encryption algorithms [224].

Algorithm	Key size (bits)	Mathematical operations/ principle	Sym-me-try	Developed by (year)
DES	56	XOR, fixed S-boxes	s	IBM (1977)
Triple DES	$122 - 168$	XOR, fixed S-boxes	s	
AES	$128 - 256$	XOR, fixed S-boxes	s	Daemen/Rijmen (1998)
RSA	variable	Prime factors	a	Rivest/Shamir/ Adleman (1977)
IDEA	128	XOR, add., mult.	s	Massey/Lai (1991)
Blowfish	< 448	XOR, add. fixed S-boxes	s	Schneider (1993)
RC5	< 2048	XOR, add., rotation	s	Rivest (1994)
CAST-128	$40 - 128$	XOR, rotation, S-boxes	s	Adams/Tavares (1997)

Therefore the 56-bit DES is no longer secure, but it is now common to use triple DES, as displayed in Fig. 7.32, or other 128-bit key systems. Table 7.18 shows that these systems seem to be secure for the next few years. The EFF cracker, for instance, today will need about 5×2^{112} days, or 7×10^{31} years, to crack the triple DES.

The first column in Table 7.18 is the commonly used abbreviations for the algorithms. The second and third columns contain the typical parameters of the algorithm. Symmetric algorithms (designated in the fourth column with an "s") are usually based on Feistel's algorithm, while asymmetric algorithms can be used in a public/private key system. The last column displays the name of the developer and the year the algorithm was first published.

7.3 Modulation and Demodulation

For a long time the goal of communications system design was to realize a fully digital receiver, consisting of only an antenna and a fully programmable circuit with digital filters, demodulators and/or decoders for error correction and cryptography on a single programmable chip. With today's FPGA gate count above one million gates this has become a reality. "FPGAs will clearly be a key technology for communication systems well into the 21st century" as predicted by Carter [225] . In this section, the design and implementation of a communication system is developed in the context of FPGAs.

7.3.1 Basic Modulation Concepts

A basic communication system transmits and receives information broadcast over a carrier frequency, say f_0. This carrier is *modulated* in amplitude, frequency or phase, proportional to the signal $x(t)$ being transmitted. Figure 7.33 shows a modulated signal for a binary transmission. For binary transmission, the modulations are called amplitude shift keying (ASK), phase shift keying (PSK), and frequency shift keying (FSK).

In general, it is more efficient to describe a (real) modulated signal with a projection of a rotating arrow on the horizontal axis, according to

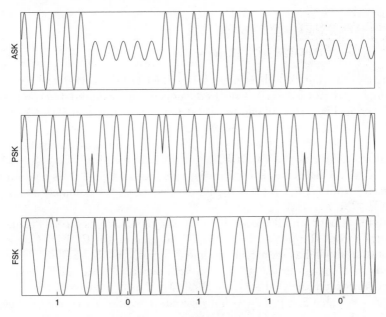

Fig. 7.33. ASK, PSK, and FSK modulation.

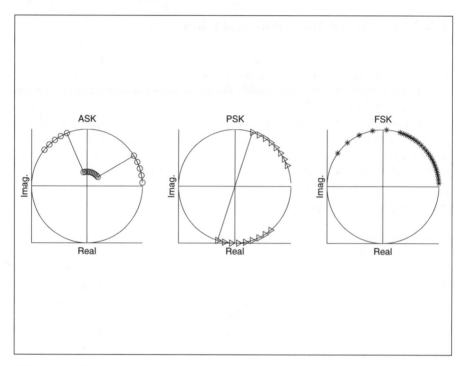

Fig. 7.34. Modulation in the complex plan.

$$s(t) = \Re\left\{ A(t)e^{j(2\pi f_0 t + \Delta\phi(t) + \phi_0)} \right\}$$
$$= A(t)\cos(2\pi f_0 t + \Delta\phi(t) + \phi_0), \tag{7.58}$$

where ϕ_0 is a (random) phase offset, $A(t)$ describes the part of the amplitude envelope, and $\Delta\phi(t)$ describes the frequency- or phase-modulated component, as shown in Fig. 7.34. As can be seen from (7.58), AM and PM/FM can be used separately to transmit different signals.

An efficient solution (that does not require large tables) for realizing universal modulator is the CORDIC algorithm discussed in Chap. 2 (p. 120). The CORDIC algorithm is used in the rotation mode, i.e., it is a coordinate converter from $(R, \theta) \rightarrow (X, Y)$. Figure 7.35 shows the complete modulator for AM, PM, and FM.

To implement amplitude modulation, the signal $A(t)$ is directly connected with the radius R input of the CORDIC. In general, the CORDIC algorithm in rotation mode has an attendant linear increase in the radius. This corresponds to a change in the gain of an amplifier and need not be taken into consideration for the AM scheme. When the linear increased radius (factor 1.6468, see Table 2.1, p. 57), is not desired, it is possible either to scale the input or the output by $1/1.6468$ with a constant coefficient multiplier.

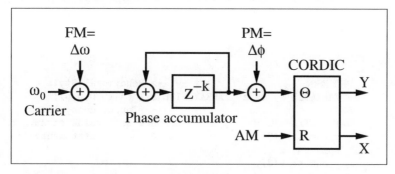

Fig. 7.35. Universal modulator using CORDIC.

The phase of the transmitted signal $\theta = 2\pi f_0 t + \Delta\phi(t)$ must also be computed. To generate the constant carrier frequency, a linearly increasing phase signal according to $2\pi f_0 t$ must be generated, which can be done with an accumulator. If FM should be generated, it is possible to modify f_0 by Δf, or to use a second accumulator to compute $2\pi \Delta f t$, and to add the results of the two accumulators. For the PM signal, a constant offset (not increasing in time) is added to the phase of the signal. These phase signals are added and applied to the angle input z or θ of the CORDIC processor. The Y register is set to zero at the beginning of the iterations.

The following example demonstrates a fully pipelined version of the CORDIC modulator.

Example 7.14: Universal Modulator using CORDIC

A universal modulator for AM, PM, and FM according to Fig. 7.35, can be designed with the following VHDL code[5] of the CORDIC part.

```
PACKAGE nine_bit_int IS    -- User-defined types
   SUBTYPE NINE_BIT IS INTEGER RANGE -256 TO 255;
   TYPE ARRAY_NINE_BIT IS ARRAY (0 TO 3) OF NINE_BIT;
END nine_bit_int;

LIBRARY work;
USE work.nine_bit_int.ALL;

LIBRARY ieee;
USE ieee.std_logic_1164.ALL;
USE ieee.std_logic_arith.ALL;

ENTITY ammod IS                        ------> Interface
          PORT (clk                : IN  STD_LOGIC;
                r_in , phi_in      : IN  NINE_BIT;
                x_out, y_out, eps  : OUT NINE_BIT);
END ammod;
```

---

[5] The equivalent Verilog code `ammod.v` for this example can be found in Appendix A on page 717. Synthesis results are shown in Appendix B on page 731.

```
ARCHITECTURE fpga OF ammod IS

BEGIN

  PROCESS                      ------> Behavioral Style
    VARIABLE x, y, z : ARRAY_NINE_BIT := (0,0,0,0);
  BEGIN                              -- Tapped delay lines
  WAIT UNTIL clk = '1';    -- Compute last value first
    x_out <= x(3);              -- in sequential statements !!
    eps   <= z(3);
    y_out <= y(3);

    IF z(2) >= 0 THEN                   -- Rotate 14 degrees
      x(3) := x(2) - y(2) /4;
      y(3) := y(2) + x(2) /4;
      z(3) := z(2) - 14;
    ELSE
      x(3) := x(2) + y(2) /4;
      y(3) := y(2) - x(2) /4;
      z(3) := z(2) + 14;
    END IF;

    IF z(1) >= 0 THEN                   -- Rotate 26 degrees
      x(2) := x(1) - y(1) /2;
      y(2) := y(1) + x(1) /2;
      z(2) := z(1) - 26;
    ELSE
      x(2) := x(1) + y(1) /2;
      y(2) := y(1) - x(1) /2;
      z(2) := z(1) + 26;
    END IF;

    IF z(0) >= 0 THEN                   -- Rotate  45 degrees
      x(1) := x(0) - y(0);
      y(1) := y(0) + x(0);
      z(1) := z(0) - 45;
    ELSE
      x(1) := x(0) + y(0);
      y(1) := y(0) - x(0);
      z(1) := z(0) + 45;
    END IF;

    IF phi_in > 90    THEN      -- Test for |phi_in| > 90
      x(0) := 0;                -- Rotate 90 degrees
      y(0) := r_in;             -- Input in register 0
      z(0) := phi_in - 90;
    ELSIF phi_in < -90 THEN
      x(0) := 0;
      y(0) := - r_in;
      z(0) := phi_in + 90;
    ELSE
      x(0) := r_in;
```

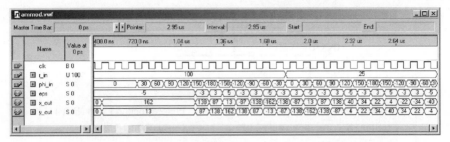

Fig. 7.36. Simulation of an AM modulator using the CORDIC algorithm.

```
       y(0) := 0;
       z(0) := phi_in;
    END IF;
    END PROCESS;

  END fpga;
```

Figure 7.36 reports the simulation of an AM signal. Note that the Altera simulation allows you to use signed data, rather then unsigned binary data (where negative values have a 512 offset). A pipeline delay of four steps is seen and the value 100 is enlarged by a factor of 1.6. A switch in radius r_in from 100 to 25 results in the maximum value x_out dropping from 163 to 42. The CORDIC modulator runs at 215.98 MHz and uses 316 LEs and no embedded multiplier.   7.14

Demodulation may be *coherent* or *incoherent*. A coherent receiver must recover the unknown carrier phase ϕ_0, while an incoherent one does not need to do so. If the receiver uses an intermediate frequency (IF) band, this type of receiver is called a superhet or double superhet (two IF bands) receiver. IF receivers are also sometimes called *heterodyne* receivers. If no IF stages are employed, a *zero IF* or *homodyne* receiver results. Figure 7.37 presents a systematic overview of the different types of receivers. Some of the receivers can only be used for one modulation scheme, while others can be used for multiple modes (e.g., AM, PM, and FM). The latter is called a universal receiver. We will first discuss the incoherent receiver, and then the coherent receiver.

All receivers use intensive filters (as discussed in Chaps. 3 and 4), in order to select only the signal components of interest. In addition, for heterodyne receivers, filters are needed to suppress the mirror frequencies, which arise from the frequency shift

$$s(t) \times \cos(2\pi f_m t) \longleftrightarrow S(f + f_m) + S(f - f_m). \tag{7.59}$$

7.3.2 Incoherent Demodulation

In an incoherent demodulation scheme, it is assumed that the exact carrier frequency is known to the receiver, but the initial phase ϕ_0 is not.

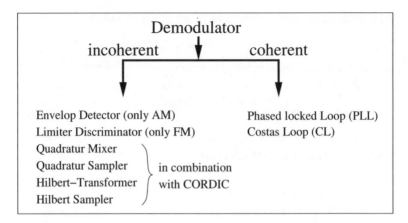

Fig. 7.37. Coherent and incoherent demodulation schemes.

If the signal component is successfully selected with digital or analog filtering, the question arises whether only one demodulation mode (e.g., AM or FM) or universal demodulator is needed. An incoherent AM demodulator can be as simple as a full or half-wave rectifier and an additional lowpass filter. For FM or PM demodulation, only the *limiter/discriminator* type of demodulator is an efficient implementation. This demodulator builds a threshold of the input signal to limit the values to ±1, and then basically "measures" the distance between the zero crossings. These receivers are easily implemented with FPGAs but sometimes produce 2π jumps in the phase signal (called "clicks" [226, 227]). There are other demodulators with better performance.

We will focus on universal receivers using in-phase and quadrature components. This type of receiver basically inverts the modulation scheme relative to (7.58) from p. 454. In a first step we have to compute, from the received cosines, components that are "in-phase" with the sender's sine components (which are in quadrature to the carrier, hence the name Q phase). These I and Q phases are used to reconstruct the arrow (rotating with the carrier frequency) in the complex plane. Now, the demodulation is just the inversion of the circuit from Fig. 7.35. It is possible to use the CORDIC algorithm in the vectoring mode, i.e., a coordinate conversion $X, Y \rightarrow R, \theta$ with $I = X$ and $Q = Y$ is used. Then the output R is directly proportional to the AM portion, and the PM/FM part can be reconstructed from the θ signal, i.e., the Z register.

A difficult part of demodulation is I/Q generation, and typically two methods are used: a quadrature scheme and a Hilbert transform.

In the quadrature scheme the input signal is multiplied by the two mixer signals, $2\cos(2\pi f_m t)$ and $-j2\sin(2\pi f_m t)$. If the signals in the IF band $f_{IF} = f_0 - f_m$ are now selected with a filter, the complex sum of these signals is then a reconstruction of the complex rotating arrow. Figure 7.38 shows this scheme while Fig. 7.39 displays an example of the I/Q generation. From the

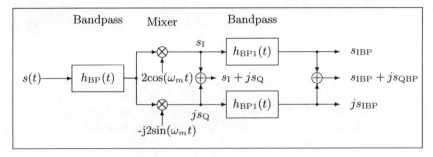

Fig. 7.38. Generation of I- and Q-phase using quadrature scheme.

spectra shown in Fig. 7.39 it can be seen that the final signal has no negative spectral components. This is typical for this type of incoherent receiver and these signals are called *analytic*.

To decrease the effort for the filters, it is desirable to have an IF frequency close to zero. In an analog scheme (especially for AM) this often introduces a new problem, that the amplifier drifts into saturation. But for a fully digital receiver, such a homodyne or zero IF receiver can be built. The bandpass filters then reduce to lowpass filters. Hogenauer's CIC filters (see Chap. 5, p. 258) are efficient realizations of these high decimation filters. Fig-

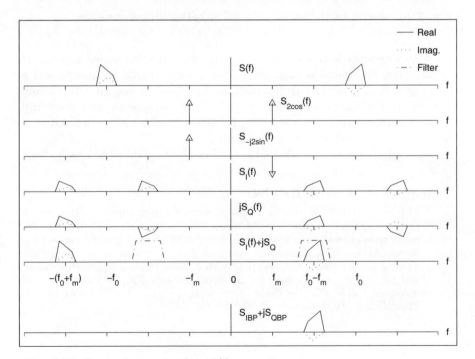

Fig. 7.39. Spectral example of the I/Q generation.

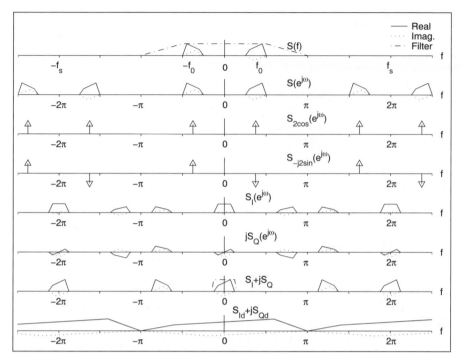

Fig. 7.40. Spectra for the zero IF receiver. Sampling frequency was 2π.

ure 7.40 shows the corresponding spectra. The real input signal is sampled at 2π. Then the signal is multiplied with a cosine signal $S_{2\cos}(e^{j\omega})$ and a sine signal $S_{-j2\sin}(e^{j\omega})$. This produces the in-phase component $S_{\mathrm{I}}(e^{j\omega})$ and the quadrature component $jS_{\mathrm{Q}}(e^{j\omega})$. These two signals are now combined into a complex analytic signal $S_{\mathrm{I}} + jS_{\mathrm{Q}}$. After the final lowpass filtering, a decimation in sampling rate can be applied.

Such a fully digital zero IF for LF has been built using FPGA technology [228].

Example 7.15: Zero IF Receiver

This receiver has an antenna, a programmable gain adjust (AGC), and a Cauer lowpass 7th-order followed by an 8-bit video A/D converter. The receiver uses eight times oversampling (0.4–1.2 MHz) for the input range from 50 to 150 kHz. The quadrature multipliers are 8×8-bit array multipliers. Two-stage CIC filters were designed with 24- and 19-bit integrator precision, and 17- and 16-bits precision for the comb sections. The final sampling rate reduction was 64. The full design could fit on a single XC3090 Xilinx FPGA. The following table shows the effort for the single units:

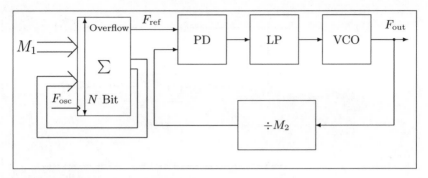

Fig. 7.41. PLL with accumulator as reference.

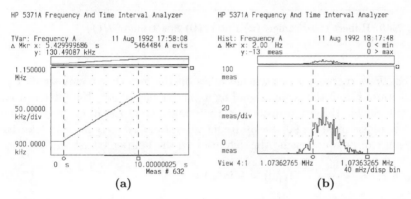

(a)   (b)

Fig. 7.42. PLL synthesizers with accumulator reference. **(a)** Behavior of the synthesizer for switching F_{out} from 900 kHz to 1.2 MHz. **(b)** Histogram of the frequency error, which is less than 2 Hz.

Design part	CLBs
Mixer with sin/cos tables	74
Two CIC filters	168
State machine and PDSP interface	18
Frequency synthesizer	32
Total	292

For the tunable frequency synthesizer an accumulator as reference for an analog phase-locked loop (PLL) was used [4]. Figure 7.41 shows this type of frequency synthesizer and Fig. 7.42 displays the measured performance of the synthesizer. The accumulator synthesizer could be clocked very high due to the fact that only the overflow is needed. A bitwise carry save adder was therefore used. The accumulator was used as a reference for the PLL that produces $F_{out} = M_2 F'_{in} = M_1 M_2 F_{in}/2^N$.

7.15

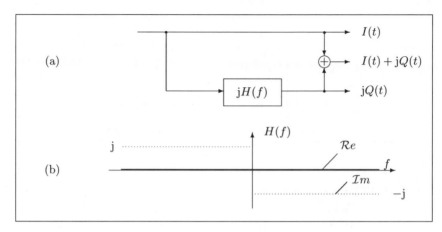

Fig. 7.43. Hilbert transformer. **(a)** Filter. **(b)** Spectrum of $H(f)$.

The *Hilbert transformer* scheme relies on the fact that a sine signal can be computed from the cosine signal by a phase delay of 90°. If a filter is used to produce this Hilbert transformer, the amplitude of the filter must be one and the phase must be 90° for all frequencies. Impulse response and transfer function can be found using the definition of the Fourier transform, i.e.,

$$h(t) = \frac{1}{\pi t} \longleftrightarrow H(j\omega) = -j\gamma(\omega) = \begin{cases} j & -\infty < \omega < 0 \\ -j & 0 < \omega < \infty, \end{cases} \quad (7.60)$$

with $\gamma(\omega) = -1 \; \forall \; \omega < 0$ and $\gamma(\omega) = 1 \; \forall \; \omega \geq 0$ as the sign function. A Hilbert filter can only be approximated by an FIR filter and resulting coefficients have been reported (see, for instance, [229, 230], [159, pp. 168–174], or [79, p. 681]).

Simplification for narrowband receivers. If the input signals are narrowband signals, i.e., the transmitted bit rate is much smaller than the carrier frequency, some simplifications in the demodulation scheme are possible. In the *input sampling scheme* it is then possible to sample at the carrier rate, or at a multiple of the period $T_0 = 1/f_0$ of the carrier, in order to ensure that the sampled signals are already free of the carrier component.

The *quadrature scheme* becomes trivial if the zero IF receiver samples at $4f_0$. In this case, the sine and cosine components are elements of $0, 1$ or -1, and the carrier phase is $0, 90°, 180° \dots$. This is sometimes referred to as "complex sampling" in the literature [231, 232]. It is possible to use *undersampling*, i.e., only every second or third carrier period is evaluated by the sampler. Then the sampled signal will still be free from the carrier frequency.

The *Hilbert transformer* can also be simplified if the signal is sampled at $T_0/4$. A Hilbert sampler of first order with $Q_{-1} = 1$ or a second-order

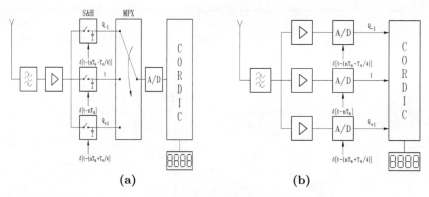

Fig. 7.44. (a) Two versions of the Hilbert sampler of first order.

type using the symmetric coefficients $Q_1 = -0.5; Q_{-1} = 0.5$ or asymmetric $Q_{-1} = 1.5; Q_{-3} = 0.5$ coefficients can be used [233].

Table 7.19. Coefficients of the Hilbert sampler.

Type	Coefficients	Bit	$\Delta f / f_0$
Zero order	$Q_{-1} = 1, 0$	8	0.005069
	$Q_{-1} = 1, 0$	12	0.000320
	$Q_{-1} = 1, 0$	16	0.000020
First- order asymmetric	$Q_{-1} = 1, 5; Q_{-3} = 0.5$	8	0.032805
	$Q_{-1} = 1, 5; Q_{-3} = 0.5$	12	0.008238
	$Q_{-1} = 1, 5; Q_{-3} = 0.5$	16	0.002069
First- order symmetric	$Q_1 = -0.5; Q_{-1} = 0.5$	8	0.056825
	$Q_1 = -0.5; Q_{-1} = 0.5$	12	0.014269
	$Q_1 = -0.5; Q_{-1} = 0.5$	16	0.003584

Table 7.19 reports the three short-term Hilbert transformer coefficients and the maximum allowed frequency offset Δf of the modulation, for the Hilbert filter providing a specified accuracy.

Figure 7.44 shows two possible realizations for the Hilbert transformer that have been used to demodulate radio control watch signals [234, 235]. The first method uses three Sample & Hold circuits and the second method uses three A/D converters to build a symmetric Hilbert sampler of first order.

Figure 7.45 shows the spectral behavior of the Hilbert sampler with a direct undersampling by two.

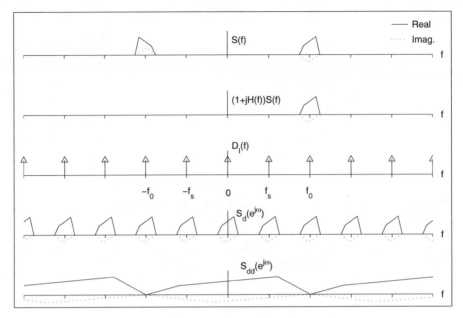

Fig. 7.45. Spectra for the Hilbert sampler with undersampling.

7.3.3 Coherent Demodulation

If the phase ϕ_0 of the receiver is known, then demodulation can be accomplished by multiplication and lowpass filtering. For AM, the received signal $s(t)$ is multiplied by $2\cos(\omega_0 t + \phi_0)$ and for PM or FM by $-2\sin(\omega_0 t + \phi_0)$. It follows that

AM:

$$A(t)\cos(2\pi f_0 t + \phi_0) \times 2\cos(2\pi f_0 t + \phi_0)$$

$$= \underbrace{A(t)}_{\text{Lowpass component}} + A(t)\cos(4\pi f_0 t + 2\phi_0) \qquad (7.61)$$

$$s_{\text{AM}}(t) = A(t) - A_0. \qquad (7.62)$$

PM:

$$-2\sin(2\pi f_0 t + \phi_0) \times \cos(2\pi f_0 t + \phi_0 + \Delta\phi(t))$$

$$= \underbrace{\sin(\Delta\phi(t))}_{\text{Lowpass component}} + \cos(4\pi f_0 t + 2\phi_0 + \Delta\phi(t)) \qquad (7.63)$$

$$\sin(\Delta\phi(t)) \approx \Delta\phi(t) \qquad (7.64)$$

$$s_{\text{PM}}(t) = \frac{1}{\eta}\Delta\phi(t). \qquad (7.65)$$

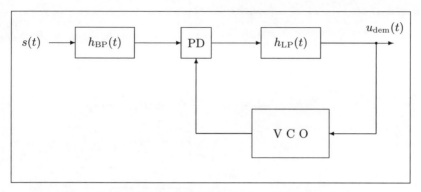

Fig. 7.46. Phase-locked loop (PLL) with (necessary) bandpass ($h_{BP}(t)$), phase detector (PD), lowpass ($h_{LP}(t)$), and voltage-controlled oscillator (VCO).

FM:

$$s_{FM}(t) = \frac{1}{\eta}\frac{d}{dt}\frac{\Delta\phi(t)}{}. \tag{7.66}$$

η is the so-called modulation index.

In the following we will discuss the types of coherent receivers that are suitable for an FPGA implementation. Typically, coherent receivers provide a 1 dB better signal-to-noise ratio than an incoherent receiver (see Fig. 7.4, p. 418). A synchronous or coherent FM receiver tracks the carrier phase of the incoming signal with a voltage-controlled oscillator (VCO) in a loop. The DC part of this voltage is directly proportional to the FM signal. PM signal demodulation requires integration of the VCO control signal, and AM demodulation requires the addition of a second mixer and a $\pi/2$ phase shifter in sequence with a lowpass filter. The risk with coherent demodulation is that for a low signal-to-noise channel, the loops may be out-of-lock, and performance will decrease tremendously.

There are two common types of coherent receiver loops: the phase-locked loop (PLL) and the Costas loop (CL). Figures 7.46 and 7.47 are block diagrams of a PLL and CL, respectively, showing the nearly doubled complexity of the CL. Each loop may be realized as an analog (linear PLL/CL) or all-digital (ADPLL, ADCL) circuit (see [236, 237, 238, 239]). The stability analysis of these loops is beyond the scope of the book and is well covered in the literature ([240, 241, 242, 243, 244]). We will discuss efficient realizations of PLLs and CLs [245, 246]. The first PLL is a direct translation of an analog PLL to FPGA technology.

Linear phase-locked loop. The difference between linear and digital loops lies in the type of input signal to be processed. A linear PLL or CL uses a fast multiplier as a phase detector, providing a possibly multilevel input signal to the loop. A digital PLL or CL can process only binary input signals. (*Digital* refers to the quality of the input signal here, not to the hardware realization!)

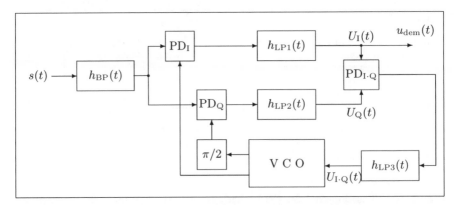

Fig. 7.47. Costas loop with (necessary) bandpass ($h_{\mathrm{BP}}(t)$), three phase detectors (PD), three lowpass filters ($h_{\mathrm{LP}}(t)$), and a voltage-controlled oscillator (VCO) with $\pi/2$ phase shifter.

As shown in Fig. 7.46, the linear PLL has three main blocks:

- Multiplier as phase detector
- Loop filter
- VCO

To keep the loop in-lock, a loop output signal-to-noise ratio larger than $4 = 6$ dB is required [239, p. 35]. Since the selection of typical antennas is not narrow enough to achieve this, an additional narrow bandpass filter has been added to Figs. 7.46 and 7.47 as a necessary addition to the demodulator [246]. The "cascaded bandpass comb" filter (see Table 5.4, p. 279) is an efficient example. However, the filter design is much easier if a fixed IF is used, as in the case of a superhet or double superhet receiver.

The VCO (or digitally controlled oscillator (DCO) for ADPLLs) oscillates with a frequency $\omega_2 = \omega_0 + K_0 \times U_f(t)$, where ω_0 is the resting point and K_0 the gain of the VCO/DCO. For sinusoidal input signals we have the signal

$$u_{\mathrm{dem}}(t) = K_d \sin\left(\Delta\phi(t)\right) \tag{7.67}$$

at the output of the lowpass, where $\Delta\phi(t)$ is the phase difference between the DCO output and the bandpass-filtered input signal. For small differences, the sine can be approximated by its argument, giving $u_{\mathrm{dem}}(t)$ proportional to $\Delta\phi(t)$ (the loop stays in-lock). If the input signal has a very sudden phase discontinuity, the loop will go out-of-lock. Figure 7.48 shows the different operation areas of the loop. The hold-in range $\omega_0 \pm \Delta\omega_H$ is the static operation limit (useful only with a frequency synthesizer). The lock-in range is the area where the PLL will lock-in within a single period of the frequency difference $\omega_1 - \omega_2$. Within the pull-in range, the loop will lock-in within the capture time T_L, which may last more than one period of $\omega_1 - \omega_2$. The pull-out

Table 7.20. Cost in CLBs of a linear PLL universal demodulator in a Xilinx XC3000 FPGA.

Function group	FM only	FM, AM, and PM
Phase detector (8×8-bit multiplier)	65	72
Loop filter (two-stage CIC)	84	168
Frequency synthesizer	34	34
DCO ($N/N + K$ divider, sin/cos table)	16	16+2
PDSP Interface	15	15
Total	214	307

range is the maximum frequency jump the loop can sustain without going out-of-lock. $\omega_0 \pm \Delta\omega_{PO}$ is the dynamic operation limit used in demodulation. There is much literature optimizing PD, loop filter and the VCO gain; see [240, 241, 242, 243, 244].

The major advantage of the linear loop over the digital PLL and CL is its noise-reduction capability, but the fast multiplier used for the PD in a linear PLL has a particularly high hardware cost, and this PD has an unstable rest point at $\pi/2$ phase shift ($-\pi/2$ is a stable point), which impedes lock-in. Table 7.20 estimates the hardware cost in CLBs of a linear PLL, using the same functional blocks as the 8-bit incoherent receiver (see Example 7.15, p. 460). Using a hardware multiplier in multiplex for AM and PM demodulation, the right-hand column reduces the circuit's cost by 58 CLBs, allowing it to fit into a 320 CLB Xilinx XC3090 device.

A comparison of these costs to those of an incoherent receiver, consuming 292 CLBs without CORDIC demodulation and 367.5 CLBs with an additional CORDIC processor [69], shows a slight improvement in the linear PLL realization. If only FM and PM demodulation are required, a digital PLL or CL, described in the next two sections, can reduce complexity dramatically.

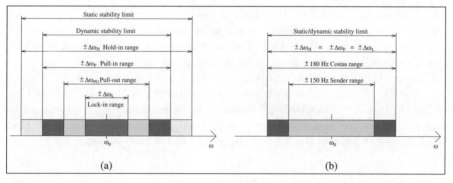

Fig. 7.48. (a) Operation area PLL/CL. **(b)** Operation area of the CL of Fig. 7.51.

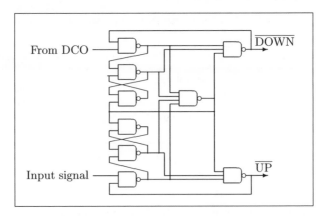

Fig. 7.49. Phase detector [19, Chap. 8 p. 127].

These designs were developed to demodulate WeatherFAX pictures, which were transmitted in central Europe by the low-frequency radio stations DCF37 and DCF54 (Carrier 117.4 kHz and 134.2 kHz; frequency modulation F1C: ± 150 Hz).

Digital PLLs. As explained in the last section, a digital PLL works with binary input signals. Phase detectors for a digital PLL are simpler than the fast multipliers used for linear PLLs; usual choices are XOR gates, edge-triggered JK flip-flops, or paired RS flip-flops with some additional gates [239, pp. 60–65]. The phase detector shown in Fig. 7.49 is the most complex, but it provides phase and frequency sensitivity and a quasi-infinite hold-in range.

Modified counters are used as loop filters for DPLLs. These may be $N/(N + K)$ counters or multistage counters, such as an N-by-M divider, where separate UP and DOWN counters are used, and a third counter measures the UP/DOWN difference. It is then possible to break off further signal processing if a certain threshold is not met. For the DCO, any typical all-digital frequency synthesizer, such as an accumulator, divider, or multiplicative generator may be used. The most frequently used synthesizer is the tunable divider, popular because of its low phase error. The low resolution of this synthesizer can be improved by using a low receiver IF [247].

One DPLL realization with very low complexity is the 74LS297 circuit, which utilizes a "pulse-stealing" design. This scheme may be improved with the phase- and frequency-sensitive J-K flip-flop, as displayed in Fig. 7.50. The PLL works as follows: the "detect flip-flop" runs with rest frequency

$$F_{\text{comp}} = \frac{F_{\text{in}}}{N} = \frac{F_{\text{osc}}}{KM}. \tag{7.68}$$

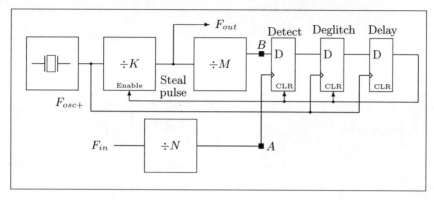

Fig. 7.50. "Pulse-stealing" PLL [248].

Table 7.21. Hardware complexity of a pulse-stealing DPLL [247].

Function group	CLBs
DCO	16
Phase detector	5
Loop filter	6
Averaging	11
PC–interface	10
Frequency synthesizer	26
Stage machine	10
Total	84

To allow tracking of incoming frequencies higher than the rest frequency, the oscillator frequency $F_{\mathrm{osc+}}$ is set slightly higher:

$$T_{\mathrm{comp}} - T_{\mathrm{comp+}} = \frac{1}{2}T_{\mathrm{osc}}, \tag{7.69}$$

such that the signal at point B oscillates half a period faster than the F_{osc} signal. After approximately two periods at the rest frequency, a one will be latched in the detector flip-flop. This signal runs through the deglitch and delay flip-flops, and then inhibits one pulse of the $\div K$ divider (thus the name "pulse-stealing"). This delays the signal at B such that the phase of signal A runs after B, and the cycle repeats. The lock-in range of the PLL has a lower bound of $F_{\mathrm{in}}|_{min}=0\,\mathrm{Hz}$. The upper bound depends on the maximum output frequency $F_{\mathrm{osc+}}/K$, so the lock-in range becomes

$$\pm\Delta\omega_L = \pm N \times F_{\mathrm{osc+}}/(K \times M). \tag{7.70}$$

A receiver can be simplified by leaving out the counters N and M. In a WeatherFAX image-decoding application the second IF of the double-

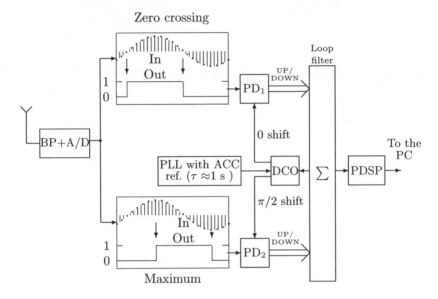

Fig. 7.51. Structure of the Costas loop.

superhet receiver is set in such a way that the frequency modulation of 300 Hz ($\delta f = 0$ Hz → white; $\delta f = 300$ Hz → black) corresponds to exactly 32 steal pulses, so that the steal pulses correspond directly to the greyscale level. We set a pixel rate of 1920 Baud, and a IF of 16.6 kHz. For each pixel, four "steal values" (number of steal pulses in an interval) are determined, so a total of $\log_2(2 \times 4) = 3$ bit shifts are used to compute the 16 gray-level values. Table 7.21 shows the hardware complexity of this PLL type.

Costas loop. This extended type of coherent loop was first proposed by John P. Costas in 1956, who used the loop for carrier recovery. As shown in Fig. 7.47, the CL has an *in-phase* and a *quadrature* path (subscripted I and Q there). With the $\pi/2$ phase shifter and the third PD and lowpass, the CL is approximately twice as complex as the PLL, but locks onto a signal twice as fast. Costas loops are very sensitive to small differences between in-phase and quadrature gain, and should therefore always be realized as all-digital circuits. The FPGA seems to be an ideal realization vehicle ([249, 250]).

For a signal $U(t) = A(t) \sin(\omega_0 t + \Delta\phi(t))$ we get, after the mixer and lowpass filters,

$$U_I(t) = K_d A(t) \cos\left(\Delta\phi(t)\right) \tag{7.71}$$
$$U_Q(t) = K_d A(t) \sin\left(\Delta\phi(t)\right), \tag{7.72}$$

where $2K_d$ is the gain of the PD. $U_I(t)$ and $U_Q(t)$ are then multiplied together in a third PD, and are lowpass filtered to get the DCO control signal:

Table 7.22. Loop filter output and DCO correction values at 32-times oversampling.

| Accumulator | | | | | |
| | | | | gray | |
Under-flow	Over-flow	Sum	DCO-IN	value	$f_{\text{carrier}} \pm \delta f$
Yes	No	$s < -(2^{13} - 1)$	3	0	$+180\,\text{Hz}$
No	No	$-(2^{13} - 1) \leq s < -2048$	2	0	$+120\,\text{Hz}$
No	No	$-2048 \leq s < -512$	1	4	$+60\,\text{Hz}$
No	No	$-512 \leq s < 512$	0	8	$+0\,\text{Hz}$
No	No	$512 \leq s < 2048$	-1	12	$-60\,\text{Hz}$
No	No	$2048 \leq s < 2^{13} - 1$	-2	15	$-120\,\text{Hz}$
No	Yes	$s \geq 2^{13} - 1$	-3	15	$-180\,\text{Hz}$

$$U_{I \times Q}(t) \sim K_d \sin\left(2\Delta\phi(t)\right). \tag{7.73}$$

A comparison of (7.67) and (7.73) shows that, for small modulations of $\Delta\phi(t)$, the slope of the control signal $U_{I \times Q}(t)$ is twice the PLL's. As in the PLL, if only FM or PM demodulation is needed, the PDs may be all digital.

Figure 7.51 shows a block diagram of a CL. The antenna signal is first filtered and amplified by a fourth-order Butterworth bandpass, then digitized by an 8-bit converter at a sampling rate 32 or 64 times the carrier base frequency. The resulting signal is split, and fed into a zero-crossing detector and a minimum/maximum detector. Two phase detectors compare the signals with a reference signal, and its $\pi/2$-shifted counterpart, synthesized by a high time-constant PLL with a reference accumulator [4, section 2]. Each phase detector has two edge detectors, which should generate a total of 4 UP and 4 DOWN signals. If more UP signals than DOWN are generated by the PDs, then the reference frequency is too low, and if more DOWN signals are generated, it is too high. The differences $\sum \text{UP} - \sum \text{DOWN}$ are accumulated for one pixel duration in a 13-bit accumulator acting as a loop filter. The loop filter data are passed to a pixel converter, which gives "correction values" to the DCO as shown in Table 7.22. The accumulated sums are also used as greyscale values for the pixel, and passed onto a PC to store and display the WeatherFAX pictures.

The smallest detectable phase offset for a 2 kBaud pixel rate is

$$f_{\text{carrier}+1} = \frac{1}{1/f_{\text{carrier}} - t_{ph37} \times 2\,\text{kBaud}/f_{\text{carrier}}} = 117.46\,\text{kHz}, \tag{7.74}$$

where $t_{ph37} = 1/(32 \times 117\,\text{kHz}) = 266\,\text{ns}$ is the sampling period at 32-times oversampling. The frequency resolution is $117.46\,\text{kHz} - (f_{\text{carrier}} = 117.4\,\text{kHz}) = 60\,\text{Hz}$. With a frequency modulation of $300\,\text{Hz}$, five greyscale values can be distinguished. Higher sampling rates for the accumulator are not possible

Table 7.23. Complexity of a fully digital Costas loop [246, p. 60].

Function group	CLBs with oversampling	
	32 times	64 times
Frequency synthesizer	33	36
Zero detection	42	42
Maximum detection	16	16
Four phase detectors	8	8
Loop filter	51	51
DCO	12	15
TMS interface	11	11
Sum	173	179

with the 3164-4ns FPGA and the limited A/D converter used. With a fast A/D converter and an Altera Flex or Xilinx XC4K FPGA, 128- and 256-times oversampling are possible.

For a maximum phase offset of π, the loop will require a maximum lock-in time T_L of $\lceil 16/3 \rceil = 6$ samples, or about 1.5 μs. Table 7.23 shows the complexity of the CL for 32 and 64 times oversampling.

Exercises

Note: If you have no prior experience with the Quartus II software, refer to the case study found in Sect. 1.4.3, p. 29. If not otherwise noted use the EP2C35F672C6 from the Cyclone II family for the Quartus II synthesis evaluations.

7.1: The following MATLAB code can be used to compute the order of an element.

```
function N = order(x,M)
% Compute the order of x modulo M
p = x; l=1;
while p ~= 1
    l = l+1; p = p * x;
    re =real(p); im = imag(p);
    p = mod(re,M) + i * mod(im,M);
end;
N=l;
```

If, for instance, the function is called with `order(2,2^25+1)` the result is 50. To compute the single factors of $2^{25} + 1$, the standard MATLAB function `factor(2^25+1)` can be used.
For
(a) $\alpha = 2$ and $M = 2^{41} + 1$
(b) $\alpha = -2$ and $M = 2^{29} - 1$

(c) $\alpha = 1 + j$ and $M = 2^{29} + 1$
(d) $\alpha = 1 + j$ and $M = 2^{26} - 1$
compute the transform length, the "bad" factors ν (i.e., order not equal order$(\alpha, 2^B \pm 1)$), all "good" prime factors M/ν, and the available input bit width $B_x = (\log_2(M/\nu) - \log_2(L))/2$.

7.2: To compute the inverse $x^{-1} \bmod M$ for $\gcd(x, M) = 1$ of the value x, we can use the fact that the following diophantic equation holds:

$$\gcd(x, M) = u \times x + v \times M \qquad \text{with} \quad u, v \in \mathbb{Z}. \tag{7.75}$$

(a) Explain how to use the MATLAB function [g u v]=gcd(x,M) to compute the multiplication inverse.
Compute the following multiplicative inverses if possible:
(b) $3^{-1} \bmod 73$;
(c) $64^{-1} \bmod 2^{32} + 1$;
(d) $31^{-1} \bmod 2^{31} - 1$;
(e) $89^{-1} \bmod 2^{11} - 1$;
(f) $641^{-1} \bmod 2^{32} + 1$.

7.3: The following MATLAB code can be used to compute Fermat NTTs for length 2, 4, 8, and 16 modulo 257.

```
function Y = ntt(x)
% Compute Fermat NTT of length 2,4,8 and 16 modulo 257
l = length(x);
switch (l)
  case 2, alpha=-1;
  case 4, alpha=16;
  case 8, alpha=4;
  case 16, alpha=2;
  otherwise, disp('NTT length not supported')
end
A=ones(l,l); A(2,2)=alpha;
%*********Computing second column
for m=3:l
 A(m,2)=mod(A(m-1,2)* alpha, 257);
end
%*********Computing rest of matrix
for m=2:l
  for n=2:l-1
    A(m,n+1)=mod(A(m,n)*A(m,2),257);
  end
end
%*********Computing NTT A*x
for k = 1:l
  C1 = 0;
  for j = 1:l
    C1 = C1 + A(k,j) * x(j);
  end
  X(k) = mod(C1, 257);
end
Y=X;
```

(a) Compute the NTT X of $x = \{1, 1, 1, 1, 0, 0, 0, 0\}$.

(b) Write the code for the appropriate INTT. Compute the INTT of X from part (a).

(c) Compute the element-by-element product $Y = X \odot X$ and $\text{INTT}(Y) = y$.

(d) Extend the code for a complex Fermat NTT and INTT for $\alpha = 1 + j$. Test your program with the identity $x = \text{INTT}(\text{NTT}(x))$.

7.4: The Walsh transform for $N = 4$ is given by:

$$W_4 = \begin{bmatrix} 1 & 1 & 1 & 1 \\ 1 & 1 & -1 & -1 \\ 1 & -1 & -1 & 1 \\ 1 & -1 & 1 & -1 \end{bmatrix}.$$

(a) Compute the scalar product of the row vectors. What property does the matrix have?

(b) Use the results from (a) to compute the inverse W_4^{-1}.

(c) Compute the 8×8 Walsh matrix W_8, by scaling the original row vector by two (i.e., $h[n/2]$) and computing an additional two "children" $h[n] + h[n-4]$ and $h[n] - h[n-4]$ from row 3 and 4. There should be no zero in the resulting W_8 matrix.

(d) Draw a function tree to construct Walsh matrices of higher order.

7.5: The Hadamard matrix can be computed using the following iteration

$$H_{2^{l+1}} = \begin{bmatrix} H_{2^l} & H_{2^l} \\ H_{2^l} & -H_{2^l} \end{bmatrix}, \tag{7.76}$$

with $H_1 = [1]$.

(a) Compute H_2, H_4, and H_8.

(b) Find the appropriate index for the rows in H_4 and H_8, compared with the Walsh matrix W_4 and W_8 from Exercise 7.4.

(c) Determine the general rule to map a Walsh matrix into a Hadamard matrix.
Hint: First compute the index in binary notation.

7.6: The following MATLAB code can be used to compute the state-space description for $p_{14} = x^{14} + x^5 + x^3 + x^1 + 1$, the nonzero elements using **nnz**, and the maximum fan-in.

```
p= input('Please define power of matrix = ')
A=zeros(14,14);
for m=1:13
  A(m,m+1)=1;
end
A(14,14)=1;
A(14,13)=1;
A(14,11)=1;
A(14,9)=1;
Ap=mod(A^p,2);
nnz(Ap)
max(sum(Ap,2))
```

(a) Compute the number of nonzero elements and fan-in for $p = 2$ to 8.

(b) Modify the code to compute the twin $p_{14} = x^{14} + x^{13} + x^{11} + x^9 + 1$. Compute the number of nonzero elements for the modified polynomial for $p = 2$ to 8.

(c) Modify the original code to compute the alternative LFSR implementation (see Fig. 7.21, p. 438) for (a) and (b) and compute the nonzero elements for $p = 2$ to 8.

7.7: (a) Compile the code for the length-6 LFSR lfsr.vhd from Example 7.12 (p. 437) using MaxPlus II.
(b) For the line

```
ff(1) <= NOT (ff(5) XOR ff(6));
```

substitute

```
ff(1) <= ff(5) XNOR ff(6);
```

and compile with MaxPlus II.
(c) Now change the Compiler settings Interfaces → VHDL Netlist Reader Settings from VHDL 1987 to VHDL 1993 and compile again. Explain the results.
Note: Using Quartus II will not produce any differences by changing the VHDL version settings from VHDL 1993 to VHDL 1987.

8. Adaptive Filters

The filters we have discussed so far had been designed for applications where the requirements for the "optimal" coefficients did not change over time, i.e., they were LTI systems. However, many real-world signals we find in typical DSP fields like speech processing, communications, radar, sonar, seismology, or biomedicine, require that the "optimal" filter or system coefficients need to be adjusted over time depending on the input signal. If the parameter changes slowly compared with the sampling frequency we can compute a "better" estimation for our optimal coefficients and adjust the filter appropriate.

In general, any filter structure, FIR or IIR, with the many architectural variations we have discussed before, may be used as an adaptive digital filter (ADF). Comparing the different structural options, we note that

- For FIR filters the direct form from Fig. 3.1 (p. 166) seems to be advantageous because the coefficient update can be done at the same time instance for all coefficients.
- For IIR filters the lattice structure shown in Fig. 4.12 (p. 227) seems to be a good choice because lattice filters possess a low fixed-point arithmetic roundoff error sensitivity and a simplified stability control of the coefficients.

From the published literature, however, it appears that FIR filters have been used more successfully than IIR filters and our focus in this chapter will therefore be efficient and fast implementation of adaptive FIR filters.

The FIR filter algorithms should converge to the optimum nonrecursive estimator solution given (originally for continuous signal) through the Wiener–Hopf equation [251]. We will then discuss the optimum recursive estimator (Kalman filter). We will compare the different options in terms of computational complexity, stability of the algorithms, initial speed of convergence, consistency of convergence, and robustness to additive noise.

Adaptive filters can now be seen to be a mature DSP field. Many books in their first edition had been published in the mid-1980s and can be used for a more in-depth study [252, 253, 254, 255, 256, 257]. More recent results may be found in textbook like [258, 259, 260]. Recent journal publications like IEEE Transactions on Signal Processing show, especially in the area of stability of LMS and its variations, essential research activity.

8.1 Application of Adaptive Filter

Although the application fields of adaptive filters are quite broad in nature, they can usually be described with one of the following four system configurations:

- Interference cancellation
- Prediction
- Inverse modeling
- Identification

We wish to discuss in the following the basic idea of these systems and present some typical successful applications for these classes. Although it may not always exactly describe the nature of the specific signals it is common to use the following notation for all systems, namely

$$x = \text{input to the adaptive filter}$$
$$y = \text{output of the adaptive filter}$$
$$d = \text{desired response (of the adaptive filter)}$$
$$e = d - y = \text{estimation error}$$

8.1.1 Interference Cancellation

In these very popular applications of the adaptive filter the incoming signal contains, beside the information-bearing signal, also an interference, which may, for example, be a random white noise or the $50/60\,\text{Hz}$ power-line hum. Figure 8.1 shows the configuration for this application. The incoming (sensor) signal $d[n]$ and the adaptive filter output response $y[n]$ to a reference signal $x[n]$ is used to compute the error signal $e[n]$, which is also the system output in the interference cancellation configuration. Thus, after convergence, the (modified) reference signal, which will represent the additive inverse of the interference is subtracted from the incoming signal.

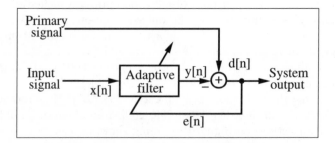

Fig. 8.1. Basic configuration for interference cancellation.

We will later study a detailed example of the interference cancellation of the power-line hum. A second popular application is the adaptive noise cancellation of echoes on telephone systems. Interference cancellation has also been used in an array of antennas (called beamformer) to adaptively remove noise interferring from unknown directions.

8.1.2 Prediction

In the prediction application the task of the adaptive filter is to provide a best prediction (usually in the least mean square sense) of a present value of a random signal. This is obviously only possible if the input signal is essential different from white noise. Prediction is illustrated in Fig. 8.2. It can be seen that the input $d[n]$ is applied over a delay to the adaptive filter input, as well as to compute the estimation error.

The predictive coding has been successfully used in image and speech signal processing. Instead of coding the signal directly, only the prediction error is encoded for transmission or storage. Other applications include the modeling of power spectra, data compression, spectrum enhancement, and event detection [253].

8.1.3 Inverse Modeling

In the inverse modeling structure the task is to provide an inverse model that represents the best fit (usually in the least squares sense) to an unknown time-varying plant. A typical communication example would be the task to estimate the multipath propagation of the signal to approximate an ideal transmission. The system shown in Fig. 8.3 illustrates this configuration. The input signal $d[n]$ enters the plant and the output of the unknown plant $x[n]$ is the input to the adaptive filter. A delayed version of the input $d[n]$ is then used to compute the error signal $e[n]$ and to adjust the filter coefficients of the adaptive filter. Thus, after convergence, the adaptive filter transfer function approximates the inverse of the transfer function of the unknown plant.

Besides the already-mentioned equalization in communication systems, inverse modeling with adaptive filters has been successfully used to improve

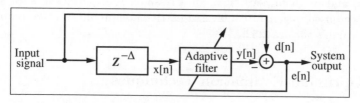

Fig. 8.2. Block diagram for prediction.

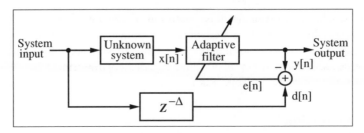

Fig. 8.3. Schematic diagram illustrating the inverse system modeling.

S/N ratio for additive narrowband noise, for adaptive control systems, in speech signal analysis, for deconvolution, and digital filter design [253].

8.1.4 Identification

In a system identification application the task is that the filter coefficients of the adaptive filter represent an unknown plant or filter. The system identification is shown in Fig. 8.4 and it can be seen that the time series, $x[n]$, is input simultaneously to the adaptive filter and another linear plant or filter with unknown transfer function. The output of the unknown plant $d[n]$ becomes the output of the entire system. After convergence the adaptive filter output $y[n]$ will approximate $d[n]$ in an optimum (usually least mean squares) sense. Provided that the order of the adaptive filter matches the order of the unknown plant and the input signal $x[n]$ is WSS the adaptive filter coefficients will converge to the same values as the unknown plant. In a practical application there will normally be an additive noise present at the output of the unknown plant (observation errors) and the filter structure will not exactly match that of the unknown plant. This will result in deviation from the perfect performance described. Due to the flexibility of this structure and the ability to individually adjust a number of input parameters independently it is one of the structures often used in the performance evaluations of adaptive filters. We will use these configurations to make a detailed comparison between LMS and RLS, the two most popular algorithms to adjust the filter coefficient of an adaptive filter.

Such system identification has been used for modeling in biology, or to model social and business systems, for adaptive control systems, digital filter design, and in geophysics [253]. In a seismology exploration, such systems have been used to generate a layered-earth model to unravel the complexities of the earth's surface [252].

8.2 Optimum Estimation Techniques

Required signal properties. In order to use successfully the adaptive filter algorithms presented in the following and to guarantee the convergence and stability of the algorithms, it is necessary to make some basic assumptions about the nature of our input signals, which from a probabilistic standpoint, can be seen as a vector of random variables. First, the input signal (i.e., the random variable vector) should be *ergodic*, i.e., statistical properties like mean

$$\eta = E\{x\} = \lim_{N \to \infty} \frac{1}{N} \sum_{n=0}^{N-1} x[n]$$

or variance

$$\sigma^2 = E\{x^2\} = \lim_{N \to \infty} \frac{1}{N} \sum_{n=0}^{N-1} (x[n] - \eta)^2$$

computed using a single input signal should show the same statistical properties like the average over an assemble of such random variables. Secondly, the signals need to be wide sense stationary (WSS), i.e., statistics measurements like average or variance measured over the assemble averages are not a function of the time, and the autocorrelation function

$$r[\tau] = E\{x[t_1]x[t_2]\} = E\{x[t + \tau]x[t]\}$$

$$= \lim_{N \to \infty} \frac{1}{N} \sum_{n=0}^{N-1} x[n]x[n + \tau]$$

depends only on the difference $\tau = t_1 - t_2$. We note in particular that

$$r[0] = E\{x[t]x[t]\} = E\{|x[t]|^2\} \tag{8.1}$$

computes the average power of the WSS process.

Definition of cost function. The definition of the cost function applied to the estimator output is a critical parameter in all adaptive filter algorithms. We need to "weight" somehow the estimation error

$$e[n] = d[n] - y[n], \tag{8.2}$$

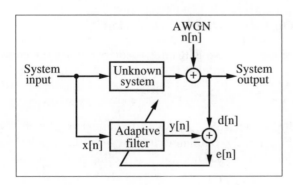

Fig. 8.4. Basic configuration for identification.

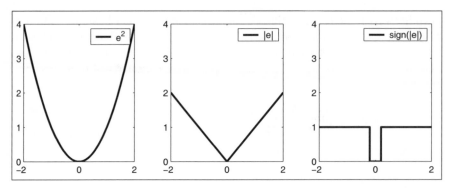

Fig. 8.5. Three possible error cost functions.

where $d[n]$ is the random variable to be estimated, and $y[n]$ is the computed estimate via the adaptive filter. The most commonly used cost function is the least-mean-squares (LMS) function given as

$$J = E\{e^2[n]\} = \overline{(d[n] - y[n])^2}. \tag{8.3}$$

It should be noted that this is not the only cost function that may be used. Alternatives are functions such as the absolute error or the nonlinear threshold functions as shown in Fig 8.5 on the right. The nonlinear threshold type may be used if a certain error level is acceptable and as we will see later can reduce the computational burden of the adaptation algorithm. It may be interesting to note that the original adaptive filter algorithms by Widrow [255] uses such a threshold function for the error.

On the other hand, the quadratic error function of the LMS method will enable us to build a stochastic gradient approach based on the Wiener–Hopf relation originally developed in the continuous signal domain. We review the Wiener–Hopf estimation in the next subsection, which will directly lead to the popular LMS adaptive filter algorithms first proposed by Widrow et al. [261, 262].

8.2.1 The Optimum Wiener Estimation

The output of the adaptive FIR filter is computed via the convolution sum

$$y[n] = \sum_{k=0}^{L-1} f_k x[n-k], \tag{8.4}$$

where the filter coefficients f_k have to be adjusted in such a way that the defined cost function J is minimum. It is, in general, more convenient to write the convolution with vector notations according to

$$y[n] = \boldsymbol{x}^T[n]\boldsymbol{f} = \boldsymbol{f}^T\boldsymbol{x}[n], \tag{8.5}$$

with $\boldsymbol{f} = [f_0 f_1 \dots f_{L-1}]^T$, $\boldsymbol{x}[n] = [x[n]x[n-1]\dots x[n-(L-1)]]^T$, are size $(L \times 1)$ vectors and T means matrix transposition or the Hermitian transposition for complex data. For $\boldsymbol{A} = [a[k,l]]$ the transposed matrix is "mirrored" at the main diagonal, i.e., $\boldsymbol{A}^T = [a[l,k]]$. Using the definition of the error function (8.2) we get

$$e[n] = d[n] - y[n] = d[n] - \boldsymbol{f}^T \boldsymbol{x}[n]. \tag{8.6}$$

The mean square error function now becomes

$$
\begin{aligned}
J &= E\{e^2[n]\} = E\{d[n] - y[n]\}^2 = E\{d[n] - \boldsymbol{f}^T \boldsymbol{x}[n]\}^2 \\
&= E\{(d[n] - \boldsymbol{f}^T \boldsymbol{x}[n])(d[n] - \boldsymbol{x}^T[n]\boldsymbol{f})\} \\
&= E\{d[n]^2 - 2d[n]\boldsymbol{f}^T \boldsymbol{x}[n] + \boldsymbol{f}^T \boldsymbol{x}[n]\boldsymbol{x}^T[n]\boldsymbol{f}\}. \tag{8.7}
\end{aligned}
$$

Note that the error is a quadratic function of the filter coefficients that can be pictured as a concave hyperparaboloidal surface, a function that never goes negative, see Fig. 8.6 for an example with two filter coefficients. Adjusting the filter weights to minimize the error involves descending along this surface with the objective of getting to the bottom of the bowl. Gradient methods are commonly used for this purpose. The choice of mean square type of cost function will enable a well-behaved quadratic error surface with a single unique minimum. The cost is minimum if we differentiate (8.7) with respect to $\boldsymbol{f}$ and set this gradient to zero, i.e.,

$$\nabla = \frac{\partial J}{\partial \boldsymbol{f}^T} = E\left\{(-2d[n]\boldsymbol{x}[n] + 2\boldsymbol{x}^T[n]\boldsymbol{x}[n]\boldsymbol{f}_{\text{opt}}\right\} = 0.$$

Assuming that the filter weight vector $\boldsymbol{f}$ and the signal vector $\boldsymbol{x}[n]$ are statistically independent (i.e., uncorrelated), it follows, that

$$E\{d[n]\boldsymbol{x}[n]\} = E\{\boldsymbol{x}[n]\boldsymbol{x}^T[n]\}\boldsymbol{f}_{\text{opt}},$$

then the optimal filter coefficient vector $\boldsymbol{f}_{\text{opt}}$ can be computed with,

$$\boldsymbol{f}_{\text{opt}} = E\{\boldsymbol{x}[n]\boldsymbol{x}^T[n]\}^{-1}E\{d[n]\boldsymbol{x}[n]\}. \tag{8.8}$$

The expectation terms are usually defined as follows:

$$
\begin{aligned}
\boldsymbol{R}_{\boldsymbol{xx}} &= E\{\boldsymbol{x}[n]\boldsymbol{x}^T[n]\} \\
&= E \begin{bmatrix} x[n]x[n] & x[n]x[n-1] & \dots & x[n]x[n-(L-1)] \\ x[n-1]x[n] & x[n-1]x[n-1] & \dots & \\ \vdots & & \ddots & \vdots \\ x[n-(L-1)]x[n] & & \dots & \end{bmatrix} \\
&= \begin{bmatrix} r[0] & r[1] & \dots & r[L-1] \\ r[1] & r[0] & \dots & r[L-2] \\ \vdots & & \ddots & \vdots \\ r[L-1] & r[L-2] & \dots & r[0] \end{bmatrix}
\end{aligned}
$$

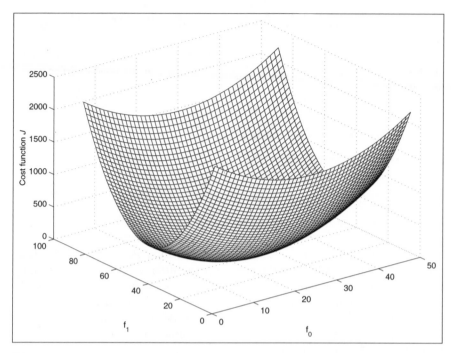

Fig. 8.6. Error cost function for the two-component case. The minimum of the cost function is at $f_0 = 25$ and $f_1 = 43.3$.

is the $(L \times L)$ autocorrelation matrix of the input signal sequence, which has the form of the Toeplitz matrix, and

$$\boldsymbol{r}_{dx} = E\{d[n]\boldsymbol{x}[n]\}$$

$$= E \begin{bmatrix} d[n]x[n] \\ d[n]x[n-1] \\ \vdots \\ d[n]x[n-(L-1)] \end{bmatrix} = \begin{bmatrix} r_{dx}[0] \\ r_{dx}[1] \\ \vdots \\ r_{dx}[L-1] \end{bmatrix}$$

is the $(L \times 1)$ cross-correlation vector between the desired signal and the reference signal. With these definitions we can now rewrite (8.8) more compactly as

$$\boldsymbol{f}_{\text{opt}} = \boldsymbol{R_{xx}}^{-1}\boldsymbol{r}_{dx}. \qquad (8.9)$$

This is commonly recognized as the Wiener–Hopf equation [251], which yields the optimum LMS solution for the filter coefficient vector $\boldsymbol{f}_{\text{opt}}$. One requirement to have a unique solution for (8.9) is that $\boldsymbol{R_{xx}}^{-1}$ exist, i.e., the autocorrelation matrix must be nonsingular, or put differently, the determinate is nonzero. Fortunately, it can be shown that for WSS signals the $\boldsymbol{R_{xx}}$ matrix is nonsingular [252, p. 41] and the inverse exists.

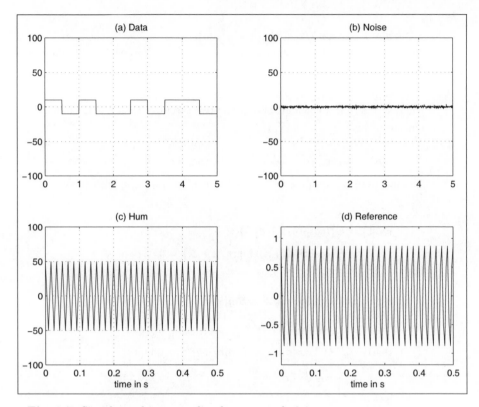

Fig. 8.7. Signals used in power-line hum example 8.1.

Using (8.7) the residue error of the optimal estimation becomes:

$$\boldsymbol{J}_{\mathrm{opt}} = E\{d[n] - \boldsymbol{f}_{\mathrm{opt}}^T \boldsymbol{x}[n]\}^2$$
$$= E\{d[n]\}^2 - 2\boldsymbol{f}_{\mathrm{opt}}^T \boldsymbol{r}_{dx} + \boldsymbol{f}_{\mathrm{opt}}^T \underbrace{\boldsymbol{R}_{xx}\boldsymbol{f}_{\mathrm{opt}}}_{\boldsymbol{r}_{dx}}$$

$$\boldsymbol{J}_{\mathrm{opt}} = \boldsymbol{r}_{dd}[0] - \boldsymbol{f}_{\mathrm{opt}}^T \boldsymbol{r}_{dx}, \tag{8.10}$$

where $\boldsymbol{r}_{dd}[0] = \sigma_{\boldsymbol{d}}^2$ is the variance of $\boldsymbol{d}$.

We now wish to demonstrate the Wiener–Hopf algorithm with the following example.

Example 8.1: Two-tap FIR Filter Interference Cancellation

Suppose we have an observed communication signal that consists of three components: The information-bearing signal, which is a Manchester encoded sensor signal $m[n]$ with amplitude $B = 10$, shown in Fig. 8.7a; an additive white Gaussian noise $n[n]$, shown in Fig. 8.7b; and a 60-Hz power-line hum interference with amplitude $A = 50$, shown in Fig. 8.7c. Assuming the sampling frequency is 4 times the power-line hum frequency, i.e., $4 \times 60 = 240\,\mathrm{Hz}$, the observed signal can therefore be formulated as follows

$$d[n] = A\cos[\pi n/2] + Bm[n] + \sigma^2 n[n].$$

The reference signal $x[n]$ (shown in Fig. 8.7d), which is applied to the adaptive filter input, is given as

$$x[n] = \cos[\pi n/2 + \phi],$$

where $\phi = \pi/6$ is a constant offset. The two-tap filter then has the following output:

$$x[n] = f_0 \cos\left[\frac{\pi}{2}n + \phi\right] + f_1 \cos\left[\frac{\pi}{2}(n-1) + \phi\right].$$

To solve (8.9) we compute first the autocorrelation for $x[n]$ with delays 0 and 1:

$$r_{xx}[0] = E\{(\cos[\pi n/2 + \phi])^2\} = \frac{1}{2}$$

$$r_{xx}[1] = E\{\cos[\pi n/2 + \phi]\sin[\pi n/2 + \phi]\} = 0.$$

For the cross-correlation we get

$$r_{dx}[0] = E\left\{(A\cos[\pi n/2] + Bm[n] + \sigma^2 n[n])\cos[\pi n/2 + \phi]\right\}$$

$$= \frac{A}{2}\cos(\phi) = 12.5\sqrt{3}$$

$$r_{dx}[1] = E\left\{(A\cos[\pi n/2] + Bm[n] + \sigma^2 n[n])\sin[\pi n/2 + \phi]\right\}$$

$$= \frac{A}{2}\cos(\phi - \pi) = \frac{50}{4} = 12.5.$$

As required for the Wiener–Hopf equation (8.9) we can now compute the (2×2) autocorrelation matrix and the (2×1) cross-correlation vector and get

$$f_{\text{opt}} = R_{xx}{}^{-1}r_{dx} \begin{bmatrix} r_{xx}[0] & r_{xx}[1] \\ r_{xx}[1] & r_{xx}[0] \end{bmatrix}^{-1} \begin{bmatrix} r_{dx}[0] \\ r_{dx}[1] \end{bmatrix}$$

$$= \begin{bmatrix} 0.5 & 0 \\ 0 & 0.5 \end{bmatrix}^{-1} \begin{bmatrix} 12.5\sqrt{3} \\ 12.5 \end{bmatrix} = \begin{bmatrix} 2 & 0 \\ 0 & 2 \end{bmatrix} \begin{bmatrix} 12.5\sqrt{3} \\ 12.5 \end{bmatrix}$$

$$= \begin{bmatrix} 25\sqrt{3} \\ 25 \end{bmatrix} = \begin{bmatrix} 43.3 \\ 25 \end{bmatrix}.$$

The simulation of these data is shown in Fig. 8.8. It shows (a) the sum of the three signals (Manchester-coded 5 bits, power-line hum of 60 Hz, and the additive white Gaussian noise) and the system output (i.e., $e[n]$) with the canceled power-line hum.   8.1

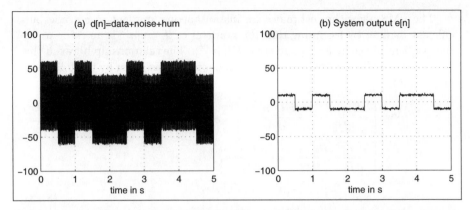

Fig. 8.8. Canceling 60-Hz power-line interference of a Manchester-coded data signal using optimum Wiener estimation.

8.3 The Widrow–Hoff Least Mean Square Algorithm

There may exist a couple of reasons why we wish to avoid a direct computation of the Wiener estimation (8.9). First, the generation of the autocorrelation matrix $\boldsymbol{R}_{xx}$ and the cross-correlation vector $\boldsymbol{r}_{dx}$ are already computationally intensive. We need to compute the autocorrelation of $\boldsymbol{x}$ and the cross-correlation between $\boldsymbol{d}$ and $\boldsymbol{x}$ and we may, for instance, not know how many data samples we need to use in order to have sufficient statistics. Secondly, if we have constructed the correlation functions we still have to compute the inverse of the autocorrelation matrix $\boldsymbol{R}_{xx}{}^{-1}$, which can be very time consuming, if the filter order gets larger. Even if a procedure is available to invert $\boldsymbol{R}_{xx}$, the precision of the result may not be sufficient because of the many computational steps involved, especially with a fixed-point arithmetic implementation.

The Widrow–Hoff least mean square (LMS) adaptive algorithm [261] is a practical method for finding a close approximation to (8.9) in real time. The algorithm does not require explicit measurement of the correlation functions, nor does it involve matrix inversion. Accuracy is limited by statistical sample size, since the filter coefficient values are based on the real-time measurements of the input signals.

The LMS algorithm is an implementation of the method of the steepest descent. According to this method, the next filter coefficient vector $\boldsymbol{f}[n+1]$ is equal to the present filter coefficient vector $\boldsymbol{f}[n]$ plus a change proportional to the negative gradient:

$$\boldsymbol{f}[n+1] = \boldsymbol{f}[n] - \frac{\mu}{2}\nabla[n]. \tag{8.11}$$

The parameter μ is the learning factor or step size that controls stability and the rate of convergence of the algorithm. During each iteration the true gradient is represented by $\nabla[n]$.

The LMS algorithm estimates an instantaneous gradient in a crude but efficient manner by assuming that the gradient of $J = e[n]^2$ is an estimate of the gradient of the mean-square error $E\{e[n]^2\}$. The relationship between the true gradient $\nabla[n]$ and the estimated gradients $\hat{\nabla}[n]$ is given by the following expression:

$$\nabla[n] = \left[\frac{\partial E\{e[n]^2\}}{\partial f_0}, \frac{\partial E\{e[n]^2\}}{\partial f_1}, \ldots, \frac{\partial E\{e[n]^2\}}{\partial f_{L-1}}\right]^T \tag{8.12}$$

$$\hat{\nabla}[n] = \left[\frac{\partial e[n]^2}{\partial f_0}, \frac{\partial e[n]^2}{\partial f_1}, \ldots, \frac{\partial e[n]^2}{\partial f_{L-1}}\right]^T$$

$$= 2e[n]\left[\frac{\partial e[n]}{\partial f_0}, \frac{\partial e[n]}{\partial f_1}, \ldots, \frac{\partial e[n]}{\partial f_{L-1}}\right]^T. \tag{8.13}$$

The estimated gradient components are related to the partial derivatives of the instantaneous error with respect to the filter coefficients, which can be obtained by differentiating (8.6), it follows that

$$\hat{\nabla}[n] = -2e[n]\frac{\partial e[n]}{\partial \boldsymbol{f}} = -2e[n]\boldsymbol{x}[n]. \tag{8.14}$$

Using this estimate in place of the true gradient in (8.11) yields:

$$\boldsymbol{f}[n+1] = \boldsymbol{f}[n] - \frac{\mu}{2}\hat{\nabla}[n] = \boldsymbol{f}[n] + \mu e[n]\boldsymbol{x}[n]. \tag{8.15}$$

Let us summarize all necessary step for the LMS algorithm[2] in the following

---

Algorithm 8.2:   Widrow–Hoff LMS Algorithm

The Widrow–Hoff LMS algorithm to adjust the L filter coefficients of an adaptive uses the following steps:

1) Initialize the $(L \times 1)$ vector $\boldsymbol{f} = \boldsymbol{x} = \boldsymbol{0} = [0, 0, \ldots, 0]^T$.
2) Accept a new pair of input samples $\{x[n], d[n]\}$ and shift $x[n]$ in the reference signal vector $\boldsymbol{x}[n]$.
3) Compute the output signal of the FIR filter, via
   $$y[n] = \boldsymbol{f}^T[n]\boldsymbol{x}[n]. \tag{8.16}$$
4) Compute the error function with
   $$e[n] = d[n] - y[n]. \tag{8.17}$$
5) Update the filter coefficients according to
   $$\boldsymbol{f}[n+1] = \boldsymbol{f}[n] + \mu e[n]\boldsymbol{x}[n]. \tag{8.18}$$
   Now continue with step 2.

Although the LMS algorithm makes use of gradients of mean-square error functions, it does not require squaring, averaging, or differentiation. The al-

---

[2] Note that in the original presentation of the algorithm [261] the update equation $\boldsymbol{f}[n+1] = \boldsymbol{f}[n] + 2\mu e[n]\boldsymbol{x}[n]$ is used because the differentiation of the gradient in (8.14) produces a factor 2. The update equation (8.15) follows the notation that is used in most of the current textbooks on adaptive filters.

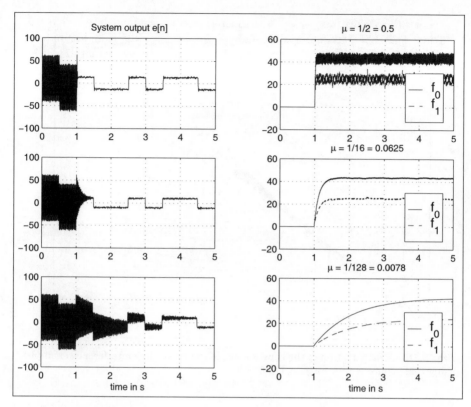

Fig. 8.9. Simulation of the power-line interference cancellation using the LMS algorithm for three different values of the step size μ. **(left)** System output $e[n]$. **(right)** Filter coefficients.

gorithm is simple and generally easy to implement in software (MATLAB code see, for instance, [260, p. 332]; C code [263], or PDSP assembler code [264]).

A simulation using the same system configuration as in Example 8.1 (p. 485) is shown in Fig. 8.9 for different values of the step size μ. Adaptation starts after 1 second. System output $e[n]$ is shown in the left column and the filter coefficient adaptation on the right. We note that depending on the value μ the optimal filter coefficients approach $f_0 = 43.3$ and $f_1 = 25$.

It has been shown that the gradient estimate used in the LMS algorithm is unbiased and that the expected value of the weight vector converges to the Wiener weight vector (8.9) when the input signals are WSS, which was anyway required in order to be able to compute the inverse of the autocorrelation matrix $\boldsymbol{R_{xx}}^{-1}$ for the Wiener estimate. Starting with an arbitrary initial filter coefficient vector, the algorithm will converge in the mean and will remain stable as long as the learning parameter μ is greater than 0 but less than an upper bound $\mu_{\max}$. Figure 8.10 shows an alternative form to

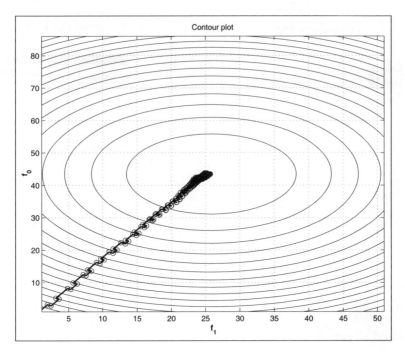

Fig. 8.10. Demonstration of the convergence of the power-line interference example using a 2D contour plot for $\mu = 1/16$.

represent the convergence of the filter coefficient adaptation by a projection of the coefficient values in a (f_0, f_1) mapping. The figure also shows the contour line with equal error. It can be seen that the LMS algorithm moves in a zigzag way towards the minimum rather than the true gradient, which would move exactly orthogonal to these error contour lines.

Although the LMS algorithm is considerably simpler than the RLS algorithm (we will discuss this later) the convergence properties of the LMS algorithm are nonetheless difficult to analysis rigorously. The simplest approach to determine an upper bound of μ makes use of the eigenvalues of $\boldsymbol{R_{xx}}$ by solving the homogeneous equation

$$0 = \det(\lambda \boldsymbol{I} - \boldsymbol{R_{xx}}), \tag{8.19}$$

where $\boldsymbol{I}$ is the $L \times L$ identity matrix. There are L eigenvalues $\lambda[k]$ that have the following properties

$$\det(\boldsymbol{R_{xx}}) = \prod_{k=0}^{L-1} \lambda[k] \qquad \text{and} \tag{8.20}$$

$$\text{trace}(\boldsymbol{R_{xx}}) = \sum_{k=0}^{L-1} \lambda[k]. \tag{8.21}$$

From this eigenvalue analysis of the autocorrelation matrix it follows that the LMS algorithm will be stable (in the mean sense) if

$$0 < \mu < \frac{2}{\lambda_{\max}}. \tag{8.22}$$

Although the filter is assumed to be stable, we will see later that this upper bound will not guarantee a finite mean square error, i.e., that $f[n]$ converges to f_{opt} and a much more stringent bound has to be used.

The simulation of the LMS algorithm in Fig. 8.9 also reveals the underlying exponential nature of the individual learning curves. Using the eigenvalue analysis we may also transform the filter coefficient in independent so-called "modes" that are no longer linear dependent. The number of natural modes is equal to the number of degrees of freedom, i.e., the number of independent components and in our case identically with the number of filter coefficients. The time constant of the kth mode is related to the k eigenvalue $\lambda[k]$ and the parameter μ by

$$\tau[k] = \frac{1}{2\mu\lambda[k]}. \tag{8.23}$$

Hence the longest time constant, $\tau_{\max}$, is associated with the smallest eigenvalue, $\lambda_{\min}$ via

$$\tau_{\max} = \frac{1}{2\mu\lambda_{\min}}. \tag{8.24}$$

Combining (8.22) and (8.24) gives

$$\tau_{\max} > \frac{\lambda_{\max}}{2\lambda_{\min}}, \tag{8.25}$$

which suggests that the larger the eigenvalue ratio (EVR), $\lambda_{\max}/\lambda_{\min}$ of the autocorrelation matrix $\boldsymbol{R_{xx}}$ the longer the LMS algorithm will take to converge. Simulation results that confirm this finding can be found for instance, in [259, p. 64] and will be discussed, in Sect. 8.3.1 (p. 493).

The results presented so far on the ADF stability can be found in most original published work by Widrow and many textbooks. However, these conditions do *not* guarantee a finite variance for the filter coefficient vector, neither do they guarantee a finite mean-square error! Hence, as many users of the algorithm realized, considerably more stringent conditions are required to ensure convergence of the algorithm. In the examples in [260, p. 130], for instance, you find the "rule of thumb" that a factor 10 smaller values for μ should be used.

More recent results indicate that the bound from (8.22) must be more restrictive. For example, the results presented by Horowitz and Senne [265] and derived in a different way by Feuer and Weinstein [266] show that the step size (assuming that the elements of the input vector $\boldsymbol{x}[n]$ are statistically independent) has to be restricted via the two conditions:

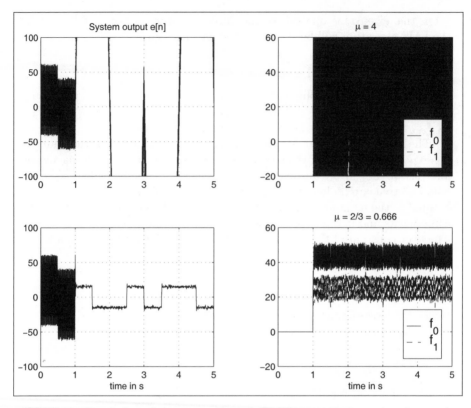

Fig. 8.11. Simulation of the power-line interference cancellation using the maximum step size values for the LMS algorithm. **(left)** System output $e[n]$. **(right)** filter coefficients.

$$0 < \mu < \frac{1}{\lambda_l} \qquad l = 0, 1, \dots, L-1 \qquad \text{and} \qquad (8.26)$$

$$\sum_{l=0}^{L-1} \frac{\mu \lambda_l}{1 - \mu \lambda_l} < 2, \qquad (8.27)$$

to ensure convergence. These conditions can not be solved analytically, but it can be shown that they are closely bounded by the following condition:

$$0 < \mu < \frac{2}{3 \times \text{trace}(\boldsymbol{R_{xx}})} = \frac{2}{3 \times L \times r_{xx}[0]}. \qquad (8.28)$$

The upper bound of (8.28) has a distinct practical advantage. Trace of $\boldsymbol{R_{xx}}$ is, by definition, (see (8.21), p. 490) the total average input signal power of the reference signal, which can easily be estimated from the reference signal $x[n]$.

Example 8.3: Bounds on Step Size

From the analysis in (8.22) we see that we first need to compute the eigenvalues of the $\boldsymbol{R_{xx}}$ matrix, i.e.

$$0 = \det(\lambda \boldsymbol{I} - \boldsymbol{R_{xx}}) = \begin{bmatrix} r_{xx}[0] - \lambda & r_{xx}[1] \\ r_{xx}[1] & r_{xx}[0] - \lambda \end{bmatrix} \tag{8.29}$$

$$= \det \begin{bmatrix} 0.5 - \lambda & 0 \\ 0 & 0.5 - \lambda \end{bmatrix} = (0.5 - \lambda)^2 \tag{8.30}$$

$$\lambda[1,2] = 0.5. \tag{8.31}$$

Using (8.22) gives

$$\mu_{\max} = \frac{2}{\lambda_{\max}} = 4. \tag{8.32}$$

Using the more restrictive bound from (8.28) yields

$$\mu_{\max} = \frac{2}{L \times 3 \times r_{xx}[0]} = \frac{2}{3 \times 2 \times 0.5} = \frac{2}{3}. \tag{8.33}$$

The simulation results in Fig. 8.11 indicate that in fact $\mu = 4$ does not show convergence, while $\mu = 2/3$ converges. $\boxed{8.3}$

We also note from the simulation shown in Fig. 8.11 that even with $\mu_{\max} = 2/3$ the convergence is much faster, but the coefficients "ripple around" essentially. Much smaller values for μ are necessary to have a smooth approach of the filter coefficient to the optimal values and to stay there.

The condition found by Horowitz and Senne [265] and Feuer and Weinstein [266] made the assumption that all inputs $\boldsymbol{x}[n]$ are statistically independent. This assumption is true if the input data come, for instance, from an antenna array of L independent sensors, however, for ADFs with the tapped delay structure, it has been shown, for instance, by Butterweck [252], that for a long filter the stability bound can be relaxed to

$$0 < \mu < \frac{2}{L \times r_{xx}[0]}, \tag{8.34}$$

i.e., compared with (8.28) the upper bound can be relaxed by a factor of 3 in the denominator. But the condition (8.34) only applies for a long filter and it may therefore saver to use (8.28).

8.3.1 Learning Curves

Learning curve, i.e., the error function J displayed over the number of iterations is an important measurement instrument when comparing the performance of different algorithms and system configurations. We wish in the following to study the LMS algorithm regarding the eigenvalue ratio $\lambda_{\max}/\lambda_{\min}$ and the sensitivity to signal-to-noise (S/N) ratio in the system to be identified.

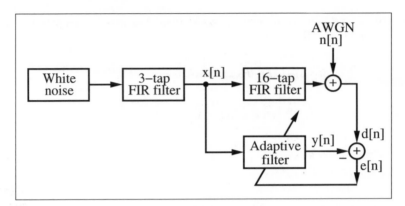

Fig. 8.12. System identification configuration for LMS learning curves.

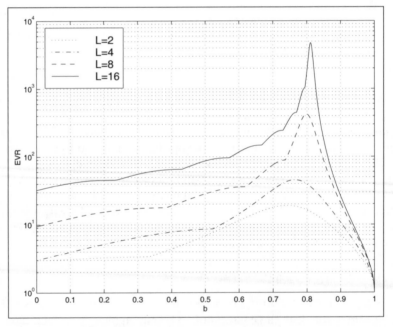

Fig. 8.13. Eigenvalues ratio for a three-tap filter system for different system size L.

A typical performance measurement of adaptive algorithms using a sys–tem-identification problem is displayed in Fig. 8.12. The adaptive filter has a length of $L = 16$ the same length as the "unknown" system, whose coefficients have to be learned. The additive noise level behind the "unknown" system has been set to two different levels -10 dB for a high-noise environment and to -48 dB for a low-noise environment equivalent to an 8-bit quantization.

Table 8.1. Four different noise-shaping FIR filters to generate power-of-ten eigenvalue ratios for $L = 16$.

No.	Impulse response	EVR
1	$0 + 1z^{-1} + 0.0z^{-2}$	1
2	$0.247665 + 0.936656z^{-1} + 0.247665z^{-2}$	10
3	$0.577582 + 0.576887z^{-1} + 0.577582z^{-2}$	100
4	$0.432663 + 0.790952z^{-1} + 0.432663z^{-2}$	1000

For the LMS algorithm the eigenvalue ratio (EVR) is the critical parameter that determines the convergence speed, see (8.25), p. 491. In order to generate a different eigenvalue ratio we use a white Gaussian noise source with $\sigma^2 = 1$ that is shaped by a digital filter. We may, for instance, use a first-order IIR filter that generates a first-order Markov process, see Exercise 8.10 (p. 533). We may alternatively filter the white noise by a three-tap symmetrical FIR filter whose coefficients are $c^T = [a, b, a]$. The FIR filter has the advantage that we can easily normalize the power. The coefficients should be normalized $\sum_k c_k^2 = 1$ in such a way that input and output sequences have the same power. This requires that

$$1 = a^2 + b^2 + a^2 \quad \text{or} \quad a = 0.5 \times \sqrt{1 - b^2}. \tag{8.35}$$

With this filter it is possible to generate different eigenvalue ratios $\lambda_{\max}/\lambda_{\min}$ as shown in Fig. 8.13 for different system size $L = 2, 4, 8$, and 16. We can now use Table 8.1 to get power-of-ten EVRs for the system of length $L = 16$.

For a white Gaussian source the R_{xx} matrix is a diagonal matrix $\sigma^2 I$ and the eigenvalues are therefore all one, i.e., $\lambda_l = 1; l = 0, 1, \ldots, L - 1$. The other EVRs can be verified with MATLAB, see Exercise 8.9 (p. 533). The impulse response of the unknown system g_k is an odd filter with coefficients $1, -2, 3, -4, \ldots, -3, 2, -1$ as shown in Fig. 8.14a. The step size for the LMS algorithm has been determined with

$$\mu_{\max} = \frac{2}{3 \times L \times E\{x^2\}} = \frac{1}{24}. \tag{8.36}$$

In order to guarantee perfect stability the step size has been chosen to be $\mu = \mu_{\max}/2 = 1/48$. The learning curve, or coefficient error is computed via the normalized error function

$$J[n] = 20 \log_{10} \left(\frac{\sum_{k=0}^{15} (g_k - f_k[n])^2}{\sum_{k=0}^{15} g_k^2} \right). \tag{8.37}$$

The coefficient adaptation for a single adaptation run with EVR=1 is shown in Fig. 8.14b. It can be seen that after 200 iterations the adaptive filter has learned the coefficient of the unknown system without an error. From the learning curves (average over 50 adaptation cycles) shown in Fig. 8.14c and d it can be seen that the LMS algorithm is very sensitive to the EVR. Many

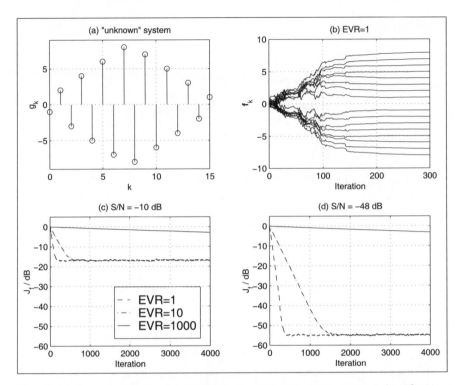

Fig. 8.14. Learning curves for the LMS algorithm using the system identification configuration shown in Fig. 8.12. (a) Impulse response g_k of the "unknown" system. (b) Coefficient learning over time. (c) Average over 50 learning curves for large system noise. (d) Average over 50 learning curves for small system noise.

iterations are necessary in the case of the high EVR. Unfortunately, many real-world signals have high EVR. Speech signals, for instance, may have EVR of 1874 [267]. On the other hand, we see from Fig. 8.14c that the LMS algorithm still adapts well in a high-noise environment.

8.3.2 Normalized LMS (NLMS)

The LMS algorithm discussed so far uses a constant step size μ proportional to the stability bound $\mu_{\max} = 2/(L \times r_{xx}[0])$. Obviously this requires knowledge of the signal statistic, i.e., $r_{xx}[0]$, and this statistic must not change over time. It is, however, possible that this statistic changes over time, and we wish to adjust μ accordingly. These can be accomplished by computing a temporary estimate for the signal power via

$$r_{xx}[0] = \frac{1}{L}\boldsymbol{x}^T[n]\boldsymbol{x}[n], \qquad (8.38)$$

and the "normalized" μ is given by

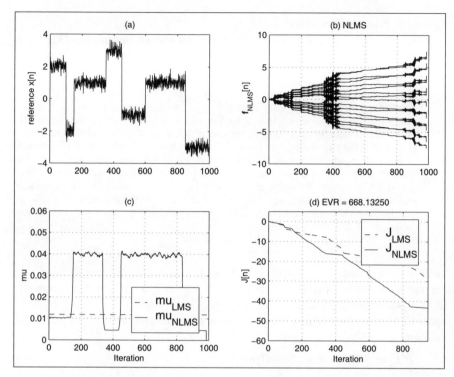

Fig. 8.15. Learning curves for the normalized LMS algorithm using the system identification configuration shown in Fig. 8.12. (a) The reference signal input $x[n]$ to the adaptive filter and the "unknown" system. (b) Coefficient learning over time for the normalized LMS. (c) Step size μ used for LMS and NLMS. (d) Average over 50 learning curves.

$$\mu_{\max}[n] = \frac{2}{\boldsymbol{x}^T[n]\boldsymbol{x}[n]}. \tag{8.39}$$

If we are concerned that the denominator can temporary become very small and μ too large, we may add a small constant δ to $\boldsymbol{x}^T[n]\boldsymbol{x}[n]$, which yields

$$\mu_{\max}[n] = \frac{2}{\delta + \boldsymbol{x}^T[n]\boldsymbol{x}[n]}. \tag{8.40}$$

To be on the safe side, we would not choose $\mu_{\max}[n]$. Instead we would use a somewhat smaller value, like $0.5 \times \mu_{\max}[n]$. The following example should demonstrate the normalized LMS algorithm.

Example 8.4: Normalized LMS

Suppose we have again the system identification configuration from Fig. 8.12 (p. 494), only this time the input signal $x[n]$ to the adaptive filter and the "unknown system" is the noisy pulse-amplitude-modulated (PAM) signal shown in Fig. 8.15a. For the conventional LMS we compute first $r_{xx}[0]$, and calculate $\mu_{\max} = 0.0118$. The step size for the normalized LMS algorithm is adjusted depending on the momentary power $\sum x[n]^2$ of the reference signal. For the computation of $\mu_{\mathrm{NLMS}}[n]$ shown in Fig. 8.15c it can be seen that at times when the absolute value of the reference signal is large the step size is reduced and for small absolute values of the reference signal, a larger step size is used. The adaptation of the coefficient displayed over time in Fig. 8.15b reflects this issue. Larger learning steps can be seen at those times when $\mu_{\mathrm{NLMS}}[n]$ is larger. An average over 50 adaptations is shown in the learning curves in Fig. 8.15d Although the EVR of the noisy PAM is larger than 600, it can be seen that the normalized LMS has a positive effect on the convergence behavior of the algorithm. $\boxed{8.4}$

The power estimation using (8.38) is a precise power snapshot of the current data vector $x[n]$. It may, however, be desired to have a longer memory in the power computation to avoid a temporary small value and a large μ value. This can be accomplished using a recursive update of the previous estimations of the power, with

$$P[n] = \beta P[n-1] + (1-\beta)|x[n]|^2, \tag{8.41}$$

with β less than but close to 1. For a nonstationary signal such as the one shown in Fig. 8.15 the choice of the parameter β must be done carefully. If we select β too small the NLMS will more and more have the performance of the original LMS algorithm, see Exercise 8.14 (p. 534).

8.4 Transform Domain LMS Algorithms

LMS algorithms that solve the filter coefficient adjustment in a transform domain have been proposed for two reasons. The goal of the *fast convolution techniques* [268] is to lower the computational effort, by using block update and transforming the convolution to compute the adaptive filter output and the filter coefficient adjustment in the transform domain with the help of a fast cyclic convolution algorithm. The second method that uses transform-domain techniques has the main goal to improve the adaptation rate of the LMS algorithm, because it is possible to find transforms that allow a "decoupling" of the modes of the adaptive filter [267, 269].

8.4.1 Fast-Convolution Techniques

Fast cyclic convolution using transforms like FFTs or NTTs can be applied to FIR filters. For the adaptive filter this leads to a block-oriented processing of the data. Although we may use any block size, the block size is usually chosen to be twice the size of the adaptive filter length so that the time delay in the coefficient update becomes not too large. It is also most often from a computational effort a good choice. In the first step a block of $2L$ input values $x[n]$ are convolved via transform with the filter coefficients $\boldsymbol{f}_L$, which produces L new filter output values $y[n]$. These results are then used to compute L error signals $e[n]$. The filter coefficient update is then done also in the transform domain, using the already transformed input sequence $x[n]$. Let us go through these block processing steps using a $L = 3$ example. We compute the three filter output signals in one block:

$$y[n] = f_0 x[n] + f_1[n]x[n-1] + f_2[n]x[n-2]$$
$$y[n+1] = f_0 x[n+1] + f_1[n]x[n] + f_2[n]x[n-1]$$
$$y[n+2] = f_0 x[n+2] + f_1[n]x[n+1] + f_2[n]x[n].$$

These can be interpreted as a cyclic convolution of

$$\{f_0, f_1, f_2, 0, 0, 0\} \circledast \{x[n+2], x[n+1], x[n], x[n-1], x[n-2], 0\}.$$

The error signals follow then with

$$e[n] = d[n] - y[n] \qquad e[n+1] = d[n+1] - y[n+1]$$
$$e[n+2] = d[n+2] - y[n+2].$$

The block processing for the filter gradient ∇ can now be written as

$$\nabla[n] = e[n]\boldsymbol{x}[n] \qquad \nabla[n+1] = e[n+1]\boldsymbol{x}[n+1]$$
$$\nabla[n+2] = e[n+2]\boldsymbol{x}[n+2].$$

The update for each individual coefficient is then computed with

$$\nabla_0 = e[n]x[n] + e[n+1]x[n+1] + e[n+2]x[n+2]$$
$$\nabla_1 = e[n]x[n-1] + e[n+1]x[n] + e[n+2]x[n-1]$$
$$\nabla_2 = e[n]x[n-2] + e[n+1]x[n+1] + e[n+2]x[n].$$

We again see that this is a cyclic convolution, only this time the input sequence $x[n]$ appears in reverse order

$$\{0, 0, 0, e[n], e[n+1], e[n+2]\}$$
$$\circledast \{0, x[n-2], x[n-1], x[n], x[n+1], x[n+2]\}.$$

In the Fourier domain the reverse order in time yields that we need to compute the conjugate transform of X. The coefficient update then becomes

$$\boldsymbol{f}[n+L] = \boldsymbol{f}[n] + \frac{\mu_B}{L}\nabla[n]. \tag{8.42}$$

Figure 8.16 shows all the necessary steps, when using the FFT for the fast convolution.

From the stability standpoint the block delay in the coefficient is not uncritical. Feuer [270] has shown that the step size has to be reduced to

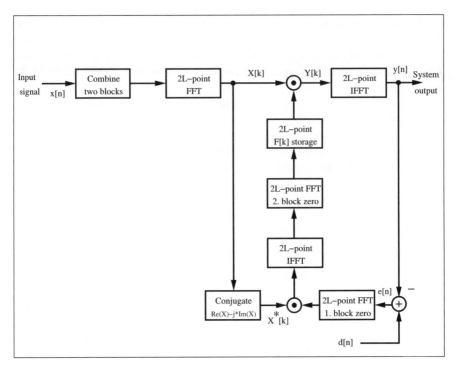

Fig. 8.16. Fast transform-domain filtering method using the FFT.

$$0 < \mu_B < \frac{2B}{(B+2) \times \text{trace}(\boldsymbol{R_{xx}})} = \frac{2}{(1+2/B)L \times r_{xx}[0]}. \qquad (8.43)$$

for a block update of B steps each. If we compare this result with the result for $\mu_{\max}$ from (8.28) page 492 we note that the values are very similar. Only for large block sizes $B >> L$ will the change in μ_B have considerable impact. This reduces to (8.28) for a block size of $B = 1$. However, the time constant is measured in blocks of L data and it follows that the largest time constant for the BLMS algorithm is L times larger then the largest time constant associated with the LMS algorithm.

8.4.2 Using Orthogonal Transforms

We have seen in Sect. 8.3.1 (p. 493) that the LMS algorithm is highly sensitive to the eigenvalue ratio (EVR). Unfortunately, many real-world signals have high EVRs. Speech signals, for instance, may have EVR of 1874 [267]. But it is also well known that the transform-domain algorithms allow a "decoupling" of the mode of the signals. The Karhunen–Loéve transform (KLT) is the optimal method in this respect, but unfortunately not a real time option, see Exercise 8.11 (p. 534). Discrete cosine transforms (DCT) and fast Fourier transform (FFT), followed by other orthogonal transforms like Walsh,

Hadamard, or Haar are the next best choice in terms of convergence speed, see Exercise 8.13, (p. 534) [271, 272].

Let us try in the following to use this concept to improve the learning rate of the identification experiment presented in Sect. 8.3.1 (p. 493), where the adaptive filter has to "learn" an impulse response of an unknown 16-tap FIR filter, as shown in Fig. 8.12 (p. 494). In order to apply the transform techniques and still to monitor the learning progress we need to compute in addition to the LMS algorithm 8.2 (p. 488) the DCT of the incoming reference signal $x[n]$ as well as the IDCT of the coefficient vector f_n. In a practical application we do not need to compute the IDCT, it is only necessary to compute it once after we reach convergence. The following MATLAB code demonstrates the transform-domain DCT-LMS algorithm.

```
for k = L:Iterations        % adapt over full length
    x = [xin;x(1:L-1)];     % get new sample
    din = g'*x + n(k);      % "unknown" filter output + AWGN
    z = dct(x);             % LxL orthogonal transform
    y = f' * z;             % transformed filter output
    err = din-y;            % error: primary - reference
    f = f + err*mu.*z;      % update weight vector
    fi = idct(f);           % filter in original domain
    J(k-L+1) = J(k-L+1) + sum((fi-g).^2); % Learning curve
end
```

The effect of a transform T on the eigenvalue spread can be computed via

$$R_{zz} = T R_{xx} T^H, \qquad (8.44)$$

where the superscript H denotes the transpose conjugate.

The only thing we have not considered so far is that the L "modes" or frequencies of the transformed input signal $z[l]$ are now more or less statistically independent input vectors and the step size μ in the original domain may no longer be appropriate to guarantee stability, or allow fast convergence. In fact, the simulations by Lee and Un [272] show that if no power normalization is used in the transform domain then the convergence did not improve compared with the time-domain LMS algorithm. It is therefore reasonable to compute for these L spectral components different step sizes according to the stability bound (8.28), p. 492, just using the power of the transform components:

$$\mu_{\max}[k] = \frac{2}{3 \times L \times r_{zz,k}[0]} \quad \text{for} \quad k = 0, 1, \ldots, L-1.$$

The additional effort is now the computation of the power normalization of all L spectral components. The MATLAB code above already includes a componentwise update via mu.*z, where the .* stands for the componentwise multiplication.

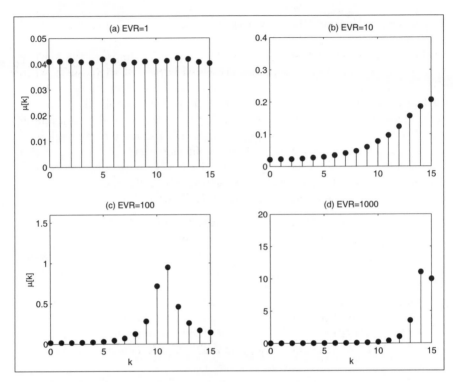

Fig. 8.17. Optimal step size for the DCT-LMS transform-domain algorithm using the system identification configuration shown in Fig. 8.12 (p. 494) for four different eigenvalue ratios. (a) Eigenvalue ratios of 1. (b) Eigenvalue ratios of 10. (c) Eigenvalue ratios of 100. (d) Eigenvalue ratios of 1000.

The adjustment in μ is somewhat similar to the normalized LMS algorithm we have discussed before. We may therefore use directly the power normalization update similar to (8.39) p. 496 for the frequency component. The effect of power normalization and transform T on the eigenvalue spread can be computed via

$$R_{zz} = \lambda^{-1} T R_{xx} T^H \lambda^{-1}, \qquad (8.45)$$

where λ^{-1} is a diagonal matrix that normalizes R_{zz} in such a way that the diagonal elements all become 1, see [271].

Figure 8.17 shows the computed step sizes for four different eigenvalue ratios of the $L = 16$ FIR filter. For a pure Gaussian input all spectral components should be equal and the step size is almost the same, as can be seen from Fig. 8.17a. The other filter shapes the noise in such a way that the power of these spectral components is increased (decreased) and the step size has to be set to a lower (higher) value.

From Fig. 8.18 the positive effect on the performance of the DCT-LMS transform-domain approach can be seen. The learning converges, even for

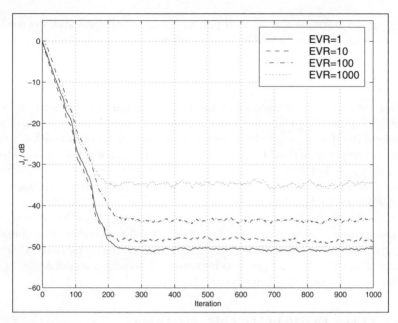

Fig. 8.18. Learning curves for the DCT transform-domain LMS algorithm using the system identification configuration shown in Fig. 8.12 (p. 494) for an average of 50 cycles using four different eigenvalue ratios.

very high eigenvalue ratios like 1000. Only the error floor and consistency of the error at −48 dB is not reached as well for high EVRs as for the lower EVRs.

One factor that must be considered in choosing the transform for real-time application algorithms is the computational complexity. In this respect, real transforms like DCT or DST transforms are superior to complex transform like the FFT, transforms with fast algorithms are better than the algorithms without. Integer transforms like Haar or Hadamard, that do not need multiplications at all, are desirable [271]. Lastly, we also need to take into account that the RLS (discussed later) is another alternative, which has, in general, a higher complexity than the LMS algorithm, but may be more efficient than a transform-domain filter approach and also yield as fast a convergence as the KLT-based LMS algorithm.

8.5 Implementation of the LMS Algorithm

We now wish to look at the task to implement the LMS algorithm with FPGAs. Before we can proceed with a HDL design, however, we need to ensure that quantization effects are tolerable. Later in this section we will then try to improve the throughput by using pipelining, and we need to ensure then also that the ADF is still stable.

8.5.1 Quantization Effects

Before we can start to implement the LMS algorithm in hardware we need to ensure that the parameter and data are well in the "green" range. This can be done if we change the software simulation from full precision to the desired integer precision. Figure 8.19 shows the simulation for 8-bit integer data and $\mu = 1/4, 1/8$ and $1/16$. Note that we can not choose μ too small, otherwise we will no longer get convergence through the large scaling of the gradient $e[n]x[n]$ with μ in the coefficient update equation (8.18), p. 488. The smaller the step size μ the more problem the algorithm has to converge to the optimal values $f_0 = 43.3$ and $f_1 = 25$. This is somehow a contrary requirement to the upper bound on μ given through the stability requirement of the algorithm. It can therefore be necessary to add fractional bits to the system to overcome these two contradictions.

8.5.2 FPGA Design of the LMS Algorithm

A possible implementation of the algorithm represented as a signal flow graph is shown in Fig. 8.20. From a hardware implementation standpoint we note that we need one scaling for μ and $2L$ general multipliers. The effort is therefore more than twice the effort of the programmable FIR filter as discussed in Chap. 3, Example 3.1 (p. 167).

We wish to study in the following the FPLD implementation of the LMS algorithm.

Example 8.5: Two-tap Adaptive LMS FIR Filter

The VHDL design[3] for a filter with two coefficients f_0 and f_1 with a step size of $\mu = 1/4$ is shown in the following listing.

```
-- This is a generic LMS FIR filter generator
-- It uses W1 bit data/coefficients bits
LIBRARY lpm;                    -- Using predefined packages
USE lpm.lpm_components.ALL;

LIBRARY ieee;
USE ieee.std_logic_1164.ALL;
USE ieee.std_logic_arith.ALL;
USE ieee.std_logic_signed.ALL;

ENTITY fir_lms IS                       ------> Interface
  GENERIC (W1 : INTEGER := 8;  -- Input bit width
           W2 : INTEGER := 16; -- Multiplier bit width 2*W1
           L  : INTEGER := 2   -- Filter length
```

---

[3] The equivalent Verilog code fir_lms.v for this example can be found in Appendix A on page 719. Synthesis results are shown in Appendix B on page 731.

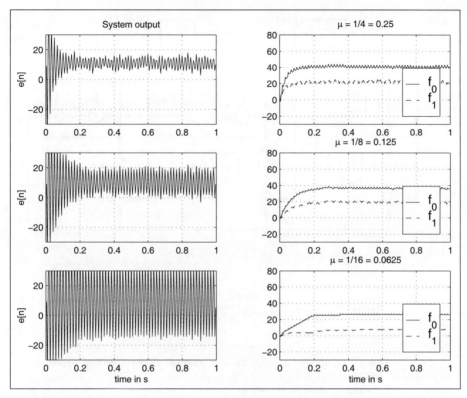

Fig. 8.19. Simulation of the power-line interference cancellation using the LMS algorithm for integer data. **(left)** System output $e[n]$. **(right)** filter coefficients.

```
                );
    PORT ( clk        : IN  STD_LOGIC;
           x_in       : IN  STD_LOGIC_VECTOR(W1-1 DOWNTO 0);
           d_in       : IN  STD_LOGIC_VECTOR(W1-1 DOWNTO 0);
    e_out, y_out      : OUT STD_LOGIC_VECTOR(W2-1 DOWNTO 0);
    f0_out, f1_out    : OUT STD_LOGIC_VECTOR(W1-1 DOWNTO 0));
END fir_lms;

ARCHITECTURE fpga OF fir_lms IS

    SUBTYPE N1BIT IS STD_LOGIC_VECTOR(W1-1 DOWNTO 0);
    SUBTYPE N2BIT IS STD_LOGIC_VECTOR(W2-1 DOWNTO 0);
    TYPE ARRAY_N1BIT IS ARRAY (0 TO L-1) OF N1BIT;
    TYPE ARRAY_N2BIT IS ARRAY (0 TO L-1) OF N2BIT;

    SIGNAL d        : N1BIT;
    SIGNAL emu      : N1BIT;
    SIGNAL y, sxty  : N2BIT;

    SIGNAL e, sxtd  : N2BIT;
```

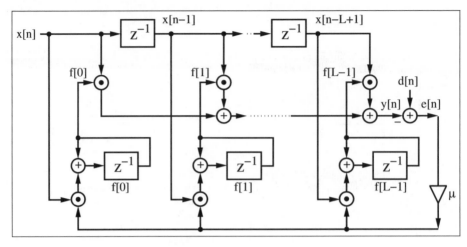

Fig. 8.20. Signal flow graph of the LMS algorithm.

```
SIGNAL  x, f    :   ARRAY_N1BIT; -- Coeff/Data arrays
SIGNAL  p, xemu :   ARRAY_N2BIT; -- Product arrays

BEGIN

  dsxt: PROCESS (d)   -- 16 bit signed extension for input d
  BEGIN
    sxtd(7 DOWNTO 0) <= d;
    FOR k IN 15 DOWNTO 8 LOOP
      sxtd(k) <= d(d'high);
    END LOOP;
  END PROCESS;

  Store: PROCESS    ------> Store these data or coefficients
  BEGIN
    WAIT UNTIL clk = '1';
        d     <= d_in;
        x(0)  <= x_in;
        x(1)  <= x(0);
        f(0)  <= f(0) + xemu(0)(15 DOWNTO 8); -- implicit
        f(1)  <= f(1) + xemu(1)(15 DOWNTO 8); -- divide by 2
  END PROCESS Store;

  MulGen1: FOR I IN 0 TO L-1 GENERATE
  FIR: lpm_mult               -- Multiply p(i) = f(i) * x(i);
       GENERIC MAP ( LPM_WIDTHA => W1, LPM_WIDTHB => W1,
                     LPM_REPRESENTATION => "SIGNED",
                     LPM_WIDTHP => W2,
                     LPM_WIDTHS => W2)
       PORT MAP ( dataa => x(I), datab => f(I),
                                       result => p(I));

       END GENERATE;
```

```
  y <= p(0) + p(1);  -- Compute ADF output

ysxt: PROCESS (y) -- Scale y by 128 because x is fraction
BEGIN
  sxty(8 DOWNTO 0) <= y(15 DOWNTO 7);
  FOR k IN 15 DOWNTO 9 LOOP
    sxty(k) <= y(y'high);
  END LOOP;
END PROCESS;


e <= sxtd - sxty;
emu <= e(8 DOWNTO 1);    -- e*mu divide by 2 and
                         -- 2 from xemu makes mu=1/4
MulGen2: FOR I IN 0 TO L-1 GENERATE
FUPDATE: lpm_mult          -- Multiply xemu(i) = emu * x(i);
         GENERIC MAP ( LPM_WIDTHA => W1, LPM_WIDTHB => W1,
                       LPM_REPRESENTATION => "SIGNED",
                       LPM_WIDTHP => W2,
                       LPM_WIDTHS => W2)
         PORT MAP ( dataa => x(I), datab => emu,
                                         result => xemu(I));
         END GENERATE;

  y_out  <= sxty;   -- Monitor some test signals
  e_out  <= e;
  f0_out <= f(0);
  f1_out <= f(1);

END fpga;
```

The design is a literal interpretation of the adaptive LMS filter architecture found in Fig. 8.20 (p. 506). The output of each tap of the tapped delay line is multiplied by the appropriate filter coefficient and the results are added. The response of the adaptive filter y and of the overall system e to a reference signal x and a desired signal d is shown in Fig. 8.21. The filter adapts after approximately 20 steps at 1 μs to the optimal values $f_0 = 43.3$ and $f_1 = 25$. The design uses 50 LEs, 4 embedded multipliers, and has a 74.59 MHz Registered Performance. $\boxed{8.5}$

The previous example also shows that the standard LMS implementation has a low Registered Performance due to the fact that two multipliers and several add operations have to be performed in one clock cycle before the filter coefficient can be updated. In the following section we wish therefore to study how to achieve a higher throughput.

8.5.3 Pipelined LMS Filters

As can be seen from Fig. 8.20 (p. 506) the original LMS adaptive filter has a long update path and hence the performance already for 8-bit data and coefficients is relatively slow. It is therefore no surprise that many attempts

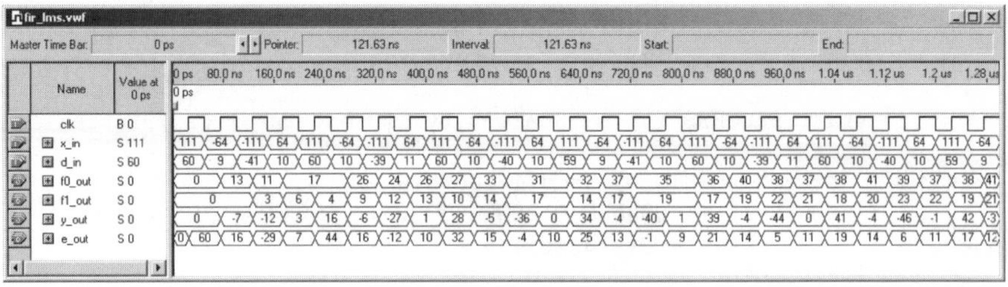

Fig. 8.21. VHDL simulation of the power-line interference cancellation using the LMS algorithm.

have been made to improve the throughput of the LMS adaptive filter. The optimal number of pipeline stages from Fig. 8.20 (p. 506) can be computed as follows: For the $(b \times b)$ multiplier f_k a total of $\log_2(b)$ stages are needed, see also (2.30) p. 85. For the adder tree an additional $\log_2(L)$ pipeline stages would be sufficient and one additional stage for the computation of the error. The coefficient update multiplication requires an additional $\log_2(b)$ pipeline stages. The total number of pipeline stages for a maximum throughput are therefore

$$D_{\mathrm{opt}} = 2\log_2(b) + \log_2(L) + 1, \tag{8.46}$$

where we have assumed that μ is a power-of-two constant and the scaling with μ can be done without the need of additional pipeline stages. If, however, the normalized LMS is used, then μ will no longer be a constant and depending on the bit width of μ additional pipeline stages will be required.

Pipelining an LMS filter is not as simple as for an FIR filter, because the LMS has, as the IIR filter, feedback. We need therefore to ensure that the coefficient of the pipelined filter still converges to the same coefficient as the adaptive filter without pipelining. Most of the ideas to pipeline IIR filters can be used to pipeline an LMS adaptive filter. The suggestion include

- Delayed LMS [263, 273, 274]
- Look-ahead transformation of the pipelined LMS [258, 275, 276]
- Transposed form LMS filter [277]
- Block transformation using FFTs [268]

We have already discussed the block transform algorithms and now wish in the following to briefly review the other techniques to improve the LMS throughput.

The Delayed LMS Algorithm. In the delayed LMS algorithm (DLMS) the assumption is that the gradient of the error $\nabla[n] = e[n]\boldsymbol{x}[n]$ does not change much if we delay the coefficient update by a couple of samples, i.e., $\nabla[n] \approx \nabla[n - D]$. It has been shown [273, 274] that as long as the delay is less than the system order, i.e., filter length, this assumption is well true and the update does not degrade the convergence speed. Long's original DLMS algorithm only considered pipelining the adder tree of the adaptive filter assuming also that multiplication and coefficient update can be done in one clock cycle (like for programmable digital signal processors [263]), but for a FPGA implementation multiplier and the coefficient update requires additional pipeline stages. If we introduce a delay of D_1 in the filter computation path and D_2 in the coefficient update path the LMS Algorithm 8.2 (p. 488) becomes:

$$e[n - D_1] = d[n - D_1] - \boldsymbol{f}^T[n - D_1]\boldsymbol{x}[n - D_1]$$
$$\boldsymbol{f}[n + 1] = \boldsymbol{f}[n - D_1 - D_2] + \mu e[n - D_1 - D_2]\boldsymbol{x}[n - D_1 - D_2].$$

The Look-ahead DLMS Algorithm. For long adaptive filters with $D = D_1 + D_2 < L$ the delayed coefficient update presented in the previous section, in general, does not change the convergence of the ADF much. It can, however, for shorter filters become necessary to reduce or even remove the change in system function completely. From the IIR pipelining method we have discussed in Chap. 4, the time domain interleaving method can always be applied. We perform just a look-ahead in coefficient computation, without alternating the overall system. Let us start with the DLMS update equations with pipelining only in the coefficient computation, i.e.,

$$e^{\mathrm{DLMS}}[n - D] = d[n - D] - \boldsymbol{x}^T[n - D]\boldsymbol{f}[n - D]$$
$$\boldsymbol{f}[n + 1] = \boldsymbol{f}[n] + \mu e[n - D]\boldsymbol{x}[n - D].$$

But the error function of the LMS would be

$$e^{\mathrm{LMS}}[n - D] = d[n - D] - \boldsymbol{x}^T[n]\boldsymbol{f}[n - D].$$

We follow the idea from Poltmann [275] and wish to compute the correction term $\Lambda[n]$, which cancels the change of the DLMS error computation compared with the LMS, i.e.,

$$\Lambda[n] = e^{\mathrm{LMS}}[n - D] - e^{\mathrm{DLMS}}[n - D].$$

The error function of the DLMS is now changed to

$$e^{\bar{\mathrm{D}}\mathrm{LMS}}[n - D] = d[n - D] - \boldsymbol{x}^T[n - D]\boldsymbol{f}[n - D] - \Lambda[n].$$

We need therefore to determine the term

$$\Lambda[n] = \boldsymbol{x}^T[n - D](\boldsymbol{f}[n] - \boldsymbol{f}[n - D]).$$

The term in brackets can be recursively determined via

$$\boldsymbol{f}[n] - \boldsymbol{f}[n - D]$$
$$= \boldsymbol{f}[n - 1] + \mu e[n - D - 1]\boldsymbol{x}[n - D_1] - \boldsymbol{f}[n - D]$$
$$= \boldsymbol{f}[n - 2] + \mu e[n - D - 2]\boldsymbol{x}[n - D - 2]$$
$$\quad + \mu e[n - D - 1]\boldsymbol{x}[n - D - 1] - \boldsymbol{f}[n - D]$$

$$= \sum_{s=1}^{D} \mu e[n - D - s]\boldsymbol{x}[n - D - s],$$

and it follows for the correction term $\varLambda[n]$ finally

$$\varLambda[n] = \boldsymbol{x}^T[n - D] \left(\sum_{s=1}^{D} \mu e[n - D - s]\boldsymbol{x}[n - D - s] \right)$$

$$e^{\bar{\text{DLMS}}}[n - D] = d[n - D] - \boldsymbol{x}^T[n - D]\boldsymbol{f}[n - D]$$

$$- \boldsymbol{x}^T[n - D] \left(\sum_{s=1}^{D} \mu e[n - D - s]\boldsymbol{x}[n - D - s] \right).$$

It can be seen that this correction term needs an additional $2D$ multiplication, which may be too expensive in some applications. It has been suggested [276] to "relax" the requirement for the correction term but some additional multipliers are still necessary.

We can, however, remove the influence of the coefficient update delay, by applying the look-ahead principle [258], i.e.,

$$\boldsymbol{f}[n + 1] = \boldsymbol{f}[n - D_1] + \mu \sum_{k=0}^{D_2-1} e[n - D_1 - k]\boldsymbol{x}[n - D_1 - k]. \tag{8.47}$$

The summation in (8.47) builds the moving average over the last D_2 gradient values, and makes it intuitively clear that the convergence will proceed more smoothly. The advantage compared with the transformation from Poltmann is that this look-ahead computation can be done without a general multiplication. The moving average in (8.47) may even be implemented with a first-order CIC filter (see Fig. 5.15, p. 260), which reduced the arithmetic effort to one adder and a subtractor.

Similar approaches to the idea from Poltmann to improve the DLMS algorithm have also been suggested [278, 279, 280].

8.5.4 Transposed Form LMS Filter

We have seen that the DLMS algorithm can be smoothed by introducing a look-ahead computation in the coefficient update, as we have used in IIR filters, but is, in general, not without additional cost. If we use, however, the transposed FIR structure (see Fig. 3.3, p. 167) instead of the direct structure, we can eliminate the delay by the adder tree completely. This will reduce the requirement for the optimal number of pipeline stages from (8.46), p. 508, by $\log_2(L)$ stages. For a LTI system both direct and transposed filters are described by the same convolution equation, but for a time-varying coefficient we need to change the filter coefficient from

$$f_k[n] \quad \text{to} \quad f_k[n - k]. \tag{8.48}$$

The equation for the estimated gradient (8.14) on page 488 now becomes

$$\hat{\nabla}[n] = -2e[n]\frac{\partial e[n-k]}{\partial \boldsymbol{f}_k[n]} \tag{8.49}$$

$$= -2e[n]\boldsymbol{x}[n-k]\frac{\boldsymbol{f}_k[n-k]}{\partial \boldsymbol{f}_k[n]}. \tag{8.50}$$

If we now assume that the coefficient update is relatively slow, i.e., $f_k[n-k] \approx f_k[n]$ the gradient becomes,

$$\hat{\nabla}[n] \approx -2e[n]\boldsymbol{x}[n], \tag{8.51}$$

and the coefficient update equation becomes:

$$f_k[n-k+1] = f_k[n-k] + \mu e[n]\boldsymbol{x}[n]. \tag{8.52}$$

The learning characteristics of the transposed-form adaptive filter algorithms have been investigated by Jones [277], who showed that we will get a somewhat slower convergence rate when compared with the original LMS algorithm. The stability bound regarding μ also needs to be determined and is found to be smaller than for the LMS algorithm.

8.5.5 Design of DLMS Algorithms

If we wish to pipeline the LMS filter from Example 8.5 (p. 504) we conclude from the discussion above (8.46) that the optimal number of pipeline stages becomes:

$$D_{\text{opt}} = 2\log_2(b) + \log_2(L) + 1 = 2 \times 3 + 1 + 1 = 8. \tag{8.53}$$

On the other hand, pipelining the multiplier can be done without additional costs and we may therefore consider only using 6 pipeline stages. Figure 8.22 shows a MATLAB simulation in 8-bit precision with a delay 6. Compared with the original LMS design from Example 8.5 (p. 504) it shows some "overswing" in the adaptation process.

Example 8.6: Two-tap Pipelined Adaptive LMS FIR Filter

The VHDL design [4] for a filter with two coefficients f_0 and f_1 with a step size of $\mu = 1/4$ is shown in the following listing.

```
-- This is a generic DLMS FIR filter generator
-- It uses W1 bit data/coefficients bits
LIBRARY lpm;                    -- Using predefined packages
USE lpm.lpm_components.ALL;

LIBRARY ieee;
USE ieee.std_logic_1164.ALL;
USE ieee.std_logic_arith.ALL;
USE ieee.std_logic_signed.ALL;
```

---

[4] The equivalent Verilog code `fir_lms.v` for this example can be found in Appendix A on page 721. Synthesis results are shown in Appendix B on page 731.

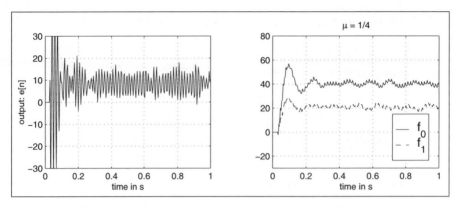

Fig. 8.22. 8-bit MATLAB simulation of the power-line interference cancellation using the DLMS algorithm with a delay of 6.

```
ENTITY fir6dlms IS                        ------> Interface
   GENERIC (W1 : INTEGER := 8; -- Input bit width
            W2 : INTEGER := 16;-- Multiplier bit width 2*W1
            L  : INTEGER := 2; -- Filter length
            Delay  : INTEGER := 3 -- Pipeline Delay
            );
   PORT ( clk    : IN STD_LOGIC;
          x_in   : IN  STD_LOGIC_VECTOR(W1-1 DOWNTO 0);
          d_in   : IN  STD_LOGIC_VECTOR(W1-1 DOWNTO 0);
        e_out, y_out : OUT STD_LOGIC_VECTOR(W2-1 DOWNTO 0);
     f0_out, f1_out  : OUT STD_LOGIC_VECTOR(W1-1 DOWNTO 0));
END fir6dlms;

ARCHITECTURE fpga OF fir6dlms IS

   SUBTYPE N1BIT IS STD_LOGIC_VECTOR(W1-1 DOWNTO 0);
   SUBTYPE N2BIT IS STD_LOGIC_VECTOR(W2-1 DOWNTO 0);
   TYPE ARRAY_N1BITF IS ARRAY (0 TO L-1) OF N1BIT;
   TYPE ARRAY_N1BITX IS ARRAY (0 TO Delay+L-1) OF N1BIT;
   TYPE ARRAY_N1BITD IS ARRAY (0 TO Delay) OF N1BIT ;
   TYPE ARRAY_N1BIT IS ARRAY (0 TO L-1) OF N1BIT;
   TYPE ARRAY_N2BIT IS ARRAY (0 TO L-1) OF N2BIT;

   SIGNAL  xemu0, xemu1        : N1BIT;
   SIGNAL  emu    : N1BIT;
   SIGNAL  y, sxty : N2BIT;

   SIGNAL  e, sxtd : N2BIT;
   SIGNAL  f       : ARRAY_N1BITF; -- Coefficient array
   SIGNAL  x       : ARRAY_N1BITX; -- Data array
   SIGNAL  d       : ARRAY_N1BITD; -- Reference array
   SIGNAL  p, xemu : ARRAY_N2BIT;  -- Product array

   BEGIN
```

```
dsxt: PROCESS (d)  -- make d a 16 bit number
BEGIN
  sxtd(7 DOWNTO 0) <= d(Delay);
  FOR k IN 15 DOWNTO 8 LOOP
    sxtd(k) <= d(3)(7);
  END LOOP;
END PROCESS;

Store: PROCESS   ------> Store these data or coefficients
BEGIN
  WAIT UNTIL clk = '1';
    d(0) <= d_in;   -- Shift register for desired data
    d(1) <= d(0);
    d(2) <= d(1);
    d(3) <= d(2);
    x(0) <= x_in;   -- Shift register for data
    x(1) <= x(0);
    x(2) <= x(1);
    x(3) <= x(2);
    x(4) <= x(3);
    f(0) <= f(0) + xemu(0)(15 DOWNTO 8); -- implicit
    f(1) <= f(1) + xemu(1)(15 DOWNTO 8); -- divide by 2
END PROCESS Store;

MulGen1: FOR I IN 0 TO L-1 GENERATE
FIR: lpm_mult             -- Multiply p(i) = f(i) * x(i);
     GENERIC MAP ( LPM_WIDTHA => W1, LPM_WIDTHB => W1,
                   LPM_REPRESENTATION => "SIGNED",
                   LPM_PIPELINE => Delay,
                   LPM_WIDTHP => W2,
                   LPM_WIDTHS => W2)
     PORT MAP ( dataa => x(I), datab => f(I),
                            result => p(I), clock => clk);
     END GENERATE;

y <= p(0) + p(1);  -- Computer ADF output

ysxt: PROCESS (y) -- scale y by 128 because x is fraction
BEGIN
  sxty(8 DOWNTO 0) <= y(15 DOWNTO 7);
  FOR k IN 15 DOWNTO 9 LOOP
    sxty(k) <= y(y'high);
  END LOOP;
END PROCESS;

e <= sxtd - sxty;        -- e*mu divide by 2 and 2
emu <= e(8 DOWNTO 1);    -- from xemu makes mu=1/4

MulGen2: FOR I IN 0 TO L-1 GENERATE
FUPDATE: lpm_mult        -- Multiply xemu(i) = emu * x(i);
     GENERIC MAP ( LPM_WIDTHA => W1, LPM_WIDTHB => W1,
                   LPM_REPRESENTATION => "SIGNED",
```

Table 8.2. Size and performance data of different pipeline options of the DLMS algorithms.

D	LEs	9 × 9-bit multipliers	MHz	Comment
0	50	4	74.59	Original LMS
1	58	4	109.28	Original DLMS
3	74	4	123.30	Pipeline of f update only
6	138	4	176.15	Pipeline all multipliers
8	179	4	368.19	Optimal number of stages

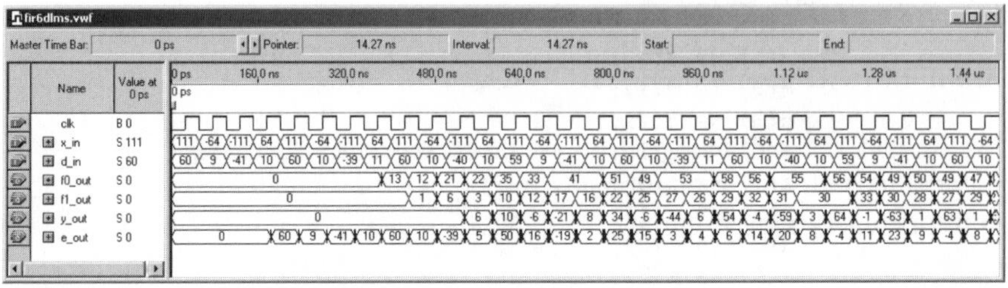

Fig. 8.23. VHDL simulation of the power-line interference cancelation using the DLMS algorithm with a delay of 6.

```
                           LPM_PIPELINE => Delay,
                           LPM_WIDTHP => W2,
                           LPM_WIDTHS => W2)
           PORT MAP ( dataa => x(I+Delay), datab => emu,
                            result => xemu(I), clock => clk);
           END GENERATE;

      y_out <= sxty;      -- Monitor some test signals
      e_out <= e;
      f0_out <= f(0);
      f1_out <= f(1);

   END fpga;
```

The design is a literal interpretation of the adaptive LMS filter architecture found in Fig. 8.20 (p. 489) with the additional delay of three pipeline stages for each multiplier. The output of each tap of the tapped delay line is multiplied by the appropriate filter coefficient and the results are added. Note the additional delays for x and d in the Store: PROCESS to make the signals coherent. The response of the adaptive filter y and of the overall system e to a reference signal x and a desired signal d is shown in the VHDL simulation in Fig. 8.23. The filter adapts after approximately 30 steps at $1.5\,\mu s$ to the optimal values $f_0 = 43.3$ and $f_1 = 25$, but it also shows some overswing in the adaptation process. The design uses 138 LEs, 4 embedded multipliers, and has a 176.15 MHz Registered Performance.   8.6

Compared with the previous example we may also consider other pipelining options. We may, for instance, use pipelining only in the coefficient update, or we may implement the optimal number of pipeline stages, i.e., 8. Table 8.2 gives an overview of the different options.

From Table 8.2 it can be seen that compared to the original LMS algorithm we may gain up to a factor of 4 speed improvement, while at the same time the additional hardware cost are only about 10%. The additional effort comes from the extra delays of the reference data $d[n]$ and the filter input $x[n]$. The limitation is just that it may become necessary for large pipeline delays to adjust μ in order to guarantee stability.

8.5.6 LMS Designs using SIGNUM Function

We saw in the previous section that the implementation cost of the LMS algorithm is already high for short filter length. The highest cost of the filter comes from the large number of general multipliers and the major goal in reducing the effort is to reduce the number of multipliers. Obviously the FIR filter part can not be reduced, but different simplifications in the computation of the coefficient update have been investigated. Given the fact that to ensure stability usually the step size is chosen much smaller than $\mu_{\max}$, the following suggestions have been made:

- Use only the sign of the reference *data* $x[n]$ not the full precision value to update the filter coefficients.
- Use only the sign of the *error* $e[n]$ not the full precision value to update the filter coefficients.
- Use both of the previous simplifications via the sign of error and data.

The three modifications can be described with the following coefficient update equations in the LMS algorithm:

$$\boldsymbol{f}[n+1] = \boldsymbol{f}[n] + \mu \times e[n] \times \text{sign}(\boldsymbol{x}[n]) \qquad \text{sign data function}$$
$$\boldsymbol{f}[n+1] = \boldsymbol{f}[n] + \mu \times \boldsymbol{x}[n] \times \text{sign}(e[n]) \qquad \text{sign error function}$$
$$\boldsymbol{f}[n+1] = \boldsymbol{f}[n] + \mu \times \text{sign}(e[n]) \times \text{sign}(\boldsymbol{x}[n]) \qquad \text{sign-sign function.}$$

We note from the simulation of the three possible simplifications shown in Fig. 8.24 that for the sign-data function almost the same result occurs as in the full precision case. This is no surprise, because our input reference signal $x[n] = \cos[\pi n/2 + \phi]$ will not be much quantized through the sign operation anyway. This is much different for the sign-error function. Here the quantization through the sign operation essentially alters the time constant of the system. But finally, after about 2.5 s the correct values are reached, although from the system output $e[n]$ we note the essential ripple in the output function even after a long simulation time. Finally, the sign-sign algorithm converges faster than the sign-error algorithm, but here also the

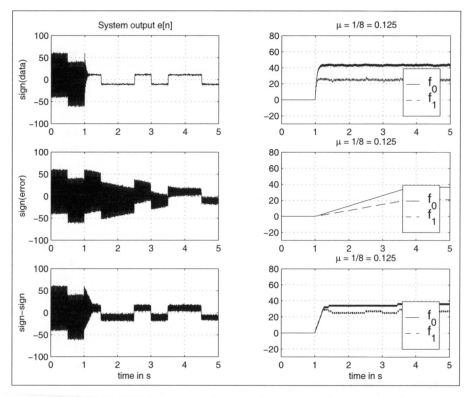

Fig. 8.24. Simulation of the power-line interference cancellation using the 3 simplified signed LMS (SLMS) algorithms. **(left)** System output $e[n]$. **(right)** filter coefficients.

system output shows essential ripple for $e[n]$. From the simulation it can be seen that the sign-function simplification (to save the L multiplications in the filter coefficient update) has to be evaluated carefully for the specific application to still guarantee a stable system and acceptable time constants of the system. In fact, it has been shown that for specific signals and application the sign algorithms does not converge, although the full precision algorithm would converge. Besides the sign effect we also need to ensure that the integer quantization through the implementation does not alter the desired system properties.

Another point to consider when using the sign function is the error floor that can be reached. This is discussed in the following example.

Example 8.7: Error Floor in Signum LMS Filters

Suppose we have a system identification configuration as discussed in Sect. 8.3.1 (p. 493), and we wish to use one of the signum-type ADF algorithms. What will then be the error floor that can be reached? Obviously through the signum operation we will lose some precision and we expect that we will not

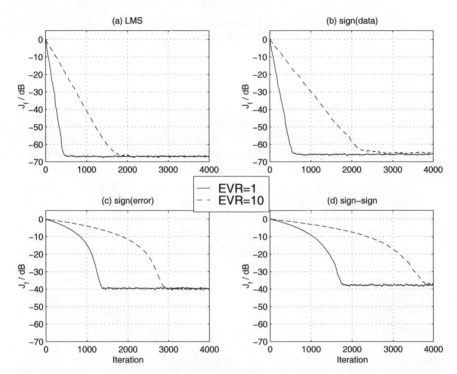

Fig. 8.25. Simulation of the system identification experiment using the 3 simplified signed LMS algorithms for an average of 50 learning curves for an error floor of −60 dB. **(a)** LMS with full precision. **(b)** signed data. **(c)** signed error algorithms. **(d)** sign-sign LMS algorithm.

reach the same low-noise level as with a full-precision LMS algorithm. We also expect that the learning rate will be somewhat decreased when compared with the full-precision LMS algorithm. This can be verified by the simulation results shown in Fig. 8.25 for an average over 50 learning curves and two different eigenvalue ratios (EVRs). The sign data algorithms shows some delay in the adaptation when compared with the full-precision LMS algorithm, but reaches the error floor, which was set to −60 dB. Signed error and sign-sign algorithms show larger delays in the adaptation and also reach only an error floor of about −40 dB. This larger error may or may not be acceptable for some applications. ⌐8.7⌐

The sign-sign algorithm is attractive from a software or hardware implementation standpoint and has been used for the International Telecommunication Union (ITU) standard for adaptive differential pulse code modulation (ADPCM) transmission. From a hardware implementation standpoint we actually do not need to implement the sign-sign algorithm, because the multiplication with μ is just a scaling with a constant and one of the single sign algorithms will already allow us to save the L multipliers we usually need for the filter coefficient update in Fig. 8.20 (p. 506).

8.6 Recursive Least Square Algorithms

In the LMS algorithm we have discussed in the previous sections the filter coefficients are gradually adjusted by a stochastic gradient method to finally approximate the Wiener–Hopf optimal solution. The recursive least square (RLS) algorithm takes another approach. Here, the estimation of the $(L \times L)$ autocorrelation matrix $\boldsymbol{R}_{xx}$ and the cross-correlation vector $\boldsymbol{r}_{dx}$ are iteratively updated with each new incoming data pair $(x[n], d[n])$. The simplest approach would be to reconstruct the Wiener–Hopf equation (8.9), i.e., $\boldsymbol{R}_{xx}\boldsymbol{f}_{\mathrm{opt}} = \boldsymbol{r}_{dx}$ and resolve it. However, this would be the equivalent of one matrix inversion as each new data point pair arrives and has the potential of being computationally expensive. The main goal of the different RLS algorithms we will discuss in the following is therefore to seek a (iterative) time recursion for the filter coefficients $\boldsymbol{f}[n+1]$ in terms of the previous least square estimate $\boldsymbol{f}[n]$ and the new data pair $(x[n], d[n])$. Each incoming new value $x[n]$ is placed in the length-L data array $\boldsymbol{x}[n] = [x[n]x[n-1]\ldots x[n-(L-1)]]^T$. We then wish to add $x[n]x[n]$ to $\boldsymbol{R}_{xx}[0,0]$, $x[n]x[n-1]$ to $\boldsymbol{R}_{xx}[0,1]$, etc. Mathematically we just compute the product $\boldsymbol{x}\boldsymbol{x}^T$ and add this $(L \times L)$ matrix to the previous estimation of the autocorrelation matrix $\boldsymbol{R}_{xx}[n]$. The recursive computation may be computed as follows:

$$\boldsymbol{R}_{xx}[n+1] = \boldsymbol{R}_{xx}[n] + \boldsymbol{x}[n]\boldsymbol{x}^T[n] = \sum_{s=0}^{n} \boldsymbol{x}[s]\boldsymbol{x}^T[s]. \tag{8.54}$$

For the cross-correlation vector $\boldsymbol{r}_{dx}[n+1]$ we also build an "improved" estimate by adding with each new pair $(x[n], d[n])$ the vector $d[n]\boldsymbol{x}[n]$ to the previous estimation of $\boldsymbol{r}_{dx}[n]$. The recursion for the cross-correlation becomes

$$\boldsymbol{r}_{dx}[n+1] = \boldsymbol{r}_{dx}[n] + d[n]\boldsymbol{x}[n], \tag{8.55}$$

we can now use the Wiener–Hopf equation in a time recursive fashion and compute

$$\boldsymbol{R}_{xx}[n+1]\boldsymbol{f}_{\mathrm{opt}}[n+1] = \boldsymbol{r}_{dx}[n+1]. \tag{8.56}$$

For the true estimates of cross- and autocorrelation matrices we would need to scale by the number of summations, which is proportional to n, but the cross- and autocorrelation matrices are scaled by the same factor, which cancel each other out in the iterative algorithm and we get for the filter coefficient update

$$\boldsymbol{f}_{\mathrm{opt}}[n+1] = \boldsymbol{R}_{xx}^{-1}[n+1]\boldsymbol{r}_{dx}[n+1]. \tag{8.57}$$

Although this first version of the RLS algorithms is computationally intensive (approximately L^3 operations are needed for the matrix inversion) it still shows the principal idea of the RLS algorithm and can be quickly programmed, for instance in MATLAB, as the following code segment shows the inner loop for length-L RLS filter algorithm:

```
x = [xin;x(1:L-1)];    % get new sample
y = f' * x;            % filter output
err = din - y;         % error: reference - filter output
Rxx = Rxx + x*x';      % update the autocorrelation matrix
rdx = rdx + din .* x;  % update the cross-correlation vector
f = Rxx^(-1) * rdx;    % compute filter coefficients
```

where Rxx is a $(L \times L)$ matrix and rdx is a $(L \times 1)$ vector. The cross-correlation vector is usually initialized with $r_{dx}[0] = 0$. The only problem with the algorithm so far arises at the first $n < L$ iterations, when $R_{xx}[n]$ only has a few nonzero entries, and consequently will be singular and no inverse exists. There are a couple of ways to tackle this problem:

• We can wait with the computation of the inverse until we find that the autocorrelation matrix is nonsingular, i.e., $\det(R_{xx}[n]) > 0$.
• We can use $R_{xx}^{+}[n] = (R_{xx}^{T}[n]R_{xx}[n])^{-1}R_{xx}^{T}[n]$ the so-called pseudoinverse, which is a standard result in linear algebra regarding the solution of an overdetermined set of linear equations.
• We can initialize the autocorrelation matrix R_{xx} with δI where δ is chosen to be a small (large) constant for high (low) S/N ratio of the input signal.

The third approach is the most popular due to the computational benefit and the possibility to set an initial "learning rate" using the constant δ. The influence of the initialization in the RLS algorithm for an experiment similar to Sect. 8.3.1 (p. 493) with an average over 5 learning curves is shown in Fig. 8.26. The upper row shows the full-length simulation over 4000 iterations, while the lower row shows the first 100 iterations only. For high S/N (-48 dB) we may use a large value for the initialization, which yields a fast convergence. For low S/N values (-10 dB) small initialization values should be used, otherwise large errors at the first iterations can occur, which may or may not be tolerable for the specific application.

A more computationally attractive approach than the first "brute force" RLS algorithm will be discussed in the following. The key idea is that we do not compute the matrix inversion at all and use a time recursion directly for $R_{xx}^{-1}[n]$, we actually will never have (or need) $R_{xx}[n]$ available. To do so, we substitute the Wiener equation for time $n + 1$, i.e., $f[n+1]R_{xx}[n+1] = r_{dx}[n+1]$ into (8.55) it follows that

$$R_{xx}[n+1]f[n+1] = R_{xx}[n]f[n] + d[n+1]x[n+1]. \tag{8.58}$$

Now we use (8.54) to get

$$R_{xx}[n+1]f[n+1] = \left(R_{xx}[n+1] - x[n+1]x^{T}[n+1]\right)f[n]$$
$$+ d[n+1]x[n+1]. \tag{8.59}$$

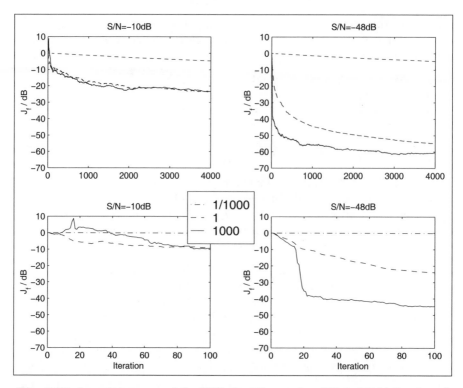

Fig. 8.26. Learning curves of the RLS algorithms using different initialization of $\boldsymbol{R}_{xx}^{-1}[0] = \delta \boldsymbol{I}$ or $\boldsymbol{R}_{xx}[0] = \delta^{-1} \boldsymbol{I}$. High S/N is -48 dB and low is -10 dB. $\delta = 1000, 1$ or $1/1000$.

We can rearrange (8.59) by multiplying by $\boldsymbol{R}_{xx}^{-1}[n+1]$ to have $\boldsymbol{f}[n+1]$ on the lefthand side of the equation:

$$\boldsymbol{f}[n+1] = \boldsymbol{f}[n] + \underbrace{\boldsymbol{R}_{xx}^{-1}[n+1]\boldsymbol{x}[n+1]}_{\boldsymbol{k}[n+1]} \underbrace{\left(d[n+1] - \boldsymbol{f}^T[n]\boldsymbol{x}[n+1]\right)}_{e[n+1]}$$
$$= \boldsymbol{f}[n] + \boldsymbol{k}[n+1]e[n+1],$$

where the *a priori error* is defined as

$$e[n+1] = d[n+1] - \boldsymbol{f}^T[n]\boldsymbol{x}[n+1],$$

and the *Kalman gain vector* is defined as

$$\boldsymbol{k}[n+1] = \boldsymbol{R}_{xx}^{-1}[n+1]\boldsymbol{x}[n+1]. \tag{8.60}$$

As mentioned above the direct computation of the matrix inversion is computationally intensive, and it is much more efficient to use again the iteration equation (8.54) to actually avoid the inversion at all. We use the so-called "matrix inversion lemma," which can be written as the following matrix identity

$$(A + BCD)^{-1}$$
$$= A^{-1} - A^{-1}B(A^{-1}BDA^{-1})(C + DA^{-1}B)^{-1},$$

which holds for all matrices A, B, C, and D, of compatible dimensions and nonsingular A. We make the following associations:

$$A = R_{xx}[n+1] \qquad\qquad B = x[n]$$
$$C = 1 \qquad\qquad\qquad D = x^T[n].$$

The iterative equation for R_{xx}^{-1} becomes:

$$R_{xx}^{-1}[n+1] = \left(R_{xx}^{-1}[n] + x[n]x^T[n] \right)^{-1}$$
$$= R_{xx}^{-1}[n] + \frac{R_{xx}^{-1}[n]x[n]x^T[n]R_{xx}^{-1}[n]}{1 + x^T[n]R_{xx}^{-1}[n]x[n]}. \qquad (8.61)$$

If we use the Kalman gain factor $k[n]$ from (8.60) we can rewrite (8.61) more compactly as:

$$R_{xx}^{-1}[n+1] = \left(R_{xx}^{-1}[n] + x[n+1]x^T[n+1] \right)^{-1}$$
$$= R_{xx}^{-1}[n] + \frac{k[n]k^T[n]}{1 + x^T[n]k[n]}.$$

This recursion is as mentioned before initialized [252] with

$$R_{xx}^{-1}[0] = \delta I \quad \text{with} \quad \delta = \begin{cases} \text{large positive constant for high SNR} \\ \text{small positive constant for low SNR.} \end{cases}$$

With this recursive computation of the inverse autocorrelation matrix the computation effort is now proportional to L^2, an essential saving for large values of L. Figure 8.27 shows a summary of the RLS adaptive filter algorithm.

8.6.1 RLS with Finite Memory

As we can see from (8.54) and (8.55) the adaptive algorithm derived so far has an infinite memory. The values of the filter coefficients are functions of all past inputs starting with time zero. As will be discussed next it is often useful to introduce a "forgetting factor" into the algorithm, so that recent data are given greater importance than older data. This not only reduces the influence of older data, it also accomplishes that through the update of the cross- and autocorrelation with each new incoming data pair no overflow in the arithmetic will occur. One way of accomplishing a finite memory is to replace the sum-of-squares cost function, by an exponentially weighted sum of the output:

$$J = \sum_{s=0}^{n} \rho^{n-s} e^2[s], \qquad (8.62)$$

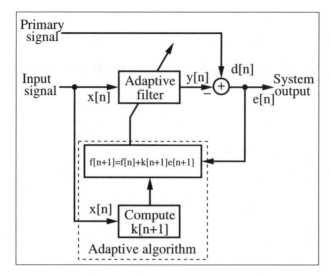

Fig. 8.27. Basic configuration for interference cancellation using the RLS algorithm.

where $0 \le \rho \le 1$ is a constant determining the effective memory of the algorithm. The case $\rho = 1$, is the infinite-memory case, as before. When $\rho < 1$ the algorithm will have an effective memory of $\tau = -1/\log(\rho) \approx 1/(1 - \rho)$ data points. The exponentially weighted RLS algorithm can now be summarized as:

Algorithm 8.8:   **RLS Algorithm**

The exponentially weighted RLS algorithm to adjust the L coefficients of an adaptive filter uses the following steps:

1) Initialize $\boldsymbol{x} = \boldsymbol{f} = [0, 0, \ldots, 0]^T$ and $\boldsymbol{R}_{xx}^{-1}[0] = \delta \boldsymbol{I}$.
2) Accept a new pair of input samples $\{x[n+1], d[n+1]\}$ and shift $x[n+1]$ input the reference signal vector $\boldsymbol{x}[n+1]$.
3) Compute the output signal of the FIR filter, via
$$y[n + 1] = \boldsymbol{f}^T[n]\boldsymbol{x}[n + 1]. \tag{8.63}$$
4) Compute the a priori error function with
$$e[n + 1] = d[n + 1] - y[n + 1]. \tag{8.64}$$
5) Compute the Kalman gain factor with
$$\boldsymbol{k}[n + 1] = \boldsymbol{R}_{xx}^{-1}[n + 1]\boldsymbol{x}[n + 1]. \tag{8.65}$$
6) Update the filter coefficient according to
$$\boldsymbol{f}[n + 1] = \boldsymbol{f}[n] + \boldsymbol{k}[n + 1]e[n + 1]. \tag{8.66}$$
7) Update the filter inverse autocorrelation matrix according to
$$\boldsymbol{R}_{xx}^{-1}[n + 1] = \frac{1}{\rho} \left(\boldsymbol{R}_{xx}^{-1}[n] + \frac{\boldsymbol{k}[n + 1]\boldsymbol{k}^T[n + 1]}{\rho + \boldsymbol{x}^T[n + 1]\boldsymbol{k}[n + 1]} \right). \tag{8.67}$$

Next continue with step 2.

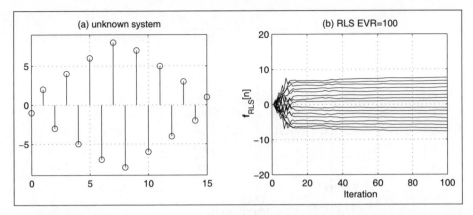

Fig. 8.28. Simulation of the $L = 16$ tap adaptive filter system identification. **(a)** Impulse response of the "unknown system." **(b)** RLS coefficient learning curves for EVR = 100.

The computational cost of the RLS are $(3L^2 + 9L)/2$ multiplications and $(3L^2 + 5L)/2$ additions or subtractions, per input sample, which is still more essential than the LMS algorithm. The advantage as we will see in the following example will be a higher rate of convergence and no need to select the step size μ, which may at times be difficult when stability of the adaptive algorithm has to be guaranteed.

Example 8.9: RLS Learning Curves

In this example we wish to evaluate a configuration called system identification to compare RLS and LMS convergence. We have used this type of performance evaluation already for LMS ADF in Sect. 8.3.1 (p. 493) The system configuration is shown in Fig. 8.12 (p. 494). The adaptive filter has a length of $L = 16$, the same length as the "unknown" system, whose coefficients have to be learned. The additive noise level behind the "unknown system" has been set to -48 dB equivalent for an 8-bit quantization. For the LMS algorithm the eigenvalue ratio (EVR) is the critical parameter that determines the convergence speed, see (8.25), p. 491. In order to generate a different eigenvalue ratio we use a white Gaussian noise source with $\sigma^2 = 1$ that is filtered by a FIR type filter shown in Table 8.1 (p. 495). The coefficients are normalized to $\sum_k h[k]^2 = 1$, so that the signal power does not change. The impulse response of the unknown system is an odd filter with coefficients $1, -2, 3, -4, \ldots, -3, 2, -1$ as shown in Fig. 8.28a. The step size for the LMS algorithm has been determined with

$$\mu_{\max} = \frac{2}{3 \times L \times E\{x^2\}} = \frac{1}{24}. \tag{8.68}$$

In order to guarantee perfect stability the step size for the LMS algorithm has been chosen to be $\mu = \mu_{\max}/2 = 1/48$. For the transform-domain DCT-LMS algorithm a power normalization for each coefficient is used, see Fig. 8.17 (p. 502). From the simulation results shown in Fig. 8.29 it can be seen that the RLS converges faster than the LMS with increased EVR. DCT-LMS converges faster than LMS and in some cases quite as fast as the RLS algo-

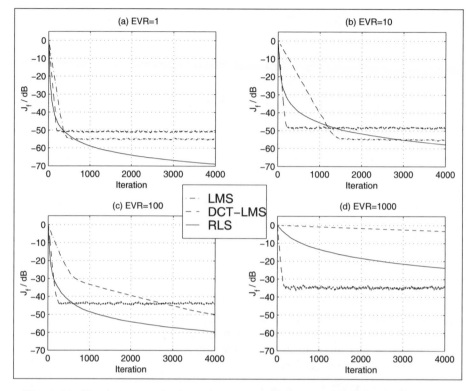

Fig. 8.29. Simulation results for a $L = 16$-tap adaptive filter system identification. Learning curve J for LMS, transform-domain DCT-LMS, and RLS with $\boldsymbol{R}_{xx}^{-1}[0] = \boldsymbol{I}$. (a) EVR = 1. (b) EVR = 10. (c) EVR = 100. (d) EVR = 1000.

rithm. The DCT-LMS algorithm has less good performance when we look at the residue-error level and consistency of convergence. For higher EVR the RLS performance is better for both level and consistency of convergence. For EVR=1 the DCT-LMS reaches the value in the 50 dB range, but for EVR = 100 only 40 dB are reached. The RLS converges below the system noise. $\boxed{8.9}$

8.6.2 Fast RLS Kalman Implementation

For the least-quare FIR fast Kalman algorithm first presented by Ljung et al. [281] the concept of single-step linear forward and backward prediction play a central role. Using these forward and backward coefficients in an all-recursive, one-dimensional Levison–Durbin type algorithm it will be possible to update the Kalman gain vector with only an $O(L)$ type effort.

A one-step *forward* predictor is presented in Fig. 8.30. The predictor estimates the present value $x[n]$ based on its L most recent past values. The *a posteriori* error in the prediction is quantified by

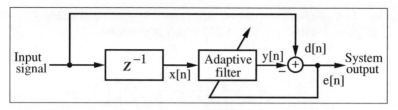

Fig. 8.30. Linear forward prediction of order L.

$$\epsilon_L^f[n] = x[n] - \hat{x}[n] = x[n] - \boldsymbol{a}^T[n]\boldsymbol{x}_L[n-1]. \tag{8.69}$$

The superscript indicates that it is the forward prediction error, while the subscript describes the order (i.e., length) of the predictor. We will drop the index L and the vector length should be L for the remainder of this section, if not otherwise noted. It is also advantageous to compute also the *a priori* error that is computed using the filter coefficient of the previous iteration, i.e.,

$$e_L^f[n] = x[n] - \boldsymbol{a}^T[n-1]\boldsymbol{x}_L[n-1]. \tag{8.70}$$

The least-quare minimum of $\epsilon_L^f[n]$ can be computed via

$$\frac{\partial(\epsilon^f[n])^2}{\partial \boldsymbol{a}^T[n]} = -E\{(x[s] - \boldsymbol{a}^T[s]\boldsymbol{x}[n])x[n-s]\} = 0 \tag{8.71}$$

$$\text{for } s = 1, 2, \ldots, L.$$

This leads again to an equation with the $(L \times L)$ autocorrelation matrix, but the right-hand side is different from the Wiener–Hopf equation:

$$\boldsymbol{R}_{\boldsymbol{xx}}[n-1]\boldsymbol{a}[n] = \boldsymbol{r}^f[n] = \sum_{s=0}^{n} \boldsymbol{x}[s-1]x[s]. \tag{8.72}$$

The minimum value of the cost function is given by

$$\alpha^f[n] = r_0^f[n] - \boldsymbol{a}^T[n]\boldsymbol{r}^f[n], \tag{8.73}$$

where $r_0^f[n] = \sum_{s=0}^{n} x[s]^2$.

The important fact about this predictor is now that the Levinson–Durbin algorithm can solve the least-quare error minimum of (8.69) in a recursive fashion, without computing a matrix inverse. To update the predictor coefficient we need the same Kalman gain factor as in (8.66) for updating the filter coefficients, namely

$$\boldsymbol{a}_L[n+1] = \boldsymbol{a}_L[n] + \boldsymbol{k}_L[n]e_L^f[n].$$

We will see later how the linear prediction coefficients can be used to iteratively update the Kalman gain factor. In order to take advantage of the fact that the data vectors from one iteration to the next only differ in the first and last element, we use an augmented-by-one version $k_{L+1}[n]$ of the Kalman gain update equation (8.65) which is given by

$$k_{L+1}[n+1] = R_{xx,L+1}^{-1}[n+1]x_{L+1}[n+1]. \tag{8.74}$$

$$= \begin{bmatrix} r_{0L}^f[n+1] & r_L^{fT}[n+1] \\ \hline r_L^f[n] & R_{xx,L}^{-1}[n] \end{bmatrix} \begin{bmatrix} x[n+1] \\ x_L[n] \end{bmatrix}. \tag{8.75}$$

In order to compute the matrix inverse of $R_{xx,L+1}^{-1}[n]$ we use a well-known theorem of matrix inversion of block matrices, i.e.,

$$M^{-1} = \begin{bmatrix} A & B \\ \hline C & D \end{bmatrix}^{-1} = \tag{8.76}$$

$$\begin{bmatrix} -(AD^{-1}C - A)^{-1} & (AD^{-1}C - A)^{-1}BD^{-1} \\ \hline D^{-1}C - (AD^{-1}C - A)^{-1} & D^{-1} - (D^{-1}CBD^{-1})(AD^{-1}C - A)^{-1} \end{bmatrix},$$

if D^{-1} is nonsingular. We now make the following associations:

$$A = r_{0L}^f[n+1] \qquad B = r_L^{fT}[n+1]$$
$$C = r_L^f[n] \qquad D = R_{xx}^{-1}[n],$$

we then get

$$D^{-1}C = R_{xx,L}^{-1}[n]r_L^f[n] = a_L[n+1]$$
$$BD^{-1} = r_L^{fT}[n+1]R_{xx}^{-1}[n] = a_L^T[n+1]$$
$$-(AD^{-1}C - A)^{-1} = -r_L^{fT}[n+1]R_{xx,L}^{-1}[n]r_L^f[n] + r_{0L}^f[n+1]$$
$$= r_{0L}^f[n+1] - a_L^T[n+1]r_L^f[n] = \alpha_L^f[n+1].$$

We can now rewrite $R_{xx,L+1}^{-1}[n+1]$ from (8.74) as

$$R_{xx,L+1}^{-1}[n+1] = \begin{bmatrix} \dfrac{1}{\alpha_L^f[n+1]} & \dfrac{a_L^T[n+1]}{\alpha_L^f[n+1]} \\ \hline \dfrac{a_L[n+1]}{\alpha_L^f[n+1]} & R_{xx,L}^{-1}[n] + \dfrac{a_L[n+1]a_L^T[n+1]}{\alpha_L^f[n+1]} \end{bmatrix}. \tag{8.77}$$

After some rearrangements (8.74) can be written as

$$k_{L+1}[n+1] = \begin{bmatrix} 0 \\ k_L[n+1] \end{bmatrix} + \frac{\epsilon_L^f[n+1]}{\alpha_L^f[n+1]} \begin{bmatrix} 1 \\ a_L[n+1] \end{bmatrix}$$
$$= \begin{bmatrix} g_L[n+1] \\ \gamma_L[n+1] \end{bmatrix}.$$

Unfortunately, we do not have a closed recursion so far. For the iterative update of the Kalman gain vector, we need besides the forward prediction coefficients, also the coefficients of the one-step *backward* predictor, whose *a posteriori* error function is

$$\epsilon^b[n] = x[n-L] - \hat{x}[n-L] = x[n-L] - b^T[n]x[n], \tag{8.78}$$

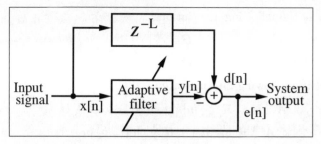

Fig. 8.31. Linear backward prediction of order L.

again all vectors are of size $(L \times 1)$. The linear backward predictor is shown in Fig. 8.31.

The *a priori* error for the backward predictor is given by

$$e_L^b[n] = x[n-L] - b^T[n-1]x_L[n],$$

The iterative equation to compute the least-quare coefficients for the backward predictor is equivalent to the forward case and given by

$$R_{xx}[n]b[n] = r^b[n] = \sum_{s=0}^{n} x[s]x[s-L], \tag{8.79}$$

and the minimum value for the total squared error becomes

$$\alpha^f[n] = r_0^b[n] - b^T[n]r^b[n],$$

where $r_0^b[n] = \sum_{s=0}^{n} x[s-L]^2$. To update the backward predictor coefficient we need again the Kalman gain factor in (8.66) as for the updating of the filter coefficients, namely

$$b_L[n+1] = b_L[n] + k_L[n+1]e_L^b[n+1].$$

Now we can again find a Levinson–Durbin type of recursive equation for the extended Kalman gain vector, only this time using the backward prediction coefficients. It follows that

$$k_{L+1}[n+1] = R_{xx,L+1}^{-1}[n+1]x_{L+1}[n+1]. \tag{8.80}$$

$$= \left[\begin{array}{c|c} R_{xx,L}[n] & r_L^b[n+1] \\ \hline r_L^{bT}[n] & r_{0L}^b[n+1] \end{array} \right]^{-1} \left[\begin{array}{c} x_L[n+1] \\ x[n-L+1] \end{array} \right]. \tag{8.81}$$

To solve the matrix inversion, we define as in (8.76) a $(L+1) \times (L+1)$ block matrix M, only this time the block A needs to be nonsingular and it follows that

$$M^{-1} = \left[\begin{array}{c|c} A & B \\ \hline C & D \end{array} \right]^{-1} =$$

$$\frac{A^{-1} - (A^{-1}BCA^{-1})(CA^{-1}B - D)^{-1} \big| A^{-1}B(CA^{-1}B - D)^{-1}}{(CA^{-1}B - D)^{-1}CA^{-1} \qquad \big| -(CA^{-1}B - D)^{-1}}.$$

We now make the following associations:

$$A = R_{xx,L}[n] \qquad\qquad B = r_L^b[n+1]$$
$$C = r_L^{bT}[n] \qquad\qquad D = r_{0L}^b[n+1],$$

we then get the following intermediate results

$$A^{-1}B = R_{xx,L}^{-1}[n]r_L^b[n+1] = b_L[n+1]$$
$$CA^{-1} = r_L^{bT}[n+1]R_{xx,L}^{-1}[n] = b_L^T[n+1]$$
$$-(CA^{-1}B - D) = -b_L^T[n+1]r_L^b[n+1] + r_{0L}^b[n+1] = \alpha_L^b[n+1].$$

Using this intermediate results in (8.78) we get

$$R_{xx,L+1}^{-1}[n] = \left[\begin{array}{c|c} R_{xx,L}^{-1}[n] + \frac{b_L[n+1]b_L^T[n+1]}{\alpha_L^b[n]} & \frac{b_L^T[n+1]}{\alpha_L^b[n]} \\ \hline \frac{b_L[n+1]}{\alpha_L^b[n]} & \frac{1}{\alpha_L^b[n]} \end{array} \right].$$

After some rearrangements (8.80) can now, using the backward prediction coefficients, be written as

$$k_{L+1}[n+1] = \left[\begin{array}{c} k_L[n+1] \\ 0 \end{array} \right] + \frac{\epsilon_L^b[n+1]}{\alpha_L^b[n+1]} \left[\begin{array}{c} b_L[n+1] \\ 1 \end{array} \right]$$
$$= \left[\begin{array}{c} g_L[n+1] \\ \gamma_L[n+1] \end{array} \right].$$

The only iterative update equation missing so far is for the minimum values of the total square errors, which is given by

$$\alpha_L^f[n+1] = \alpha_L^f[n] + \epsilon_L^f[n+1]e_L^f[n+1] \qquad\qquad (8.82)$$
$$\alpha_L^b[n+1] = \alpha_L^b[n] + \epsilon_L^b[n+1]e_L^b[n+1]. \qquad\qquad (8.83)$$

We now have all iterative equations available to define the

Algorithm 8.10:   **Fast Kalman RLS Algorithm**

The prewindowed fast Kalman RLS algorithm to adjust the L filter coefficients of an adaptive filter uses the following steps:

1) Initialize $\boldsymbol{x} = \boldsymbol{a} = \boldsymbol{b} = \boldsymbol{f} = \boldsymbol{k} = [0, 0, \ldots, 0]^T$ and $\alpha^f = \alpha^b = \delta$

2) Accept a new pair of input samples $\{x[n+1], d[n+1]\}$.

3) Compute now the following equations to update $\boldsymbol{a}, \boldsymbol{b}$, and $\boldsymbol{k}$ in sequential order

$$e_L^f[n+1] = x[n+1] - \boldsymbol{a}^T[n]\boldsymbol{x}_L[n]$$

$$\boldsymbol{a}_L[n+1] = \boldsymbol{a}_L[n] + \boldsymbol{k}_L[n]e_L^f[n+1]$$

$$\epsilon_L^f[n+1] = x[n+1] - \boldsymbol{a}^T[n+1]\boldsymbol{x}_L[n]$$

$$\alpha_L^f[n+1] = \alpha_L^f[n] + \epsilon_L^f[n+1]e_L^f[n+1]$$

$$\boldsymbol{k}_{L+1}[n+1] = \begin{bmatrix} 0 \\ \boldsymbol{k}_L[n+1] \end{bmatrix} = \frac{\epsilon_L^f[n+1]}{\alpha_L^f[n+1]} \begin{bmatrix} 1 \\ \boldsymbol{a}_L[n+1] \end{bmatrix}$$

$$= \begin{bmatrix} \boldsymbol{g}_L[n+1] \\ \gamma_L[n+1] \end{bmatrix}$$

$$e_L^b[n+1] = x[n+1-L] - \boldsymbol{b}^T[n]\boldsymbol{x}_L[n+1]$$

$$\boldsymbol{k}_L[n+1] = \frac{\boldsymbol{g}_L[n+1] - \gamma_L[n+1]\boldsymbol{b}^T[n]}{1 + \gamma_L[n+1]e_L^b[n+1]}$$

$$\boldsymbol{b}_L[n+1] = \boldsymbol{b}_L[n] + \boldsymbol{k}_L[n+1]e_L^b[n+1].$$

4) Shift the $x[n+1]$ in the reference signal vector $\boldsymbol{x}[n+1]$ and compute the following two equations in order to update the adaptive filter coefficients:

$$e_L[n+1] = d[n+1] - \boldsymbol{f}_L^T[n]\boldsymbol{x}_L[n+1]$$

$$\boldsymbol{f}_L[n+1] = \boldsymbol{f}_L[n] + \boldsymbol{k}_L[n+1]e_L[n+1].$$

Next continue with step 2.

Counting the computational effort we find that step 3 needs 2 divisions, $8L+2$ multiplications, and $7L+2$ add or subtract operations. The coefficient update in step 4 uses an additional $2L$ multiply and add/subtract operations, that the total computational effort is $10L+2$ multiplications, $9L+2$ add/subtract operations and 2 divisions.

8.6.3 The Fast a Posteriori Kalman RLS Algorithm

A careful inspection of Algorithm 8.10 reveals that the original fast Kalman algorithm as introduced by Ljung et al. [281] is mainly based on the *a priori* error equations. In the fast a posteriori error sequential technique (FAEST) introduced by Carayannis et al. [282] to a greater extent the *a posteriori* error is used. The algorithm explores even more the iterative nature of the different parameters in the fast Kalman algorithm, which will reduce the computational effort by an additional $2L$ multiplications. Otherwise, the original fast Kalman and the FAEST use mainly the same ideas, i.e., extended by one length Kalman gain, and the use of the forward and backward predictions a and b. We also introduce the forgetting factor ρ. The following listing shows the inner loop of the FAEST algorithm in MatLab:

```
%********* FAEST Update of k, a, and b
ef=xin - a'*x;           % a priori forward prediction error
ediva=ef/(rho*af);       % a priori forward error/minimal error
ke(1)=-ediva;            % extended Kalman gain vector update
ke(2:l+1)=k - ediva*a;%  split the l+1 length vector
epsf=ef*psi;             % a posteriori forward error
a=a+epsf*k;                  % update forward coefficients
k=ke(1:l) + ke(l+1).*b;  % Kalman gain vector update
eb=-rho*alphab*ke(l1);   % a priori backward error
alphaf=rho*alphaf+ef*epsf;  % forward minimal error
alpha=alpha+ke(l+1)*eb+ediva*ef; % prediction crosspower
psi=1.0/alpha;           % psi makes it a 2 div algorithm
epsb=eb*psi;             % a posteriori backward error update
alphab=rho*alphab+eb*epsb;  % minimum backward error
b=b-k*epsb;              % update backward prediction coefficients
x=[xin;x(1:l-1)];        % shift new value into filter taps
%******** Time updating of the LS FIR filter
e=din-f'*x;              % error: reference - filter output
eps=-e*psi;              % a posteriori error of adaptive filter
f=f+w*eps;               % coefficient update
```

The total effort (not counting the exponential weight with ρ) is 2 divisions, $7L + 8$ multiplications and $7L + 4$ additions or subtractions.

8.7 Comparison of LMS and RLS Parameters

Finally, Table 8.3 compares the algorithms we have introduced in this chapter. The table shows a comparison in terms of computation complexity for the basic stochastic gradient (SG) methods like signed LMS (SLMS), normalized LMS (NLMS) or block LMS (BLMS) algorithm using a FFT. Transform-domain algorithms are listed next, but the effort does not include the power normalization, i.e., L normalizations in the transform domain. From the RLS algorithms we have discussed the (fast) Kalman algorithm and the FAEST algorithm. Lattice algorithm (not discussed) in general, require a large number of division and square root computations and it has been suggested to use the logarithmic number system (see Chap. 2, p. 65) in this case [283].

The data in Table 8.3 are based on the discussion in Chap. 6 of DCT and DFT and their implementation using fast DIF or DIT algorithms. For DCT or DFT of length 8 and 16 more efficient (Winograd-type) algorithms have been developed using even fewer operations. A length-8 DCT (see Fig. 6.23, p. 389),

Table 8.3. Complexity comparison for LMS and RLS algorithms for length-L adaptive filter. TDLMS without normalization. Add L multiplications and $2L$ add/subtract and L divide, if normalization is used in the TDLMS algorithms.

Algorithm	Implementation	Computational load		
		mult	add/sub	div
SG	LMS	$2L$	$2L$	-
	SLMS	L	$2L$	-
	NLMS	$2L + 1$	$2L + 2$	1
	BLMS (FFT)	$10\log_2(L) + 8$	$15\log_2)L) + 30$	
SG	Hadamard	$2L$	$4L - 2$	-
TDLMS	Haar	$2L$	$2L + 2\log_2(L)$	-
	DCT	$2L + \frac{3L}{2}\log_2(L) + L$	$2L + \frac{3L}{2}\log_2(L)$	-
	DFT	$2L + \frac{3L}{2}\log_2(L)$	$2L + \frac{3L}{2}\log_2(L)$	-
	KLT	$2L + L^2 + L$	$2L + 2L$	-
RLS	direct	$2L^2 + 4L$	$2L^2 + 2L - 2$	2
	fast Kalman	$10L + 2$	$9L + 2$	2
	lattice	$8L$	$8L$	$6L$
	FAEST	$7L + 8$	$7L + 4$	2

for instance, uses 12 multiplications and a DCT transform-domain algorithm can then be implemented with $2 \times 8 + 12 = 28$ multiplications, which compares to the FAEST algorithms $7 \times 8 + 8 = 64$. But this calculation does not take into account that a power normalization is mandatory for all TDLMS (otherwise there is no fast convergence compared with the standard LMS algorithm [271, 272]). The effort for the division may be larger than the multiplication effort. When the power normalization factor can be determined beforehand it may be possible to implement the division with hardwired scaling operations. FAEST needs only 2 divisions, independent of the ADF length.

A comparison of the RLS and LMS adaptation speed was presented in Example 8.9 (p. 523), which shows that RLS-type algorithms adapt much faster than the LMS algorithm, but the LMS algorithm can be improved essentially with transform-domain algorithms, like the DCT-LMS. Also, error floor and consistency of the error is, in general, better for the RLS algorithm, when compared with LMS or TDLMS algorithms. But none of the RLS-type algorithms can be implemented without division operations, which will require usually a larger overall system bit width, at least a fractional number representation, or even a floating-point representation [283]. The LMS algorithm on the other hand, can be implemented with only a few bits as presented in Example 8.5 (p. 504).

Exercises

Note: If you have no prior experience with the Quartus II software, refer to the case study found in Sect. 1.4.3, p. 29. If not otherwise noted use the EP2C35F672C6 from the Cyclone II family for the Quartus II synthesis evaluations.

8.1: Suppose the following signal is given

$$x[n] = A\cos[2\pi n/T + \phi].$$

(a) Determine the power or variance σ^2.
(b) Determine the autocorrelation function $r_{xx}[\tau]$.
(c) What is the period of $r_{xx}[\tau]$?

8.2: Suppose the following signal is given

$$x[n] = A\sin[2\pi n/T + \phi] + n[n],$$

where $n[n]$ is a white Gaussian noise with variance σ_n^2.
(a) Determine the power or variance σ^2 of the signal $x[n]$
(b) Determine the autocorrelation function $r_{xx}[\tau]$.
(c) What is the period of $r_{xx}[\tau]$?

8.3: Suppose the following two signals are given:

$$x[n] = \cos[2\pi n/T_0] \qquad y[n] = \cos[2\pi n/T_1].$$

(a) Determine the cross-correlation function $r_{xy}[\tau]$.
(b) What is the condition for T_0 and T_1 that $r_{xy}[\tau] = 0$?

8.4: Suppose the following signal statistics have been determined:

$$\boldsymbol{R_{xx}} = \begin{bmatrix} 2 & 1 \\ 1 & 2 \end{bmatrix} \qquad \boldsymbol{r}_{dx} = \begin{bmatrix} 4 \\ 5 \end{bmatrix} \qquad \boldsymbol{R}_{dd}[0] = 20.$$

Compute
(a) Compute $\boldsymbol{R}_{xx}^{-1}$.
(b) The optimal Wiener filter weight.
(c) The error for the optimal filter weight.
(d) The eigenvalues and the eigenvalue ratio.

8.5: Suppose the following signal statistics for a second-order system are given:

$$\boldsymbol{R_{xx}} = \begin{bmatrix} r_0 & r_1 \\ r_1 & r_0 \end{bmatrix} \qquad \boldsymbol{r}_{dx} = \begin{bmatrix} c_0 \\ c_1 \end{bmatrix} \qquad \boldsymbol{R}_{dd}[0] = \sigma_d^2.$$

The optimal filter with coefficient should be f_0 and f_1.
(a) Compute $\boldsymbol{R}_{xx}^{-1}$.
(b) Determine the optimal filter weight error as a function of f_0 and f_1.
(c) Determine f_0 and f_1 as a function of r and c.
(d) Assume now that $r_1 = 0$. What are the optimal filter coefficients f_0 and f_1?

8.6: Suppose the desired signal is given as:

$$d[n] = \cos[2\pi n/T_0].$$

The reference signal $x[n]$ that is applied to the adaptive filter input is given as

$$x[n] = \sin[2\pi n/T_0] + 0.5\cos[2\pi n/T_1],$$

where $T_0 = 5$ and $T_1 = 3$. Compute for a second-order system:
(a) $\boldsymbol{R}_{xx}, \boldsymbol{r}_{dx}$. and $\boldsymbol{R}_{dd}[0]$.
(b) The optimal Wiener filter weight.
(c) The error for the optimal filter weight.
(d) The eigenvalues and the eigenvalue ratio.
(e) Repeat (a)–(d) for a third-order system.

8.7: Suppose the desired signal is given as:

$$d[n] = \cos[2\pi n/T_0] + n[n],$$

where $n[n]$ is a white Gaussion noise with variance 1. The reference signal $x[n]$ that is applied to the adaptive filter input is given as

$$x[n] = \sin[2\pi n/T_0],$$

where $T_0 = 5$. Compute for a second-order system:
(a) $\boldsymbol{R}_{xx}, \boldsymbol{r}_{dx}$. and $\boldsymbol{R}_{dd}[0]$.
(b) The optimal Wiener filter weight.
(c) The error for the optimal filter weight.
(d) The eigenvalues and the eigenvalue ratio.
(e) Repeat (a)–(d) for a third-order system.

8.8: Suppose the desired signal is given as:

$$d[n] = \cos[4\pi n/T_0]$$

where $n[n]$ is a white Gaussian noise with variance 1. The reference signal $x[n]$, which is applied to the adaptive filter input, is given as

$$x[n] = \sin[2\pi n/T_0] - \cos[4\pi n/T_0],$$

with $T_0 = 5$. Compute for a second-order system:
(a) $\boldsymbol{R}_{xx}, \boldsymbol{r}_{dx}$. and $\boldsymbol{R}_{dd}[0]$.
(b) The optimal Wiener filter weight.
(c) The error for the optimal filter weight.
(d) The eigenvalues and the eigenvalue ratio.
(e) Repeat (a)–(d) for a third-order system.

8.9: Using the 4 FIR filters given in Sect. 8.3.1 (p. 493) use C or MATLAB to compute the autocorrelation function and the eigenvalue ratio using the autocorrelation for of a (filtered) sequence of 10 000 white noise samples. For the following system length (i.e., size of autocorrelation matrix):
(a) $L = 2$.
(b) $L = 4$.
(c) $L = 8$.
(d) $L = 16$.
Hint: The MATLAB functions: `randn`, `filter`, `xcorr`, `toeplitz`, `eig` are helpful.

8.10: Using an IIR filter with one pole $0 < \rho < 1$ use C or MATLAB to compute the autocorrelation function and plot the eigenvalue ratio using the autocorrelation for

a (filtered) sequence of 10 000 white noise samples. For the following system length
(i.e., size of autocorrelation matrix):
(a) $L = 2$.
(b) $L = 4$.
(c) $L = 8$.
(d) $L = 16$.
(e) Compare the results from (a) to (d) with the theoretical value EVR $= (1 + \rho)/(1 - \rho))^2$ of Markov-1 processes [269].
Hint: The MATLAB functions: `randn, filter, xcorr, toeplitz, eig` are helpful.

8.11: Using the FIR filter for EVR $= 1000$ given in Sect. 8.3.1 (p. 493) use C or
MATLAB to compute the eigenvectors of the autocorrelation for $L = 16$. Compare
the eigenvectors with the DCT basis vectors.

8.12: Using the FIR filter for EVR $= 1000$ given in Sect. 8.3.1 (p. 493) use C or
MATLAB to compute the eigenvalue ratios of the transformed power normalized
autocorrelation matrices from (8.45) on page 502 for $L = 16$ using the following
transforms:
(a) Identity transform (i.e., no transform).
(b) DCT.
(c) Hadamard.
(d) Haar.
(e) Karhunen–Loéve.
(f) Build a ranking of the transform from (a)–(e).

8.13: Using the one pole IIR filter from Exercise 8.10 use C or MATLAB to compute
for 10 values of ρ in the range 0.5 to 0.95 the eigenvalue ratios of the transformed
power normalized autocorrelation matrices from (8.45) on page 502 for $L = 16$
using the following transforms:
(a) Identity transform (i.e., no transform).
(b) DCT.
(c) Hadamard.
(d) Haar.
(e) Karhunen–Loéve.
(f) Build a ranking of the transform from (a)–(e).

8.14: Use C or MATLAB to rebuild the power estimation shown for the nonstationary
signal shown in Fig. 8.15 (p. 497). For the power estimation use
(a) Equation (8.38) page 496.
(b) Equation (8.41) page 498 with $\beta = 0.5$.
(c) Equation (8.41) page 498 with $\beta = 0.9$.

8.15: Use C or MATLAB to rebuild the simulation shown in Example 8.1 (p. 485)
for the following filter length:
(a) L=2.
(b) L=3.
(c) L=4.
(d) Compute the exact Wiener solution for L=3.
(e) Compute the exact Wiener solution for L=4.

8.16: Use C or MATLAB to rebuild the simulation shown in Example 8.3 (p. 492)
for the following filter length:
(a) L=2.
(b) L=3.
(c) L=4.

8.17: Use C or MATLAB to rebuild the simulation shown in Example 8.6 (p. 511) for the following pipeline configuration:
(a) DLMS with 1 pipeline stages.
(b) DLMS with 3 pipeline stages.
(c) DLMS with 6 pipeline stages.
(d) DLMS with 8 pipeline stages.

8.18: (a) Change the filter length of the adaptive filter in Example 8.5 (p. 504) to three.
(b) Make a functional compilation (with the Quartus II compiler) of the HDL code for the filter.
(c) Perform a functional simulation of the filter with the inputs $d[n]$ and $x[n]$.
(d) Compare the results with the simulation in Exercise 8.15b and d.

8.19: (a) Change the filter length of the adaptive filter in Example 8.5 (p. 504) to four.
(b) Make a functional compilation (with the Quartus II compiler) of the HDL code for the filter.
(c) Perform a functional simulation of the filter with the inputs $d[n]$ and $x[n]$.
(d) Compare the results with the simulation in Exercise 8.15c and e.

8.20: (a) Change the DLMS filter design from Example 8.6 (p. 511) pipeline of $e[n]$ only, i.e. DLMS with 1 pipeline stage.
(b) Make a functional compilation (with the Quartus II compiler) of the HDL code for the filter.
(c) Perform a functional simulation of the filter with the inputs $d[n]$ and $x[n]$.
(d) Compare the results with the simulation in Exercise 8.17a.
(e) Determine the **Registered Performance** and the used resources (LEs, multipliers, and M4Ks) of your D=1 design using the device EP2C35F672C6 from the Cyclone II family.
(f) Repeat (e) for the EPF10K70RC240-4 from the Flex 10K family.

8.21: (a) Change the DLMS filter design from Example 8.6 (p. 511) pipeline of f update only, i.e. DLMS with 3 pipeline stages.
(b) Make a functional compilation (with the Quartus II compiler) of the HDL code for the filter.
(c) Perform a functional simulation of the filter with the inputs $d[n]$ and $x[n]$.
(d) Compare the results with the simulation in Exercise 8.17b.
(e) Determine the **Registered Performance** and the used resources (LEs, multipliers, and M4Ks) of your D=3 design using the device EP2C35F672C6 from the Cyclone II family.
(f) Repeat (e) for the EPF10K70RC240-4 from the Flex 10K family.

8.22: (a) Change the DLMS filter design from Example 8.6 (p. 511) pipeline with an optimal number of stages, i.e. DLMS with 8 pipeline stages, 3 for each multiplier and one stage each for $e[n]$ and $y[n]$.
(b) Make a functional compilation (with the Quartus II compiler) of the HDL code for the filter.
(c) Perform a functional simulation the filter with the inputs $d[n]$ and $x[n]$.
(d) Compare the results with the simulation in Exercise 8.17d.
(e) Determine the **Registered Performance** and the used resources (LEs, multipliers, and M4Ks) of your D=8 design using the device EP2C35F672C6 from the Cyclone II family.
(f) Repeat (e) for the EPF10K70RC240-4 from the Flex 10K family.

9. Microprocessor Design

Introduction

When you think of current microprocessors (µPs), the Intel Itanium processor with 592 million transistors may come to mind. Designing this kind of microprocessor with an FPGA, you may ask? Now the author has become overconfident with the capabilities of today's FPGAs. Clearly today's FPGAs will not be able to implement such a top-of-the-range µP with a single FPGA. But there are many applications where a less-powerful µP can be quite helpful. Remember a µP trades performance of the hardwired solution with gate efficiency. In software, or when designed as an FSM, an algorithm like an FFT may ran slower, but usually needs much less resources. So the µP we build with FPGAs are more of the microcontroller type than a fully featured modern Intel Pentium or TI VLIW PDSP. A typical application we discuss later would be the implementation of a controller for a radix-2 FFT. Now you may argue that this can be done with an FSM. And yes that is true and we basically consider our FPGA µP design nothing else then an FSM augmented by a program memory that includes the operation the FSM will perform, see Fig. 9.1. In fact the early versions of the Xilinx PicoBlaze processor were called Ken Chapman programmable state machine (KCPSM) [284]. A complete µP design usually involves several steps, such as the architecture exploration phase, the instruction set design, and the development tools. We will discuss these steps in the following in more details. You are encouraged to study in addition a computer architecture book; there are many available today as this is a standard topic in most undergraduate computer engineering curricula [285, 286, 287, 288, 289, 290, 291]. But before we go into details of µP design let us first have a look back at the begin of the µP era.

9.1 History of Microprocessors

Usually microprocessor are classified into three major classes: the general-purpose or CISC processor, reduced instruction set processors (RISC), and programmable digital signal processors (PDSP). Let us now have a brief look how these classes of microprocessor have developed.

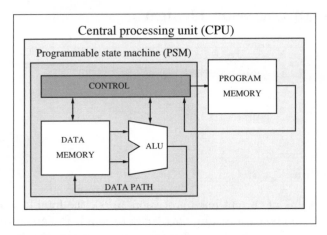

Fig. 9.1. The Xilinx KCPSM a.k.a. PicoBlaze.

9.1.1 Brief History of General-Purpose Microprocessors

By 1968 the typical general-purpose minicomputers in use were 16-bit architectures using about 200 MSI chips on a single circuit board. The MSI had about 100 transistors each per chip. A popular question [292] then was: can we also build a single CPU with (only) 150, 80, or 25 chips?

At about the same time Robert Noyce and Gordon Moore, formerly with Fairchild Corp., started a new company first called NM Electronics and later renamed Intel, whose main product was memory chips. In 1969 Busicom, a Japanese calculator manufacture, asked Intel to design a set of chips for their new family of programmable calculators. Intel did not have the manpower to build the 12 different custom chips requested by Busicom, since good IC designers at that time were hard to find. Instead Intel's engineer Ted Hoff suggested building a more-general four-chip set that would access its instructions from a memory chip. A programmable state machine (PSM) with memory was born, which is what we today call a microprocessor. After nine months with the help of F. Faggin the group of Hoffs delivered the Intel 4004, a 4-bit CPU that could be used for the BCD arithmetic used in the Busicom calculators. The 4004 used 12-bit program addresses and 8-bit instructions and it took five clock cycles for the execution of one instruction. A minimum working system (with I/O chips) could be build out of two chips only: the 4004 CPU and a program ROM. The 1 MHz clock allowed one to add multidigit BCD numbers at a rate of 80 ns per digit [293].

Hoff's vision was now to extend the use of the 4004 beyond calculators to digital scales, taxi meters, gas pumps, elevator control, medical instruments, vending machines, etc. Therefore he convinced the Intel management to obtain the rights from Busicom to sell the chips to others too. By May 1971 Intel gave a price concession to Busicom and obtained in exchange the rights to sell the 4004 chips for applications other then calculators.

Table 9.1. The family of Intel microprocessors [294] (IA = instruction set architecture).

Name	Year intro.	MHz	IA	Process technology	#Trans- istors
4004	1971	0.108	4	10μ	2300
8008	1972	0.2	8	10μ	3500
8080	1974	2	8	6μ	4500
8086	1978	5-10	16	3μ	29K
80286	1982	6-12.5	16	1.5μ	134K
80386	1985	16-33	32	1μ	275K
80486	1989	25-50	32	0.8μ	1.2M
Pentium	1993	60-66	32	0.8μ	3.1M
Pentium II	1997	200-300	32	0.25μ	7.5M
Pentium 3	1999	650-1400	32	0.25μ	9.5M
Pentium 4	2000	1300-3800	32	0.18μ	42M
Xeon	2003	1400-3600	64	0.09μ	178M
Itanium 2	2004	1000-1600	64	0.13μ	592M

One of the concerns was that the performance of the 4004 could not compete with state-of-the-art minicomputers at the time. But another invention, the EPROM, also from Intel by Dov Frohamn-Bentchkovsky helped to market a 4004 development system. Now programs did not need an IC factory to generate the ROM, with the associated long delays in the development. EPROMs could be programmed and reprogrammed by the developer many times if necessary.

Some of Intel's customer asked for a more-powerful CPU, and an 8-bit CPU was designed that could handle the 4-bit BCD arithmetic of the 4004. Intel decided to build a 8008 and it also supported standard RAM and ROM devices, custom memory was no longer required as for the 4004 design. Some of the shortcoming of the 8008 design were fixed by the 8080 design in 1974 that used now about 4500 transistors. In 1978 the first 16-bit µP was introduced, the 8086. In 1982 the 80286 followed: a 16-bit µP but with about six times the performance of the 8086. The 80386, introduced in 1985, was the first µP to support multitasking. In the 80387 a mathematics coprocessor was added to speed up floating-point operations. Then in 1989 the 80486 was introduced with an instruction cache and instruction pipelining as well as a mathematics coprocessor for floating-point operations. In 1993 the Pentium family was introduced, with now two pipelines for execution, i.e., a super-scalar architecture. The next generation of Pentium II introduced in 1997 added multimedia extension (MMX) instructions that could perform some parallel vector-like MAC operations of up to four operands. The Pentium 3 and 4 followed with even more advanced features, like hyperthreading and SSE instructions to speed up audio and video processing. The largest processor as of 2006 is the Intel Itanium with two processor cores and a whopping

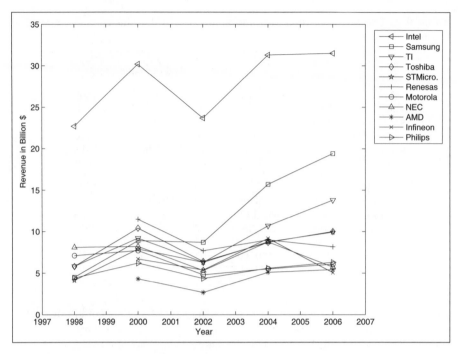

Fig. 9.2. Top semiconductor company revenue.

592 million transistors. It includes 9 MB of L3 cache alone. The whole family of Intel processors is shown in Table 9.1.

Looking at the revenue of semiconductor companies it is still impressive that Intel has maintained its lead over many years mainly with just one product, the microprocessor. Other microprocessor-dominated companies like Texas Instruments, Motorola/Freescale or AMD have much lower revenue. Other top companies such as Samsung or Toshiba are dominated by memory technology, but still do not have Intel's revenue, which has been in the lead for many years now.

9.1.2 Brief History of RISC Microprocessors

The Intel architecture discussed in the last section is sometimes called a complex instruction set computer (CSIC). Starting from the early CPUs, subsequent designs tried to be compatible, i.e., being able to run the same programs. As the bitwidth of data and programs expanded you can imagine that this compatibility came at a price: performance, although Moore's law allowed this CISC architecture to be quite successful by adding new components and features like numeric coprocessors, data and program caches, MMX, and SSE instructions and the appropriate instructions that support these additions to improve performance. Intel μPs are characterized by having many instructions and supporting many addressing modes. The μP manuals are usually 700 or more pages thick.

Around 1980 research from CA University of Berkeley (Prof. Patterson), IBM (later called PowerPC) and at Stanford University (Prof. Henessy, on what later become the microprocessor without interlocked pipeline stages, i.e., MIPS μP family) analyzed μPs and came to the conclusion that, from a performance standpoint, CISC machines had several problems. In Berkeley this new generation of μPs were called RISC-1 and RISC-2 since one of their most important feature was a limited number of instructions and addressing modes, leading to the name reduced instruction set computer (RISC).

Let us now briefly review some of the most important features by which (at least early) RISC and CISC machines could be characterized.

- The *instruction set* of a CISC machines is rich, while a RISC machine typically supports fewer than 100 instructions.
- The *word length* of RISC instructions is fixed, and typically 32 bits long. In a CISC machine the word length is variable. In Intel machines we find instructions from 1 to 15 bytes long.
- The *addressing modes* supported by a CISC machine are rich, while in a RISC machine only very few modes are supported. A typically RISC machine supports just immediate and register base addressing.
- The *operands* for ALU operations in a CISC machine can come from instruction words, registers, or memory. In a RISC machine no direct memory operands are used. Only memory load/store to or from registers is allowed and RISC machine are therefore called load/store architectures.
- In *subroutines* parameters and data are usually linked via a stack in CISC machines. A RISC machine has a substantial number of registers, which are used to link parameter and data to subroutines.

In the early 1990s it become apparent that neither RISC nor CISC architectures in their purest form were better for all applications. CISC machines, although still supporting many instruction and addressing modes, today take advantage of a larger number of CPU registers and deep pipelining. RISC machines today like the MIPS, PowerPC, SUN Sparc or DEC alpha, have hundreds of instructions, and some need multiple cycles and hardly fit the title reduced instruction set computer.

9.1.3 Brief History of PDSPs

In 1980 Intel [295] introduced the 2930, an analog signal processor that was used in control systems and included an ADC and DAC on chip[1] to implement the algorithms in a DSP-like fashion, see Fig. 9.3.

---

[1] We would call today such a μP a microcontroller, since I/O functions are integrated on the chip.

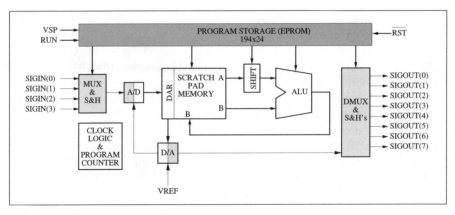

Fig. 9.3. The 2920 functional block diagram [295].

The 2920 had a 40×25 bit scratch path memory, a 192-word program EPROM, a shifter, and an ALU and could implement multiplications with a series shift-and-add/subtracts.

Example 9.1: A multiplication in the 2920 with a constant $C = 1.88184_{10} = 1.1110000111_2$ was coded first in CSD as $C = 10.00\bar{1}0001001\bar{1}_2$ and was then implemented with the following instructions:

```
ADD  Y,X,L01;
SUB  Y,X,R03;
ADD  Y,X,R07;
SUB  Y,X,R10;
```

where the last operand Rxx (Lxx) describes the right (left) shift by xx of the second operand before the adding or subtracting.   9.1

There are a few building block macrofunctions that can be used for the 2920, they are listed in the following table:

Function	# Instructions
Constant multiply	1-5
Variable multiply	10-26
Triangle generator	6-10
Threshold detector	2-4
Sine wave generator	8-12
Single real pole	2-6

although we will not call the 2920 a PDSP it shares same important features with the PDSP, which was improved with the first-generation PDSPs introduced at the start of the 1980s. The first-generation PDSPs like TI's TSM320C10 and NEC's μPD 7720 were characterized by a *Harvard architecture*, i.e., the program and data were located in separate memory. The first

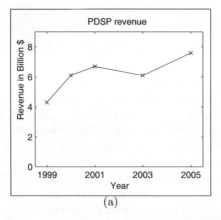
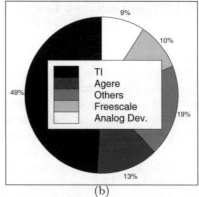

Fig. 9.4. PDSPs (a) processor revenue. (b) Market share (Agere formerly Lucent/AT&T; Freescale formerly Motorola).

generation also had a hardwired multiplier that consumed most of the chip area and allowed a single multiply in one clock cycle. The second-generation PDSPs introduced around 1985 with TI's TMS 320C25, Motorola's MC56000, or the DSP16 from AT&T allowed a MAC operation in one clock cycle and zero-overhead loops were introduced. The third generation of PDSPs, with μPs like the TMS320C30, Motorola's 96002, and the DSP32C, introduced after 1988, now supported a single clock cycle 32-bit floating-point multiplication typically using the IEEE single-precision floating-point standard, see Sect. 2.2.3, p. 71. The forth generation with the TMS320C40 and TMS320C80 introduced around 1992 now included multi-core MACs. The newest and most powerful PDSPs introduced after 1997 like the TMS320C60x, Philips Trimedia, or Motorola Starcore, are very long instruction word (VLIW) machines. Other architectures in today's PDSPs are SIMD architectures (e.g., ADSP-2126x SHARC), superscalar machines (e.g., Renesas SH77xxx), matrix math engine (e.g., Intrinsity, FastMath), or a combination of these techniques. PDSPs architectures today are more diverse than ever.

PDSPs have done very well in the last 20 years, as can be seen from Fig. 9.4a and are expected to deliver three trillion instructions per second by 2010 [296]. The market share of PDSPs has not changed much over the years and is shown in Fig. 9.4b, with Texas Instruments (TI) leading the field by a large margin. Recently some PDSP cores have also become available for FPGA designs such as Cast Inc.'s (www.cast.com) version of the TI TMS32025 and Motorola/Freescale's 56000. We can assume that most of the core instructions and architecture features are supported, while some of the I/O units such as the timer, DMA, or UARTs included on the PDSPs are usually not implemented in the FPGA cores.

9.2 Instruction Set Design

The instruction set of a microprocessor (μP) describes the collection of actions a μP can perform. The designer usually asks first, which kind of *arithmetic* operation do I need in the applications for which I will use my μP. For DSP application, for instance, special concern would be applied to support for (fast) add and multiply. A multiply done by a series of smaller shift-adds is probably not a good choice in heavy DSP processing. As well as these ALU operations we also need some *data move* instructions and some *program flow*-type instructions such as branch or goto.

The design of an instruction set also depends however on the underlying μP architecture. Without considering the hardware elements we can not fully define our instruction set. Because instruction set design is a complex task, it is a good idea to break up the development into several steps. We will proceed by answering the following questions:

1) What are the addressing modes the μP supports?
2) What is the underlying data flow architecture, i.e., how many operands are involved in an instruction?
3) Where can we find each of these operands (e.g., register, memory, ports)?
4) What type of operations are supported?
5) Where can the next instruction be found?

9.2.1 Addressing Modes

Addressing modes describe how the operands for an operation are located. A CISC machine may support many different modes, while a design for performance like in RISC or PDSPs requires a limitation to the most often used addressing modes. Let us now have a look at the most frequently supported modes in RISC and PDSPs.

Implied addressing. In implied addressing operands come from or go to a location that is implicitly and not explicitly defined by the instruction, see Fig. 9.5. An example would be the ADD operation (no operands listed) used in a stack machine. All arithmetic operations in a stack machine are performed using the two top elements of the stack. Another example is the ZAC operation of a PDSP, which clears the accumulator in a TMS320 PDSP [297]. The following listing shows some examples of implied addressing for different microprocessors:

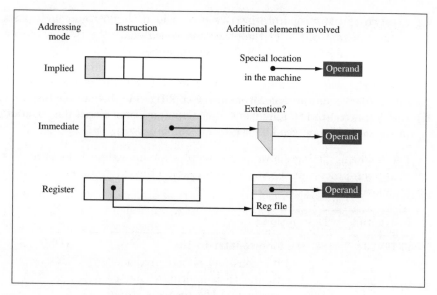

Fig. 9.5. Implied, immediate and register addressing.

Instruction	Description	µP
ZAT	Clear accumulator and T register.	TMS320C50
APAC	The contents of the P register is added to the accumulator register and replaces the accumulator register value.	TMS320C50
RET	The PC is loaded with the value of the ra register.	Nios II
BRET	Copies the b status into the status register and loads the PC with the ba register value.	Nios II

Immediate addressing. In the immediate addressing mode the operand (i.e., constant) is included in the instruction itself. This is shown in Fig. 9.5. A small problem arises due to the fact that the constant provided within one instruction word is usually shorter in terms of number of bits than the (full) data words used in the µP. There are several approaches to solve this problem:

a) Use a sign extension since most constants (like increments or loop counters) used in programs are small and do not require full length anyway. We should use sign extension (i.e., MSB is copied to the high word) rather then zero extension so that the value −1 is extended correctly (see the MPY -5 example below).

b) Use two or more separate instructions to load the low and high parts of the constant one after the other and concatenate the two parts to a full-length word (see the `LPH DAT0` example below).

c) Use a double-length instruction format to access long constants. If we extend the default size by a second word, this usually provides enough bits to load a full-precision constant (see `RPT #1111h` example below).

d) Use a (barrel) `shift` alignment of the operand that aligns the constant in the desired way (see `ADD #11h,2` example below).

The following listing shows five examples for immediate addressing for different microprocessors. Examples 2-5 describe the extension method mentioned above.

Instruction	Description	μP
CNTR=10;	Set the loop counter to 10.	ADSP
MPY -5	The 13-bit constant is sign extended to 16 bits and then multiplied by the TREG0 register and the result is stored in the P register.	TMS320C50
LPH DAT0	Load the upper half of the 32-bit product register with the data found in memory location DAT0.	TMS320C50
RPT #1111h	Repeat the next instruction $1111_{16} + 1 = 4370_{10}$ times. This is a two-word instruction and the constant is 16 bits long.	TMS320C50
ADD #11h,2	Add 11h shifted by two bits to the accumulator.	TMS320C50

To avoid having always two memory accesses, we can also combine method (a) sign extension and (b) high/low addressing. We then only need to make sure that the load high is done after the sign extension of the low word.

Register addressing. In the register addressing mode operands are accessed from registers within the CPU, and no external memory access takes place, see Fig. 9.5. The following listing shows some examples of register addressing for different microprocessors:

Instruction	Description	μP
MR=MX0*MY0(RND)	Multiply the MX0 and MY0 registers with round and store in the MR register.	ADSP
SUB sA,sB	Subtract the sB from the sA register and store in the sA register.	PicoBlaze
XOR r6,r7,r8	Compute the logical exclusive XOR of the registers r7 and r8 and store the result in the register r6.	Nios II
LAR AR3,#05h	Load the auxiliary register AR3 with the value 5.	TMS320C50
OR %i1,%i2	OR the registers i1 and i2 and replace i1.	Nios
SWAP %g2	Swap the 16-bit half-word values in the 32-bit register g2 and put it back in g2.	Nios

Since in most machines register access is much faster and consumes less power than regular memory access this is a frequently used mode in RISC machines. In fact all arithmetic operations in a RISC machine are usually done with CPU registers only; memory access is only allowed via separate load/store operations.

Memory addressing. To access external memory direct, indirect, and combination modes are typically used. In the *direct* addressing mode part of the instruction word specifies the memory address to be accessed, see Fig. 9.6, and the FETCH example below. Here the same problem as for immediate addressing occurs: the bits provided in the instruction word to access the memory operand is too small to specify the full memory address. The full address length can be constructed by using an auxiliary register that may be ex- or implicitly specified. If the auxiliary register is added to the direct memory address this is called *based* addressing, see the LDBU example below. If the auxiliary register is just used to provide the missing MSB this is called *page-wise* addressing, since the auxiliary register allows us to specify a page within which we can access our data, see the AND example below. If we need to access data outside the page, we need to update the page pointer first. Since the register in the based addressing mode represents a full-length address we can use the register without a direct memory address. This is then called *indirect* addressing, see the PM(I6,M6) example below.

The following listing shows four examples for typical memory addressing modes for different microprocessors.

Instruction	Description	μP
FETCH sX,ss	Read the scratch pad RAM location **ss** into register **sX**.	PicoBlaze
LDBU r6,100(r5)	Compute the sum of 100 and the register **r5** and load the data from this address into the register **r6**.	Nios II
AND DAT16	Assuming that the 9-bit data page register points to page 4, this will access the data word from memory location $4 \times 128 + 16$ and perform an **AND** operation with the accumulator.	TMS320C50
PM(I6,M6)=AR	The **AR** register is stored in the program memory by using the register **I6** as the address. After the memory access the address register **I6** is incremented by the value from the modify register **M6**.	ADSP

Since the indirect address mode can usually only point to a limited number of index registers this usually shortens the instruction words and it is the most popular addressing mode in PDSPs. In RISC machines based addressing is preferred, since it allows easy access to an array by specifying the base and the array element via an offset, see the LDBU example above.

PDSP specific addressing modes. PDSPs are known to process DSP algorithms much more efficiently than standard GPP or RISC machines. We now want to discuss three addressing modes typical used in PDSPs that have shown to be particular helpful in DSP algorithms:

- Auto de/increment addressing
- Circular addressing
- Bitreversed addressing

The auto and circular addressing modes let us compute convolutions and correlations more efficiently, while bitreversed addressing is used in FFT algorithms.

The *auto increment* or *decrement* of the address pointer is a typical address modification that is used in PDSPs for convolutions or correlations. Since convolutions or correlations can be described by a repeated multiply-accumulate (MAC) operation, i.e., the inner (scalar) product of two vectors [see (3.2), p. 166] it can be argued that, after each MAC operation, an address modification increment or decrement depending on the ordering of the data and coefficient is necessary to update the address pointer for the

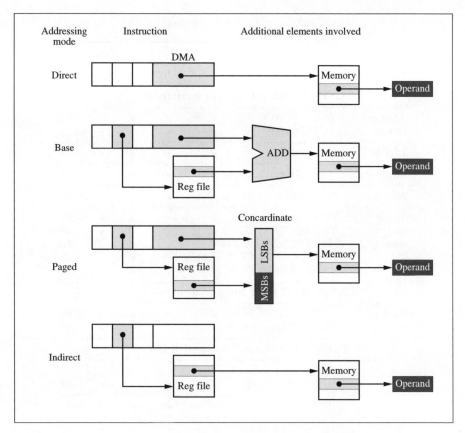

Fig. 9.6. Memory addressing: direct, based, paged, and indirect.

next data/coefficient pair. After each memory access a de/increment of the data/coefficient read pointer is therefore performed. This allow us to design a single-cycle MAC – the address update is not done by the CPU as in RISC machines; a separate address register file is used for this purpose, see Fig. 9.7. An FIR filtering using the ADSP assembler coding can be done via the following few steps:

```
        CNTR = 256;
        MR=0, MX0=DM(I0,M1), MY0(I4,M5)
        DO FIR UNTIL CE;
FIR: MR=MR+MX0*MY0(SS), MX0=DM(I0,M1), MY0=PM(I4,M5)
```

After initializing the loop counter `CNTR` and the pointer for data x coming from the data memory `DM` and y coming from the program memory `PM`, the `DO UNTIL` loop allows one to compute one MAC instruction and two data loads in each clock cycle. The data pointers `I0` and `I4` are then updated via `M1` and `M5`, respectively. This example also shows another feature of PDSPs,

Table 9.2. Properties of PDSPs useful in convolutions (©1999 Springer Press [5]).

Vendor	Type	Accu bits	Super Harvard	Modulo address	Bit-re-verse	Hard-ware loops	MAC rate MHz
PDSPs 16×16 bit integer multiplier							
Analog	ADSP-2187	40	✓	✓	✓	✓	52
Device	ADSP-21csp01	40	✓	✓	–	–	50
Lucent	DSP1620	36	✓	–	✓	✓	120
Motorola	DSP56166	36	✓	✓	✓	✓	60
NEC	µPD77015	40	✓	✓	✓	✓	33
Texas	TMS320F206	32	✓	–	✓	✓	80
Instrum-	TMS320C51	32	✓	✓	✓	✓	100
ents	TMS320C549	40	✓	✓	✓	✓	200
	TMS320C601	40	2 MAC	✓	✓	✓	200
	TMS320C80	32	4 MAC	✓	✓	✓	50
PDSPs 24×24 bit integer multiplier							
Motorola	DSP56011	56	✓	✓	✓	✓	80
	DSP56305	56	✓	✓	✓	✓	100
PDSPs 24×24 bit float multiplier							
Analog Device	SHARC 21061	80	✓	✓	✓	✓	40
Motorola	DSP96002	96	✓	✓	✓	✓	60
Texas	TMS320C31	40	✓	✓	✓	✓	60
Instru-ments	TMS320C40	40	✓	✓	✓	✓	60

the so-called zero-overhead loop. The update and loop counter check done at the end of the loop do not require any additional cycles as in GPP or RISC machines.

To motivate *circular addressing* let us revisit how data and coefficient typically are arranged in a continuous data processing scheme. In the first instance to compute $y[0]$ the data $x[k]$ and coefficient $f[k]$ are aligned as follows:

$f[L-1]$	$f[L-2]$	$f[L-3]$	$\cdots$	$f[1]$	$f[0]$
$x[0]$	$x[1]$	$x[2]$	$\cdots$	$x[L-2]$	$x[L-1]$

where the coefficient and data in the same column are multiplied and all products are accumulated to compute $y[0]$. In the next step to compute $y[1]$ we need the following data

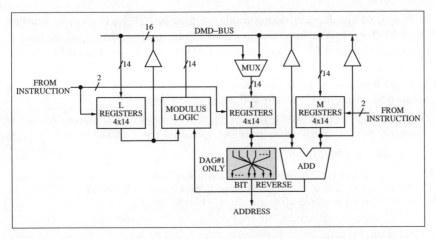

Fig. 9.7. Address generation in the ADSP programmable digital signal processor.

$f[L-1]$	$f[L-2]$	$f[L-3]$	$\cdots$	$f[1]$	$f[0]$
$x[1]$	$x[2]$	$x[3]$	$\cdots$	$x[L-1]$	$x[L]$

We can solve this problem by shifting every data word after a MAC operation. The TMS320 family, for instance, provides such a `MACD`, i.e., a MAC plus data move instruction [298]. After N `MACD` operations the whole vector x is shifted by one position. Alternatively we can look at the data that are used in the computation and we see that all data can stay in the same place; only the oldest element $x[0]$ needs to be replaced by $x[L]$ – the other coefficients keep their place. We therefore have the following memory arrangement for the x data vector:

$x[L]$	$x[1]$	$x[2]$	$\cdots$	$x[L-2]$	$x[L-1]$

If we now start our data pointer with $x[1]$ the processing works fine as before; only when we reach the end of the data buffer $x[L-1]$ we do need to reset the address pointer. Assuming the address buffer can be described by the following four parameters:

- L = buffer length
- I = current address
- M = modify value (signed)
- B = base address of buffer

we can describe the required address computation by the following equation

$$\text{new address} = (I + M - B) \bmod L + B. \tag{9.1}$$

This is called circular addressing, since after each memory modification we check if the resulting address is still in the valid range. Figure 9.7 shows the address generator that is used by the ADSP PDSPs, using (9.1).

The third special addressing mode we find in PDSPs is used to simplify the radix-2 FFT. We have discussed in Chap. 6 that the input or output sequence in radix-2 FFTs appear in bitreverse order, see Fig. 6.14, p. 367. The bitreverse index computation usually take many clock cycles in software since the location of each bit must be reversed. PDSPs support this by a special addressing mode as the following example shows.

Example 9.2: The ADSP [299] and TMS320C50 [298] both support bitreverse addressing. Here is an assembler code example from the TMS320 family:

```
ADD  * BRO-,8
```

The content of the **INDX** register is first subtracted from the current auxiliary register. Then a bitreverse of the address value is performed to locate the operand. The loaded data word is added after a shift of 8 bits to the accumulator.   9.2

Table 9.2 shows an overview of PDSPs and the supported addressing modes. All support auto de/increment. Circular buffers and bitreverse is also supported by most PDSPs.

9.2.2 Data Flow: Zero-,One-, Two- or Three-Address Design

A typical assembler coding of an instruction lists first the operation code followed by the operand(s). A typical ALU operation requires two operands and, if we also want to specify a separate result location, a natural way that makes assembler easy for the programmer would be to allow that the instruction word has an operation code followed by three operands. However, a three-operand choice can require a long instruction word. Assume our embedded FPGA memory has 1K words then at least 30 bits, not counting the operation code, are required in the direct addressing mode. In a modern CPU that address 4 GB requires a 32-bit address, three operands in direct addressing would require at least 96-bit instruction words. As a result limiting the number of operands will reduce the instruction word length and save resources. A zero-address or stack machine would be perfect in this regard. Another way would be to use a register file instead of direct memory access and only allow load/store of single operands as is typical in RISC machines. For a CPU with eight registers we would only need 9 bits to specify the three operands. But we then need extra operation code to load/store data memory data in the register file.

In the following sections we will discuss the implications the hardware and instruction sets when we allow zero to three operands in the instruction word.

Stack machine: a zero-address CPU. A zero-address machine, you may ask, how can this work? We need to recall from the addressing modes, see Sect. 9.2.1, p. 544, that operands can be specified implicitly in the instruction. For instance in a TI PDSP all products are stored in the product register P, and this does not need to be specified in the multiply instruction since all multiply results will go there. Similarly in a zero-address or stack machine, all two-operand arithmetic operations are performed with the two top elements of a stack [300]. A stack by the way can be seen as a last-in first-out (LIFO). The element we put on the stack with the instruction PUSH will be the first that comes out when we use a POP operation. Let us briefly analyze how an expression like

$$d = 5 - a + b * c \tag{9.2}$$

is computed by a stack machine. The left side shows the instruction and the right side the contents of the stack with four entries. The top of the stack is to the left.

Instruction	Stack			
	TOP	2	3	4
PUSH 5	5	$-$	$-$	$-$
PUSH a	a	5	$-$	$-$
SUB	$5 - a$	$-$	$-$	$-$
PUSH b	b	$5 - a$	$-$	$-$
PUSH c	c	b	$5 - a$	$-$
MUL	$c \times b$	$5 - a$	$-$	$-$
ADD	$c \times b + 5 - a$	$-$	$-$	$-$
POP d	$-$	$-$	$-$	$-$

It can be seen that all arithmetic operations (ADD, SUB, MUL) use the implicitly specified operands top-of-stack and second-of stack, and are in fact zero-address operations. The memory operations PUSH and POP however require one operand.

The code for the stack machine is called postfix (or reverse Polish) operation, since first the operands are specified and then the operations. The standard arithmetic as in (9.2) is called infix notation, e.g., we have the two congruent representations:

$$\underbrace{5 - a + b * c}_{\text{Infix}} \longleftrightarrow \underbrace{5a - bc * +}_{\text{Postfix}} \tag{9.3}$$

In Exercises 9.8-9.11 (p. 636) some more examples of these two different arithmetic modes are shown. Some may recall that the postfix notation is exactly the same coding the HP41C pocket calculator requires. The HC41C too used a stack with four values. Figure 9.8a shows the machine architecture.

Accumulator machine: a one-address CPU. Let use now add a single accumulator to the CPU and use this accumulator both as the source for one operand and as the destination for the result. The arithmetic operations are of the form

$$\text{acc} \leftarrow \text{acc} \,\square\, \text{op1}, \tag{9.4}$$

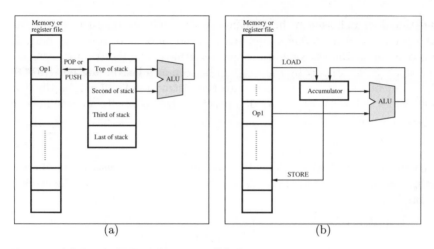

Fig. 9.8. (a) Stack CPU architecture. (b) Accumulator machine architecture.

where □ describes an ALU operation like ADD, MUL, or AND. The underlying architecture of the TI TMS320 [297] family of PDSPs is of this type and is shown in Fig. 9.8b. In ADD or SUB operations, for instance, a single operand is specified. The example equation (9.2) from the last section would be coded in the TMS320C50 [298] assembler code as follows:

Instruction	Description
ZAP	; Clear accu and P register
ADD #5h	; Add 5 to the accu
SUB DAT1	; Subtract DAT1 from the accu
LT DAT2	; Load DAT2 in the T register
MPY DAT3	; Multiply T and DAT3 and store in P register
APAC	; Add the P register to the accu
SACL DAT4	; Store accu at the address DAT4

The example assumes that the variables a-d have been mapped to data memory words DAT1-DAT4. Comparing the stack machine with the accumulator machine we can make the following conclusions:

- The size of the instruction word has not changed, since the stack machine also requires POP and PUSH operations that include an operand
- The number of instructions to code an algebraic expression is not essentially reduced (seven for an accumulator machine; eight for a stack machine)

A more-substantial reduction in the number of instructions required to code an algebraic expression is expected when we use a two-operand machine, as discussed next.

The two-address CPU. In a two-address machine we have arithmetic operations that allows us to specify the two operands independently, and the destination operand is equal to the first operand, i.e., the operations are of the form

$$\text{op1} \leftarrow \text{op1} \ \Box \ \text{op2}, \tag{9.5}$$

where $\Box$ describes an ALU operation like SUB, DIV, or NAND. The PicoBlaze from Xilinx [301, 284] and the Nios processor [302] from Altera use this kind of data flow, which is shown in Fig. 9.9a. The limitation to two operands allows in these cases the use of a 16-bit instruction word[2] format. The coding of our algebraic example equation (9.2) would be coded in assembler for the PicoBlaze as follows:

Instruction	Description
LOAD sD,sB	; Store register B in register D
MUL sD,sC	; Multiply D with register C
ADD sD,5	; Add 5 to D
SUB sD,sA	; Subtract A from D

In order to avoid an intermediate result for the product a rearrangement of the operation was necessary. Note that PicoBlaze does not have a separate MUL operation and the code is therefore for demonstration of the two-operand principle only. The PicoBlaze uses 16 registers, each 8 bits wide. With two operands and 8-bit constant values this allows us to fit the operation code and operands or constant in one 16-bit data word.

We can also see that two-operand coding reduces the number of operations essentially compared with stack or accumulator machines.

The three-address CPU. The three-address machine is the most flexible of all. The two operands and the destination operand can come or go into different registers or memory locations, i.e., the operations are of the form

$$\text{op1} \leftarrow \text{op2} \ \Box \ \text{op3}. \tag{9.6}$$

Most modern RISC machine like the PowerPC, MicroBlaze or Nios II favor this type of coding [303, 304, 305]. The operands however are usually register operands or no more than one operand can come from data memory. The data flow is shown in Fig. 9.9b.

Programming in assembler language with the three-operand machine is a straightforward task. The coding of our arithmetic example equation (9.2) will look for a Nios II machine as follows:

---

[2] Some recent PicoBlaze coding now use 18-bit instruction words since this is the memory width of the Xilinx block RAMs [284, 301].

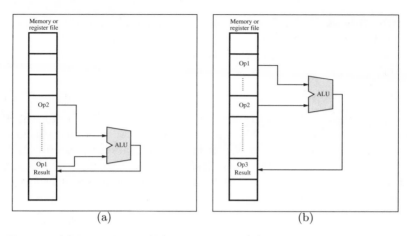

Fig. 9.9. (a) Two address CPU architecture. (b) Three address machine architecture.

Instruction	Description
SUBI r4,r1,5	; Subtract 5 from r1 register and store in r4
MUL r5,r2,r3	; Multiply registers r2 and r3 and store in r5
ADD r4,r4,r5	; Add registers r4 and r5 and store in r4

assuming that the registers r1-r4 hold the values for the variables a through d. This is the shortest code of all four machines we have discussed so far. The price to pay is the larger instruction word. In terms of hardware implementation we will not see much difference between two- and three-operand machines, since the register files need separate multiplexer and demultiplexer anyway.

Comparison of Zero-,One-, Two- and Three-Address CPUs

Let us summarize our findings:

- The stack machine has the longest program and the shortest individual instructions.
- Even a stack machine needs a one-address instruction to access memory.
- The three-address machine has the shortest code but requires the largest number of bits per instruction.
- A register file can reduce the size of the instruction words. Typically in three-address machines two registers and one memory operand are allowed.
- A load/store machine only allows data moves between memory and registers. Any ALU operation is done with the register file.
- Most designs make the assumption that register access is faster than memory access. While this is true in CBIC or FPGAs that use external memory,

Table 9.3. Comparison of different design goal in zero- to three-operand CPUs.

Goal	# of operands			
	0	1	2	3
Ease of assembler	worst	...	...	best
Simple C compiler	best	...	...	worst
# of code words	worst	...	...	best
Instruction length	best	...	...	worst
Range of immediate	worst	...	...	best
Fast operand fetch and decode	best	...	...	worst
Hardware size	best	...	...	worst

inside the FPGA register file access and embedded memory access times are in the same range, providing the option to realize the register file with embedded (three-port) memories.

The above finding seems unsatisfying. There seems no best choice and as a result each style has been used in practice as our coding examples show. The question then is: why hasn't one particular data flow type emerged as optimal? An answer to this question is not trivial since many factors, like ease of programming, size of code, speed of processing, and hardware requirements need to be considered. Let us compare the different designs based on this different design goals. The summary is shown in Table 9.3.

The *ease of assembler* coding is proportional to the complexity of the instruction. A three-address assembler code is much easier to read and code than the many PUSH and POP operations we find in stack machine assembler coding. The design of a *simple C compiler* on the other hand is much simpler for the stack machine since it easily employs the postfix operation that can be much more simply analyzed by a parser. Managing a register file in an efficient way is a very hard task for a compiler. The number of *code words* in arithmetic operation is much shorter for two- and three-address operation, since intermediate results can easily be computed. The *instruction length* is directly proportional to the number of operands. This can be simplified by using registers instead of direct memory access, but the instruction length still is much shorter with less operands. The size of the *immediate* operand that can be stored depends on the instruction length. With shorter instruction words the constants that can be embedded in the instructions are shorter and we may need multiple load or double word length instructions, see Fig. 9.6 (memory addressing). The operand *fetch and decode* is faster if fewer operands are involved. As a stack machine always uses the two top elements of a stack, no long MUX or DEMUX delays from register files occur. The *hardware size* mainly depends on the register file. A three-operand CPU has the highest requirements, a stack machine the smallest; ALU and control unit are similar in size.

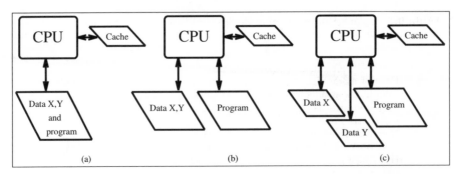

Fig. 9.10. Memory architectures (**a**) Von-Neuman machine (GPP) [15, 16]. (**b**) Harvard architecture with separate program and data bus. (**c**) Super Harvard architecture with two data busses.

In conclusion we can say that each particular architecture has its strengths and weaknesses and must also match the designer tools, skills, design goal in terms of size/speed/power, and development tools like the assembler, instruction set simulator, or C compiler.

9.2.3 Register File and Memory Architecture

In the early days of computers when memory was expensive *von Neuman* suggested a new highly celebrated innovation: to place the data and program in the same memory, see Fig. 9.10a. At that time computer programs were often hardwired in an FSM and only data memory used RAM. Nowadays the image is different: memory is cheap, but access speed is still for a typical RISC machine much slower than for CPU registers. In a three-address machine we therefore need to think about where the three operands should come from. Should all operands be allowed to come from main memory, or only two or one, or should we implement a *load/store architecture* that only allows single transfer between register and memory, but require that all ALU operations are done with CPU registers? The VAX PDP-11 is always quoted as the champion in this regard and allows multiple memory as well as multiple register operations. For an FPGA design we have the additional limitation that the number of instruction words is typically in the kilo range and the von Neuman approach is not a good choice. All the requirements for multiplexing data and program words would waste time and can be avoided if we use separate program and data memory. This is what is called a *Harvard architecture*, see Fig. 9.10b. For PDSPs designs it would be even better (think of an FIR filter application) if we can use three different memory ports: the coefficient and data come from two separate data memory locations x and y, while the accumulated results are held in CPU registers. The third memory is required for the program memory. Since many DSP algorithms are short, some PDSPs like the ADSP try to save the third bus by implementing a small cache. After the first run through the loop the instructions are in the cache and the program memory can be used as a second data memory. This three-bus architecture is shown in Fig. 9.10c and is usually called a *super Harvard architecture*.

A GPP machine like Intel's Pentium or RISC machines usually use a memory hierarchy to provide the CPU with a continuous data stream but also allow one to use cheaper memory for the major data and programs. Such a memory hierarchy starts with very fast CPU registers, followed by level-1, level-2 data and/or program caches to main DRAM memory, and external media like CD-ROMs or tapes. The design of such a memory system is much more sophisticated then what we can design inside our FPGA.

From a hardware implementation standpoint the design of a CPU can be split into three main parts:

- Control path, i.e., a finite state machine,
- ALU
- Register file

of the three items, although not difficult to design, the register file seems to be the block with the highest cost when implemented with LEs. From these high implementation costs it appears that we need to compromise between more registers that make a μP easy to use and the high implementation costs of a larger file, such as 32 registers. The following example shows the coding of a typical RISC register file.

Example 9.3: A RISC Register File

When designing a RISC register file we usually have a larger number of registers to implement. In order to avoid additional instructions for indirect addressing (offset zero) or to clear a register (both operands zero) or register move instructions, usually the first register is set permanently to zero. This may appear to be a large waste for a machine with few registers, but simplifies the assembler coding essential, as the following examples show:

Instruction	Description
ADD r3,r0,r0	; Set register r3 to zero.
ADD r4,r2,r0	; Move register r2 to register r4.
LDBU r5,100(r0)	Compute the sum of 100 and register r0=0 ; and load data from this address into register r5.

Note that the pseudo-instruction above only work under the assumption that the first register r0 is zero.

The following VHDL code[3] shows the generic specification for a 16-register file with 8-bit width.

```
-- Desciption: This is a W x L bit register file.
LIBRARY ieee;
USE ieee.std_logic_1164.ALL;
```

---

[3] The equivalent Verilog code reg_file.v for this example can be found in Appendix A on page 723. Synthesis results are shown in Appendix B on page 731.

```
ENTITY reg_file IS
  GENERIC(W : INTEGER := 7; -- Bit width-1
          N : INTEGER := 15); -- Number of regs-1
  PORT ( clk, reg_ena : IN std_logic;
         data  : IN STD_LOGIC_VECTOR(W DOWNTO 0);
         rd, rs, rt : IN integer RANGE 0 TO 15;
         s, t : OUT STD_LOGIC_VECTOR(W DOWNTO 0));
END;

ARCHITECTURE fpga OF reg_file IS

  SUBTYPE bitw IS STD_LOGIC_VECTOR(W DOWNTO 0);
  TYPE SLV_NxW IS ARRAY (0 TO N) OF bitw;
  SIGNAL r : SLV_NxW;

BEGIN

  MUX: PROCESS    -- Input mux inferring registers
  BEGIN
    WAIT UNTIL clk = '1';
    IF reg_ena = '1' AND rd > 0 THEN
      r(rd) <= data;
    END IF;
  END PROCESS MUX;

  DEMUX: PROCESS (r, rs, rt) --  2 output demux
  BEGIN                      --  without registers
    IF rs > 0 THEN -- First source
      s <= r(rs);
    ELSE
      s <= (OTHERS => '0');
    END IF;
    IF rt > 0 THEN -- Second source
      t <= r(rt);
    ELSE
      t <= (OTHERS => '0');
    END IF;
  END PROCESS DEMUX;

  END fpga;
```

The first process, MUX, is used to store the incoming data in the register file.
Note that the register zero is not overwritten since it should be zero all the
time. The second process, DEMUX, hosts the two decoders to read out the two
operands for the ALU operations. Here again access to register 0 is answered
with a value of zero. The design uses 211 LEs, no embedded multiplier, and
no M4Ks. A Registered Performance can not be measured since there is no
register-to-register path.

We check the register file with the simulation shown in Fig. 9.11. The input
data (address rd) is written continuously as data into the file. The output s
is set via the rs to register 2, while output t is set to register 3 using rt. We
note that the register enable low signal between 500 and 700 ns means that
register 2 is not overwritten with the data value 12. But for register 3 the

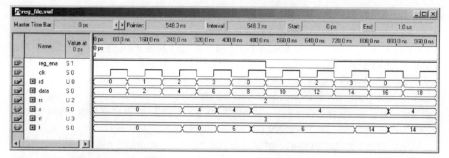

Fig. 9.11. Simulation of the register file.

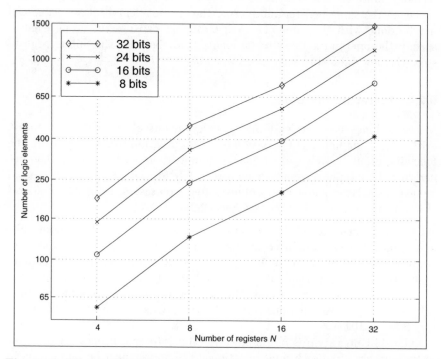

Fig. 9.12. LEs for different register file configurations.

enable is again high at 750 ns and the new value 14 is written into register 3, as can be seen from the t signal.

9.3

Figure 9.12 shows the LEs data for register number in the range 4 to 32 and bitwidth 8, 16, 24, and 32.

With Xilinx FPGAs it would also be possible to use the LEs as 16×1 dual-port memory, see Chap. 1, Table 1.4, p. 11. With Altera FPGAs the only option to save the substantial number of LEs used for the register file would be to use a three-port memory, or two embedded dual-port memory blocks as the register file. We would write the same data in both memories and can read the two sources from the other port of the memory. This principle has been used in the Nios µP and can greatly reduce the LE count. We may then only use the lower 16 or 32 registers or offer (as it is done in the Nios µP) a window of registers. The window can be moved, for instance, in a subroutine call and the basic register do not need to be saved on a stack or in memory as would otherwise be necessary.

From the timing requirement however we now have the problem that BlockRAMs are synchronous memory blocks, and we can not load and store memory addresses and data with the same clock edge from both ports, i.e., replacing the same register value using the current demultiplexer value can not be done with the same clock edge. But we can use the rising edge to specify the operand address to be loaded and then use the falling edge to store the new value and set the write enable.

9.2.4 Operation Support

Most machines have at least one instruction out of the three categories: arithemtic/logic unit (ALU), data move, and program control. Let us in the following briefly review some typical examples from each category. The underlying data type is usually a multiple of bytes, i.e., 8, 16, or 32 bits of integer data type; some more-sophisticated processors use a 32- or 64-bit IEEE floating-point data type, see Sect. 2.2.3, p. 71.

ALU instructions. ALU instruction include arithmetic, logic, and shift operations. Typical supported arithmetic instructions for two operands are addition (ADD), subtraction (SUB), multiply (MUL) or multiply-and-accumulate (MAC). For a single operand, absolute (ABS) and sign inversion (NEG) are part of a minimum set. Division operation is typically done by a series of shift-subtract-compare instructions since an array divider can be quite large, see Fig. 2.28, p. 104.

The shift operation is useful since in b-bit integer arithmetic a bit grow to $2 \times b$ occurs after each multiplications. The shifter may be implicit as in the TMS320 PDSP from TI or provided as separate instructions. Logical and arithmetic (i.e., correct sign extension) as well as rotations are typical supported. In a block floating-point data format exponent detection (i.e., determining the number of sign bits) is also a required operation.

The following listing shows arithmetic and shift operations for different microprocessors.

Instruction	Description	µP
ADD *,8,AR3	The * indicates that the auxiliary memory point ARP points to one of the eight address registers that is used for the memory access. The word from that location is left shifted by eight bits before being added to the accumulator. After the instruction the ARP points to AR3, which is used for the next instruction.	TMS320C50
MACD Coeff,Y	Multiply the coefficient and Y and store the result in the product register P. Then move the register Y by one location.	TMS320C50
NABS r3, r4	Store the negative absolute value of r4 in r3.	PowerPC
DIV r3,r2,r1	This instruction divides the register r2 by r1 and stores the quotient in the register r3.	Nios II
SR=SRr OR ASHIFT 5	Right shift the SR register by five bits and use sign extension.	ADSP

Although logic operations are less often used in basic DSP algorithms like filtering or FFT, some more-complex systems that use cryptography or error correction algorithms need basic logic operations such as AND, OR, and NOT. For error correction EXOR and EQUIV are also useful. If the instruction number is critical we can also use a single NAND or NOR operation and all other Boolean operations can be derived from these universal functions, see Exercise 1.1, p. 42.

Data move instructions. Due to the large address space and performance concerns most machines are closer to the typical RISC load/store architecture than the universal approach of the VAX PDP-11 that allows all operands of an instruction to come from memory. In the load/store philosophy we only allow data move instructions between memory and CPU registers, or different registers – a memory location can *not* be part of an ALU operation. In PDSP designs a slightly different approach is taken. Most data access is done with indirect addressing, since a typical PDSP like the ADSP or TMS320 has separate memory address generation units that allow auto de/increment and modulo addressing, see Fig. 9.7, p. 551. These address computations are done in parallel to the CPU and do not require additional CPU clock cycles.

The following listing shows data move instructions for different microprocessors.

Instruction	Description	μP
st [%fp],%g1	Store register g1 at the memory location specified in the fp register.	Nios
LWZ R5,DMA	Move the 32-bit data from the memory location specified by DMA to register R5	PowerPC
MX0=DM(I2,M1)	Load from data memory into register MX0 the word pointed to by the address register I2 and post-increment I2 by M1.	ADSP
IN STAT, DA3	Read the word from the peripheral on port address 3 and store the data in the new location STAT.	TMS320

Program flow instructions. Under control flow we group instructions that allow us to implement loops, call subroutines, or jump to a specific program location. We may also set the μP to idle, waiting for an interrupt to occur, which indicates new data arrival that need to be processed.

One specific type of hardware support in PDSPs that is worth mentioning is the so-called zero-overhead loops. Usually at the end of a loop the μP decrements the loop counter and checks if the end of the loop is reached. If not, the program flow continues at the begin of the loop. This check would require about four instructions, and with typical PDSP algorithms (e.g., FIR filter) with a loop length of instruction 1, i.e., a single MAC, 80% of the time would be spent on the loop control. The loops in PDSP are in fact so short that the TMS320C10 provides a RPT #imm instruction such that the next instruction is repeated #imm+1 times. Newer PDSPs like the ADSP or TMS320C50 also allow longer loops and nested loops of several levels. In most RISC machine applications the loops are usually not as short as for PDSPs and the loop overhead is not so critical. In addition RISC machines use delay branch slots to avoid NOPs in pipeline machines.

The following listing shows program flow instructions for different microprocessors.

Instruction	Description	µP
CALL FIR	Call the subroutine that starts at the label FIR.	TMS32010
BUN r1	Branch to the location stored in the register r1 if in the previous floating-point operation one or more values were NANs.	PowerPC
RET	The end of a subroutine is reached and the PC is loaded with the value stored in the register ra.	Nios II
RPT #7h	Repeat the next instruction $7 + 1 = 8$ times. This is a one-word instruction due to the small constant value.	TMS320C50
CNTR=10; DO L UNTIL CE;	Repeat the loop from the next instruction on up to the label L until the counter CNTR expires.	ADSP

Additional hardware logic is provided by PDSPs to enable these short loops without requiring additional clock cycles at the end of the loop. The initialization of zero-overhead loops usually consists of specifying the number of loops and the end-of-loop label or the number of instructions in the loop; see the ADSP example above. Concurrently to the operation execution the control unit checks if the next instruction is still in the loop range, otherwise it loads the next instruction into the instruction register and continues with the instruction from the start of the loop. From the overview in Table 9.2, p. 550 it can be concluded that all second-generation PSDPs support this feature.

9.2.5 Next Operation Location

In theory we can simplify the next operation computation by providing a fourth operand that includes the address of the next instruction word. But since almost all instructions are executed one after the other (except for jump-type instructions), this is mainly redundant information and we find no commercial microprocessor today that uses this concept.

Only if we design an ultimate RISC machine (see Exercise 9,12, p. 637) that contains only one instruction we do need to include the next address or (better) the offset compared to the current instruction in the instruction word [306].

9.3 Software Tools

According to Altera's Nios online net seminar [307] one of the main reasons why Altera's Nios development systems have been a huge success[4] is based on the fact that, besides a fully functional microprocessor, also all the necessary software tools including a GCC-based C compiler is generated at the same time when the IP block parametrization takes place. You can find many free μP cores on the web, see for instance:

- http://www.opencores.org/   OPENCORES.ORG
- http://www.free-ip.com/   The free IP project
- http://www.fpgacpu.org/   FPGA CPU

but most of them lack a full set of development tools and are therefore less useful. A set of development tools (best case) should include

- Assembler, linker, and loader/basic terminal program
- Instruction set simulator
- C compiler

Figure 9.13 explains the different levels of abstraction in the development tools. In the following we will briefly describe the main programs used to develop these tools. You may also consider using the language for instruction set architecture (LISA) system originally developed at ISS, RWTH Aachen [308], and now a commercial product of CoWare Inc., which automatically generates an assembler and instruction set simulator, and the C compiler with a few additional specifications in a semiautomatic way. In Sect. 9.5.2 (p. 610) we will review this type of design flow.

Writing a compiler can be a time-consuming project. A good C compiler, for instance, requires up to 50 man-years of work [309, 310, 311].

Nowadays we can benefit from the program developed in the GNU project that provides several useful utilities that speed up compiler development:

- The GNU tool Flex [312] is a scanner or lexical analyzer that recognizes patterns in text, similar to what the UNIX utility grep or the line editor sed can do for single pattern.
- The GNU tool Bison [313] a, YACC-compatible parser generator [314] allows us to describe a grammar in Bakus–Naur form (BNF), which can initiate actions if expressions are found in the text.
- For the GNU C compiler gcc we can take advantage of the tutorial written by R. Stallman [315] to adapt the C compiler to the actual μP we have or plan to build.

All three tools are freely available under the terms of the GNU public licence and for all three tools we have included documentation on the book CD under book3e/uP, which also includes many useful examples.

---

[4] 10,000 systems were sold in the first four years after introduction.

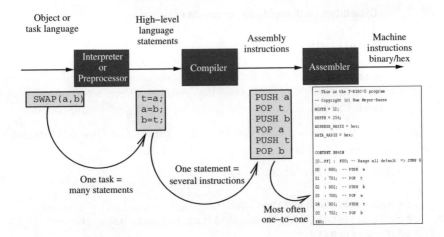

Fig. 9.13. Models and tools in programming.

9.3.1 Lexical Analysis

A program that can recognize lexical patterns in a text is called a *scanner*. Flex, compatible with the original AT&T Lex, is a tool that can generate such a scanner [316]. Flex uses an input file (usual extension *.l) and produces C source code that can then be compiled on the same or a different system. A typical scenario is that you generate the parser under UNIX or Linux with the GNU tools and, since most Altera tools run on a PC, we compile the scanner under MS-DOS so that we can use it together with the Quartus II software. The default UNIX file that is produced by Flex uses a lex.yy.c file name. We can change this by using the option -oNAME.C to generate the output NAME.C instead. Note that there is no space between -o and the new name under UNIX. Assume we have a Flex input file simple.l then we use the two steps:

```
flex -osimple.c simple.l
gcc -o simple.exe simple.c
```

to generate a scanner called simple.exe under UNIX. Even a very short input file produces about 1,500 lines of C-code and has a size of 35 KB. We can already see from these data the great help this utility can be. We can also FTP the C-code simple.c to an MS-DOS PC and then compile it with a C compiler of our choice. The question now is how do we specify the pattern in Flex for our scanner, i.e., how does a typical Flex input file look? Let use first have a look at the formal arrangement in the Flex input file. The file consists of three parts:

```
%{
```

```
 C header and defines come here
%}
definitions ...
%%
rules ...
%%
user C code ...
```

The three sections are separated by two %% symbols. Here is a short example
of an input file.

```
/* A simple flex example */
%{
/* C-header and definitions */
#include <stdlib.h> /* needed for malloc, exit etc **/
#define YY_MAIN 1
%}
%%
.|\n            ECHO; /* Rule section */
%%
/* User code here */
int yywrap(void) { return 1; }
```

The YY_MAIN define is used in case you want **Flex** to provide the main routine
for the scanner. We could also have provided a basic main routine as follows

```
    main() { lex(); }
```

The most important part is the rule section. There we specify the pattern
followed by the actions. The pattern . is any character except new line,
and \n is new line. The bar | stands for the or combination. You can see
that most coding has a heavy C-code flavor. The associated action ECHO will
forward each character to standard output. So our scanner works similarly
to the **more**, **type**, or **cat** utility you may have used before. Note that **Flex**
is column sensitive. Only patterns in the rule section are allowed to start in
the first column; not even a comment is allowed to start here. Between the
pattern and the actions, or between multiple actions combined in parenthesis
{}, a space is needed.

We have already discussed two special symbols used by **Flex**: the dot .
that describes any character and the new line symbol \n. Table 9.4 shows
the most often used symbols. Note that these are the same kinds of symbols
used by the utility **grep** or the line editor **sed** to specify regular expressions.
Here are some examples of how to specify a pattern:

Pattern	Matches
a	the character a.
a{1,3}	One to three a's, i.e., a|aa|aaa.
a|b|c	any single character from a, b, or c.
[a-c]	any single character from a, b, or c, i.e., a|b|c.
ab*	a and zero or more b's, i.e., a|ab|abb|abbb...
ab+	a and one or more b's, i.e., ab|abb|abbb...
a\+b	string a+b .
[\t\n]+	one or more space, tab or new lines.
^L	Begin of line must be an L.
[^a-b]	any character except a, b, or c.

Using these pattern we now build a more-useful scanner that performs a lexical analysis of a VHDL file and reports the types of items he finds. Here is the Flex file vhdlex.l

```
/* Lexical analysis for a toy VHDL-like language */

%{
#include <stdio.h>
#include <stdlib.h>
%}
DIGIT          [0-9]
ID             [a-z][a-z0-9_]*
ASSIGNMENT     [(<=)|(:=)]
GENERIC        [A-Z]
DELIMITER      [;,)(':]
COMMENT        "--"[^\n]*
LABEL          [a-zA-Z][a-zA-Z0-9]*[:]
%%
{DIGIT}+       { printf( "An integer: %s (%d)\n", yytext,
                 atoi( yytext ) ); }
IN|OUT|ENTITY|IS|END|PORT|ARCHITECTURE|OF|WAIT|UNTIL {
                 printf( "A keyword: %s\n", yytext ); }
BEGIN|PROCESS { printf( "A keyword: %s\n", yytext ); }

{ID}           printf( "An identifier: %s\n", yytext );
"<="           printf( "An assignment: %s\n", yytext );
"="            printf( "Equal condition: %s\n", yytext );
{DELIMITER} printf( "A delimiter: %s\n", yytext );
{LABEL}        printf( "A label: %s\n", yytext );

"+"|"-"|"*"|"/"    printf( "An operator: %s\n", yytext );
{COMMENT}          printf( "A comment: %s\n", yytext );
[ \t\n]+           /* eat up whitespace */
```

Table 9.4. Special symbols used by `Flex`.

Symbol	Meaning
.	Any single character except new line
\n	New line
*	Zero or more copies of the preceding expression
+	One or more copies of the preceding expression
?	Zero or one of the preceding expression
^	Begin of line or negated character class
$	End of line symbol
|	Alternate, i.e., or expressions
()	Group of expressions
"+"	Literal use of expression within quotes
[]	Character class
{}	How many times an expression is used
\	Escape sequence to use a special symbol as a character only

```
.          printf( "Unrecognized character: %s\n", yytext );

%%

int yywrap(void) { return 1; }

main( argc, argv )
int argc;
char **argv;
{
  ++argv, --argc; /* skip over program name */
  if ( argc > 0 )
    yyin = fopen( argv[0], "r" );
  else
    yyin = stdin;
  yylex();
}
```

We compile the files with the following step under UNIX:

```
flex -ovhdlex.c vhdlex.l
gcc -o vhdlex.exe vhdlex.c
```

Assume we have the following small VHDL example:

```
ENTITY d_ff IS -- Example flip-flop
PORT(clk, d    :IN bit;
     q              :OUT bit);
END;
```

```
ARCHITECTURE fpga OF d_ff IS
BEGIN
P1: PROCESS (clk)
   BEGIN
      WAIT UNTIL clk='1';  --> gives always FF
         q <= d;
   END PROCESS;
END fpga;
```

then calling our scanner with vhdlex.exe < d_ff.vhd will produce the following output:

```
A keyword: ENTITY
An identifier: d_ff
A keyword: IS
A comment: -- Example flip-flop
A keyword: PORT
A delimiter: (
An identifier: clk
A delimiter: ,
An identifier: d
A delimiter: :
A keyword: IN
An identifier: bit
A delimiter: ;
An identifier: q
A delimiter: :
A keyword: OUT
An identifier: bit
A delimiter: )
A delimiter: ;
A keyword: END
A delimiter: ;
A keyword: ARCHITECTURE
An identifier: fpga
A keyword: OF
An identifier: d_ff
...
```

After the two introductory examples we can now take on a more-challenging task. Let us build an asm2mif converter that reads in assembler code and outputs an MIF file that can be loaded into the block memory as used by the Quartus II software. To keep things simple let us use the assembler code of a stack machine with the following 16 operations (sorted by their operation code):

ADD, NEG, SUB, OPAND, OPOR, INV, MUL, POP,

```
        PUSHI, PUSH, SCAN, PRINT, CNE, CEQ, CJP, JMP
```

Since we should also allow forward referencing labels in the assembler code
we need to have a two-pass analysis. In the first pass we make a list of all
variables and labels and their code lines. We also assign a memory location
to variables when we find one. In the second run we can then translate our
assembler code line-by-line into MIF code. We start with the MIF file header
that includes the data formats and then each line will have a address, an
operation and possibly an operand. We can display the original assembler
code at the end of each line by preceding -- comment symbols as in VHDL.
Here is the `Flex` input file for our two-pass scanner:

```
/* Scanner for assembler to MIF file converter */
%{
#include <stdio.h>
#include <string.h>
#include <math.h>
#include <errno.h>
#include <stdlib.h>
#include <time.h>
#include <ctype.h>
#define DEBUG 0
int state =0; /* end of line prints out IW */
int    icount =0; /* number of instructions */
int    vcount =0; /* number of variables */
int    pp =1; /** preprocessor flag **/
char  opis[6], lblis[4], immis[4];
struct inst {int adr; char opc; int imm; char *txt;} iw;
struct init  {char *name; char code;} op_table[20] = {
   "ADD"    , '0', "NEG"    , '1', "SUB"    , '2',
   "OPAND"  , '3', "OPOR"   , '4', "INV"    , '5',
   "MUL"    , '6', "POP"    , '7', "PUSHI"  , '8',
   "PUSH"   , '9', "SCAN"   , 'a', "PRINT"  , 'b',
   "CNE"    , 'c', "CEQ"    , 'd', "CJP"    , 'e',
   "JMP"    , 'f', 0,0 };
FILE      *fid;
int add_symbol(int value, char *symbol);
int lookup_symbol(char *symbol);
void list_symbols();
void conv2hex(int value, int Width);
char lookup_opc(char *opc);
%}
DIGIT      [0-9]
VAR        [a-z][a-z0-9_]*
COMMENT    "--"[^\n]*
LABEL      L[0-9]+[:]
```

```
GOTO              L[0-9]+
%%
\n                    {if (pp) printf( "-- end of line \n");
                      else { if ((state==2) && (pp==0))
                    /* print out an instruction at end of line */
                    {conv2hex(iw.adr,8); printf("  : %c",iw.opc);
                    conv2hex(iw.imm,8);
                    printf(";  -- %s  %s\n",opis,immis); }
                    state=0;iw.imm=0;
                    }}
{DIGIT}+          { if (pp) printf( "-- An integer: %s (%d)\n",
                                    yytext, atoi( yytext ) );
                        else   {iw.imm=atoi( yytext ); state=2;
                                    strcpy(immis,yytext);}}
POP|PUSH|PUSHI|CJP|JMP {
                    if (pp)
              printf( "-- %d) Instruction with operand: %s\n",
                                      icount++, yytext);
                    else {state=1; iw.adr=icount++;
                              iw.opc=lookup_opc(yytext);}}
CNE|CEQ|SCAN|PRINT|ADD|NEG|SUB|OPAND|OPOR|INV|MUL {
                    if (pp)  printf( "-- %d) ALU Instruction: %s\n",
                                      icount++, yytext );
                      else  {  state=2; iw.opc=lookup_opc(yytext);
                          iw.adr=icount++; strcpy(immis,"   ");}}
{VAR}             { if (pp)  {printf( "-- An identifier: %s\n",
                          yytext ); add_symbol(vcount, yytext);}
                      else {state=2;iw.imm=lookup_symbol(yytext);
                                    strcpy(immis,yytext);}}
{LABEL}     { if (pp) {printf( "-- A label: %s lenth=%d
                        Icount=%d\n", yytext , yyleng, icount);
                                  add_symbol(icount, yytext);}}
{GOTO}   {if (pp) printf( "-- A goto label: %s\n", yytext );
              else {state=2;sprintf(lblis,"%s:",yytext);
            iw.imm=lookup_symbol(lblis);strcpy(immis,yytext);}}
{COMMENT} {if (pp) printf( "-- A comment: %s\n", yytext );}
[ \t]+      /* eat up whitespace */
.           printf( "Unrecognized character: %s\n", yytext );
%%

int yywrap(void) { return 1; }

int main( argc, argv )
int argc;
```

```
char **argv;
{
  ++argv, --argc; /* skip over program name */
  if ( argc > 0 )
    yyin = fopen( argv[0], "r" );
  else
{ printf("No input file -> EXIT\n"); exit(1);}
  printf("--- First path though file ---\n");
  yylex();
  if (yyin != NULL) fclose(yyin);
  pp=0;
  printf("\n-- This is the T-RISC program with ");
  printf("%d lines and %d variables\n",icount,vcount);
  icount=0;
  printf("-- for the book DSP with FPGAs 3/e\n");
  printf("-- Copyright (c) Uwe Meyer-Baese\n");
  printf("-- WIDTH = 12; DEPTH = 256;\n");
  if (DEBUG) list_symbols();
  printf("ADDRESS_RADIX = hex; DATA_RADIX = hex;\n\n");
  printf("CONTENT BEGIN\n");
  printf("[0..FF] :  F00; -- ");
  printf("Set all address from 0 to 255 => JUMP 0\n");
  if (DEBUG) printf("--- Second path through file ---\n");
  yyin = fopen( argv[0], "r" );
  yylex();
  printf("END;\n");
}


/* define a linked list of symbols */
struct symbol { char *symbol_name; int symbol_value;
                struct symbol *next; };

struct symbol *symbol_list;/*first element in symbol list*/

extern void *malloc();

int add_symbol(int value, char *symbol)
{
    struct symbol *wp;

    if(lookup_symbol(symbol) >= 0 ) {
printf("-- Warning: symbol %s already defined \n", symbol);
        return 0;
    }
```

```
    wp = (struct symbol *) malloc(sizeof(struct symbol));
    wp->next = symbol_list;
    wp->symbol_name = (char *) malloc(strlen(symbol)+1);
    strcpy(wp->symbol_name, symbol);
        if (symbol[0]!='L') vcount++;
    wp->symbol_value = value;
    symbol_list = wp;
    return 1;    /* it worked */
}

int lookup_symbol(char *symbol)
{
    struct symbol *wp = symbol_list;
    for(; wp; wp = wp->next) {
        if(strcmp(wp->symbol_name, symbol) == 0)
        {if (DEBUG)
                printf("-- Found symbol %s value is: %d\n",
                                  symbol, wp->symbol_value);
            return wp->symbol_value;}
    }
    if (DEBUG) printf("-- Symbol %s not found!!\n",symbol);
    return -1;   /* not found */
}

char lookup_opc(char *opc)
{ int k;
  strcpy(opis,opc);
  for (k=0; op_table[k].name != 0; k++)
    if (strcmp(opc,op_table[k].name) == 0)
                                return (op_table[k].code);
  printf("******* Ups, no opcode for: %s --> exit \n",opc);
  exit(1);
}

void list_symbols()
{
    struct symbol *wp = symbol_list;
      printf("--- Print the Symbol list: ---\n");
    for(; wp; wp = wp->next)
      if (wp->symbol_name[0]=='L') {
    printf("-- Label     : %s  line = %d\n",
                        wp->symbol_name, wp->symbol_value);
        } else {
    printf("-- Variable : %s  memory @ %d\n",
```

```
                              wp->symbol_name, wp->symbol_value);
        }

}

/*********** CONV_STD_LOGIC_VECTOR(value, bits) *********/
void
conv2hex(int value, int Width)
{
    int            W, k, t;
    extern FILE    *fid;
    t = value;
    for (k = Width - 4; k >= 0; k-=4) {
        W = (t >> k) % 16; printf( "%1x", W);
    }
}
```

The variable **pp** is used to decide if the preprocessing phase or the second phase is running. Labels and variables are stored in a symbol table using the functions `add_symbol` and `lookup_symbol`. For labels we store the instruction line the label occurs, while for variables we assign a running number as we go through the code. An output of the symbol table for two labels and two variables will look as follows:

```
...
--- Print the Symbol list: ---
-- Label     : L01:  line = 17
-- Label     : L00:  line = 4
-- Variable : k   memory @ 1
-- Variable : x   memory @ 0
...
```

This will be displayed in the debug mode (set `#define DEBUG 1`) of the scanner, to display the symbol list. Here are the UNIX instructions to compile and run the code:

```
flex -oasm2mif.c asm2mif.l
gcc -o asm2mif.exe asm2mif.c
asm2mif.exe factorial.asm
```

The factorial program `factorial.asm` for the stack machine looks as follows.

```
            PUSHI   1
            POP     x
            SCAN
            POP     k
LOO:        PUSH    k
```

```
            PUSHI    1
            CNE
            CJP      L01
            PUSH     x
            PUSH     k
            MUL
            POP      x
            PUSH     k
            PUSHI    1
            SUB
            POP      k
            JMP      L00
L01:        PUSH     x
            PRINT
```

The output generated by asm2mif is shown next.

```
-- This is the T-RISC program with 19 lines and 2 variables
-- for the book DSP with FPGAs 3/e
-- Copyright (c) Uwe Meyer-Baese
WIDTH = 12;
DEPTH = 256;
ADDRESS_RADIX = hex;
DATA_RADIX = hex;

CONTENT BEGIN
[0..FF] :  F00; -- Set address from 0 to 255 => JUMP 0
00  : 801;  -- PUSHI  1
01  : 700;  -- POP  x
02  : a00;  -- SCAN
03  : 701;  -- POP  k
04  : 901;  -- PUSH  k
05  : 801;  -- PUSHI  1
06  : c00;  -- CNE
07  : e11;  -- CJP  L01
08  : 900;  -- PUSH  x
09  : 901;  -- PUSH  k
0a  : 600;  -- MUL
0b  : 700;  -- POP  x
0c  : 901;  -- PUSH  k
0d  : 801;  -- PUSHI  1
0e  : 200;  -- SUB
0f  : 701;  -- POP  k
10  : f04;  -- JMP  L00
11  : 900;  -- PUSH  x
12  : b00;  -- PRINT
```

END;

9.3.2 Parser Development

From the program name YACC, i.e., yet another compiler-compiler [314], we see that at the time YACC was developed it was an often performed task to write a parser for each new μP. With the popular GNU UNIX equivalent Bison, we have a tool that allows us to define a grammar. Why not use Flex to do the job, you may ask? In a grammar we allow recursive expressions like $a + b, a + b + c, a + b + c + d$, and if we use Flex then for each algebraic expression it would be necessary to define the patterns and actions, which would be a large number even for a small number of operations and operands.

YACC or Bison both use the Bakus–Naur form or BNF that was developed to specify the language Algol 60. The grammar rules in Bison use terminals and nonterminals. Terminals are specified with the keyword %token, while nonterminals are declared through their definition. YACC assigns a numerical code to each token and it expects these codes to be supplied by a lexical analyzer such as Flex. The grammar rule use a look-ahead left recursive parsing (LALR) technique. A typical rule is written like

```
expression  :  NUMBER '+' NUMBER    { $$ = $1 + $3; }
```

We see that an expression consisting of a number followed by the add symbol and a second number can be reduced to a single expression. The associated action is written in {} parenthesis. Say in this case that we add element 1 and 3 from the operand stack (element 2 is the add sign) and push back the result on the value stack. Internally the parser uses an FSM to analyze the code. As the parser reads tokens, each time it reads a token it recognizes, it pushes the token onto an internal stack and switches to the next state. This is called a *shift*. When it has found all symbols of a rule it can *reduce* the stack by applying the action to the value stack and the reduction to the parse stack. This is the reason why this type of parser is sometimes called a shift-reduce parser.

Let us now build a complete Bison specification around this simple add rule. To do so we first need the formal structure of the Bison input file, which typically has the extension *.y. The Bison file has three major parts:

```
%{
 C header and declarations come here
%}
Bison definitions ...
%%
Grammar rules ...
%%
User C code ...
```

It is not an accident that this looks very similar to the Flex format. Both original programs Lex and YACC were developed by colleagues at AT&T [314, 316] and the two programs work nicely together as we will see later. Now we are ready to specify our first Bison example add2.y

```
/* Infix notation add two calculator */
%{
#define YYSTYPE double
#include <math.h>
void yyerror(char *);
%}

/* BISON declarations */
%token NUMBER
%left '+'

%% /* Grammar rules and actions follows */
program :     /* empty */
          | program exp '\n'      { printf("    %lf\n",$2); }
          ;

exp       : NUMBER                { $$ = $1;}
          | NUMBER '+' NUMBER     { $$ = $1 + $3; }
          ;

%% /* Additional C-code goes here */

#include <ctype.h>
int yylex(void)
{ int c;
  /* skip white space and tabs */
  while ((c = getchar()) == ' '|| c == '\t');
  /* process numbers */
  if (c == '.' || isdigit(c)) {
    ungetc(c,stdin);
    scanf("%lf", &yylval);
    return NUMBER;
  }
  /* Return end-of-file */
  if (c==EOF) return(0);
  /* Return single chars */
  return(c);
}
```

```
/* Called by yyparse on error */
void yyerror(char *s) { printf("%s\n", s); }

int main(void)  { return yyparse(); }
```

We have added the token NUMBER to our rule to allow us to use a single number as a valid expression. The other addition is the program rule so that the parser can accept a list of statements, rather than just one statement. In the C-code section we have added a little lexical analysis that reads in operands and the operation and skips over whitespace. Bison calls the routine yylex every time it needs a token. Bison also requires an error routine yyerror that is called in case there is a parse error. The main routine for Bison can be short, a return yyparse() is all that is needed. Let use now compile and run our first Bison example.

```
bison -o -v add2.c add2.y
gcc -o add2.exe add2.c -lm
```

If we now start the program, we can add two floating-point numbers at a time and our program will return the sum, e.g.,

```
user: add2.exe
user: 2+3
add2:    5.000000
user: 3.4+5.7
add2:    9.100000
```

Let us now have a closer look at how Bison performs the parsing. Since we have turned on the -v option we also get an output file that has the listing of all rules, the FSM machine information, and any shift-reduce problems or ambiguities. Here is the output file add2.output

```
Grammar
rule 1    program ->/* empty */
rule 2    program -> program exp '\n'
rule 3    exp -> NUMBER
rule 4    exp -> NUMBER '+' NUMBER

Terminals, with rules where they appear

$ (-1)
'\n' (10) 2
'+' (43) 4
error (256)
NUMBER (257) 3 4

Nonterminals, with rules where they appear
```

```
program (6)
    on left: 1 2, on right: 2
exp (7)
    on left: 3 4, on right: 2


state 0
    $default reduce using rule 1 (program)
    program go to state 1

state 1
    program  ->  program . exp '\n'   (rule 2)
    $     go to state 7
    NUMBER shift, and go to state 2
    exp  go to state 3

state 2
    exp  ->  NUMBER .   (rule 3)
    exp  ->  NUMBER . '+' NUMBER   (rule 4)
    '+'  shift, and go to state 4
    $default reduce using rule 3 (exp)

state 3
    program  ->  program exp . '\n'   (rule 2)
    '\n'shift, and go to state 5

state 4
    exp  ->  NUMBER '+' . NUMBER   (rule 4)
    NUMBER shift, and go to state 6

state 5
    program  ->  program exp '\n' .   (rule 2)
    $default reduce using rule 2 (program)

state 6
    exp  ->  NUMBER '+' NUMBER .   (rule 4)
    $default reduce using rule 4 (exp)

state 7
    $     go to state 8

state 8
    $default accept
```

At the start of the output file we see our rules are listed with separate rule values. Then the list of terminals follow. The terminal NUMBER, for instance, was assigned the token value 257. These first lines are very useful if you want to debug the input file, for instance, if you have ambiguities in your grammar rules this would be the place to check what went wrong. More about ambiguities a little later. We can see that in normal operation of the FSM the shifts are done in states 1, 2, and 4, for the first number, the add operation, and the second number, respectively. The reduction is done in state 6, and the FSM has a total of eight states.

Our little calculator has many limitations, it can, for instance, not do any subtraction. If we try to subtract we get the following message

```
user: add2.exe
add2: 7-2
parse error
```

Not only is our repertoire limited to adds, but also the number of operands is limited to two. If we try to add three operands our grammar does not yet allow it, e.g.,

```
user: 2+3+4
add2: parse error
```

As we see the basic calculator can only add two numbers not three. To have a more-useful version we add a recursive grammar rule, the operations $*, /, -, \hat{}$, and a symbol table that allows us to specify variables. The C-code for the symbol table can be found in examples in the literature, see, for instance, [313, p. 23], [317, p. 15], or [318, p. 65]. The lexical analysis for Flex is shown next.

```
/* Inifix calculator with symbol table, error recovery and
                                             power-of */
%{
    #include "ytab.h"
    #include <stdlib.h>
    void yyerror(char *);
%}

%%

[a-z]        { yylval = *yytext - 'a';
               return VARIABLE; }

[0-9]+       { yylval = atoi(yytext);
               return INTEGER; }

[-+()^=/*\n]    { return *yytext; }
```

```
[ \t]    ;        /* skip whitespace */

.                 yyerror("Unknown character");

%%

int yywrap(void) { return 1; }
```

We see that we now also have VARIABLEs using the small single characters
a to z besides the integer NUMBER tokens. yytext and yylval are the text
and value associated with each token. Table 9.5 shows the variables used
in the Flex ↔Bison communication. The grammar for our more-advanced
calculator calc.y now looks as follows.

```
%{
        #include <stdio.h>
        #include <math.h>
        #define YYSTYPE int
        void yyerror(char *);
        int yylex(void);
        int symtable[26];
%}

%token INTEGER VARIABLE
%left '+' '-'
%left '*' '/'
%left NEG      /* Negation, i.e. unary minus */
%right '^'    /* exponentiation             */

%%

program:
        program statement '\n'
        | /* NULL */
        ;

statement:
        expression                  { printf("%d\n", $1);}
        | VARIABLE '=' expression   { symtable[$1] = $3; }
        ;

expression:
        INTEGER
        | VARIABLE                  { $$ = symtable[$1];}
        | expression '+' expression   { $$ = $1 + $3; }
```

Table 9.5. Special functions and variables used in the `Flex`↔`Bison` communication, see Appendix A [313] for a full list.

Item	Meaning
`char *yytext`	Token text
`file *yyin`	Flex input file
`file *yyout`	Flex file destination for ECHO
`int yylength`	Token length
`int yylex(void)`	Routine called by parser to request tokens
`int yylval`	Token value
`int yywrap(void)`	Routine called by the `Flex` when end of file is reached
`void yyparse();`	The main parser routine
`void yyerror(char *s)`	Called by `yyparse` on error

```
| expression '-' expression      { $$ = $1 - $3; }
| expression '*' expression      { $$ = $1 * $3; }
| expression '/' expression      {
   if ($3) $$ = $1 / $3;
   else { $$=1; yyerror("Division by zero !\n");}}
/* Exponentiation */
| expression '^' expression   { $$ = pow($1, $3); }
/* Unary minus */
| '-' expression %prec NEG       { $$ = -$2; }
| '(' expression ')'             { $$ = $2; }
;


%%

void yyerror(char *s) { fprintf(stderr, "%s\n", s); }

int main(void) { yyparse(); }
```

There are several new things in this grammar specification we need to discuss. The specification of the terminals using `%left` and `%right` ensures the right associativity. We prefer that $2 - 3 - 5$ is computed as $(2 - 3) - 5 = -6$ (i.e., left associative) and not as $2 - (3 - 5) = 4$, i.e., right associativity. For the exponentiation ^ we use right associativity, since 2^2^2 should be grouped as $2^{(2^2)}$. The operands listed later in the token list have a higher precedence. Since * is listed after + we assign a higher precedence to multiply compared with add, e.g., $2 + 3 * 5$ is computed as $2 + (3 * 5)$ rather than $(2 + 3) * 5$. If we do not specify this precedence the grammar will report many reduce-shift conflicts, since it does not know if it should reduce or shift if an item like $2 + 3 * 5$ is found.

For the divide grammar rule we have introduced error handling for divide by zero. If we had not used this kind of error handling a divide by zero would terminate the calculator, like

```
user: 10/0
calc: Floating exception (core dumped)
```

producing a large core dump file. With the error handling we get much smoother behavior, e.g.,

```
user: 30/3
calc: 10
user: 10/0
calc: Division by zero !
```

but calc would allow us to continue operation.

The grammar rules are now written in a recursive fashion, i.e., an expression can consist of expression operation expression terms. Here are a few more examples that also show the use of left and right associativity.

```
user: 2+3*5
calc: 17
user: 1-2-5
calc: -6
user: x=3*10
user: y=2*5-9
user: x+y
calc: 31
user: #
calc: Unknown character
calc: parse error
```

Any special unknown character will stop the calculator. The book CD contains the C source code as well as an executable you can play with.

Let us now briefly look at the compilation steps and the Flex↔Bison communication. Since we first need to know from Bison which kind of token it expects from Flex we run this program first, i.e.,

```
bison -y -d -o ytab.c calc.y
```

This will generate the files ytab.c, ytab.output, and ytab.h. The header file ytab.h contains the token values:

```
#ifndef YYSTYPE
#define YYSTYPE int
#endif
#define INTEGER 257
#define VARIABLE      258
#define NEG     259
extern YYSTYPE yylval;
```

Now we can run `Flex` to generate the lexical analyzer. With

```
flex -olexyy.c calc.l
```

we will get the file `lexyy.c`. Finally we compile both `C` source files and link them together in `calc.exe`

```
gcc -c ytab.c lexyy.c
gcc ytab.o lexyy.o -o calc.exe -lm
```

The `-lm` option for the math library was used since we have some more-advanced math function like `pow` in our calculator.

We should now have the knowledge to write a more-challenging task. One would be a program c2asm that generates assembler code from a simple C-like language. In [319] we find such code for a three-address machine. In [317] all the steps to produce assembler code for a C-like language for a stack machine are given. [320, 315, 309, 321] describe more-advanced C compiler designs. For a stack-like machine we would allow as input a file for a factorial code as

```
x=1;
scan k;
while (k != 1) {
    x = x * k;
    k = k - 1;
}
print x;
```

The program reads data from the inport (e.g., 8-pin DIP switch) calculates the factorial and outputs the data to a two-digit seven-segment display, as we have on the UP2 or Cyclone II boards. Using the `c2asm.exe` program from the book CD, see `book3e/uP`, we would get

```
        PUSHI   1
        POP     x
        SCAN
        POP     k
L00:    PUSH    k
        PUSHI   1
        CNE
        CJP     L01
        PUSH    x
        PUSH    k
        MUL
        POP     x
        PUSH    k
        PUSHI   1
        SUB
        POP     k
```

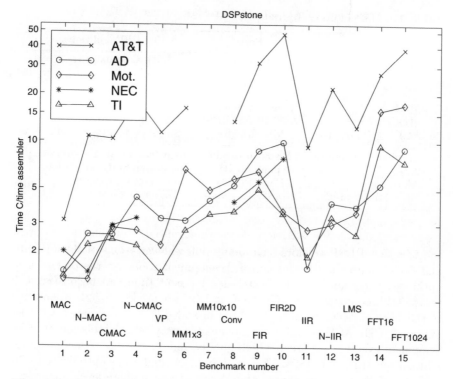

Fig. 9.14. The 15 benchmarks from the DSPstone project.

```
           JMP     LOO
L01:       PUSH    x
           PRINT
```

We can then use our `asm2mif` utility (see p. 578) to produce a program that can be used to generate an MIF file that can be used in a stack machine. All that is needed is to design a stack machine, which will be discussed in the next section.

A challenge that remains is the design of a good C compiler for a PDSP due to the dedicate registers and computational units like the address generators in PDSPs. The DSPstone benchmark developed at ISS, RWTH Aachen [322, 323] uses 15 typical PDSP routines ranging from simple MACs to complex FFT to evaluate the code produced by C compilers in comparison with hand-written assembler code. As can be seen from the DSPstone benchmark shown in Fig. 9.14, the GCC-based compiler for AT&T, Motorola, and Analog PDSPs produce on average 9.58 times less-efficient code than optimized assembler code. The reason the GCC retargetable compiler was used is that the development of a high-performance compiler usually requires an effort of 20 to 50 man-years, which is prohibitively large for most projects, given

Table 9.6. DSP additions to traditional μP [324].

Company	Part	Key DSP additions
ARM	ARM9E	Single-cycle MAC
Fujitsu	SPARClite family	Integer MAC and multimedia assistance
IBM	PowerPC family	Integer MAC
IDT	79RC4650 (MIPS)	Integer MAC
MIPS Technologies	MIPS64 5Kc	Single-cycle integer MAC
Hewlett-Packard	PA_8000 family	Registers for MPEG decode
Intel	Pentium III	Streaming SIMD extensions
Motorola	PowerPC G4	Vector processor
SUN Microsystems	UltraSPARC family	VIZ imaging instruction set

the fact that PDSP vendors have many different variations of their PDSPs and each requires a different set of development tools. At the cost of less-optimized C compiler much shorter development time can be observed, if retargetable compilers like GNU GCC or LCC are used [315, 321]. A solution to this critical design problem has been provided by the associated compiler experts (ACE). ACE provides a highly flexible, easy-retargetable compiler development system that creates high-quality, high-performance compilers for a broad spectrum of PDSPs. They have implemented many optimizations for the intermediate code representation derived from the high-level language.

9.4 FPGA Microprocessor Cores

In recent years we have seen that traditional μPs have been enhanced to enable DSP operations more efficiently. In addition for some processor like Intel Pentiums or SUN SPARC we have seen the addition of MMX multimedia instruction extensions and VIZ instruction set extensions, that allows graphics and matrix multiplication more efficiently. Table 9.6 gives an overview of enhancements of traditional μPs.

The traditional RISC μP has been the basis for FPGA designs for hardcore as well as softcore processors. CISC processors are usually not used in embedded FPGA applications. Xilinx used the PowerPC as a basis for their very successful Virtex II PRO FPGA family that includes 1-4 PowerPC RISC processors. Altera decided to use the ARM processor family in their Excalibur FPGA series, which is one of the most popular RISC processors in embedded applications like mobile phones. Although these Altera FPGAs are still available they are no longer recommended for new designs. Since the performance of the processor also depends on the process technology used and that ARM-based FPGAs are not produced in the new technology, we have witnessed that Altera's softcore Nios II processor achieves about the same performance

Table 9.7. Dhrystone µP performance.

µP name	Device used	Speed (MHz)	D-MIPS measured
ARM922T	Excalibur	200	210
MicroBlaze	Virtex-II PRO-7	150	125
MicroBlaze	Spartan-3(-4)	85	68
Nios II	Cyclone 2	90	105
PPC405	Virtex-4 FX	450	700

as ARM-based hardcore µPs. Xilinx also offer a 32-bit RISC softcore, called MicroBlaze and an 8-bit PicoBlaze processor. No Xilinx 16-bit softcore processor is available and since 16 bit is a good bitwidth for DSP algorithms we design such a machine in Sect. 9.5.2, p. 610.

Table 9.7 gives an overview of the performance of FPGA hard- and softcore µPs measured in Dhrystone MIPS (D-MIPS). D-MIPS is a collection of rather short benchmark compared with the SPEC benchmark most often used in computer architecture literature. Some of the SPEC benchmarks would probably not run, especially on softcore processors.

9.4.1 Hardcore Microprocessors

Hardcore microprocessor, although not as flexible as softcore processors, are still attractive due to their relative small die sizes and the higher possible clock rate and Dhrystone MIPS rates. Xilinx favor the PowerPC series from IBM and Motorola, while Altera has used the ARM922T core, a standard core used in many embedded applications like mobile phones. Let us have in the following a brief look at both architectures.

Xilinx PowerPC. The Xilinx hardcore processor used in Virtex II PRO devices is a fully featured PowerPC 405 core. The RISC µP has a 32-bit Harvard architecture and consists of the following functional unit, shown in Fig. 9.15:

- Instruction and data cache, 16 KB each
- Memory management unit with 64-entry translation lookaside buffer (TLB)
- Fetch & decode unit
- Execution unit with thirty two 32-bit general-purpose registers, ALU, and MAC
- Timers
- Debug logic

The PPC405 *instruction and data caches* are both 16 KB in size while they are organized in 256 lines each with 32 bytes, i.e., $2^8 \times 32 \times 8 = 2^{16}$.

Since data and program are separated this should give a similar performance as a four-way set associate cache in a von Neuman machine. The cache organization along with higher speed are the major reasons that the PPC405 outperforms a softcore, although here the cache size can be adjusted to the specific application, but are usually organized as direct mapped caches, see Exercise 9.29, p. 640.

The key feature of the 405 core can be summarized as follows:

- Embedded 450+ MHz Harvard architecture core
- Five-stage data path pipeline
- 16 KB two-way set associative instruction cache
- 16 KB two-way set associative data cache
- Hardware multiply/divide unit

The instruction cache unit (ICU) delivers one or two instructions per clock cycle over a 64-bit bus. The data cache unit (DCU) transfers 1,2,3,4, or 8 bytes per clock cycle. Another difference to most softcores is the branch prediction, which usually assumes that branches with negative displacements are taken. The execution unit (EXU) of the PPC405 is a single issue unit that contains a register file with 32 32-bit general-purpose registers (GPRs), an ALU, and a MAC unit that performs all integer instructions in hardware. As with typical RISC processors a load/store method is used, i.e., reading and writing of ALU or MAC operation is done with GPRs only. Another features usually not found in softcores is the MMU that allows the PPC405 to address a 4 GB address space. To avoid access to the page table a cache that keeps track of recently used address mappings, called the translation look-aside buffer (TLB) with 64 entries, is used. The PPC also contains three 64-bit timers: the programmable interval timer (PIT), the fixed interval timer (FIT), and a watchdog timer (WDT). All resources of the PPC405 core can be accessed through the debug logic. The ROM monitor, JTAG debugger, and instruction trace tool are supported. For more details on the PPC405 core, see [325].

The auxiliary processor unit (APU) is a key embedded processing feature that has been added to the PowerPC for Virtex-4 FX devices,[5] see Table 1.4, p. 11. Here is a summary of the most interesting APU features [326]:

- Supports user defined instructions
- Supports up to four 32-bit word data transfers in a single instruction
- Allows to build a floating-point or general-purpose coprocessor
- Supports autonomous instructions, i.e., no pipeline stalls
- 32-bit instruction width and 64-bit data
- four cycle cache line transfer

The APU is hooked up directly to the PowerPC pipeline and assembler or C-code instructions allow the access of this unit. For floating-point FIR filters

---

[5] This section was suggested by A. Vera from UNM.

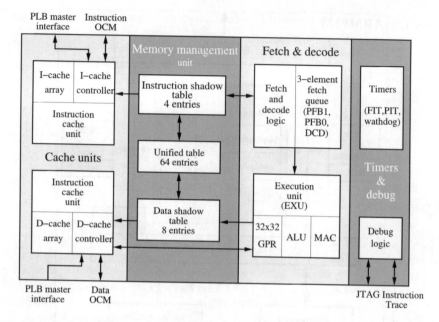

Fig. 9.15. The PPC405 core used by Xilinx Virtex II PRO devices.

improvements by a factor of 20 over software emulations have been reported
[326]. A 16-bit integer 8×8 pixel 2D-IDCT block interfaced via the APU
results also in a speed-up factor of about 20. By comparison, if the IDCT
hardware is connected via the PowerPC local bus and not through the APU
the system performance will be reduced. This is caused by the PowerPC
local bus arbitration overhead and the large number of 32-bit load/store
instructions required in the 8×8 pixel 2D-IDCT block [326].

Altera's ARM. Altera has included in the Excalibur FPGA family the
ARM922T hardcore processor, which includes the ARM9TDMI core, instruc-
tion and data caches, a memory management unit (MMU), debug logic, an
AMBA bus interface, and a coprocessor interface. The key features of the
ARM922T core can be summarized as follows:

- Embedded 200 MHz (210 Dhrystone MIPS) Harvard architecture core
- Five-stage data path pipeline
- 8 KB 64-way set associative instruction cache
- 8 KB 64-way set associative data cache
- Hardware multiply unit
- Low-power 0.8 mW/MHz; small size 6.55 mm^2
- Three-operand 32-bit instructions or
- Two-operand 16-bit thumb instructions

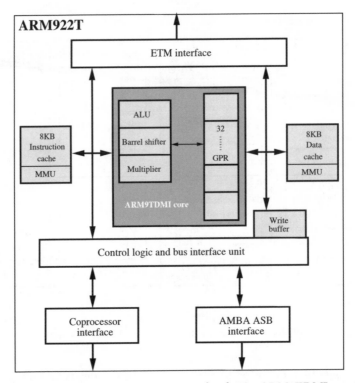

Fig. 9.16. The ARM922T overall architecture [327]. The ARM9TDMI core internal architecture is shown in dark gray [328].

The MMU and cache architecture used in these embedded processor are usually more sophisticated and complex and are the reason the Dhrystone MIPS rate is higher than for a softcore running with the same clock rate. The ARM922T uses an 8-word-per-line architecture. Both data and instruction cache a 64-way set-associate. In addition the ARM922T included a write buffer with 16 data words and four addresses to avoid stalling in case of cache miss. The MMU can map memory of sizes as small as 1 KB to 1 MB pages. The MMU uses two separate 64-entry translation lookaside buffers (TLBs) for the data and instructions.

The advanced microprocessor bus architecture (AMBA) is an often used bus architecture in embedded systems and is supported by the ARM922T.

The ARM9TDMI core is shown in Fig. 9.16 as the grey area. The core can operate on standard 32-bit instructions as well on the shorter thumb set that uses 16 bits only and allows one to pack two instructions in one 32-bit memory word. The core uses a five-stage pipeline that has the following sequence: (1) fetch, (2) decode and register read, (3) execute, (4) memory access, and multiply completion, (5) write register. The CPU contains 31 general-purpose registers, with 16 registers visible at a time, while the others are reserved for context switching. Register 15 is used as the PC, register 14 holds return address for subroutine calls, and register 13 is usually used as the stack pointer. Besides the registers the core includes an ALU, a barrel shifter, and a hardware multiplier. In the following we will briefly study the thumb instruction coding.

Example 9.4: Thumb Instruction Coding

The instructions show first the regular 32-bit coding and then the thumb coding of the same instruction. The first instruction ADDS an 8-bit immediate value to a register and stores the result in the same register, i.e., Rd = Rd + immed$_8$

1110	00101001	Rd$_4$	Rd$_4$	0000	immed$_8$

The thumb instruction keeps the immediate length at 8 bits, but the two register must be the same, and the register selection is reduced from 16 to 8 registers to fit in the 16-bit instruction:

00110	Rd$_3$	immed$_8$

The arithmetic shift right (ASR) instruction allows one to divide a signed number by a power-of-2 value. The shift amount is specified as 5-bit immediate. Rd is the destination register and Rm the source, i.e., Rd = Rm >>> immed$_5$. For the standard 32-bit instruction all 16 register can be used, i.e.,

1110	00011011	SBZ$_4$	Rd$_4$	immed$_5$	100	Rm$_4$

while for the thumb encoding only the first eight can be used as the source and destination to meet the 16-bit instruction length, i.e.,

00010	immed$_5$	Rm$_3$	Rd$_3$

The multiply operation produces a 32-bit result only, and our source and destination must be the same in the thumb instruction, i.e., Rd = (Rm * Rd)$_{32}$. The thumb encoding has the following format:

010000	1101	Rm$_3$	Rd$_3$

while the equivalent 32-bit instruction has the following format:

1110	00000001	Rd$_4$	SBZ$_4$	Rd$_4$	1001	Rd$_4$

9.4

As we can see from the previous example the 16-bit thumb instruction set preserves most of the feature of the 32-bit ISA, however most operations are now of the two-operand form, i.e., have to share one operand, and the number of register is reduced from 16 to 8.

Some of the more-complex instructions like the multiply-accumulate instruction MLA, which is actually a four-operand operation in the ARM922T, has no equivalent in the thumb instruction set, as the following example shows.

Example 9.5:
The 32-bit instruction MLA to compute Rd = (Rm * Rs) + Rn$_{32}$ is coded as follows

cond$_4$	0000001	S	Rd$_4$	Rn$_4$	Rs$_4$	1001	Rm$_4$

It is interesting to notice that it is a four-operand operation and therefore will not fit in the thumb ISA.

9.5

The fact the the MLA (a.k.a. MAC) is not included in the 16-bit thumb set is particular unfortunate if you think of DSP operations that have many MACs.

9.4.2 Softcore Microprocessors

Altera and Xilinx provide their own proprietary µP softcores. These processors do not try to reproduce an industry-standard processor but rather take advantage of the special hardware elements available with the FPGAs. The Xilinx PicoBlaze, for instance, makes use of the feature that the LEs can be used as dual-port RAMs, resulting in very small area requirements; Altera's Nios processor replaces the register file with M4K memory blocks that allow one to save a large number of LEs.

The most popular FPGA-based industry-standard processors are offered by third-party vendors through the FPGA vendor partner programs. We find here popular embedded µP softcores such as the Motorola's 68HC11, Microchip's PIC, or the TMS320C25 from Texas Instruments. Let us now have a closer look at these FPGA-based softcore processors.

An 8-bit processor: the Xilinx PicoBlaze. 8-bit FPGA softcores are available for many instruction sets like Intel's 8080 or 8051, Zilog's Z80, Microchip's PIC family, MOS Technology's 6502 (popular in early Apple and Atari computer), Motorola/Freescales 68HC11, or Atmel AVR microprocessors. At www.edn.com/microprocessor a full list of current controllers is provided. The 8-bitters have become the favorite controllers. Sales are about 3 billion controllers per year, compared with 1 billion 4- or 16/32-bit controllers. 4-bit processors usually do not have the required performance, while 16- or 32-bit controllers are usually too expensive.

One of the most important driving forces in the microcontroller market has been the automotive and home appliance market. In cars, for instance, only a few high-performance microcontrollers are needed for audio, or engine control; the other more than 50 microcontrollers are used in such functions as electric mirrors, air bags, the speedometer, and door locking, to name just a few. Xilinx PicoBlaze fits right in these popular 8-bit applications and provide a nice and free-of-charge development platform. The assembler/link/loader and VHDL code for the core are available royalty free. Optimized for Xilinx devices (the low-level LUT implementation of many functions like the ALU and register file using dual-port memories would make it hard to use an Altera device) the core is very small, and characterized by the following key features and performance (depending on the device family used) data, see Fig. 9.17.

- 16-byte-wide general-purpose data registers
- 256-1024 instruction words
- Byte-wide ALU operation with carry and zero flags
- 64-byte internal scratchpad RAM
- 256 input/output ports
- Four to 31 locations of CALL/RETURN stack

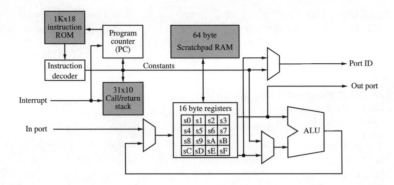

Fig. 9.17. The PicoBlaze a.k.a. KCPSM core from Xilinx.

- Each instruction takes two clock cycles
- Twenty-one MIPS for CoolRunner II to 100 MIPS for Virtex-4
- Instruction size 16 to 18-bits
- Eight to 16 8-bit registers
- Size 76-96 slices (Virtex/Spartan) or 212 macrocells in CoolRunner-II

A free C compiler is also available written by Francesco Poderico; it is royalty free and available for download, see www.xilinx.com.

Example 9.6: The following C-code segment

```
// DSPstone benchmark 1
char  a, b, c, d;
void main()
{ d = c + a * b; }
```

will be translated into the following assembler code for the PicoBlaze:

```
;**********************************************************
;             Picoblaze Small C Compiler for Xilinx PicoBlaze
;   Picoblaze C Compiler for PicoBlaze, Version alpha 1.7.7
;**********************************************************

NAMEREG sf , XL
NAMEREG se , YL
NAMEREG sd , ZL
NAMEREG sc , XH
NAMEREG sa , ZH
NAMEREG sb , TMP
NAMEREG s9 , SH
NAMEREG s8 , SL
NAMEREG s7 , KH
NAMEREG s6 , KL
NAMEREG s5 , TMP2
CONSTANT _a ,ff
CONSTANT _b ,fe
CONSTANT _c ,fd
```

```
CONSTANT _d ,fc
LOAD YL , fc
JUMP _main


;// DSPstone benchmark 1
;char   a, b, c, d;
;void main(){
_main:
;d = c + a*b;
INPUT ZL ,_c
SUB YL , 01
OUTPUT ZL,(YL)
INPUT ZL ,_a
SUB YL , 01
OUTPUT ZL,(YL)
INPUT ZL ,_b
INPUT XL,(YL)
ADD YL , 01
LOAD XH,XL
AND XH,80
JUMP Z,L2
LOAD XH,ff
L2:
LOAD ZH,ZL
AND ZH,80
JUMP Z,L3
LOAD ZH,ff
L3:
call _sign_mult
INPUT XL,(YL)
ADD YL , 01
ADD XL , ZL
OUTPUT XL,_d
;}
_end_main: jump _end_main; end of program!

; MULT SUBROUTINE
_mult:
LOAD TMP , 0f
LOAD  SL, XL
LOAD  SH, XH
LOAD  XL, 00
LOAD  XH, 00
_m1: SR0 ZH
SRA ZL
JUMP NC , _m2
ADD XL , SL
ADDCY XH , SH
_m2: SL0 SL
SLA SH
SUB TMP , 01
```

```
JUMP NZ , _m1
LOAD ZL,XL
LOAD ZH,XH
RETURN
_sign_mult:
LOAD TMP2,00
LOAD TMP,XH
AND TMP,80
JUMP Z,_check_member2
LOAD TMP2,01
XOR XL,ff
XOR XH,ff
ADD XL,01
ADDCY XH,00
_check_member2:
LOAD TMP,ZH
AND TMP,80
JUMP Z,_do_mult
XOR TMP2,01
XOR ZL,ff
XOR ZH,ff
ADD ZL,01
ADDCY ZH,00
_do_mult:
CALL _mult
AND TMP2,01
JUMP NZ,_invert_mult
RETURN
_invert_mult:
XOR XL,ff
XOR XH,ff
ADD XL,01
ADDCY XH,00
RETURN

;0 error(s) in compilation
```

9.6

As can be seen from the example the C compiler uses many instructions for this short DSP code sequence; even worse because the PicoBlaze, as most 8-bitter, does not have a hardware multiply, which is done by a series of shift and adds that slows down the program further.

Although Altera does not promote its own 8-bitter, the AMPP partner supports several instruction sets, like 8081, Z80, 68HC11, PIC and 8051, see Table 9.8. For Xilinx devices we find besides the PicoBlaze also support for 8051, 68HC11, and PIC ISA.

Notice how the low-level hardware optimization of the PicoBlaze with only 177 four-input LUTs makes it the smallest and fastest 8-bitter for Xilinx devices.

Table 9.8. FPGA 8-bitter ISA support. Vendors: DI= Dolphin Integration (France); CI= CAST Inc.(NJ, USA); DCD=Digital core design (Poland); N/A= Information not available

μP name	Device used	LE / Slices	BRAM/ M4Ks	Speed (MHz)	Vendor
C8081	EP1S10-5	2061	3	108	CI
CZ80CPU	EP1C6-6	3897	–	82	CI
DF6811CPU	Stratix-7	2220	4	73	DCD
DFPIC1655X	Cyclone-II-6	663	N/A	91	DCD
DR8051	Cyclone-II-6	2250	N/A	93	DCD
Flip8051	Xc2VP4-7	1034	N/A	62	DI
DP8051	Spartan-III-5	1100	N/A	73	DCD
DF6811CPU	Spartan-III-5	1312	N/A	73	DCD
DFPIC1655X	Spartan-III-5	386	3	52	DCD
PicoBlaze	Spartan-III	96	1	88	Xilinx

An 16-bit processor: the Altera Nios. The Nios embedded processor is a configurable RISC processor with a 16- or 32-bit datapath. Nios embedded systems can be created with any number of peripherals. Figure 9.18 shows the SOPC builder 32-bit standard configuration of the Nios processor.

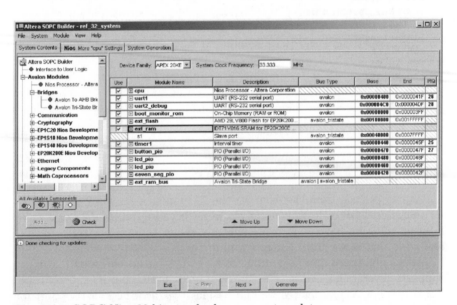

Fig. 9.18. SOPC Nios 32-bit standard processor template.

Table 9.9 shows the base core sizes for the Nios embedded processor and some of the IP core peripherals that integrate with the standard Nios em-

Table 9.9. Nios core and peripheral sizes, logic elements (LE) count, and embedded array blocks (M4K).

Unit	LE	M4K
16-bit data path Nios	950	2
32-bit datapath Nios	1250	3
UART, fixed baud rate	170	
Timer	244	
Serial peripheral interface(SPI):		
8-bit master, one slave	103	
SPI: 8-bit master, two slaves	108	
General-purpose I/O: 32-bit, tristate	138	
SDRAM controller	380	
External memory/peripheral: 32 bit	110	
External memory/peripheral: 16 bit	85	

bedded processor to form complete microprocessing units. Most peripherals can be parameterized to fit the specific application and can be instantiated multiple times within a single µP. In addition, customer-designed logic and peripherals can be integrated with the Nios processor to deliver a unique µP. The creation of these custom µPs can be done in minutes using the Altera SOPC builder tool, and synthesized to run on any Altera FPGA. In addition to the IP cores listed in Fig. 9.18, SOPC builder features additional IP cores available from Altera and Altera's megafunction partners program (AMPP).

Nios processors lower than version 2.0 have a three-stage pipeline (load, decode, execute) and each instruction takes a predictable amount of time. For versions later than 2.0 the three-stage pipeline is replaced by a five-stage pipeline with sophisticated prefetch logic, interlocking, and hazard management. The pipeline logic hides these details from the programmer and makes it more difficult to analyze the execution time just via instruction count only, since the latency of the instruction depends on many factors, like the pre- or post instruction, operands, and memory location, to name just a few. Altera provide a best-case estimate for each instruction the actual latency however maybe longer. The minimum clock-cycle estimate for some typical instructions is shown in Table 9.10.

The Altera Nios differs from other softcore processor solutions in the market by including custom instruction features, see Fig. 9.19. Custom instruction design is a process of implementing a complex sequence of standard instructions in hardware in order to reduce them to a single-instruction macro that can be accessed by software. The custom instructions can be used to implement complex processing tasks in single-cycle (combinatorial) and multi-cycle (sequential) operations. In addition, these user-added custom instructions can access memory as well as logic outside the Nios system. As an example design in the case study section, we will see a radix-2 FFT for custom implementation due to its wide range of possible transform lengths

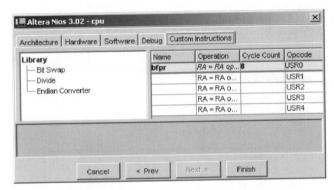

Fig. 9.19. Custom instruction features of the Nios processor.

Table 9.10. Altera clock cycles in 5 stage pipeline Nios processors [329].

Function name	memory location	Clock cycle	Comment
ASR,ASRI,LSL,LSLI,LSR,LSRI	–	1	Shift operations
MUL	–	2	$16 \times 16 \rightarrow$ 32-bit
JMP, CALL	–	2	control flow
LD, ST	on-chip	2	Load and store
TRET	–	3	Return function
TRAP	–	4	Hold processor
LD, ST	off-chip	4	Load and store

(all power-of-two transform lengths), DFT calculations, decreased memory requirement, and easy hardware implementation of the small butterfly processor. The butterfly processor is implemented as a custom logic block and its software macro generated is then used in the software code for the radix-2 FFT. The performance of the software code with custom instructions for the butterfly processor with different multiplier optimizations available with the Nios processor is then compared with software-only code.

With the wide range of densities available in FPGA devices and the small sizes of Nios embedded systems, system designers can divide complex problems into smaller tasks and use multiple Nios embedded processors. These Nios processors can be customized with a wide selection of peripherals, defining very simple to very complex µP systems. By targeting low-cost devices, powerful, customized embedded systems can be realized at the best cost in the industry.

The Nios processor shown in Fig. 9.20 has a pipelined general-purpose RISC architecture [330, 331, 332, 302, 333]. The 32-bit processor has a separate 16-bit instruction bus and 32-bit data bus. The register file is configurable to have 128, 256 or 512 registers but at one time only 32 of these registers are accessible as general-purpose registers through software using

Table 9.11. Nios processor core multiplier options.

Multiplication option	Clock cycles 32-bit product	Hardware effort
Software	80	0
MSTEP	18	14-24 LEs
MUL	3	427-462 LEs

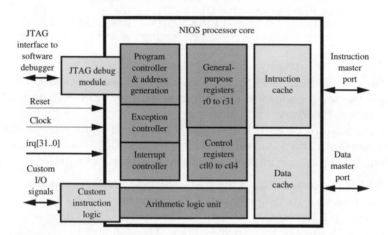

Fig. 9.20. Nios processor core.

a sliding window. These large numbers of internal registers are used to accelerate subroutine calls and local variable access. The CPU template is in general configurable with instruction and data cache memory, which increases its performance, but the APEX device with its limited M4K memory block architecture, used by Altera in the Nios development board, does not support this feature. The Nios instruction set can be configured to increase software performance and can be modified either by adding custom instructions or by using predefined instruction set extensions provided with the processor template. The three predefined multiplier optimizations for the Nios processor are listed in Table 9.11, giving the number of clock cycles and size required for each of the multiplier options:

- The MUL instruction includes a hardware 16×16-bit integer multiplier.
- The MSTEP instruction provides the hardware to execute one step of a 16×16-bit multiply in one clock cycle.
- Software multiplication uses the C runtime libraries to implement integer multiplication with sequences of shift and add instructions.

Table 9.12. FPGA 16/24-bit ISA support. Vendors: CI= CAST Inc.(NJ, USA); DCD=Digital core design (Poland)

μP name	Device used	LE / slices	BRAM/ M4Ks	Speed (MHz)	Vendor
C56000 PDSP	EP1C25-6	13,531	12	51	CI
C68000	EP1C12C-6	6152	–	57	CI
D68000	Cyclone-6	6604	n/a	44	DCD
C32025 PDSP	Spartan-III-5	2211	19	44	CI
D68000	Virtex-II PRO-7	3415	n/a	65	DCD

Any of these three multiplier options could be used to implement 16×16-bit multiplication in software. Depending on the overall processor architecture the additional hardware effort may vary.

Alternative 16/24-bit microprocessors are offered by IP vendors, that rebuild standard PDSPs (Motorola 56000; TI TMS320C25) or GPP (Motorola 68000). Table 9.12 give an overview of available core and required resources.

An 32-bit processor: the Xilinx MicroBlaze. As an example of a 32-bit softcore processor let us in the following study the Xilinx MicroBlaze. The MicroBlaze is a Harvard 32-bit data and instruction RISC processor, that is available with three or five pipeline stages. The standard key features of the MicroBlaze core can be summarized as follows:

- MicroBlaze v4 is a three-pipeline-stage core with 0.92 DMIPS/MHz
- MicroBlaze v5 is a five-pipeline-stage core with 1.15 DMIPS/MHz
- The ALU, shifter, and 32×32 register file are standard

Optional items that can be included at configuration time are:

- Barrel shifter
- Array multiplier
- Divider
- Floating-point unit for add, subtraction, multiply, divide, and comparison
- Data cache from 2-64 KB
- Instruction cache from 2 to 64 KB

The five-stage pipeline is executed in the following steps: (1) fetch, (2) decode, (3) execute, (4) memory access, and (5) writeback.

The data and instruction caches have a direct mapped cache architecture. The cache can be access in blocks of four or eight words. One or more Block-RAMs are used to store the data, while an additional BlockRAM is used to store the tag data. Let us study a typical cache configuration.

Example 9.7: MicroBlaze Cache Configuration

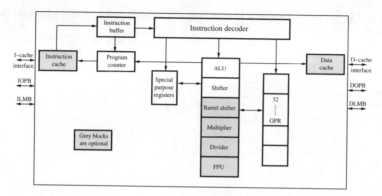

Fig. 9.21. The MicroBlaze softcore architecture core from Xilinx.

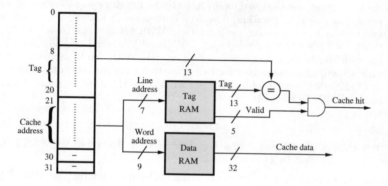

Fig. 9.22. The MicroBlaze cache configuration for 16 MB cache.

Let us design in the following a cache that uses only two BlockRAMs. Since BlockRAMs in Spartan-III are of size 16 Kbits, we conclude, for 32-bit word size, that we can store 512×32 words in a single BlockRAM. Using a cache size of less than 16 Kbits or 2 KB will not really save any resources and that is why this is the smallest available cache size. Now we have do decide if we want to use the four or eight words-per-line configuration. Usually with eight words per line we can address a larger external memory, while four words per line gives a faster decoder, so let us start with the four word per line and see what the maximum external memory we can address is.

With 512 words in our cache, and four words per line we conclude that we have to store 128 tags in our tag memory. Then our tag BlockRAM needs to be configured as 128×32 memory. Each line now needs a valid bit and four valid bits for each word, which leaves up to 27 bits for the tags. Therefore the external memory will be limited to an address space of 27 tag bits plus 11 LSBs used to address 2 KB by the cache, i.e., a 38-bit word can be addressed, i.e., $2^{38} = 256$ GB memory. This is probably much larger than the actual main memory and more than the 32-bit address space of the MicroBlaze. This configuration (with 13 tag bits to address the 16 MB of the Nexsys board) is shown in Fig. 9.22.

Table 9.13. DMIPS comparison for different memory organizations (D=data; I=program memory).

External SRAM		Internal Cache		BRAM		LCs	DMIPS
D	I	D	I	D	I		
✓	✓	–	–	–	–	8718	7.139
✓	✓	✓	✓	–	–	9076	29.13
–	✓	–	–	✓	–	8812	47.81
–	–	–	–	✓	✓	8718	59.78

At the upper end Xilinx allows one to use a 64 KB cache memory. We then need 32 BlockRAMs for the cache data alone. If we now use one more Block-RAM to store the tags, we again have to decide if we use the four or eight words-per-line configuration. The 64 KB requires 16 K words, and 4 K lines for the four words-per-line configuration. With 4 K lines we then need to use the $2^{12} \times 4$ BlockRAM configuration. We need four valid and one line bit, so one BlockRAM for the tags will not be enough. With two BlockRAMs for the tags we will have three bits for tags, i.e., the main memory can have a size of 512 KB. With each additional BlockRAM we can then increase the main memory by a factor of 16.   9.7

Other example configurations are discussed in Exercises 9.33 and 9.34, p. 641.

Given the option that we can now use a cache memory for the MicroBlaze one question is still open: what is the best memory configuration for maximum performance? This question has been evaluated by Fletcher [334] and the results are shown in Table 9.13. For almost all embedded microprocessor, when the FPGA just hosts the processor and the program and data are kept in external SRAM memory, the performance improves if we add an on-chip data and/or program cache, see rows 4 and 5 in Table 9.13. However, if we are able to keep the whole main memory inside the FPGA this will be better than a design with external memory plus caches, as Table 9.13 shows for the Dhrystone benchmarks, which are much smaller then the SPEC benchmarks in a conventional computer and fit inside the FPGA BlockRAMs.

Alternative 32-bit microprocessors are offered by Altera and IP vendors, which rebuild custom or standard GPP (e.g., Motorola 68000). The Altera Nios II is available in three different versions: A fast six-stage pipeline version that provides 1.16 DMIPS/MHz; a standard version with five pipeline stages that delivers 0.74 DMIPS/MHz; and an economy version with minimal core size and a one-stage pipeline that provides 0.15 DMIPS/MHz. Table 9.14 gives an overview of available core and required resources.

Table 9.14. FPGA 16/24-bit ISA support. Vendors: CI= CAST Inc.(NJ, USA); N/A = data not available

μP name	Device used	LE / slices	BRAM/ M4Ks	Speed	Vendor
C68000-AHB	EP1C6F256C6	5822	4	59 MHz	CI
Nios-II fast	EP2C20F484C6	1595	N/A	105 DMIPS	Altera
Nios-II std	EP2C20F484C6	1033	N/A	57 DMIPS	Altera
Nios-II eco.	EP2C20F484C6	542	N/A	22 DMIPS	Altera
C68000-AHB	Spartan-III	2923	–	40 MHz	CI

9.5 Case Studies

Finally let use have a look at three more-detailed design projects. The first is the HDL design of a complete zero-address, i.e., stack machine, that uses the assembler and C compiler tools we developed in Sect. 9.3, p. 566. The second case study is a LISA-based DWT processor design that shows the wide variety (simple μP to true vector processor) that can be built with a few LISA instructions [335]. The final case study shows how a custom DSP block can be tightly couple with Altera's Nios processor. The FFT butterfly processor is chosen and hardware as well as software optimizations are discussed [336, 337].

9.5.1 T-RISC Stack Microprocessors

Let us start our design explorations with a simple stack machine. Although the stack machine is called a zero-address machine, we still need a couple of bits in the instruction to define immediate operands. If we select 8-bit data and 4-bit instructions then we can define 16 instructions and have a 12-bit data word that is easily represented in the simulation by three half bytes. We can choose seven ALU instructions (0-6), five data move instructions (7-11), and four control flow instructions. More specifically, our instructions set then becomes:

- The ALU instructions use the top of the stack (TOS) and the second of the stack (if applicable) and can be further split into:
  - Four arithmetic operations: ADD, SUB, MUL, and INV
  - Three logic operations: OPAND, OPOR, and OPNOT
- Data move instructions move data from/to the top of the stack:
  - POP <var> stores the TOS data word memory location var.
  - PUSH <var> loads a data word var from memory and places it on TOS.
  - PUSHI <imm> puts the immediate value imm on TOS.
  - SCAN reads 8 bits from the input port and puts them on TOS.
  - PRINT writes 8 bits from the TOS to the output port.

- The four program control instructions are:
  - CNE and CEQ compare the two top elements of the stack and set the jump control register JC accordingly.
  - CJP <imm> loads the PC with the value imm if the JC is set to true.
  - JMP <imm> loads the PC with the value imm.

Since with Cyclone II devices we can only implement synchronous memory, we have to use a register for the input data or address. This makes it impossible to have a PC update, program, data, and TOS update in one clock cycle and a minimum of two clock cycles must be used. Figure 9.23a shows the implemented timing. The PC is update on the first falling edge. This update is used as the input for the address of the program memory. The output data of the program memory is stored by the next rising edge in the PROM output register. The instructions are then decoded, and only for the POP operation do we store TOS with the next falling edge in the data memory. For any ALU or PUSH operation the memory is routed through the ALU and stored in the TOS register with the next rising edge. At the same time we update the values in the stack. A four-value stack as used in the popular HP41 pocket calculators should provide enough stack depth. Since such a short stack is used, it is much easier to use registers than a LIFO M4K memory block. The proposed timing works for all instructions except the control flow instructions. Since the comparison values has to be stored in the JC register we need an extra clock cycle to implement a conditional jump operation. In the first step we update the JC register and in the next clock cycle the PC is updated according to the JC register, see Fig. 9.23b.

Example 9.8: A Stack Machine

The following VHDL code[6] shows the four-entry stack machine. As a test program, factorial computation (C-code see p. 586; MIF code see p. 578) is shown in the simulation.

```
-- Title: T-RISC stack machine
-- Description: This is the top control path/FSM of the
-- T-RISC, with a single three-phase clock cycle design
-- It has a stack machine/0-address-type instruction word
-- The stack has only four words.

LIBRARY lpm; USE lpm.lpm_components.ALL;

LIBRARY ieee;
USE ieee.STD_LOGIC_1164.ALL;
USE ieee.STD_LOGIC_arith.ALL;
USE ieee.STD_LOGIC_signed.ALL;

ENTITY trisc0 IS
  GENERIC (WA : INTEGER := 7; -- Address bit width -1
           WD : INTEGER := 7); -- Data bit width -1
```

---

[6] The equivalent Verilog code trisc0.v for this example can be found in Appendix A on page 724. Synthesis results are shown in Appendix B on page 731.

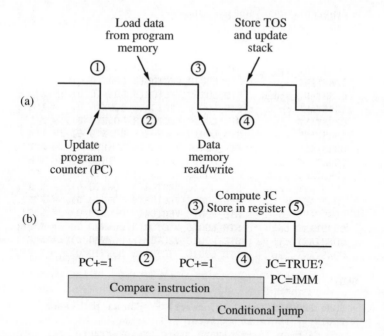

Fig. 9.23. Timing for T-RISC operations. (a) Usual four clock edges used in most instructions. (b) Two-instruction sequence timing used for conditional jump instructions.

```
PORT(reset, clk : IN  STD_LOGIC;
     jc_OUT      : OUT BOOLEAN;
     me_ena      : OUT STD_LOGIC;
     iport       : IN  STD_LOGIC_VECTOR(WD DOWNTO 0);
     oport       : OUT STD_LOGIC_VECTOR(WD DOWNTO 0);
     s0_OUT, s1_OUT, dmd_IN, dmd_OUT : OUT
                            STD_LOGIC_VECTOR(WD DOWNTO 0);
     pc_OUT, dma_OUT, dma_IN : OUT
                            STD_LOGIC_VECTOR(WA DOWNTO 0);
     ir_imm      : OUT STD_LOGIC_VECTOR(7 DOWNTO 0);
     op_code     : OUT STD_LOGIC_VECTOR(3 DOWNTO 0));
END;

ARCHITECTURE fpga OF trisc0 IS

  TYPE state_type IS (ifetch, load, store, incpc);
  SIGNAL state    : state_type;
  SIGNAL op    : STD_LOGIC_VECTOR(3 DOWNTO 0);
  SIGNAL imm, s0, s1, s2, s3, dmd
                            : STD_LOGIC_VECTOR(wd DOWNTO 0);
  SIGNAL pc, dma : STD_LOGIC_VECTOR(wa DOWNTO 0);
  SIGNAL pmd, ir   : STD_LOGIC_VECTOR(11 DOWNTO 0);
  SIGNAL eq, ne, mem_ena, not_clk : STD_LOGIC;
```

```
    SIGNAL jc        :  boolean;

-- OP Code of instructions:
  CONSTANT add   : STD_LOGIC_VECTOR(3 DOWNTO 0) := X"0";
  CONSTANT neg   : STD_LOGIC_VECTOR(3 DOWNTO 0) := X"1";
  CONSTANT sub   : STD_LOGIC_VECTOR(3 DOWNTO 0) := X"2";
  CONSTANT opand : STD_LOGIC_VECTOR(3 DOWNTO 0) := X"3";
  CONSTANT opor  : STD_LOGIC_VECTOR(3 DOWNTO 0) := X"4";
  CONSTANT inv   : STD_LOGIC_VECTOR(3 DOWNTO 0) := X"5";
  CONSTANT mul   : STD_LOGIC_VECTOR(3 DOWNTO 0) := X"6";
  CONSTANT pop   : STD_LOGIC_VECTOR(3 DOWNTO 0) := X"7";
  CONSTANT pushi : STD_LOGIC_VECTOR(3 DOWNTO 0) := X"8";
  CONSTANT push  : STD_LOGIC_VECTOR(3 DOWNTO 0) := X"9";
  CONSTANT scan  : STD_LOGIC_VECTOR(3 DOWNTO 0) := X"A";
  CONSTANT print : STD_LOGIC_VECTOR(3 DOWNTO 0) := X"B";
  CONSTANT cne   : STD_LOGIC_VECTOR(3 DOWNTO 0) := X"C";
  CONSTANT ceq   : STD_LOGIC_VECTOR(3 DOWNTO 0) := X"D";
  CONSTANT cjp   : STD_LOGIC_VECTOR(3 DOWNTO 0) := X"E";
  CONSTANT jmp   : STD_LOGIC_VECTOR(3 DOWNTO 0) := X"F";

BEGIN

    FSM: PROCESS (op, clk, reset) -- FSM of processor
    BEGIN -- store in register ?
        CASE op IS -- always store except Branch
          WHEN pop    => mem_ena <= '1';
          WHEN OTHERS => mem_ena <= '0';
        END CASE;
        IF reset = '1' THEN
          pc <= (OTHERS => '0');
        ELSIF FALLING_EDGE(clk) THEN
          IF ((op=cjp) AND NOT jc ) OR  (op=jmp) THEN
            pc <= imm;
          ELSE
            pc <= pc + "00000001";
          END IF;
        END IF;
        IF reset = '1' THEN
          jc <= false;
        ELSIF rising_edge(clk) THEN
          jc <= (op=ceq AND s0=s1) OR (op=cne AND s0/=s1);
        END IF;
    END PROCESS FSM;

    -- Mapping of the instruction, i.e., decode instruction
    op   <= ir(11 DOWNTO 8);   -- Operation code
    dma  <= ir(7 DOWNTO 0);    -- Data memory address
    imm  <= ir(7 DOWNTO 0);    -- Immidiate operand

    prog_rom: lpm_rom
    GENERIC MAP ( lpm_width => 12,
                  lpm_widthad => 8,
                  lpm_outdata => "registered",
```

```
                    lpm_address_control => "unregistered",
                    lpm_file => "TRISCOFAC.MIF")
PORT MAP ( outclock => clk, address => pc, q => pmd);
not_clk <= NOT clk;

data_ram: lpm_ram_dq
GENERIC MAP ( lpm_width => 8,
              lpm_widthad => 8,
              lpm_indata => "registered",
              lpm_outdata => "unregistered",
              lpm_address_control => "registered")
PORT MAP ( data => s0, we => mem_ena, inclock => not_clk,
           address => dma, q => dmd);

ALU: PROCESS (op, clk)
VARIABLE temp: STD_LOGIC_VECTOR(2*WD+1 DOWNTO 0);
BEGIN
  IF rising_edge(clk) THEN
    CASE op IS
      WHEN add    =>   s0  <= s0 + s1;
      WHEN neg    =>   s0  <= -s0;
      WHEN sub    =>   s0  <= s1 - s0;
      WHEN opand  =>   s0  <= s0 AND s1;
      WHEN opor   =>   s0  <= s0 OR s1;
      WHEN inv    =>   s0  <= NOT s0;
      WHEN mul    =>   temp := s0 * s1;
                       s0  <= temp(WD DOWNTO 0);
      WHEN pop    =>   s0  <= s1;
      WHEN push   =>   s0  <= dmd;
      WHEN pushi  =>   s0  <= imm;
      WHEN scan   =>   s0 <= iport;
      WHEN print  =>   oport <= s0; s0<=s1;
      WHEN OTHERS =>   s0 <= (OTHERS => '0');
    END CASE;
    CASE op IS    -- Specify the stack operations
      WHEN pushi | push | scan => s3<=s2; s2<=s1; s1<=s0;
                                              -- Push type
      WHEN cjp | jmp | inv | neg => NULL;
                                  -- Do nothing for branch
      WHEN OTHERS =>   s1<=s2; s2<=s3; s3<=(OTHERS=>'0');
                                  -- Pop all others
    END CASE;
  END IF;
END PROCESS ALU;

-- Extra test pins:
dmd_OUT <= dmd; dma_OUT <= dma; -- Data memory I/O
dma_IN <= dma; dmd_IN  <= s0;
pc_OUT <= pc; ir <= pmd; ir_imm <= imm; op_code <= op;
                                            -- Program
jc_OUT <= jc; me_ena <= mem_ena; -- Control signals
s0_OUT <= s0; s1_OUT <= s1;    -- Two top stack elements
```

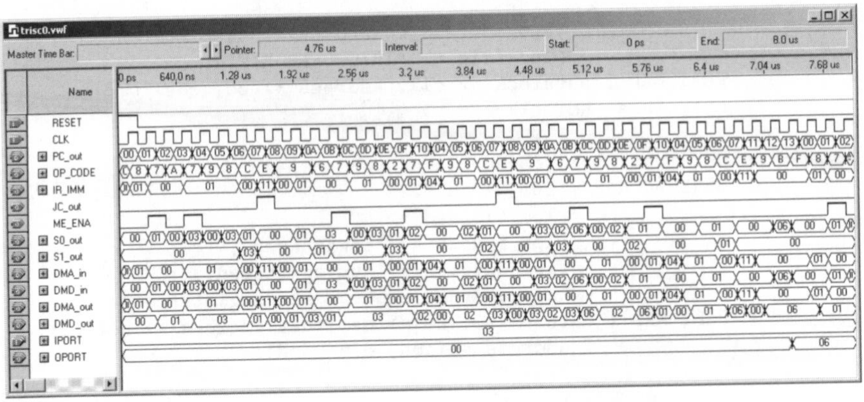

Fig. 9.24. T-RISC simulation of factorial example.

```
END fpga;
```
We see in the coding first the generic definition followed by the entity that includes the ports and the test pins. The architecture part starts with general-purpose signals and then the op-code for all 16 instructions are listed as constant values. The first process in the architecture body **FSM** hosts the finite state machine that is used to control the microprocessor. Then program and data memory are instantiated via ROM and RAM blocks. All operations that include an update of the stack are included in **ALU**. The previous constant definitions allow very intuitive coding of the actions. Finally some extra test pins are assigned that are visible as output ports. The design uses 198 LEs, two M4Ks, 1 embedded multiplier and has a **Registered Performance** of 115.65 MHz.   9.8

The simulation from the previous design for a factorial program is shown in Fig. 9.24. The program starts by loading the input value from the IPORT. Then the computation of the factorial starts. First the loop variable is evaluated; if it is larger than 1, then x is multiplied by k. Then k is decremented and the program jumps to the start of the loop. After two runs through the loop the program is finished and the factorial result $(2 \times 3 = 6)$ is transported to OPORT.

9.5.2 LISA Wavelet Processor Design

A microprocessor is a much more-efficient way of using FPGA resources than a direct hardware implementation of an algorithm and has become in recent years one of the most important IP blocks for FPGA vendors. Altera, for instance, reported that they sold 10,000 systems of the Nios microprocessor development systems in the first three years alone. Xilinx reported an even larger number of downloads of their MicroBlaze microprocessors.

A new generation of design tools is empowering software developers to take their algorithmic expressions straight into FPGA hardware without hav-

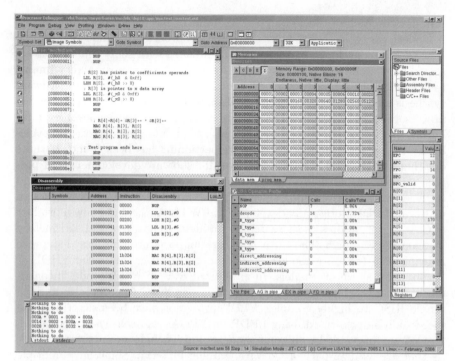

Fig. 9.25. LISA development tools: (left) disassembler, (center) memory monitor and pipeline profiles, (right) file and register window.

ing to learn traditional hardware design techniques. These tools and associated design methodologies are classified collectively as electronic system-level (ESL) design, broadly referring to system design and verification methodologies that begin at a higher level of abstraction than the current mainstream HDL. The language for instruction set architecture (LISA), for instance, allows us to specify a processor instruction or cycle accurately using a few LISA operations, then explore architecture using a tool generator and profiler (see Fig. 9.25), and finally determine speed/size/power parameters via automatically synthesized VHDL or Verilog code. ESL tools have been around for a while, and many perceive that these tools are predominantly focused on ASIC design flows. But with ASIC mask charges of $4 million in 65 nm technology the number of designs using FPGAs is rapidly increasing. In reality, an increasing number of ESL tool providers (e.g., Celoxica, Codetronix, CoWare, Binachip, Impulse Accelerated, Mimosys) are focusing on programmable logic.

Today the majority of microprocessors are employed in embedded systems. This number is not surprising because a typical home today may have a Laptop/PCs with a high-performance microprocessor but probably dozens of embedded systems, including electronic entertainment, household, and telecom devices, each of them equipped with one or more embedded processors. A modern car typically has more than 50 microprocessors. Embedded processors are often developed by relatively small teams with short time-to-market requirements, and the processor design automation is clearly a very important issue. Once a model of a new processor is available, existing hardware synthesis tools enable the path to FPGA implementation. However embedded processor design typically begins at a much higher abstraction level, even far beyond an instruction set and involves several architecture exploration cycles before the optimum hardware/software partioning has been found. It turns out that this requires a number of tools for software development and profiling. These are normally written manually – a major source of cost and inefficiency in embedded processor design so far. The LISA processor design platform (LPDP), originally developed at RWTH Aachen and now a product of CoWare Inc. addresses these issues in a highly innovative and satisfactory manner, see Fig. 9.25. The LISA language supports profiling-based stepwise refinement of processor models down to cycle accuracy and even VHDL or Verilog RTL synthesis models. In an elegant way, it avoids model inconsistencies otherwise inevitable in traditional design flows. Microprocessors from simple RISC to highly complex VLIW processor have been described and successfully implemented using LPDP for FPGAs and cell-based ASICs.

CoWare provides 14 different models. This include seven tutorial models that are used as part of CoWare training material. Some have multiple versions like the more than ten different designs in the QSIP_X models. Four starting point models are provided and used as a skeleton for starting a new architecture. Three different IP models for classic architectures are also included. All models are instruction accurate and most of the models are Harvard-type RISC models that are also cycle accurate. Pipeline stages vary from three to five. Provided are all types of modern processor from simple RISC (QSIP), over PDSP like LT_DSP_32p3 to VLIW LT_VLIW_32p4 to special processors like a 16- to 4096-point FFT processor LT_FFT_48p3. Table 9.15 shows the properties of some example models.

LISA 18-bit instruction word RISC processor. Xilinx offers a 32-bit MicroBlaze and a 8-bit PicoBlaze RISC processor but no processor with 16 or 24-bit, as typically for DSP algorithms used, is offered. Let us in the following design such a 16-bit RISC machine with the LPDP. Since a 16-bit processor fits in between the Micro- and PicoBlaze we will call our RISC processor NanoBlaze.

For an FPGA design we can start with the three-pipeline RISC tutorial design of the LISA 2.0 QSIP_12 model and extend the ISA to make it more useful for the FPGA design. The BlockRAMs in Xilinx FPGAs are 18 bits wide and the instruction words should therefore also be 18 bits wide. When using BlockRAM there will be no gain when using instruction words short than 18 bits. The byte-wide access in QSIP model should be changed to flat 18-bit instruction and data access. Changes then have to be included in the

Table 9.15. LISA example models (CC=C compiler generated).

Name	CC	Pipeline stages	Description
QSIP_X	–	3	• Harvard RISC architecture • 12 different tutorial versions • Single cycle ALU • Pipeline and zero pipeline version
LT_DSP_32p3	✓	3	• Single cycle ALU with MAC • Zero overhead loops • 32-bit instruction • 24-bit data path • 48-bit accumulator
LT_VLIW_p4	–	4	• QSIP like ISA • Parallel load/store • Parallel arithmetic instructions

instruction counter, memory configuration *.cmd file, step_cyle, and the data memory instructions LDL, LDH, and LDR. The following listing show the supported instructions of the designed NanoBlaze.

- Arithemtic/logic unit (ALU) instructions:
  - ADD: three-operand add operation with two source operands and a third destination operand.
  - MUL: three-operand multiply operation with two source operands and a third destination operand. Only the lower 16 bits of the product are preserved.
- Data move instructions:
  - LDL: load the lower 8 bits of the data word with a constant value.
  - LDH: load the upper 8 bits of the data word with a constant value.
  - LDR: load register from memory. The memory location can be specified explicitly as a constant or indirectly via a general-purpose register.
  - STR: store register content to memory. The memory location can be specified explicitly or indirectly via a general-purpose register.
- Program control instructions:
  - BC: the condition branch checks a (loop) register for zero and not zero.
  - B: an unconditional branch.
  - BDS: the delay branch is a condition BC with the feature that the next instruction after the BDS instruction is also executed.

The basic instruction set of the DWT RISC processor consists of nine instructions that were designed using 28 LISA operations. The instruction coding of the instruction in the execution pipeline stage is shown in Fig. 9.26.

The NanoBlaze processor can now be synthesized and implemented in an FPGA. Depending on the type of memory used (i.e., CLB- or BlockRAM-based) we get the synthesis results shown in Table 9.16.

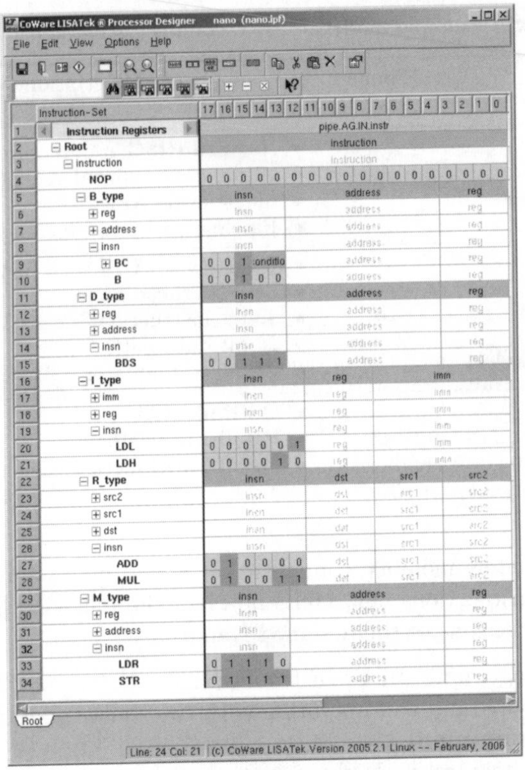

Fig. 9.26. NanoBlaze instruction set architecture.

Table 9.16. NanoBlaze synthesis result for the Xilinx device XC3S1000-4ft256.

Parameter	NanoBlaze with CLB-based RAM	NanoBlaze with BlockRAM
Slices	1896	1893
4-input LUT	3443	3602
Multiplier	1	1
BlockRAMs	0	2
Total gates	32,986	162,471
Clock period	13.293 ns	13.538 ns
F_{max}	75.2 MHz	73.9 MHz

Example 9.9: If we now use the RISC processor to implement a length-8 DWT processor as shown in Fig. 5.55, p. 314, we need two length-8 filter $g[n]$ and $h[n]$ and for each output sample pair 16 multiply and 14 add operations are necessary. For 100 samples with an output downsampling by 2 the arithmetic requirements for the DWT filter band would therefore be $8 \times 100 = 800$ multiplications and $7 \times 100 = 700$ additions. From the Calls shown in the instruction profile in Fig. 9.27 we see that the number of multiplication is in fact 800, however the number of add instructions

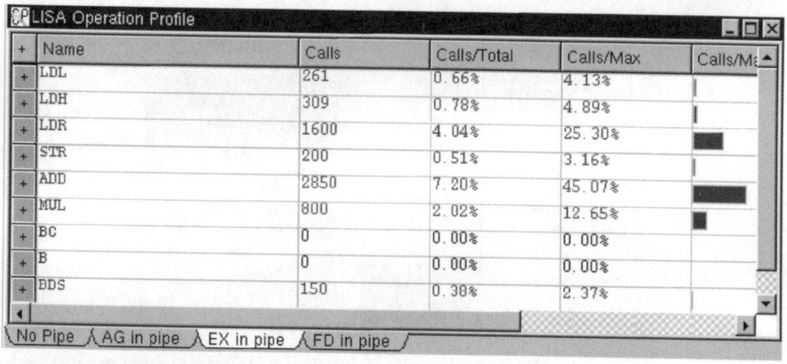

Name	Calls	Calls/Total	Calls/Max	Calls/Ma
LDL	261	0.66%	4.13%	
LDH	309	0.78%	4.89%	
LDR	1600	4.04%	25.30%	
STR	200	0.51%	3.16%	
ADD	2850	7.20%	45.07%	
MUL	800	2.02%	12.65%	
BC	0	0.00%	0.00%	
B	0	0.00%	0.00%	
BDS	150	0.38%	2.37%	

Fig. 9.27. NanoBlaze operation profile for 100-point length-8 dual-channel DWT including memory initialization, i.e., 300 instructions.

was more than four times higher as expected. This is due to the fact that the register updates for the memory access are also computed with the general-purpose ALU. 9.9

In addition to the large number of add operations to update the memory register pointer also 1 600 `LDR` load operations were performed. This can be substantially improved if we use auto-increment, indirect memory access as used for the `MAC` operation in PDSPs.

LISA programmable digital signal processor. From the DWT processor discussed in the last section we have seen the large required arithmetic count for updating the memory pointer and the memory access itself. A single multiply accumulate instruction in NanoBlaze requires the following operations:

```
; use pointer R[2] and R[3] to load operands
LDR R[8], R[2]
LDR R[9], R[3]
; increment register pointer using R[1]=1
; multiply and add result in R[4] and avoid data hazards
ADD R[2], R[2], R[1]
MUL R[7], R[8], R[9]
ADD R[3], R[3], R[1]
ADD R[4], R[4], R[7]
```

DSP algorithms typically operate on linear data arrays (i.e., vectors) and post-auto-increments or decrements in the memory pointer are therefore frequently used. In addition a fused add and multiply, usually called `MAC`, allows the previous six instructions to be combined into one single instruction, i.e.,

```
; load and multiply the values from pointer R[2] and R[3],
; and add the product to register R[4]
MAC R[4],R[3],R[2]
```

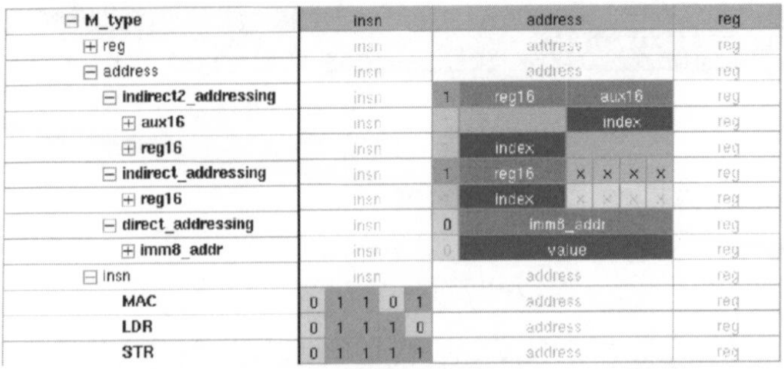

M_type	insn	address						reg
reg	insn	address						reg
address	insn	address						reg
indirect2_addressing	insn	1	reg16		aux16			reg
aux16	insn			index				reg
reg16	insn		index					reg
indirect_addressing	insn	1	reg16		x	x	x	x	reg
reg16	insn		index		x	x	x	reg
direct_addressing	insn	0	imm8_addr					reg
imm8_addr	insn	0	value					reg
insn	insn	address						reg
MAC	0 1 0 1	address						reg
LDR	0 1 1 0	address						reg
STR	0 1 1 1	address						reg

Fig. 9.28. Programmable digital signal processor (DSP18) instruction set additions.

The ISA additions to the NanoBlaze is shown in Fig. 9.28. The addition of such a MAC operation to the instruction set requires two major modification. First we need to provide a LISA operation that allows two indirect memory accesses. In hardware this results in a more-complex address generation unit and a dual-output-port data memory that supports two reads in one clock cycle. Secondly, we need to add the LISA operation for the MAC instructions.

The LISA operation to implement the MAC instruction can be implemented as follows:

```
/* This LISA operation implements the instruction MAC.    */
/* It accumulates the product of two register and stores  */
/* the result in a destination register.                  */
OPERATION MAC IN pipe.EX
{
  DECLARE
    {
      REFERENCE address;
      REFERENCE reg;
    }
  CODING { 0b01101 }
  SYNTAX { "MAC" }
  BEHAVIOR
    {
      short tmp1, tmp2, s1, s2; /* Temporary */
      short tmp_reg;
      short res;

      tmp_reg = reg;

      s1 = (data_mem[EX.IN.ar] & (char)0xffff);
      s2 = (data_mem[EX.IN.ar1] & (char)0xffff);
      res = tmp_reg + s1 * s2;
```

```
#pragma analyze(off)
      printf ("%04X * %04X + %04X = %04X\n",
                                      s1, s2, tmp_reg, res);
#pragma analyze(on)

      reg = res;

   }
}
```

The MAC LISA operation starts with the DECLARE section that refers to elements that are defined in other LISA operations. The CODING section that describes the OP code follows. The assembler syntax would be MAC, and finally in the BEHAVIOR section the implementation of the instruction is shown. We have used an additional printf inside the operation to monitor progress. While this does not change the hardware, we can monitor the output of our MAC instruction directly in the debugger window; see the lower window in Fig. 9.25.

We can then go ahead and synthesize the new processor that we want to call DSP18 due to the PDSP-like added features in the instruction set and perform a testbench simulation in the ModelSim simulator.

Example 9.10: To verify the functionality of the generated VHDL code we use the ModelSim simulator. LPDP generated all the required HDL code (VHDL or Verilog) and all the required simulation scripts (i.e., ModelTech *.do-files). As the test values we use $x = [1, 2, 3]; g = [10, 20, 40]$; and the MAC operation should progress as follows:

1. MAC = 1*10=10
2. MAC = 2*20=40 => 40+10=50
3. MAC = 3*40=120 =>120+50=170

The correct function can be seen from the ModelSim simulation from reg_r_4, shown in Fig 9.29, which shows the contents of register $r[4]$.   9.10

If we now write the program for the same 100-point length-8 DWT as in the last section, we will see the large impact the MAC operation has on the overall instruction count. Now the 800 MAC operations are the dominate operations and much fewer explicit add or memory operations are required. The total instruction count improves from 5,870 for the NanoBlaze to 1,968 for the DSP18. The operation profile (including memory initialization, i.e., 300 operations) for the DWT example is shown in Fig. 9.30. Since the DSP18 is larger and the addressing modes are more sophisticated than in the NanoBlaze, the overall registered performance decreases to 39 MHz when using CLB-based RAM and 51 MHz when using BlockRAM. Table 9.17 shows the implementation results for two different external memory configurations.

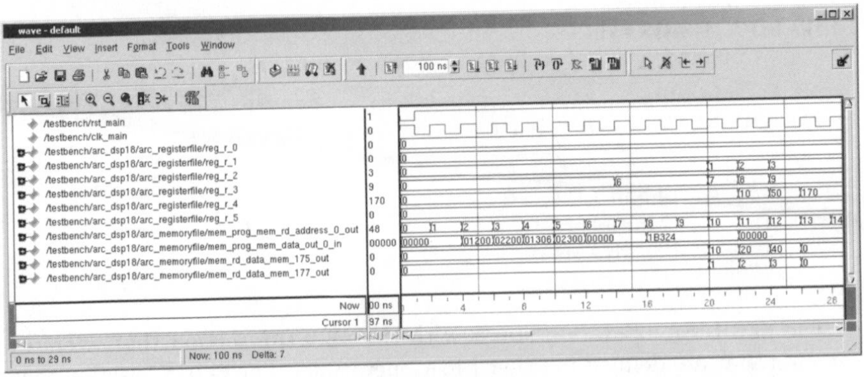

Fig. 9.29. DSP18 testbench for MAC operation.

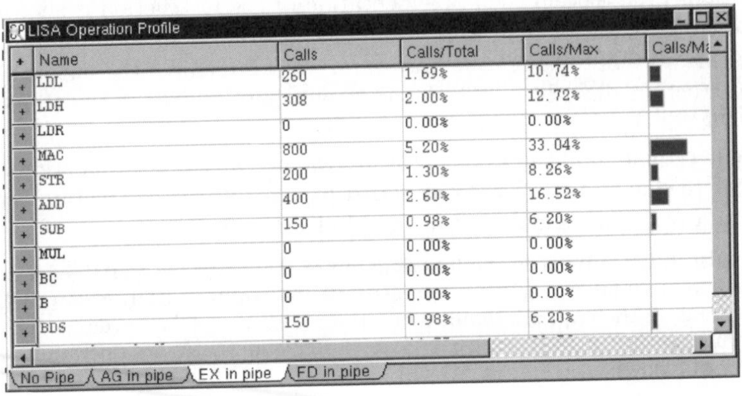

Name	Calls	Calls/Total	Calls/Max
LDL	260	1.69%	10.74%
LDH	308	2.00%	12.72%
LDR	0	0.00%	0.00%
MAC	800	5.20%	33.04%
STR	200	1.30%	8.26%
ADD	400	2.60%	16.52%
SUB	150	0.98%	6.20%
MUL	0	0.00%	0.00%
BC	0	0.00%	0.00%
B	0	0.00%	0.00%
BDS	150	0.98%	6.20%

Fig. 9.30. DSP18 operation profile for 100-points length-8 dual-channel DWT including memory initialization, i.e., 300 instructions.

LISA true vector processor. General-purpose CPUs have improved in recent years by exploring instruction-level parallelism (ILP), adding on-chip cache and floating-point units, speculative branch execution, and improved speed etc. One particular problem that occurs now is that the logic to track dependencies between all in-flight instructions grows quadratically with the number of instructions [287]. As a result these improvements have considerably slowed down since 2002 and the use of multiple CPUs on the same die is now favored instead of increasing clock speed. This requires code to be written for parallel processors, which may be less efficient than using a vector processor to start with. Vector processors were successfully commercialized long before ILP machines and use an alternative approach to controlling multiple function units with deep pipelines. Vector processors like Cray, NEC, or Fujitsu VP100 provide high-level instructions that work on vectors, i.e., a linear array of numbers. Usually vector processor are characterized by using:

- a vector array with dedicated load/store unit

Table 9.17. DSP18 synthesis result for the XC3S1000-4ft256 Spartan-3 Xilinx device.

Parameter	DSP18 with CLB-based RAM	DSP18 with BlockRAM
Slices	3145	2679
4-input LUT	6053	5183
Multiplier	2	2
BlockRAMs	0	2
Total gates	81,509	177,203
Clock period	25.542 ns	19.565 ns
F_{max}	39.15 MHz	51.11 MHz

- functional unit that are highly pipelined
- hazard control is minimized
- support of vector instruction, that replace a complete loop by a single instruction

Figure 9.31 shows an experimental vector processor that is a vector extension of the popular MIPS machine called VMIPS. VMIPS was introduced in 2001 has eight vector registers each with 64 elements, one load/store, and five arithmetic units, one lane, and runs at 500 MHz.

The second DSPstone benchmark: $d[k] = a[k] \times b[k]; 0 \le k \le N$, for instance, would be implemented in VMIPS by

```
MULV. D        V1,V2,V3
```

i.e., multiply the elements of the vectors V2 and V3 and put each result in the vector register V1.

However, as can be seen from Fig. 9.31, the typically implemented vector processor architecture only looks to a programmer as a vector machine. Inside the vector processor we may typically find eight vector registers where each vectors has 32 to 1024 (Fujitsu VP100) elements but usually only one floating-point arithmetic unit for each operation. A vector multiply or add, like in DSPstone2, still requires N clock cycles (not counting the initialization). Multiple lanes that allow more than one floating-point operation per clock cycle are limited. In the last 30 years of vector processor history only two machines (the NEC SX/5 from 1998 and the Fujitsu VPP5000 introduced in 1999) have had over 10 lanes, but the quotient of register elements to lanes is still only 3% for the 512-elements-per-vector NEC SX/5 with 16 lanes. The reason that typically VPs do not have more than one lane is that the floating-point units in 64 bits need a large die area.

Another weakness of current vector processors is their limited usefulness for DSP operations. In DSPs we not only need a vector multiply, instead more often we need a inner product computation, i.e.,

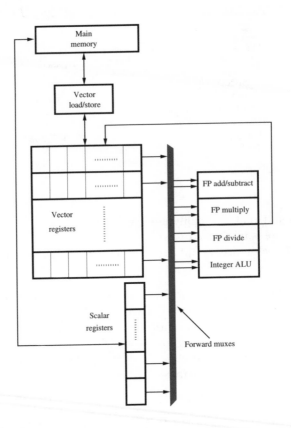

Fig. 9.31. The VMIPS vector processor.

$$\mathbf{X} \times \mathbf{Y} = \sum_{k=0}^{N-1} X[k] \times Y[k]. \tag{9.7}$$

While the multiplication can be done in a vector element-by-element parallel fashion the summation requires the addition of all products in an adder tree. This is usually not supported with vector instructions. A third operation that is not supported in most vector processors is the (cyclic) shift of the vector register elements. For instance, if an FIR application requires the vector elements $x[0] \ldots x[N-1]$, then in the next step the elements $x[1] \ldots x[N]$ are needed. A PDSP uses cyclic addressing to address this issue. In a vector processor it is usually necessary to reload the complete vector.

In an FPGA design we can therefore improve the processing by

- Adding the vector shift instructions VSXY and VSXY to our instruction set to load two words from data memory, shift the two vector registers of the data or coefficients, and place the two new values in the first location.

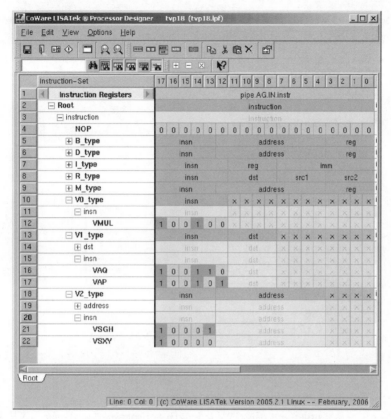

Fig. 9.32. The true vector processor (TVP) instruction set additions.

- Since modern FPGA can have up to 512 embedded multipliers, we can implement as many multipliers as vector elements are in a vector. A VMUL instruction will perform 2×8 multiplications and place the products in the two product vector registers P and Q.
- Implement (inner product) vector sum instructions VAP and VAQ that adds up all elements in a (product) register vector.

We call such a machine a true vector processor (TVP) since vector operations like vector multiply are no longer translated into a sequence of single multiplies – all operations are done in parallel. For a two-channel length-8 wavelet processor we would therefore require 16 embedded multipliers. The Spartan-3 device XC3S1000-4ft256 that is used in the low-cost Nexys Digilent university boards, for instance, has 24 embedded 18×18-bit multipliers available, more than enough for our TVP.

The inner product sum may be a concern in terms of speed since here horizontal $L - 1$ additions need to be performed for a vector register with L

elements. But we can perform the additions on a binary adder tree, as the following LISA code examples shows for the VAP instruction

```
/* Vector scalar add of all P register */
  OPERATION VAP IN pipe.EX {
  DECLARE
    {
      REFERENCE dst;
    }
  CODING { 0b100101 }
  SYNTAX { "VAP" }
  BEHAVIOR
    {short t1,t2,t3,t4,t5,t6,t7;
      t1 = P[0] + P[1];
      t2 = P[2] + P[3];
      t3 = P[4] + P[5];
      t4 = P[6] + P[7];
      t5 = t1 + t2;
      t6 = t3 + t4;
      t7 = t5 + t6;
      dst = t7;
    }
}
```

which reduces the worst-case delay from seven adds to three.

If we now implement the DWT length-8 processor with the TVP ISA we find that the inner loop is much shorter, i.e., only nine instructions. The downsampling by 2 of the DWT requires two shifts of the vectors X and Y for each new output sample.

```
_loop:
          VSXY  R[2],R[3]
          VSXY  R[2],R[3]
          VMUL
          VAP  R[4]
          VAQ  R[5]
          STR R[4], R[6]
          STR R[5], R[6]
          BDS @_loop, R[7]
          ; next instruction is in the branch delay slot
          SUB R[7],R[7],R[1]
```

The overall instruction count when compared with the DSP18 is further decreased, as can be seen from the profile shown in Fig. 9.34. The total instruction count for TVP is only 479 instructions.

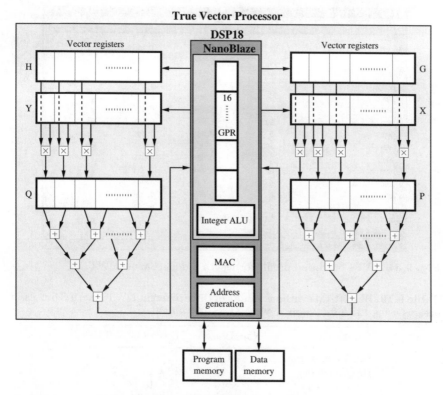

Fig. 9.33. The true vector processor (TVP) architecture.

Table 9.18. TVP synthesis result for the Xilinx device XC3S1000-4ft256.

Parameter	μP only	with BlockRAM
Slices	4907	4993
4-input LUT	8850	9226
Multiplier	18	18
BlockRAMs	0	2
Total gates	141,158	274,463
Clock Period	22.082 ns	20.799
F_{max}	45.3 MHz	48.1 MHz

From Table 9.18 we notice the larger resource requirements and lower maximum operation frequency.

Final LISA processor design comparison. Let us finally compare all three designs in terms of size, speed, and overall throughput in MSPS for a DWT length-8 example. The key synthesis properties are summarized in

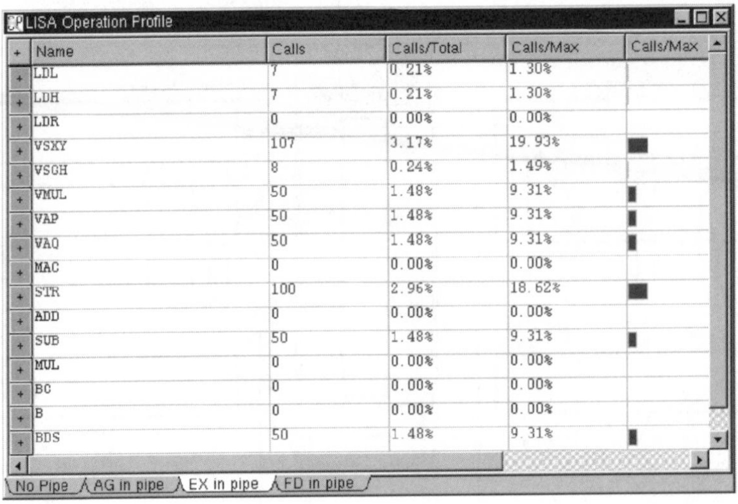

Fig. 9.34. TVP1 operation profile for length-8 dual-channel DWT.

Table 9.19. BlockRAM synthesis results of three different DWT length-8 processor designs using LISA for Xilinx device: XC3S1000-4ft256

Parameter	NanoBlaze	DSP18	TVP
LISA operations	28	32	40
Prog. memory	$2^7 \times 18$	$2^7 \times 18$	$2^7 \times 18$
Data memory	$2^8 \times 16$	$2^8 \times 16$	$2^8 \times 16$
BRAMs	2	2	2
Gates	162,471	177,203	274,463
MHz	73.9	51.11	48.1

Table 9.19. The device used is a Spartan-3 XC3S1000-4ft256 as in the Nexys Digilent university boards (see http://www.digilentinc.com/), with 7680 slices, 15360 four-input LUTs, 24 embedded multiplier, and 24 BlockRAMs with 18 Kbit each, see Fig. 1.11, p. 18. When data or program memory are implemented with CLB-based RAM (called distributed RAM by Xilinx) then about 800 four-input LUTs are required for a $2^8 \times 16$ and about 120 four-input LUTs for a $2^7 \times 18$ memory.

The overall performance of the three processors is measured by the mega samples per second (MSPS) throughput when implementing a length-8 DWT as shown in Fig. 5.55, p. 314. We need two length-8 filter $g[n]$ and $h[n]$ and for each output sample pair 16 multiply and 14 add operations are computed. For 100 samples with an output downsampling by 2 the arithmetic requirements for the DWT filter band would therefore be $8 \times 100 = 800$ multiplications and $7 \times 100 = 700$ additions, or 800 MAC calls. From the NanoBlaze instruction profile however we see that many addition cycles for LDR and ADD are required,

Table 9.20. Comparison of three different DWT length-8 processor designs using LISA for the Xilinx Spartan-3 device XC3S1000-4ft256.

Parameter	NanoBlaze	DSP18	TVP
LDL	261	260	7
LDH	309	308	7
LDR	1600	0	0
VSXY	–	–	107
VSGH	–	–	8
VMUL	–	–	50
VAP	–	–	50
VAQ	–	–	50
MAC	–	800	0
STR	100	100	100
ADD	2750	400	0
SUB	–	50	50
MUL	800	0	0
BC	0	0	0
BC	0	0	0
BDS	50	50	50
Total	5870	1968	479
Clock	73.9	51.1	48.1
MSPS	1.26	2.60	10.0

due to the fact that the register updates for memory access are also computed with the general-purpose ALU. The DSP18 processor reduces the LDR and ADD essentially, and although the maximum clock frequency is decreased, the overall throughput is improved by a factor of 2. If we use a true vector processor, then an inner product can be computed in two clock cycles and the overall throughput is improved by a factor of 8 compared with NanoBlaze and a factor of 4 when compared with a single-core MAC DSP18 design.

Finally, the performance data of the three LISA based processors and a direct RNS polyphase implementation (4217 LEs, 155 MSPS, [338]) are compared in Fig. 9.35. We see the large improvement from NanoBlaze to TVP, but still a direct mapping into hardware is another magnitude faster than any microprocessor solution. The hardware architecture however can only implement one configuration, while the TVP software architecture can implement many different algorithms.

9.5.3 Nios FFT Design

As a Nios design example we will now study a DIF radix-2 FFT [164] using the custom instruction feature for the butterfly processor implementation. The discrete Fourier transform for an N-point input signal $x[n]$ has been discussed in Chap. 6, see (6.2), p. 344, i.e.,

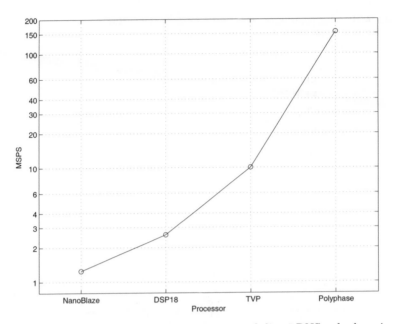

Fig. 9.35. Comparison of LISA-based processor and direct RNS polyphase implementation.

$$X[k] = \sum_{n=0}^{N-1} x[n] \mathrm{e}^{-\mathrm{j}2\pi kn/N} = \sum_{n=0}^{N-1} x[n] W_N^{kn}. \qquad (9.8)$$

For the DIF radix-2 FFT algorithm, decimation is used to decompose the N-point DFT into successively smaller DFT. The decimation process for an N-point input sequence is carried out $\log_2(N)$ times. Each decimation step rearranges the input sequence into even and odd indexed sequences. The total number of complex multiplications is therefore reduced to $(N/2)\log_2(N)$. Fig. 9.36a illustrates the signal flow graph of the eight-point decimation-in-frequency radix-2 FFT algorithm, showing the different stages, groups, and butterflies. The basic computation performed at each stage is called a butterfly, as shown in Fig. 9.36b. In this algorithm, the input is in normal order while the DFT output is in bitreverse order.

To evaluate the custom instruction feature of the Nios processor, the DIF radix-2 FFT algorithm described above is implemented in two steps:

- Software implementation
- Software implementation with custom instruction for butterfly processor

Software implementation. Software implementation uses the characteristics of the algorithm illustrated in Table 9.21. As the Gnupro compiler provided with the Nios development board, APEX edition, supports C programs,

Table 9.21. Characteristics of the DIF radix-2 algorithm.

Stage number k	Stage 0	Stage 1	Stage 2	...	Stage $\log_2(N) - 1$
Number of groups per stage $(p = 2^k)$	1	2	4	...	$N/2$
Butterflies per group $(t = N/2p)$	$N/2$	$N/4$	$N/8$	...	1
Increment exponent twiddle-factors	1	2	4	...	$N/2$

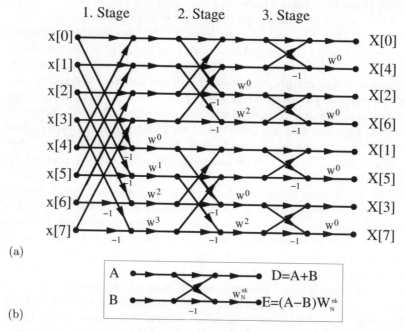

(a)

(b)

Fig. 9.36. DIF FFT. (a) Signal flow graph length-8. (b) Butterfly computation.

the software code for the algorithm is written in the C language [332, 329, 302]. The first row gives the stage number for a length-N FFT and is designed as the outermost loop. The second row gives the number of groups based on the stage number and is considered the second loop in the algorithm implementation. The third row gives the number of butterflies per group based on the group number and forms the innermost loop. The final row shows how the twiddle-factors increment based on a particular stage. The hardware effort for the Nios processor depending on the multiplier implemented without a custom logic block ranges from 2701 to 3163 logic cells is shown in the second column of Table 9.22.

Table 9.22. Hardware effort of Nios processor with different multiplier options and butterfly custom logic.

Multiplier option	Standard processor	Custom processor
Software	2701 LCs	4414 LCs
MSTEP	2716 LCs	4433 LCs
MUL	3163 LCs	4841 LCs

Creation of custom logic block. For the custom implementation of a butterfly processor, the custom instruction features of the Nios processor must be considered [333]. The custom logic block in the Nios processor connects directly to the ALU of the Nios processor as a reconfigurable functional unit and therefore provides an interface with predefined ports and names present in the processor. The predefined physical ports and names that could be used for custom logic design are shown in Fig. 9.37a. The number of custom logic blocks that can be implemented in the Nios embedded processor system is restricted to five, but with the presence of an 11-bit prefix port up to 2048 functions for each block can be performed. In the butterfly processor custom instruction design, the custom logic block for the butterfly processor is written in VHDL [339].

Chapter 6 describes an efficient butterfly processor code `bfproc.vhd` for 8-bit data values, see Example 6.12, p. 370. In this code, the processor is implemented with one adder, one subtraction, and a component instantiation for twiddle-factor multiplier. The twiddle-factor multiplication is efficiently computed using component instantiations of three `lpm_mult` and three `lpm_add_sub` modules. The output of the twiddle-factor multiplier is scaled such that it has the same data format as the input. The algorithm used for the twiddle-factor multiplier uses three coefficients of the twiddle-factor, $C, C + S$, and $C - S$, where C and S are the real and imaginary coefficients of the twiddle-factor. The complex twiddle-factor multiplication $R + j \times I = (X + j \times Y)(C + j \times S)$ is efficiently performed when these coefficients are used. The real and imaginary parts of the complex twiddle-factor multiplication using these three coefficients can be computed as $R = (C - S) \times Y + C \times (X - Y)$ and $I = (C + S) \times X - C \times (X - Y)$, respectively. To ensure short latency for in-place FFT computation, the complex multiplier is implemented without pipeline stages. The butterfly processor is designed to compute scaled outputs where the output produced by the design is equal to half the actual output value. The butterfly processor de-

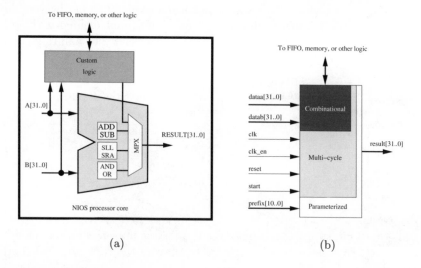

(a)   (b)

Fig. 9.37. (a) Adding custom logic to the Nios ALU. (b) Physical ports for the custom logic block.

sign uses flip-flops for the input, coefficient, and the output data to have a single-input/output registered design.

For custom implementation, this design is modified to use the predefined physical ports for multicycle logic shown in Fig. 9.37b. The parameterized prefix port is used in the design to define various read and write functions required to read and write the complex input and output data present in the butterfly computation. A total of eight prefix-defined functions are required for proper implementation of the design. The design is compiled with a EPF20k200EFC484-2X (APEX device) present in the Nios development board using the Quartus II software. It requires about 1700 LEs on the APEX device. The performance ($f_{\max}$) of the design is found to be 30.82 MHz. The simulation of the design shows that it requires two clock cycles for valid implementation of the design.

Instantiation of custom logic block. The custom logic block is instantiated with the aid of the Nios CPU configuration wizard, which is a part of SOPC builder (system integration tool) present in the Quartus II software available with the Nios development kit. The only input for this instantiation process is the number of clock cycles required for valid implementation of the custom logic block created. For the butterfly processor custom logic block, the number of clock cycles required is eight, based on the simulation result. The instantiation process includes the custom logic block with the Nios ALU and generates a software macro for this block known as a custom instruction. The new Nios processor with the custom logic block is then recompiled and downloaded to the APEX FPLD device in the Nios development board. The

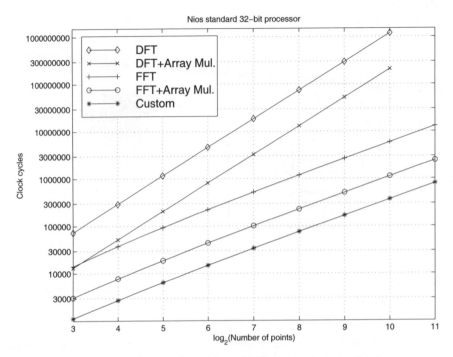

Fig. 9.38. Number of clock cycles with and without an array multiplier.

entire design requires, depending on the multiplier used, between 4414 and 4841 LEs of the APEX device, see the third column in Table 9.22. The downloaded design acts as a platform for software implementation using custom instruction. The butterfly computation in software could then be performed using the software macro functions where the function of the software macro is defined by the prefix port values. Finally the software implementation of the DIF radix-2 FFT is modified by using custom instructions for the butterfly computation.

Code optimization for custom logic design. A first direct approach to code the DIF FFT is to use a standard program like the FORTRAN code by Burrus and Parks [164], convert the code to the language C and introduce the custom instructions. Such a code [336] will look like:

```
1: dwStartTick=GetTickCount();        // Record Start Time
2: S = log10(N)/log10(2);             // Number of stages
3: Stages for (k=0; k<S; k++)
4: Loop
5:   { p = 1<<k;                      // Number of groups
6:     t = N/(p << 1);            // Number of butterflies
7:     f = t << 1;                         // in each Group
8:     for (j=0; j<p; j++)                   // Group Loop
```

```
 9:   { I = f * j;           // Jump to each group in a stage
10:       for (i=0; i<t; i++)                // Butterfly Loop
11:       { a = I + t;                 // Butterfly calculation
12:         Are = (*(xr+I) << 1); Aim = (*(xi+I) << 1);
13:         /***Custom Instructions**/
14:         nm_bfpr_pfx(1,Are,Aim);    // Read Are and Aim
15:         Bre = (*(xr+a) << 1); Bim = (*(xi+a) << 1);
16:         nm_bfpr_pfx(2,Bre,Bim);    // Read Bre and Bim
17:         Wr = *(wnr+(p*i)); Wi = *(wni+(p*i));
18:         nm_bfpr_pfx(3,Wr,0);               // Read C
19:         W1 = Wr + Wi; W2 = Wr - Wi;
20:         nm_bfpr_pfx(4,W1,W2);      // Read C+S and C-S
21:         *(xr+I) = nm_bfpr_pfx(5,0,0);      // Write real
22:         *(xi+I) = nm_bfpr_pfx(6,0,0); // Write imaginary
23:         *(xr+a) = nm_bfpr_pfx(7,0,0); // Write the real
24:         *(xi+a) = nm_bfpr_pfx(8,0,0); // Write imaginary
25:         I = I + 1;                 // Jump to each butterfly
26:   } } }                           // in a Group
27: lTicksUsed=GetTickCount();        // Record end time
```

Several different DIF Radix-2 FFT versions were tested and the following list shows the most successful changes:

0) Initial version using direct C conversion from FORTRAN [336], but with $\log_{10}()$ computation for the number of stages outside the FFT routine, i.e., a LUT. Now the FFTs with custom instructions should not depend on the Nios multiplier type.

1) Improved software code: no multiplications or divides in the index computations and moving code to avoid multiplies or divides, see, for instance, listing line 5.

2) COS/SIN load outside the butterfly, i.e., computes all groups with the same twiddle-factor, rather then running through a whole group.

3) Uses a complex data type, i.e., real and imaginary parts are stored in memory next to each other. This improves memory access, which was verified by analyzing the generated assembler code.

These different methods can improve the number of clock cycles, depending on the multiplier type, between 17% and 61% for 256-point FFTs, as shown in Table 9.23.

On the VHDL hardware side additional improvements were implemented as follows:

0) Original version from the 2/e Springer book DSP with FPGAs [57] adjusted in MS project [336]:
   • change the I/O ports (dataa, datab result, etc.) as required by Nios
   • change to 32-bit-width I/O and internal 18×18-bit multipliers

1) Removal of second pipeline from CCMUL

Table 9.23. Speed increase for 256-point FFT with software-only improvements for different multiplier options.

Multiplier option	Clock cycles software multiplier	Clock cycles MSTEP multiplier	Clock cycles array multiplier
Original 0	1345627	771295	331836
Improved 1	1236345	695673	281765
Improved 2	1228978	688306	268772
Improved 3	518953	643423	227663
Gain	61%	17%	31%

- CMUL included in the `bfp2.vhd` code
- No `lpm`, only `STD_LOGIC` functions for add and multiply
- Move COS+/−SIN to the FPGA side from software (removing one custom instruction)
- No scaling by 2 in the hardware → remove (`<<1`) in software

2) Further VHDL optimizations:
   - simplify result sign extension
   - Reset is asynchronous
   - Start is used as the enable for the flip-flops

Table 9.24. Speed increase for 256-point FFT with software and hardware improvements.

	Clock cycles array multiplier	Gain
Original software	331836	
Custom instructions	181747	45%
Software only	135615	59%
Hardware only	113054	65%
Software+hardware	75516	77%

The overall gain using the custom instruction and further optimization is in the range of 45% − 65%. Together with a careful software development a gain of 77% for a 256-point FFT is observed, see Table 9.24. The following code shows the optimized C-code. Both hardware as well as software modifications are present in the code.

```
1: dwStartTick=GetTickCount();
2: k2 = N; dw = 1;
3: for (l = 1; l <= S; l++)
4: {k1 = k2; k2 >>= 1; w = 0;
```

```
 6:  for (k = 0; k < k2; k++) {
 7:    Wr = coef[w].r;
 8:    Wi = coef[w].i;
 9:    nm_bfp3_pfx(3,Wr,Wi);              // Read COS+SIN
10:    w += dw;
11:    for (i1 = k; i1 < N; i1 += k1) {
12:      i2 = i1 + k2;
13:      /***Custom Instructions**/
14:      tr = x[i1].r; ti = x[i1].i;
15:      nm_bfp3_pfx(1,tr,ti);           // Read Are and Aim
16:      tr = x[i2].r; ti = x[i2].i;
17:      nm_bfp3_pfx(2,tr,ti);           // Read Bre and Bim
18:      x[i1].r = nm_bfp3_pfx(4,0,0);  // Write real
19:      x[i1].i = nm_bfp3_pfx(5,0,0);  // Write imaginary
20:      x[i2].r = nm_bfp3_pfx(6,0,0);  // Write real
21:      x[i2].i = nm_bfp3_pfx(7,0,0);  // Write imaginary
22:    }
23:  }
24:  dw <<= 1;
25: }
```

Nios FFT performance results. By measuring the number of clock cycles required for custom implementation and software-only implementation of the butterfly processor, the speed relation between the two can be measured with different multiplier optimizations, as shown in Table 9.25. The clock cycle measurements are taken based on the Gnupro C/C++ compiler available with the Nios development board. The increase in speed from custom implementation for a single butterfly computation is given by

$$\text{Speed increase} = \frac{\text{Clock cycles } \textit{software} \text{ only design}}{\text{Clock cycles for custom design}} \tag{9.9}$$

Table 9.25. Speed increase for a single butterfly computation.

Multiplier option	Clock cycles with software	Clock cycles with custom instruction	Improvement factor
Software	1227	119	10.3
MSTEP	698	119	5.8
MUL	295	119	2.47

Furthermore, the performance of FFTs of different length using custom implementation and software-only implementation of the butterfly processor can be compared. Figures 9.38 and 9.39 show this comparison along with reference DFT data using the direct computation as in Equation (9.8), p. 626.

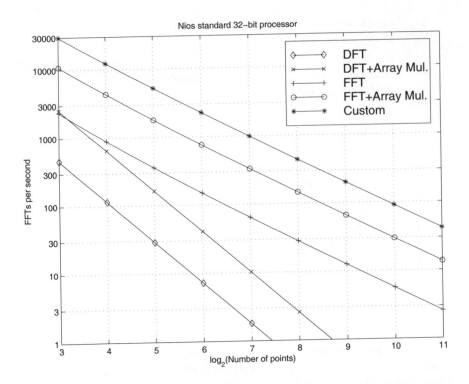

Fig. 9.39. Number of DFT/FFT per second with and without array multiplier.

It is observed that the overall performance decreases with increasing length of the FFT for each implementation while the performance of the custom implementation for each of the FFTs computed is higher than the software-only implementation. Of all the multiplier optimizations the MUL optimization gives the best overall performance for both implementations. Apart from the performance calculations, it is observed that custom implementation results in a small additional quantization error in the output values, which increases with increasing length of the FFT.

Exercises

Note: If you have no prior experience with the Quartus II software, refer to the case study found in Sect. 1.4.3, p. 29. If not otherwise noted use the EP2C35F672C6 from the Cyclone II family for the Quartus II synthesis evaluations.

9.1: Using the following categories

(A) Applications software
(B) Systems software
(C) High-level programming language
(D) Personal computer
(E) Minicomputer/workstation

(F) Output device
(G) Input device
(H) Semiconductor
(I) Integrated circuit
(J) Supercomputer

classify the following examples by putting the answers in the provided brackets []
to the right.

Device driver [   ]
Wafer [   ]
MS Word [   ]
Microphone [   ]
CD-ROM [   ]

GCC [   ]
Pentium Pro [   ]
SRAM [   ]
SUN [   ]
Metal layer [   ]

LCD display [   ]
copy (DOS) or cp (UNIX) [   ]
Internet browser [   ]
Loudspeaker [   ]
dir (DOS) or ls (UNIX)[   ]

9.2: Repeat Exercise 9.1 for the following items:

Spreadsheet [   ]
DEC Alpha [   ]
Operating system [   ]
Microprocessor [   ]
Cathode ray tube display [   ]

Mouse [   ]
Cray-1 [   ]
DRAM [   ]
PowerPC [   ]
Printer [   ]

Silicon [   ]
Macintosh [   ]
Pascal [   ]
Compiler [   ]
Assembler [   ]

9.3: Consider a machine with three instructions (ADD/MUL/DIV) with the following
clock cycles per instruction (CPI) data:

Instruction class CPI for this instruction class
ADD = 1, MUL = 1, DIV= 1.

We have measured the code for the same program for two different compilers
and obtained the following data:

Code from	Instruction counts for each class		
	ADD	MUL	DIV
Compiler 1	25×10^9	5×10^9	5×10^9
Compiler 2	47×10^9	18×10^9	5×10^9

Assume that the machine's clock rate is 200 MHz. Compute execution time and
MIPS rate = instruction count /(execution time $\times 10^6$) for the two compilers using
the following steps. Compute:
(a) CPU clock cycles for compiler 1 =
(b) Execution time using compiler 1 =
(c) MIPS rate using compiler 1 =
(d) CPU clock cycles for compiler 2 =
(e) Execution time using compiler 2 =
(f) MIPS rate using compiler 2 =
(g) Which compiler is better in terms of MIPS?
(h) Which compiler is better in terms of execution time?

9.4: Repeat Exercise 9.3 for the following CPI data: ADD=1 MUL=5 DIV=8 and a machine clock rate of 250 MHz. We have measured the code for the same program for
two different compilers and obtained the following data:

Code from	Instruction counts for each class		
	ADD	MUL	DIV
Compiler 1	35×10^9	10×10^9	5×10^9
Compiler 2	60×10^9	5×10^9	5×10^9

Answer the questions (a)-(h).

9.5: Compare zero- to three-address machines by writing programs to compute

$$g = (a * b - c)/(d * e - f) \tag{9.10}$$

for the following instruction sets:
(a) Stack: PUSH Op1, POP Op1, ADD, SUB, MUL, DIV.
(b) Accumulator LA Op1, STA Op1, ADD Op1, SUB Op1, MUL Op1, DIV Op1. All arithmetic operation use as the second operand the accumulator and the result is stored in the accumulator, see (9.4), p. 553
(c) Two-operand LT Op1, M; ST Op1, M; ADD Op1, Op2; SUB Op1, Op2; MUL Op1, Op2; DIV Op1, Op2. All arithmetic operations use two operands according to (9.5), p. 555.
(c) Three-operand LT Op1, M; ST Op1, M; ADD Op1, Op2, Op3; SUB Op1, Op2, Op3; MUL Op1, Op2, Op3; DIV Op1, Op2, Op3. All arithmetic operations use three operands according to (9.6), p. 555.

9.6: Repeat Exercise 9.5 for the following arithmetic expression:

$$f = (a - b)/(c + d \times e). \tag{9.11}$$

You may rearrange the expression if necessary.

9.7: Repeat Exercise 9.5 for the following arithmetic expression:

$$h = (a - b)/((c + d) * (f - g)). \tag{9.12}$$

You may use additional temporary variables if necessary.

9.8: Convert the following infix arithmetic expressions to postfix expressions.
(a) $a + b - c - d + e$
(b) $(a + b) * c$
(c) $a + b * c$
(d) $(a + b)^{\hat{}}(c - d)$ ($\hat{}$ power-of sign)
(e) $(a - b) \times (c + d) + e$
(f) $a \times b + c \times d - e \times f$
(g) $(a - b) \times (((c - d \times e) \times f)/g) \times h$

9.9: For the arithmetic expression from Exercise 9.8 generate the assembler code for a stack machine using the program c2asm.exe from the book CD.

9.10: Convert the following postfix notation to infix.
(a) $abc + /$
(b) $ab + cd - /$
(c) $ab - c + d\times$
(d) $ab/cd \times -$
(e) $abcde + \times \times /$
(f) $ab + c^{\hat{}}$ (with $\hat{}$ power-of sign)
(g) $abcde/f \times -g + h/ \times +$

9.11: Which of the following pairs of postfix expression are equivalent?
(a) $ab + c+$ and $abc + +$
(b) $ab - c-$ and $abc - -$
(c) $ab * c*$ and $abc * *$
(d) $ab\hat{}c\hat{}$ and $abc\hat{}\,\hat{}$ (with $\hat{}$ power-of sign)
(e) $ab \times c+$ and $cab \times +$
(f) $ab \times c+$ and $abc + \times$
(g) $abc + \times$ and $ab \times bc \times +$

9.12: The URISC machine suggested by Parhami [306] has a single instruction only, which performs the following: subtract operand 1 from operand 2 and replace operand 2 with the result, then jump to the target address if the result was negative. The instructions are of the form `urisc op1,op2,offset`, where offset specifies the next instruction offset, which is 1 most of the time. Develop URISC code for the following functions:
(a) `clear (src)`.
(b) `dest = (src1)+(src2)`.
(c) `exchange (src1) and (src2)`.
(d) `goto label`.
(e) `if (src1>=src2), goto label`.
(f) `if (src1=src2), goto label`.

9.13: Extend the lexical VHDL analysis `vhdlex.l` (see p. 571) to include pattern matching for:
(a) Bits `'0'` and `'1'`
(b) Bit vectors `"0/1..0/1"`
(c) Data type definition like `BIT` and `STD_LOGIC`
(d) All keywords found in HDL code `example`
(e) All keywords found in HDL code `sqrt`

9.14: Write a lexical VHDL analysis `float.l` that recognizes floating-point numbers of the type
(a) 1.5
(b) 1.5e10
(c) 1.5e10 and integers, e.g., 5

9.15: Write a lexical VHDL analysis `vhdlwc.l` that counts character, VHDL key words, other words, and lines, similar to the UNIX commands `wc`.

9.16: Write a simple lexical natural-language analyzer. Your scanner should classify each word into verbs and others. Example session:

```
user:  do you like vhdl
lexer: do: is a verb
lexer: you: is not a verb
lexer: like: is a verb
lexer: vhdl: is not a verb
```

9.17: Extend the natural-language lexical analyzer from Exercise 9.16 so that the scanner classifies the items into verb, adverb, preposition, conjunction, adjective, and pronoun. Examples: is, very, to, if, their, and I, respectively.

9.18: Extend the natural-language lexical analyzer from Exercise 9.16 by adding a symbol table that allows you to add new words and its type to your dictionary. The type definition has to start in the first column. Example session:

```
user:   verb is am run
user:   noun dog cat
user:   dog run
lexer:  dog: noun
lexer:  run: verb
```

9.19: Develop a Bison grammar nlp that finds a valid sentence of the form subject
VERB object, where subject can be of type NOUN or PRONOUN and object is of type
NOUN. For the lexical analysis use a similar scanner to that developed in Exercise
9.18. If you run the sentence parser, print out a statement if the sentence is valid
or invalid. Example session:

```
user:   verb like enjoy hate
user:   noun vhdl verilog
user:   pronoun I you she he
user:   I like vhdl
nlp:    Sentence is valid
user:   you like she
nlp:    Sentence is invalid
```

9.20: Add to the Bison grammar of add2.y for float number type the grammar
rules for
(a) Basic math -,*,/
(b) Trigonometric functions, e.g., sin, cos, arctan
(c) The operations sqrt, ln, and log.
Example session:

```
user: sqrt(4*4+3*3)
calc: 5.0
user: arctan(1)
calc: 0.7853
user: log(1000)
calc: 3.0
```

9.21: Rewrite the Bison grammar of calc.y (keep calc.1) to have a reverse Polish or postfix rpcalc.y calculator. The calculator should support the operations
+,-,*,/, and ^. Verify your calculator with the following example session:

```
user:   1 3 +
rpcalc: 4
user:   1 3 + 5 * 2 ^
rpcalc: 256
user:   (5*9)/(6*5-3*3*3)
rpcalc: 15
```

9.22: Write a Bison grammar to analyze arithmetic expression and output three-
address code. Use temporary variables to store intermediate results. At the end
print the operation code and the symbol table. Example session:

```
r=(a*b-c)/(d*e-f);
-- Result from quads:

Intermediate code:
Quadruples          3AC
Op Dst Op1 Op2
(*,  4,   2,   3)       T1 <= a * b
```

```
(-,   6,   4,   5)      T2 <= T1 - c
(*,   9,   7,   8)      T3 <= d * e
(-,  11,   9,  10)      T4 <= T3 - f
(/,  12,   6,  11)      T5 <= T2 / T4
(=,   1,  12,  --)      r  <= T5
```

Symbol table:
```
 1 : r
 2 : a
 3 : b
 4 : T1
 5 : c
 6 : T2
 7 : d
 8 : e
 9 : T3
10 : f
11 : T4
12 : T5
```

9.23: Determine the equivalent assembler code and the values of the registers $t0, $t1, and $t2 for the C program below. Assume that the C language variables A, B, and C have the following register assignments: A=$t0, B=$t1, and C=$t2. Use − if the register value is unknown. Do not leave blank fields in the table. Use ADD, SUB, ADDI, or the SRL (shift right logical) instructions. Do not use SLL.

Step	C-code	Assembler instruction	$t0	$t1	$t2
1	A = 48;				
2	B= 2 * A;				
3	C = 10;				
4	C = C / 4;				
5	A = A / 16;				

Instruction formats:

Addition (with overflow)	ADD	rd, rs, rt
Subtract (with overflow)	SUB	rd, rs, rt
Addition immediate (with overflow)	ADDI	rd, rs, imm
Shift right logical	SRL	rd, rs, shamt

9.24: Repeat Exercise 9.23 for the following instructions:

Step	C-code	Assembler instruction	$t0	$t1	$t2
1	B = 96;				
1	A= 2*B;				
3	C = 20;				
4	C = C / 8;				
5	B = B / 32;				

9.25: Many RISC machines do not have a special instruction to reset a register. Show using add, addi, or substract instructions how to reset the register $t0. Register $zero and the immediate value 0 should be used. Fill in the blanks in the following code.

(a) ADD   # Compute t0 = register with zero + register with zero
(b) ADDI   # Compute t0 = register with zero + 0
(c) SUB   # Compute t0 = t0 - t0;

9.26: Many RISC machines do not have a special move instruction. Show using add, addi, or substract instructions how to move the register $s0 to $t0, i.e., set $t0 to $s0. Register $zero and the immediate value 0 should be used. Fill in the blanks in the following code.

(a) ADD   # Compute t0 = s0 + register with zero
(b) ADDI   # Compute t0 = s0 + 0
(c) SUB   # Compute t0 = s0 - 0

9.27: Determine the equivalent C-code and the values of the registers $t0, $t1, and $t2 for the assembler program below. Assume that the C language variables a, b, and c have the following register assignments: a=$t0, b=$t1, and c=$t2. Use − if the register value is unknown. Do not leave blank fields in the table.

Step	Instruction	C-code	$t0	$t1	$t2
1	ADDI $t2, $zero, 32				
2	SRL $t1, $t2, 3				
3	ADDI $t0, $zero, 2				
4	SUB $t2, $zero, $t0				
5	SLL $t0, $t0, 5				

9.28: Repeat Exercise 9.27 for the following instructions:

Step	Instruction	C-code	$t0	$t1	$t2
1	ORI $t0, $zero, 4				
2	SLL $t0, $t0, 2				
3	SUB $t1, $zero, $t0				
4	ORI $t2, $zero, 64				
5	SRL $t1, $t2, 3				

9.29: Determine the cache contents for the memory access sequence 2,1,2,5,1 for the following three caches with four blocks.
(a) Direct mapped cache

Address	Cache contents			
	Hit or miss	0	1	2	3
2				
1				
2				
5				
1				

(b) Fully associative (starting with first unused location)

Address	Cache contents			
	Hit or miss 0 1 2 3			
2				
1				
2				
5				
1				

(c) Two-way set associative (replacing the least recently used)

Addr.	Cache contents							
Hit or miss	Set 0	Flag	Set 0	Flag	Set 1	Flag	Set 1	Flag
2								
1								
2								
5								
1								

9.30: Repeat Exercise 9.29 for the following memory access sequence: 2,6,3,2,3.

9.31: For a cache with 8 KB of data and with a 32-bit data/address word width compute the total memory size using the following steps:
(a) How many words are there in the cache?
(b) How many tag bits are there in the cache?
(c) What is the total size of the cache?
(d) Compute the overhead in percentage of the cache.

9.32: Repeat Exercise 9.31 with 4 KB of data and a 32-bit data/address word width.

9.33: In Example 9.7 (p. 603) the MicroBlaze cache was discussed. Determine the number of BlockRAMs for the following MicroBlaze data: main memory 64 KB; cache size 4 KB; words per line=8.
(a) The number of BlockRAMs to store the data
(b) The number of BlockRAMs to store the tags
(c) Using the data from (a) and (b), determine the maximum main memory size that can be addressed with this configuration.

9.34: In Example 9.7 (p. 603) the MicroBlaze cache was discussed. Determine the number of BlockRAMs for the following MicroBlaze data: main memory 16 KB; cache size 2 kB; words per line=4.
(a) The number of BlockRAMs to store the data
(b) The number of BlockRAMs to store the tags
(c) Using the data from (a) and (b), determine the maximum main memory size that can be addressed with this configuration.

9.35: (a) Design the PREP benchmark 7 (which is equivalent to benchmark 8) shown in Fig. 9.40a with the Quartus II software. The design is a 16-bit binary up-counter. It has an asynchronous reset **rst**, an active-high clock enable **ce**, an active-high load signal **fd**, and 16-bit data input **d[15..0]**. The registers are positive-edge triggered via `clk`, see the simulation in Fig. 9.40c for the function test. The following table summarizes the functions:

clk	rst	ld	ce	q[15..0]
X	0	X	X	0000
∫	1	1	X	d[15..0]
∫	1	0	0	No change
∫	1	0	1	Increment

(b) Determine the `Registered Performance` and the used resources (LEs, multipliers, and M4Ks) for a single copy. Compile the HDL file with the synthesis optimization technique set to `Speed`, `Balanced` or `Area` as found in the `Analysis &` `Synthesis Settings` section under `EDA Tool Settings` in the `Assignments` menu. Which synthesis options are optimal in terms of size and `Registered Performance`? Select one of the following devices:

(b1) EP2C35F672C6 from the Cyclone II family
(b2) EPF10K70RC240-4 from the Flex 10K family
(b3) EPM7128LC84-7 from the MAX7000S family

(c) Design the multiple instantiation for benchmark 7, as shown in Fig. 9.40b.

(d) Determine the `Registered Performance` and the used resources (LEs, multipliers, and M2Ks/M4Ks) for the design with the maximum number of instantiations of PREP benchmark 7. Use the optimal synthesis option you found in (b) for the following devices:

(d1) EP2C35F672C6 from the Cyclone II family
(d2) EPF10K70RC240-4 from the Flex 10K family
(d3) EPM7128LC84-7 from the MAX7000S family

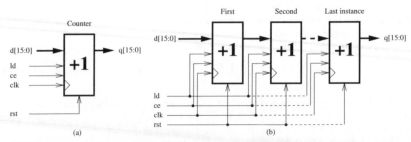

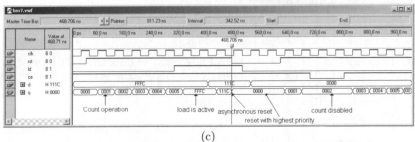

Fig. 9.40. PREP benchmark 7 and 8. **(a)** Single design. **(b)** Multiple instantiation. **(c)** Testbench to check the function.

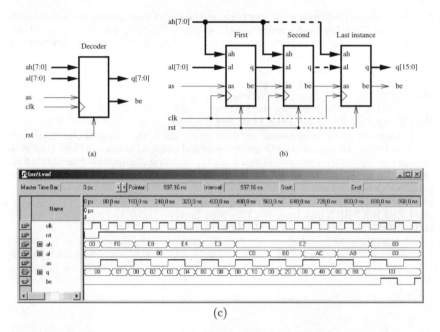

Fig. 9.41. PREP benchmark 9. **(a)** Single design. **(b)** Multiple instantiation. **(c)** Testbench to check the function.

9.36: **(a)** Design the PREP benchmark 9 shown in Fig. 9.41a with the Quartus II software. The design is a memory decoder common in microprocessor systems. The addresses are decoded only when the address strobe **as** is active. Addresses that fall outside the decoder activate a bus error **be** signal. The design has a 16-bit input **a[15..0]**, an asynchronous active-low reset **rst** and all flip-flops are positive-edge triggered via **clk**. The following table summarizes the behavior:

rst	as	clk	A (hex)	q[7..0] (binary)	be
0	X	X	X	00000000	0
1	0	∫	X	00000000	0
1	1	0	X	q[7..0]	be
1	1	∫	FFFF to F000	00000001	0
1	1	∫	EFFF to E800	00000010	0
1	1	∫	E7FF to E400	00000100	0
1	1	∫	E3FF to E300	00001000	0
1	1	∫	E2FF to E2C0	00010000	0
1	1	∫	E2BF to E2B0	00100000	0
1	1	∫	E2AF to E2AC	01000000	0
1	1	∫	E2AA	10000000	0
1	1	∫	E2AA to 0000	00000000	1

Where X is don't care. Try to match the simulation in Fig. 9.41c for the function test. Note that the original **be** definition requires be stored **be** for **as=f**, but the simulation shows **be** differently. The coding matches the simulation rather then the original truth table.

(b) Determine the `Registered Performance` and the used resources (LEs, multipliers, and M2Ks/M4Ks) for a single copy. Compile the HDL file with the synthesis optimization technique set to `Speed`, `Balanced` or `Area` as found in the `Analysis & Synthesis Settings` section under `EDA Tool Settings` in the `Assignments` menu. Which synthesis options are optimal in terms of size and `Registered Performance`? Select one of the following devices:

(b1) EP2C35F672C6 from the Cyclone II family

(b2) EPF10K70RC240-4 from the Flex 10K family

(b3) EPM7128LC84-7 from the MAX7000S family

(c) Design the multiple instantiation for benchmark 9, as shown in Fig. 9.41b.

(d) Determine the `Registered Performance` and the used resources (LEs, multipliers, and M2Ks/M4Ks) for the design with the maximum number of instantiations of PREP benchmark 9. Use the optimal synthesis option you found in (b) for the following devices:

(d1) EP2C35F672C6 from the Cyclone II family

(d2) EPF10K70RC240-4 from the Flex 10K family

(d3) EPM7128LC84-7 from the MAX7000S family

References

1. B. Dipert: "EDN's first annual PLD directory," *EDN*, pp. 54–84 (2000). http://www.ednmag.com/ednmag/reg/2000/08172000/17cs.htm
2. S. Brown, Z. Vranesic: *Fundamentals of Digital Logic with VHDL Design* (McGraw-Hill, New York, 1999)
3. D. Smith: *HDL Chip Design* (Doone Publications, Madison, Alabama, USA, 1996)
4. U. Meyer-Bäse: *The Use of Complex Algorithm in the Realization of Universal Sampling Receiver using FPGAs (in German)* (VDI/Springer, Düsseldorf, 1995), Vol. 10, no. 404, 215 pages
5. U. Meyer-Bäse: *Fast Digital Signal Processing (in German)* (Springer, Heidelberg, 1999), 370 pages
6. P. Lapsley, J. Bier, A. Shoham, E. Lee: *DSP Processor Fundamentals* (IEEE Press, New York, 1997)
7. D. Shear: "EDN's DSP Benchmarks," *EDN* **33**, 126–148 (1988)
8. J. Donovan: (2002), "The truth about 300 mm," http://www.eet.com
9. Plessey: (1990), "Datasheet," ERA60100
10. J. Greene, E. Hamdy, S. Beal: "Antifuse Field Programmable Gate Arrays," *Proceedings of the IEEE*, pp. 1042–56 (1993)
11. J. Rose, A. Gamal, A. Sangiovanni-Vincentelli: "Architecture of Field-Programmable Gate Arrays," *Proceedings of the IEEE*, pp. 1013–29 (1993)
12. Xilinx: "PREP Benchmark Observations," in *Xilinx-Seminar*, San Jose (1993)
13. Altera: "PREP Benchmarks Reveal FLEX 8000 is Biggest, MAX 7000 is Fastest," in *Altera News & Views*, San Jose (1993)
14. Actel: "PREP Benchmarks Confirm Cost Effectiveness of Field Programmable Gate Arrays," in *Actel-Seminar* (1993)
15. E. Lee: "Programmable DSP Architectures: Part I," *IEEE Transactions on Acoustics, Speech and Signal Processing Magazine*, pp. 4–19 (1988)
16. E. Lee: "Programmable DSP Architectures: Part II," *IEEE Transactions on Acoustics, Speech and Signal Processing Magazine* pp. 4–14 (1989)
17. R. Petersen, B. Hutchings: "An Assessment of the Suitability of FPGA-Based Systems for Use in Digital Signal Processing," *Lecture Notes in Computer Science* **975**, 293–302 (1995)
18. J. Villasenor, B. Hutchings: "The Flexibility of Configurable Computing," *IEEE Signal Processing Magazine* pp. 67–84 (1998)
19. Xilinx: (1993), "Data book," XC2000, XC3000 and XC4000
20. Altera: (1996), "Datasheet," FLEX 10K CPLD family
21. Altera: (2005), "Cyclone II Device Handbook," Vol. 1
22. F. Vahid: *Embedded System Design* (Prentice Hall, Englewood Cliffs, New Jersey, 1990)
23. J. Hakewill: "Gainin Control over Silicon IP," *Communication Design*, online (2000)

24. E. Castillo, U. Meyer-Baese, L. Parrilla, A. Garcia, A. Lloris: "Watermarking Strategies for RNS-Based System Intellectual Property Protection," in *Proc. of 2005 IEEE Workshop on Signal Processing Systems SiPS'05* Athens (2005), pp. 160–165

25. O. Spaniol: *Computer Arithmetic: Logic and Design* (John Wiley & Sons, New York, 1981)

26. I. Koren: *Computer Arithmetic Algorithms* (Prentice Hall, Englewood Cliffs, New Jersey, 1993)

27. E.E. Swartzlander: *Computer Arithmetic, Vol. I* (Dowden, Hutchingon and Ross, Inc., Stroudsburg, Pennsylvania, 1980), also reprinted by IEEE Computer Society Press 1990

28. E. Swartzlander: *Computer Arithmetic, Vol. II* (IEEE Computer Society Press, Stroudsburg, Pennsylvania, 1990)

29. K. Hwang: *Computer Arithmetic: Principles, Architecture and Design* (John Wiley & Sons, New York, 1979)

30. N. Takagi, H. Yasuura, S. Yajima: "High Speed VLSI multiplication algorithm with a redundant binary addition tree," *IEEE Transactions on Computers* **34** (2) (1985)

31. D. Bull, D. Horrocks: "Reduced-Complexity Digital Filtering Structures using Primitive Operations," *Electronics Letters* pp. 769–771 (1987)

32. D. Bull, D. Horrocks: "Primitive operator digital filters," *IEE Proceedings-G* **138**, 401–411 (1991)

33. A. Dempster, M. Macleod: "Use of Minimum-Adder Multiplier Blocks in FIR Digital Filters," *IEEE Transactions on Circuits and Systems II* **42**, 569–577 (1995)

34. A. Dempster, M. Macleod: "Comments on "Minimum Number of Adders for Implementing a Multiplier and Its Application to the Design of Multiplierless Digital Filters"," *IEEE Transactions on Circuits and Systems II* **45**, 242–243 (1998)

35. F. Taylor, R. Gill, J. Joseph, J. Radke: "A 20 Bit Logarithmic Number System Processor," *IEEE Transactions on Computers* **37** (2) (1988)

36. P. Lee: "An FPGA Prototype for a Multiplierless FIR Filter Built Using the Logarithmic Number System," *Lecture Notes in Computer Science* **975**, 303–310 (1995)

37. J. Mitchell: "Computer multiplication and division using binary logarithms," *IRE Transactions on Electronic Computers* **EC-11**, 512–517 (1962)

38. N. Szabo, R. Tanaka: *Residue Arithmetic and its Applications to Computer Technology* (McGraw-Hill, New York, 1967)

39. M. Soderstrand, W. Jenkins, G. Jullien, F. Taylor: *Residue Number System Arithmetic: Modern Applications in Digital Signal Processing*, IEEE Press Reprint Series (IEEE Press, New York, 1986)

40. U. Meyer-Bäse, A. Meyer-Bäse, J. Mellott, F. Taylor: "A Fast Modified CORDIC-Implementation of Radial Basis Neural Networks," *Journal of VLSI Signal Processing* pp. 211–218 (1998)

41. V. Hamann, M. Sprachmann: "Fast Residual Arithmetics with FPGAs," in *Proceedings of the Workshop on Design Methodologies for Microelectronics* Smolenice Castle, Slovakia (1995), pp. 253–255

42. G. Jullien: "Residue Number Scaling and Other Operations Using ROM Arrays," *IEEE Transactions on Communications* **27**, 325–336 (1978)

43. M. Griffin, M. Sousa, F. Taylor: "Efficient Scaling in the Residue Number System," in *IEEE International Conference on Acoustics, Speech, and Signal Processing* (1989), pp. 1075–1078

44. G. Zelniker, F. Taylor: "A Reduced-Complexity Finite Field ALU," *IEEE Transactions on Circuits and Systems* **38** (12), 1571–1573 (1991)
45. IEEE: "Standard for Binary Floating-Point Arithmetic," *IEEE Std 754-1985* pp. 1–14 (1985)
46. IEEE: "A Proposed Standard for Binary Floating-Point Arithmetic," *IEEE Transactions on Computers* **14** (12), 51–62 (1981). Task P754
47. N. Shirazi, P. Athanas, A. Abbott: "Implementation of a 2-D Fast Fourier Transform on an FPGA-Based Custom Computing Machine," *Lecture Notes in Computer Science* **975**, 282–292 (1995)
48. M. Bayoumi, G. Jullien, W. Miller: "A VLSI Implementation of Residue Adders," *IEEE Transactions on Circuits and Systems* pp. 284–288 (1987)
49. A. Garcia, U. Meyer-Bäse, F. Taylor: "Pipelined Hogenauer CIC Filters using Field-Programmable Logic and Residue Number System," in *IEEE International Conference on Acoustics, Speech, and Signal Processing* Vol. 5 (1998), pp. 3085–3088
50. L. Turner, P. Graumann, S. Gibb: "Bit-serial FIR Filters with CSD Coefficients for FPGAs," *Lecture Notes in Computer Science* **975**, 311–320 (1995)
51. J. Logan: "A Square-Summing, High-Speed Multiplier," *Computer Design* pp. 67–70 (1971)
52. Leibowitz: "A Simplified Binary Arithmetic for the Fermat Number Transform," *IEEE Transactions on Acoustics, Speech and Signal Processing* **24**, 356–359 (1976)
53. T. Chen: "A Binary Multiplication Scheme Based on Squaring," *IEEE Transactions on Computers* pp. 678–680 (1971)
54. E. Johnson: "A Digital Quarter Square Multiplier," *IEEE Transactions on Computers* pp. 258–260 (1980)
55. Altera: (2004), "Implementing Multipliers in FPGA Devices," application note 306, Ver. 3.0
56. D. Anderson, J. Earle, R. GOldschmidt, D. Powers: "The IBM System/360 Model 91: Floating-Point Execution Unit," *IBM Journal of Research and Development* **11**, 34–53 (1967)
57. U. Meyer-Baese: *Digital Signal Processing with Field Programmable Gate Arrays*, 2nd edn. (Springer-Verlag, Berlin, 2004), 527 pages
58. A. Croisier, D. Esteban, M. Levilion, V. Rizo: (1973), "Digital Filter for PCM Encoded Signals," US patent no. 3777130
59. A. Peled, B. Liu: "A New Realization of Digital Filters," *IEEE Transactions on Acoustics, Speech and Signal Processing* **22** (6), 456–462 (1974)
60. K. Yiu: "On Sign-Bit Assignment for a Vector Multiplier," *Proceedings of the IEEE* **64**, 372–373 (1976)
61. K. Kammeyer: "Quantization Error on the Distributed Arithmetic," *IEEE Transactions on Circuits and Systems* **24** (12), 681–689 (1981)
62. F. Taylor: "An Analysis of the Distributed-Arithmetic Digital Filter," *IEEE Transactions on Acoustics, Speech and Signal Processing* **35** (5), 1165–1170 (1986)
63. S. White: "Applications of Distributed Arithmetic to Digital Signal Processing: A Tutorial Review," *IEEE Transactions on Acoustics, Speech and Signal Processing Magazine*, 4–19 (1989)
64. K. Kammeyer: "Digital Filter Realization in Distributed Arithmetic," in *Proc. European Conf. on Circuit Theory and Design* (1976), Genoa, Italy
65. F. Taylor: *Digital Filter Design Handbook* (Marcel Dekker, New York, 1983)
66. H. Nussbaumer: *Fast Fourier Transform and Convolution Algorithms* (Springer, Heidelberg, 1990)
67. H. Schmid: *Decimal Computation* (John Wiley & Sons, New York, 1974)

68. Y. Hu: "CORDIC-Based VLSI Architectures for Digital Signal Processing," *IEEE Signal Processing Magazine* pp. 16–35 (1992)
69. U. Meyer-Bäse, A. Meyer-Bäse, W. Hilberg: "**CO**ordinate **R**otation **DI**gital Computer (CORDIC) Synthesis for FPGA," *Lecture Notes in Computer Science* **849**, 397–408 (1994)
70. J.E. Volder: "The CORDIC Trigonometric computing technique," *IRE Transactions on Electronics Computers* **8** (3), 330–4 (1959)
71. J. Walther: "A Unified algorithm for elementary functions," *Spring Joint Computer Conference* pp. 379–385 (1971)
72. X. Hu, R. Huber, S. Bass: "Expanding the Range of Convergence of the CORDIC Algorithm," *IEEE Transactions on Computers* **40** (1), 13–21 (1991)
73. D. Timmermann (1990): "CORDIC-Algorithmen, Architekturen und monolithische Realisierungen mit Anwendungen in der Bildverarbeitung," Ph.D. thesis, VDI Press, Serie 10, No. 152
74. H. Hahn (1991): "Untersuchung und Integration von Berechnungsverfahren elementarer Funktionen auf CORDIC-Basis mit Anwendungen in der adaptiven Signalverarbeitung," Ph.D. thesis, VDI Press, Serie 9, No. 125
75. G. Ma (1989): "A Systolic Distributed Arithmetic Computing Machine for Digital Signal Processing and Linear Algebra Applications," Ph.D. thesis, University of Florida, Gainesville
76. Y.H. Hu: "The Quantization Effects of the CORDIC-Algorithm," *IEEE Transactions on Signal Processing* pp. 834–844 (1992)
77. M. Abramowitz, A. Stegun: *Handbook of Mathematical Functions*, 9th edn. (Dover Publications, Inc., New York, 1970)
78. W. Press, W. Teukolsky, W. Vetterling, B. Flannery: *Numerical Recipes in C*, 2nd edn. (Cambridge University Press, Cambrige, 1992)
79. A.V. Oppenheim, R.W. Schafer: *Discrete-Time Signal Processing* (Prentice Hall, Englewood Cliffs, New Jersey, 1992)
80. D.J. Goodman, M.J. Carey: "Nine Digital Filters for Decimation and Interpolation," *IEEE Transactions on Acoustics, Speech and Signal Processing* pp. 121–126 (1977)
81. U. Meyer-Baese, J. Chen, C. Chang, A. Dempster: "A Comparison of Pipelined RAG-n and DA FPGA-Based Multiplierless Filters," in *IEEE Asia Pacific Conference on Circuits and Systems, APCCAS 2006.* (2006), pp. 1555–1558
82. O. Gustafsson, A. Dempster, L. Wanhammer: "Extended Results for Minimum-Adder Constant Integer Multipliers," in *IEEE International Conference on Acoustics, Speech, and Signal Processing* Phoenix (2002), pp. 73–76
83. Y. Wang, K. Roy: "CSDC: A New Complexity Reduction Technique for Multiplierless Implementation of Digital FIR Filters," *IEEE Transactions on Circuits and Systems I* **52** (0), 1845–1852 (2005)
84. H. Samueli: "An Improved Search Algorithm for the Design of Multiplierless FIR Filters with Powers-of-Two Coefficients," *IEEE Transactions on Circuits and Systems* **36** (7), 1044-1047 (1989)
85. Y. Lim, S. Parker: "Discrete Coefficient FIR Digital Filter Design Based Upon an LMS Criteria," *IEEE Transactions on Circuits and Systems* **36** (10), 723–739 (1983)
86. Altera: (2004), "FIR Compiler: MegaCore Function User Guide," ver. 3.1.0
87. R. Hartley: "Subexpression Sharing in Filters Using Canonic Signed Digital Multiplier," *IEEE Transactions on Circuits and Systems II* **30** (10), 677–688 (1996)
88. R. Saal: *Handbook of Filter Design* (AEG-Telefunken, Frankfurt, Germany, 1979)

89. C. Barnes, A. Fam: "Minimum Norm Recursive Digital Filters that Are Free of Overflow Limit Cycles," *IEEE Transactions on Circuits and Systems* pp. 569–574 (1977)
90. A. Fettweis: "Wave Digital Filters: Theorie and Practice," *Proceedings of the IEEE* pp. 270–327 (1986)
91. R. Crochiere, A. Oppenheim: "Analysis of Linear Digital Networks," *Proceedings of the IEEE* **63** (4), 581–595 (1995)
92. A. Dempster, M. Macleod: "Multiplier blocks and complexity of IIR structures," *Electronics Letters* **30** (22), 1841–1842 (1994)
93. A. Dempster, M. Macleod: "IIR Digital Filter Design Using Minimum Adder Multiplier Blocks," *IEEE Transactions on Circuits and Systems II* **45**, 761–763 (1998)
94. A. Dempster, M. Macleod: "Constant Integer Multiplication using Minimum Adders," *IEE Proceedings - Circuits, Devices & Systems* **141**, 407–413 (1994)
95. K. Parhi, D. Messerschmidt: "Pipeline Interleaving and Parallelism in Recursive Digital Filters - Part I: Pipelining Using Scattered Look-Ahead and Decomposition," *IEEE Transactions on Acoustics, Speech and Signal Processing* **37** (7), 1099–1117 (1989)
96. H. Loomis, B. Sinha: "High Speed Recursive Digital Filter Realization," *Circuits, Systems, Signal Processing* **3** (3), 267–294 (1984)
97. M. Soderstrand, A. de la Serna, H. Loomis: "New Approach to Clustered Look-ahead Pipelined IIR Digital Filters," *IEEE Transactions on Circuits and Systems II* **42** (4), 269–274 (1995)
98. J. Living, B. Al-Hashimi: "Mixed Arithmetic Architecture: A Solution to the Iteration Bound for Resource Efficient FPGA and CPLD Recursive Digital Filters," in *IEEE International Symposium on Circuits and Systems* Vol. I (1999), pp. 478–481
99. H. Martinez, T. Parks: "A Class of Infinite-Duration Impulse Response Digital Filters for Sampling Rate Reduction," *IEEE Transactions on Acoustics, Speech and Signal Processing* **26** (4), 154–162 (1979)
100. K. Parhi, D. Messerschmidt: "Pipeline Interleaving and Parallelism in Recursive Digital Filters - Part II: Pipelined Incremental Block Filtering," *IEEE Transactions on Acoustics, Speech and Signal Processing* **37** (7), 1118–1134 (1989)
101. M. Shajaan, J. Sorensen: "Time-Area Efficient Multiplier-Free Recursive Filter Architectures for FPGA Implementation," in *IEEE International Conference on Acoustics, Speech, and Signal Processing* (1996), pp. 3269–3272
102. P. Vaidyanathan: *Multirate Systems and Filter Banks* (Prentice Hall, Englewood Cliffs, New Jersey, 1993)
103. S. Winograd: "On Computing the Discrete Fourier Transform," *Mathematics of Computation* **32**, 175–199 (1978)
104. Z. Mou, P. Duhamel: "Short-Length FIR Filters and Their Use in Fast Nonrecursive Filtering," *IEEE Transactions on Signal Processing* **39**, 1322–1332 (1991)
105. P. Balla, A. Antoniou, S. Morgera: "Higher Radix Aperiodic-Convolution Algorithms," *IEEE Transactions on Acoustics, Speech and Signal Processing* **34** (1), 60–68 (1986)
106. E.B. Hogenauer: "An Economical Class of Digital Filters for Decimation and Interpolation," *IEEE Transactions on Acoustics, Speech and Signal Processing* **29** (2), 155–162 (1981)
107. Harris: (1992), "Datasheet," HSP43220 Decimating Digital Filter
108. Motorola: (1989), "Datasheet," DSPADC16 16–Bit Sigma–Delta Analog-to-Digital Converter

109. O. Six (1996): "Design and Implementation of a Xilinx universal XC-4000 FPGAs board," Master's thesis, Institute for Data Technics, Darmstadt University of Technology
110. S. Dworak (1996): "Design and Realization of a new Class of Frequency Sampling Filters for Speech Processing using FPGAs," Master's thesis, Institute for Data Technics, Darmstadt University of Technology
111. L. Wang, W. Hsieh, T. Truong: "A Fast Computation of 2-D Cubic-Spline Interpolation," *IEEE Signal Processing Letters* **11** (9), 768–771 (2004)
112. T. Laakso, V. Valimaki, M. Karjalainen, U. Laine: "Splitting the Unit Delay," *IEEE Signal Processing Magazine* **13** (1), 30–60 (1996)
113. M. Unser: "Splines: a Perfect Fit for Signal and Image Processing," *IEEE Signal Processing Magazine* **16** (6), 22–38 (1999)
114. S. Cucchi, F. Desinan, G. Parladori, G. Sicuranza: "DSP Implementation of Arbitrary Sampling Frequency Conversion for High Quality Sound Application," in *IEEE International Symposium on Circuits and Systems* Vol. 5 (1991), pp. 3609–3612
115. C. Farrow: "A Continuously Variable Digital Delay Element," in *IEEE International Symposium on Circuits and Systems* Vol. 3 (1988), pp. 2641–2645
116. S. Mitra: *Digital Signal Processing: A Computer-Based Approach*, 3rd edn. (McGraw Hill, Boston, 2006)
117. S. Dooley, R. Stewart, T. Durrani: "Fast On-line B-spline Interpolation," *IEE Electronics Letters* **35** (14), 1130–1131 (1999)
118. Altera: "Farrow-Based Decimating Sample Rate Converter," in *Altera Application Note AN-347* San Jose (2004)
119. F. Harris: "Performance and Design Considerations of the Farrow Filter when used for Arbitrary Resampling of Sampled Time Series," in *Conference Record of the Thirty-First Asilomar Conference on Signals, Systems & Computers* Vol. 2 (1997), pp. 1745–1749
120. M. Unser, A. Aldroubi, M. Eden: "B-spline Signal Processing: I – Theory," *IEEE Transactions on Signal Processing* **41** (2), 821–833 (1993)
121. P. Vaidyanathan: "Generalizations of the Sampling Theorem: Seven Decades after Nyquist," *Circuits and Systems I: Fundamental Theory and Applications* **48** (9), 1094–1109 (2001)
122. Z. Mihajlovic, A. Goluban, M. Zagar: "Frequency Domain Analysis of B-spline Interpolation," in *Proceedings of the IEEE International Symposium on Industrial Electronics* Vol. 1 (1999), pp. 193–198
123. M. Unser, A. Aldroubi, M. Eden: "Fast B-spline Transforms for Continuous Image Representation and Interpolation," *IEEE Transactions on Pattern Analysis and Machine Intelligence* **13** (3), 277–285 (1991)
124. M. Unser, A. Aldroubi, M. Eden: "B-spline Signal Processing: II – Efficiency Design and Applications," *IEEE Transactions on Signal Processing* **41** (2), 834–848 (1993)
125. M. Unser, M. Eden: "FIR Approximations of Inverse Filters and Perfect Reconstruction Filter Banks," *Signal Processing* **36** (2), 163–174 (1994)
126. T. Blu, P. Thévenaz, M. Unser: "MOMS: Maximal-Order Interpolation of Minimal Support," *IEEE Transactions on Image Processing* **10** (7), 1069–1080 (2001)
127. T. Blu, P. Thévenaz, M. Unser: "High-Quality Causal Interpolation for On-line Unidimenional Signal Processing," in *Proceedings of the Twelfth European Signal Processing Conference (EUSIPCO'04)* (2004), pp. 1417–1420
128. A. Gotchev, J. Vesma, T. Saramäki, K. Egiazarian: "Modified B-Spline Functions for Efficient Image Interpolation," in *First IEEE Balkan Conference on*

Signal Processing, Communications, Circuits, and Systems (2000), pp. 241–244

129. W. Hawkins: "FFT Interpolation for Arbitrary Factors: a Comparison to Cubic Spline Interpolation and Linear Interpolation," in *Proceedings IEEE Nuclear Science Symposium and Medical Imaging Conference* Vol. 3 (1994), pp. 1433–1437

130. A. Haar: "Zur Theorie der orthogonalen Funktionensysteme," *Mathematische Annalen* **69**, 331–371 (1910). Dissertation Göttingen 1909

131. W. Sweldens: "The Lifting Scheme: A New Philosophy in Biorthogonal Wavelet Constructions," in *SPIE, Wavelet Applications in Signal and Image Processing III* (1995), pp. 68–79

132. C. Herley, M. Vetterli: "Wavelets and Recursive Filter Banks," *IEEE Transactions on Signal Processing* **41**, 2536–2556 (1993)

133. I. Daubechies: *Ten Lectures on Wavelets* (Society for Industrial and Applied Mathematics (SIAM), Philadelphia, 1992)

134. I. Daubechies, W. Sweldens: "Factoring Wavelet Transforms into Lifting Steps," *The Journal of Fourier Analysis and Applications* **4**, 365–374 (1998)

135. G. Strang, T. Nguyen: *Wavelets and Filter Banks* (Wellesley-Cambridge Press, Wellesley MA, 1996)

136. D. Esteban, C. Galand: "Applications of Quadrature Mirror Filters to Split Band Voice Coding Schemes," in *IEEE International Conference on Acoustics, Speech, and Signal Processing* (1977), pp. 191–195

137. M. Smith, T. Barnwell: "Exact Reconstruction Techniques for Tree-Structured Subband Coders," *IEEE Transactions on Acoustics, Speech and Signal Processing* pp. 434–441 (1986)

138. M. Vetterli, J. Kovacevic: *Wavelets and Subband Coding* (Prentice Hall, Englewood Cliffs, New Jersey, 1995)

139. R. Crochiere, L. Rabiner: *Multirate Digital Signal Processing* (Prentice Hall, Englewood Cliffs, New Jersey, 1983)

140. M. Acheroy, J.M. Mangen, Y. Buhler.: "Progressive Wavelet Algorithm versus JPEG for the Compression of METEOSAT Data," in *SPIE, San Diego* (1995)

141. T. Ebrahimi, M. Kunt: "Image Compression by Gabor Expansion," *Optical Engineering* **30**, 873–880 (1991)

142. D. Gabor: "Theory of communication," *J. Inst. Elect. Eng. (London)* **93**, 429–457 (1946)

143. A. Grossmann, J. Morlet: "Decomposition of Hardy Functions into Square Integrable Wavelets of Constant Shape," *SIAM J. Math. Anal.* **15**, 723–736 (1984)

144. U. Meyer-Bäse: "High Speed Implementation of Gabor and Morlet Wavelet Filterbanks using RNS Frequency Sampling Filters," in *Aerosense 98 SPIE, Orlando* (1998), Vol. 3391, pp. 522–533

145. U. Meyer-Bäse: "Die Hutlets – eine biorthogonale Wavelet-Familie: Effiziente Realisierung durch multipliziererfreie, perfekt rekonstruierende Quadratur Mirror Filter," *Frequenz* pp. 39–49 (1997)

146. U. Meyer-Bäse, F. Taylor: "The Hutlets - a Biorthogonal Wavelet Family and their High Speed Implementation with RNS, Multiplier-free, Perfect Reconstruction QMF," in *Aerosense 97 SPIE, Orlando* (1997), Vol. 3078, pp. 670–681

147. M. Heideman, D. Johnson, C. Burrus: "Gauss and the History of the Fast Fourier Transform," *IEEE Transactions on Acoustics, Speech and Signal Processing Magazine* **34**, 265–267 (1985)

148. C. Burrus: "Index Mappings for Multidimensional Formulation of the DFT and Convolution," *IEEE Transactions on Acoustics, Speech and Signal Processing* **25**, 239–242 (1977)
149. B. Baas: (1998), "SPIFFEE an energy-efficient single-chip 1024-point FFT processor," http://www-star.stanford.edu/ bbaas/fftinfo.html
150. G. Sunada, J. Jin, M. Berzins, T. Chen: "COBRA: An 1.2 Million Transistor Exandable Column FFT Chip," in *Proceedings of the International Conference on Computer Design: VLSI in Computers and Processors* (IEEE Computer Society Press, Los Alamitos, CA, USA, 1994), pp. 546–550
151. TMS: (1996), "TM-66 swiFFT Chip," Texas Memory Systems
152. SHARP: (1997), "BDSP9124 Digital Signal Processor," http://www.butterflydsp.com
153. J. Mellott (1997): "Long Instruction Word Computer," Ph.D. thesis, University of Florida, Gainesville
154. P. Lavoie: "A High-Speed CMOS Implementation of the Winograd Fourier Transform Algorithm," *IEEE Transactions on Signal Processing* **44** (8), 2121–2126 (1996)
155. G. Panneerselvam, P. Graumann, L. Turner: "Implementation of Fast Fourier Transforms and Discrete Cosine Transforms in FPGAs," in *Lecture Notes in Computer Science* Vol. 1142 (1996), pp. 1142:272–281
156. Altera: "Fast Fourier Transform," in *Solution Brief 12*, Altera Corparation (1997)
157. G. Goslin: "Using Xilinx FPGAs to Design Custom Digital Signal Processing Devices," in *Proceedings of the DSP^X* (1995), pp. 595–604
158. C. Dick: "Computing 2-D DFTs Using FPGAs," *Lecture Notes in Computer Science: Field-Programmable Logic* pp. 96–105 (1996)
159. S.D. Stearns, D.R. Hush: *Digital Signal Analysis* (Prentice Hall, Englewood Cliffs, New Jersey, 1990)
160. K. Kammeyer, K. Kroschel: *Digitale Signalverarbeitung* (Teubner Studienbücher, Stuttgart, 1989)
161. E. Brigham: *FFT*, 3rd edn. (Oldenbourg Verlag, München Wien, 1987)
162. R. Ramirez: *The FFT: Fundamentals and Concepts* (Prentice Hall, Englewood Cliffs, New Jersey, 1985)
163. R.E. Blahut: *Theory and practice of error control codes* (Addison-Wesley, Melo Park, California, 1984)
164. C. Burrus, T. Parks: *DFT/FFT and Convolution Algorithms* (John Wiley & Sons, New York, 1985)
165. D. Elliott, K. Rao: *Fast Transforms: Algorithms, Analyses, Applications* (Academic Press, New York, 1982)
166. A. Nuttall: "Some Windows with Very Good Sidelobe Behavior," *IEEE Transactions on Acoustics, Speech and Signal Processing* **ASSP-29** (1), 84–91 (1981)
167. U. Meyer-Bäse, K. Damm (1988): "Fast Fourier Transform using Signal Processor," Master's thesis, Department of Information Science, Darmstadt University of Technology
168. M. Narasimha, K. Shenoi, A. Peterson: "Quadratic Residues: Application to Chirp Filters and Discrete Fourier Transforms," in *IEEE International Conference on Acoustics, Speech, and Signal Processing* (1976), pp. 12–14
169. C. Rader: "Discrete Fourier Transform when the Number of Data Samples is Prime," *Proceedings of the IEEE* **56**, 1107–8 (1968)
170. J. McClellan, C. Rader: *Number Theory in Digital Signal Processing* (Prentice Hall, Englewood Cliffs, New Jersey, 1979)

171. I. Good: "The Relationship between Two Fast Fourier Transforms," *IEEE Transactions on Computers* **20**, 310–317 (1971)

172. L. Thomas: "Using a Computer to Solve Problems in Physics," in *Applications of Digital Computers* (Ginn, Dordrecht, 1963)

173. A. Dandalis, V. Prasanna: "Fast Parallel Implementation of DFT Using Configurable Devices," *Lecture Notes in Computer Science* **1304**, 314–323 (1997)

174. U. Meyer-Bäse, S. Wolf, J. Mellott, F. Taylor: "High Performance Implementation of Convolution on a Multi FPGA Board using NTT's defined over the Eisenstein Residuen Number System," in *Aerosense 97 SPIE, Orlando* (1997), Vol. 3068, pp. 431–442

175. Xilinx: (2000), "High-Performance 256-Point Complex FFT/IFFT," product specification

176. Altera: (2004), "FFT: MegaCore Function User Guide," ver. 2.1.3

177. Z. Wang: "Fast Algorithms for the Discrete W transfrom and for the discrete Fourier Transform," *IEEE Transactions on Acoustics, Speech and Signal Processing* pp. 803–816 (1984)

178. M. Narasimha, A. Peterson: "On the Computation of the Discrete Cosine Transform," *IEEE Transaction on Communications* **26** (6), 934–936 (1978)

179. K. Rao, P. Yip: *Discrete Cosine Transform* (Academic Press, San Diego, CA, 1990)

180. B. Lee: "A New Algorithm to Compute the Discrete Cosine Transform," *IEEE Transactions on Acoustics, Speech and Signal Processing* **32** (6), 1243–1245 (1984)

181. S. Ramachandran, S. Srinivasan, R. Chen: "EPLD-Based Architecture of Real Time 2D-discrete Cosine Transform and Qunatization for Image Compression," in *IEEE International Symposium on Circuits and Systems* Vol. III (1999), pp. 375–378

182. C. Burrus, P. Eschenbacher: "An In-Place, In-Order Prime Factor FFT Algorithm," *IEEE Transactions on Acoustics, Speech and Signal Processing* **29** (4), 806–817 (1981)

183. J. Pollard: "The Fast Fourier Transform in a Finite Field," *Mathematics of Computation* **25**, 365–374 (1971)

184. F. Taylor: "An RNS Discrete Fourier Transform Implementation," *IEEE Transactions on Acoustics, Speech and Signal Processing* **38**, 1386–1394 (1990)

185. C. Rader: "Discrete Convolutions via Mersenne Transforms," *IEEE Transactions on Computers* **C-21**, 1269–1273 (1972)

186. N. Bloch: *Abstract Algebra with Applications* (Prentice Hall, Englewood Cliffs, New Jersey, 1987)

187. J. Lipson: *Elements of Algebra and Algebraic Computing* (Addison-Wesley, London, 1981)

188. R. Agrawal, C. Burrus: "Fast Convolution Using Fermat Number Transforms with Applications to Digital Filtering," *IEEE Transactions on Acoustics, Speech and Signal Processing* **22**, 87–97 (1974)

189. W. Siu, A. Constantinides: "On the Computation of Discrete Fourier Transform using Fermat Number Transform," *Proceedings F IEE* **131**, 7–14 (1984)

190. J. McClellan: "Hardware Realization of the Fermat Number Transform," *IEEE Transactions on Acoustics, Speech and Signal Processing* **24** (3), 216–225 (1976)

191. TI: (1993), "User's Guide," TMS320C50, Texas Instruments

192. I. Reed, D. Tufts, X. Yu, T. Truong, M.T. Shih, X. Yin: "Fourier Analysis and Signal Processing by Use of the Möbius Inversion Formula," *IEEE Transactions on Acoustics, Speech and Signal Processing* **38** (3), 458–470 (1990)

193. H. Park, V. Prasanna: "Modular VLSI Arichitectures for Computing the Arithmetic Fourier Transform," *IEEE Transactions on Signal Processing* **41** (6), 2236–2246 (1993)
194. H. Lüke: *Signalübertragung* (Springer, Heidelberg, 1988)
195. D. Herold, R. Huthmann (1990): "Decoder for the Radio Data System (RDS) using Signal Processor TMS320C25," Master's thesis, Institute for Data Technics, Darmstadt University of Technology
196. U. Meyer-Bäse, R. Watzel: "A comparison of DES and LFSR based FPGA Implementable Cryptography Algorithms," in *3rd International Symposium on Communication Theory & Applications* (1995), pp. 291–298
197. U. Meyer-Bäse, R. Watzel: "An Optimized Format for Long Frequency Paging Systems," in *3rd International Symposium on Communication Theory & Applications* (1995), pp. 78–79
198. U. Meyer-Bäse: "Convolutional Error Decoding with FPGAs," *Lecture Notes in Computer Science* **1142**, 376–175 (1996)
199. R. Watzel (1993): "Design of Paging Scheme and Implementation of the Suitable Cryto-Controller using FPGAs," Master's thesis, Institute for Data Technics, Darmstadt University of Technology
200. J. Maier, T. Schubert (1993): "Design of Convolutional Decoders using FPGAs for Error Correction in a Paging System," Master's thesis, Institute for Data Technics, Darmstadt University of Technology
201. U. Meyer-Bäse et al.: "Zum bestehenden Übertragungsprotokoll kompatible Fehlerkorrektur," in *Funkuhren Zeitsignale Normalfrequenzen* (1993), pp. 99–112
202. D. Herold (1991): "Investigation of Error Corrections Steps for DCF77 Signals using Programmable Gate Arrays," Master's thesis, Institute for Data Technics, Darmstadt University of Technology
203. P. Sweeney: *Error Control Coding* (Prentice Hall, New York, 1991)
204. D. Wiggert: *Error-Control Coding and Applications* (Artech House, Dedham, Mass., 1988)
205. G. Clark, J. Cain: *Error-Correction Coding for Digital Communications* (Plenum Press, New York, 1988)
206. W. Stahnke: "Primitive Binary Polynomials," *Mathematics of Computation* pp. 977–980 (1973)
207. W. Fumy, H. Riess: *Kryptographie* (R. Oldenbourg Verlag, München, 1988)
208. B. Schneier: *Applied Cryptography* (John Wiley & Sons, New York, 1996)
209. M. Langhammer: "Reed-Solomon Codec Design in Programmable Logic," *Communication System Design (www.csdmag.com)* pp. 31–37 (1998)
210. B. Akers: "Binary Decusion Diagrams," *IEEE Transactions on Computers* pp. 509–516 (1978)
211. R. Bryant: "Graph-Based Algorithms for Boolean Function Manipulation," *IEEE Transactions on Computers* pp. 677–691 (1986)
212. A. Sangiovanni-Vincentelli, A. Gamal, J. Rose: "Synthesis Methods for Field Programmable Gate Arrays," *Proceedings of the IEEE* pp. 1057–83 (1993)
213. R. del Rio (1993): "Synthesis of boolean Functions for Field Programmable Gate Arrays," Master's thesis, Univerity of Frankfurt, FB Informatik
214. U. Meyer-Bäse: "Optimal Strategies for Incoherent Demodulation of Narrow Band FM Signals," in *3rd International Symposium on Communication Theory & Applications* (1995), pp. 30–31
215. J. Proakis: *Digital Communications* (McGraw-Hill, New York, 1983)
216. R. Johannesson: "Robustly Optimal One-Half Binary Convolutional Codes," *IEEE Transactions on Information Theory* pp. 464–8 (1975)

217. J. Massey, D. Costello: "Nonsystematic Convolutional Codes for Sequential Decoding in Space Applications," *IEEE Transactions on Communications* pp. 806–813 (1971)

218. F. MacWilliams, J. Sloane: "Pseudo-Random Sequences and Arrays," *Proceedings of the IEEE* pp. 1715–29 (1976)

219. T. Lewis, W. Payne: "Generalized Feedback Shift Register Pseudorandom Number Algorithm," *Journal of the Association for Computing Machinery* pp. 456–458 (1973)

220. P. Bratley, B. Fox, L. Schrage: *A Guide to Simulation* (Springer-Lehrbuch, Heidelberg, 1983), pp. 186–190

221. M. Schroeder: *Number Theory in Science and Communication* (Springer, Heidelberg, 1990)

222. P. Kocher, J. Jaffe, B.Jun: "Differential Power Analysis," in *Lecture Notes in Computer Science* (1999), pp. 388–397, www.cryptography.com

223. EFF: *Cracking DES* (O'Reilly & Associates, Sebastopol, 1998), Electronic Frontier Foundation

224. W. Stallings: "Encryption Choices Beyond DES," *Communication System Design (www.csdmag.com)* pp. 37–43 (1998)

225. W. Carter: "FPGAs: Go reconfigure," *Communication System Design (www.csdmag.com)* p. 56 (1998)

226. J. Anderson, T. Aulin, C.E. Sundberg: *Digital Phase Modulation* (Plenum Press, New York, 1986)

227. U. Meyer-Bäse (1989): "Investigation of Thresholdimproving Limiter/Discriminator Demodulator for FM Signals through Computer simulations," Master's thesis, Department of Information Science, Darmstadt University of Technology

228. E. Allmann, T. Wolf (1991): "Design and Implementation of a full digital zero IF Receiver using programmable Gate Arrays and Floatingpoint DSPs," Master's thesis, Institute for Data Technics, Darmstadt University of Technology

229. O. Herrmann: "Quadraturfilter mit rationalem Übertragungsfaktor," *Archiv der elektrischen Übertragung (AEÜ)* pp. 77–84 (1969)

230. O. Herrmann: "Transversalfilter zur Hilbert-Transformation," *Archiv der elektrischen Übertragung (AEÜ)* pp. 581–587 (1969)

231. V. Considine: "Digital Complex Sampling," *Electronics Letters* pp. 608–609 (1983)

232. T.E. Thiel, G.J. Saulnier: "Simplified Complex Digital Sampling Demodulator," *Electronics Letters* pp. 419–421 (1990)

233. U. Meyer-Bäse, W. Hilberg: (1992), "Schmalbandempfänger für Digitalsignale," German patent no. 4219417.2-31

234. B. Schlanske (1992): "Design and Implementation of a Universal Hilbert Sampling Receiver with CORDIC Demodulation for LF FAX Signals using Digital Signal Processor," Master's thesis, Institute for Data Technics, Darmstadt University of Technology

235. A. Dietrich (1992): "Realisation of a Hilbert Sampling Receiver with CORDIC Demodulation for DCF77 Signals using Floatingpoint Signal Processors," Master's thesis, Institute for Data Technics, Darmstadt University of Technology

236. A. Viterbi: *Principles of Coherent Communication* (McGraw-Hill, New York, 1966)

237. F. Gardner: *Phaselock Techniques* (John Wiley & Sons, New York, 1979)

238. H. Geschwinde: *Einführung in die PLL-Technik* (Vieweg, Braunschweig, 1984)

239. R. Best: *Theorie und Anwendung des Phase-locked Loops* (AT Press, Schwitzerland, 1987)

240. W. Lindsey, C. Chie: "A Survey of Digital Phase-Locked Loops," *Proceedings of the IEEE* pp. 410–431 (1981)

241. R. Sanneman, J. Rowbotham: "Unlock Characteristics of the Optimum Type II Phase-Locked Loop," *IEEE Transactions on Aerospace and Navigational Electronics* pp. 15–24 (1964)

242. J. Stensby: "False Lock in Costas Loops," *Proceedings of the 20th Southeastern Symposium on System Theory*, pp. 75–79 (1988)

243. A. Mararios, T. Tozer: "False-Lock Performance Improvement in Costas Loops," *IEEE Transactions on Communications* pp. 2285–88 (1982)

244. A. Makarios, T. Tozer: "False-Look Avoidance Scheme for Costas Loops," *Electronics Letters* pp. 490–2 (1981)

245. U. Meyer-Bäse: "Coherent Demodulation with FPGAs," *Lecture Notes in Computer Science* **1142**, 166–175 (1996)

246. J. Guyot, H. Schmitt (1993): "Design of a full digital Costas Loop using programmable Gate Arrays for coherent Demodulation of Low Frequency Signals," Master's thesis, Institute for Data Technics, Darmstadt University of Technology

247. R. Resch, P. Schreiner (1993): "Design of Full Digital Phase Locked Loops using programmable Gate Arrys for a low Frequency Reciever," Master's thesis, Institute for Data Technics, Darmstadt University of Technology

248. D. McCarty: "Digital PLL Suits FPGAs," *Elektronic Design* p. 81 (1992)

249. J. Holmes: "Tracking-Loop Bias Due to Costas Loop Arm Filter Imbalance," *IEEE Transactions on Communications* pp. 2271–3 (1982)

250. H. Choi: "Effect of Gain and Phase Imbalance on the Performance of Lock Detector of Costas Loop," *IEEE International Conference on Communications, Seattle* pp. 218–222 (1987)

251. N. Wiener: *Extrapolation, Interpolation and Smoothing of Stationary Time Series* (John Wiley & Sons, New York, 1949)

252. S. Haykin: *Adaptive Filter Theory* (Prentice Hall, Englewood Cliffs, New Jersey, 1986)

253. B. Widrow, S. Stearns: *Adaptive Signal Processing* (Prentice Hall, Englewood Cliffs, New Jersey, 1985)

254. C. Cowan, P. Grant: *Adaptive Filters* (Prentice Hall, Englewood Cliffs, New Jersey, 1985)

255. A. Papoulis: *Probability, Random Variables, and Stochastic Processes* (McGraw–Hill, Singapore, 1986)

256. M. Honig, D. Messerschmitt: *Adaptive Filters: Structures, Algorithms, and Applications* (Kluwer Academic Publishers, Norwell, 1984)

257. S. Alexander: *Adaptive Signal Processing: Theory and Application* (Springer, Heidelberg, 1986)

258. N. Shanbhag, K. Parhi: *Pipelined Adaptive Digital Filters* (Kluwer Academic Publishers, Norwell, 1994)

259. B. Mulgrew, C. Cowan: *Adaptive Filters and Equalisers* (Kluwer Academic Publishers, Norwell, 1988)

260. J. Treichler, C. Johnson, M. Larimore: *Theory and Design of Adaptive Filters* (Prentice Hall, Upper Saddle River, New Jersey, 2001)

261. B. Widrow, J. Glover, J. McCool, J. Kaunitz, C. Williams, R. Hearn, J. Zeidler, E. Dong, R. Goodlin: "Adaptive Noise Cancelling: Principles and Applications," *Proceedings of the IEEE* **63**, 1692–1716 (1975)

262. B. Widrow, J. McCool, M. Larimore, C. Johnson: "Stationary and Nonstationary Learning Characteristics of the LMS Adaptive Filter," *Proceedings of the IEEE* **64**, 1151–1162 (1976)

263. T. Kummura, M. Ikekawa, M. Yoshida, I. Kuroda: "VLIW DSP for Mobile Applications," *IEEE Signal Processing Magazine* **19**, 10–21 (2002)

264. Analog Device: "Application Handbook," 1987

265. L. Horowitz, K. Senne: "Performance Advantage of Complex LMS for Controlling Narrow-Band Adaptive Arrays," *IEEE Transactions on Acoustics, Speech and Signal Processing* **29**, 722–736 (1981)

266. A. Feuer, E. Weinstein: "Convergence Analysis of LMS Filters with Uncorrelated Gaussian Data," *IEEE Transactions on Acoustics, Speech and Signal Processing* **33**, 222–230 (1985)

267. S. Narayan, A. Peterson, M. Narasimha: "Transform Domain LMS Algorithm," *IEEE Transactions on Acoustics, Speech and Signal Processing* **31**, 609–615 (1983)

268. G. Clark, S. Parker, S. Mitra: "A Unified Approach to Time- and Frequency-Domain Realization of FIR Adaptive Digital Filters," *IEEE Transactions on Acoustics, Speech and Signal Processing* **31**, 1073–1083 (1983)

269. F. Beaufays (1995): "Two-Layer Structures for Fast Adaptive Filtering," Ph.D. thesis, Stanford University

270. A. Feuer: "Performance Analysis of Block Least Mean Square Algorithm," *IEEE Transactions on Circuits and Systems* **32**, 960–963 (1985)

271. D. Marshall, W. Jenkins, J. Murphy: "The use of Orthogonal Transforms for Improving Performance of Adaptive Filters," *IEEE Transactions on Circuits and Systems* **36** (4), 499–510 (1989)

272. J. Lee, C. Un: "Performance of Transform-Domain LMS Adaptive Digital Filters," *IEEE Transactions on Acoustics, Speech and Signal Processing* **34** (3), 499–510 (1986)

273. G. Long, F. Ling, J. Proakis: "The LMS Algorithm with Delayed Coefficient Adaption," *IEEE Transactions on Acoustics, Speech and Signal Processing* **37**, 1397–1405 (1989)

274. G. Long, F. Ling, J. Proakis: "Corrections to "The LMS Algorithm with Delayed Coefficient Adaption"," *IEEE Transactions on Signal Processing* **40**, 230–232 (1992)

275. R. Poltmann: "Conversion of the Delayed LMS Algorithm into the LMS Algorithm," *IEEE Signal Processing Letters* **2**, 223 (1995)

276. T. Kimijima, K. Nishikawa, H. Kiya: "An Effective Architecture of Pipelined LMS Adaptive Filters," *IEICE Transactions Fundamentals* **E82-A**, 1428–1434 (1999)

277. D. Jones: "Learning Characteristics of Transpose-Form LMS Adaptive Filters," *IEEE Transactions on Circuits and Systems II* **39** (10), 745–749 (1992)

278. M. Rupp, R. Frenzel: "Analysis of LMS and NLMS Algorithms with Delayed Coefficient Update Under the Presence of Spherically Invariant Processsess," *IEEE Transactions on Signal Processing* **42**, 668–672 (1994)

279. M. Rupp: "Saving Complexity of Modified Filtered-X-LMS abd Delayed Update LMS," *IEEE Transactions on Circuits and Systems II* **44**, 57–60 (1997)

280. M. Rupp, A. Sayed: "Robust FxLMS Algorithms with Improved Convergence Performance," *IEEE Transactions on Speech and Audio Processing* **6**, 78–85 (1998)

281. L. Ljung, M. Morf, D. Falconer: "Fast Calculation of Gain Matrices for Recursive Estimation Schemes," *International Journal of Control* **27**, 1–19 (1978)

282. G. Carayannis, D. Manolakis, N. Kalouptsidis: "A Fast Sequential Algorithm for Least-Squares Filtering and Prediction," *IEEE Transactions on Acoustics, Speech and Signal Processing* **31**, 1394–1402 (1983)

283. F. Albu, J. Kadlec, C. Softley, R. Matousek, A. Hermanek, N. Coleman, A. Fagan: "Implementation of (Normalised RLS Lattice on Virtex," *Lecture Notes in Computer Science* **2147**, 91–100 (2001)

284. Xilinx: (2005), "PicoBlaze 8-bit EMbedded Microcontroller User Guide," www.xilinx.com

285. V. Heuring, H. Jordan: *Computer Systems Design and Architecture*, 2nd edn. (Prentice Hall, Upper Saddle RIver, New Jersey, 2004), contribution by M. Murdocca

286. D. Patterson, J. Hennessy: *Computer Organization & Design: The Hardware/Software Interface*, 2nd edn. (Morgan Kaufman Publishers, Inc., San Mateo, CA, 1998)

287. J. Hennessy, D. Patterson: *Computer Architecture: A Quantitative Approach*, 3rd edn. (Morgan Kaufman Publishers, Inc., San Mateo, CA, 2003)

288. M. Murdocca, V. Heuring: *Principles of Computer Architecture*, 1st edn. (Prentice Hall, Upper Saddle River, NJ, 2000), jAVA machine overview

289. W. Stallings: *Computer Organization & Architecture*, 6th edn. (Prentice Hall, Upper Saddle River, NJ, 2002)

290. R. Bryant, D. O'Hallaron: *Computer Systems: A Programmer's Perspective*, 1st edn. (Prentice Hall, Upper Saddle River, NJ, 2003)

291. C. Rowen: *Engineering the Complex SOC*, 1st edn. (Prentice Hall, Upper Saddle River, NJ, 2004)

292. S. Mazor: "The History of the Microcomputer – Invention and Evolution," *Proceedings of the IEEE* **83** (12), 1601–8 (1995)

293. H. Faggin, M. Hoff, S. Mazor, M. Shima: "The History of the 4004," *IEEE Micro Magazine* **16**, 10–20 (1996)

294. Intel: (2006), "Microprocessor Hall of Fame," http://www.intel.com/museum

295. Intel: (1980), "2920 Analog Signal Processor," design handbook

296. TI: (2000), "Technology Inovation," www.ti.com/sc/techinnovations

297. TI: (1983), "TMS3210 Assembly Language Programmer's Guide," digital signal processor products

298. TI: (1993), "TMS320C5x User's Guide," digital signal processing products

299. Analog Device: (1993), "ADSP-2103," 3-Volt DSP Microcomputer

300. P. Koopman: *Stack Computers: The New Wave*, 1st edn. (Mountain View Press, La Honda, CA, 1989)

301. Xilinx: (2002), "Creating Embedded Microcontrollers," www.xilinx.com, Part 1-5

302. Altera: (2003), "Nios-32 Bit Programmer's Reference Manual," Nios embedded processor, Ver. 3.1

303. Xilinx: (2002), "Virtex-II Pro," documentation

304. Xilinx: (2005), "MicroBlaze – The Low-Cost and Flexible Processing Solution," www.xilinx.com

305. Altera: (2003), "Nios II Processor Reference Handbook," NII5V-1-5.0

306. B. Parhami: *Computer Architecture: From Microprocessor to Supercomputers*, 1st edn. (Oxford University Press, New York, 2005)

307. Altera: (2004), "Netseminar Nios processor," http://www.altera.com

308. A. Hoffmann, H. Meyr, R. Leupers: *Architecture Exploration for Embedded Processors with LISA*, 1st edn. (Kluwer Academic Publishers, Boston, 2002)

309. A. Aho, R. Sethi, J. Ullman: *Compilers: Principles, Techniques, and Tools*, 1st edn. (Addison Wesley Longman, Reading, Massachusetts, 1988)

310. R. Leupers: *Code Optimization Techniques for Embedded Processors*, 2nd edn. (Kluwer Academic Publishers, Boston, 2002)

311. R. Leupers, P. Marwedel: *Retargetable Compiler Technology for Embedded Systems*, 1st edn. (Kluwer Academic Publishers, Boston, 2001)

312. V. Paxson: (1995), "Flex, Version 2.5: A Fast Scanner Generator," http://www.gnu.org

313. C. Donnelly, R. Stallman: (2002), "Bison: The YACC-Compatible Parser Generator," http://www.gnu.org

314. S. Johnson: (1975), "YACC – Yet Another Compiler-Compiler," technical report no. 32, AT&T

315. R. Stallman: (1990), "Using and Porting GNU CC," http://www.gnu.org

316. W. Lesk, E. Schmidt: (1975), "LEX – a Lexical Analyzer Generator," technical report no. 39, AT&T

317. T. Niemann: (2004), "A Compact Guide to LEX & YACC," http://www.epaperpress.com

318. J. Levine, T. Mason, D. Brown: *lex & yacc*, 2nd edn. (O'Reilly Media Inc., Beijing, 1995)

319. T. Parsons: *Intorduction to Compiler Construction*, 1st edn. (Computer Science Press, New York, 1992)

320. A. Schreiner, H. Friedman: *Introduction to Compiler Construction with UNIX*, 1st edn. (Prentice-Hall, Inc, Englewood Cliffs, New Jersey, 1985)

321. C. Fraser, D. Hanson: *A Retargetable C Compilers: Design and Implementation*, 1st edn. (Addison-Wesley, Boston, 2003)

322. V. Zivojnovic, J. Velarde, C. Schläger, H. Meyr: "DSPSTONE: A DSP-orioented Benchmarking Methodology," in *International Conference* (19), pp. 1–6

323. Institute for Integrated Systems for Signal Processing: (1994), "DSPstone," final report

324. W. Strauss: "Digital Signal Processing: The New Semiconductor Industry Technology Driver," *IEEE Signal Processing Magazine* pp. 52–56 (2000)

325. Xilinx: (2002), "Virtex-II Pro Platform FPGA," handbook

326. Xilinx: "Accelerated System Performance with APU-Enhanced Processing," *Xcell Journal* pp. 1-4, first quarter (2005)
Xilinx: (2007), "Virtex-4 Online Documentation," http://www.xilinx.com

327. ARM: (2001), "ARM922T with AHB: Product Overview," http://www.arm.com

328. ARM: (2000), "ARM9TDMI Technical Reference Manual," http://www.arm.com

329. Altera: (2004), "Nios Software Development Reference Manual," http://www.altera.com

330. Altera: (2004), "Nios Development Kit, APEX Edition," Getting Started User Guide

331. Altera: (2004), "Nios Development Board Document," http://www.altera.com

332. Altera: (2004), "Nios Software Development Tutorial," http://www.altera.com

333. Altera: (2004), "Custom Instruction Tutorial," http://www.altera.com

334. B. Fletcher: "FPGA Embedded Processors," in *Embedded Systems Conference* San Francisco, CA (2005), p. 18, www.memec.com

335. U. Meyer-Baese, A. Vera, S. Rao, K. Lenk, M. Pattichis: "FPGA Wavelet Processor Design using Language for Instruction-set Architectures (LISA)," in *Proc. SPIE Int. Soc. Opt. Eng.* Orlando (2007), Vol. 6576, pp. 6576U1–U12

336. D. Sunkara (2004): "Design of Custom Instruction Set for FFT using FPGA-Based Nios processors," Master's thesis, FSU

337. U. Meyer-Baese, D. Sunkara, E. Castillo, E.A. Garcia: "Custom Instruction Set NIOS-Based OFDM Processor for FPGAs," in *Proc. SPIE Int. Soc. Opt. Eng.* Orlando (2006), Vol. 6248, pp. 6248o01–15

338. J. Ramirez, U. Meyer-Baese, A. Garcia: "Efficient Wavelet Architectures using Field- Programmable Logic and Residue Number System Arithmetic," in *Proc. SPIE Int. Soc. Opt. Eng.* Orlando (2004), Vol. 5439, pp. 222–232

339. J. Bhasker: *Verilog HDL Synthesis* (Start Galaxy Publishing, Allentown, PA, 1998)

340. IEEE: (1995), "Standard Hardware Description Language Based on the Verilog Hardware Description Language," language reference manual std 1364-1995

341. IEEE: (2001), "Standard Verilog Hardware Description Language," language reference manual std 1364-2001

342. S. Sutherland: "The IEEE Verilog 1364-2001 Standard: Whats New, and Why You Need It," in *Proceedings 9th Annual International HDL Conference and Exhibition* Santa Clara, CA (2000), p. 8, http://www.sutherland-hdl.com

343. Xilinx: (2004), "Verilog-2001 Support in XST," XST version 6.1 help

344. Altera: (2004), "Quartus II Support for Verilog 2001," Quartus II version 4.2 help

345. Synopsys: (2003), "Common VCS and HDL Compiler (Presto Verilog) 2001 Constructs," SolvNet doc id: 002232

346. J. Ousterhout: *Tcl and the Tk Toolkit*, 1st edn. (Addison-Wesley, Boston, 1994)

347. M. Harrison, M. McLennan: *Effective Tcl/Tk Programming*, 1st edn. (Addison-Wesley, Reading, Massachusetts, 1998)

348. B. Welch, K. Jones, H. J: *Practical Programming in Tcl and Tk*, 1st edn. (Prentice Hall, Upper Saddle River, NJ, 2003)

A. Verilog Source Code 2001

The first and second editions of the book include Verilog using the IEEE 1364-1995 standard [340]. This third edition takes advantage of several improvements that are documented in the IEEE 1364-2001 standard, which is available for about $100 from the IEEE bookstore [341, 342]. The Verilog 2001 improvements have now found implementations in all major design tools, [343, 344, 345] and we want to briefly to review the most important new features that are used. We only review the implemented new features; for all the new features see [341].

- The entity description in the 1364-1995 standard requires that (similar to the Kernighan and Ritchie C coding) all ports appear twice, first in the port list and then in the port data-type description, e.g.,

  ```
  module iir_pipe (x_in, y_out, clk); //----> Interface

      parameter W = 14; // Bit width - 1
      input         clk;
      input  [W:0]  x_in;   // Input
      output [W:0]  y_out;  // Result
      ...
  ```

  Note that all ports (x_in, y_out, and clk) are defined twice. In the 1364-2001 standard (see Sect. 12.3.4 LRM) this duplication is no longer required, i.e., the new coding is done as follows:

  ```
  module iir_pipe       //----> Interface
    #(parameter W = 14) // Bit width - 1
     (input          clk,
       input  signed [W:0]  x_in,   // Input
       output signed [W:0]  y_out); // Result
  ```

- Signed data types are available in the 1364-2001 standard, which allows arithmetic signed operations to be simplified. In the signal definition line the signed keyword is introduced after the input, output, reg or wire keywords, e.g.,

  ```
  reg signed [W:0] x, y;
  ```

Conversion between signed and unsigned type can be accomplished via the $signed or $unsigned conversion functions, see Sect. 4.5 LRM. For signed constants we introduce a small s *or* capital S between the hyphen and the base, e.g., `'sd90` for a signed constant 90. Signed arithmetic operations can be done using the conventional divide / or multiply * operators. For power-of-2 factors we can use the new arithmetic left <<< or right shift >>> operations, see Sect. 4.1.23 LRM. Note the three shift operations symbols used to distinguished to the unsigned shifts that use two shift symbols. Signed or zero extension is automatically done depending on the data type. From the IIR filter examples we can now replace the old-style operation:

```
...
y   <= x + {y[W],y[W:1]} + {{2{y[W]}},y[W:2]};
...                        // i.e., x + y / 2 + y / 4;
```

with the 1364-2001 Verilog style operations using the divide operator:

```
y <= x + y / 'sd2 + y / 'sd4;  // div with / uses 92 LEs.
```

Note the definition as signed divide for the constants, the code

```
y <= x + y / 2 + y / 4;
```

will show incorrect simulation results in Quartus II in versions 4.0, 4.2, 5.0, 5.1 but works fine in our web edition 6.0. Alteratively we can use the arithmetic right shift operator to implement the divide by power-of-2 values.

```
y <= x + (y >>> 1) + (y >>> 2);// div with >>> uses 60 LEs
```

It is evident that this notation makes the arithmetic operation much more readable than the old-style coding. Although both operations are functional equivalent, the Quartus synthesis results reveals that the divide is mapped to a different architecture and needs therefore more LEs (92 compared with 60 LEs) than used by the arithmetic shift operations. From the comparison to the VHDL synthesis data we conclude that 60 LEs is the expect result. This is the reason we will use the arithmetic left and right shift throughout the examples.

- The implicit `event_expression` list allows one to add automatically all right-hand-side variables to be added to the event expression, i.e., from the `bfproc` examples

```
...
always @(Are or Bre or Aim or Bim)
begin
...
```

we now simply use

```
    ...
    always @(*)
    begin
    ...
```

This reduces essentially the RTL simulation errors due to missing variables in the event listing, see Sect. 9.7.5 LRM for details.

- The generate statement introduced in the 1364-2001 Verilog standard allows one to instantiate several components using a single generate loop construct. The LRM Sects. 12.1.3.2-4 show eight different generate examples. We use the generate in the fir_gen, fir_lms, and fir6dlms files. Note that you also need to define a genvar as a loop variable used for the generate statement, see Sect. 12.1.3.1 LRM for details.

The 1364-2001 Verilog standard introduces 21 new keywords. We use the new keywords endgenerate, generate, genvar, signed, and unsigned. We have not used the new keywords:

```
automatic, cell, config, design, endconfig, incdir, include,
instance, liblist, library, localparam, noshowcancelled,
pulsestyle_onevent, pulsestyle_ondetect, showcancelled, use
```

The next pages contain the Verilog 1364-2001 code of all design examples. The old style Verilog 1364-1995 code can be found in [57]. The synthesis results for the examples are listed on page 731.

```
//**********************************************************
// IEEE STD 1364-2001 Verilog file: example.v
// Author-EMAIL: Uwe.Meyer-Baese@ieee.org
//**********************************************************
//'include "220model.v"    // Using predefined components

module example    //----> Interface
  #(parameter WIDTH =8)    // Bit width
  (input   clk,
   input   [WIDTH-1:0] a, b, op1,
   output  [WIDTH-1:0] sum, d);

   wire [WIDTH-1:0]   c;        // Auxiliary variables
   reg  [WIDTH-1:0]   s;        // Infer FF with always
   wire [WIDTH-1:0]   op2, op3;

   wire   clkena, ADD, ena, aset, sclr, sset, aload, sload,
                   aclr, ovf1, cin1; // Auxiliary lpm signals

// Default for add:
   assign cin1=0; assign aclr=0; assign ADD=1;
```

```
   assign ena=1; assign aclr=0; assign aset=0;
   assign sclr=0; assign sset=0; assign aload=0;
   assign sload=0; assign clkena=0; // Default for FF

   assign op2 = b;        // Only one vector type in Verilog;
               // no conversion int -> logic vector necessary

// Note when using 220model.v ALL component's signals
// must be defined, default values can only be used for
// the parameters.

   lpm_add_sub add1          //----> Component instantiation
   ( .result(op3), .dataa(op1), .datab(op2)); // Used ports
// .cin(cin1),.cout(cr1), .add_sub(ADD), .clken(clkena),
// .clock(clk), .overflow(ovl1), .aclr(aclr)); // Unused
     defparam add1.lpm_width = WIDTH;
     defparam add1.lpm_representation = "SIGNED";

   lpm_ff reg1
   ( .data(op3), .q(sum), .clock(clk));  // Used ports
// .enable(ena), .aclr(aclr), .aset(aset), .sclr(sclr),
// .sset(sset), .aload(aload), .sload(sload)); // Unused
     defparam reg1.lpm_width = WIDTH;

   assign c = a + b; //----> Continuous assignment statement

   always @(posedge clk)   //----> Behavioral style
   begin : p1              // Infer register
     s = c + s;            // Signal assignment statement
   end
   assign d = s;

endmodule

//*********************************************************
// IEEE STD 1364-2001 Verilog file: fun_text.v
// Author-EMAIL: Uwe.Meyer-Baese@ieee.org
//*********************************************************
// A 32-bit function generator using accumulator and ROM
//'include "220model.v"

module fun_text              //----> Interface
  #(parameter WIDTH = 32)    // Bit width
 (input          clk,
```

```verilog
    input   [WIDTH-1:0]  M,
    output  [7:0]  sin, acc);

    wire [WIDTH-1:0] s, acc32;
    wire [7:0]    msbs;                  // Auxiliary vectors
    wire  ADD, ena, aset, sclr, sset; // Auxiliary signals
    wire aload, sload, aclr, ovf1, cin1, clkena;

    // Default for add:
    assign clkena=0; assign cin1=0; assign ADD=1;
    //default for FF:
    assign ena=1; assign aclr=0; assign aset=0;
    assign sclr=0; assign sset=0; assign aload=0;
    assign sload=0;

    lpm_add_sub add_1                           // Add M to acc32
    ( .result(s), .dataa(acc32), .datab(M)); // Used ports
//   .cout(cr1), .add_sub(ADD), .overflow(ovl1),  // Unused
//   .clock(clk),.cin(cin1), .clken(clkena), .aclr(aclr));
//
        defparam add_1.lpm_width = WIDTH;
        defparam add_1.lpm_representation = "UNSIGNED";


    lpm_ff reg_1                                 // Save accu
    ( .data(s), .q(acc32), .clock(clk));        // Used ports
//   .enable(ena), .aclr(aclr), .aset(aset), // Unused ports
//   .sset(sset), .aload(aload), .sload(sload),.sclr(sclr));
        defparam reg_1.lpm_width = WIDTH;

    assign msbs = acc32[WIDTH-1:WIDTH-8];
    assign acc  = msbs;

    lpm_rom rom1
    ( .q(sin), .inclock(clk), .outclock(clk),
                              .address(msbs)); // Used ports
//                   .memenab(ena) ) ;        // Unused port
        defparam rom1.lpm_width = 8;
        defparam rom1.lpm_widthad = 8;
        defparam rom1.lpm_file = "sine.mif";

endmodule

//***********************************************************
// IEEE STD 1364-2001 Verilog file: cmul7p8.v
```

```
// Author-EMAIL: Uwe.Meyer-Baese@ieee.org
//*************************************************************
module cmul7p8                        // ------> Interface
        (input  signed [4:0] x,
          output signed [4:0] y0, y1, y2, y3);

  assign y0 = 7 * x / 8;
  assign y1 = x / 8 * 7;
  assign y2 = x/2 + x/4 + x/8;
  assign y3 = x - x/8;

endmodule

//*************************************************************
// IEEE STD 1364-2001 Verilog file: add1p.v
// Author-EMAIL: Uwe.Meyer-Baese@ieee.org
//*************************************************************
//'include "220model.v"

module add1p
#(parameter WIDTH   = 19, // Total bit width
            WIDTH1  = 9,  // Bit width of LSBs
            WIDTH2  = 10) // Bit width of MSBs
  (input  [WIDTH-1:0] x, y,  // Inputs
   output [WIDTH-1:0] sum,   // Result
   input              clk,   // Clock
   output          LSBs_Carry); // Test port

  reg [WIDTH1-1:0] l1, l2, s1; // LSBs of inputs
  reg [WIDTH1:0] r1;           // LSBs of inputs
  reg [WIDTH2-1:0] l3, l4, r2, s2; // MSBs of input

  always @(posedge clk) begin
    // Split in MSBs and LSBs and store in registers
    // Split LSBs from input x,y
    l1[WIDTH1-1:0] <= x[WIDTH1-1:0];
    l2[WIDTH1-1:0] <= y[WIDTH1-1:0];
    // Split MSBs from input x,y
    l3[WIDTH2-1:0] <= x[WIDTH2-1+WIDTH1:WIDTH1];
    l4[WIDTH2-1:0] <= y[WIDTH2-1+WIDTH1:WIDTH1];
/************* First stage of the adder  ***************/
    r1 <= {1'b0, l1} + {1'b0, l2};
    r2 <= l3 + l4;
/************* Second stage of the adder ***************/
    s1 <= r1[WIDTH1-1:0];
```

```
  // Add MSBs (x+y) and carry from LSBs
     s2 <= r1[WIDTH1] + r2;
  end

  assign LSBs_Carry = r1[WIDTH1]; // Add a test signal

 // Build a single output word of WIDTH = WIDTH1 + WIDTH2
  assign sum = {s2, s1};

endmodule

//***********************************************************
// IEEE STD 1364-2001 Verilog file: add2p.v
// Author-EMAIL: Uwe.Meyer-Baese@ieee.org
//***********************************************************
// 22-bit adder with two pipeline stages
// uses no components

//'include "220model.v"

module add2p
#(parameter WIDTH    = 28,    // Total bit width
            WIDTH1  = 9,     // Bit width of LSBs
            WIDTH2  = 9,     // Bit width of middle
            WIDTH12 = 18,    // Sum WIDTH1+WIDTH2
            WIDTH3  =  10)   // Bit width of MSBs
 (input   [WIDTH-1:0] x, y,  // Inputs
  output  [WIDTH-1:0] sum,   // Result
  output LSBs_Carry, MSBs_Carry,  // Single bits
  input   clk);              // Clock

  reg [WIDTH1-1:0] l1, l2, v1, s1; // LSBs of inputs
  reg [WIDTH1:0]   q1;             // LSBs of inputs
  reg [WIDTH2-1:0] l3, l4, s2;     // Middle bits
  reg [WIDTH2:0]   q2, v2;         // Middle bits
  reg [WIDTH3-1:0] l5, l6, q3, v3, s3; // MSBs of input

  // Split in MSBs and LSBs and store in registers
  always @(posedge clk) begin
    // Split LSBs from input x,y
    l1[WIDTH1-1:0] <= x[WIDTH1-1:0];
    l2[WIDTH1-1:0] <= y[WIDTH1-1:0];
    // Split middle bits from input x,y
    l3[WIDTH2-1:0] <= x[WIDTH2-1+WIDTH1:WIDTH1];
    l4[WIDTH2-1:0] <= y[WIDTH2-1+WIDTH1:WIDTH1];
```

```
      // Split MSBs from input x,y
      l5[WIDTH3-1:0] <= x[WIDTH3-1+WIDTH12:WIDTH12];
      l6[WIDTH3-1:0] <= y[WIDTH3-1+WIDTH12:WIDTH12];
//************** First stage of the adder  ****************
      q1 <= {1'b0, l1} + {1'b0, l2};  // Add LSBs of x and y
      q2 <= {1'b0, l3} + {1'b0, l4};  // Add LSBs of x and y
      q3 <= l5 + l6;                  // Add MSBs of x and y
//************** Second stage of the adder ****************
      v1 <= q1[WIDTH1-1:0];           // Save q1
// Add result from middle bits (x+y) and carry from LSBs
      v2 <= q1[WIDTH1] + {1'b0,q2[WIDTH2-1:0]};
// Add result from MSBs bits (x+y) and carry from middle
      v3 <= q2[WIDTH2] + q3;
//************** Third stage of the adder ****************
      s1 <= v1;                       // Save v1
      s2 <= v2[WIDTH2-1:0];           // Save v2
// Add result from MSBs bits (x+y) and 2. carry from middle
      s3 <= v2[WIDTH2] + v3;
   end

  assign LSBs_Carry = q1[WIDTH1]; // Provide test signals
  assign MSBs_Carry = v2[WIDTH2];

// Build a single output word of WIDTH=WIDTH1+WIDTH2+WIDTH3
  assign sum ={s3, s2, s1};   // Connect sum to output pins

endmodule


//*************************************************************
// IEEE STD 1364-2001 Verilog file: add3p.v
// Author-EMAIL: Uwe.Meyer-Baese@ieee.org
//*************************************************************
// 37-bit adder with three pipeline stages
// uses no components

//'include "220model.v"

module add3p
#(parameter WIDTH   = 37, // Total bit width
            WIDTH0  = 9,  // Bit width of LSBs
            WIDTH1  = 9,  // Bit width of 2. LSBs
            WIDTH01 = 18, // Sum WIDTH0+WIDTH1
            WIDTH2  = 9,  // Bit width of 2. MSBs
            WIDTH012 = 27,  // Sum WIDTH0+WIDTH1+WIDTH2
```

```
                WIDTH3   =  10)   // Bit width of MSBs
 (input   [WIDTH-1:0] x, y,   // Inputs
  output  [WIDTH-1:0] sum,   // Result
  output LSBs_Carry, Middle_Carry, MSBs_Carry, // Test pins
  input                clk);  // Clock

  reg [WIDTH0-1:0] l0, l1, r0, v0, s0;    // LSBs of inputs
  reg [WIDTH0:0] q0;                      // LSBs of inputs
  reg [WIDTH1-1:0] l2, l3, r1, s1;       // 2. LSBs of input
  reg [WIDTH1:0] v1, q1;                 // 2. LSBs of input
  reg [WIDTH2-1:0] l4, l5, s2, h7;       // 2. MSBs bits
  reg [WIDTH2:0] q2, v2, r2;             // 2. MSBs bits
  reg [WIDTH3-1:0] l6, l7, q3, v3, r3, s3, h8;
                                          // MSBs of input


always @(posedge clk) begin
// Split in MSBs and LSBs and store in registers
  // Split LSBs from input x,y
  l0[WIDTH0-1:0] <= x[WIDTH0-1:0];
  l1[WIDTH0-1:0] <= y[WIDTH0-1:0];
  // Split 2. LSBs from input x,y
  l2[WIDTH1-1:0] <= x[WIDTH1-1+WIDTH0:WIDTH0];
  l3[WIDTH1-1:0] <= y[WIDTH1-1+WIDTH0:WIDTH0];
  // Split 2. MSBs from input x,y
  l4[WIDTH2-1:0] <= x[WIDTH2-1+WIDTH01:WIDTH01];
  l5[WIDTH2-1:0] <= y[WIDTH2-1+WIDTH01:WIDTH01];
  // Split MSBs from input x,y
  l6[WIDTH3-1:0] <= x[WIDTH3-1+WIDTH012:WIDTH012];
  l7[WIDTH3-1:0] <= y[WIDTH3-1+WIDTH012:WIDTH012];

//************ First stage of the adder  *****************
  q0 <= {1'b0, l0} + {1'b0, l1};  // Add LSBs of x and y
  q1 <= {1'b0, l2} + {1'b0, l3};  // Add 2. LSBs of x / y
  q2 <= {1'b0, l4} + {1'b0, l5};  // Add 2. MSBs of x/y
  q3 <= l6 + l7;                  // Add MSBs of x and y
//************ Second stage of the adder *****************
  v0 <= q0[WIDTH0-1:0];           // Save q0
// Add result from 2. LSBs (x+y) and carry from LSBs
  v1 <= q0[WIDTH0] + {1'b0, q1[WIDTH1-1:0]};
// Add result from 2. MSBs (x+y) and carry from 2. LSBs
  v2 <= q1[WIDTH1] + {1'b0, q2[WIDTH2-1:0]};
// Add result from MSBs (x+y) and carry from 2. MSBs
  v3 <= q2[WIDTH2] + q3;
```

```
//************** Third stage of the adder *****************
  r0 <= v0;  // Delay for LSBs
  r1 <= v1[WIDTH1-1:0];  // Delay for 2. LSBs
// Add result from 2. MSBs (x+y) and carry from 2. LSBs
  r2 <= v1[WIDTH1] + {1'b0, v2[WIDTH2-1:0]};
// Add result from MSBs (x+y) and carry from 2. MSBs
  r3 <= v2[WIDTH2] + v3;
//************ Fourth stage of the adder ****************
  s0 <= r0;                // Delay for LSBs
  s1 <= r1;                // Delay for 2. LSBs
  s2 <= r2[WIDTH2-1:0];    // Delay for 2. MSBs
// Add result from MSBs (x+y) and carry from 2. MSBs
  s3 <= r2[WIDTH2] + r3;
end

assign LSBs_Carry   = q0[WIDTH1];  // Provide test signals
assign Middle_Carry = v1[WIDTH1];
assign MSBs_Carry   = r2[WIDTH2];

// Build a single output word of
// WIDTH = WIDTH0 + WIDTH1 + WIDTH2 + WIDTH3
assign sum = {s3, s2, s1, s0}; // Connect sum to output

endmodule


//************************************************************
// IEEE STD 1364-2001 Verilog file: mul_ser.v
// Author-EMAIL: Uwe.Meyer-Baese@ieee.org
//************************************************************
module mul_ser        //----> Interface
  (input          clk, reset,
   input  signed [7:0]  x,
   input         [7:0]  a,
   output reg signed [15:0] y);

   always @(posedge clk) //-> Multiplier in behavioral style
   begin : States
     parameter s0=0, s1=1, s2=2;
     reg [2:0] count;
     reg [1:0] s;                 // FSM state register
     reg signed [15:0] p, t;      // Double bit width
     reg   [7:0] a_reg;

     if (reset)                   // Asynchronous reset
```

```
        s <= s0;
      else
        case (s)
          s0 : begin            // Initialization step
            a_reg <= a;
            s <= s1;
            count = 0;
            p <= 0;         // Product register reset
            t <= x;         // Set temporary shift register to x
          end
          s1 : begin          // Processing step
            if (count == 7)   // Multiplication ready
              s <= s2;
            else
              begin
              if (a_reg[0] == 1) // Use LSB for bit select
                p <= p + t;      // Add 2^k
              a_reg <= a_reg >>> 1;
              t <= t <<< 1;
              count = count + 1;
              s <= s1;
            end
          end
          s2 : begin          // Output of result to y and
            y <= p;         // start next multiplication
            s <= s0;
          end
        endcase
    end

endmodule

//****************************************************************
// IEEE STD 1364-2001 Verilog file: div_res.v
// Author-EMAIL: Uwe.Meyer-Baese@ieee.org
//****************************************************************
// Restoring Division
// Bit width:  WN          WD          WN          WD
//          Numerator / Denominator = Quotient and Remainder
// OR:        Numerator = Quotient * Denominator + Remainder

module div_res(
  input        clk, reset,
  input  [7:0] n_in,
  input  [5:0] d_in,
```

```verilog
output reg [5:0] r_out,
output reg [7:0] q_out);

parameter s0=0, s1=1, s2=2, s3=3; // State assignments

// Divider in behavioral style
always @(posedge clk or posedge reset)
begin : F // Finite state machine
  reg [3:0] count;
  reg [1:0] s;            // FSM state
  reg [13:0] d;           // Double bit width unsigned
  reg signed [13:0] r;    // Double bit width signed
  reg [7:0] q;

  if (reset)              // Asynchronous reset
    s <= s0;
  else
    case (s)
      s0 : begin          // Initialization step
        s <= s1;
        count = 0;
        q <= 0;           // Reset quotient register
        d <= d_in << 7;   // Load aligned denominator
        r <= n_in;        // Remainder = numerator
      end
      s1 : begin          // Processing step
        r <= r - d;       // Subtract denominator
        s <= s2;
      end
      s2 : begin          // Restoring step
        if (r < 0) begin  // Check r < 0
          r <= r + d;     // Restore previous remainder
          q <= q << 1;    // LSB = 0 and SLL
        end
        else
          q <= (q << 1) + 1; // LSB = 1 and SLL
        count = count + 1;
        d <= d >> 1;

        if (count == 8)   // Division ready ?
          s <= s3;
        else
          s <= s1;
      end
```

```
        s3 : begin          // Output of result
          q_out <= q[7:0];
          r_out <= r[5:0];
          s <= s0;   // Start next division
        end
      endcase
  end

endmodule

//**********************************************************
// IEEE STD 1364-2001 Verilog file: div_aegp.v
// Author-EMAIL: Uwe.Meyer-Baese@ieee.org
//**********************************************************
// Convergence division after
//                    Anderson, Earle, Goldschmidt, and Powers
// Bit width:  WN          WD           WN           WD
//        Numerator / Denominator = Quotient and Remainder
// OR:       Numerator = Quotient * Denominator + Remainder

module div_aegp
  (input          clk, reset,
   input  [8:0] n_in,
   input  [8:0] d_in,
   output reg [8:0] q_out);

   always @(posedge clk or posedge reset) //-> Divider in
   begin : States                         // behavioral style
     parameter s0=0, s1=1, s2=2;
     reg [1:0] count;
     reg [1:0] state;
     reg [9:0] x, t, f;        // one guard bit
     reg [17:0] tempx, tempt;

     if (reset)               // Asynchronous reset
       state <= s0;
     else
       case (state)
         s0 : begin           // Initialization step
           state <= s1;
           count = 0;
           t <= {1'b0, d_in};   // Load denominator
           x <= {1'b0, n_in};   // Load numerator
         end
         s1 : begin              // Processing step
```

```
          f = 512 - t;        // TWO - t
          tempx = (x * f);  // Product in full
          tempt = (t * f);  // bitwidth
          x <= tempx >> 8;  // Factional f
          t <= tempt >> 8;  // Scale by 256
          count = count + 1;
          if (count == 2)     // Division ready ?
            state <= s2;
          else
            state <= s1;
        end
        s2 : begin          // Output of result
          q_out <= x[8:0];
          state <= s0;   // Start next division
        end
      endcase
  end

endmodule

//*************************************************************
// IEEE STD 1364-2001 Verilog file: cordic.v
// Author-EMAIL: Uwe.Meyer-Baese@ieee.org
//*************************************************************
module cordic #(parameter W = 7)  // Bit width - 1
(input        clk,
 input  signed [W:0] x_in, y_in,
 output reg signed [W:0] r, phi, eps);


// There is bit access in Quartus array types
// in Verilog 2001, therefore use single vectors
// but use a separate line for each array!
  reg signed [W:0] x [0:3];
  reg signed [W:0] y [0:3];
  reg signed [W:0] z [0:3];

  always @(posedge clk) begin  //----> Infer registers
    if (x_in >= 0)             // Test for x_in < 0 rotate
      begin                    // 0, +90, or -90 degrees
      x[0] <= x_in; // Input in register 0
      y[0] <= y_in;
      z[0] <= 0;
      end
    else if (y_in >= 0)
```

```
      begin
      x[0] <= y_in;
      y[0] <= - x_in;
      z[0] <= 90;
      end
   else
      begin
      x[0] <= - y_in;
      y[0] <= x_in;
      z[0] <= -90;
      end

   if (y[0] >= 0)                       // Rotate 45 degrees
      begin
      x[1] <= x[0] + y[0];
      y[1] <= y[0] - x[0];
      z[1] <= z[0] + 45;
      end
   else
      begin
      x[1] <= x[0] - y[0];
      y[1] <= y[0] + x[0];
      z[1] <= z[0] - 45;
      end

   if (y[1] >= 0)                       // Rotate 26 degrees
      begin
      x[2] <= x[1] + (y[1] >>> 1); // i.e. x[1] + y[1] /2
      y[2] <= y[1] - (x[1] >>> 1); // i.e. y[1] - x[1] /2
      z[2] <= z[1] + 26;
      end
   else
      begin
      x[2] <= x[1] - (y[1] >>> 1); // i.e. x[1] - y[1] /2
      y[2] <= y[1] + (x[1] >>> 1); // i.e. y[1] + x[1] /2
      z[2] <= z[1] - 26;
      end

   if (y[2] >= 0)                       // Rotate 14 degrees
      begin
         x[3] <= x[2] + (y[2] >>> 2); // i.e. x[2] + y[2]/4
         y[3] <= y[2] - (x[2] >>> 2); // i.e. y[2] - x[2]/4
         z[3] <= z[2] + 14;
      end
```

```
      else
        begin
          x[3] <= x[2] - (y[2] >>> 2); // i.e. x[2] - y[2]/4
          y[3] <= y[2] + (x[2] >>> 2); // i.e. y[2] + x[2]/4
          z[3] <= z[2] - 14;
        end

      r   <= x[3];
      phi <= z[3];
      eps <= y[3];
    end

endmodule

//************************************************************
// IEEE STD 1364-2001 Verilog file: arctan.v
// Author-EMAIL: Uwe.Meyer-Baese@ieee.org
//************************************************************
module arctan  #(parameter W = 9,    // Bit width
                           L  = 5)    // Array size
  (input clk,
   input signed [W-1:0] x_in,
   //output reg signed [W-1:0] d_o [1:L],
   output wire signed [W-1:0] d_o1, d_o2 ,d_o3, d_o4 ,d_o5,
   output reg signed [W-1:0] f_out);

  reg signed [W-1:0] x;    // Auxilary signals
  wire signed [W-1:0] f;
  wire signed [W-1:0] d [1:L]; // Auxilary array
  // Chebychev coefficients c1, c2, c3 for 8-bit precision
  // c1 = 212; c3 = -12; c5 = 1;

  always @(posedge clk) begin
    x <= x_in;       // FF for input and output
    f_out <= f;
  end

    // Compute sum-of-products with
    // Clenshaw's recurrence formula
  assign d[5] = 'sd1;    // c5=1
  assign d[4] = (x * d[5]) / 128;
  assign d[3] = ((x * d[4]) / 128) - d[5] - 12; // c3=-12
  assign d[2] = ((x * d[3]) / 128) - d[4];
  assign d[1] = ((x * d[2]) / 128) - d[3] + 212; // c1=212
  assign f    = ((x * d[1]) / 256) - d[2];
```

```
                                     // last step is different

   assign d_o1 = d[1];   // Provide test signals as outputs
   assign d_o2 = d[2];
   assign d_o3 = d[3];
   assign d_o4 = d[4];
   assign d_o5 = d[5];
endmodule


//**********************************************************
// IEEE STD 1364-2001 Verilog file: ln.v
// Author-EMAIL: Uwe.Meyer-Baese@ieee.org
//**********************************************************
module ln #(parameter N = 5, // -- Number of coeffcients-1
            parameter W= 17) // -- Bitwidth -1
       (input clk,
        input signed [W:0] x_in,
        output reg signed [W:0] f_out);

   reg signed [W:0] x, f;   // Auxilary register
   wire signed [W:0] p [0:5];
   reg signed [W:0] s [0:5];

// Polynomial coefficients for 16-bit precision:
// f(x) = (1  + 65481 x -32093 x^2 + 18601 x^3
//                       -8517 x^4 + 1954 x^5)/65536
   assign p[0] = 18'sd1;
   assign p[1] = 18'sd65481;
   assign p[2] = -18'sd32093;
   assign p[3] = 18'sd18601;
   assign p[4] = -18'sd8517;
   assign p[5] = 18'sd1954;

   always @(posedge clk)
   begin : Store
     x <= x_in;      // Store input in register
   end

   always @(posedge clk)            // Compute sum-of-products
   begin :  SOP
     integer k; // define the loop variable
     reg signed [35:0] slv;

     s[N] = p[N];
```

```
// Polynomial Approximation from Chebyshev coefficients
   for (k=N-1; k>=0; k=k-1)
   begin
     slv   = x * s[k+1]; // no FFs for slv
     s[k]  = (slv >>> 16) + p[k];
   end       // x*s/65536 problem 32 bits
   f_out  <= s[0];       // make visable outside
 end

endmodule

//***********************************************************
// IEEE STD 1364-2001 Verilog file: sqrt.v
// Author-EMAIL: Uwe.Meyer-Baese@ieee.org
//***********************************************************

module sqrt  ////> Interface
 (input         clk, reset,
  input  [16:0] x_in,
  output [16:0] a_o, imm_o, f_o,
  output reg [2:0]  ind_o,
  output reg [1:0]  count_o,
  output [16:0] x_o,pre_o,post_o,
  output reg [16:0] f_out);

  // Define the operation modes:
  parameter load=0, mac=1, scale=2, denorm=3, nop=4;
  //   Assign the FSM states:
  parameter start=0, leftshift=1, sop=2,
                  rightshift=3,  done=4;

  reg [2:0] s, op;
  reg [16:0] x; // Auxilary
  reg signed [16:0] a, b, f, imm; // ALU data
  reg [16:0] pre, post;
  // Chebychev poly coefficients for 16-bit precision:
  wire signed [16:0] p [0:4];

  assign p[0] = 7563;
  assign p[1] = 42299;
  assign p[2] = -29129;
  assign p[3] = 15813;
  assign p[4] = -3778;

  always @(posedge reset or posedge clk) //------> SQRT FSM
```

```
begin : States                        // sample at clk rate
  reg signed [3:0] ind;
  reg [1:0] count;

  if (reset)                          // Asynchronous reset
    s <= start;
  else begin
    case (s)                          // Next State assignments
      start : begin                   // Initialization step
        s <= leftshift; ind = 4;
        imm <= x_in;                  // Load argument in ALU
        op <= load; count = 0;
      end
      leftshift : begin               // Normalize to 0.5 .. 1.0
        count = count + 1; a <= pre; op <= scale;
        imm <= p[4];
        if (count == 3) begin         // Normalize ready ?
          s <= sop; op <= load; x <= f;
        end
      end
      sop :  begin                    // Processing step
        ind = ind - 1; a <= x;
        if (ind == -1) begin          // SOP ready ?
          s <= rightshift; op <= denorm; a <= post;
        end else begin
          imm <= p[ind]; op <= mac;
        end
      end
      rightshift : begin // Denormalize to original range
        s <= done; op <= nop;
      end
      done :  begin                   // Output of results
        f_out <= f;                   // I/O store in register
        op<=nop;
        s <= start;
      end                             // start next cycle
    endcase
  end
  ind_o <= ind;
  count_o <= count;
end

always @(posedge clk) // Define the ALU operations
begin : ALU
```

```
    case (op)
        load    : f  <= imm;
        mac     : f  <= (a * f / 32768) + imm;
        scale   : f  <= a * f;
        denorm  : f  <= (a * f /32768);
        nop     : f  <= f;
        default : f  <= f;
    endcase
  end

  always @*
  begin : EXP
    reg [16:0] slv;
    reg [16:0] po, pr;
    integer K, L;

    slv = x_in;
    // Compute pre-scaling:
    for (K=0; K <= 15; K= K+1)
      if (slv[K] == 1)
        L <= K;
    pre = 1 << (14-L);
    // Compute post scaling:
    po = 1;
    for (K=0; K <= 7; K= K+1) begin
      if (slv[2*K] == 1)    // even 2^k gets 2^k/2
        po = 1 << (K+8);
// sqrt(2): CSD Error = 0.0000208 = 15.55 effective bits
// +1 +0. -1 +0 -1 +0 +1 +0 +1 +0 +0 +0 +0 +1
// 9     7    5    3    1                  -5
      if (slv[2*K+1] == 1) // odd k has sqrt(2) factor
        po = (1<<(K+9)) - (1<<(K+7)) - (1<<(K+5))
             + (1<<(K+3)) + (1<<(K+1)) + (1<<(K-5));
    end
    post <= po;
  end

  assign a_o = a;    // Provide some test signals as outputs
  assign imm_o = imm;
  assign f_o = f;
  assign pre_o = pre;
  assign post_o = post;
  assign x_o = x;
```

```
endmodule

//************************************************************
// IEEE STD 1364-2001 Verilog file: fir_gen.v
// Author-EMAIL: Uwe.Meyer-Baese@ieee.org
//************************************************************
// This is a generic FIR filter generator
// It uses W1 bit data/coefficients bits
module fir_gen
#(parameter W1 = 9,    // Input bit width
             W2 = 18,   // Multiplier bit width 2*W1
             W3 = 19,   // Adder width = W2+log2(L)-1
             W4 = 11,   // Output bit width
             L  = 4,    // Filter length
             Mpipe = 3) // Pipeline steps of multiplier
 (input clk, Load_x,  // std_logic
  input signed [W1-1:0] x_in, c_in,  // Inputs
  output signed [W4-1:0] y_out);  // Results

  reg signed [W1-1:0]  x;
  wire signed [W3-1:0]  y;
// 1D array types i.e. memories supported by Quartus
// in Verilog 2001; first bit then vector size
  reg  signed [W1-1:0] c [0:3]; // Coefficient array
  wire signed [W2-1:0] p [0:3]; // Product array
  reg  signed [W3-1:0] a [0:3]; // Adder array

  wire  signed [W2-1:0] sum;  // Auxilary signals
  wire  clken, aclr;

  assign sum=0; assign aclr=0; // Default for mult
  assign clken=0;

//----> Load Data or Coefficient
  always @(posedge clk)
    begin: Load
    if (! Load_x) begin
      c[3] <= c_in; // Store coefficient in register
      c[2] <= c[3];   // Coefficients shift one
      c[1] <= c[2];
      c[0] <= c[1];
      end
    else begin
      x <= x_in; // Get one data sample at a time
      end
```

```
      end

//----> Compute sum-of-products
   always @(posedge clk)
     begin: SOP
   // Compute the transposed filter additions
     a[0] <= p[0] + a[1];
     a[1] <= p[1] + a[2];
     a[2] <= p[2] + a[3];
     a[3] <= p[3]; // First TAP has only a register
   end
   assign y = a[0];

   genvar I; //Define loop variable for generate statement
   generate
   for (I=0; I<L; I=I+1) begin: MulGen
// Instantiate L pipelined multiplier
   lpm_mult mul_I               // Multiply x*c[I] = p[I]
     (.clock(clk), .dataa(x), .datab(c[I]), .result(p[I]));
// .sum(sum), .clken(clken), .aclr(aclr)); // Unused ports
     defparam mul_I.lpm_widtha = W1;
     defparam mul_I.lpm_widthb = W1;
     defparam mul_I.lpm_widthp = W2;
     defparam mul_I.lpm_widths = W2;
     defparam mul_I.lpm_pipeline = Mpipe;
     defparam mul_I.lpm_representation = "SIGNED";
   end
   endgenerate

   assign y_out = y[W3-1:W3-W4];

endmodule

//**********************************************************
// IEEE STD 1364-2001 Verilog file: fir_srg.v
// Author-EMAIL: Uwe.Meyer-Baese@ieee.org
//**********************************************************
module fir_srg              //----> Interface
 (input          clk,
   input signed [7:0] x,
   output reg signed [7:0] y);

// Tapped delay line array of bytes
   reg  signed  [7:0] tap [0:3];
// For bit access use single vectors in Verilog
```

```
    integer I;

    always @(posedge clk)  //----> Behavioral style
    begin : p1
     // Compute output y with the filter coefficients weight.
     // The coefficients are [-1  3.75  3.75  -1].
     // Multiplication and division for Altera MaxPlusII can
     // be done in Verilog 2001 with signed shifts !
      y <= (tap[1] <<< 1) + tap[1] + (tap[1] >>> 1) - tap[0]
          + ( tap[1] >>> 2) + (tap[2] <<< 1) + tap[2]
          + (tap[2] >>> 1) + (tap[2] >>> 2) - tap[3];

      for (I=3; I>0; I=I-1) begin
        tap[I] <= tap[I-1];  // Tapped delay line: shift one
      end
      tap[0] <= x;    // Input in register 0
    end

endmodule

//**********************************************************
// IEEE STD 1364-2001 Verilog file: dafsm.v
// Author-EMAIL: Uwe.Meyer-Baese@ieee.org
//**********************************************************
'include "case3.v" // User-defined component

module dafsm           //--> Interface
  (input          clk, reset,
   input  [2:0]   x_in0, x_in1, x_in2,
   output [2:0]   lut,
   output reg [5:0]  y);

   reg    [2:0]   x0, x1, x2;
   wire   [2:0]   table_in, table_out;

   reg [5:0] p;  // temporary register

   assign table_in[0] = x0[0];
   assign table_in[1] = x1[0];
   assign table_in[2] = x2[0];

   always @(posedge clk or posedge reset)
   begin : DA                    //----> DA in behavioral style
     parameter s0=0, s1=1;
     reg [0:0] state;
```

```verilog
    reg [1:0] count;   // Counts the shifts

  if (reset)                  // Asynchronous reset
    state <= s0;
  else
  case (state)
    s0 : begin         // Initialization
      state <= s1;
      count = 0;
      p   <= 0;
      x0 <= x_in0;
      x1 <= x_in1;
      x2 <= x_in2;
    end
    s1 : begin                    // Processing step
      if (count == 3) begin    // Is sum of product done?
        y <= p;                  // Output of result to y and
        state <= s0;             // start next sum of product
      end
      else begin
        p <= (p >> 1) + (table_out << 2); // p/2+table*4
        x0[0] <= x0[1];
        x0[1] <= x0[2];
        x1[0] <= x1[1];
        x1[1] <= x1[2];
        x2[0] <= x2[1];
        x2[1] <= x2[2];
        count = count + 1;
        state <= s1;
      end
    end
  endcase
end

  case3 LC_Table0
  ( .table_in(table_in), .table_out(table_out));

  assign lut = table_out; // Provide test signal

endmodule

//**********************************************************
// IEEE STD 1364-2001 Verilog file: case3.v
// Author-EMAIL: Uwe.Meyer-Baese@ieee.org
//**********************************************************
```

```verilog
module case3
  (input  [2:0] table_in,  // Three bit
   output reg [2:0] table_out); // Range 0 to 6

// This is the DA CASE table for
// the 3 coefficients: 2, 3, 1

  always @(table_in)
  begin
    case (table_in)
      0 :      table_out =  0;
      1 :      table_out =  2;
      2 :      table_out =  3;
      3 :      table_out =  5;
      4 :      table_out =  1;
      5 :      table_out =  3;
      6 :      table_out =  4;
      7 :      table_out =  6;
      default : ;
    endcase
  end

endmodule

//*************************************************************
// IEEE STD 1364-2001 Verilog file: case5p.v
// Author-EMAIL: Uwe.Meyer-Baese@ieee.org
//*************************************************************
module case5p
 (input         clk,
  input  [4:0] table_in,
  output reg [4:0] table_out);       // range 0 to 25

  reg [3:0] lsbs;
  reg [1:0] msbs0;
  reg [4:0] table0out00, table0out01;

// These are the distributed arithmetic CASE tables for
// the 5 coefficients: 1, 3, 5, 7, 9

  always @(posedge clk) begin
    lsbs[0] = table_in[0];
    lsbs[1] = table_in[1];
    lsbs[2] = table_in[2];
    lsbs[3] = table_in[3];
```

```
      msbs0[0] = table_in[4];
      msbs0[1] = msbs0[0];
    end

// This is the final DA MPX stage.
  always @(posedge clk) begin
    case (msbs0[1])
      0 : table_out <= table0out00;
      1 : table_out <= table0out01;
      default : ;
    endcase
  end

// This is the DA CASE table 00 out of 1.
  always @(posedge clk) begin
    case (lsbs)
      0  : table0out00 = 0;
      1  : table0out00 = 1;
      2  : table0out00 = 3;
      3  : table0out00 = 4;
      4  : table0out00 = 5;
      5  : table0out00 = 6;
      6  : table0out00 = 8;
      7  : table0out00 = 9;
      8  : table0out00 = 7;
      9  : table0out00 = 8;
      10 : table0out00 = 10;
      11 : table0out00 = 11;
      12 : table0out00 = 12;
      13 : table0out00 = 13;
      14 : table0out00 = 15;
      15 : table0out00 = 16;
      default ;
    endcase
  end

// This is the DA CASE table 01 out of 1.
  always @(posedge clk) begin
    case (lsbs)
      0  : table0out01 = 9;
      1  : table0out01 = 10;
      2  : table0out01 = 12;
      3  : table0out01 = 13;
      4  : table0out01 = 14;
```

```
      5  : table0out01 = 15;
      6  : table0out01 = 17;
      7  : table0out01 = 18;
      8  : table0out01 = 16;
      9  : table0out01 = 17;
     10 : table0out01 = 19;
     11 : table0out01 = 20;
     12 : table0out01 = 21;
     13 : table0out01 = 22;
     14 : table0out01 = 24;
     15 : table0out01 = 25;
     default ;
   endcase
 end

endmodule

//**********************************************************
// IEEE STD 1364-2001 Verilog file: darom.v
// Author-EMAIL: Uwe.Meyer-Baese@ieee.org
//**********************************************************
//'include "220model.v"

module darom  //--> Interface
 (input          clk, reset,
  input  [2:0]   x_in0, x_in1, x_in2,
  output [2:0]   lut,
  output reg [5:0]  y);

  reg    [2:0]   x0, x1, x2;
  wire   [2:0]   table_in, table_out;

  reg [5:0] p;  // Temporary register
  wire ena;

  assign ena=1;

  assign table_in[0] = x0[0];
  assign table_in[1] = x1[0];
  assign table_in[2] = x2[0];

  always @(posedge clk or posedge reset)
  begin : DA                    //----> DA in behavioral style
    parameter s0=0, s1=1;
    reg [0:0] state;
```

```verilog
    reg [1:0] count;   // Counts the shifts

    if (reset)              // Asynchronous reset
      state <= s0;
    else
      case (state)
        s0 : begin    // Initialization
          state <= s1;
          count = 0;
          p   <= 0;
          x0 <= x_in0;
          x1 <= x_in1;
          x2 <= x_in2;
        end
        s1 : begin                    // Processing step
          if (count == 3) begin  // Is sum of product done?
            y <= (p >> 1) + (table_out << 2);// Output to y
            state <= s0;   // and start next sum of product
          end
          else begin
            p <= (p >> 1) + (table_out << 2);
            x0[0] <= x0[1];
            x0[1] <= x0[2];
            x1[0] <= x1[1];
            x1[1] <= x1[2];
            x2[0] <= x2[1];
            x2[1] <= x2[2];
            count = count + 1;
            state <= s1;
          end
        end
        default : ;
      endcase
  end

  lpm_rom rom_1     // Used ports:
  ( .outclock(clk),.address(table_in), .q(table_out));
// .inclock(clk),   .memenab(ena)); // Unused
    defparam rom_1.lpm_width = 3;
    defparam rom_1.lpm_widthad = 3;
    defparam rom_1.lpm_outdata = "REGISTERED";
    defparam rom_1.lpm_address_control = "UNREGISTERED";
    defparam rom_1.lpm_file = "darom3.mif";
```

```verilog
    assign lut = table_out; // Provide test signal

endmodule

//*********************************************************
// IEEE STD 1364-2001 Verilog file: dasign.v
// Author-EMAIL: Uwe.Meyer-Baese@ieee.org
//*********************************************************
'include "case3s.v" // User-defined component

module dasign                   //-> Interface
 (input          clk, reset,
  input signed  [3:0] x_in0, x_in1, x_in2,
  output [3:0]  lut,
  output reg signed [6:0]  y);

  reg   signed  [3:0] x0, x1, x2;
  wire signed   [2:0] table_in;
  wire signed   [3:0] table_out;

  reg [6:0] p;  // Temporary register

  assign table_in[0] = x0[0];
  assign table_in[1] = x1[0];
  assign table_in[2] = x2[0];

  always @(posedge clk or posedge reset)// DA in behavioral
  begin : DA                                      // style
    parameter s0=0, s1=1;
    integer k;
    reg [0:0] state;
    reg [2:0] count;            // Counts the shifts

    if (reset)                  // Asynchronous reset
      state <= s0;
    else
      case (state)
        s0 : begin              // Initialization step
          state <= s1;
          count = 0;
          p   <= 0;
          x0 <= x_in0;
          x1 <= x_in1;
          x2 <= x_in2;
        end
```

```
        s1 : begin                 // Processing step
          if (count == 4) begin// Is sum of product done?
            y <= p;                // Output of result to y and
            state <= s0;           // start next sum of product
          end else begin //Subtract for last accumulator step
            if (count ==3)   // i.e. p/2 +/- table_out * 8
              p <= (p >>> 1) - (table_out <<< 3);
            else             // Accumulation for all other steps
              p <= (p >>> 1) + (table_out <<< 3);
            for (k=0; k<=2; k= k+1) begin      // Shift bits
              x0[k] <= x0[k+1];
              x1[k] <= x1[k+1];
              x2[k] <= x2[k+1];
            end
            count = count + 1;
            state <= s1;
          end
        end
      endcase
  end

  case3s LC_Table0
  ( .table_in(table_in), .table_out(table_out));

  assign lut = table_out; // Provide test signal

endmodule

//***********************************************************
// IEEE STD 1364-2001 Verilog file: case3s.v
// Author-EMAIL: Uwe.Meyer-Baese@ieee.org
//***********************************************************
module case3s
  (input  [2:0] table_in,  // Three bit
   output reg [3:0] table_out); // Range -2 to 4 -> 4 bits

// This is the DA CASE table for
// the 3 coefficients: -2, 3, 1

  always @(table_in)
  begin
    case (table_in)
      0 :     table_out =  0;
      1 :     table_out = -2;
      2 :     table_out =  3;
```

```
        3 :        table_out =  1;
        4 :        table_out =  1;
        5 :        table_out = -1;
        6 :        table_out =  4;
        7 :        table_out =  2;
      default : ;
    endcase
  end

endmodule

//*********************************************************
// IEEE STD 1364-2001 Verilog file: dapara.v
// Author-EMAIL: Uwe.Meyer-Baese@ieee.org
//*********************************************************
'include "case3s.v" // User-defined component

module dapara                    //----> Interface
  (input          clk,
   input signed [3:0]  x_in,
   output reg signed[6:0]  y);

  reg signed   [2:0] x [0:3];
  wire signed [3:0] h [0:3];
  reg  signed [4:0] s0, s1;
  reg  signed [3:0] t0, t1, t2, t3;

  always @(posedge clk)  //----> DA in behavioral style
  begin : DA
    integer k,l;
    for (l=0; l<=3; l=l+1) begin      // For all 4 vectors
      for (k=0; k<=1; k=k+1) begin    // shift all bits
       x[l][k] <= x[l][k+1];
      end
    end
    for (k=0; k<=3; k=k+1) begin     // Load x_in in the
      x[k][2] <= x_in[k];            // MSBs of the registers
    end
//   y <= h[0] + (h[1] <<< 1) + (h[2] <<< 2) - (h[3] <<< 3);
// Sign extensions, pipeline register, and adder tree:
    t0 <= h[0]; t1 <= h[1]; t2 <= h[2]; t3 <= h[3];
    s0 <= t0 + (t1 <<< 1);
    s1 <= t2 - (t3 <<< 1);
    y  <= s0 + (s1 <<< 2);
    end
```

```
      genvar i;// Need to declare loop variable in Verilog 2001
      generate      //    One table for each bit in x_in
        for (i=0; i<=3; i=i+1) begin:LC_Tables
          case3s LC_Table0 ( .table_in(x[i]), .table_out(h[i]));
        end
      endgenerate

endmodule

//**********************************************************
// IEEE STD 1364-2001 Verilog file: iir.v
// Author-EMAIL: Uwe.Meyer-Baese@ieee.org
//**********************************************************
module iir #(parameter W = 14)      // Bit width - 1
      ( input  signed [W:0] x_in,   // Input
        output signed [W:0] y_out,  // Result
        input           clk);

  reg signed [W:0] x, y;

// initial begin
//   y=0;
//   x=0;
// end

// Use FFs for input and recursive part
always @(posedge clk) begin      // Note: there is a signed
  x  <= x_in;                    // integer in Verilog 2001
  y  <= x + (y >>> 1) + (y >>> 2); // >>> uses fewer LEs

//y  <= x + y / 2 + y / 4; // div with / uses more LEs

end

assign  y_out = y;              // Connect y to output pins

endmodule

//**********************************************************
// IEEE STD 1364-2001 Verilog file: iir_pipe.v
// Author-EMAIL: Uwe.Meyer-Baese@ieee.org
//**********************************************************
module iir_pipe      //----> Interface
```

```
#(parameter W = 14) // Bit width - 1
(input          clk,
 input signed [W:0]  x_in,     // Input
 output signed [W:0]  y_out);  // Result

 reg signed [W:0] x, x3, sx;
 reg signed [W:0] y, y9;

 always @(posedge clk)  // Infer FFs for input, output and
 begin                  // pipeline stages;
   x   <= x_in;         // use nonblocking FF assignments
   x3  <= (x >>> 1) + (x >>> 2);
                        // i.e. x / 2 + x / 4 = x*3/4
   sx  <= x + x3; // Sum of x element i.e. output FIR part
   y9  <= (y >>> 1) + (y >>> 4);
                        // i.e. y / 2 + y / 16 = y*9/16
   y   <= sx + y9;                    // Compute output
 end

 assign y_out = y ;    // Connect register y to output pins

endmodule

//**********************************************************
// IEEE STD 1364-2001 Verilog file: iir_par.v
// Author-EMAIL: Uwe.Meyer-Baese@ieee.org
//**********************************************************
module iir_par              //----> Interface
#(parameter W = 14) // bit width - 1
 (input          clk, reset,
  input signed [W:0]  x_in,
  output signed [W:0]  y_out,
  output          clk2);

 reg signed [W:0] x_even, xd_even, x_odd, xd_odd, x_wait;
 reg signed [W:0] y_even, y_odd, y_wait, y;
 reg signed [W:0] sum_x_even, sum_x_odd;
 reg         clk_div2;

 always @(posedge clk)            // Clock divider by 2
 begin : clk_divider              // for input clk
   clk_div2 <= ! clk_div2;
 end

 always @(posedge clk)            // Split x into even
```

```
  begin : Multiplex            // and odd samples;
    parameter even=0, odd=1;   // recombine y at clk rate
    reg [0:0] state;

    if (reset)                 // Asynchronous reset
      state <= even;
    else
      case (state)
        even : begin
            x_even <= x_in;
            x_odd <= x_wait;
            y <= y_wait;
            state <= odd;
        end
        odd : begin
          x_wait <= x_in;
          y <= y_odd;
          y_wait <= y_even;
          state <= even;
        end
      endcase
  end

  assign y_out = y;
  assign clk2  = clk_div2;

  always @(negedge clk_div2)
  begin: Arithmetic
    xd_even <= x_even;
    sum_x_even <= x_odd+ (xd_even >>> 1) + (xd_even >>> 2);
                  // i.e. x_odd + x_even / 2 + x_even /4
    y_even <= sum_x_even + (y_even >>> 1) + (y_even >>> 4);
              // i.e. sum_x_even + y_even / 2 + y_even /16
    xd_odd <= x_odd;
    sum_x_odd <= xd_even + (xd_odd >>> 1) + (xd_odd >>> 4);
                  // i.e. x_even + xd_odd / 2 + xd_odd /4
    y_odd  <= sum_x_odd + (y_odd >>> 1)+ (y_odd >>> 4);
              // i.e. sum_x_odd + y_odd / 2 + y_odd / 16
  end

endmodule

//*************************************************************
// IEEE STD 1364-2001 Verilog file: cic3r32.v
// Author-EMAIL: Uwe.Meyer-Baese@ieee.org
```

```verilog
//**********************************************************
module cic3r32  //----> Interface
 (input         clk, reset,
  input signed [7:0] x_in,
  output signed [9:0] y_out,
  output reg clk2);

  parameter hold=0, sample=1;
  reg [1:0] state;
  reg [4:0]  count;
  reg signed [7:0] x;        // Registered input
  reg signed [25:0] i0, i1 , i2; // I section  0, 1, and 2
  reg signed [25:0] i2d1, i2d2, c1, c0;       // I + COMB 0
  reg signed [25:0] c1d1, c1d2, c2;      // COMB section 1
  reg signed [25:0] c2d1, c2d2, c3;      // COMB section 2

  always @(posedge clk or posedge reset)
  begin : FSM
    if (reset) begin          // Asynchronous reset
      count <= 0;
      state <= hold;
      clk2  <= 0;
    end else begin
      if (count == 31) begin
          count <= 0;
          state <= sample;
          clk2  <= 1;
      end else begin
        count <= count + 1;
        state <= hold;
        clk2  <= 0;
      end
    end
  end

  always @(posedge clk) // 3 integrator sections
  begin : Int
      x     <= x_in;
      i0    <= i0 + x;
      i1    <= i1 + i0 ;
      i2    <= i2 + i1 ;
   end

  always @(posedge clk) // 3 comb sections
```

```
    begin : Comb
      if (state == sample) begin
        c0    <= i2;
        i2d1 <= c0;
        i2d2 <= i2d1;
        c1    <= c0 - i2d2;
        c1d1 <= c1;
        c1d2 <= c1d1;
        c2    <= c1  - c1d2;
        c2d1 <= c2;
        c2d2 <= c2d1;
        c3    <= c2  - c2d2;
      end
    end

    assign y_out = c3[25:16];

endmodule

//*************************************************************
// IEEE STD 1364-2001 Verilog file: cic3s32.v
// Author-EMAIL: Uwe.Meyer-Baese@ieee.org
//*************************************************************
module cic3s32              //----> Interface
 (input        clk, reset,
  output reg clk2,
  input  signed [7:0] x_in,
  output signed [9:0] y_out);

  parameter hold=0, sample=1;
  reg [1:0] state;
  reg [4:0]  count;
  reg signed [7:0]  x;                    // Registered input
  reg signed [25:0] i0;                   // I section 0
  reg signed [20:0] i1;                   // I section 1
  reg signed [15:0] i2;                   // I section 2
  reg signed [13:0] i2d1, i2d2, c1, c0;   // I+C0
  reg signed [12:0] c1d1, c1d2, c2;       // COMB 1
  reg signed [11:0] c2d1, c2d2, c3;       // COMB 2

  always @(posedge clk or posedge reset)
  begin : FSM
    if (reset) begin            // Asynchronous reset
      count <= 0;
      state <= hold;
```

```
          clk2  <= 0;
      end else begin
        if (count == 31) begin
          count <= 0;
          state <= sample;
          clk2  <= 1;
        end
        else begin
          count <= count + 1;
          state <= hold;
          clk2  <= 0;
        end
      end
    end


    always @(posedge clk) // 3 integrator sections
    begin : Int
        x   <= x_in;
        i0  <= i0 + x;
        i1  <= i1 + i0[25:5];
        i2  <= i2 + i1[20:5];
    end


    always @(posedge clk) // 3 comb sections
    begin : Comb
      if (state == sample) begin
        c0   <= i2[15:2];
        i2d1 <= c0;
        i2d2 <= i2d1;
        c1   <= c0 - i2d2;
        c1d1 <= c1[13:1];
        c1d2 <= c1d1;
        c2   <= c1[13:1] - c1d2;
        c2d1 <= c2[12:1];
        c2d2 <= c2d1;
        c3   <= c2[12:1] - c2d2;
      end
    end

  assign y_out = c3[11:2];

endmodule

//***********************************************************
// IEEE STD 1364-2001 Verilog file: db4poly.v
```

```verilog
// Author-EMAIL: Uwe.Meyer-Baese@ieee.org
//**********************************************************
module db4poly      //----> Interface
 (input           clk, reset,
  output          clk2,
  input signed [7:0]   x_in,
  output signed [16:0]  x_e, x_o, g0, g1, // Test signals
  output signed [8:0]   y_out);

  reg signed [7:0] x_odd, x_even, x_wait;
  reg  clk_div2;

// Register for multiplier, coefficients, and taps
  reg signed [16:0] m0, m1, m2, m3, r0, r1, r2, r3;
  reg signed [16:0] x33, x99, x107;
  reg signed [16:0] y;

  always @(posedge clk or posedge reset) // Split into even
  begin : Multiplex             // and odd samples at clk rate
    parameter even=0, odd=1;
    reg [0:0] state;

    if (reset)                  // Asynchronous reset
      state <= even;
    else
      case (state)
        even : begin
          x_even <= x_in;
          x_odd  <= x_wait;
          clk_div2 = 1;
          state <= odd;
        end
        odd : begin
          x_wait <= x_in;
          clk_div2 = 0;
          state <= even;
        end
      endcase
  end

  always @(x_odd, x_even)
  begin : RAG
// Compute auxiliary multiplications of the filter
    x33  = (x_odd <<< 5) + x_odd;
```

```
    x99  = (x33 <<< 1) + x33;
    x107 = x99 + (x_odd << 3);
// Compute all coefficients for the transposed filter
    m0 = (x_even <<< 7) - (x_even <<< 2); // m0 = 124
    m1 = x107 <<< 1;                      // m1 = 214
    m2 = (x_even <<< 6) - (x_even <<< 3)
                                + x_even; // m2 =  57
    m3 = x33;                             // m3 = -33
  end

  always @(negedge clk_div2) // Infer registers;
  begin : AddPolyphase      // use nonblocking assignments
//---------- Compute filter G0
    r0 <=  r2 + m0;         // g0 = 128
    r2 <=  m2;              // g2 = 57
//---------- Compute filter G1
    r1 <=  -r3 + m1;        // g1 = 214
    r3 <=  m3;              // g3 = -33
// Add the polyphase components
    y <= r0 + r1;
  end

// Provide some test signals as outputs
  assign x_e = x_even;
  assign x_o = x_odd;
  assign clk2 = clk_div2;
  assign g0 = r0;
  assign g1 = r1;

  assign y_out = y >>> 8; // Connect y / 256 to output

endmodule

//**********************************************************
// IEEE STD 1364-2001 Verilog file: rc_sinc.v
// Author-EMAIL: Uwe.Meyer-Baese@ieee.org
//**********************************************************
module rc_sinc #(parameter OL = 2, //Output buffer length-1
                           IL = 3,  //Input buffer length -1
                           L  = 10) // Filter length -1
  (input clk, reset,  // Clock + reset for the registers
   input signed [7:0] x_in,
   output [3:0] count_o,
   output ena_in_o, ena_out_o, ena_io_o,
   output signed [8:0] f0_o, f1_o, f2_o,
```

```verilog
output signed [8:0] y_out);

reg [3:0] count; // Cycle R_1*R_2
reg ena_in, ena_out, ena_io; // FSM enables
reg signed [7:0] x [0:10]; // TAP registers for 3 filters
reg signed [7:0] ibuf [0:3]; // TAP in registers
reg signed [7:0] obuf [0:2]; // TAP out registers
reg signed [8:0] f0, f1, f2; // Filter outputs

// Constant arrays for multiplier and taps:
wire signed [8:0] c0 [0:10];
wire signed [8:0] c2 [0:10];

// filter coefficients for filter c0
assign c0[0] = -19; assign c0[1] = 26;  assign c0[2]=-42;
assign c0[3] = 106; assign c0[4] = 212; assign c0[5]=-53;
assign c0[6] = 29;  assign c0[7] = -21; assign c0[8]=16;
assign c0[9] = -13; assign c0[10] = 11;

// filter coefficients for filter c2
assign c2[0] = 11; assign c2[1] = -13;assign c2[2] = 16;
assign c2[3] = -21;assign c2[4] = 29; assign c2[5] = -53;
assign c2[6] = 212;assign c2[7] = 106;assign c2[8] = -42;
assign c2[9] = 26; assign c2[10] = -19;

always @(posedge reset or posedge clk)
begin : FSM  // Control the system and sample at clk rate
  if (reset)                 // Asynchronous reset
    count <= 0;
  else begin
    if (count == 11)
      count <= 0;
    else
      count <= count + 1;
  end
end

always @(posedge clk)
begin           // set the enable signal for the TAP lines
    case (count)
      2, 5, 8, 11 : ena_in <= 1;
      default     : ena_in <= 0;
    endcase
```

```verilog
    case (count)
      4, 8    : ena_out <= 1;
      default : ena_out <= 0;
    endcase

    if (count == 0)
      ena_io <= 1;
    else
      ena_io <= 0;
end

always @(posedge clk)        //----> Input delay line
begin :  INPUTMUX
  integer I;     // loop variable

  if (ena_in) begin
    for (I=IL; I>=1; I=I-1)
      ibuf[I] <= ibuf[I-1];        // shift one
    ibuf[0] <= x_in;               // Input in register 0
  end
end

always @(posedge clk)        //----> Output delay line
begin  : OUPUTMUX
  integer I;     // loop variable

  if (ena_io) begin  // store 3 samples in output buffer
    obuf[0] <= f0;
    obuf[1] <= f1;
    obuf[2] <= f2;
  end
  else if (ena_out) begin
    for (I=OL; I>=1; I=I-1)
      obuf[I] <= obuf[I-1];        // shift one
  end
end

always @(posedge clk)        //----> One tapped delay line
begin : TAP                  // get 4 samples at one time
  integer I;     // loop variable

  if (ena_io) begin
  for (I=0; I<=3; I=I+1)
    x[I] <= ibuf[I];   // take over input buffer
```

```
      for (I=4; I<=10; I=I+1) // 0->4; 4->8 etc.
        x[I] <= x[I-4];        // shift 4 taps

    end
  end

  always @(posedge clk) // Compute sum-of-products for f0
  begin : SOP0
    reg signed [16:0] sum; // temp sum
    reg signed [16:0] p [0:10]; // temp products
    integer I;

    for (I=0; I<=L; I=I+1) // Infer L+1  multiplier
      p[I] = c0[I] * x[I];

    sum = p[0];
    for (I=1; I<=L; I=I+1)        // Compute the direct
      sum = sum + p[I];          // filter adds

    f0 <= sum >>> 8;
  end

always @(posedge clk) // Compute sum-of-products for f1
 begin : SOP1
   f1 <= x[5];  // No scaling, i.e., unit inpulse
 end

  always @(posedge clk) // Compute sum-of-products for f2
  begin : SOP2
    reg signed[16:0] sum; // temp sum
    reg signed [16:0] p [0:10]; // temp products
    integer I;

    for (I=0; I<=L; I=I+1) // Infer L+1  multiplier
      p[I] = c2[I] * x[I];

    sum = p[0];
    for (I=1; I<=L; I=I+1)        // Compute the direct
      sum = sum + p[I];          // filter adds

      f2 <= sum >>> 8;
  end
```

```verilog
   // Provide some test signals as outputs
   assign f0_o = f0;
   assign f1_o = f1;
   assign f2_o = f2;
   assign count_o = count;
   assign ena_in_o = ena_in;
   assign ena_out_o = ena_out;
   assign ena_io_o = ena_io;

   assign y_out = obuf[OL]; // Connect to output

endmodule

//**********************************************************
// IEEE STD 1364-2001 Verilog file: farrow.v
// Author-EMAIL: Uwe.Meyer-Baese@ieee.org
//**********************************************************
module farrow #(parameter IL = 3) // Input buffer length -1
      (input clk, reset,  // Clock/reset for the registers
       input signed [7:0] x_in,
       output [3:0] count_o,
       output ena_in_o, ena_out_o,
       output signed [8:0] c0_o, c1_o, c2_o, c3_o,
       output [8:0] d_out,
       output reg signed [8:0] y_out);

  reg [3:0] count; // Cycle R_1*R_2
  wire [6:0] delta; // Increment d
  reg ena_in, ena_out; // FSM enables
  reg signed [7:0] x [0:3];
  reg signed [7:0] ibuf [0:3]; // TAP registers
  reg [8:0]  d; // Fractional Delay scaled to 8 bits
  // Lagrange matrix outputs:
  reg signed [8:0] c0, c1, c2, c3;

  assign delta = 85;

  always @(posedge reset or posedge clk)     // Control the
  begin : FSM                  // system and sample at clk rate
    reg [8:0] dnew;
    if (reset) begin                 // Asynchronous reset
      count <= 0;
      d <= delta;
    end else begin
      if (count == 11)
```

```verilog
        count <= 0;
      else
        count <= count + 1;
      if (ena_out) begin        // Compute phase delay
       dnew = d + delta;
         if (dnew >= 255)
           d <= 0;
         else
           d <= dnew;
      end
   end
end

always @(posedge clk)
begin            // Set the enable signals for the TAP lines
    case (count)
      2, 5, 8, 11 : ena_in <= 1;
      default     : ena_in <= 0;
    endcase

    case (count)
      3, 7, 11 : ena_out <= 1;
      default  : ena_out <= 0;
    endcase
end

always @(posedge clk)        //----> One tapped delay line
begin : TAP
  integer I;    // loop variable

  if (ena_in) begin
    for (I=1; I<=IL; I=I+1)
      ibuf[I-1] <= ibuf[I];    // Shift one

    ibuf[IL] <= x_in;                // Input in register IL

  end
end

always @(posedge clk)
begin : GET                    // Get 4 samples at one time
  integer I;    // loop variable

  if (ena_out) begin
```

```
    for (I=0; I<=IL; I=I+1)
      x[I] <= ibuf[I];    // take over input buffer
    end
  end

  // Compute sum-of-products:
  always @(posedge clk) // Compute sum-of-products for f0
  begin : SOP
    reg signed [8:0] y; // temp's

// Matrix multiplier iV=inv(Vandermonde) c=iV*x(n-1:n+2)'
//       x(0)    x(1)         x(2)       x(3)
// iV=     0    1.0000         0          0
//    -0.3333  -0.5000     1.0000    -0.1667
//     0.5000  -1.0000     0.5000         0
//    -0.1667   0.5000    -0.5000     0.1667
    if (ena_out) begin

      c0 <= x[1];
      c1 <= (-85 * x[0] >>> 8) - (x[1]/2) + x[2] -
                                 (43 * x[3] >>> 8);
      c2 <= ((x[0] + x[2]) >>> 1) - x[1] ;
      c3 <= ((x[1] - x[2]) >>> 1) +
                           (43 * (x[3] - x[0]) >>> 8);

// Farrow structure = Lagrange with Horner schema
// for u=0:3, y=y+f(u)*d^u; end;
      y = c2 + ((c3 * d) >>> 8); // d is scale by 256
      y = ((y * d) >>> 8) + c1;
      y = ((y * d) >>> 8) + c0;

      y_out <= y; // Connect to output + store in register
    end
  end

  assign c0_o = c0; // Provide test signals as outputs
  assign c1_o = c1;
  assign c2_o = c2;
  assign c3_o = c3;
  assign count_o = count;
  assign ena_in_o = ena_in;
  assign ena_out_o = ena_out;
  assign d_out = d;
```

```
        endmodule

        //**********************************************************
        // IEEE STD 1364-2001 Verilog file: cmoms.v
        // Author-EMAIL: Uwe.Meyer-Baese@ieee.org
        //**********************************************************
        module cmoms #(parameter IL = 3)  // Input buffer length -1
                (input clk, reset,  // Clock/reset for registers
                 input signed [7:0] x_in,
                 output [3:0] count_o,
                 output ena_in_o, ena_out_o,
                 output signed [8:0] c0_o, c1_o, c2_o, c3_o, xiir_o,
                 output signed [8:0] y_out);


           reg [3:0] count; // Cycle R_1*R_2
           reg [1:0] t;
           reg ena_in, ena_out; // FSM enables
           reg signed [7:0] x [0:3];
           reg signed [7:0] ibuf [0:3]; // TAP registers
           reg signed [8:0] xiir; // iir filter output

           reg signed [16:0] y, y0, y1, y2, y3, h0, h1; // temp's

           // Spline matrix output:
           reg signed [8:0] c0, c1, c2, c3;

           // Precomputed value for d**k :
           wire signed [8:0] d1 [0:2];
           wire signed [8:0] d2 [0:2];
           wire signed [8:0] d3 [0:2];

           assign d1[0] = 0; assign d1[1] = 85; assign d1[2] = 171;
           assign d2[0] = 0; assign d2[1] = 28; assign d2[2] = 114;
           assign d3[0] = 0; assign d3[1] =  9; assign d3[2] =  76;


           always @(posedge reset or posedge clk) // Control the
           begin : FSM                    // system sample at clk rate
             if (reset) begin             // Asynchronous reset
               count <= 0;
               t <= 1;
             end else begin
               if (count == 11)
                 count <= 0;
```

```
        else
          count <= count + 1;
        if (ena_out)
          if (t>=2)     // Compute phase delay
            t <= 0;
          else
            t <= t + 1;
      end
    end
    assign t_out = t;

    always @(posedge clk) // set the enable signal
    begin                 // for the TAP lines
        case (count)
          2, 5, 8, 11 : ena_in <= 1;
          default     : ena_in <= 0;
        endcase

        case (count)
          3, 7, 11    : ena_out <= 1;
          default : ena_out <= 0;
        endcase
    end

//   Coeffs: H(z)=1.5/(1+0.5z^-1)
    always @(posedge clk)       //----> Behavioral Style
    begin : IIR  // Compute iir coefficients first
      reg signed [8:0] x1;    // x * 1

      if (ena_in) begin
        xiir <= (3 * x1 >>> 1) - (xiir >>> 1);
        x1 = x_in;
      end
    end

    always @(posedge clk)       //----> One tapped delay line
    begin : TAP
      integer I;    // loop variable

      if (ena_in) begin
        for (I=1; I<=IL; I=I+1)
          ibuf[I-1] <= ibuf[I];    // Shift one

        ibuf[IL] <= xiir;          // Input in register IL
```

```
      end
    end

    always @(posedge clk)        //----> One tapped delay line
    begin : GET                  // get 4 samples at one time
      integer I;    // loop variable

      if (ena_out) begin
      for (I=0; I<=IL; I=I+1)
        x[I] <= ibuf[I];    // take over input buffer
      end
    end

    // Compute sum-of-products:
    always @(posedge clk) // Compute sum-of-products for f0
    begin :  SOP
// Matrix multiplier C-MOMS matrix:
//    x(0)        x(1)        x(2)        x(3)
//    0.3333      0.6667      0           0
//   -0.8333      0.6667      0.1667      0
//    0.6667     -1.5         1.0        -0.1667
//   -0.1667      0.5        -0.5         0.1667
      if (ena_out) begin

        c0 <= (85 * x[0] + 171 * x[1]) >>> 8;
        c1 <= (171 * x[1] - 213 * x[0] + 43 * x[2]) >>> 8;
        c2 <= (171 * x[0] - (43 * x[3]) >>> 8)
                              - (3 * x[1] >>> 1) + x[2];
        c3 <= (43 * (x[3] - x[0]) >>> 8)
                              + ((x[1] - x[2]) >>> 1);

// No Farrow structure, parallel LUT for delays
// for u=0:3, y=y+f(u)*d^u; end;

        y0 <= c0 * 256; // Use pipelined adder tree
        y1 <= c1 * d1[t];
        y2 <= c2 * d2[t];
        y3 <= c3 * d3[t];
        h0 <= y0 + y1;
        h1 <= y2 + y3;
        y  <= h0 + h1;
      end
    end
    assign y_out = y >>> 8; // Connect to output
```

```
      assign c0_o = c0; // Provide some test signals as outputs
      assign c1_o = c1;
      assign c2_o = c2;
      assign c3_o = c3;
      assign count_o = count;
      assign ena_in_o = ena_in;
      assign ena_out_o = ena_out;
      assign xiir_o = xiir;

endmodule

//*********************************************************
// IEEE STD 1364-2001 Verilog file: db4latti.v
// Author-EMAIL: Uwe.Meyer-Baese@ieee.org
//*********************************************************
module db4latti
  (input          clk, reset,
   output         clk2,
   input signed   [7:0]  x_in,
   output signed [16:0] x_e, x_o,
   output reg signed [8:0]   g, h);

      reg signed   [7:0]  x_wait;
      reg signed   [16:0] sx_up, sx_low;
      reg  clk_div2;
      wire signed [16:0] sxa0_up, sxa0_low;
      wire signed [16:0] up0, up1, low1;
      reg signed   [16:0] low0;

      always @(posedge clk or posedge reset) // Split into even
      begin : Multiplex            // and odd samples at clk rate
        parameter even=0, odd=1;
        reg [0:0] state;

        if (reset)                  // Asynchronous reset
          state <= even;
        else
          case (state)
            even : begin
              // Multiply with 256*s=124
              sx_up   <= (x_in <<< 7) - (x_in <<< 2);
              sx_low  <= (x_wait <<< 7) - (x_wait <<< 2);
              clk_div2 <= 1;
              state <= odd;
```

```
              end
              odd : begin
                x_wait <= x_in;
                clk_div2 <= 0;
                state <= even;
              end
           endcase
      end


//******** Multipy a[0] = 1.7321
// Compute: (2*sx_up  - sx_up /4)-(sx_up /64 + sx_up /256)
   assign sxa0_up  = ((sx_up <<< 1)  - (sx_up >>> 2))
                      - ((sx_up >>> 6) + (sx_up >>> 8));
// Compute: (2*sx_low - sx_low/4)-(sx_low/64 + sx_low/256)
   assign sxa0_low = ((sx_low <<< 1) - (sx_low >>> 2))
                      - ((sx_low >>> 6) + (sx_low >>> 8));


//******** First stage -- FF in lower tree
   assign up0 = sxa0_low + sx_up;
   always @(negedge clk_div2)
   begin: LowerTreeFF
       low0 <= sx_low - sxa0_up;
   end


//******** Second stage: a[1]=-0.2679
// Compute:   (up0 - low0/4) - (low0/64 + low0/256);
   assign up1  = (up0 - (low0 >>> 2))
                   - ((low0 >>> 6) + (low0 >>> 8));
// Compute: (low0 + up0/4) + (up0/64  +  up0/256)
   assign low1 = (low0 + (up0 >>> 2))
                        + ((up0 >>> 6) + (up0 >>> 8));

   assign x_e  = sx_up;        // Provide some extra
   assign x_o  = sx_low;       // test signals
   assign clk2 = clk_div2;

   always @(negedge clk_div2)
   begin: OutputScale
     g <= up1 >>> 8;        // i.e. up1 / 256
     h <= low1 >>> 8;       // i.e. low1 / 256;
   end

endmodule

//*********************************************************
```

```
// IEEE STD 1364-2001 Verilog file: rader7.v
// Author-EMAIL: Uwe.Meyer-Baese@ieee.org
//*********************************************************
module rader7              //---> Interface
 (input          clk, reset,
  input   [7:0]  x_in,
  output reg signed [10:0] y_real, y_imag);

  reg signed [10:0]  accu;                  // Signal for X[0]
// Direct bit access of 2D vector in Quartus Verilog 2001
// possible no auxiliary signal for this purpose necessary
  reg signed [18:0]  im [0:5];
  reg signed [18:0]  re [0:5];
// real is keyword in Verilog and can not be an identifier
                                  // Tapped delay line array
  reg signed [18:0]  x57, x111, x160, x200, x231, x250 ;
                                  // The filter coefficients
  reg signed [18:0]  x5, x25, x110, x125, x256;
                          // Auxiliary filter coefficients
  reg signed [7:0]   x, x_0;             // Signals for x[0]

  always @(posedge clk or posedge reset)    // State machine
  begin : States                          // for RADER filter
    parameter Start=0, Load=1, Run=2;
    reg [1:0] state;
    reg [4:0] count;

    if (reset)                  // Asynchronous reset
      state <= Start;
    else
      case (state)
        Start : begin         // Initialization step
          state <= Load;
          count <= 1;
          x_0 <= x_in;        // Save x[0]
          accu <= 0 ;         // Reset accumulator for X[0]
          y_real  <= 0;
          y_imag  <= 0;
        end
        Load : begin // Apply x[5],x[4],x[6],x[2],x[3],x[1]
          if (count == 8)     // Load phase done ?
            state <= Run;
          else begin
            state <= Load;
```

```verilog
          accu <= accu + x;
        end
        count <= count + 1;
      end
      Run : begin // Apply again x[5],x[4],x[6],x[2],x[3]
        if (count == 15) begin // Run phase done ?
          y_real  <= accu;        // X[0]
          y_imag  <= 0;  // Only re inputs => Im(X[0])=0
          state <= Start;        // Output of result
        end                      // and start again
        else begin
          y_real  <= (re[0] >>> 8) + x_0;
                                  // i.e. re[0]/256+x[0]
          y_imag  <= (im[0] >>> 8);    // i.e. im[0]/256
          state <= Run;
        end
        count <= count + 1;
      end
    endcase
end

always @(posedge clk)    // Structure of the two FIR
begin : Structure        // filters in transposed form
  x <= x_in;
  // Real part of FIR filter in transposed form
  re[0] <= re[1] + x160  ;   // W^1
  re[1] <= re[2] - x231  ;   // W^3
  re[2] <= re[3] - x57   ;   // W^2
  re[3] <= re[4] + x160  ;   // W^6
  re[4] <= re[5] - x231  ;   // W^4
  re[5] <= -x57;             // W^5

  // Imaginary part of FIR filter in transposed form
  im[0] <= im[1] - x200  ;   // W^1
  im[1] <= im[2] - x111  ;   // W^3
  im[2] <= im[3] - x250  ;   // W^2
  im[3] <= im[4] + x200  ;   // W^6
  im[4] <= im[5] + x111  ;   // W^4
  im[5] <= x250;             // W^5
end

always @(posedge clk)    //  Note that all signals
begin : Coeffs           //  are globally defined
// Compute the filter coefficients and use FFs
```

```
    x160   <= x5 <<< 5;         // i.e. 160 = 5 * 32;
    x200   <= x25 <<< 3;        // i.e. 200 = 25 * 8;
    x250   <= x125 <<< 1;       // i.e. 250 = 125 * 2;
    x57    <= x25 + (x <<< 5);  // i.e. 57 = 25 + 32;
    x111   <= x110 + x;         // i.e. 111 = 110 + 1;
    x231   <= x256 - x25;       // i.e. 231 = 256 - 25;
  end

  always @*                  // Note that all signals
  begin : Factors            // are globally defined
  // Compute the auxiliary factor for RAG without an FF
    x5   = (x <<< 2) + x;  // i.e. 5 = 4 + 1;
    x25  = (x5 <<< 2) + x5;       // i.e. 25 = 5*4 + 5;
    x110 = (x25 <<< 2) + (x5 <<< 2);// i.e. 110 = 25*4+5*4;
    x125 = (x25 <<< 2) + x25;     // i.e. 125 = 25*4+25;
    x256 = x <<< 8;               // i.e. 256 = 2 ** 8;
  end

endmodule

//**********************************************************
// IEEE STD 1364-2001 Verilog file: ccmul.v
// Author-EMAIL: Uwe.Meyer-Baese@ieee.org
//**********************************************************
//'include "220model.v"

module ccmul #(parameter W2 = 17,    // Multiplier bit width
                         W1 = 9,     // Bit width c+s sum
                         W  = 8)     // Input bit width
  (input clk,  // Clock for the output register
   input signed [W-1:0] x_in, y_in, c_in,  // Inputs
   input signed [W1-1:0] cps_in, cms_in,  // Inputs
   output reg signed [W-1:0]    r_out, i_out); // Results


   wire signed [W-1:0] x, y, c ;       // Inputs and outputs
   wire signed [W2-1:0] r, i, cmsy, cpsx, xmyc, sum; //Prod.
   wire signed [W1-1:0] xmy, cps, cms, sxtx, sxty;//x-y etc.


   wire  clken, cr1, ovl1, cin1, aclr, ADD, SUB;
                                       // Auxiliary signals
   assign cin1=0; assign aclr=0; assign ADD=1; assign SUB=0;
   assign cr1=0; assign sum=0; assign clken=0;
                                       // Default for add
```

```
    assign x   = x_in;   // x
    assign y   = y_in;   // j * y
    assign c   = c_in;   // cos
    assign cps = cps_in; // cos + sin
    assign cms = cms_in; // cos - sin

    always @(posedge clk) begin
      r_out <= r[W2-3:W-1];        // Scaling and FF for output
      i_out <= i[W2-3:W-1];
    end

//********* ccmul with 3 mul. and 3 add/sub  **************
    assign sxtx = x;      // Possible growth for
    assign sxty = y;      // sub_1 -> sign extension

    lpm_add_sub sub_1                   // Sub: x - y
    ( .result(xmy), .dataa(sxtx), .datab(sxty));// Used ports
//   .add_sub(SUB), .cout(cr1), .overflow(ovl1), .cin(cin1),
//   .clken(clken), .clock(clk), .aclr(aclr));  // Unused
      defparam sub_1.lpm_width = W1;
      defparam sub_1.lpm_representation = "SIGNED";
      defparam sub_1.lpm_direction = "sub";

    lpm_mult mul_1                      // Multiply  (x-y)*c = xmyc
    ( .dataa(xmy), .datab(c), .result(xmyc)); // Used ports
//   .sum(sum), .clock(clk), .clken(clken), .aclr(aclr));
                                        // Unused ports
      defparam mul_1.lpm_widtha = W1;
      defparam mul_1.lpm_widthb = W;
      defparam mul_1.lpm_widthp = W2;
      defparam mul_1.lpm_widths = W2;
      defparam mul_1.lpm_representation = "SIGNED";

    lpm_mult mul_2                      // Multiply (c-s)*y = cmsy
    ( .dataa(cms), .datab(y), .result(cmsy)); // Used ports
//   .sum(sum), .clock(clk), .clken(clken), .aclr(aclr));
                                        // Unused ports
      defparam mul_2.lpm_widtha = W1;
      defparam mul_2.lpm_widthb = W;
      defparam mul_2.lpm_widthp = W2;
      defparam mul_2.lpm_widths = W2;
      defparam mul_2.lpm_representation = "SIGNED";

    lpm_mult mul_3                      // Multiply (c+s)*x = cpsx
```

```
   ( .dataa(cps), .datab(x), .result(cpsx)); // Used ports
//   .sum(sum), .clock(clk), .clken(clken), .aclr(aclr));
                                              // Unused ports
     defparam mul_3.lpm_widtha= W1;
     defparam mul_3.lpm_widthb = W;
     defparam mul_3.lpm_widthp = W2;
     defparam mul_3.lpm_widths = W2;
     defparam mul_3.lpm_representation = "SIGNED";

  lpm_add_sub add_1         // Add:  r <= (x-y)*c + (c-s)*y
   ( .dataa(cmsy), .datab(xmyc), .result(r));  // Used ports
//   .add_sub(ADD), .cout(cr1), .overflow(ovl1), .cin(cin1),
//   .clken(clken), .clock(clk), .aclr(aclr));  // Unused
     defparam add_1.lpm_width = W2;
     defparam add_1.lpm_representation = "SIGNED";
     defparam add_1.lpm_direction = "add";

  lpm_add_sub sub_2         // Sub: i <= (c+s)*x - (x-y)*c
   ( .dataa(cpsx), .datab(xmyc), .result(i)); // Used ports
//  .add_sub(SUB), .cout(cr1), .overflow(ovl1), .clock(clk),
//   .cin(cin1),  .clken(clken), .aclr(aclr));  // Unused
     defparam sub_2.lpm_width = W2;
     defparam sub_2.lpm_representation = "SIGNED";
     defparam sub_2.lpm_direction = "sub";

endmodule

//*************************************************************
// IEEE STD 1364-2001 Verilog file: bfproc.v
// Author-EMAIL: Uwe.Meyer-Baese@ieee.org
//*************************************************************
//'include "220model.v"
//'include "ccmul.v"

module bfproc #(parameter W2 = 17,  // Multiplier bit width
                W1 = 9,     // Bit width c+s sum
                W  = 8)     // Input bit width
 (input clk,  // Clock for the output register
  input signed [W-1:0] Are_in, Aim_in,      // 8-bit inputs
  input signed [W-1:0] Bre_in, Bim_in, c_in,// 8-bit inputs
  input signed [W1-1:0] cps_in, cms_in,  // coefficients
  output reg signed [W-1:0] Dre_out, Dim_out,// registered
  output signed   [W-1:0] Ere_out, Eim_out); // results

  reg signed [W-1:0] dif_re, dif_im;       // Bf out
```

```verilog
  reg signed [W-1:0] Are, Aim, Bre, Bim;   // Inputs integer
  reg signed [W-1:0] c;                     // Input
  reg signed [W1-1:0] cps, cms;             // Coefficient in

  always @(posedge clk)    // Compute the additions of the
  begin                    // butterfly using integers
    Are     <= Are_in;     // and store inputs
    Aim     <= Aim_in;     // in flip-flops
    Bre     <= Bre_in;
    Bim     <= Bim_in;
    c       <= c_in;             // Load from memory cos
    cps     <= cps_in;           // Load from memory cos+sin
    cms     <= cms_in;           // Load from memory cos-sin
    Dre_out <= (Are >>> 1) + (Bre >>> 1); // Are/2 + Bre/2
    Dim_out <= (Aim >>> 1) + (Bim >>> 1); // Aim/2 + Bim/2
  end

    // No FF because butterfly difference "diff" is not an
  always @(*)                              // output port
  begin
    dif_re = (Are >>> 1) - (Bre >>> 1);//i.e. Are/2 - Bre/2
    dif_im = (Aim >>> 1) - (Bim >>> 1);//i.e. Aim/2 - Bim/2
  end

//*** Instantiate the complex twiddle factor multiplier
ccmul ccmul_1                     // Multiply (x+jy)(c+js)
( .clk(clk), .x_in(dif_re), .y_in(dif_im),  .c_in(c),
  .cps_in(cps), .cms_in(cms), .r_out(Ere_out),
                                  .i_out(Eim_out));

endmodule

//*********************************************************
// IEEE STD 1364-2001 Verilog file: lfsr.v
// Author-EMAIL: Uwe.Meyer-Baese@ieee.org
//*********************************************************
module lfsr            //----> Interface
  (input     clk,
  output [6:1] y);  // Result

  reg [6:1] ff; // Note that reg is keyword in Verilog and
                          // can not be variable name
  integer i;

  always @(posedge clk) begin // Length-6 LFSR with xnor
```

```
    ff[1] <= ff[5] ~^ ff[6]; // Use nonblocking assignment
    for (i=6; i>=2 ; i=i-1) // Tapped delay line: shift one
      ff[i] <= ff[i-1];
  end

  assign   y = ff;            // Connect to I/O pins

endmodule

//***********************************************************
// IEEE STD 1364-2001 Verilog file: lfsr6s3.v
// Author-EMAIL: Uwe.Meyer-Baese@ieee.org
//***********************************************************
module lfsr6s3             //----> Interface
  (input         clk,
   output [6:1]  y);  // Result

  reg [6:1] ff; // Note that reg is keyword in Verilog and
                            // can not be variable name

  always @(posedge clk) begin // Implement three-step
    ff[6] <= ff[3];            // length-6 LFSR with xnor;
    ff[5] <= ff[2];            // use nonblocking assignments
    ff[4] <= ff[1];
    ff[3] <= ff[5] ~^ ff[6];
    ff[2] <= ff[4] ~^ ff[5];
    ff[1] <= ff[3] ~^ ff[4];
  end

  assign  y = ff;

endmodule

//***********************************************************
// IEEE STD 1364-2001 Verilog file: ammod.v
// Author-EMAIL: Uwe.Meyer-Baese@ieee.org
//***********************************************************
module ammod #(parameter W = 8)  // Bit width - 1
  (input         clk,                  //----> Interface
   input signed [W:0] r_in,
   input signed [W:0] phi_in,
   output reg signed [W:0] x_out, y_out, eps);

  reg signed [W:0] x [0:3]; // There is bit access in 2D
```

```verilog
reg signed [W:0] y [0:3]; // array types in
reg signed [W:0] z [0:3]; //  Quartus Verilog 2001

always @(posedge clk) begin //----> Infer register
  if (phi_in > 90)                // Test for |phi_in| > 90
    begin                         // Rotate 90 degrees
    x[0] <= 0;
    y[0] <= r_in;                        // Input in register 0
    z[0] <= phi_in - 'sd90;
    end
  else
    if (phi_in < - 90)
      begin
      x[0] <= 0;
      y[0] <= - r_in;
      z[0] <= phi_in + 'sd90;
      end
    else
      begin
      x[0] <= r_in;
      y[0] <= 0;
      z[0] <= phi_in;
      end

  if (z[0] >= 0)                   // Rotate 45 degrees
    begin
    x[1] <= x[0] - y[0];
    y[1] <= y[0] + x[0];
    z[1] <= z[0] - 'sd45;
    end
  else
    begin
    x[1] <= x[0] + y[0];
    y[1] <= y[0] - x[0];
    z[1] <= z[0] + 'sd45;
    end

  if (z[1] >= 0)                   // Rotate 26 degrees
    begin
    x[2] <= x[1] - (y[1] >>> 1); // i.e. x[1] - y[1] /2
    y[2] <= y[1] + (x[1] >>> 1); // i.e. y[1] + x[1] /2
    z[2] <= z[1] - 'sd26;
    end
  else
```

```
      begin
      x[2] <= x[1] + (y[1] >>> 1); // i.e. x[1] + y[1] /2
      y[2] <= y[1] - (x[1] >>> 1); // i.e. y[1] - x[1] /2
      z[2] <= z[1] + 'sd26;
      end

    if (z[2] >= 0)                    // Rotate 14 degrees
      begin
        x[3] <= x[2] - (y[2] >>> 2); // i.e. x[2] - y[2]/4
        y[3] <= y[2] + (x[2] >>> 2); // i.e. y[2] + x[2]/4
        z[3] <= z[2] - 'sd14;
      end
    else
      begin
        x[3] <= x[2] + (y[2] >>> 2); // i.e. x[2] + y[2]/4
        y[3] <= y[2] - (x[2] >>> 2); // i.e. y[2] - x[2]/4
        z[3] <= z[2] + 'sd14;
      end

    x_out <= x[3];
    eps   <= z[3];
    y_out <= y[3];
  end

endmodule

//************************************************************
// IEEE STD 1364-2001 Verilog file: fir_lms.v
// Author-EMAIL: Uwe.Meyer-Baese@ieee.org
//************************************************************
// This is a generic LMS FIR filter generator
// It uses W1 bit data/coefficients bits

module fir_lms          //----> Interface
 #(parameter W1 = 8,    // Input bit width
             W2 = 16,   // Multiplier bit width 2*W1
             L  = 2,    // Filter length
             Delay = 3) // Pipeline steps of multiplier
 (input clk,  // 1 bit input
  input signed [W1-1:0] x_in, d_in,  // Inputs
  output signed [W2-1:0] e_out, y_out,  // Results
  output signed [W1-1:0] f0_out, f1_out);  // Results

// Signed data types are supported in 2001
// Verilog, and used whenever possible
```

```
reg   signed [W1-1:0] x [0:1]; // Data array
reg   signed [W1-1:0] f [0:1]; // Coefficient array
reg   signed [W1-1:0] d;
wire signed [W1-1:0] emu;
wire signed [W2-1:0] p [0:1]; // 1. Product array
wire signed [W2-1:0] xemu [0:1]; // 2. Product array
wire signed [W2-1:0]  y, sxty, e, sxtd;


wire  clken, aclr;
wire  signed [W2-1:0] sum;  // Auxilary signals


assign sum=0; assign aclr=0; // Default for mult
assign clken=0;

always @(posedge clk) // Store these data or coefficients
  begin: Store
    d <= d_in; // Store desired signal in register
    x[0] <= x_in; // Get one data sample at a time
    x[1] <= x[0];   // shift 1
    f[0] <= f[0] + xemu[0][15:8]; // implicit divide by 2
    f[1] <= f[1] + xemu[1][15:8];
  end

// Instantiate L pipelined multiplier
  genvar I;
  generate
    for (I=0; I<L; I=I+1) begin: Mul_fx
  lpm_mult mul_xf               // Multiply  x[I]*f[I] = p[I]
    ( .dataa(x[I]), .datab(f[I]), .result(p[I]));
//    .clock(clk), .sum(sum),
//    .clken(clken), .aclr(aclr)); // Unused ports
    defparam mul_xf.lpm_widtha = W1;
    defparam mul_xf.lpm_widthb = W1;
    defparam mul_xf.lpm_widthp = W2;
    defparam mul_xf.lpm_widths = W2;
//    defparam mul_xf.lpm_pipeline = Delay;
    defparam mul_xf.lpm_representation = "SIGNED";
    end // for loop
  endgenerate


assign y = p[0] + p[1];  // Compute ADF output
```

```
    // Scale y by 128 because x is fraction
    assign e = d - (y >>> 7) ;
    assign emu = e >>> 1;  // e*mu divide by 2 and
                           // 2 from xemu makes mu=1/4

// Instantiate L pipelined multiplier
  generate
    for (I=0; I<L; I=I+1) begin: Mul_xemu
    lpm_mult mul_I              // Multiply xemu[I] = emu * x[I];
      ( .dataa(x[I]), .datab(emu), .result(xemu[I]));
//    .clock(clk), .sum(sum),
//    .clken(clken), .aclr(aclr)); // Unused ports
      defparam mul_I.lpm_widtha = W1;
      defparam mul_I.lpm_widthb = W1;
      defparam mul_I.lpm_widthp = W2;
      defparam mul_I.lpm_widths = W2;
//    defparam mul_I.lpm_pipeline = Delay;
      defparam mul_I.lpm_representation = "SIGNED";
      end // for loop
  endgenerate

  assign  y_out  = y;     // Monitor some test signals
  assign  e_out  = e;
  assign  f0_out = f[0];
  assign  f1_out = f[1];

endmodule

//*************************************************************
// IEEE STD 1364-2001 Verilog file: fir6dlms.v
// Author-EMAIL: Uwe.Meyer-Baese@ieee.org
//*************************************************************
// This is a generic DLMS FIR filter generator
// It uses W1 bit data/coefficients bits

module fir6dlms          //----> Interface
 #(parameter W1 = 8,    // Input bit width
             W2 = 16,   // Multiplier bit width 2*W1
             L  = 2,    // Filter length
             Delay = 3) // Pipeline steps of multiplier
  (input clk,  // 1 bit input
   input signed [W1-1:0] x_in, d_in,  // Inputs
   output signed [W2-1:0] e_out, y_out,  // Results
   output signed [W1-1:0] f0_out, f1_out); // Results
```

```verilog
// 2D array types memories are supported by Quartus II
// in Verilog, use therefore single vectors
  reg signed [W1-1:0] x [0:4], f0, f1;
  reg signed [W1-1:0] f[0:1];
  reg  signed [W1-1:0] d[0:3]; // Desired signal array
  wire signed [W1-1:0] emu;
  wire signed [W2-1:0] xemu[0:1]; // Product array
  wire signed [W2-1:0] p[0:1]; // Product array
  wire  signed [W2-1:0]  y, sxty, e, sxtd;

  wire  clken, aclr;
  wire  signed [W2-1:0] sum;  // Auxilary signals


  assign sum=0; assign aclr=0; // Default for mult
  assign clken=0;

  always @(posedge clk) // Store these data or coefficients
    begin: Store
      d[0] <= d_in; // Shift register for desired data
      d[1] <= d[0];
      d[2] <= d[1];
      d[3] <= d[2];
      x[0] <= x_in; // Shift register for data
      x[1] <= x[0];
      x[2] <= x[1];
      x[3] <= x[2];
      x[4] <= x[3];
      f[0] <= f[0] + xemu[0][15:8]; // implicit divide by 2
      f[1] <= f[1] + xemu[1][15:8];
    end

// Instantiate L pipelined multiplier
  genvar I;
  generate
    for (I=0; I<L; I=I+1) begin: Mul_fx
  lpm_mult mul_xf              // Multiply  x[I]*f[I] = p[I]
  (.clock(clk), .dataa(x[I]), .datab(f[I]), .result(p[I]));
//   .sum(sum), .clken(clken), .aclr(aclr)); // Unused ports
    defparam mul_xf.lpm_widtha = W1;
    defparam mul_xf.lpm_widthb = W1;
    defparam mul_xf.lpm_widthp = W2;
    defparam mul_xf.lpm_widths = W2;
```

```
      defparam mul_xf.lpm_pipeline = Delay;
      defparam mul_xf.lpm_representation = "SIGNED";
      end // for loop
   endgenerate

   assign y = p[0] + p[1];  // Compute ADF output

   // Scale y by 128 because x is fraction
   assign e = d[3] - (y >>> 7);
   assign emu = e >>> 1;  // e*mu divide by 2 and
                          // 2 from xemu makes mu=1/4

// Instantiate L pipelined multiplier
   generate
      for (I=0; I<L; I=I+1) begin: Mul_xemu
   lpm_mult mul_I          // Multiply xemu[I] = emu * x[I];
      (.clock(clk), .dataa(x[I+Delay]), .datab(emu),
                                    .result(xemu[I]));
//    .sum(sum), .clken(clken), .aclr(aclr)); // Unused ports
      defparam mul_I.lpm_widtha = W1;
      defparam mul_I.lpm_widthb = W1;
      defparam mul_I.lpm_widthp = W2;
      defparam mul_I.lpm_widths = W2;
      defparam mul_I.lpm_pipeline = Delay;
      defparam mul_I.lpm_representation = "SIGNED";
      end // for loop
   endgenerate

   assign  y_out  = y;   // Monitor some test signals
   assign  e_out  = e;
   assign  f0_out = f[0];
   assign  f1_out = f[1];

endmodule

// Desciption: This is a W x L bit register file.
//*****************************************************
// IEEE STD 1364-2001 Verilog file: reg_file.v
// Author-EMAIL: Uwe.Meyer-Baese@ieee.org
//*****************************************************
module reg_file  #(parameter W = 7, // Bit width -1
                             N = 15) //Number of register - 1
       (input clk, reg_ena,
        input [W:0] data,
        input [3:0]  rd, rs, rt ,
```

```
      output reg [W:0] s, t);

  reg [W:0] r [0:N];

  always @(posedge clk) // Input mux inferring registers
  begin : MUX
    if ((reg_ena == 1) & (rd > 0))
      r[rd] <= data;
  end

  // 2 output demux without registers
  always @*
  begin : DEMUX
    if (rs > 0) // First source
      s = r[rs];
    else
      s = 0;
    if (rt > 0) // Second source
      t = r[rt];
    else
      t = 0;
  end

endmodule

//*************************************************************
// IEEE STD 1364-2001 Verilog file: trisc0.v
// Author-EMAIL: Uwe.Meyer-Baese@ieee.org
//*************************************************************
// Title: T-RISC stack machine
// Description: This is the top control path/FSM of the
// T-RISC, with a single three-phase clock cycle design
// It has a stack machine/0-address-type instruction word
// The stack has only four words.
//'include "220model.v"

module trisc0 #(parameter WA = 7,  // Address bit width -1
                          WD = 7)   // Data bit width -1
  (input reset, clk,  // Clock for the output register
   output  jc_OUT, me_ena,
   input [WD:0] iport,
   output reg [WD:0] oport,
   output [WD:0] s0_OUT, s1_OUT, dmd_IN, dmd_OUT,
   output [WA:0] pc_OUT, dma_OUT, dma_IN,
   output [7:0]  ir_imm,
```

```
output [3:0] op_code);

//parameter ifetch=0, load=1, store=2, incpc=3;
reg [1:0] state;

wire [3:0] op;
wire [WD:0] imm, dmd;
reg [WD:0] s0, s1, s2, s3;
reg [WA:0] pc;
wire [WA:0] dma;
wire [11:0] pmd, ir;
wire eq, ne, not_clk;
reg mem_ena, jc;

// OP Code of instructions:
  parameter
  add  = 0,  neg  = 1, sub  = 2, opand = 3, opor = 4,
  inv  = 5,  mul  = 6, pop  = 7, pushi = 8, push = 9,
  scan = 10, print = 11, cne = 12, ceq  = 13, cjp = 14,
  jmp  = 15;

// Code of FSM:
  always @(op) // Sequential FSM of processor
               // Check store in register ?
      case (op)  // always store except Branch
        pop     : mem_ena <= 1;
        default : mem_ena <= 0;
      endcase

  always @(negedge clk or posedge reset)
      if (reset == 1)  // update the program counter
        pc <= 0;
      else begin     // use falling edge
        if (((op==cjp) & (jc==0)) | (op==jmp))
          pc <= imm;
        else
          pc <= pc + 1;
      end

  always @(posedge clk or posedge reset)
    if (reset)            // compute jump flag and store in FF
      jc <= 0;
    else
      jc <= ((op == ceq) & (s0 == s1)) |
```

```verilog
                                        ((op == cne) & (s0 != s1));

    // Mapping of the instruction, i.e., decode instruction
    assign op  = ir[11:8];   // Operation code
    assign dma = ir[7:0];    // Data memory address
    assign imm = ir[7:0];    // Immidiate operand

    lpm_rom prog_rom
    ( .outclock(clk),.address(pc), .q(pmd));  // Used ports
 // .inclock(clk),  .memenab(ena)); // Unused
       defparam prog_rom.lpm_width = 12;
       defparam prog_rom.lpm_widthad = 8;
       defparam prog_rom.lpm_outdata = "REGISTERED";
       defparam prog_rom.lpm_address_control = "UNREGISTERED";
       defparam prog_rom.lpm_file = "TRISCOFAC.MIF";

    assign not_clk = ~clk;

    lpm_ram_dq data_ram
    ( .inclock(not_clk),.address(dma), .q(dmd),
      .data(s0), .we(mem_ena));  // Used ports
 // .outclock(clk)); // Unused
       defparam data_ram.lpm_width = 8;
       defparam data_ram.lpm_widthad = 8;
       defparam data_ram.lpm_indata = "REGISTERED";
       defparam data_ram.lpm_outdata = "UNREGISTERED";
       defparam data_ram.lpm_address_control = "REGISTERED";


    always @(posedge clk)
    begin : P3
      integer temp;

      case (op)
         add   :   s0  <= s0 + s1;
         neg   :   s0  <= -s0;
         sub   :   s0  <= s1 - s0;
         opand :   s0  <= s0 & s1;
         opor  :   s0  <= s0 | s1;
         inv   :   s0  <= ~ s0;
         mul   :   begin temp = s0 * s1;  // double width
                   s0  <= temp[WD:0]; end  // product
         pop   :   s0  <= s1;
         push  :   s0  <= dmd;
```

```
      pushi  :    s0  <= imm;
      scan   :    s0 <= iport;
      print  :    begin oport <= s0; s0<=s1; end
      default:    s0 <= 0;
    endcase
    case (op) // SPECIFY THE STACK OPERATIONS
      pushi, push, scan : begin s3<=s2; s2<=s1; s1<=s0; end
                                              // Push type
      cjp, jmp,  inv | neg : ;    // Do nothing for branch
      default :  begin s1<=s2; s2<=s3; s3<=0; end
                                        // Pop all others
    endcase
  end

  // Extra test pins:
  assign dmd_OUT = dmd; assign dma_OUT = dma; //Data memory
  assign dma_IN = dma; assign dmd_IN  = s0;
  assign pc_OUT = pc; assign ir = pmd; assign ir_imm = imm;
  assign op_code = op;  // Program control
  // Control signals:
  assign jc_OUT = jc; assign me_ena = mem_ena;
  // Two top stack elements:
  assign s0_OUT = s0; assign s1_OUT = s1;

endmodule
```

B. VHDL and Verilog Coding

Unfortunately, today we find *two* HDL languages are popular. The US west coast and Asia prefer Verilog, while the US east coast and Europe more frequently use VHDL. For digital signal processing with FPGAs, both languages seem to be well suited, but some VHDL examples were in the past a little easier to read because of the supported signed arithmetic and multiply/divide operations in the IEEE VHDL 1076-1987 and 1076-1993 standards. This gap has disappeared with the introduction of the Verilog IEEE standard 1364-2001, as it also includes signed arithmetic. Other constraints may include personal preferences, EDA library and tool availability, data types, readability, capability, and language extensions using PLIs, as well as commercial, business and marketing issues, to name just a few. A detailed comparison can be found in the book by Smith [3]. Tool providers acknowledge today that both languages need to be supported.

It is therefore a good idea to use an HDL code style that can easily be translated into either language. An important rule is to avoid any keyword in *both* languages in the HDL code when naming variables, labels, constants, user types, etc. The IEEE standard VHDL 1076-1987 uses 77 keywords and an extra 19 keywords are used in VHDL 1076-1993 (see VHDL 1076-1993 Language Reference Manual (LRM) on p. 179). New in VHDL 1076-1993 are:

```
GROUP, IMPURE, INERTIAL, LITERAL, POSTPONED, PURE, REJECT
ROL, ROR, SHARED, SLA, SLL, SRA, SRL, UNAFFECTED, XNOR,
```

which are unfortunately *not* highlighted in the MaxPlus II editor but with the Quartus II. The IEEE standard Verilog 1364-1995, on the other hand, has 102 keywords (see LRM, p. 604). Together, both HDL languages have 201 keywords, including 19 in common. Table B.1 shows VHDL 1076-1993 keywords in capital letters, Verilog 1364-2001 keywords in small letters, and the common keywords with a capital first letter. New in Verilog 1076-2001 are:

```
automatic, cell, config, design, endconfig, endgenerate,
generate, genvar, incdir, include, instance, liblist,
library, localparam, noshowcancelled, pulsestyle_onevent,
pulsestyle_ondetect, showcancelled, signed, unsigned, use
```

Table B.1. VHDL 1076-1993 and Verilog 1364-2001 keywords.

ABS	event	notif0	
ACCESS	EXIT	notif1	SIGNAL
AFTER	FILE	NULL	signed
ALIAS	For	OF	OF
ALL	force	ON	SLA
always	forever	OPEN	SLL
And	fork	Or	small
ARCHITECTURE	Function	OTHERS	specify
ARRAY	Generate	OUT	specparam
ASSERT	GENERIC	output	SRA
assign	genvar	PACKAGE	SRL
ATTRIBUTE	GROUP	parameter	strong0
automatic	GUARDED	pmos	strong1
Begin	highz0	PORT	SUBTYPE
BLOCK	highz1	posedge	supply0
BODY	If	POSTPONED	supply1
buf	ifnone	primitive	table
BUFFER	IMPURE	PROCEDURE	task
bufif0	IN	PROCESS	THEN
bufif1	incdir	pull0	time
BUS	include	pull1	TO
Case	INERTIAL	pulldown	tran
casex	initial	pullup	tranif0
casez	Inout	pulsestyle_onevent	tranif1
cell	input	pulsestyle_ondetect	TRANSPORT
cmos	instance	PURE	tri
config	integer	RANGE	tri0
COMPONENT	IS	rcmos	tri1
CONFIGURATION	join	real	triand
CONSTANT	LABEL	realtime	trior
deassign	large	RECORD	trireg
default	liblist	reg	TYPE
defparam	Library	REGISTER	UNAFFECTED
design	LINKAGE	REJECT	UNITS
disable	LITERAL	release	unsigned
DISCONNECT	LOOP	REM	UNTIL
DOWNTO	localparam	repeat	Use
edge	macromodule	REPORT	VARIABLE
Else	MAP	RETURN	vectored
ELSIF	medium	rnmos	Wait
End	MOD	ROL	wand
endcase	module	ROR	weak0
endconfig	Nand	rpmos	weak1
endfunction	negedge	rtran	WHEN
endgenerate	NEW	rtranif0	While
endmodule	NEXT	rtranif1	wire
endprimitive	nmos	scalared	WITH
endspecify	Nor	SELECT	wor
endtable	noshowcancelled	SEVERITY	Xnor
endtask	Not	SHARED	Xor
ENTITY		showcancelled	

B.1 List of Examples

These synthesis results for all examples can be easily reproduced by using the scripts qvhdl.tcl in the VHDL or Verilog directories of the CD-ROM. Run the TCL script with

```
quartus_sh -t qvhdl.tcl > qvhdl.txt
```

The script produces for each design four parameters. For the trisc0.vhd, for instance, we get:

```
. . . .
_____
trisc0 fmax: 115.65 MHz ( period = 8.647 ns )
trisc0 LEs: 198 / 33,216 ( < 1 % )
trisc0 M4K bits: 5,120 / 483,840 ( 1 % )
trisc0 DSP blocks: 1 / 70 ( 1 % )
_____
. . . .
```

then grep through the report qvhdl.txt file using fmax:, LEs: etc.

From the script you will notice that the following special options of Quartus II web edition 6.0 were used:

- Device set Family to Cyclone II and then under Available devices select EP2C35F672C6.
- For Timing Analysis Settings set Default required fmax: to 3 ns.
- For Analysis & Synthesis Settings from the Assignments menu
  - set Optimization Technique to Speed
  - Deselect Power-Up Don't Care
- In the Fitter Settings select as Fitter effort Standard Fit (highest effort)

The table below displays the results for all VHDL and Verilog examples given in this book. The table is structured as follows. The first column shows the entity or module name of the design. Columns 2 to 6 are data for the VHDL designs: the number of LEs shown in the report file; the number of 9×9-bit multipliers; the number of M4K memory blocks; the Registered Performance; and the page with the source code. The same data are provided for the Verilog design examples, shown in columns 7 to 9. Note that VHDL and Verilog produce the same data for number of 9×9-bit multiplier and number of M4K memory blocks, but the LEs and Registered Performance do not always match.

Design	LEs	VHDL 9 × 9 Mult.	M4Ks	f_{MAX} MHz	Page	LEs	Verilog f_{MAX} MHz	Page
add_1p	125	no	0	316.46	78	77	390.63	666
add_2p	234	no	0	229.04	78	144	283.85	667
add_3p	372	no	0	215.84	78	229	270.42	668
ammod	316	no	0	215.98	455	277	288.85	717
arctan	100	4	0	32.09	134	99	32.45	676
bfproc	131	3	0	95.73	370	83	116.09	715
ccmul	39	3	0	–	368	39	–	713
cic3r32	337	no	0	282.17	262	337	269.69	694
cic3s32	205	no	0	284.58	269	205	284.50	696
cmoms	372	10	0	85.94	303	239	107.48	705
cmul7p8	48	no	0	-	59	48	–	665
cordic	235	no	0	222.67	126	197	317.16	674
dafsm	32	no	0	420.17	189	30	420.17	683
dapara	33	no	0	214.96	202	45	420.17	691
darom	27	no	1	218.29	196	27	218.96	687
dasign	56	no	0	236.91	199	47	328.19	688
db4latti	418	no	0	58.81	324	248	74.69	709
db4poly	173	no	0	136.65	250	158	136.31	697
div_aegp	64	4	0	134.63	94	64	134.63	671
div_res	127	no	0	265.32	100	115	257.86	673
example	24	no	0	420.17	15	24	420.17	663
farrow	279	6	0	43.91	292	175	65.77	703
fir6dlms	138	4	0	176.15	511	138	174.52	721
fir_gen	184	4	0	329.06	167	184	329.06	680
fir_lms	50	4	0	74.59	504	50	74.03	719
fir_srg	114	no	0	97.21	179	70	106.15	682
fun_text	32	no	1	264.20	30	32	264.20	664
iir	62	no	0	160.69	217	30	234.85	692
iir_par	268	no	0	168.12	237	199	136.87	693
iir_pipe	124	no	0	207.08	231	75	354.48	692
lfsr	6	no	0	420.17	437	6	420.17	716
lfsr6s3	6	no	0	420.17	440	6	420.17	717
ln	88	10	0	32.76	145	88	32.76	677
mul_ser	121	no	0	256.15	82	140	245.34	670
rader7	443	no	0	137.06	355	404	159.41	710
rc_sinc	448	19	0	61.93	285	416	81.47	699
reg_file	211	no	0	-	559	211	–	723
sqrt	336	2	0	82.16	150	317	82.73	678
tris0	198	1	2	115.65	606	166	71.94	724

B.2 Library of Parameterized Modules (LPM)

Throughout the book we use six different LPM megafunctions (see Fig. B.1), namely:

- lpm_ff, the flip-flop megafunction
- lpm_add_sub, the adder/subtractor megafunction
- lpm_ram_dq, the RAM megafunction
- lpm_rom, the ROM megafunction
- lpm_divide, the divider megafunction, and
- lpm_mult, the multiplier megafunction

These megafunctions are explained in the following, along with their port definitions, parameters, and resource usage. This information is also available using the Quartus II help under megafunctions/LPM.

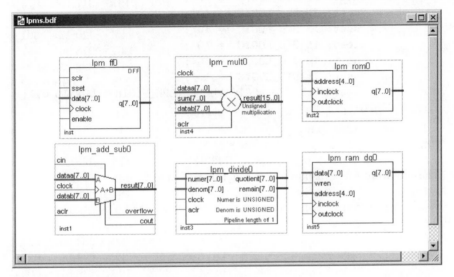

Fig. B.1. Six LPM megafunction used.

B.2.1 The Parameterized Flip-Flop Megafunction (lpm_ff)

The lpm_ff function is useful if features are needed that are not available in the DFF, DFFE, TFF, and TFFE primitives, such as synchronous or asynchronous set, clear, and load inputs. We have used this megafunction for the following designs: example, p. 15 and fun_text, p. 30.

Altera recommends instantiating this function as described in "Using the MegaWizard Plug-In Manager" in the Quartus II help.

The port names and order for Verilog HDL prototypes are:

```
module lpm_ff ( q, data, clock, enable, aclr,
                aset, sclr, sset, aload, sload) ;
```

The VHDL component declaration is shown below:

```
COMPONENT lpm_ff
   GENERIC (LPM_WIDTH: POSITIVE;
            LPM_AVALUE: STRING := "UNUSED";
            LPM_FFTYPE: STRING := "FFTYPE_DFF";
            LPM_TYPE: STRING := "L_FF";
            LPM_SVALUE: STRING := "UNUSED";
            LPM_HINT: STRING := "UNUSED");
   PORT (data: IN STD_LOGIC_VECTOR(LPM_WIDTH-1 DOWNTO 0);
        clock: IN STD_LOGIC;
       enable: IN STD_LOGIC := '1';
        sload: IN STD_LOGIC := '0';
         sclr: IN STD_LOGIC := '0';
         sset: IN STD_LOGIC := '0';
        aload: IN STD_LOGIC := '0';
         aclr: IN STD_LOGIC := '0';
         aset: IN STD_LOGIC := '0';
            q: OUT STD_LOGIC_VECTOR(LPM_WIDTH-1 DOWNTO 0));
END COMPONENT;
```

Ports

The following table displays all input ports of lpm_ff:

Port name	Required	Description	Comments
data	No	T-type flip-flop: Toggle enable D-type flip-flop: Data input	Input port LPM_WIDTH wide. If the data input is not used, at least one of the aset, aclr, sset, or sclr ports must be used. Unused data inputs default to GND.
clock	Yes	Positive-edge triggered clock	
enable	No	Clock Enable input	Default = 1
sclr	No	Synchronous clear input	If both sset and sclr are used and both are asserted, sclr is dominant. The sclr signal affects the output q values before polarity is applied to the ports.
sset	No	Synchronous set input	Sets q outputs to the value specified by LPM_SVALUE, if that value is present, or sets the q outputs to all 1s. If both sset and sclr are used and both are asserted, sclr is dominant. The sset signal affects the output q values before polarity is applied to the ports.
sload	No	Synchronous load input. Loads the flipflop with the value on the data input on the next active clock edge.	Default = 0. If sload is used, data must be used. For load operation, sload must be high (1) and enable must be high (1) or unconnected. The sload port is ignored when the LPM_FFTYPE parameter is set to DFF.
aclr	No	Asynchronous clear input	If both aset and aclr are used and both are asserted, aclr is dominant. The aclr signal affects the output q values before polarity is applied to the ports.
aset	No	Asynchronous set input	Sets q outputs to the value specified by LPM_AVALUE, if that value is present, or sets the q outputs to all 1s.
aload	No	Asynchronous load input. Asynchronously loads the flip-flop with the value on the data input.	Default = 0. If aload is used, data must be used.

The following table displays all OUTPUT ports of lpm_ff:

Port Name	Required	Description	Comments
q	Yes	Data output from D or T flip-flops	Output port LPM_WIDTH wide

Parameters

The following table shows the parameters of the `lpm_ff` component:

Parameter	Type	Required	Description
LPM_WIDTH	Integer	Yes	Width of the data and q ports
LPM_AVALUE	Integer	No	Constant value that is loaded when aset is high. If omitted, defaults to all 1s. The LPM_AVALUE parameter is limited to a maximum of 32 bits.
LPM_SVALUE	Integer	No	Constant value that is loaded on the rising edge of clock when sset is high. If omitted, defaults to all 1s.
LPM_FFTYPE	String	No	Values are DFF, TFF, and UNUSED. Type of flip-flop. If omitted, the default is DFF. When the LPM_FFTYPE parameter is set to DFF, the sload port is ignored.
LPM_HINT	String	No	Allows you to specify Altera-specific parameters in VHDL design files. The default is UNUSED.
LPM_TYPE	String	No	Identifies the LPM entity name in the VHDL design files.

Note that for Verilog LPM 220 synthesizable code (i.e., `220model.v`) the following parameter ordering applies: `lpm_type`, `lpm_width`, `lpm_avalue`, `lpm_svalue`, `lpm_pvalue`, `lpm_fftype`, `lpm_hint`.

Function

The following table is an example of the T-type flip-flop behavior in `lpm_ff`:

aclr	aset	enable	clock	sclr	sset	sload	Q[LPM_WIDTH-1..0]
1	X	X	X	X	X	X	000...
0	1	X	X	X	X	X	111... or LPM_AVALUE
0	0	0	X	X	X	X	q[LPM_WIDTH-1..0]
0	0	1	∫	1	X	X	000...
0	0	1	∫	0	1	X	111... or LPM_SVALUE
0	0	1	∫	0	0	1	data[LPM_WIDTH-1..0]
0	0	1	∫	0	0	0	q[LPM_WIDTH-1..0] xor data[LPM_WIDTH-1..0]

The table header spans: Inputs (aclr aset enable clock sclr sset sload) | Outputs (Q[LPM_WIDTH-1..0])

Resource Usage

The megafunction `lpm_ff` uses one logic cell per bit.

B.2.2 The Parameterized Adder/Subtractor Megafunction (lpm_add_sub)

Altera recommends using the `lpm_add_sub` function to replace all other types of adder/subtractor functions, including old-style adder/subtractor macrofunctions. We have used this megafunction for the following designs: `example`, p. 15, `fun_text`, p. 30, `ccmul`, p. 368.

Altera recommends instantiating this function as described in "Using the MegaWizard Plug-In Manager" in the Quartus II help.

The port names and order for Verilog HDL prototypes are:

```
module lpm_add_sub ( cin,
                     dataa, datab,
                     add_sub, clock, aclr,
                     result, cout, overflow);
```

The VHDL component declaration is shown below:

```
COMPONENT lpm_add_sub
  GENERIC (LPM_WIDTH: POSITIVE;
           LPM_REPRESENTATION: STRING := "SIGNED";
           LPM_DIRECTION: STRING := "UNUSED";
           LPM_HINT: STRING := "UNUSED";
           LPM_PIPELINE: INTEGER := 0;
           LPM_TYPE: STRING := "L_ADD_SUB");
  PORT (dataa, datab
               : IN STD_LOGIC_VECTOR(LPM_WIDTH-1 DOWNTO 0);
        aclr, clken, clock, cin : IN STD_LOGIC := '0';
        add_sub                 : IN STD_LOGIC := '1';
        result : OUT STD_LOGIC_VECTOR(LPM_WIDTH-1 DOWNTO 0);
        cout, overflow          : OUT STD_LOGIC);
END COMPONENT;
```

Ports

The following table displays all input ports of `lpm_add_sub`:

Port name	Required	Description	Comments
cin	No	Carry-in to the low-order bit. If the operation is ADD, low = 0 and high = +1. If the operation is SUB, low = -1 and high = 0.	If omitted, the default is 0 (i.e., low if the operation is ADD and high if the operation is SUB).
dataa	Yes	Augend/Minuend	Input port LPM_WIDTH wide
datab	Yes	Addend/Subtrahend	Input port LPM_WIDTH wide
add_sub	No	If the signal is high, the operation = dataa + datab. If the signal is low, the operation = dataa − datab.	If the LPM_DIRECTION parameter is used, add_sub cannot be used. If omitted, the default is ADD. Altera recommends that you use the LPM_DIRECTION parameter to specify the operation of the lpm_add_sub function, rather than assigning a constant to the add_sub port.
clock	No	Clock for pipelined usage	The clock port provides pipelined operation for the lpm_add_sub function. For LPM_PIPELINE values other than 0 (default value), the clock port must be connected.
clken	No	Clock enable for pipelined usage	Available for VHDL only
aclr	No	Asynchronous clear for pipelined usage	The pipeline initializes to an undefined (X) logic level. The aclr port can be used at any time to reset the pipeline to all 0s, asynchronously to the clock signal.

The following table displays all output ports of `lpm_add_sub`:

Port Name	Required	Description	Comments
result	Yes	dataa + or − datab + or − cin	Output port LPM_WIDTH wide
cout	No	Carry-out (borrow-in) of the MSB	If overflow is used, cout cannot be used. The cout port has a physical interpretation as the carry-out (borrow-in) of the MSB. cout is most meaningful for detecting overflow in UNSIGNED operations.
overflow	No	Result exceeds available precision.	If overflow is used, cout cannot be used. The overflow port has a physical interpretation as the XOR of the carry-in to the MSB with the carry-out of the MSB. overflow is meaningful only when the LPM_REPRESENTATION parameter value is SIGNED.

Parameters

The following table shows the parameters of the `lpm_add_sub` component:

Parameter	Type	Re-quired	Description
LPM_WIDTH	Integer	Yes	Width of the dataa, datab, and result ports.
LPM_DIRECTION	String	No	Values are ADD, SUB, and UNUSED. If omitted, the default is DEFAULT, which directs the parameter to take its value from the add_sub port. The add_sub port cannot be used if LPM_DIRECTION is used. Altera recommends that you use the LPM_DIRECTION parameter to specify the operation of the lpm_add_sub function, rather than assigning a constant to the add_sub port.
LPM_-REPRESEN-TATION	String	No	Type of addition performed: SIGNED, UNSIGNED, or UNUSED. If omitted, the default is SIGNED.
LPM_PIPELINE	Integer	No	Specifies the number of clock cycles of latency associated with the result output. A value of zero (0) indicates that no latency exists, and that a purely combinatorial function will be instantiated. If omitted, the default is 0 (nonpipelined).
LPM_HINT	String	No	Allows you to specify Altera-specific parameters in VHDL design files. The default is UNUSED.
LPM_TYPE	String	No	Identifies the LPM entity name in VHDL design files.
ONE_INPUT_-IS_CONSTANT	String	No	Altera-specific parameter. Values are YES, NO, and UNUSED. Provides greater optimization, if one input is constant. If omitted, the default is NO.
MAXIMIZE_-SPEED	Integer	No	Altera-specific parameter. You can specify a value between 0 and 10. If used, MaxPlus II attempts to optimize a specific instance of the lpm_add_sub function for speed rather than area, and overrides the setting of the Optimize option in the Global Project Logic Synthesis dialog box (Assign menu). If MAXIMIZE_SPEED is unused, the value of the Optimize option is used instead. If the setting for MAXIMIZE_SPEED is 6 or higher, the compiler will optimize lpm_add_sub megafunctions for higher speed; if the setting is 5 or less, the compiler will optimize for smaller area.

Note that for Verilog LPM 220 synthesizable code (i.e., 220model.v) the following parameter ordering applies: lpm_type, lpm_width, lpm_direction, lpm_representation, lpm_pipeline, lpm_hint.

Function

The following table is an example of the UNSIGNED behavior in
`lpm_add_sub`:

Inputs			Outputs	
add_sub	dataa	datab	cout,result	overflow
1	a	b	a + b + cin	cout
0	a	b	a - b - cin	!cout

The following table is an example of the SIGNED behavior in `lpm_add_sub`:

Inputs			Outputs	
add_sub	dataa	datab	cout,sum	overflow
1	a	b	$a + b$+cin	$a \geq 0$ and $b \geq 0$ and $sum < 0$ or $a < 0$ and $b < 0$ and $sum \geq 0$
0	a	b	$a - b$−cin	$a >= 0$ and $b < 0$ and $sum < 0$ or $a < 0$ and $b \geq 0$ and $sum \geq 0$

Resource Usage

The following table summarizes the resource usage for an `lpm_add_sub` mega-function used to implement a 16-bit unsigned adder with a carry-in input and a carry-out output. Logic cell usage scales linearly in proportion to adder width.

Design goals		Design results		
Device family	Optimization	LEs	Speed (ns)	Notes
FLEX 6K, 8K, and 10K	Routability Speed	45 18	53 17	Speed for EPF8282A-2
MAX 5K, 7K, and 9K	Routability	28 (22)	23	Speed for EPM7128E-7

Numbers of shared expanders used are shown in parentheses.

B.2.3 The Parameterized Multiplier Megafunction (lpm_mult)

Altera recommends that you use `lpm_mult` to replace all other types of multiplier functions, including old-style multiplier macrofunctions. We have used this megafunction for the designs fir_gen, p. 167, ccmul, p. 368. fir_lms, p. 504, and fir6dlms, p. 511.

Altera recommends instantiating this function as described in "Using the MegaWizard Plug-In Manager" in the Quartus II help.

The port names and order for Verilog HDL prototype are:

```
module lpm_mult ( dataa, datab, sum, aclr, clock,
                  result);
```

The VHDL component declaration is shown below:

```
COMPONENT lpm_mult
  GENERIC (LPM_WIDTHA: POSITIVE;
           LPM_WIDTHB: POSITIVE;
           LPM_WIDTHS: POSITIVE;
           LPM_WIDTHP: POSITIVE;
           LPM_REPRESENTATION: STRING := "UNSIGNED";
           LPM_PIPELINE: INTEGER := 0;
           LPM_TYPE: STRING := "L_MULT";
           LPM_HINT : STRING := "UNUSED");
  PORT (dataa : IN STD_LOGIC_VECTOR(LPM_WIDTHA-1 DOWNTO 0);
        datab : IN STD_LOGIC_VECTOR(LPM_WIDTHB-1 DOWNTO 0);
        aclr, clken, clock : IN STD_LOGIC := '0';
        sum   : IN STD_LOGIC_VECTOR(LPM_WIDTHS-1 DOWNTO 0)
                                        := (OTHERS => '0');
        result: OUT STD_LOGIC_VECTOR(LPM_WIDTHP-1 DOWNTO 0)
        );
  END COMPONENT;
```

Ports

The following table displays all input ports of lpm_mult:

Port name	Required	Description	Comments
dataa	Yes	Multiplicand	Input port LPM_WIDTHA wide
datab	Yes	Multiplier	Input port LPM_WIDTHB wide
sum	No	Partial sum	Input port LPM_WIDTHS wide
clock	No	Clock for pipelined usage	The clock port provides pipelined operation for the lpm_mult function. For LPM_PIPELINE values other than 0 (default value), the clock port must be connected.
clken	No	Clock enable for pipelined usage	Available for VHDL only.
aclr	No	Asynchronous clear for pipelined usage	The pipeline initializes to an undefined (X) logic level. The aclr port can be used at any time to reset the pipeline to all 0s, asynchronously to the clock signal.

The following table displays all output ports of lpm_mult:

Port Name	Required	Description	Comments
result	Yes	result = dataa * datab + sum. The product LSB is aligned with the sum LSB.	Output port LPM_WIDTHP wide. If LPM_WIDTHP < max (LPM_WIDTHA + LPM_WIDTHB, LPM_WIDTHS) or (LPM_WIDTHA + LPM_WIDTHS), only the LPM_WIDTHP MSBs are present.

Parameters

The following table shows the parameters of the lpm_mult component:

Parameter	Type	Re-quired	Description
LPM_WIDTHA	Integer	Yes	Width of the dataa port
LPM_WIDTHB	Integer	Yes	Width of the datab port
LPM_WIDTHP	Integer	Yes	Width of the result port
LPM_WIDTHS	Integer	Yes	Width of the sum port. Required even if the sum port is not used.
LPM_REPRESENTATION	String	No	Type of multiplication performed: SIGNED, UNSIGNED, or UNUSED. If omitted, the default is UNSIGNED.
LPM_PIPELINE	Integer	No	Specifies the number of clock cycles of latency associated with the result output. A value of zero (0) indicates that no latency exists, and that a purely combinatorial function will be instantiated. If omitted, the default is 0 (non-pipelined).
LPM_HINT	String	No	Allows you to assign Altera-specific parameters in VHDL design files. The default is UNUSED.
LPM_TYPE	String	No	Identifies the LPM entity name in VHDL design files.
INPUT_A_IS_CONSTANT	String	No	Altera-specific parameter. Values are YES, NO, and UNUSED. If dataa is connected to a constant value, setting INPUT_A_IS_CONSTANT to YES optimizes the multiplier for resource usage and speed. If omitted, the default is NO.
INPUT_B_IS_CONSTANT	String	No	Altera-specific parameter. Values are YES, NO, and UNUSED. If datab is connected to a constant value, setting INPUT_B_IS_CONSTANT to YES optimizes the multiplier for resource usage and speed. The default is NO.

Parameter	Type	Re-quired	Description
USE_EAB	String	No	Altera-specific parameter. Values are ON, OFF, and UNUSED. Setting the USE_EAB parameter to ON allows Quartus II to use EABs to implement 4 × 4 or (8 × constant value) building blocks in FLEX 10K devices. Altera recommends that you set USE_EAB to ON only when LCELLS are in short supply. If you wish to use this parameter, when you instantiate the function in a GDF, you must specify it by entering the parameter name and value manually with the Edit Ports/Parameters dialog box (Symbol menu). You can also use this parameter name in a TDF or a Verilog design file. You must use the LPM_HINT parameter to specify the USE_EAB parameter in VHDL design files.
DEDICATED_ _MULTIPLIER_ _CIRCUITRY	String	No	Altera-specific parameter. You must use the LPM_HINT parameter to specify the DEDICATED_MULTIPLIER_CIRCUITRY parameter in VHDL design files. Specifies whether to use dedicated multiplier circuitry. Values are 'AUTO, YES, and 'NO,. If omitted, the default is AUTO.
LATENCY	Integer	No	Altera-specific parameter. Same as LPM_PIPELINE. (This parameter is provided only for backward compatibility with MaxPlus II pre-version 7.0 designs. For all new designs, you should use the LPM_PIPELINE parameter instead.)
MAXIMIZE_ _SPEED	Integer	No	Altera-specific parameter. You can specify a value between 0 and 10. If used, MaxPlus II attempts to optimize a specific instance of the lpm_mult function for speed rather than area, and overrides the setting of the Optimize option in the Global Project Logic Synthesis dialog box (Assign menu). If MAXIMIZE_SPEED is unused, the value of the Optimize option is used instead. If the setting for MAXIMIZE_SPEED is 6 or higher, the compiler will optimize lpm_mult megafunctions for higher speed; if the setting is 5 or less, the compiler will optimize for smaller area.
LPM_HINT	String	No	Allows you to specify Altera-specific parameters in VHDL design files. The default is UNUSED.

Note that specifying a value for MAXIMIZE_SPEED has an effect only if LPM_REPRESENTATION is set to SIGNED.

Note that for Verilog LPM 220 synthesizable code (i.e., 220model.v) the following parameter ordering applies: lpm_type, lpm_widtha, lpm_widthb, lpm_widths, lpm_widthp, lpm_representation, lpm_pipeline, lpm_hint.

Function

The following table is an example of the UNSIGNED behavior in lpm_mult:

Inputs dataa datab sum			Outputs product
a	b	s	LPM_WIDTHP most significant bits of $a * b + s$

Resource Usage

The following table summarizes the resource usage for an lpm_mult function used to implement 4-bit and 8-bit multipliers with LPM_PIPELINE = 0 and without the optional sum input. Logic cell usage scales linearly in proportion to the square of the input width.

Design goals		Design results			
Device family	Optimization	Width	LEs	Speed (ns)	Notes
FLEX 6K, 8K, and 10K	Routability Speed	8 8	121 163	80 52	Speed for EPF8282A-2
FLEX 6K, 8K, and 10K	Routability Speed	4 4	29 41	34 27	Speed for EPF8282A-2
MAX 5K, 7K, and 9K	Routability Speed	4 4	26 (11) 27 (4)	23 19	Speed for EPM7128E-7

Numbers of shared expanders used are shown in parentheses. In the FLEX 10K device family, the 4-bit by 4-bit multiplier example shown above can be implemented in a single EAB.

B.2.4 The Parameterized ROM Megafunction (lpm_rom)

The lpm_rom block is parameterized ROM with separate input and output clocks. We have used this megafunction for the designs fun_text, p. 30 and darom, p. 196.

The lpm_rom block can also be used with older device families, e.g., Flex 10K. Altera translates for newer devices like Cyclone II the lpm_rom in the altsyncram megafunction block. But the altsyncram is not supported for Flex 10K that is used on the popular UP2 boards.

Altera recommends instantiating this function as described in "Using the MegaWizard Plug-In Manager" in the Quartus II help.

The port names and order for Verilog HDL prototype are:

```
module lpm_rom ( address, inclock, outclock, memenab,
                 q);
```

The VHDL component declaration is shown below:

```
COMPONENT lpm_rom
  GENERIC (LPM_WIDTH    : POSITIVE;
           LPM_TYPE     : STRING := "L_ROM";
           LPM_WIDTHAD  : POSITIVE;
           LPM_NUMWORDS : POSITIVE;
           LPM_FILE     : STRING;
           LPM_ADDRESS_CONTROL : STRING := "REGISTERED";
           LPM_OUTDATA  : STRING := "REGISTERED";
           LPM_HINT     : STRING := "UNUSED");
  PORT(address : IN STD_LOGIC_VECTOR(LPM_WIDTHAD-1 DOWNTO 0);
       inclock : IN STD_LOGIC := '1';
       outclock : IN STD_LOGIC := '1';
       memenab : IN STD_LOGIC := '1';
       q        : OUT STD_LOGIC_VECTOR(LPM_WIDTH-1 DOWNTO 0)
       );
  END COMPONENT;
```

Ports

The following table displays all input ports of lpm_rom:

Port name	Required	Description	Comments
address	Yes	Address input to the memory	Input port LPM_WIDTHAD wide
inclock	No	Clock for input registers	The address port is synchronous (registered) when the inclock port is connected, and is asynchronous (registered) when the inclock port is not connected.
outclock	No	Clock for output registers	The addressed memory content-to-q response is synchronous when the outclock port is connected, and is asynchronous when it is not connected.
memenab	No	Memory enable input	High = data output on q, Low = high-impedance outputs

The following table displays all output ports of lpm_rom:

Port Name	Required	Description	Comments
q	Yes	Output of memory	Output port LPM_WIDTH wide

Parameters

The following table shows the parameters of the `lpm_rom` component:

Parameter	Type	Re-quired	Description
LPM_WIDTH	Integer	Yes	Width of the q port.
LPM_WIDTHAD	Integer	Yes	Width of the address port. LPM_WIDTHAD should be (but is not required to be) equal to $\log_2(\text{LPM\_NUMWORDS})$. If LPM_WIDTHAD is too small, some memory locations will not be addressable. If it is too large, addresses that are too high will return undefined logic levels.
LPM_NUMWORDS	Integer	Yes	Number of words stored in memory. In general, this value should be (but is not required to be) $2^{\text{LPM\_WIDTHAD}} - 1 < \text{LPM\_NUMWORDS} \leq 2^{\text{LPM\_WIDTHAD}}$. If omitted, the default is $2^{\text{LPM\_WIDTHAD}}$.
LPM_FILE	String	No	Name of the Memory Initialization File (*.mif) or Hexadecimal (Intel-format) File (*.hex) containing ROM initialization data (<filename>), or UNUSED.
LPM_ADDRESS_CONTROL	String	No	Values are REGISTERED, UNREGISTERED, and UNUSED. Indicates whether the address port is registered. If omitted, the default is REGISTERED.
LPM_OUTDATA	String	No	Values are REGISTERED, UNREGISTERED, and UNUSED. Indicates whether the q and eq ports are registered. If omitted, the default is REGISTERED.
LPM_HINT	String	No	Allows you to specify Altera-specific parameters in VHDL design files. The default is UNUSED.
LPM_TYPE	String	No	Identifies the LPM entity name in VHDL design files.

Note that for Verilog LPM 220 synthesizable code (i.e., `220model.v`) the following parameter ordering applies: `lpm_type`, `lpm_width`, `lpm_widthad`, `lpm_numwords`, `lpm_address_control`, `lpm_outdata`, `lpm_file`, `lpm_hint`.

Function

The following table shows the *synchronous* read from memory behavior of `lpm_rom`:

OUTCLOCK	MEMENAB	Function
X	L	q output is high impedance (memory not enabled)
⌐	H	No change in output
⌠	H	The output register is loaded with the contents of the memory location pointed to by address. q outputs the contents of the output register.

The output q is asynchronous and reflects the data in the memory to which address points. The following table shows the *asynchronous* memory operations behavior of `lpm_rom`:

MEMENAB	Function
L	q output is high-impedance (memory not enabled)
H	The memory location pointed to by address is read

Totally asynchronous memory operations occur when neither `inclock` nor `outclock` is connected. The output q is asynchronous and reflects the memory location pointed to by address. Since this totally asynchronous memory operation is only available with Flex 10K devices, but not with Cyclone II, we do not use this mode in our designs. Either input or output is registered in all of our designs that use memory blocks.

Resource Usage

The Megafunction `lpm_rom` uses one embedded cell per memory bit.

B.2.5 The Parameterized Divider Megafunction (lpm_divide)

Altera recommends that you use `lpm_divide` to replace all other types of divider functions, including old-style divide macrofunction. We have used this megafunction for the array divider designs p. 103.

Altera recommends instantiating this function as described in "Using the MegaWizard Plug-In Manager" in the Quartus II help.

The port names and order for Verilog HDL prototype are:

```
module lpm_divide ( quotient, remain, numer, denom,
                            clock, clken, aclr )
```

The VHDL component declaration is shown below:

```
COMPONENT lpm_divide
  GENERIC ( LPM_WIDTHN: POSITIVE;
            LPM_WIDTHD: POSITIVE;
            LPM_NREPRESENTATION: STRING: = "UNSIGNED";
            LPM_DREPRESENTATION: STRING: = "UNSIGNED";
            LPM_TYPE: STRING :="LPM_DIVIDE";
```

```
                LPM_PIPELINE: INTEGER := 0;
                LPM_HINT: STRING:= "UNUSED";
            );
    PORT    ( numer: IN STD_LOGIC_VECTOR(LPM_WIDTHN-1 DOWNTO 0);
              denom: IN STD_LOGIC_VECTOR(LPM_WIDTHD-1 DOWNTO 0);
              clock, aclr: IN STD_LOGIC := '0';
              clken: IN STD_LOGIC := '1';
              quotient: OUT STD_LOGIC_VECTOR(LPM_WIDTHN-1 DOWNTO 0);
              remain: OUT STD_LOGIC_VECTOR(LPM_WIDTHD-1 DOWNTO 0)
            );
    END COMPONENT;
```

Ports

The following table displays all input ports of `lpm_divide`:

Port name	Required	Description	Comments
numer	Yes	Numerator	Input port LPM_WIDTHN wide.
denom	Yes	Denominator	Input port LPM_WIDTHD wide.
clock	No	Clock input for pipelined usage.	You must connect the clock input if you set LPM_PIPELINE to a value other than 0.
clken	No	Clock enable for pipelined usage.	
aclr	No	Asynchronous clear signal.	The aclr port may be used at any time to reset the pipeline to all 0s asynchronously to the clock input.

The following table displays all output ports of `lpm_divide`:

Port Name	Required	Description	Comments
quotient	Yes	Output port LPM_WIDTHN wide.	You must use either the quotient or the remain ports.
remain	Yes	Output port LPM_WIDTHD wide.	You must use either the quotient or the remain ports.

Parameters

The following table shows the parameters of the `lpm_divide` component:

Parameter	Type	Re-quired	Description
LPM_WIDTHN	Integer	Yes	Width of the **numer** and **quotient** port
LPM_WIDTHD	Integer	Yes	Width of the **denom** and **remain** port
LPM_NREPRESENTATION	String	No	Specifies whether the numerator is SIGNED or UNSIGNED. Only UNSIGNED is supported for now.
LPM_DREPRESENTATION	String	No	Specifies whether the denominator is SIGNED or UNSIGNED. Only UNSIGNED is supported for now.
LPM_PIPELINE	Integer	No	Specifies the number of clock cycles of latency associated with the **quotient** and **remain** outputs. A value of zero (0) indicates that no latency exists, and that a purely combinatorial function will be instantiated. If omitted, the default is 0 (nonpipelined). You cannot specify a value for the LPM_PIPELINE parameter that is higher than LPM_WIDTHN.
LPM_TYPE	String	No	Identifies the LPM entity name in VHDL design files.
LPM_HINT	String	No	Allows you to assign Altera-specific parameters in VHDL design files. The default is UNUSED.

You can pipeline a design by connecting the `clock` input and specifying the number of clock cycles of latency with the `LPM_PIPELINE` parameter.

Note that for Verilog LPM 220 synthesizable code (i.e., 220model.v) the following parameter ordering applies: `lpm_type`, `lpm_widthn`, `lpm_widthd`, `lpm_nrepresentation`, `lpm_drepresentation`, `lpm_pipeline`.

B.2.6 The Parameterized RAM Megafunction (lpm_ram_dq)

The `lpm_ram_dq` block is parameterized RAM with separate input and output ports. The `lpm_ram_dq` block can also be used with older device families, e.g., Flex10K. Altera translates for newer devices like Cyclone II the `lpm_ram_dq` in the `altsyncram` megafunction block. But the `altsyncram` is not supported for Flex10K that is used on the popular UP2 boards.

We have used this megafunction for the design `trisc0`, p. 606.

Altera recommends instantiating this function as described in "Using the MegaWizard Plug-In Manager" in the Quartus II help.

The port names and order for Verilog HDL prototype are:

```
module lpm_ram_dq ( q, data, inclock, outclock, we, address);
```

The VHDL component declaration is shown below:

```
COMPONENT lpm_ram_dq
    GENERIC (LPM_WIDTH            : POSITIVE;
             LPM_WIDTHAD          : POSITIVE;
             LPM_NUMWORDS         : NATURAL    := 0;
             LPM_INDATA           : STRING     := "REGISTERED";
             LPM_ADDRESS_CONTROL  : STRING     := "REGISTERED";
             LPM_OUTDATA          : STRING     := "REGISTERED";
             LPM_FILE             : STRING     := "UNUSED";
             LPM_TYPE             : STRING     := "LPM_RAM_DQ";
             LPM_HINT             : STRING     := "UNUSED" );
    PORT (data : IN STD_LOGIC_VECTOR(LPM_WIDTH-1 DOWNTO 0);
       address : IN STD_LOGIC_VECTOR(LPM_WIDTHAD-1 DOWNTO 0);
      inclock, outclock          : IN STD_LOGIC := '0';
          we : IN STD_LOGIC;
          q  : OUT STD_LOGIC_VECTOR(LPM_WIDTH-1 DOWNTO 0)
          );
END COMPONENT;
```

Ports

The following table displays all input ports of `lpm_ram_dq`:

Port name	Required	Description	Comments
address	Yes	Address input to the memory	Input port LPM_WIDTHAD wide
data	Yes	Data input to the memory	Input port LPM_WIDTHAD wide
inclock	No	Clock for input registers	The address port is synchronous (registered) when the inclock port is connected, and is asynchronous (registered) when the inclock port is not connected.
outclock	No	Clock for output registers	The addressed memory content-to-q response is synchronous when the outclock port is connected, and is asynchronous when it is not connected.
we	Yes	Memory enable input	Write enable input. Enables write operations to the memory when high.

The following table displays all output ports of `lpm_ram_dq`:

Port Name	Required	Description	Comments
q	Yes	Output of memory	Output port LPM_WIDTH wide

Parameters

The following table shows the parameters of the `lpm_ram_dq` component:

Parameter	Type	Re-quired	Description
LPM_WIDTH	Integer	Yes	Width of the q port.
LPM_WIDTHAD	Integer	Yes	Width of the **address** port. **LPM_WIDTHAD** should be (but is not required to be) equal to $\log_2$(LPM_NUMWORDS). If LPM_WIDTHAD is too small, some memory locations will not be addressable. If it is too large, addresses that are too high will return undefined logic levels.
LPM_NUMWORDS	Integer	Yes	Number of words stored in memory. In general, this value should be (but is not required to be) $2^{\text{LPM\_WIDTHAD}} - 1 < \text{LPM\_NUMWORDS} \leq 2^{\text{LPM\_WIDTHAD}}$. If omitted, the default is $2^{\text{LPM\_WIDTHAD}}$.
LPM_FILE	String	No	Name of the **Memory Initialization File** (*.mif) or **Hexadecimal** (Intel-format) File (*.hex) containing ROM initialization data (<filename>), or **UNUSED**.
LPM_ADDRESS_CONTROL	String	No	Values are **REGISTERED**, **UNREGISTERED**, and **UNUSED**. Indicates whether the address port is registered. If omitted, the default is **REGISTERED**.
LPM_OUTDATA	String	No	Values are **REGISTERED**, **UNREGISTERED**, and **UNUSED**. Indicates whether the q and eq ports are registered. If omitted, the default is **REGISTERED**.
LPM_HINT	String	No	Allows you to specify Altera-specific parameters in VHDL design files. The default is **UNUSED**.
LPM_TYPE	String	No	Identifies the LPM entity name in the VHDL design files.

Note that for Verilog LPM 220 synthesizable code (i.e., `220model.v`) the following parameter ordering applies: `lpm_type`, `lpm_width`, `lpm_widthad`, `lpm_numwords`, `lpm_address_control`, `lpm_outdata`, `lpm_file`, `lpm_hint`.

Function

The following table shows the *synchronous* read and write memory behavior of
`lpm_ram_dq`:

inclock	we	Function
X	-	No change (requires rising clock edge).
∫	H	The memory location pointed to by address is loaded with data.
∫	L	The memory location pointed to by address is read from the array. If outclock is not used, the read data appears at the outputs.

The following table shows the synchronous read from memory from memory
operations behavior of `lpm_ram_dq`:

outclock	Function
—	No change
∫	The memory location pointed to by address is read and written into the output register.

Totally asynchronous memory operations occur when neither `inclock` nor
`outclock` is connected. The output `q` is asynchronous and reflects the memory
location pointed to by `address`. Since this totally asynchronous memory operation
is only available with Flex 10K devices, but not with Cyclone II, we do not use this
mode in our designs. Either the input or output is registered in all of our designs
that use memory blocks.

Resource Usage

The Megafunction `lpm_ram_dq` uses one embedded cell per memory bit.

C. Glossary

ACC	Accumulator
ACT	Actel FPGA family
ADC	Analog-to-digital converter
ADCL	All-digital CL
ADF	Adaptive digital filter
ADPCM	Adaptive differential pulse code modulation
ADPLL	All-digital PLL
ADSP	Analog Devices digital signal processor family
AES	Advanced encryption standard
AFT	Arithmetic Fourier transform
AHDL	Altera HDL
AHSM	Additive half square multiplier
ALU	Arithmetic logic unit
AM	Amplitude modulation
AMBA	Advanced microprocessor bus architecture
AMD	Advanced Micro Devices, Inc.
ASCII	American standard code for information interchange
ASIC	Application-specific IC
AWGN	Additive white Gaussian noise
BCD	Binary coded decimal
BDD	Binary decision diagram
BLMS	Block LMS
BP	Bandpass
BRS	Base removal scaling
BS	Barrelshifter
CAE	Computer-aided engineering
CAM	Content addressable memory
CAST	Carlisle Adams and Stafford Tavares
CBC	Cipher block chaining
CBIC	Cell-based IC
CD	Compact disc
CFA	Common factor algorithm
CFB	Cipher feedback
CIC	Cascaded integrator comb
CISC	Complex instruction set computer
CL	Costas loop
CLB	Configurable logic block
C-MOMS	Causal MOMS

CMOS	Complementary metal oxide semiconductor
CODEC	Coder/decoder
CORDIC	Coordinate rotation digital computer
COTS	Commercial off-the-shelf technology
CPLD	Complex PLD
CPU	Central processing unit
CQF	Conjugate quadrature filter
CRNS	Complex RNS
CRT	Chinese remainder theorem
CSOC	Canonical self-orthogonal code
CSD	Canonical signed digit
CWT	Continuous wavelet transform
CZT	Chirp-z transform
DA	Distributed arithmetic
DAC	Digital-to-analog converter
DAT	Digital audio tap
DB	Daubechies filter
DC	Direct current
DCO	Digital controlled oscillator
DCT	Discrete cosine transform
DCU	Data cache unit
DES	Data encryption standard
DFT	Discrete Fourier transform
DHT	Discrete Hartley transform
DIF	Decimation in frequency
DIT	Decimation in time
DLMS	Delayed LMS
DMA	Direct memory access
DMIPS	Dhrystone MIPS
DMT	Discrete Morlet transform
DPLL	Digital PLL
DSP	Digital signal processing
DST	Discrete sine transform
DWT	Discrete wavelet transform
EAB	Embedded array block
ECB	Electronic code book
ECL	Emitter coupled logic
EDIF	Electronic design interchange format
EFF	Electronic Frontier Foundation
EPF	Altera FPGA family
EPROM	Electrically programmable ROM
ERA	Plessey FPGA family
ERNS	Eisenstein RNS
ESA	European Space Agency
EVR	Eigenvalue ratio
EXU	Execution unit
FAEST	Fast a posteriori error sequential technique
FCT	Fast cosine transform
FC2	FPGA compiler II
FF	Flip-flop

FFT	Fast Fourier transform
FIFO	First-in first-out
FIR	Finite impulse response
FIT	Fused internal timer
FLEX	Altera FPGA family
FM	Frequency modulation
FNT	Fermat NTT
FPGA	Field-programmable gate array
FPL	Field-programmable logic (combines CPLD and FPGA)
FPLD	FPL device
FSF	Frequency sampling filter
FSK	Frequency shift keying
FSM	Finite state machine
GAL	Generic array logic
GF	Galois field
GNU	GNU's not Unix
GPP	General-purpose processor
GPR	General-purpose register
HB	Half-band filter
HI	High frequency
HDL	Hardware description language
HSP	Harris Semiconductor DSP ICs
IBM	International Business Machines (corporation)
IC	Integrated circuit
ICU	Instruction cache unit
IDCT	Inverse DCT
IDEA	International data encryption algorithm
IDFT	Inverse discrete Fourier transform
IEEE	Institute of Electrical and Electronics Engineers
IF	Inter frequency
IFFT	Inverse fast Fourier transform
IIR	Infinite impulse response
I-MOMS	Interpolating MOMS
INTT	Inverse NTT
IP	Intellectual property
I/Q	In-/Quadrature phase
ISA	Instruction set architecture
ITU	International Telecommunication Union
JPEG	Joint photographic experts group
JTAG	Joint test action group
KCPSM	Ken Chapman PSM
KLT	Karhunen–Loeve transform
LAB	Logic array block
LAN	Local area network
LC	Logic cell
LE	Logic element
LIFO	Last-in first-out

LISA	Language for instruction set architecture
LF	Low frequency
LFSR	Linear feedback shift register
LMS	Least-mean-square
LNS	Logarithmic number system
LO	Low frequency
LP	Lowpass
LPM	Library of parameterized modules
LRS	Serial left right shifter
LS	Least-square
LSB	Least-significant bit
LSI	Large scale integration
LTI	Linear time-invariant
LUT	Look-up table
MAC	Multiplication and accumulate
MACH	AMD/Vantis FPGA family
MAG	Multiplier adder graph
MAX	Altera CPLD family
MIF	Memory initialization file
MIPS	Microprocessor without interlocked pipeline
MIPS	Million instructions per second
MLSE	Maximum-likelihood sequence estimator
MMU	Memory management unit
MMX	Multimedia extension
MNT	Mersenne NTT
MOMS	Maximum order minimum support
μP	Microprocessor
MPEG	Moving Picture Experts Group
MPX	Multiplexer
MSPS	Millions of sample per second
MRC	Mixed radix conversion
MSB	Most significant bit
MUL	Multiplication
NCO	Numeric controlled oscillators
NLMS	Normalized LMS
NP	Nonpolynomial complex problem
NRE	Nonreccurring engineering costs
NTT	Number theoretic transform
OFB	Open feedback (mode)
O-MOMS	Optimal MOMS
PAM	Pulse-amplitude modulated
PC	Personal computer
PCI	Peripheral component interconnect
PD	Phase detector
PDSP	Programmable digital signal processor
PFA	Prime factor algorithm
PIT	Programmable interval timer
PLA	Programmable logic array
PLD	Programmable logic device

PLL	Phase-locked loop
PM	Phase modulation
PREP	Programmable Electronic Performance (cooperation)
PRNS	Polynomial RNS
PROM	Programmable ROM
PSK	Phase shift keying
PSM	Programmable state machine
QDFT	Quantized DFT
QLI	Quick look-in
QFFT	Quantized FFT
QMF	Quadrature mirror filter
QRNS	Quadratic RNS
QSM	Quarter square multiplier
RAM	Random-access memory
RC	Resistor/capacity
RF	Radio frequency
RISC	Reduced instruction set computer
RLS	Recursive least square
RNS	Residue number system
ROM	Read-only memory
RPFA	Rader prime factor algorithm
RS	Serial right shifter
RSA	Rivest, Shamir, and Adelman
SD	Signed digit
SG	Stochastic gradient
SIMD	Single instruction multiple data
SLMS	Signed LMS
SM	Signed magnitude
SNR	Signal-to-noise ratio
SPEC	System performance evaluation cooperation
SPLD	Simple PLD
SPT	Signed power-of-two
SR	Shift register
SRAM	Static random-access memory
SSE	Streaming SIMD extension
STFT	Short-term Fourier transform
TDLMS	Transform-domain LMS
TLB	Translation look-aside buffer
TLU	Table look-up
TMS	Texas Instruments DSP family
TI	Texas Instruments
TOS	Top of stack
TTL	Transistor transistor logic
TVP	True vector processor
UART	Universal asynchronous receiver/transmitter
VCO	Voltage-control oscillator
VHDL	VHSIC hardware description language

VHSIC	Very-high-speed integrated circuit
VLIW	Very long instruction word
VLSI	Very large integrated ICs
WDT	Watchdog timer
WFTA	Winograd Fourier transform algorithm
WSS	Wide sense stationary
XC	Xilinx FPGA family
XNOR	Exclusive NOR gate
YACC	Yet another compiler-compiler

D. CD-ROM File: "1readme.ps"

The accompanying CD-ROM includes

- A full version of the Quartus II software
- Altera datasheets for Cyclone II devices
- All VHDL/Verilog design examples and utility programs and files

To install the Quartus II 6.0 web edition software first read the licence agreement carefully. Since the Quartus II 6.0 web edition software uses many other tools (e.g., GNU, Berkeley Tools, SUN microsystems tool, etc.) you need to agree to their licence agreements too before installing the software. To install the software start the self-extracting file `quartusii_60_web_edition.exe` on the CD-ROM in the `Altera` folder. After the installation the user must register the software through Altera's web page at `www.altera.com` in order to get a permanent licence key. Otherwise the temporary licence key expires after the 30-day grace period and the software will no longer run. Altera frequently update the Quartus II software to support new devices and you may consider downloading the latest Quartus II version from the Altera webpage directly, but keep in mind that the files are large and that the synthesis results will differ slightly for another version. Altera's University program now delivers the files via download, which can take long time with a 56 Kbit/s MODEM.

The design examples for the book are located in the directories `book3e/vhdl` and `book3e/verilog` for the VHDL and Verilog examples, respectively. These directories contain, for each example, the following four files:

- The VHDL or Verilog source code (`*.vhd` and `*.v`)
- The Quartus project files (`*.qpf`)
- The Quartus setting files (`*.qsf`)
- The Simulator wave form file (`*.vwf`)

For the design `fun_graf`, the block design file (`*.bdf`) is included in `book3e/vhdl`. For the examples that utilize M4Ks (i.e., `fun_text`, `darom`, and `trisc0`), the memory initialization file (`*.mif`) can be found on the CD-ROM. To simplify the compilation and postprocessing, the source code directories include the additional (`*.bat`) files and Tcl scripts shown below:

File	Comment
qvhdl.tcl	Tcl script to compile all design examples. Note that the device can be changed from Cyclone II to Flex, Apex or Stratix just by changing the comment sign # in column 1 of the script.
qclean.bat	Cleans all temporary Quartus II compiler files, but not the report files (*.map.rpt), the timing analyzer output files (*.tan.rpt), and the project files *.qpf and *.qsf.
qveryclean.bat	Cleans all temporary compiler files, *including* all report files (*.rep) and project files.

Use the DOS prompt and type

```
quartus_sh -t qvhdl.tcl > qvhdl.txt
```

to compile all design examples and then qclean.bat to remove the unnecessary files. The Tcl script qvhdl.tcl is included on the CD. The Tcl script language developed by the Berkeley Professor John Ousterhout [346, 347, 348] (used by most modern CAD tools: Altera Quartus, Xilinx ISE, ModelTech, etc.) allows a comfortable scripting language to define setting, specify functions, etc. Given the fact that many tools also use the graphic toolbox Tcl/Tk we have witnessed that many tools now also looks almost the same.

Two search procedures (show_fmax and show_resources) are used within the Tcl script qvhdl.tcl to display resources and Registered Performance. The script includes all settings and also alternative device definitions. The protocol file qvhdl.txt has all the useful synthesis data. For the trisc0 processor, for instance, the list for the Cyclone II device EP2C35F672C6 is:

```
....
--------------------------------------------------------
trisc0 fmax: 115.65 MHz ( period = 8.647 ns )
trisc0 LEs: 198 / 33,216 ( < 1 % )
trisc0 M4K bits: 5,120 / 483,840 ( 1 % )
trisc0 DSP blocks: 1 / 70 ( 1 % )
--------------------------------------------------------
....
```

The results for all examples are summarized in Table B.1, p. 731.

Other devices are prespecified and include the EPF10K20RC240-4 and EPF10K70RC240-4 from the UP1 and UP2 University boards, the EP20K200EFC484-2X from the Nios development boards, and three devices from other DSP boards available from Altera, i.e., the EP1S10F484C5, EP1S25F780C5, and EP2S60F1020C4ES.

Using Compilers Other Then Quartus II

Synopsys FPGA_CompilerII

The main advantage of using the FPGA_CompilerII (FC2) from Synopsys was that it was possible to synthesize examples for other devices like Xilinx, Vantis, Actel, or QuickLogic with the same tool. The Tcl scripts vhdl.fc2, and verilog.fc2, respectively, were provided the necessary commands for the shell mode of FC2, i.e., fc2_shell in the second edition of the book [57]. Synopsys, however, since 2006 no longer supports the FPGA_CompilerII and it is therefore not a good idea to use the compiler anymore since the newer devices can not be selected.

Model Technology

By using the synthesizable public-domain models provided by the EDIF organization (at www.edif.org), it is also possible to use other VHDL/Verilog simulators then Quartus II.

Using MTI and VHDL. For VHDL, the two files 220pack.vhd and 220model.vhd must first be compiled. For the ModelSim simulator vsim from Model Technology Inc., the script mti_vhdl.do can be used for device-independent compilation and simulation of the design examples. The script is shown below:

```
#---------------------------------------------------------------
# Model Technology VHDL compiler script for the book
# Digital Signal Processing with FPGAs (3.edition)
# Author-EMAIL: Uwe.Meyer-Baese@ieee.org
#---------------------------------------------------------------

echo Create Library directory lpm
vlib lpm

echo Compile lpm package.
vcom -work lpm -explicit -quiet 220pack.vhd 220model.vhd

echo Compile chapter 1 entitys.
vcom -work lpm -quiet example.vhd fun_text.vhd

echo Compile chapter 2 entitys.
vcom -work lpm -explicit -quiet add1p.vhd add2p.vhd
vcom -work lpm -explicit -quiet add3p.vhd mul_ser.vhd
vcom -work lpm -explicit -quiet cordic.vhd

echo Compile chapter 3 components.
vcom -work lpm -explicit -quiet case3.vhd case5p.vhd
vcom -work lpm -explicit -quiet case3s.vhd
echo Compile chapter 3 entitys.
vcom -work lpm -explicit -quiet fir_gen.vhd fir_srg.vhd
vcom -work lpm -explicit -quiet dafsm.vhd darom.vhd
vcom -work lpm -explicit -quiet dasign.vhd dapara.vhd

echo Compile chapter 4 entitys.
vcom -work lpm -explicit -quiet iir.vhd iir_pipe.vhd
vcom -work lpm -explicit -quiet iir_par.vhd

echo Compile chapter 5 entitys.
vcom -work lpm -explicit -quiet cic3r32.vhd cic3s32.vhd
vcom -work lpm -explicit -quiet db4poly.vhd db4latti.vhd

echo Compile chapter 6 entitys.
```

```
vcom -work lpm -explicit -quiet rader7.vhd ccmul.vhd
vcom -work lpm -explicit -quiet bfproc.vhd

echo Compile chapter 7 entitys.
vcom -work lpm -explicit -quiet rader7.vhd ccmul.vhd
vcom -work lpm -explicit -quiet bfproc.vhd

echo Compile 2. edition entitys.
vcom -work lpm -explicit -quiet div_res.vhd div_aegp.vhd
vcom -work lpm -explicit -quiet fir_lms.vhd fir6dlms.vhd

echo Compile 3. edition entitys from chapter 2.
vcom -work lpm -explicit -quiet cmul7p8.vhd arctan.vhd
vcom -work lpm -explicit -quiet ln.vhd sqrt.vhd

echo Compile 3. edition entitys from chapter 5.
vcom -work lpm -explicit -quiet rc_sinc.vhd farrow.vhd
vcom -work lpm -explicit -quiet cmoms.vhd

echo Compile 3. edition entitys from chapter 9.
vcom -work lpm -explicit -quiet reg_file.vhd trisc0.vhd
```

Start the ModelSim simulator and then type

```
do mti_vhdl.do
```

to execute the script.

Using MTI and Verilog. Using the Verilog interface with the lpm library from EDIF, i.e., 220model.v, needs some additional effort. When using 220model.v it is necessary to specify *all* ports in the Verilog lpm components. There is an extra directory book3e/verilog/mti, which provides the design examples with a full set of lpm port specifications. The designs use

```
'\include "220model.v"
```

at the beginning of each Verilog file to include the lpm components, if necessary. Use the script mti_v1.csh and mti_v2.csh to compile all Verilog design examples with Model Technology's vcom compiler.

In order to load the memory initialization file (*.mif), it is required to be familiar with the programming language interface (PLI) of the Verilog 1364-1995 IEEE standard (see LRM Sect. 17, p. 228 ff). With this powerful PLI interface, conventional C programs can be dynamically loaded into the Verilog compiler. In order to generate a dynamically loaded object from the program convert_hex2ver.c, the path for the include files veriuser.h and acc_user.h must be specified. Use -I when using the gcc or cc compiler under SUN Solaris. Using, for instance, the gcc compiler under SUN Solaris for the Model Technology Compiler, the following commands are used to produce the shared object:

```
gcc -c -I/<install_dir>/modeltech/include convert_hex2ver.c
ld -G -B symbolic -o convert_hex2ver.sl convert_hex2ver.o
```

By doing so, ld will generate a warning "Symbol referencing errors," because all symbols are first resolved within the shared library at link time, but these warnings can be ignored.

It is then possible to use these shared objects, for instance, with Model Technology's vsim in the first design example fun_text.v, with

```
vsim  -pli convert_hex2verl.sl  lpm.fun_text
```

To learn more about PLIs, check out the Verilog IEEE standard 1364-1995, or the vendor's user manual of your Verilog compiler.

We can use the script `mti_v1.do` to compile all Verilog examples with MTI's `vlog`. Just type

```
do mti_v1.do
```

in the ModelTech command line. But `vlog` does not perform a check of the correct component port instantiations or shared objects. A second script, `mti_v2.do`, can be used for this purpose. Start the `vsim` simulator (without loading a design) and execute the `DO` file with

```
do mti_v2.do
```

to perform the check for all designs.

Using Xilinx ISE The conversion of designs from Altera Quartus II to Xilinx ISE seems to be easy if we use standard HDL. Unfortunately there a couple of issues that needs to be addressed. We assume that the ModelTech simulation environment and the web version (i.e., no core generation) is used. We like to discuss in the following a couple of items that address the ISE/ModelTech design entry. We describe the Xilinx ISE 6.2 web edition and ModelTech 5.7g version.

1) The Xilinx simulation with timing ("Post-Place & Route") uses a bitwise simulation model on the LUT level. Back annotations are only done for the I/O ports, and are ALL from type `standard_logic` or `standard_logic_vector`. In order to match the behavior and the simulation with timing we therefore need to use only the `standard_logic` or `standard_logic_vector` data type for I/O. As a consequence no integers, generic, or custom I/O data type, (e.g., like the subtype byte see `cordic.vhd`) can be used.

2) The ISE software supports the development of testbenches with the "Test Bench Waveform." Use `New Source...` under the `Project` menu. This waveform will give you a quick way to generate a testbench that is used by ModelTech, both for behavior as well as simulation with timing. There are some benefits and drawbacks with the testbencher. For instance, you can not assign negative integers in the waveforms, you need to build the two's complement, i.e., equivalent unsigned number by hand.

3) If you have feedback, you need to initialize the register to zero in your HDL code. You can not do this with the testbencher: for instance, ModelTech initialize all integer signals to the smallest value, i.e., -128 for a 8-bit number, if you add two integers, the result will be $-128 - 128 = -256 < -128$ and ModelTech will stop and report an overflow. Some designs, e.g., `cordic`, `cic3r32`, `cic3s32`, only work correctly in behavior simulation if all integers are changed to `standard_logic_vector` data type. Changing I/O ports alone and using the conversion function does not always guarantee correct simulation results.

4) Simulation with timing usually needs one clock cycle more than behavior code until all logic is settled. The input stimuli should therefore be zero in the first clock cycle (ca. 100 ns) and, if you want to match behavior and timing simulation, and the design uses a (small) FSM for control, you need to add a synchronous or asynchronous reset. You need to do this for the following 2/e designs: `dafsm`, `dadrom`, `dasign`, `db4latti`, `db4poly`, `div_aegp`, `div_res`, `iir_par`, `mul_ser`, `rader7`.
Just add a control part for the FSM like this:

```
-- IF rising_edge(clk) THEN          -- Synchronous reset
--     IF reset = '1' THEN
--         state <= s0;
--     ELSE
     IF reset = '1' THEN             -- Asynchronous reset
       state <= s0;
     ELSIF rising_edge(clk) THEN
       CASE state IS
       WHEN s0 =>                    -- Initialization step
   . . .
```

Although at first glance this synchronous or asynchronous control seems to be cheap because the FSM is small, we need to keep in mind that, if the reset is active, all signals that are assigned in the state s0 of the FSM need to be preserved with their initial state value. The following table shows the synthesis results for the three different reset styles for the design file dafsm.vhd (small distribute arithmetic state machine):

Reset style	Performance/ns	4-input LUT	Gates
No reset (original code)	3.542	20	339
synchronous	3.287	29	393
asynchronous	3.554	29	393

Designs with reset usually have a higher LUT and gate count. Depending on the design, synchronous or asynchronous reset can also have a (small) influence on performance.

5) Back annotation is only done for I/O ports. If we want to monitor internal nets, we can try to find the appropriate net name in the *_timesim.vhd file, but that is quite complicated and may change in the next compiler run. A better idea is to introduce additional test outputs, see, for instance, fir_lms.vhd for f1_out and f2_out. In the behavioral (but not in the timing) simulation internal test signals and variables can be monitored. Modify the *.udo file and add, for instance for the fir_srg_tb.vhd file, add wave /fir_srg_tb/uut/tap to the testbench.

3) There are a couple of nice features in the Xilinx ISE package too: there is no need for special lpm blocks to use the internal resources for multiplier, shifter, RAMs or ROMs. Some other features are:

a) ISE converts a shift register in a single CLB-based shift register. This can save some resources.

b) Multipliers can be implemented with LUTs only, including block multipliers (if available) or even pipelined LUTs, which is done via pipeline retiming. Just right click on the Synthesize-XST menu in the Processes, select HDL Options under Process Properties and the last entry is the multiplier style. But note that for pipelined LUT design the additional register must be placed at the *output* of the multiplier. Pipeline retiming is not done if the additional registers are at the inputs. You need about $\log_2(B)$ additional registers to have good timing (see Chap. 2 on pipeline multiplier). This has an impact on the Registered Performance, LUT usage, and gates as the following table shows for the fir_gen.vhd example, i.e., length 4 programmable FIR filter (from Chap. 3):

Synthesis style	Speed in ns	4-input LUT	mul. blocks	Gates
Block multiplier	9.838	57	4	17552
LUT (no pipeline)	15.341	433	0	6114
LUT (3 stage pipeline)	6.762	448	0	9748

For this multiplier size (9 bit) the pipelined LUT seems to be attractive, both for speed as well as gate count. If the number of LUTs is limited, the block multiplier provides the next best alternative.

c) If you follow the recommended style the Xilinx software synthesis tool (see XST manual and ISE help "Inferring BlockRAM in VHDL") maps your HDL code to the block RAM (see, fun_text.vhd). If the table is small, the ISE auto option selects the LUT-based implementation for a ROM table (see darom.vhd). You can also initialize the table in the HDL code and use it as a ROM. Please see the XST manual Chap. 3, "FPGA Optimization" for details on ROM implementation. There are some limitations that apply to the initialization of BlockRAMs (see, XST Chap. 2)

Utility Programs and Files

A couple of extra utility programs are also included on the CD-ROM[1] and can be found in the directory book3e/util:

File	Description
sine3e.exe	Program to generate the MIF files for the function generator in Chap. 1
csd3e.exe	Program to find the canonical signed digit representation of integers or fractions as used in Chap. 2
fpinv3e.exe	Program to compute the floating-point tables for reciprocals as used in Chap. 2
dagen.exe	Program to generate the VHDL code for the distributed arithmetic files used in Chap. 3
ragopt.exe	Program to compute the reduced adder graph for constant-coefficient filters as used in Chap. 3. It has 10 predefined lowpass and half-band filters. The program uses a MAG cost table stored in the file mag14.dat
cic.exe	Program to compute the parameters for a CIC filter as used in Chap. 5

The programs are compiled using the author's MS Visual C++ standard edition software (available for $50–100 at all major retailers) for DOS window applications and should therefore run on Windows 95 or higher. The DOS script Testall.bat produces the examples used in the book.

Also under book3e/util we find the following utility files:

---

[1] You need to copy the programs to your harddrive first; you can not start them from the CD directly since the program write out the results in text files.

File	Description
quickver.pdf	Quick reference card for Verilog HDL from QUALIS
quickvhd.pdf	Quick reference card for VHDL from QUALIS
quicklog.pdf	Quick reference card for the IEEE 1164 logic package from QUALIS
93vhdl.vhd	The IEEE VHDL 1076-1993 keywords
95key.v	The IEEE Verilog 1364-1995 keywords
01key.v	The IEEE Verilog 1364-2001 keywords
95direct.v	The IEEE Verilog 1364-1995 compiler directives
95tasks.v	The IEEE Verilog 1364-1995 system tasks and functions

In addition, the CD-ROM includes a collection of useful Internet links (see file dsp4fpga.htm under book3e/util), such as device vendors, software tools, VHDL and Verilog resources, and links to online available HDL introductions, e.g., the "Verilog Handbook" by Dr. D. Hyde and "The VHDL Handbook Cookbook" by Dr. P. Ashenden.

Microprocessor Project Files and Programs

All microprocessor-related tools and documents can be found in the book3e/uP folder. Six software Flex/Bison projects along with their compiler scripts are included:

- build1.bat and simple.l are used for a simple Flex example.
- build2.bat, d_ff.vhd, and vhdlcheck.l are a basic VHDL lexical analysis.
- build3.bat, asm2mif.l, and add2.txt are used for a simple Flex example.
- build4.bat, add2.y, and add2.txt are used for a simple Bison example.
- build5.bat, calc.l, calc.y and calc.txt is an infix calculator and are used to demonstrate the Bison/Flex communication.
- build6.bat, c2asm.h, c2asm.h, c2asm.c, lc2asm.c, yc2asm.c and factorial.c are used for a C-to-assembler compiler for a stack computer.

The *.txt files are used as input files for the programs. The buildx.bat can be used to compile each project separately; alternatively you can use the uPrunall.bat under Unix to compile and run all files in one step. The compiled files that run under SunOS UNIX end with *.exe while the DOS programs end with *.com.

Here is a short description of the other supporting files in the book3e/uP directory: Bison.pdf contains the Bison compiler, i.e., the YACC-compatible parser generator, written by Charles Donnelly and Richard Stallman; Flex.pdf is the description of the fast scanner generator written by Vern Paxson.

Index

Accumulator 10, 257
- μP 553, 557
Actel 9
Adaptive filter 477–535
Adder
- binary 75
- fast carry 76
- floating-point 110, 114
- LPM 15, 30, 78, 368, 733, 737
- pipelined 78
- size 77
- speed 77
Agarwal–Burrus NTT 410
Algorithms
- Bluestein 350
- chirp-z 350
- Cooley–Tukey 367
- CORDIC 120–130
- common factor (CFA) 362
- Goertzel 350
- Good–Thomas 363
- fast RLS 10
- LMS 488, 531
- prime factor (PFA) 362
- Rader 353, 413
- Radix-r 366
- RLS 522, 531
- Widrow–Hoff LMS 488
- Winograd DFT 360
- Winograd FFT 376
Altera 9, 20
AMD 9
Arbitrary rate conversion 280–308
Arctan approximation 132
ARM922T μP 592

Bartlett window 175, 345
Bijective 259
Bison μP tool 567, 578
Bitreverse 390

Blackman window 175, 345
Blowfish 452
B-spline rate conversion 296
Butterfly 366, 370

CAST 452
C compiler 586, 587
Chebyshev series 131
Chirp-z algorithm 350
CIC filter 258–273
- RNS design 260
- interpolator 340
Coding bounds 423
Codes
- block
-- decoders 426
-- encoder 425
- convolutional
-- comparison 436
-- complexity 435
-- decoder 430, 434
-- encoder 430, 434
- tree codes 429
Contour plot 490
Convergence 490, 491, 492
- time constant 491
Convolution
- Bluestein 391
- cyclic 391
- linear 116, 165
Cooley–Tukey
- FFT 367
- NTT 409
CORDIC algorithm 120–130
cosine approximation 137
Costas loop
- architecture 470
- demodulation 470
- implementation 472
CPLD 6, 5

Cryptography 436–452
Cypress 9

Daubechies 314, 319, 330, 337, 337
Data encryption standard (DES) 446–452
DCT
- definition 387
- fast implementation 389
- 2D 387
- JPEG 387
Decimation 245
Decimator
- CIC 261
- IIR 236
Demodulator 458
- Costas loop 470
- I/Q generation 459
- zero IF 460
- PLL 465
DFT
- computation using
-- NTT 417
-- Walsh–Hadamard transformation 417
-- AFT 417
- definition 344 - inverse 344
- filter bank 309
- Rader 363
- real 347
- Winograd 360
Digital signal processing (DSP) 2, 116
Discrete
- Cosine transform, *see DCT 390*
- Fourier transform, *see DFT 344*
- Hartley transform 393
- Sine transform (DST) 387
- Wavelet transform (DWT) 332–337
-- LISA µP 610
Distributed arithmetic 116–122
- Optimization
-- Size 121
-- Speed 122
- signed 199
Divider 91–104
- array
-- performance 103
-- size 104
- convergence 100
- fast 99
- LPM 103, 733, 749
- nonperforming 96, 157
- nonrestoring 98, 157
- restoring 94

- types 92
Dyadic DWT 332

Eigenfrequency 259
Eigenvalues ratio 494, 495, 501, 502, 524
Encoder 425, 430, 434
Error
- control 418–436
- cost functions 482
- residue 485
Exponential approximation 141

Farrow rate conversion 292
Fast RLS algorithm 10
Fermat NTT 407
FFT
- comparison 380
- Good–Thomas 363
- group 366
- Cooley–Tukey 367
- in-place 381
- IP core 383
- index map 362
- Nios co-processor 627
- Radix-r 366
- rate conversion 282
- stage 366
- Winograd 376
Filter 165–239
- cascaded integrator comb (CIC) 258–273
- causal 170
- CSD code 179 - conjugate mirror 323
- distributed arithmetic (DA) 189
- finite impulse response (FIR) 165–204
- frequency sampling 277
- infinite impulse response (IIR) 216–210
- IP core 205
- lattice 324
- polyphase implementation 250
- signed DA 199
- symmetric 172
- transposed 167
- recursive 280
Filter bank
- constant
-- bandwidth 329
-- Q 329
- DFT 309
- two-channel 314–328
-- aliasing free 317
-- Haar 316
-- lattice 324

-- linear-phase 327
-- lifting 321
-- QMF 314
-- orthogonal 323
-- perfect reconstruction 316
-- polyphase 323
-- mirror frequency 314
-- comparison 328
Filter design
- Butterworth 222
- Chebyshev 223
- Comparison of FIR to IIR 216
- elliptic 222
- equiripple 177
- frequency sampling 277
- Kaiser window 174
- Parks–McClellan 177
Finite impulse response (FIR), *see*
Filter 165–204
Flex µP tool 567, 571
Flip-flop
- LPM 15, 30, 78, 78, 78, 733
Floating-point
- addition 110
- arithmetic 104
- conversion to fixed-point 106
- division 111
- multiplication 108
- numbers 71
- reciprocal 113
- synthesis results 114
FPGA
- Altera's Cyclone II 22
- architecture 6
- benchmark 10
- Compiler II 762
- design compilation 33
- floor plan 33
- graphical design entry 30
- performance analysis 36
- power dissipation 13
- registered performance 36
- routing 5, 23
- simulation 34
- size 20, 22
- technology 9
- timing 26
- waveform files 43
- Xilinx Spartan-3 20
FPL, *see FPGA and CPLD*
Fractal 336
Fractional delay rate conversion 284
Frequency sampling filter 277

Function approximation
- arctan 132
- cosine 137
- Chebyshev series 131
- exponential 141
- logarithmic 145
- sine 137
- square root 150
- Taylor series 121

Galois Field 423
Gauss primes 69
General-purpose µP 538, 588
Generator 67, 68
Gibb's phenomenon 174
Good–Thomas
- FFT 363
- NTT 409
Goodman/Carey half-band filter 275,
318, 337
Gradient 487

Hadamard 474
half-band filter
- decimator 276
- factorization 317
- Goodman and Carey 275, 318
- definition 274
Hamming window 175, 345
Hann window 175, 345
Harvard µP 558
Hogenauer filter, *see CIC*
Homomorphism 258

IDEA 452
Identification 480, 494, 503
Isomorphism 258
Image compression 387
Index 67
- multiplier 68
- maps
-- in FFTs 362
-- in NTTs 409
Infinite impulse response (IIR) filter
216–239
- finite wordlength effects 228
- fast filtering using
-- time-domain interleaving 231
-- clustered look-ahead pipelining 233
-- scattered look-ahead pipelining 234
-- decimator design 235
-- parallel processing 237
-- RNS design 239

In-place 381
Instruction set design 544
Intel 539
Interference cancellation 478, 522
Interpolation
- CIC 340
- *see rate conversion*
Inverse
- multiplicative 363
- additive 404
- system modeling 480
IP core
- FFT 383
- FIR filter 205
- NCO 35

JPEG, *see Image compression*

Kaiser
- window 345
- window filter design 175
Kalman gain 520, 522, 526
Kronecker product 376

Learning curves 493
- RLS 520, 523
Lexical analysis *(see Flex)*
LISA µP 567, 610–626
LPM
- add_sub 15, 30, 78, 368, 733, 737
- divider 103, 733, 749
- flip-flop 15, 30, 78, 78, 78, 733
- multiplier 167, 504, 511, 368, 733, 741
- RAM 606
- ROM 30, 196, 733, 746
Lifting 321
Linear feedback shift register 438
LMS algorithm 488, 531
- normalized 496, 498
- design 506,
- pipelined 508
-- delayed 508
-- design 511, 514
-- look-ahead 510
-- transposed 511
-- block FFT 500
- simplified 516, 517
-- error floor 516
Logarithmic approximation 145

MAC 78
Mersenne NTT 408
MicroBlaze µP 603

Microprocessor
- Accumulator 553, 557
- Bison tool 567, 578
- C compiler 586, 587
- DWT 610
- GPP 538, 588
- Instruction set design 544
-- Profile 611, 615, 618, 624
- Intel 539
- FFT co-processor 627
- Flex tool 567, 571
- Lexical analysis *(see Flex)*
- LISA 567, 610–626
- Hardcore
-- PowerPC 591
-- ARM922T 592
- Harvard 558
- Softcore
-- MicroBlaze 603
-- Nios 598
-- PicoBlaze 538, 595
- Parser *(see Bison)*
- PDSP 2, 12, 114, 550, 616
- RISC 540
-- register file 559
- Stack 553, 557, 606
- Super Harvard 558
- Three address 555, 557
- Two address 555, 557
- Vector 620
- Von-Neuman 558
Möbius function 416
MOMS rate conversion 301
Multiplier
- adder graph 184, 229
- array 84
- block 91
- Booth 154
- complex 156, 368
- FPGA array 85
- floating-point 108, 114
- half-square 88
- index 68
- LPM 167, 504, 511, 368, 733, 741
- performance 86
- QRNS 69
- quarter square 90, 239
- serial/parallel 83
- size 87
Modulation 453
- using CORDIC 457
Modulo
- adder 68

- multiplier 68,
- reconstruction 273

NAND 5, 42
NCO IP core 35
Nios µP 598
Number representation
- canonical signed digit (CSD) 58, 229
- diminished by one (D1) 57, 405
- floating-point 74
- fractional 59, 178
- one's complement (1C) 57, 405
- two's complement (2C) 57
- sign magnitude (SM) 57
Number theoretic transform 401–417
- Agarwal–Burrus 410
- convolution 405
- definition 401
- Fermat 407
- Mersenne 408
- wordlength 408

Order
- filter 166
- for NTTs 408
Ordering, see index map
Orthogonal
- wavelet transform 319
- filter bank 323

Parser (see Bison)
Perfect reconstruction 316
Phase-locked loop (PLL)
- with accumulator reference 461
- demodulator 466
- digital 468
- implementation 467, 469
- linear 465
PicoBlaze µP 538, 595
Plessey ERA 5
Pole/zero diagram 236, 323
Polynomial rate conversion 290
Polyphase representation 250, 320
Power
- dissipation 13
- estimation 496, 498
- line hum 485, 487, 490, 492, 505, 516
PowerPC µP 591
Prediction 479
- forward 525
- backward 527
Prime number
- Fermat 403

- Mersenne 403
Primitive element 67
Programmable signal processor 2, 12, 114, 550, 616
- addressing generation 551
Public key systems 452

Quadratic RNS (QRNS) 69
Quadrature Mirror Filter (QMF) 314

Rader
- DFT 363
- NTT 413
Rate conversion
- arbitrary 280–308
- B-spline 296
- Farrow 292
- FFT-based 282
- fractional delay 284
- MOMS 301
- polynomial 290
- rational 249
Rational rate conversion 249
RC5 452
Rectangular window 175, 345
Reduced adder graph 184, 229
RISC µP 540
- register file 559
RLS algorithm 518, 522, 529
RNS
- CIC filter 260
- complex 70
- IIR filter 239
- Quadratic 69
- scaling 273
ROM
- LPM 30, 196, 733, 746
RSA 452

Sampling
- Frequency 345
- Time 345
- see rate conversion
Sea of gates Plessey ERA 5
Self-similar 332
Sine approximation 137
Simulator
- ModelTechnology 618, 763
Square root approximation 150
Stack µP 553, 557, 606
Step size 492, 492, 493, 502
Subband filter 309
Super Harvard µP 558

Symmetry
- in filter 172
- in cryptographic algorithms 452
Synthesizer
- accumulator 29
- PLL with accumulator 461

Taylor series 121
Theorem
- Chinese remainder 67
Three address µP 555, 557
Two-channel filter bank 314–328
- comparison 328
- lifting 321
- orthogonal 323
- QMF 323
- polyphase 320
Transformation
- arithmetic Fourier 417
- continuous Wavelet 332
- discrete cosine 390
- discrete Fourier 344
-- inverse (IDFT) 344
- discrete Hartley 393
- discrete Wavelet 332–337
- domain LMS 500
- Fourier 345
- Fermat NTT 407
- pseudo-NTT 409
- short-time Fourier (STFT) 329
- discrete sine 387
- Mersenne NTT 408
- number theoretic 401–417
- Walsh–Hadamard 417
Triple DES 451
Two address µP 555, 557

Vector µP 620
Verilog
- key words 729
VHDL
- styles 15
- key words 729
Von-Neuman µP 558

Walsh 473
Wavelets 332–337
- continuous 332
- linear-phase 327
- LISA processor 610–626
- orthogonal 319
Widrow–Hoff LMS algorithm 488
Wiener–Hopf equation 484

Windows 175, 345
Winograd DFT algorithm 360
Winograd FFT algorithm 376
Wordlength
- IIR filter 228
- NTT 408

Zech logarithm 68